DRAFTING
FOR
INDUSTRY

By
WALTER C. BROWN

Professor, Division of Technology
Arizona State University, Tempe

South Holland, Illinois
THE GOODHEART-WILLCOX COMPANY, INC.
Publishers

Library of Congress Catalog Card Number 89-33831
International Standard Book Number 0-87006-767-2

1234567890-90-8765432109

Library of Congress Cataloging in Publication Data

Brown, Walter Charles,
  Drafting for industry / by Walter C. Brown.

    p. cm.
  Bibliography: p.
  Includes index.
  ISBN 0-87006-767-2
  1. Mechanical drawing. I. Title.
T353.B873 1990
604.2--dc20   89-33831
                       CIP

INTRODUCTION

DRAFTING FOR INDUSTRY provides instruction and information on technical drafting techniques from fundamentals skills and processes through computer-aided drafting and design. It is intended to help you develop basic technical drafting skills as well as the advanced techniques used by industry. This book is intended to teach you to communicate and express ideas in an understandable, efficient and accurate manner.

Drafting problems of an ''exercise'' nature have been kept to a minimum, and actual problems encountered in the drafting rooms of industry have been selected to enrich the study of drafting. The creative approach to problem solving, so essential in all technical careers today, is emphasized throughout the text.

Relevant career information is presented on drafting and related occupational areas. Content is based on an extensive study of drafting practices in over 200 modern industries.

DRAFTING FOR INDUSTRY is a truly comprehensive text presented in an easy-to-understand and well-illustrated style.

Walter C. Brown

Industry photo. Technical drafting techniques enable ideas to be communicated and expressed in an understandable, efficient and accurate manner.   (Standard Oil of California)

CONTENTS

PART I
DRAFTING FUNDAMENTALS

Drafting fundamentals encompasses all of the basic skills required to produce drawings that convey necessary and specific information to the reader. This part of the text covers INSTRUMENT DRAFTING, TECHNICAL SKETCHING and LETTERING, GEOMETRY, THE DESIGN METHOD and MULTIVIEW, SECTIONAL and PICTORIAL DRAWING.

First, however, consider the many CAREER OPPORTUNITIES.

The industrial drawing represents the most effective means of communicating ideas about complex mechanisms. (Dow Chemical U.S.A.)

7

Chapter 1
CAREERS IN DRAFTING

Never in our history has the world been so technically oriented. With all the many new developments in industry, science, medicine, technology and space exploration, the language of drafting has become essential to the communication of ideas. See Fig. 1-1.

Drafting is a graphic language used by industry to communicate ideas and plans from the creative-design stage through production, Fig. 1-2. Anyone associated with the processes of production (and most of us are, either as consumers or technical workers) will benefit from a study of drafting.

DRAFTING IN TODAY'S WORLD

The purpose of this text is to aid you in a systematic study of this graphic means of communication. Whether your goal is to develop an understanding of drafting in a technical world, or enter one of the career fields related to drafting, you will find this a challenging area of study.

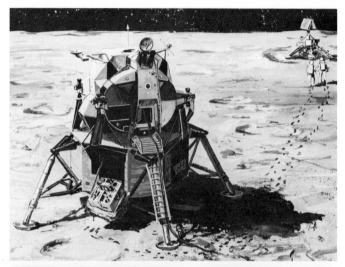

Fig. 1-1. Advances in industrial technology depend heavily on drawings for communication.   (3M Company)

Fig. 1-2. From creative design to reality, drafting had a part in exploration of the lunar surface.   (NASA)

Today, drafting is a highly developed science and effective means of meeting the challenge of an industrial nation. It accomplishes this, in keeping with improvements made in methods, materials and standards, by following the long used drafting system — third-angle projection, Fig. 1-3.

8

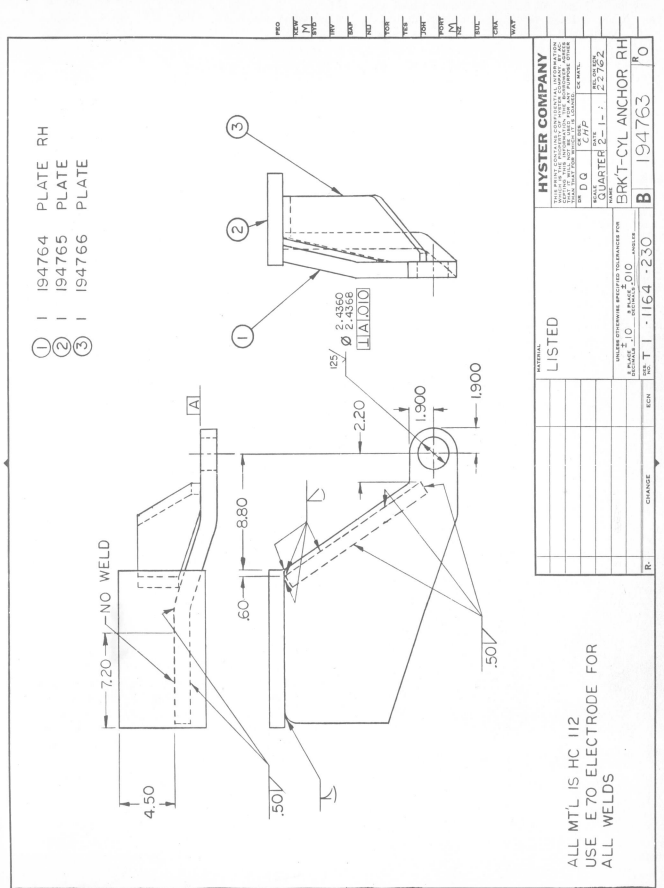

Fig. 1-3. Drawings in the United States are third-angle projections.

Drafting for Industry

The fundamentals of drafting are the same regardless of the type of drawing. Symbols used, style of lettering and general arrangement of the drawing may vary from mechanical to electronic or architectural to map drafting. However, the fundamental process of producing the finished drawings, standards, checking procedures and methods of reproducing drawings are much the same. Once the basics of drafting are understood, those with a continuing interest in drafting may select an area of specialization to gain additional experience.

CAREERS IN DRAFTING

Job titles and duties in the field of drafting vary with each industry, and the nature of activities may vary between industries under the same general classification. However, the following career activities in drafting are typical for the job levels and industries discussed. These should be supplemented with career information concerning specific industries of your choice.

The following career fields do not require a college degree, although preparation in a technical school or community college would be advantageous.

DRAFTING TRAINEES

Drafting trainees are expected to have, at the time of employment, a basic understanding and skill in the use of drafting instruments, a knowledge of procedures for representing views of objects, and the ability to produce neat freehand lettering. Generally, they will work under close supervision of senior drafting personnel, Fig. 1-4.

Trainees' duties include the revision of drawings, redrawing or repairing damaged drawings, and gathering information from reference sources necessary for detailing components (for themselves or other drafters). The work-training program will involve drawing detail views, sectional views, dimensioning and preparing tables and working drawings. Trainees must become familiar with the company's drafting standards expressed in their drafting room manuals.

Courses taken in drafting, mathematics, science, electronics, metals and woods are recommended.

DETAIL DRAFTERS

Detail drafters are well informed in the fundamentals of drafting and have gained proficiency and speed in handling instruments. They usually work as detailers in the preparation of working drawings for manufacturing or construction. Primarily, they will prepare detail drawings, revise drawings and bills of material, work on simple assembly drawings, wiring or circuit diagrams, charts and graphs.

Detail drafters should be thoroughly familiar with drafting standards and symbols and be able to make basic calculations in their area of drafting. They should also have a practical knowledge of engineering or architectural materials and procedures.

LAYOUT DRAFTERS

It is the job of layout drafters to prove out the product design, using sketches and models and a scaled layout drawing. Design layout serves to determine the manufacturing feasibility of the product.

This is an exacting type of drafting and requires a knowledge of the field and products being drawn. It may include preparation of some original layouts and studies to determine proper fits or clearances. It also could involve some changes in the design after consulting with the engineer in charge.

Since layout drafters may be required to make dimensional computations and allowances, a knowledge of machine shop practices and materials is essential, as well as an ability to research reference manuals.

DESIGN DRAFTERS

Design drafters are senior level drafters and represent the highest level of drafters on the board. After having acquired considerable experience in the drafting field, they do layout work and prepare complex detail and assembly drawings of machines, equipment, structures, wiring diagrams, piping diagrams and construction drawings. They work from basic data supplied by architects, engineers or industrial designers.

Design drafters must possess a sound knowledge of good engineering and drafting procedures, shop practices, mathematics and science. They may make design changes where required in consultation with the design engineer or architect, Fig. 1-5. Then they follow through on changes made to see that these are corrected on other drawings involved. They may prepare cost estimates based on the materials and parts list, according to the design problem. And they usually instruct and supervise other drafters in the support activities of assigned task or other related drawing problems.

Fig. 1-4. The drafting trainee will experience a variety of drawing activities. (Honeywell, Inc.)

Fig. 1-5. The manufacturing of this Tracking and Data Relay Satellite System User Transponder is a team effort of engineer-designer, drafter and manufacturing personnel. (Motorola, Inc.)

TECHNICAL ILLUSTRATOR

Technical illustration is the drawing of objects, usually machine parts, assemblies or mechanisms, in pictorial form. These illustrations may be enhanced by line work, shading or colors to give them a more realistic appearance.

Technical illustrators should have an understanding of drafting fundamentals, including the construction of pictorial views. They should be able to read and interpret blueprints and must have some background in art, industrial design and manufacturing processes.

Technical illustrations are used in manufacturing and production to assist workers in interpreting blueprints. It also proves to be of value to those who are unable to read blueprints by helping them visualize the object and its construction, Fig. 1-7. A more "artistic" type of technical illustration is used in marketing and advertising literature.

CHECKER

When a drawing is finished by a drafter, it must be checked for accuracy, completeness, clarity and manufacturing feasibility, Fig. 1-6. This check is made by an experienced drafter called a "checker."

Checkers are persons who understand manufacturing processes and are thoroughly familiar with the American National Drafting Standards as well as the drafting practices in the particular industry. Usually, they have reached the level of design drafters and may suggest modifications in design or specifications and other changes to facilitate production. Approval signature of a checker normally will appear in the title block of the drawing.

Fig. 1-6. The checking of a drawing requires the skills and knowledge of an experienced drafter. (Honeywell, Inc.)

Fig. 1-7. The technical illustrator's presentation of a section of the Bay Area Rapid Transit tube being lowered into place on the floor of San Francisco Bay. (Bay Area Rapid Transit District)

CAREERS RELATED TO DRAFTING

There are a number of related career opportunities for students who have talent in the drafting field. All require a background in drafting, which is used on the job. Many of these career fields require preparation at the college level.

11

Fig. 1-8. An architectural model of a major business development.
(Urban Investment and Development Co.)

ARCHITECTS

Architects plan, design and oversee the construction of projects such as buildings, housing developments, city planning and landscape architecture (parks, golf courses, etc.), Fig. 1-8. Because the field is so broad, architects tend to specialize in one phase of work such as residences, churches, schools or landscape projects.

Architects' duties include a study of clients' needs and desires. They make preliminary plans and sketches, finished drawings and renderings, cost estimates, prepare specifications and supervise construction of the project.

Fig. 1-9. A group of industrial designers discussing the design features of a telephone.   (Western Electric)

Training usually consists of four or five years of college, which includes on-the-board drafting experience. An architect should have a good understanding of mathematics and science as well as art and the humanities.

INDUSTRIAL DESIGNER

The profession of industrial design is concerned with the development of solutions to three-dimensional problems involving esthetics, materials, manufacturing processes, human factors and creativity. The industrial designer's task is working with scientific ideas and discoveries, endeavoring to develop these into useful products and services for humans.

Products such as office machines, furniture, systems for controlling forest fires and special equipment to enable a handicapped child live a fuller and more productive life are goals of the industrial designer's efforts, Fig. 1-9.

There are two areas of emphasis in the four-year college preparatory program for this career. One is product design; the other is mechanical design.

The product designer works primarily with those problems where the user interacts with the product (design of a telephone, automobile steering wheel or furniture for a library).

The mechanical designer's field includes problems where there is a machine-to-machine relationship, and no direct human interaction is involved (design of an automobile transmission, a machine tool and die or an improved conveyor system for materials handling).

Although industrial designers may not be working as drafters, they should have a good background in drafting, mathematics and science. They should be creative and have a thorough understanding of design problem-solving techniques.

The design drafting process is not a job for one person. It involves a team of engineers, designers and drafters, Fig. 1-10.

Careers in Drafting

FLOW DIAGRAM OF DESIGN-DRAFTING PROCESS IN A LARGE INDUSTRY

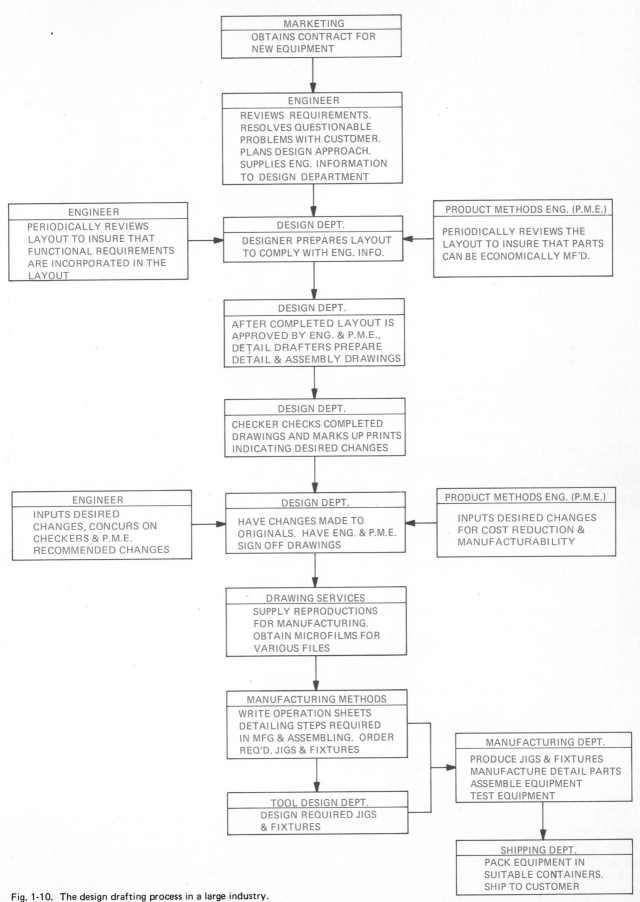

Fig. 1-10. The design drafting process in a large industry.
(Sperry Flight Systems Div.)

ENGINEER

Professional engineers are also concerned with creative design solutions. They usually possess an extensive background in science and mathematics since they must be able to apply these principles in searching out practical problem solutions, Fig. 1-11.

Fig. 1-11. A design engineer checking calculations for a bridge design. (Virginia Department of Highways and Transportation)

fiber for the benefit of man. More specifically, the agricultural engineer's specialty is the design and development of farm machinery, farm structures, processing equipment and the control and conservation of water, Fig. 1-12.

Fig. 1-12. This cantaloupe harvester collects the melons from trenches and moves them on a conveyor belt along each side of the tractor into a padded bin behind the driver. (University of Arizona)

The professional engineer's field of interest and knowledge is broad, even though she or he may specialize in one area of engineering. A good understanding of drafting procedures is necessary so that there is communication with other members of the technical team. The chief means of expressing ideas to others is by original freehand sketches and in reviewing the drawings prepared by drafters and making suggestions for their alteration.

There are a number of areas of specialization in the broad field of engineering, all of which require four to five years of college education. Some of the better known areas are presented here.

AEROSPACE ENGINEERING deals with the design and development of all types of conventional and experimental aircraft and aerospace vehicles. Persons working in aerospace engineering tend to specialize in one phase of activity, either aerodynamics, propulsion systems, structures, instrumentation or manufacturing.

AGRICULTURAL ENGINEERING is concerned with the problems of production, handling, and processing of food and

CERAMIC ENGINEERING is the research, design and development of nonmetallic materials into useful products for man. Examples include glassware, electrical insulators and the fusing of refractory materials as a protective coating for metals. The latter process was responsible for designing and developing the protective shield for the nose cone of space ships reentering the earth's atmosphere. It also applies to the production of metal signs, cooking utensils and sinks.

CHEMICAL ENGINEERING is that branch of engineering that processes materials to undergo chemical change. Chemical engineers design and develop the processes and equipment that change raw materials into useful products such as petro-chemicals, plastics, synthetic fibers and medicines.

CIVIL ENGINEERING involves the design and development of transportation systems — highways, railroads and airports. This field of engineering also includes the design and construction of water systems, waste disposal systems, marine harbors, pipelines, buildings, dams and bridges, Fig. 1-13.

ELECTRICAL ENGINEERING has two major branches, electrical power and electronics. The electrical power area involves the generation, transmission and utilization of electrical energy. Engineers specializing in this field design major

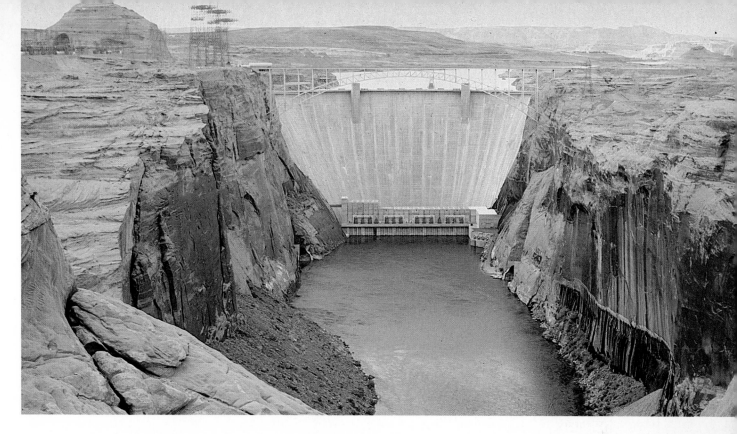

Fig. 1-13. This hydroelectric dam and bridge at Glen Canyon Dam, Page, Arizona was designed in part by civil engineers. (U. S. Department of the Interior)

projects — electrical distribution systems for large buildings or industry. They are also involved in the design of equipment — electrical generators and motors for various purposes.

Engineers working in the area of electronics are concerned with communication systems — radio and television broadcasting. They also work with industrial electronics in designing computer equipment and automated control systems for numerically controlled machines or processing equipment.

INDUSTRIAL ENGINEERING is a new form of engineering concerned with the design, operation and management of systems. Industrial engineers design plant layout and devise improved methods of manufacturing and processing. They also work with quality control, production control and cost analysis. In these activities, they work with engineers in other areas of specialization and with personnel managers in developing and coordinating these systems, Fig. 1-14.

Fig. 1-14. BART, The Bay Area Rapid Transit system operating in the San Francisco Bay area, is the product of many engineering fields. (Bay Area Rapid Transit District)

MECHANICAL ENGINEERING is concerned with the design and development of mechanical devices ranging from extremely small components to large earth moving equipment, Fig. 1-15. Mechanical engineers usually specialize in a given area such as machinery, automobiles, ships, turbines, jet engines or manufacturing facilities.

Fig. 1-16. This gas processing plant was designed and its construction supervised by petroleum engineers. (Cities Service Oil Co.)

Fig. 1-15. Earth moving equipment must be designed and engineered to withstand rough and heavy work. (Hyster Co.)

MINING AND METALLURGICAL ENGINEERING involves the location, extraction and refining of metals. The metallurgical engineer may specialize in one general area of the total field — mining and extraction, refining or welding of metals. His or her responsibilities, for example, could include altering of the structure of a metal through alloying or other processes to produce a metal with certain characteristics to perform a special purpose.

NUCLEAR ENGINEERING is devoted to research, design and development in the field of nuclear energy. This includes the design and operation of electrical generating plants (powered by nuclear energy) which may be used in ships, submarines or locomotives. This new field of engineering offers many challenges in the area of power systems, as well as chemistry, biology and medicine.

PETROLEUM ENGINEERING is concerned with the location and recovery of petroleum resources and the development and transportation of petroleum products, Fig. 1-16. Considerable research has been done on the utilization of petroleum resources, yet much must be done to conserve these resources which are in greater and greater demand each year. New sources of petroleum and gases need to be located while the search for substitute materials continues.

WOMEN IN DRAFTING

More women are employed today in various types of drafting and related careers than ever before. The success of women who have entered drafting, design, architectural and engineering positions has demonstrated their potential in all fields of drafting, Fig. 1-17.

PROBLEMS AND ACTIVITIES

The following problems and activities are designed to help you understand the nature of drafting and its opportunities.
1. Select an object around the home or school, such as the molding around a door or window trim, and describe it using only written words.
2. Make a freehand sketch of the object described above and compare the clarity and ease with which each description was prepared.
3. Review current issues of magazines such as Aztlan, Ebony, Jet, Popular Mechanics or Scientific American and prepare a report for presentation in class on an individual who has distinguished himself or herself in drafting or a related career.
4. From magazines, newspapers and brochures available, find as many different types and uses made of drafting as you can. These may include drawings, sketches, graphs, charts or diagrams in which drafting procedures were used. Mount your collection on notebook paper and be prepared to show them in class. Preserve the collection for later reference.
5. Interview an architect, industrial designer, drafter or

Fig. 1-17. Increasing opportunities are available to women in all phases of drafting.   (International Harvester)

engineer on the nature of the work and its rewards. Find out as much as you can about educational requirements and specialized training needed, the nature of the drafting work, pay compared to other types of work and opportunities for employment and advancement.

6. Interview several persons not directly engaged in technical work. This could include your parents, neighbors and others in the community. Obtain their opinion on the value of drafting to the average citizen. Find out what activities they have done in which a knowledge of drafting was (or would have been) helpful.

7. Bring an actual blueprint (or whiteprint) of a house plan or some manufactured machine part to class. Discuss it with other members of the class.

8. Select a person who has become well known in a technical field related to drafting (your school or public librarian can help you find material). Prepare a report to be made in class on how that person became involved in this work and the contributions he or she has made.

SELECTED ADDITIONAL READING

1. U. S. Bureau of Labor Statistics, Occupational Outlook Handbook, Government Printing Office, Washington, D.C., latest edition.

Careers in Engineering Design Drafting will involve Computer Graphics.   (McDonnell-Douglas, St. Louis)

Chapter 2
INSTRUMENT DRAFTING

The purpose of an engineering or technical drawing is to convey information about a machine part or assembly as clearly and simply as possible. The process of making technical drawings with the use of instruments, templates, scales and other mechanical equipment is called "instrument drafting."

Because of the need for accuracy in design and clarity in communicating necessary information, most drawings used in technical work are instrument drawings, Fig. 2-1. Today, the production and servicing of industrial components requires careful preparation of the drawings, plus systematic checking prior to the start of actual work on the parts described.

The purpose of this chapter is to enable you to gain a knowledge of basic drawing instruments and to instruct you in their proper use and care. After you have become familiar with these basic "tools," you can easily acquire skill with more specialized instruments used in drafting. Some of the instruments needed for technical drawing are shown in Fig. 2-2.

DRAWING EQUIPMENT

Some schools furnish drafting equipment for their students. Others require that students purchase their own. Good instruments are expensive, but it is wise to invest in quality merchandise. Unless you are familiar with drafting equipment,

Fig. 2-1. The modern engineering design drafter makes use of numerous drafting instruments and machines.
(Sperry Flight Systems Div.)

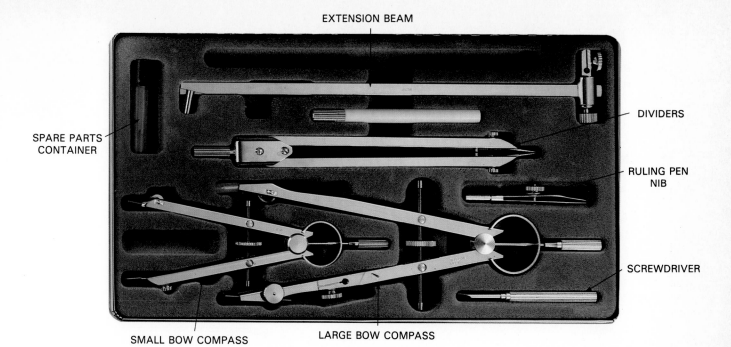

EXTENSION BEAM

SPARE PARTS
CONTAINER

DIVIDERS

RULING PEN
NIB

SCREWDRIVER

SMALL BOW COMPASS

LARGE BOW COMPASS

Fig. 2-2. A basic set of drafting instruments.
(Teledyne Post)

get the advice of your instructor before making a purchase.

The equipment and supplies listed below are adequate for most drafting work:

INSTRUMENTS

Small compass with pen attachment
Large compass with pen attachment
Lengthening bar
Friction divider
Ruling pen
Box of leads
Screwdriver

OTHER EQUIPMENT

Drawing board (approximately 18'' x 24'')
T-square 24'' plastic edge
30-60 deg. triangle — 10''
45 deg. triangle — 8''
Architect's and engineer's scales
Lettering instrument, Ames or Braddock
Protractor
Irregular curve
Circle and ellipse templates
Erasing shield
Soft rubber eraser
Vinyl eraser
Dusting brush
Sketch pad
Drawing paper, tracing paper or vellum
Drafting tape
Drafting pencils or mechanical lead holder
Sandpaper pad, file or pencil pointer

DRAWING BOARDS

Drawing boards usually are made of selected softwoods with a smooth, firm surface and cleated at the ends to prevent warping, Fig. 2-3. Care should be exercised to avoid marring the surface.

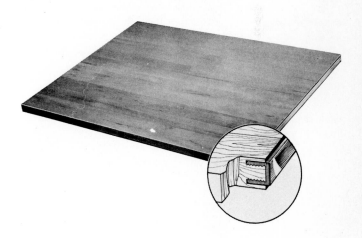

Fig. 2-3. Drawing board with wood and metal cleated ends to restrain warping. (Teledyne Post)

Also available is a vinyl board-top material with excellent resiliency. It mounts over a regular drawing board or drafting table top to improve the drawing surface.

In addition to a smooth surface, the drawing board must have a straight working edge to serve as a base for the T-square and reference line for drawing. Check the working edge for

19

trueness with a framing square or other true straight-line tool, Fig. 2-4. Correct a wooden edge which is not true by planing with a hand plane or jointer.

Most drafting tables come with a drafting board surface, which eliminates the need for a separate board. Generally, these tables are vinyl covered.

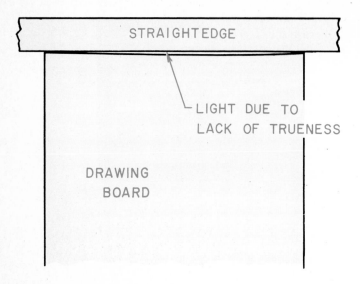

Fig. 2-4. Checking the working edge of a drawing board for trueness.

T-SQUARE

The T-square consists of two parts, a head and a blade fastened together rigidly, Fig. 2-5. Some T-squares have transparent plastic relieved edges that permit lines to be viewed under the edge of the blade. This also avoids the tendency for ink to run under the blade when inking a drawing.

T-squares range in length from 18 to 60 inches in 6 inch intervals. The length should be approximately the same as the width of the drawing board being used.

The three principal uses of the T-square are: (1) to align the paper on the board, (2) serve as a guide in drawing horizontal lines, (3) serve as a base for triangles. In addition, the T-square is used in conjunction with lettering instruments and various drafting templates.

Occasionally check the T-square for rigidity of the head and for straightness of the blade. If the head is loose, remove the screws and clean off the old glue. Add new glue and firmly reset the screws.

To check the straightness of the blade, mark two points on a sheet of paper at a distance equal to the approximate length of the T-square blade, Fig. 2-6. Join these two points with a sharp line along the edge of the T-square.

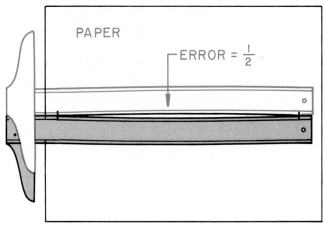

Fig. 2-6. Checking the T-square for trueness.

Next, revolve the paper 180 degrees so that the points are reversed. Then join the points with a sharp line. If the lines coincide, the blade is true. If they do not, the error is one-half the difference between the lines.

This error can be corrected by lightly sanding the blade at the high points. Use a piece of fine abrasive paper wrapped around a block of wood, and check your progress frequently to insure an accurate edge.

Be careful in handling the T-square so that the edge does not strike other instruments or sharp edges that may dent or nick the ruling surface. Never use the T-square as a guide for a knife when cutting material. This could nick the straightedge.

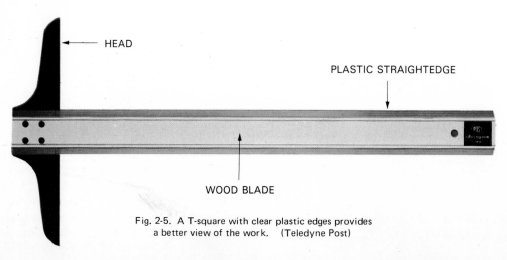

Fig. 2-5. A T-square with clear plastic edges provides a better view of the work. (Teledyne Post)

PARALLEL STRAIGHTEDGE

Drawing boards and tables may be equipped with a parallel-ruling straightedge, Fig. 2-7. The straightedge is preferred for large drawings (lengths over four feet) and for vertical board work. It is easily manipulated and retains its parallel position.

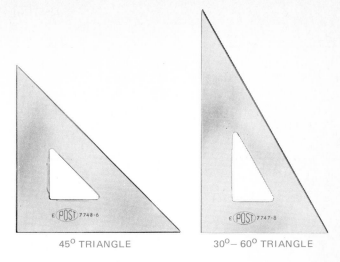

45° TRIANGLE   30° – 60° TRIANGLE

Fig. 2-8. Transparent plastic triangles. (Teledyne Post)

Fig. 2-7. Drafting board equipped with parallel-ruling straightedge.

The straightedge may be moved up or down the board by holding it at any point along its length. It is supported at both ends by a cable which operates over a series of pulleys, keeping the straightedge parallel.

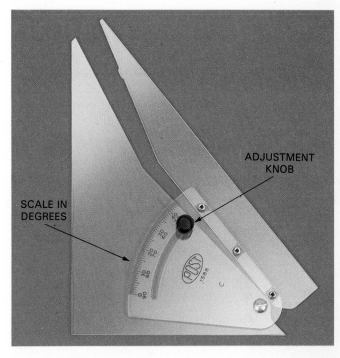

ADJUSTMENT KNOB

SCALE IN DEGREES

Fig. 2-9. Special purpose triangles. Above. An adjustable triangle. Below. Lettering guides with special alignment angles.

TRIANGLES

The most commonly used drafting triangles are 45 degree and 30-60 degree types, Fig. 2-8. They are made of transparent plastic, and their size is designated by the height of the triangle. Most popular sizes are the 8 inch in the 45 degree type and the 10 inch in the 30-60 degree. Both may be purchased in sizes from 4 inches to 18 inches.

Special purpose triangles are made for drawing guide lines for lettering, laying out roof pitches and intricate angular work, Fig. 2-9. An adjustable triangle permits the laying off of any angle from 0 degrees to 90 degrees.

Occasionally check all triangles for nicks by running your fingernail along the edges. Such defects are caused by hitting the triangles against sharp edges, dropping them, or using them as a guide for a knife in cutting or trimming.

Minor defects, nicks or misalignment can be corrected by sanding the edge with fine abrasive paper wrapped around a block of wood.

Test triangles for accuracy by drawing a vertical line;

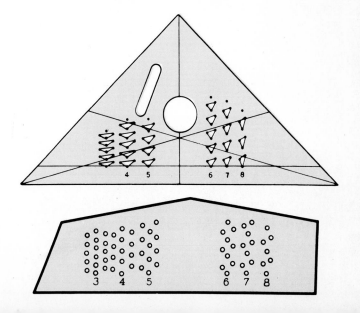

reverse the triangle and draw a second line, starting at the same point, Fig. 2-10. If the lines coincide, the triangles are true. If not, the error is equal to one-half the distance between the lines.

Fig. 2-10. Checking triangles for trueness.

DRAFTING MACHINES

Manual drafting machines have become very popular in industry as well as in many school drafting rooms, Fig. 2-11. The manual drafting machine combines the functions of the T-square, straightedge, triangles, protractor and scales into one machine.

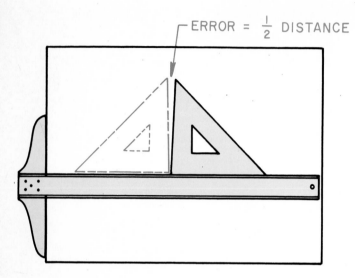

Fig. 2-11. A typical industrial track-type drafting machine. (VEMCO)

There are two types of manual drafting machines in use — the X-Y coordinate or track-type, Fig. 2-11, and the arm-type, Fig. 2-12. The arm-type is the least expensive. The track-type has the advantage of being more versatile and less troublesome in maintaining accuracy.

Fig. 2-12. An arm-type drafting machine in use by a design drafter. (Keuffel & Esser Co.)

Essentially, basic operating principles are the same for the two types of drafting machines. This is true for the various makes of machines, although control mechanisms may vary with each. If an instruction manual is available, review it to become familiar with the controls and adjustments of the machine.

Alignment of the drafting machine should be checked periodically by comparing results with previously drawn horizontal and vertical lines. When differences occur, check board mounting clamps, scale chuck clamps and base-line protractor clamp. If these are in order, check the maintenance manual for the machine or with your instructor.

When using the drafting machine — regardless of make — observe the following procedures:

BASE-LINE ALIGNMENT

To establish a base line in the desired position:

1. Set protractor indexing mechanism at "zero," then release wing nut or lever securing protractor, Fig. 2-13.

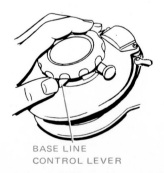

BASE LINE
CONTROL LEVER

Fig. 2-13. Release the base line control lever to align base line.

2. Adjust horizontal scale to desired base line. (This may be horizontal or any other angle for necessity or comfort.)

3. Tighten protractor wing nut or screw to "fix" zero setting.

VERTICAL SCALE ALIGNMENT

To align the vertical scale at 90 degrees from the horizontal:

1. Make certain screws holding scale chucks are tight, then firmly insert scales in base plate.

2. With protractor head at zero, draw a reference line along edge of horizontal scale, Fig. 2-14. Hold scale with one hand so pencil pressure does not deflect scale at free end.

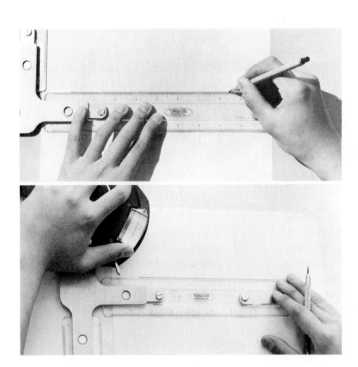

Fig. 2-14. Aligning the vertical scale. Above. Drawing reference line with head set at zero. Below. Checking alignment with head set at 90 degrees.

3. Index protractor head 90 degrees clockwise to check alignment along reference line, Fig. 2-14.

4. If necessary, align scale by releasing scale chuck clamping screws, adjusting scale and tightening screws.

READING THE VERNIER SCALE

To read or set an angle on the vernier scale:

1. Read number of full degrees vernier zero line shows on protractor. In Fig. 2-15, reading is 12 degrees plus a fraction more counterclockwise from zero.

2. Locate line on counterclockwise side of vernier that aligns exactly with a line on protractor. In Fig. 2-15, 15 minute line aligns.

3. Reading of this angle, then, is 12 degrees 15 minutes.

4. To read a negative angle (below protractor zero), read number of full degrees between zero on the protractor and zero on the vernier. Then locate line clockwise from zero on vernier that exactly aligns with a line on the protractor. Note that this is 5 degrees 35 minutes in Fig. 2-16.

5. To set scales for a certain angle, release protractor clamp, revolve scales until protractor head and vernier align with required number of degrees and minutes.

Fig. 2-15. Vernier scale reading in degrees and minutes above horizontal (counterclockwise).

Fig. 2-16. Vernier scale reading on negative (below horizontal or clockwise) side of protractor zero.

DRAFTING MEDIA

There are several types of drafting media (materials) available for use in the preparation of drawings or tracings. These may be classified in three groups — paper, cloth and film. They differ in qualities of strength, translucency, erasability, permanence and stability. The characteristics of each is covered here to assist you in selecting the proper material for a particular drawing.

ENGLISH SYSTEM

METRIC SYSTEM

8 1/2 x 11 MULTIPLES		9 x 12 MULTIPLES	
LETTER DESIGNATION	SHEET SIZE	LETTER DESIGNATION	SHEET SIZE
A	8 1/2 x 11	A	9 x 12
B	11 x 17	B	12 x 18
C	17 x 22	C	18 x 24
D	22 x 34	D	24 x 36
E	34 x 44	E	36 x 48

(a)

(b)

DESIGNATION	MILLIMETERS	INCHES
A4	210 x 297	8.27 x 11.69
A3	297 x 420	11.69 x 16.54
A2	420 x 594	16.54 x 23.39
A1	594 x 841	23.39 x 33.11
A0	841 x 1189	33.11 x 46.81

(c)

Fig. 2-17. Standard flat sheet sizes.

PAPER

Industrial drafting practices have changed in recent years. Now the use of opaque drawing papers is limited to high grade permanent papers for the preparation of maps and master drawings intended to be photographed. Less expensive papers are used for beginning drafting classes in schools. Even here, however, the tendency is to use inexpensive types of translucent materials.

Opaque drawing papers are available in cream, light green, blue tint and white. White papers are preferable for drawings to be photographed later. Light colored papers reduce eye strain and are less likely to soil.

Media size is important too. Because of their widespread use in industry, two series of drawing sheet sizes are recognized as standard. The sizes are 8 1/2 x 11 and 9 x 12 inches and multiples of each, designated by letter sizes, Fig. 2-17 (a) and (b). Drawing sheet sizes in the metric system are shown at (c).

TRACING PAPER

Tracing paper is a translucent drawing paper. Its name was derived from the practice of first making a drawing in pencil on opaque paper, then "tracing" it in ink on an overlay sheet of translucent paper.

Today, however, the practice in industry is to develop the master drawing in pencil directly on tracing paper from which reproductions can be made (eliminating time and expense of preparing a "tracing"). Inking is reserved primarily for permanent drawings or for photographic reproduction work.

Tracing papers (as differentiated from vellums described below) are natural papers which have no transparentizing agents added. Natural papers made fairly strong and durable are not very transparent. Papers with high transparency are only moderately durable.

VELLUM

Vellums are referred to as "transparentized" or "prepared" tracing papers. They provide strength, transparency, durability (handling and folding) and erasability without "ghosting" (erased lines showing through).

Vellum papers are made of 100 percent rag content and impregnated with a synthetic resin to provide high transparency. They are available in white or blue tint and are highly resistive to discoloration and brittleness due to age. The working qualities of vellum make it the most commonly used in industry even though it is more expensive than tracing paper.

TRACING CLOTH

Tracing cloth is considerably more expensive than paper media. However, tracings are made on cloth when greater transparency and permanence are desired.

Tracing cloth is a finely woven cotton fiber material that has been sized with starch to provide a surface that takes pencil and ink. It comes in white for pencil tracings and blue tint for ink. The working side is dull or frosted. You can erase pencil and ink lines without damaging the surface by using a soft rubber or vinyl eraser.

Tracing cloth is subject to expansion and shrinkage due to changes in moisture content of the air (unstable). Therefore, one section or part of a drawing should be completed at a time.

POLYESTER FILM

The latest development in the drafting media field is polyester film. With the exception of cost, it has the best qualifications for a drawing medium. It provides dimensional stability, great resistance to tearing, easy erasing (with soft eraser or erasing fluid) and high transparency. It is waterproof and will not discolor or become brittle with age.

Polyester film has an excellent working surface for pencil, ink or typewriter. Many industries feel that the added cost of this medium is offset by its advantages, plus the ease with which changes can be made on the tracing from time to time.

FASTENING DRAWING SHEET TO BOARD

The drawing sheet should be positioned on the drawing board about 2 inches from the working edge (left edge for right handers and right edge for left handers). It should be placed up from the bottom edge approximately 4 to 6 inches to allow for the manipulation of the T-square and triangles.

Next, place the T-square near the bottom of the board and position the drawing sheet on its blade near the working edge, Fig. 2-18. Fasten the upper corners of the sheet with drafting tape and remove the T-square. Smooth the sheet out to the corners and fasten the lower corners.

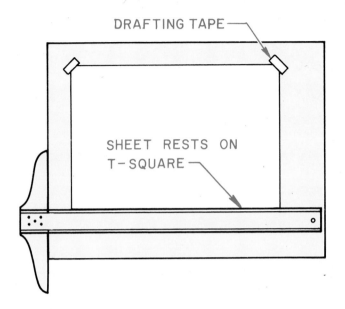

Fig. 2-18. Fastening the drawing sheet to the board.

Transparent cellophane tape should not be used to attach the drawing sheet. It frequently tears the sheet upon removal.

When a drafting machine is being used, the drawing sheet may be placed at a slight angle. This provides a more "natural" drawing and lettering position than the horizontal position necessary when using a T-square or a parallel rule.

DRAWING PENCILS

Drawing pencils are manufactured in a variety of types, Fig. 2-19:
(a) Standard wood-case.
(b) Rectangular leads for drawing long lines with a wedge point.
(c) Mechanical lead holders in standard lead sizes.
(d) Fractional millimeter lead sizes called "thin leads."

Although more expensive, refill pencils are convenient to use, remain the same length and save the time required in sharpening wood-case pencils.

Leads used in drawing pencils are manufactured by a special process designed to make them strong and capable of producing sharp, even density lines. Drawing leads are graded in eighteen degress of hardness from 7B (very soft) to 9H (very hard).

The softer grades of pencil leads (2H, 3H and 4H) deposit more graphite on the paper and produce more opaque lines. However, many drafters continue to use the harder grades because they produce sharper lines and do not smudge as readily during the drafting process.

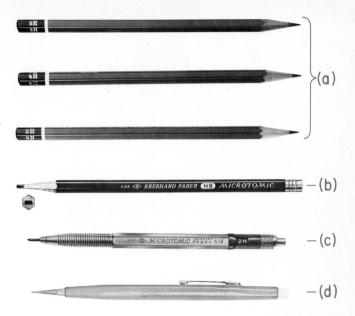

Fig. 2-19. Wood-case drafting pencils and mechanical lead holders.

Special pencil leads with a plastic base are manufactured for use on polyester drafting films. They come in five grades of hardness from K1 (very soft) to K5 (very hard).

SHARPENING THE PENCIL

When sharpening wood-case pencils, first remove enough wood to expose 3/8 inch of lead on the end opposite the grade marking, Fig. 2-20 (a). Use a knife or a drafter's pencil sharpener with special cutters which will remove wood only, as

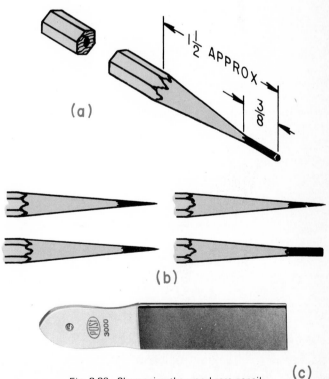

Fig. 2-20. Sharpening the wood-case pencil.

Fig. 2-21. Left. A drafter's pencil sharpener. Fig. 2-22. Right. Pencil leads can be shaped to a conical point in this lead pointer. (Teledyne Post)

shown in Fig. 2-21.

If a knife is used, exercise care to prevent nicking the lead, causing it to break under pressure of use. Lead in standard size lead holders should be extended slightly beyond normal use position for pointing. Thin leads do not require pointing.

Two types of points are used on drafting pencils — conical and wedge, Fig. 2-20 (b). The conical point is used for general line work and lettering. It is shaped in a lead pointer, Fig.

2-22, on a sandpaper pad or file, Fig. 2-20 (c).

For final shaping, finish the point on a piece of scrap paper. Remove all excess graphite dust from the pencil point by wiping it on a felt pad or soft cloth.

The wedge point is used for drawing long straight lines because it holds its point (edge) longer than the conical point. Shape the point on a sandpaper pad or file by dressing the two sides to produce a sharp edge. Finish on scrap paper and

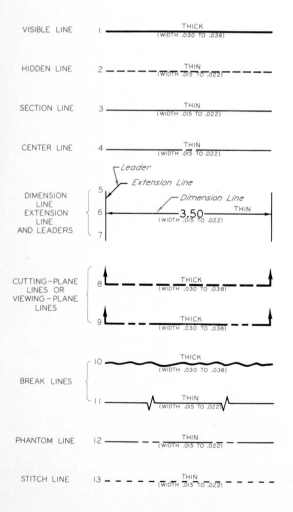

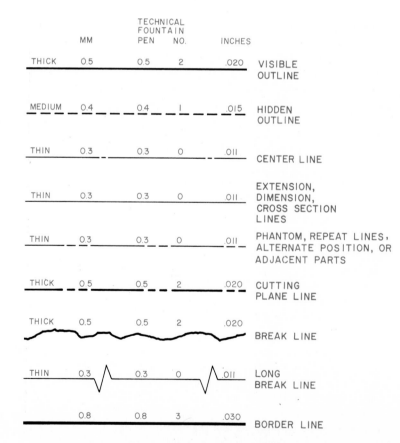

Fig. 2-23. Left. Alphabet of lines using two line weights and recommended for drawings to be reproduced by available photographic methods. (ANSI) Above. Alphabet of lines using three line weights.

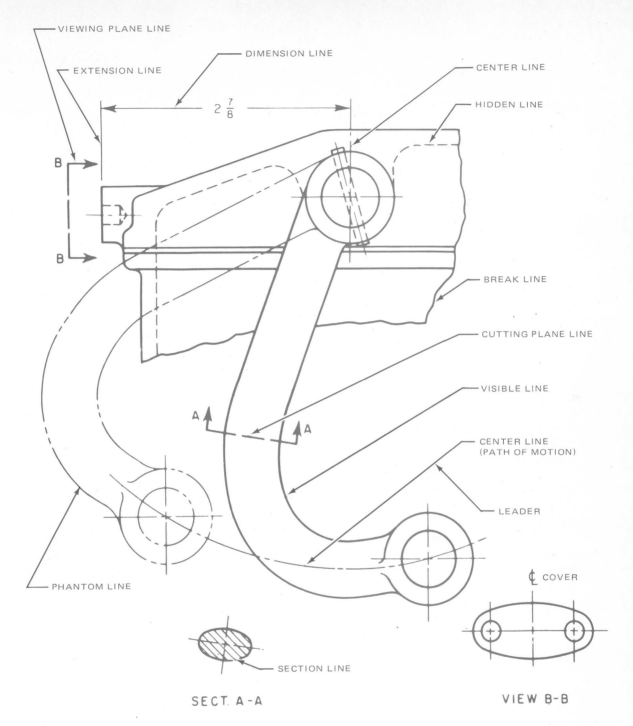

VIEWING PLANE LINE

EXTENSION LINE

DIMENSION LINE

CENTER LINE

HIDDEN LINE

$2\frac{7}{8}$

B

B

BREAK LINE

CUTTING PLANE LINE

VISIBLE LINE

CENTER LINE
(PATH OF MOTION)

A

A

LEADER

PHANTOM LINE

⊄ COVER

SECTION LINE

SECT. A-A

VIEW B-B

Fig. 2-24. Application of line symbols (ANSI).

remove the excess graphite dust.

To maintain a neat work area and produce clean drawings, do not sharpen the pencil over your drawing or instruments. Before storing, remove the graphite dust from the sandpaper pad or file by tapping it lightly against the inside of a waste basket.

ALPHABET OF LINES

The American Society of Mechanical Engineers (ASME) has developed a standard for lines which is accepted throughout the industry. Known as the Alphabet of Lines, Fig. 2-23, this standard is designed to give life and meaning to a drawing.

The Alphabet of Lines reveals shape, size, hidden surfaces, interior detail, alternate positions of parts, etc. Lines shown in Fig. 2-23 should be studied carefully, and each drawing produced should conform to the standard.

Note that the lines differ in width (sometimes referred to as thickness or weight). They also differ in character so that each is easily distinguishable from another and each conveys a particular meaning on the drawing. The use of these lines on a drawing is illustrated in Fig. 2-24.

Different widths of lines are used on drawings, as shown in Fig. 2-23. All lines should be dense black, regardless of width, as thin pencil lines tend to "burn out" during the reproduction process of making prints. Any variation in lines should be in width and character only. Pencil lines are likely to be slightly thinner than corresponding ink lines. However, pencil lines should be as thick as practical to provide acceptable reproductions.

VISIBLE LINES

Visible lines are used to outline the visible edges or contours of the object that can be seen by an observer. Visible object lines should stand out sharply when contrasted with other lines on the drawing.

HIDDEN LINES

Hidden lines indicate edges, surfaces and corners of an object that are concealed from the view of the observer. They are thin or medium lines made up of short dashes, evenly spaced. The dashes are approximately 1/8 inch long and the spaces approximately 1/32 inch. They may vary sightly with the size of a drawing.

Hidden lines start and end with a dash. They make contact with visible or hidden lines from which they start or end, Fig. 2-25 (a). If the hidden line is a continuation of a visible line,

then a gap is shown, (b). A gap is also shown when a hidden line crosses but does not intersect another line, (c). Dashes should join at corners, (d), and arcs start with dashes at tangent points, (e).

Hidden lines should be omitted wherever they are not needed for clarity. However, until you gain experience in drafting, it is well that you include all required lines unless otherwise directed by your instructor.

SECTION LINES

Sections lines are sometimes called "cross-hatching." These lines represent surfaces exposed by a cutting plane (an invisible plane passing through an object). Section lines, Fig. 2-26, are usually drawn at an angle of 45 degrees with a sharp 2H pencil. Draw the lines dark and thin to contrast with the heavier object lines. On average size drawings, space the lines by eye about 1/16 inch apart (small drawings, 1/32 inch; large drawings, 1/8 inch). Spacing of section lines should be uniform.

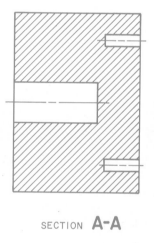

SECTION **A-A**

Fig. 2-26. Section lines are uniformly spaced, thin, dark lines.

The Ames lettering instrument shown in Fig. 4-8 can be used along your triangle as an aid in obtaining uniform section lines. The section lining shown in Fig. 2-26 is for the cast iron and malleable iron, and it is recommended for general sectioning of all materials on detail and assembly drawings. The section lining for other materials is shown on page 182.

CENTER LINES

Center lines are thin lines composed of long and short dashes alternately spaced with a long dash at each end. They indicate axes of symmetrical parts, circles and paths of motion, Fig. 2-24. Depending on the size of the drawing, the long dashes vary from approximately 3/4 to 1 1/2 inches (or longer). The short dash is approximately 1/16 to 1/8 inch in length. Center lines should intersect at the short dashes if possible. They should extend only a short distance beyond the object of the drawing unless needed for dimensioning or other functions.

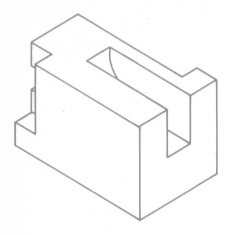

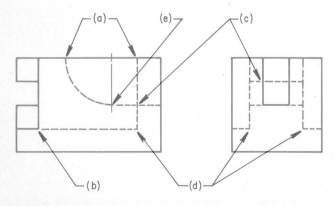

Fig. 2-25. Method of representation for the use of hidden lines.

DIMENSION LINES, EXTENSION LINES AND LEADERS

Dimension lines indicate the extent and direction of dimensions and are terminated by arrowheads, Fig. 2-24. Extension lines indicate the termination of a dimension. The extension line extends from approximately 1/16 inch from the object to 1/8 inch beyond the last arrowhead. Leaders are drawn to notes or identification symbols used on the drawing.

CUTTING PLANE LINES

Cutting plane lines indicate the location of the edge view of the cutting plane, Fig. 2-24. Two forms of lines are approved for general use. The first is composed of alternating long dashes (approximately 3/4 to 1 1/2 inches or longer, depending on drawing size) and pairs of short dashes (approximately 1/8 inch with 1/16 inch spaces).

The second form is composed of equal dashes, approximately 1/4 inch (or more) in length. This form contrasts well and is quite effective on complicated drawings. Both forms have ends bent at 90 degrees and terminated by arrowheads to indicate the direction of viewing the section. If the heavyweight line will obscure some of the detail, portions of the line overlaying the object may be omitted.

BREAK LINES

Break lines are used to limit a partial view of a broken section. For short breaks, a thick line is drawn freehand, Fig. 2-23. A long break line is drawn with long, thin, ruled dashes joined by freehand "zig-zags." It is used for long breaks, particularly in structural drawing.

PHANTOM LINES

Phantom lines show alternate positions, repeated details and paths of motion, Fig. 2-24. They consist of thin, long dashes, approximately 3/4 to 1 1/2 inches in length, alternated with pairs of short dashes, 1/8 inch in length with 1/16 inch spaces.

DATUM DIMENSIONS

Datums are lines, points and surfaces that are assumed to be accurate. They are placed on drawings as datum dimensions since they may be used for exact reference and location purposes. The word datum is used in the dimension line for such revealing information. These are lightweight lines with the same characteristics as phantom lines.

CONSTRUCTION LINES

Construction lines are very light, gray lines used to lay out all work. They should be light enough on a drawing so they will not reproduce when making a print. On drawings for display or photoreproduction, they should not be visible beyond an arm's length.

BORDER LINES

Border lines, while actually not a part of the standard, are used as a "frame" for the drawing and are the heaviest of all lines, Fig. 2-23.

ERASING AND ERASING TOOLS

It would be desirable to make a drawing without erasing. However, mistakes do happen and changes in existing drawings must frequently be made.

Erasing is a technique which the drafter must perfect if he or she wants to do good work. Much erasing time and damage to drawings may be saved by drawing all lines first as construction lines. They can be "heavied-in" for final finish.

Two types of erasers are useful in drafting: the firm textured rubber eraser for erasing ink lines, and the soft vinyl eraser for erasing pencil lines and cleaning drawings, Fig. 2-27. Erasers containing gritty abrasives (such as ink erasers) should not be used. These will damage the drawing surface and produce "ghosting" on reproductions.

Fig. 2-27. The red rubber and soft vinyl erasers are recommended for drafting erasers. (Eugene Dietzgen Co.)

Steel erasing knives should not be used for general line erasing. They are useful in removing ink lines that have overrun, are too wide, or have been made by error, Fig. 2-28. They must be kept sharp. Use them with a light sideways stroke. Special nonabrasive vinyl erasers and erasing fluids are used on tracing cloth and polyester film.

Fig. 2-28. Steel erasing knife for removing overruns in inking. (Keuffel & Esser Co.)

ERASING PROCEDURE

When making erasures, follow this procedure:
1. Clean eraser by rubbing it on a scrap of paper.
2. With your free hand, hold drawing securely to avoid wrinkling.
3. Rub soft vinyl eraser lightly back and forth to erase detail or line.
4. For erasing deeply grooved pencil or ink lines, place a triangle under the paper for backing.
5. If necessary to protect details close by, use an erasing shield, Fig. 2-29. (A piece of stiff paper will serve if an erasing shield is not available.)

Fig. 2-29. Using an erasing shield to protect parts of a drawing which are not to be removed.

6. Clean drawing with vinyl eraser before final finishing of lines.
7. Remove erasure dust with dust brush, Fig. 2-30, or soft cloth.

Fig. 2-30. Frequent use of the dust brush will help keep your drawing clean.

8. After cleaning front of drawing, turn drawing over and inspect back for dirt that may have been transferred from the drawing board to the back of the drawing.

ELECTRIC ERASERS

Most industries today, because of frequent alterations in product design, find it necessary to make changes on existing drawings. Improvements in drafting media, as well as the electric erasing machine, Fig. 2-31, make changes a simple matter. Exercise care in using the electric erasing machine not to press too hard or to remain in one spot too long. This could mar or distort the drawing surface or cause "ghosting" in reproduction of prints.

Only soft-rubber or vinyl erasers should be used in electric erasing machines. A very gentle pressure avoids overheating the drawing surface. Use of a piece of thin gage copper, brass or aluminum sheet under the area to be erased will dissipate the heat and reduce the possibility of damage to the drawing.

Fig. 2-31. The electric eraser is a time saver for drafters. (Teledyne Post)

NEATNESS IN DRAFTING

The first impression is a lasting one. Remember that the appearance of your drawing is the first reflection of your ability as a drafter. People have a tendency, and rightly so, to associate NEATNESS and ABILITY in drafting.

Practice cleanliness from the start, and make it a habit to guard against soiling a drawing in any way. The primary source of dirt on a drawing is pencil graphite smeared from lines by the sliding of T-square, triangles, shirt sleeves and hands across the drawing.

Another practice that detracts from the neatness of a drawing is making lines too dark and heavy at the start and later discovering that changes need to be made. The line and its deep "groove" are hard to remove and will appear as a "ghost" line on the reproduction.

The following suggestions will help you keep a drawing clean:

1. Wash your hands before starting to draw and occasionally during the drawing period if your hands tend to be oily. Since you are continually working over the drawing, clean hands will do much to keep the drawing clean.

2. Always wipe the dust and dirt from your instruments with a soft cloth before starting to draw and frequently during use. A thorough cleaning occasionally with a soft eraser or erasing solvent keeps instruments in good condition.

3. Lay out all views with light lines using a hard pencil. "Heavy-in" lines only when you are sure all parts are correct.

4. Remove dust as soon as it collects. After each line is drawn, blow particles of loose graphite from the sheet. Remove erasure dust immediately with a soft cloth or dusting brush.

5. Do not slide instruments across drawing. Tilt the T-square by pressing down on the head before sliding. Lift the straightedge and triangles to prevent the smear of graphite dust from lines already drawn.

6. Sharpen pencil away from drawing and enclose the sandpaper pad or file in an envelope before storing in the drawer with drawings or instruments.

7. Keep an orderly drawing area. Have only the tools and equipment needed on top of the desk. This will prevent crowding and avoid the possibility of instruments falling on the floor, Fig. 2-32.

8. Use a paper overlay when lettering or working over parts of a drawing already completed.

9. Cover drawing at night or store it in a drawer to prevent dust from gathering.

10. Store completed drawings in a portfolio to prevent damage from sliding in the drawer.

Fig. 2-32. An orderly work area will contribute to cleaner drawings and better reproductions. (Fisher Body Division, GMC)

SCALES

The scale is one of the most frequently used drawing instruments. In addition to laying off measurements, it is the device used to reduce or enlarge a drawing of an object to a suitable size.

A standard scale is made in one of two basic shapes, flat or triangular. The flat scale is available in three bevel shapes, regular two-bevel, opposite two-bevel and four-bevel, Fig. 2-33. The triangular scale is available in two styles, regular and concave, Fig. 2-33.

Fig. 2-33. Shapes of standard drafting scales.

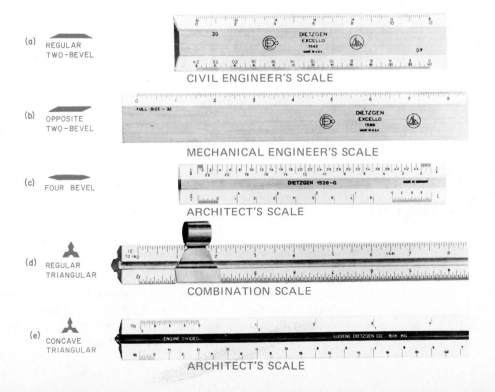

(a) REGULAR TWO-BEVEL

CIVIL ENGINEER'S SCALE

(b) OPPOSITE TWO-BEVEL

MECHANICAL ENGINEER'S SCALE

(c) FOUR BEVEL

ARCHITECT'S SCALE

(d) REGULAR TRIANGULAR

COMBINATION SCALE

(e) CONCAVE TRIANGULAR

ARCHITECT'S SCALE

The triangular scale has the advantage of having more scales available on one instrument (6 to 11). A special guard aids in keeping proper scale in position, Fig. 2-33 (d). Many drafters prefer a flat-type scale because it is easier to manipulate.

Scales are made of boxwood, boxwood with plastic faces, all plastic and aluminum. They may have engine (machine) divided, precision molded or die-engraved graduations. The better scales are engine divided. Most are made of boxwood with white plastic overlayed on the faces to make the divisions easy to read, Fig. 2-33 (a) and (b).

The divisions on scales are either "open divided" or "fully divided."

Open divided means that the main units are numbered along the entire length of each scale. Also, only the first major unit is subdivided into fractional or decimal segments of the major unit, Fig. 2-33 (c) and (e).

Some open divided scales have two compatible scales on the same face reading from opposite ends. The larger of the two scales is twice the size of the smaller, Fig. 2-33 (c).

The fully divided scale has all units along the entire scale subdivided, Fig. 2-33 (a), (b) and (d). This has the advantage of allowing the drafter to lay off several values from the same origin without resetting.

A large variety of scales is required to draw objects ranging from small machine parts to large area maps. Therefore, scales are classified according to their use. The following are the most common:

ARCHITECT'S SCALE

Architect's scales, available in all five shapes shown in Fig. 2-33, are most commonly used in making drawings for the building and structural industry. They are also used by many mechanical drafters since the major units are divided into feet and inches.

The various scales usually represented on an architects's triangular scale are shown in Fig. 2-34. These scales are arranged in pairs on each of 5 faces, plus a full size scale marked "16" to indicate the inch is subdivided into 16ths on the sixth face.

Architect's scales may be purchased with scales 2 and 4 inches to the foot. In this case the 3/32 and 3/16 inch scales are omitted from the triangular scale.

Architect's scales are open divided (except on full size scale). The end unit beyond the zero is subdivided into 12 parts representing inches of the foot. On the two smaller scales, however, the subdivisions represent 2 inches. On the four larger scales, the finest division represents fractional parts of an inch.

CIVIL ENGINEER'S OR DECIMAL SCALE

The civil engineer's scale is commonly referred to as the decimal scale because of its increasing use in manufacturing and other industries. This fully divided scale has long been used in civil engineering where large reductions are required for drawings such as maps and charts.

Scales on a civil engineer's scale are divided into units representing decimal parts of the inch. Each one inch unit on the 10 scale is subdivided into 10 parts, each .10 inch in size. The 50 scale has 50 parts to the inch, each 1/50 or .02 inch, Fig. 2-35.

The divisions may represent feet, pounds, bushels, time or any other quantity. The units may be expanded to represent any proportional number (for example, 50 scale could represent 50 feet, 500 feet, 5000 feet, etc.). The subdivisions of the major units would have corresponding values.

MECHANICAL ENGINEER'S SCALE

The mechanical engineer's scale is useful in drawing machine parts where the dimensions are in inches or fractional parts of an inch. Common graduations for mechanical engineer's scales represent one inch and are shown in Fig. 2-36.

1″	= 1″	(full size)
3″	= 1'–0″	(quarter size)
1 1/2″	= 1'–0″	(eighth size)
1″	= 1'–0″	(twelfth size)
3/4″	= 1'–0″	(1/16 size)
1/2″	= 1'–0″	(1/24 size)
3/8″	= 1'–0″	(1/32 size)
1/4″	= 1'–0″	(1/48 size)
3/16″	= 1'–0″	(1/64 size)
1/8″	= 1'–0″	(1/96 size)
3/32″	= 1'–0″	(1/128 size)

Fig. 2-34. Scales usually included on the architect's scale.

1″	= 1″	(full size)
1/2″	= 1″	(half size)
1/4″	= 1″	(quarter size)
1/8″	= 1″	(eighth size)

Fig. 2-36. Mechanical Engineer's Scales.

Fig. 2-35. A decimal scale with the 10 and 50 scales shown.
(Keuffel & Esser Co.)

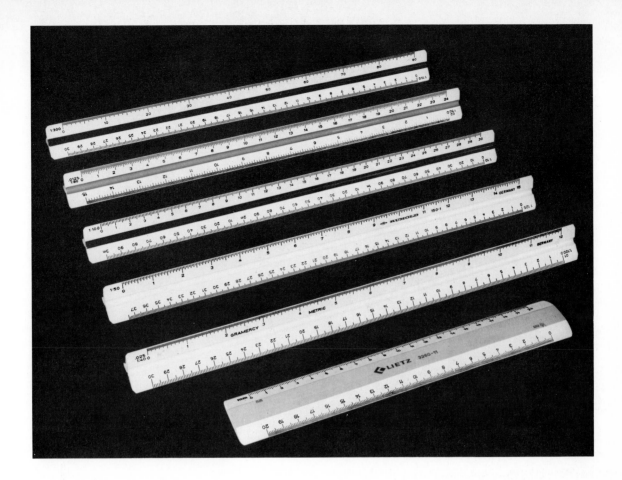

Fig. 2-37. Some examples of metric scales.

These scales are often called the "size" scales. For example, a size scale of "1/8" represents a drawing one eighth the size of the object. The "1" represents the full-size scale.

COMBINATION SCALES

A triangular scale combining selected scales from the architect's (1/8, 1/4, 1/2, 1 in. to the foot), civil engineer's (50 parts to the inch), and mechanical engineer's scales (1/4, 1/2, 3/8, 3/4 and full size) is available, Fig. 2-33 (d). Special scales are also available for such uses as map, aerial photography and statistical work.

METRIC SCALES

Metric scales have a variety of shapes and sizes, Fig. 2-37. Metric scales are "size" scales. For example, 1:20 means the drawing is 1/20th the size of the actual object.

A scale of 1:20 may be referenced on the drawing as 1:20 or 5 cm = 1 m, Fig. 2-38 (a). A scale of 1:100 means the drawing is 1/100th the size of the actual object and may be referenced as 1:100 or as 1 cm = 1 m, (b). Fig. 2-37 shows the 1:100 scale. When this scale is used as a reduction scale of 1:100, the 1, 2, 3 . . . represent 1 meter (1 cm = 100 cm or 1 meter). The 1:100 scale may also be used as a full-size scale since the smallest division is actually 1 millimeter (mm). The 1, 2, 3 . . . are actually full-size centimeters (10 millimeters).

Since the metric scales are decimal scales (base-ten number system) their ratios may be changed by multiplying (or dividing) by a multiple of 10. For example, the 1:100 scale may be changed to 1:1000 by multiplying by 10 — each of the numerals, 1, 2, 3 . . . multiplied by 10 — now represent 10 m, 20 m, 30 m . . .

Most industries using metrics have metric scales in the sizes normally used so that multiplying or dividing will not be necessary to change scales.

The metric system of measurement originated in France in 1793 and has gradually been adopted by most countries throughout the world. In the United States a sizable number of industries are going to full metric. Every industry will one day go to metric or find the economics of remaining outside prohibitive. Some industries will use an interim system called "dual dimensioning" (decimal inch and metric) which is discussed in Chapter 8.

(a) SCALE 1:20 (5 cm = 1 m)

(b) SCALE 1:100 (1 cm = 1 m)

(c) SCALE 1:100,000 (1 cm = 1 km)

Fig. 2-38. Metric scales are referenced on a drawing to indicate the units of the ratio.

The metric system was founded on a measurement unit from nature, one/ten millionth of the distance from the North Pole to the equator. The unit was called "meter." More

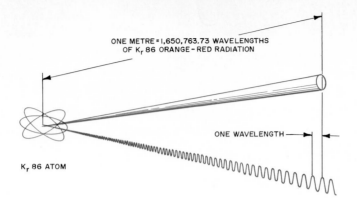

ONE METRE = 1,650,763.73 WAVELENGTHS
OF Kᵣ 86 ORANGE - RED RADIATION

ONE WAVELENGTH

Kᵣ 86 ATOM

Fig. 2-39. The length of the metre scientifically and accurately established.

accurate scientific studies in 1961 established the length of the meter as equal to 1,650,763.73 wavelengths of the orange-red light given off by krypton 86, Fig. 2-39. This determination is accurate to 1 part in 100,000,000 and can be reproduced in scientific laboratories throughout the world.

INTERNATIONAL SYSTEM OF UNITS (SI)

In 1954, and again in 1960, the metric system and its units were redefined in order to correct some deficiencies that had developed over the years. The 1960 General Conference on Weights and Measures formally gave the revised metric system its new title "Systeme International d' Unites" with the universal abbreviation "SI Units."

SI units of measure in metric system conform to reason, are consistent and fit together. The authorized translation from French definition of the seven basic and two supplementary units of the International System are shown in Fig. 2-40.

The metric system of numbers works in the same way as the decimal-inch system. Only the size of the units and terms vary. Both are on the base-ten number system which makes it easy to shift from one multiple or submultiple to another.

Each of the powers of 10 in the metric system is given a special name, Fig. 2-40. Those multiple and submultiple factors from power 10^3 to 10^{-6} set off in red, Fig. 2-41, are the units most commonly used.

MULTIPLICATION FACTORS	NAME	SI SYMBOL
1 000 000 000 000 = 10^{12}	terameter	Tm
1 000 000 000 = 10^9	gigameter	Gm
1 000 000 = 10^6	megameter	Mm
1 000 = 10^3	kilometer	km
100 = 10^2	hectometer	hm
10 = 10^1	dekameter	dam
1 = 10^0	meter	m
0.1 = 10^{-1}	decimeter	dm
0.01 = 10^{-2}	centimeter	cm
0.001 = 10^{-3}	millimeter	mm
0.000 001 = 10^{-6}	micrometer	μm
0.000 000 001 = 10^{-9}	nanometer	nm
0.000 000 000 001 = 10^{-12}	picometer	pm
0.000 000 000 000 001 = 10^{-15}	femtometer	fm
0.000 000 000 000 000 001 = 10^{-18}	attometer	am

Fig. 2-41. Metric and SI multiple and submultiple units.

Any drafter using the metric system of linear measure should learn all common units and their relation to other factors, Fig. 2-42. Actually, changing from kilometers to meters is a simple shift of the decimal point, Fig. 2-43.

Kilometer	=	1 000	meters (thousands)
Hectometer	=	100	meters (hundreds)
Dekameter	=	10	meters (tens)
Meter	=	1	meter (unit of linear measure)
Decimeter	=	0.1	meter (tenths)
Centimeter	=	0.01	meter (hundredths)
Millimeter	=	0.001	meter (thousandths)
Micrometer	=	0.000 001	meter (millionths)

Fig. 2-42. Common linear units in the metric system.

BASIC SI UNITS

QUANTITY	NAME OF UNIT	SYMBOL
Length	meter	m
Mass (weight)	kilogram	kg
Time	second	s
Electric current	ampere	A
Temperature	kelvin	K
Luminous intensity	candela	cd
Amount of substance	mole	mol

SUPPLEMENTARY UNITS

UNIT	NAME OF UNIT	SYMBOL
Plane angle	radian	rad
Solid angle	steradian	sr

Fig.2-40. The seven basic and two supplementary units of the International Systems of Units.

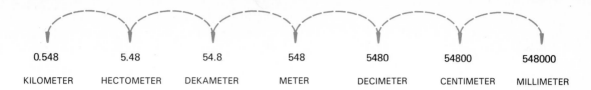

0.548	5.48	54.8	548	5480	54800	548000
KILOMETER	HECTOMETER	DEKAMETER	METER	DECIMETER	CENTIMETER	MILLIMETER

Fig. 2-43. Numbers in the metric system are added, subtracted, multiplied and divided in the same way as decimal-inch numbers.

METRIC UNITS USED ON INDUSTRIAL DRAWINGS

The basic unit of the metric system is the meter. However, the following derived-decimal units are used in various industrial fields as the official unit of measure on metric drawings, Fig. 2-44. On drawings in the topographical field where the distance is great, the kilometer is used and a sizable scale reduction results.

INDUSTRY	UNIT OF MEASURE	INTERNATIONAL SYMBOL	MULTIPLE FACTOR
Topographical	kilometer	km	$10^3 = 1\,000$ m
Building, Construction	meter	m	$10^0 = 1$ m
Lumber, Cabinet	centimeter	cm	$10^{-2} = 0.01$ m
Mechanical Design, Manufacturing	millimeter	mm	$10^{-3} = 0.001$ m

Fig. 2-44. Metric units of measure on industrial drawings by type of industry.

The meter serves as the official unit of measure in the building and construction fields, whereas the centimeter (1/100 m) is the official unit in the lumber and cabinet industries. In the precision manufacturing industries (aerospace, automotive, computer and machine parts) where close tolerances must be maintained, the official unit is the millimeter (1/1000 m).

Some typical metric scale designations for drawings which have been reduced or enlarged are as follows:

REDUCTIONS		ENLARGEMENTS
1:2	1:50	2:1
1:2.5	1:100	5:1
1:5	1:200	10:1
1:10	1:500	20:1
1:20	1:1000	50:1

DRAWING TO SCALE

All drawings should be drawn to scale except schematics and tables. Drawing to scale refers to a drawing that has been reduced proportionally from actual size so that it can be placed on the drawing sheet, Fig. 2-52. Or a drawing may have to be enlarged proportionally over the actual object size for clarity and detail, Fig. 2-45.

Drawing to scale is also used in referring to drawings of objects drawn full size. In this case, the scale would be 1″ = 1″.

The scale to be selected for a particular drawing depends upon the size of the object to be drawn. In general, the drawing of the object should be as large as possible to nicely fit a standard sheet size.

As indicated in previous sections, there are numerous scale sizes from which to select. The first reduction on the architect's scale is to quarter size, Fig. 2-34. The first reduction on the mechanical engineer's scale is half size, Fig. 2-36. If these reductions are insufficient, then a smaller scale must be selected.

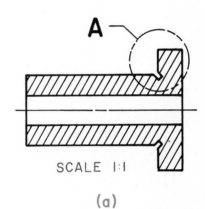

SCALE 1:1

(a)

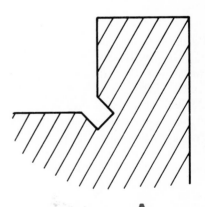

DETAIL **A**
SCALE 4X SIZE

(b)

Fig. 2-45. A machine part drawn to a reduced scale and a feature of the part drawn to an enlarged scale to show detail.

READING THE SCALE

Here is how to read an open divided architect's scale for a mixed number. To read 2 feet and 3 1/2 inches on the eighth-size scale (1 1/2″ = 1′ − 0″), start with the numeral 2 in the open divided section. Then move to the fully divided unit to locate the 3 1/2 inches, Fig. 2-46.

on the drawing represents 100 on the object. The measurement in Fig. 2-49 would then be 4.35 metres.

The difference in the meaning of the words scale and size as used in scale drawings should be noted. The scale 1/4″ = 1′ − 0″ is a common scale for drawing house plans (often referred to as "quarter scale"). Note in Fig. 2-34 that this scale is actually 1/48 size since the quarter inch on the

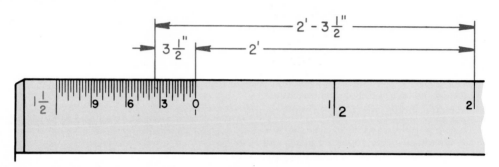

Fig. 2-46. Measuring with the architect's scale.

A reading of 3 5/8 inches on an open divided mechanical engineer's half-size scale would be read in a similar manner, Fig. 2-47. Remember that units on this scale represent inches. Start with the numeral 3 in the open divided section and move to 5/8 inch in the fully divided end unit.

drawing represents one foot on the house plan. The word scale refers to the name of a particular scale and not to the size of the drawing.

The word size does refer to the size of the drawing in relation to the size of the object. Hence, the "quarter size"

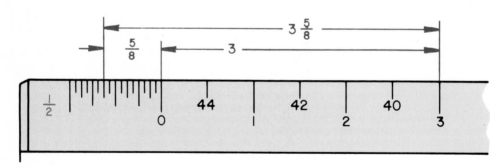

Fig. 2-47. Measuring with the mechanical engineer's scale.

To read a decimal dimension of 2.125 inches on the civil engineer's fully divided 10-scale, Fig. 2-48 (a), start from the 0 and move past the 2. Continue past the first tenth (.10) to one-fourth (.025) of the next tenth (judged by eye). This represents the decimal .125, Fig. 2-48 (a).

The same reading can be made with greater accuracy on the 50-scale, where each subdivision equals 1/50 or .02 of an inch, Fig. 2-48 (b). Move from 0 to 10 (5 major divisions to the inch, 10 = 2 inches) and on to 6 subdivisions (6 x .02 = 12). Continue to one-fourth of the next subdivision (1/4 of .02 = .005) for a measurement of 2.125 inches. The 50-scale is the one most commonly used in the machine-parts manufacturing industries where decimal dimensioning is standard practice.

Measuring with the metric scale is shown in Fig. 2-49, using the .01 scale as a full size scale. The measurement reads 43.5 mm. To use the same scale as a reducing scale of 1:100, let each numbered unit (actually a centimetre) represent a metre — a reduction of 1 to 100, Fig. 2-49. That is, one unit

scale on the mechanical engineer's scale, where 1/4″ = 1″, will produce a drawing one-quarter the size of the object.

The scale of a drawing is usually indicated in the title block of the drawing in a manner similar to that shown below:

Full size — 1/1, or FULL SIZE
Enlarged — 2/1, 4/1, 10/1, 10X, TWICE SIZE
Reduced — 1/4″ = 1′ 0″, 1/2, 1/4, 1/10, 1/50, HALF SIZE,
        QUARTER SIZE

Views which have been drawn to a scale other than that indicated in the title block should have the scale noted below the view, Fig. 2-45 (b).

LAYING OFF MEASUREMENTS

Position the scale on the sheet with the particular scale to be used, face up and away from you, Fig. 2-50. Eye the scale

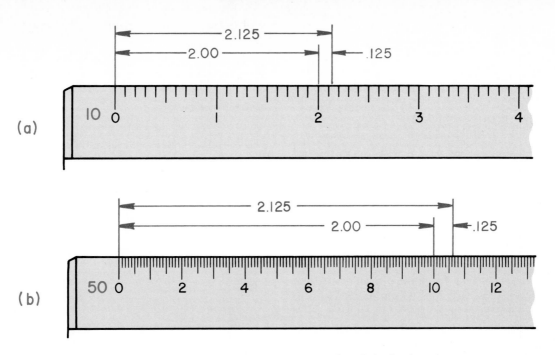

Fig. 2-48. Measuring with the civil engineer's or decimal scale.

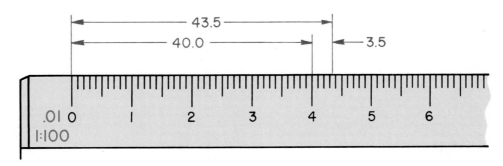

Fig. 2-49. Measuring with the metric scale.

directly from above. Then, with a sharp conical pencil, mark the desired distance lightly with a short dash at right angles to the scale.

Successive distances on the same line should be laid off without shifting the scale, Fig. 2-51. Double check your additions of these successive distances by measuring each

Fig. 2-50. Eye the scale directly from above when laying off a measurement.

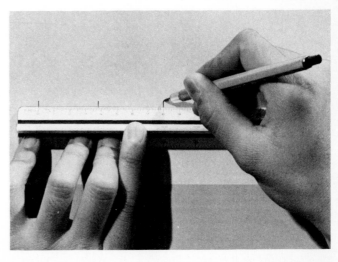

Fig. 2-51. Laying off successive distances.

individually after they have been marked off.

The dividers or compass should never be used to take distances directly from the scale. This procedure is harmful to the scale. Mark distances on the sheet, then set the dividers or compass to these marks.

SHEET FORMAT

Recommendations are made by the American Society of Mechanical Engineers for drawing sheet borders and basic format data. However, these vary somewhat by the needs of particular industries. Included here are the practices found in most industries.

SHEET MARGINS

The recommended margins for drawing sheets vary from 1/4 inch on A and B size sheets to 1/2 inch for size D and E. Up to one inch may be used on the left edge if the sheet is to be bound. If sheet is to be rolled, 4 to 8 inches should be left beyond the margin for protection.

TITLE BLOCK

A "title block" is included on a drawing to provide certain recorded data pertinent to the drawing and to provide supplementary information, Fig. 2-52. The title block is usually located in the lower right hand corner of the drawing just above the border line.

Fig. 2-52. An industry title block.

Sometimes a title strip containing the same information is used. The title strip extends partially or completely across the lower portion of the sheet. Suggested title block layouts are given on page 603 for use on your drawings.

DRAFTING INSTRUMENT PROCEDURES

Basic drafting procedures in instrument usage are presented in this section to assist the student in forming good habits. Study the material carefully and refer to it as needed in actual use of the various instruments.

DRAWING HORIZONTAL LINES

Horizontal lines are drawn along the upper edge of the T-square (parallel or drafting machine straightedge) with the T-square in position along the working edge of the drawing board. Press down on the head, Fig. 2-53 (a), to prevent the

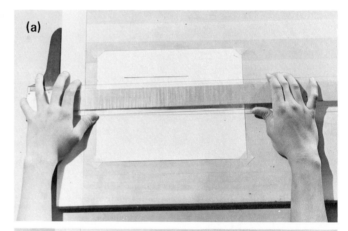

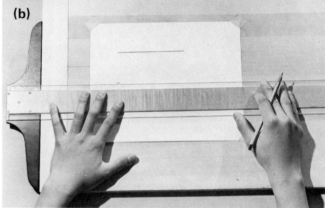

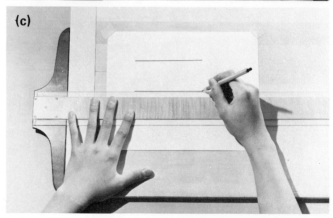

Fig. 2-53. Positioning the T-square and drawing a horizontal line.

blade from sliding over the drawing (rubbing graphite across the drawing from existing pencil lines), while bringing it into approximate position for the line to be drawn.

Let the left hand slide from the head to the blade with the fingers resting on the blade and the thumb on the drawing board, Fig. 2-53 (b). Your fingers are now in position to make the final adjustment to bring the T-square blade in perfect alignment where the line is to be drawn. Hold the T-square in this position and draw a light line from the left to the right, Fig. 2-53 (c).

Note that the pencil is inclined in the direction the line is being drawn. It is tilted slightly away from the drafter to cause the pencil point to follow accurately along the T-square edge. Let the little finger glide along the T-square blade to help steady your hand. Rotate the pencil between thumb and fingers slowly to retain a conical point. Draw horizontal lines at the top of the sheet first and work down.

If you are left handed, use the T-square along the right hand working edge of the drawing board and draw horizontal lines from right to left.

DRAWING VERTICAL LINES

Vertical lines are drawn with the vertical edge of either the 30-60 degree or 45 degree triangle supported on the upper edge of the T-square. Position the T-square below the starting point of the vertical line and place the triangle on the T-square, holding it and the T-square firmly with the palm and fingers, Fig. 2-54.

Draw the lines upward, away from the body. Hold the pencil at a 60 degree angle with the paper and tilt the pencil away from the triangle so the point will follow accurately along the edge of the triangle.

To maintain accuracy, never draw lines too close to ends of the triangle. Rotate the pencil as you draw to retain a fine point. Draw the vertical lines at the left side of sheet first.

Left handers should reverse the T-square and place the triangle with the working edge facing the right side of the board.

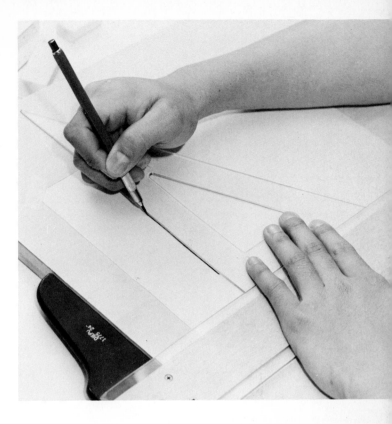

Fig. 2-54. Drawing a vertical line.

DRAWING INCLINED LINES

Inclined lines at 30, 45 and 60 degrees may be drawn by using one of the triangles with the T-square, Fig. 2-55. By using the two triangles in combination, lines at 15 and 75 degrees may be drawn, Fig. 2-56.

By using the two triangles individually or in combination, a complete circle may be divided into 24 sectors of 15 degrees each, Fig. 2-57. Inclined lines on degree settings other than these 15 degree sectors may be set off with the protractor, Fig. 2-60, or the drafting machine, Fig. 2-15.

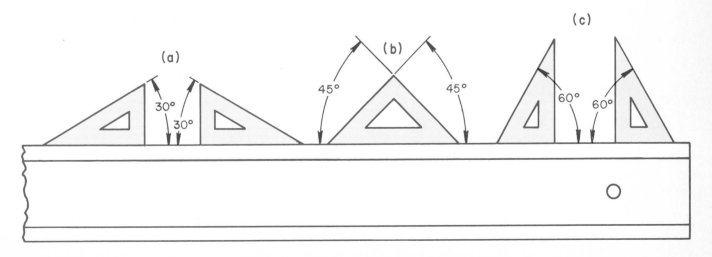

Fig. 2-55. Drawing lines inclined at 30, 45 and 60 degrees.

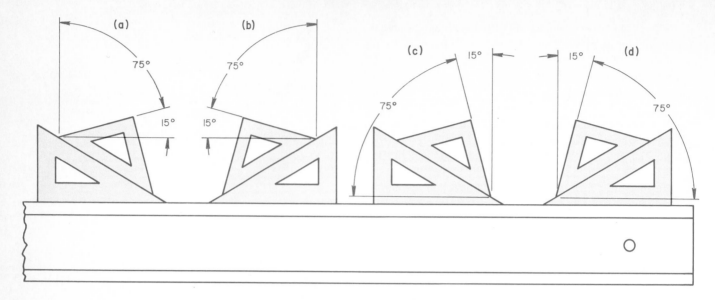

Fig. 2-56. Combining triangles to draw lines at angles of 15 and 75 degrees.

DRAWING PARALLEL INCLINED LINES

Lines parallel to inclined lines may be drawn by the use of the T-square and one triangle, as shown in Fig. 2-58 (a). Adjust the T-square to align the triangle with the given line A-B. Hold the T-square firm, slide the triangle to the desired location and draw the parallel line. The process of drawing lines parallel to inclined lines may be performed by the use of two triangles (eliminating T-square) if the lines are not widely separated, as illustrated in Fig. 2-58 (b).

DRAWING A PERPENDICULAR TO AN INCLINED LINE

Place the hypotenuse of a triangle along the edge of the T-square and adjust until one side of the triangle is aligned with the given line, Fig. 2-59 (a). Hold T-square firmly, slide triangle until second side is in the desired location and draw the perpendicular line.

When a longer perpendicular line is required, place the side of one triangle against the T-square and adjust until the

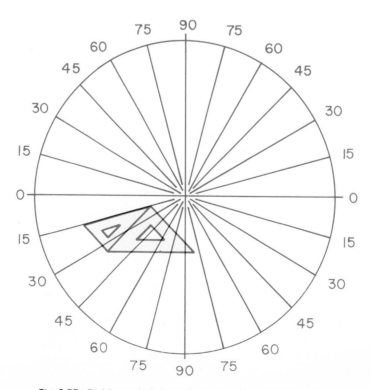

Fig. 2-57. Dividing a circle into 15 degree sectors with triangles.

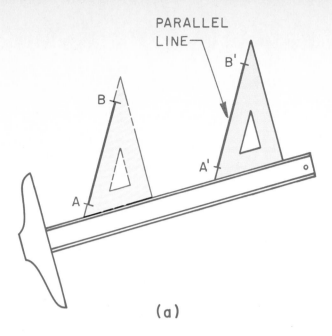

PARALLEL
LINE

(a)

(b)

Fig. 2-58. Drawing parallel inclined lines.

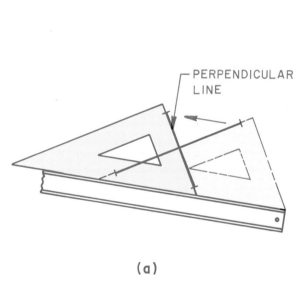

PERPENDICULAR
LINE

(a)

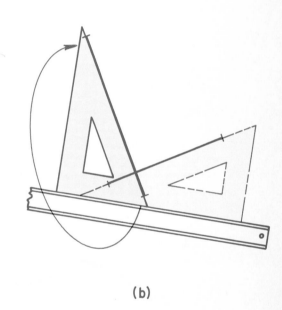

(b)

Fig. 2-59. Drawing lines perpendicular to inclined lines.

hypotenuse is aligned with the given line. Hold the T-square firmly and revolve the triangle until its other side is against the T-square. Then slide to desired location and draw line, Fig. 2-59 (b). This process may be performed with the T-square and either triangle or with the two triangles in combination.

THE PROTRACTOR

The protractor is used to measure and to set off angles which are not obtainable with the T-square and triangles. Protractors are available in several designs, including the simple semicircular type, Fig. 2-60. An adjustable triangle provides more accuracy. The protractor with a vernier attachment, Fig. 2-61, is used for extreme accuracy in laying out angles in map work.

To lay out an angle with the semicircular protractor, set the vertex indicator at the location of the vertex of the angle to be

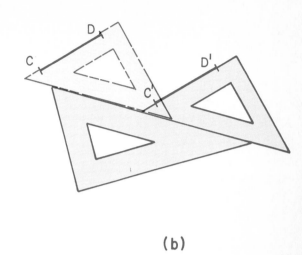

Fig. 2-60. Semicircular protractor for use in laying out angles. (Keuffel & Esser Co.)

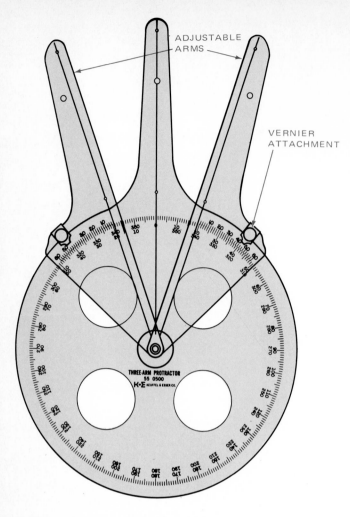

Fig. 2-61. Protractor with vernier attachment for more accuracy in laying out angles. (Keuffel & Esser Co.)

drawn, Fig. 2-62 (a). Mark the desired angle with a fine point, (b). Align the triangle with two points by placing your pencil on one point and revolving the triangle in line, (c). Then draw a line between the vertex and the point marking the required angle, (d).

Protractors are also available in graduations other than degrees. A percentage protractor, shown in Fig. 2-63, is useful in laying out circle graphs.

PENCIL TECHNIQUE WITH INSTRUMENTS

To produce accurate and clean drawings, it is important that you develop proper pencil techniques when using instruments. Careful observation, practice and following these suggestions will aid you in developing this technique:

1. All layout work on drawings should first be done with light construction lines. Use a 4H pencil and rotate it slowly when drawing to help retain the conical point.
2. Before the drawing is "heavied-in," erase all unnecessary lines.
3. "Heavy-in" all lines to their proper line weight, Fig. 2-23, using the proper grade pencil to give you the best results with the paper being used.
4. For accuracy and to minimize working over finished lines, pencil them in the following order: (1) arcs and circles, (2) horizontal lines starting at the top of the drawing and working down, (3) vertical lines from left to right, (4) inclined lines from top down and from left to right.
5. To prevent smearing of lines, dust loose graphite from the drawing after each line is drawn, and avoid sliding the T-square, triangles and other instruments across the drawing.

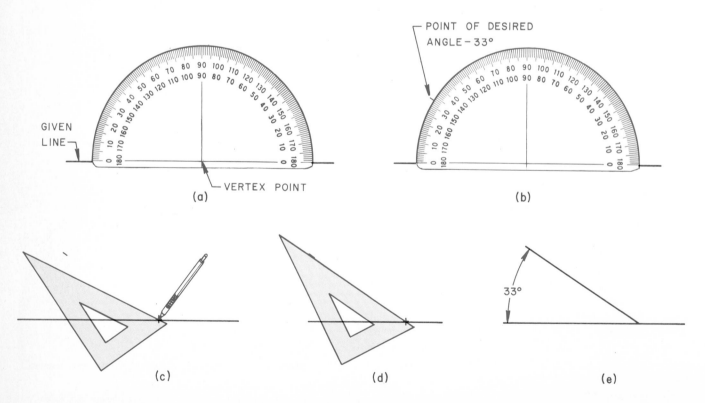

Fig. 2-62. Laying out an angle with the semicircular protractor.

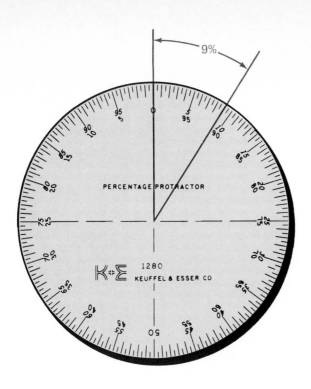

Fig. 2-63. The percentage protractor is useful in constructing circle graphs.

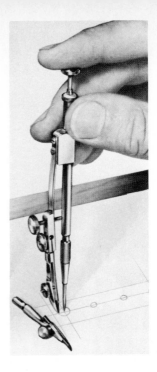

Fig. 2-65. Drop bow compass is used to draw very small arcs and circles. (Eugene Dietzgen Co.)

THE COMPASS

Large and small bow pencil compasses are used frequently in drafting. The bow instrument has a steel ringhead and a side or center adjusting screw, Fig. 2-64. The small bow instrument is used for drawing smaller circles with radii of approximately one inch or less. A large bow compass is used for circles with radii up to 5 or 6 inches.

The drop bow compass, Fig. 2-65, which revolves around a center shaft, is used for drawing very small arcs and circles.

When drawing an arc or circle with a large radius, bend the legs of a friction compass at the knees and adjust them to meet the paper in a vertical position, Fig. 2-66. A lengthening bar, Fig. 2-67, may be employed for drawing circles with larger diameters.

Fig. 2-66. Adjust legs of a friction compass to meet the drawing surface vertically.

Fig. 2-64. The bow compass. (Teledyne Post)

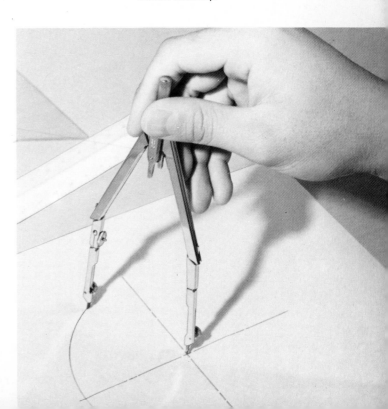

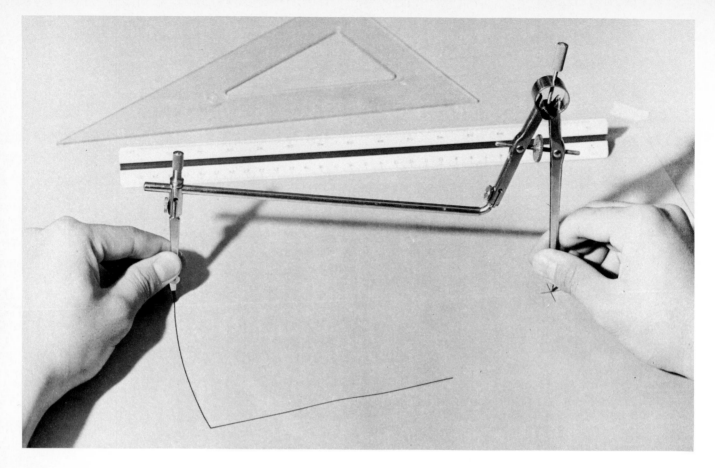

Fig. 2-67. Using the bow compass with lengthening bar.

BEAM COMPASS

The beam compass is used for drawing large arcs and circles. It consists of two sets of points, one a needle pivot point and the other a holder for a pen or pencil point, and a beam to which the points clamp, Fig. 2-68. To draw arcs and circles with the beam compass, hold the pivot point steady with one hand and swing the pen or pencil point with the other.

SHARPENING THE COMPASS LEAD

The compass is used with both pencil and pen attachments. Lead used in the compass should be about one grade softer (F or H) than that used in your pencil. This will allow you to exert less pressure on the compass. The lead should extend approximately 3/8 inch; sharpen it to a chisel point as shown in Fig. 2-69. After sharpening, adjust the lead to a length of 1/32 inch shorter than needle point.

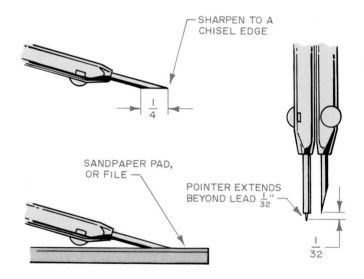

SHARPEN TO A CHISEL EDGE

$\frac{1}{4}$

SANDPAPER PAD, OR FILE

POINTER EXTENDS BEYOND LEAD $\frac{1}{32}$"

$\frac{1}{32}$

Fig. 2-69. Sharpening and adjusting the compass lead.

Fig. 2-68. The beam compass is used to draw extremely large circles.   (Teledyne Post)

DRAWING ARCS AND CIRCLES

The bow instrument is adjusted to a radius setting by twisting the adjusting screw between the thumb and forefinger, Fig. 2-70.

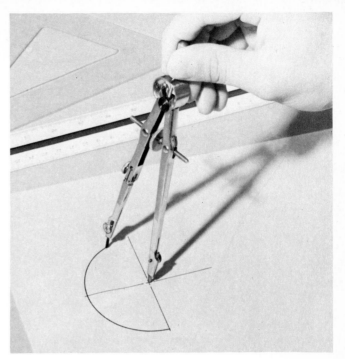

Fig. 2-71. Drawing a circle with the compass.

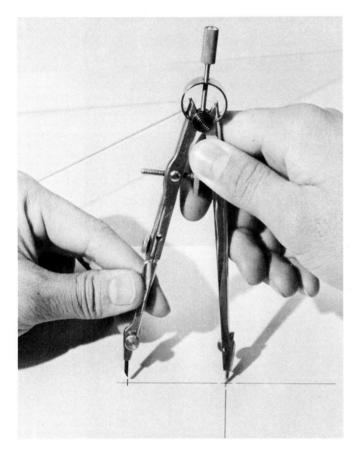

Fig. 2-70. Setting the bow compass with the adjusting screw.

To set the compass, measure off the radius on a scrap of paper (or lightly on your drawing) and adjust the compass accordingly. Test the setting by drawing the circle lightly on the drawing or scrap paper, then measure the diameter.

To draw a circle, hold the compass in one hand and start in a clockwise direction, lean the compass slightly forward and revolve the handle between the thumb and forefinger, Fig. 2-71 (left handers, counterclockwise). Draw the circle or arc lightly at first. When you are ready to "heavy-in," make repeated turns to darken the line.

Arcs and circles to be joined by straight lines should be drawn first. When a number of concentric circles are to be drawn with the compass, draw the smaller circles first since there is a tendency for the needle point hole to become enlarged. A center tack, Fig. 2-72, is sometimes used to prevent this.

DRAFTING TEMPLATES AND THEIR USE

Industrial drafting rooms make extensive use of templates. They provide economy and consistency when drawing com-

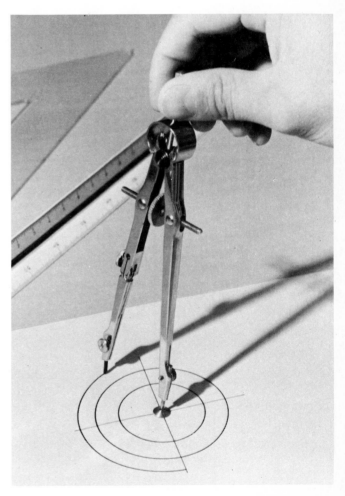

Fig. 2-72. A center tack prevents damage to sheet from compass needle.

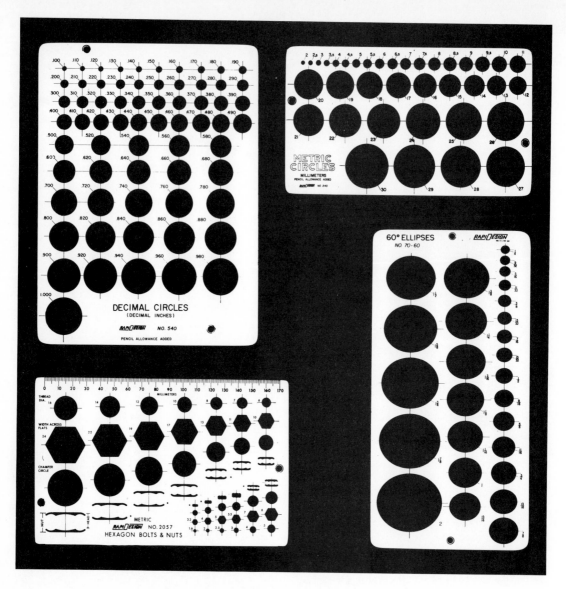

Fig. 2-73. A wide variety of templates are used in drafting.

ELECTRICAL SYMBOLS

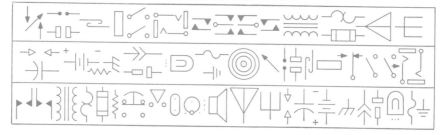

WELDING SYMBOLS

Fig. 2-74. Symbol templates for use with lettering equipment.
(Keuffel & Esser Co.)

monly used characters and symbols. Templates are available for nearly all standard size circles in fractional, decimal and metric graduations, Fig. 2-73. Templates for ellipses and bolt heads, and symbols for nearly every field of drafting are available to speed the drafter's work.

Templates are also available for use with lettering equipment for drawing many types of symbols, Fig. 2-74.

To draw circles, arcs or ellipses with the aid of a template, first lay out the center lines on the drawing, Fig. 2-75. Then align the center lines of the template and draw the figure. Other types of templates are aligned with features on the drawing.

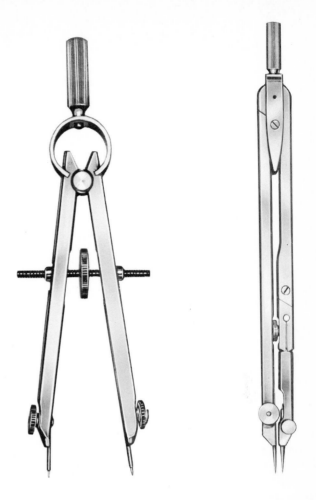

Fig. 2-76. Bow divider and friction joint divider. (Teledyne Post)

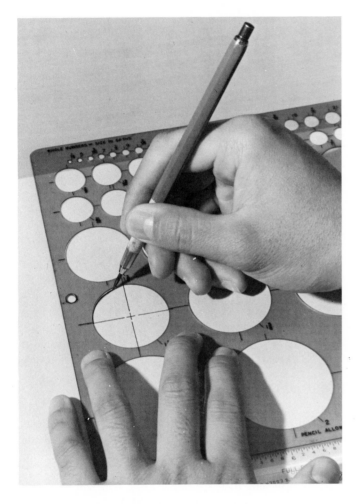

Fig. 2-75. Drawing a circle with the aid of a template.

DIVIDERS

Two types of dividers are used extensively by draftsmen, bow dividers and friction joint dividers, Fig. 2-76. Dividers are used to transfer distances and to divide straight and circular lines into equal parts.

Dividers are adjusted in the same manner as the compass. To transfer or step off distances, hold the knurled handle and place the dividers in position, Fig. 2-77. Mark the distances by making a slight dent in the paper with the divider point. Mark this dent with a light pencil mark or circle the dent.

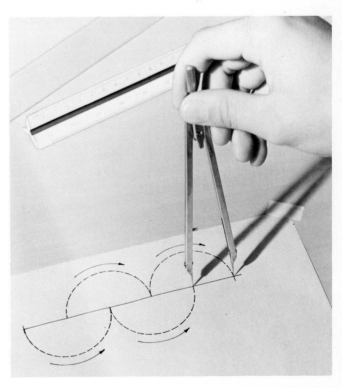

Fig. 2-77. Stepping off distances with the dividers.

Fig. 2-78. Dividing a line into three equal parts with the dividers.

To divide a line into three equal spaces, for example, set the dividers for an estimated 1/3 of the distance and step off, Fig. 2-78. Correct any error in estimation by decreasing or increasing the divider setting by 1/3 of the error and making another trial. Careful estimation of the distance and adjustment after the first trial should enable you to complete the division in 2 or 3 trials. Avoid puncturing the paper with the divider points.

PROPORTIONAL DIVIDERS

A proportional divider is a special instrument used for dividing linear and circular measurements into equal parts. It is also used to lay off measurements in a given proportion.

The instrument consists of two legs held together by a sliding pivot. It can be adjusted to obtain various ratios between the two sets of points on the ends of the legs, Fig. 2-79.

Graduations on the dividers vary from less expensive proportional dividers (with ratios for division of lines) to dividers with vernier graduations. Settings for any desired ratio between 1:1 and 1:10 and ratios for circles, squares and area may be made.

Examples of the use of proportional dividers include:
1. Dividing straight lines into any number of equal parts.
2. Lengthening straight lines to any given proportion.
3. Dividing the circumference of a circle into any number of equal parts.
4. Laying off the circumference of a circle from the diameter of that circle.
5. Laying out a square equal in area to a circle based on the diameter of that circle.

With the more expensive instruments, a table of settings is provided for use in setting the dividers for various proportions.

IRREGULAR CURVES

All curves which do not follow a circular arc are known as irregular curves. These curves are common in sheet metal developments, cam diagrams, aerospace drawings and various charts.

The instrument used for drawing the final smooth curve through plotted points is called an irregular curve or French curve. These curves are available in many shapes, a few of

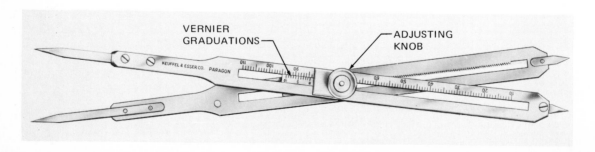

Fig. 2-79. Proportional dividers with vernier. (Keuffel & Esser Co.)

Instrument Drafting

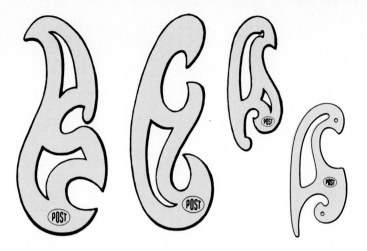

Fig. 2-80. These instruments are called irregular curves or French curves and are available in many shapes and sizes.

If the curve is symmetrical in repeated phases (development of a cam diagram, for example), the same segment of the irregular curve should be used, (f). Marking the curve with a pencil when the first symmetrical segment is drawn will aid in locating that segment for successive phases of the curve.

Flexible curve rules, Fig. 2-82 (a), and splines with lead weights, (b), are useful in ruling a smooth curve through a number of points.

Fig. 2-82. Flexible rules (a) and splines (b) for drawing irregular curves. (Keuffel & Esser Co.)

which are shown in Fig. 2-80. The instruments are made up of a series of geometric curves in various combinations.

To draw irregular curves, plot a series of points to accurately establish the curve, Fig. 2-81 (a). Then lightly sketch a freehand line to join the points in a smooth curve, (b). Draw in the final smooth line with the irregular curve, matching it with three or more points on the curve. Draw a segment at a time until the line is complete, (c).

Check to see that the general curvature of the irregular curve is placed in the same direction as the curve of the line to be drawn. Do not draw the full distance matched by the irregular curve. Stop short and make the next setting of the irregular curve flow out of the previous one, (d). When the plotted points of the curve reverse direction, watch for the point of tangency where the irregular curve (instrument) should be reversed and overlap the previous setting with a smooth flow, free of "humps", (e).

INKING

Inking in industrial drafting for the production of working drawings is used very little today, but inking drawings for technical publications is quite common. The difficulty of

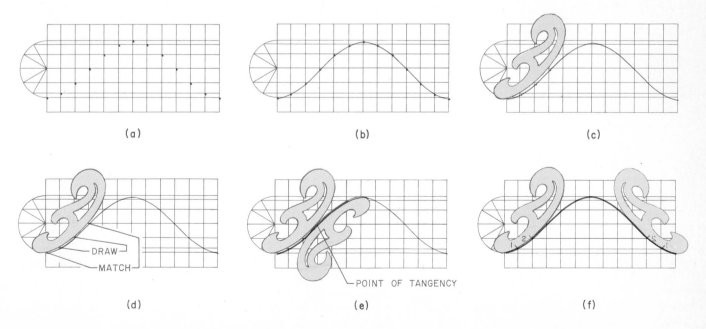

Fig. 2-81. Drawing a smooth irregular curve with the aid of an irregular (French) curve instrument.

inking has been greatly reduced with the introduction of improved instruments and drafting media (papers and polyester films).

Inked drawings provide a sharper line definition and make cleaning of the finished drawing much easier. It is possible to erase right over inked lines with a soft eraser and remove penciled construction lines.

INSTRUMENTS FOR INKING

The ruling pen is included in most drafting sets for inking lines. It has two, adjustable, sharp nibs that permit lines of varying widths to be drawn, Fig. 2-83. Add the ink to the pen between the nibs directly from a bottle designed for this purpose, Fig. 2-84. Do not fill the pen too full as the weight of the ink may cause a line wider at the start than at the finish. Usually 1/4 inch in the nib is sufficient. Practice will help in arriving at the right amount.

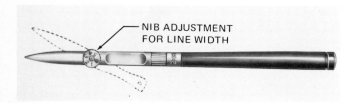

Fig. 2-83. A ruling pen used for inking lines of varying widths. (Keuffel & Esser Co.)

Fig. 2-84. Loading ink into a ruling pen with bulb top and plastic pipette.

The ruling pen is used to draw a line along the straightedge, triangle or irregular curve. A ruling pen attachment for the compass is usually included in a set of instruments. Clean ruling pens frequently during use and when the drawing is completed. Use a damp tissue, wiping both the inside and outside of the nibs.

The technical pen, Fig. 2-85, has largely replaced the use of ruling pens in industrial and technical illustration drafting rooms. Pens are available in a series of point sizes, assuring uniformity of line weight throughout a single drawing or several drawings.

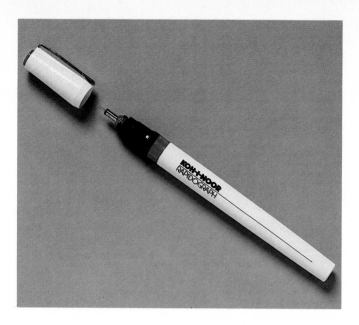

Fig. 2-85. The technical pen is a versatile instrument for inking drawings. (Koh-i-noor Rapidograph, Inc.)

The technical pen has a supply reservoir of ink and does not require filling for each use. Today's technical pens offer instant start-up in inking, even after weeks of storage. The technical pen is most versatile and can be used with straightedges, irregular curves, compasses, templates of all sorts and in instrument lettering devices.

PROCEDURE FOR INKING

The following procedures tend to produce neater inked drawings:
1. Lay out the drawing, using light construction lines with a 2H or harder pencil.
2. Ink arcs from tangent point to tangent point (points where an arc is joined by a straight line). These points are located by drawing a light circle and drawing the tangent line lightly so that it just touches the circle. This point is the point of tangency.
3. Ink full circles and ellipses.
4. Ink irregular curves.
5. Ink all straight lines of one line weight from top to bottom, then from left to right (left-handers opposite). Continue inking remaining straight lines of different weight.
6. Ink notes, dimensions, arrowheads and title block.

PROBLEMS AND ACTIVITIES

The problems in Figs. 2-86, 2-87 and 2-88 are one-view drawing problems to provide you an opportunity to become familiar with the use of basic drafting instruments. Use size A drawing sheets and draw the objects assigned by your instructor. You may use the inch or metric scale in drawing the objects. Select a scale size to make good use of available drawing space without crowding. Do not dimension.

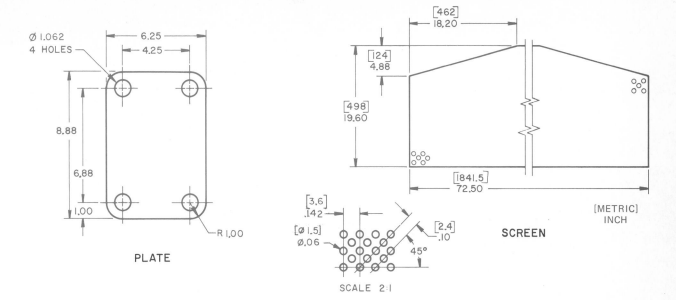

PLATE

SCREEN

SCALE 2:1

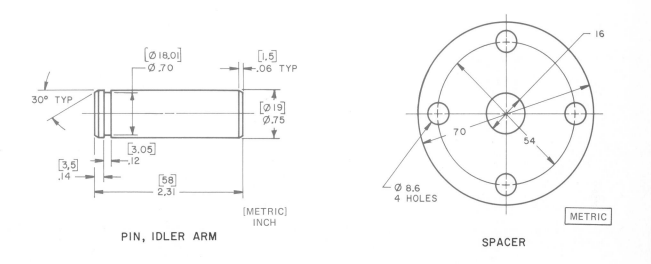

PIN, IDLER ARM

SPACER

[METRIC]

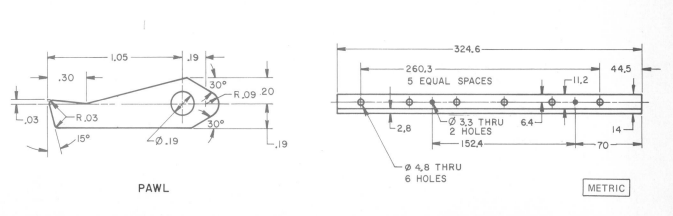

PAWL

BACK WEAR PLATE

METRIC

Fig. 2-86. Beginning one-view drawing problems.

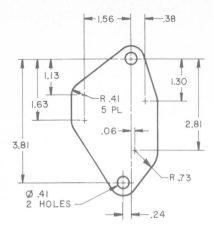

PLATE, COVER

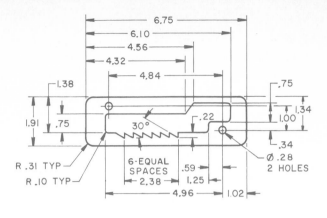

GATE, THROTTLE GUIDE

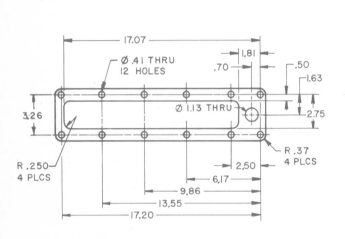

GASKET, WATER INLET CONN

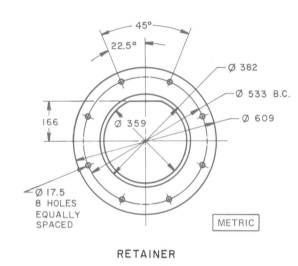

RETAINER

METRIC

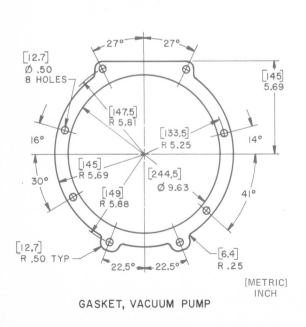

GASKET, VACUUM PUMP

[METRIC]
INCH

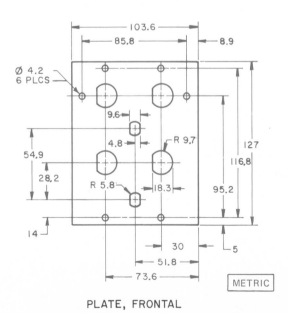

PLATE, FRONTAL

METRIC

Fig. 2-87. Intermediate one-view drawing problems.

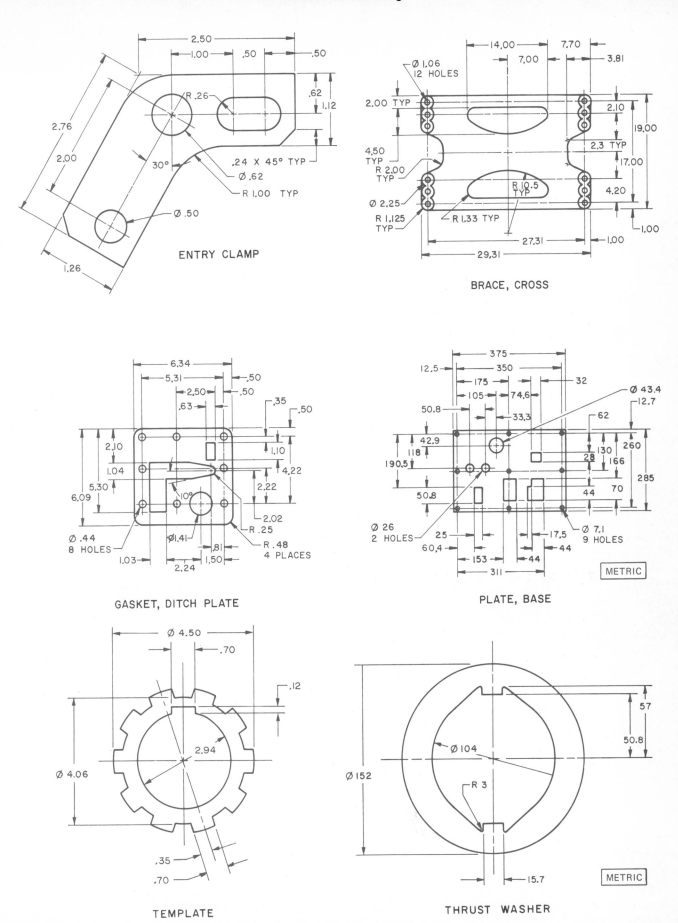

ENTRY CLAMP

BRACE, CROSS

GASKET, DITCH PLATE

PLATE, BASE

TEMPLATE

THRUST WASHER

Fig. 2-88. Advanced one-view drawing problems.

Chapter 3
TECHNICAL SKETCHING

Freehand technical sketching is a method of making a drawing without the use of instruments. It is a technique essential to all who work in a technical field. A good description of the process is "thinking and drafting," since the person sketching should be able to concentrate on the solution to the problem without being encumbered with manipulation of instruments.

Most drafters and engineers use freehand sketching to "think through" solutions to drafting problems before starting an instrument drawing. Sketching also permits them to quickly convey ideas to others, especially in the area of design improvement. Once the design or problem solution has been sketched, it is given to a drafter to prepare an accurate instrument drawing.

This chapter is presented to acquaint you with the skills and procedures necessary to do freehand technical sketching. The techniques you learn will be particularly useful when combined with material presented in the chapters on instrument drafting, working drawings, pictorial drawings and dimensioning.

PURPOSES OF A SKETCH

Freehand sketches are used for numerous purposes including the following:
1. To transmit ideas graphically to others.
2. Preliminary planning of a problem prior to making an instrument drawing. This includes the selection and blocking out of views.
3. To provide a clearer picture of an object to be drawn in orthographic projection.
4. For recording technical notes and descriptions of parts in the shop for later use in the drafting room.
5. To "think through" the solution to a design or construction problem.
6. To substitute for an instrument drawing when the use of a sketch will do the job.

SKETCHING EQUIPMENT

Freehand sketching requires very little equipment or material. It readily lends itself to use by the drafter in the field or shop, away from the drafting room. A pencil (F or HB grade), soft eraser and some paper are all that is needed.

Several types of papers are suitable for sketching, depending on the nature of the job. You can use plain, cross section or isometric (a type of pictorial drawing discussed in Chapter 11) grid paper. Also available are bond typing paper, drafting paper and tracing paper.

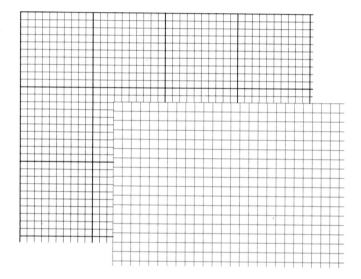

Fig. 3-1. Cross section papers for freehand sketching.

The cross section papers, Fig. 3-1, come in varying grid sizes and are helpful in line work and proportions. Isometric grid paper also aids in these areas and in obtaining the proper position of the axes, Fig. 3-2 (a). Cross section tracing papers may be purchased with fade-out grid lines. When the sketch is completed and prints are made, the grid lines do not reproduce, Fig. 3-2 (b).

You may want to start with cross sectioned paper but it is well for you to learn to sketch on plain paper as soon as possible. This will help develop your skill and accuracy in freehand sketching without the use of aids.

SKETCHING TECHNIQUE

When sketching, hold the pencil with a grip firm enough to control the strokes but not so tight as to stiffen your strokes or cramp your hand. Your arm and hand should have a free and easy movement. The point of the pencil should extend

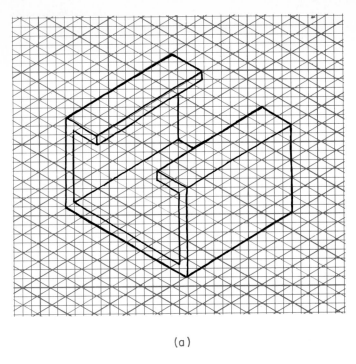

(a)

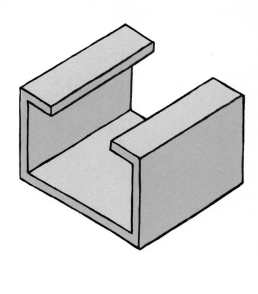

(b)

Fig. 3-2. Isometric grid papers with nonreproducible grid.

approximately 1 1/2 inches beyond your fingertips, a little farther than in normal drawing or lettering, Fig. 3-3. This will permit better observation of your work and provide a more relaxed position.

SKETCHED LINE

INSTRUMENT LINE

Fig. 3-4. Comparison of freehand sketch and instrument lines.

various kinds of freehand lines will be discussed in sections that follow.

While you should strive for neatness and good technique in freehand sketching, you should expect that freehand lines will look different than those drawn with instruments. Good freehand sketches have character all their own, Fig. 3-4.

Fig. 3-3. Correct position for holding the pencil when sketching.

Rotate the pencil slightly between strokes to retain the point longer and produce sharper lines. Initial lines should be firm and light, but not fuzzy. Avoid making grooves in your paper caused by too much pressure since they are difficult to remove and may result in a messy drawing.

When sketching straight lines, your eye should be on the point of termination of the line, and use a series of short strokes to reach that point. Specific steps in sketching the

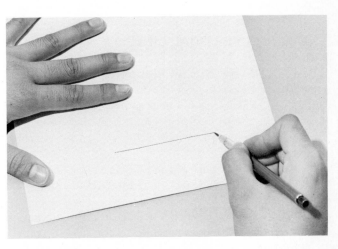

Fig. 3-5. Pencil position for sketching horizontal lines.

55

SKETCHING HORIZONTAL LINES

Horizontal lines, Fig. 3-5, are sketched with a movement of the forearm approximately perpendicular to the line being sketched. You will find that four steps are essential in sketching horizontal lines:

1. Locate and mark the end points of the line to be sketched, Fig. 3-6 (a).

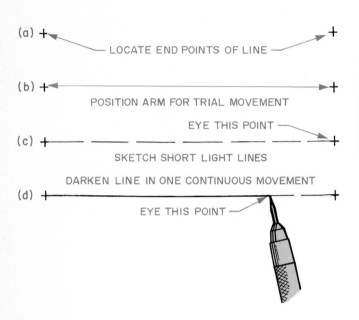

Fig. 3-6. Steps in sketching horizontal lines from left to right.

2. Position arm by making trial movements from left to right without marking paper, (b).
3. Sketch short, light lines between points, (c). Keep your eye on the point where the line is to end. Do not permit your eye to follow the pencil.
4. Remove unnecessary lines with a soft eraser and darken line to form one continuous line of uniform weight. In this step, eye should lead point of pencil along light sketch line, (d). With practice, you will soon develop the hand and eye coordination necessary for making good freehand lines.

SKETCHING VERTICAL LINES

Vertical lines are sketched from top to bottom, using the same short strokes in series as for horizontal lines. When making the strokes, position your arm comfortably about 15 degrees with the vertical line, Fig. 3-7. A finger and wrist movement, or pulling arm movement, are best for sketching vertical lines.

Follow these four steps when sketching vertical lines:

1. Locate and mark end points of line to be sketched, Fig. 3-8 (a).
2. Position arm by making trial movements from top to bottom without marking paper, (b).
3. Sketch short, light lines between points, (c). Keep your eye on point where line is to end.

4. Remove unnecessary lines with a soft eraser and darken line to form one continuous line of uniform weight. In this step, eye should lead point of pencil along light sketch line, (d).

Fig. 3-7. How to hold the pencil when sketching vertical lines.

You may find it easier to sketch vertical or horizontal lines if the paper is rotated counterclockwise to form a slight angle, Fig. 3-7. Straight lines that are parallel to the edge of the drafting board, such as border lines, may be drawn by letting the third and fourth fingers slide along the edge of the board as a guide, Fig. 3-9.

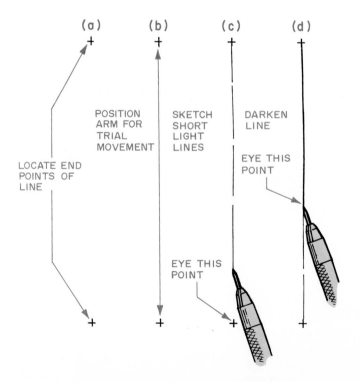

Fig. 3-8. Steps in sketching vertical lines.

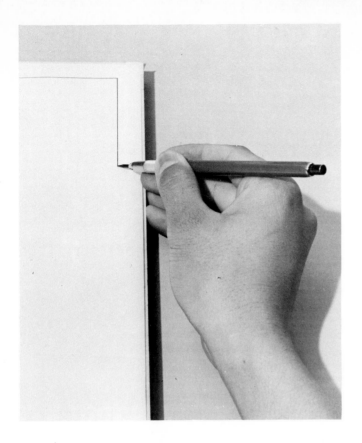

Fig. 3-9. Using fingers along edge of drawing board as a guide for sketching straight lines.

sketching horizontal and vertical lines, Fig. 3-10 (a). If you prefer, rotate the paper to sketch these lines as if they are horizontal or vertical, (b).

Angles can be estimated quite accurately by first sketching a right angle (90 degrees), then subdividing it to get the desired angle. Fig. 3-11 illustrates how to obtain an angle of 30 degrees.

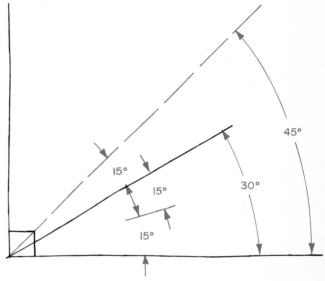

Fig. 3-11. Estimating an angle of 30 degrees by freehand sketching.

SKETCHING INCLINED LINES AND ANGLES

All lines that are not horizontal or vertical are called inclined lines. Sketch between two points or at a designated angle, using the same strokes and techniques employed in

SKETCHING CIRCLES AND ARCS

There are several methods of sketching circles and arcs. All are sufficiently accurate. Familiarize yourself with various techniques to use method best suited to a particular problem.

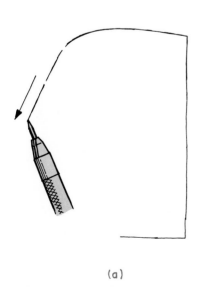

(a)

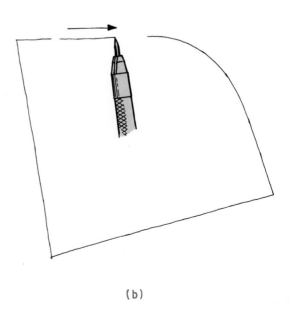

(b)

Fig. 3-10. Methods for sketching inclined lines.

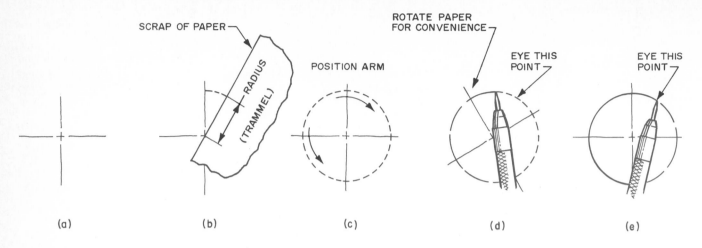

Fig. 3-12. Center-line method of sketching a circle.

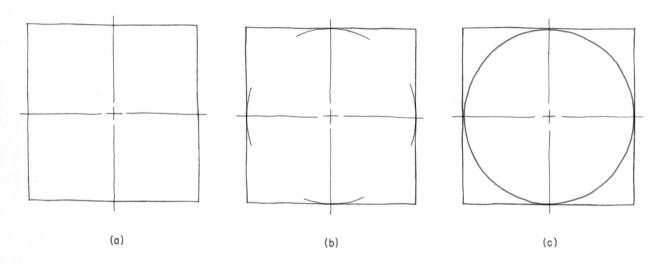

Fig. 3-13. Enclosing square method of sketching a circle.

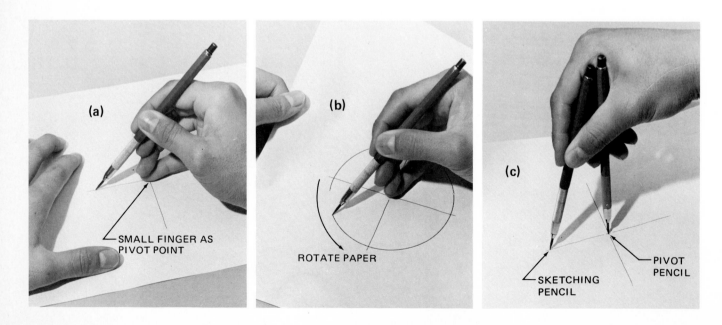

Fig. 3-14. Hand-pivot methods of sketching a circle.

CENTER-LINE METHOD

Six steps are used in the center-line method of freehand sketching:

1. Locate center of circle and draw horizontal and vertical center lines, Fig. 3-12 (a).
2. With a scrap of paper used as a trammel (dividers), locate points in radius, (b).
3. Position arm for sketching in a downward movement to the right, (c). Include at least three points in each movement.
4. Keep an eye on next point, (d). Rotate paper for convenience.
5. Lightly sketch complete circle.
6. Darken circle with a uniform dense line by letting eye lead pencil along light construction line, (e).

ENCLOSING-SQUARE METHOD

The following steps are necessary when sketching by the enclosing-square method:

1. Lightly sketch a square equal in size to diameter of desired circle, Fig. 3-13 (a).
2. Sketch center lines.
3. Sketch circle, starting with tangent arcs at midpoints of square, (b).
4. Complete circle, (c).

HAND-PIVOT METHOD

The hand-pivot method is a quick and easy method of sketching circles.

1. With pencil in hand, let finger pivot on center of circle and rotate paper slowly, Fig. 3-14, (a) and (b).
2. Sketch a light line by this method and darken later. Some draftsmen use two pencils with second one serving as the pivot, (c).

FREE-CIRCLE METHOD

The free-circle method of freehand sketching involves more skill in performance but can be developed with practice. The following steps are used:

1. Decide location of circle. With a forearm movement, lightly describe several full circles, Fig. 3-15 (a).
2. Select and darken circle with best form and erase extra lines, (b).

It may be seen that this is the fastest and most useful of all methods once skill is achieved.

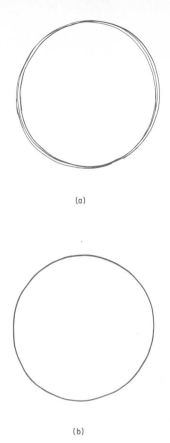

(a)

(b)

Fig. 3-15. Free-circle method of sketching a circle.

SKETCHING ELLIPSES

Occasionally it is necessary to sketch an ellipse. Three methods are presented here to aid you in producing a good sketch.

RECTANGULAR METHOD

The rectangular method is similar to sketching a circle in the enclosing square.

1. Sketch a rectangle with a length equal to major axis of ellipse and width equal to minor axis, Fig. 3-16 (a).
2. Sketch ellipse with tangent arcs at midpoints of sides, (b).
3. Complete ellipse and darken, (c).

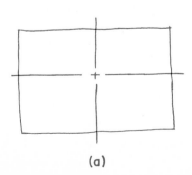

(a)

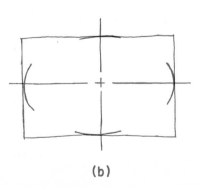

(b)

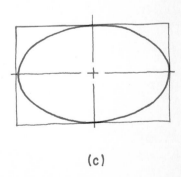

(c)

Fig. 3-16. Rectangular method of sketching an ellipse.

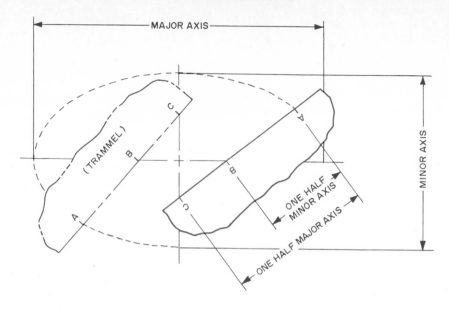

Fig. 3-17. Trammel method of sketching an ellipse.

TRAMMEL METHOD

Four steps are used to do freehand sketches by the trammel method:

1. Sketch major and minor axes, Fig. 3-17.
2. On a scrap of paper (the trammel), lay off three points A, B and C as shown, with AB equal to one-half of minor axis and AC equal to one-half of major axis.
3. With B and C on axes of ellipse, plot arc of ellipse.
4. Sketch ellipse through these points and darken.

FREE-ELLIPSE METHOD

The free-ellipse method is the same technique as used for sketching by the free-circle method. Observe the following steps:

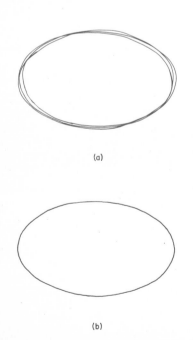

(a)

(b)

Fig.3-18. Free-ellipse method of sketching an ellipse.

1. With a forearm movement, lightly describe two or three full ellipses, Fig. 3-18 (a).
2. Select and darken ellipse with best form and erase extra lines, (b).

SKETCHING IRREGULAR CURVES

An irregular curve may be sketched freehand by connecting a series of points at intervals of 1/4 to 1/2 inch along its path, Fig. 3-19. Include at least three points in each stroke. Lead out of the previous curve into the next.

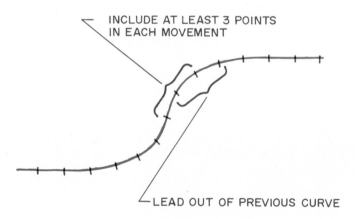

Fig. 3-19. Sketching an irregular curve.

PROPORTION IN SKETCHING

There is more to sketching than making straight or curved lines. Sketches must contain correct proportions.

Proportion is the relation of one part to another or to the whole object. You must keep the width, height and depth of the object in your sketch in the same proportion to that of the

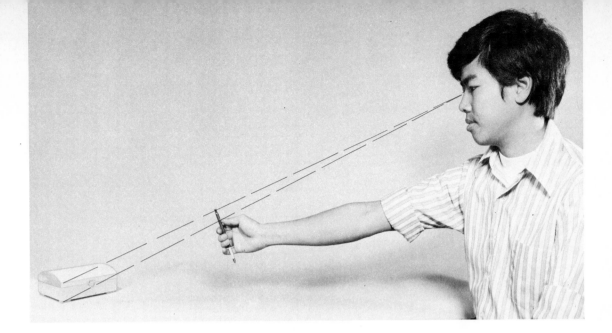

Fig. 3-20. Gaging proportions in sketching by pencil-sight method.

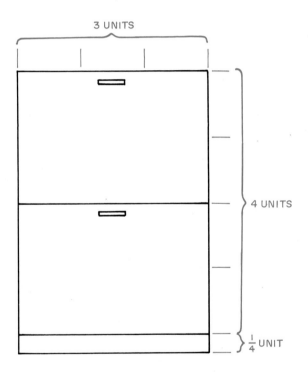

3 UNITS

4 UNITS

$\frac{1}{4}$ UNIT

TWO DRAWER FILE

Fig. 3-21. Unit method of gaging proportions.

object itself. If not, the sketch may not convey an accurate description.

There are several techniques which may be used by the drafter to obtain good proportions. One of these is the ''pencil-sight'' method. With pencil in hand, extend your arm forward in a stiff arm position and use your thumb on the pencil to gage the proportions of an object, Fig. 3-20. These distances may be laid off directly on your sketch.

The size of your sketch may be varied by the distance you stand from the object. To increase the size of the sketch, stand closer to the object; to decrease, stand farther away. However, in estimating the proprtions of an object for a particular sketch, always stand in the same position. The pencil-sight technique is particularly useful in making sketches of an actual object rather than from a picture of the object.

Another useful technique in estimating proportions is the ''unit'' method. This method involves establishing a relationship between distances on an object by breaking each distance into units. Compare the width to the height and select a unit that will fit each distance, Fig. 3-21. Distances laid off on your sketch should be the same proportion although the units may vary in size. This method is useful when making a sketch from a picture of the object.

You will find it helpful in sketching to first enclose the object in a rectangle, square or other appropriate geometric form of the correct proportion. Then subdivide this form to

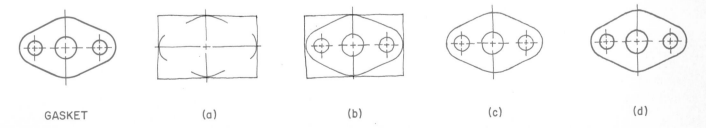

GASKET   (a)   (b)   (c)   (d)

Fig. 3-22. Steps involved in sketching the gasket shown at the left.

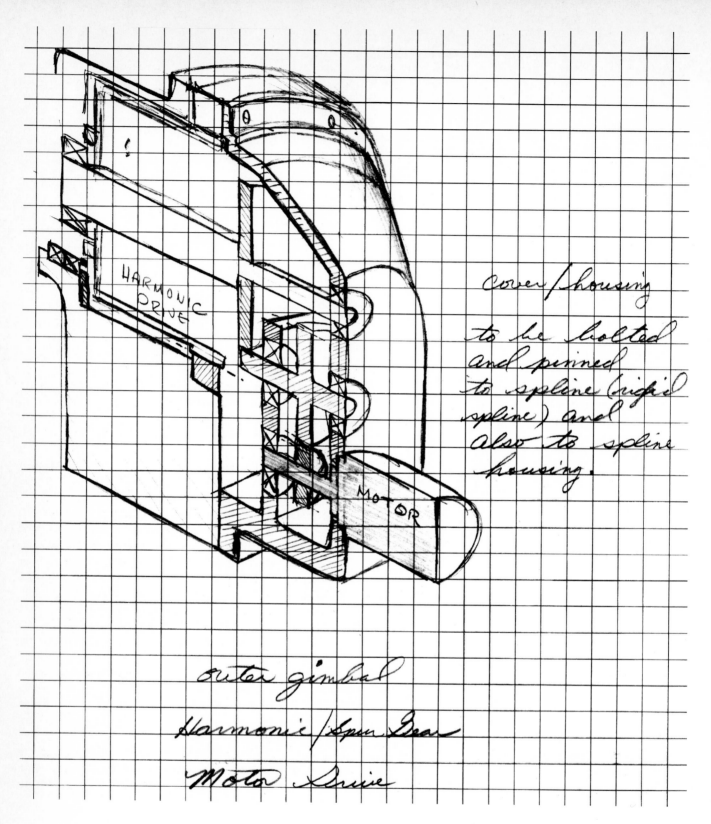

HARMONIC DRIVE

Cover / housing to be bolted and pinned to spline (rigid spline) and also to spline housing.

MOTOR

outer gimbal

Harmonic / Spur Gear

Motor Drive

Fig. 3-23. A preliminary design sketch on grid paper.

obtain the parts of the object. Once the outside proportions are established, the smaller parts are easily divided.

Proportioning is a matter of estimating distances and locating these correctly in your sketch. This takes skill acquired through continuous practice. Try laying off distances two, three or four times the length of another. Also practice dividing a distance into halves, thirds, fourths and fifths. Check your estimated proportions with a scale.

STEPS IN SKETCHING AN OBJECT

The following steps will help you lay out and complete freehand technical sketches. Care should be taken to make the sketch large enough to show the smallest details of the object.

1. Enclose the object in a rectangle, square, etc., of the correct proportion, Fig. 3-22 (a).
2. Sketch major subdivisions and details of the object, (b).

3. Remove unnecessary lines and clean sketch with a soft vinyl eraser, (c).
4. Darken finish lines of the object, (d).

AIDS TO FREEHAND SKETCHING

Several aids to freehand sketching have been discussed: scraps of paper for measuring, cross sectioned and isometric ruled paper, enclosing square for sketching circles, etc. The aim in freehand sketching, however, should be to develop your skill to a point where aids are no longer necessary.

Sketching is a means for a drafter to quickly communicate a shape and size description of an object to another person. It is also used to record information in the field for use at a later time. Being able to do freehand sketching without aids saves time and gives you the freedom to think while sketching a design problem.

DESIGN SKETCHING

Sketching is a valuable tool in the design stages of product development. Most engineers and designers use sketches of ideas in the early steps of design leading to detail drawings, Fig. 3-23. You will have an opportunity in Chapter 9 to use your sketching technique in the solution of design problems.

PROBLEMS AND ACTIVITIES

The problems presented here have been carefully selected to provide meaningful practice in freehand technical sketching. Use 8 1/2 x 11 in. (size A) plain paper — unless directed otherwise by your instructor — plus a pencil and an eraser. Scales and straightedges are not to be used.

Practice sketching strokes on scrap paper as you review each section prior to doing the assigned problems. Do those problems assigned by your instructor.

SKETCHING STRAIGHT LINES AND ANGLES

Sketch a border (see Fig. 3-9), then divide your sheet into four rectangles (estimate the dividing point, do not measure), Fig. 3-24. Proceed as follows:
1. Sketch horizontal lines in rectangle No. 1. Allow 1/2 inch space between lines and strive for straight sharp lines.

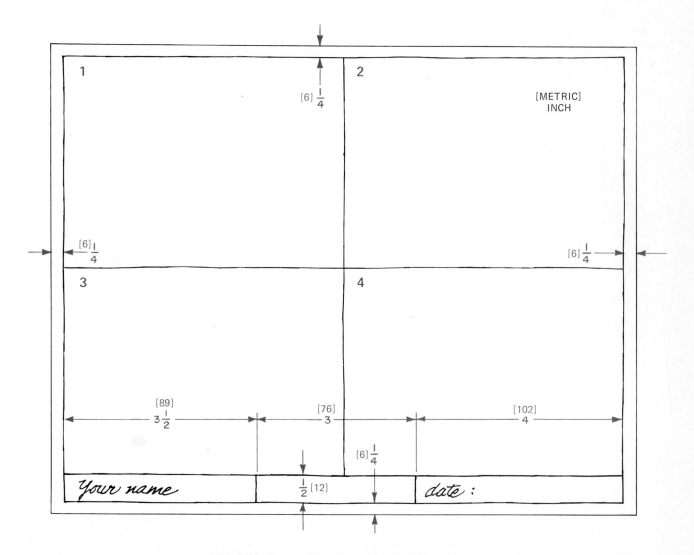

Fig. 3-24. Suggested sheet layout for sketching problems.

2. Sketch a series of vertical lines in rectangle No. 2. Allow 1/2 inch between lines and work to achieve true vertical lines.

3. Sketch inclined lines in rectangle No. 3. Your angle of inclination will be determined by the first line. Sketch this line as a diagonal between opposite corners of the rectangle. Space additional parallel lines 1/2 inch from this line. Your lines should be straight, sharp, parallel and uniformly spaced.

4. Use the bottom line of rectangle No. 4 as a reference line. Starting 1/4 inch from the left end of this line, sketch the following angles at 1/4 inch intervals (about 2 inches in length) starting upward and to the right: 75, 45 and 20 degrees. Sketch a second series of angles in the same manner from the right end of the reference line. Slant these upward and to the left at 60, 35 and 15 degrees. Sign your name and date the sketch.

SKETCHING CIRCLES AND ARCS

Divide your sheet into four rectangles as shown in Fig. 3-24 and sketch the following circles and arcs:

5. Sketch, by the center-line method, a 2 1/2 inch circle in rectangle No. 1. Center the circle in the rectangle. Your finished circle should appear as one sharp, freehand line.

6. Sketch two circles in rectangle No. 2, using the enclosing square and hand-pivot methods. Select the size of the circles so that space is well used but not crowded. Darken the finished circles, but retain the light construction lines for review by your instructor.

7. In rectangle No. 3, sketch a 2 inch diameter circle, using the free circle method. Locate the circle in the center of the rectangle. Erase your light "trial" circle and darken the finished circle.

8. Lightly sketch a rectangle inside rectangle No. 4 at a distance of 1/2 inch inside the border lines. Sketch an arc in each of the corners starting at the lower left hand corner and working clockwise around the rectangle. The arcs have radii of: 1/2, 1 1/2, 1 and 3/4 inch. Darken the finished rectangle and arcs. Sign your name and date the sketch.

SKETCHING ELLIPSES AND IRREGULAR CURVES

Divide your sheet into four rectangles as shown in Fig. 3-24. Sketch the following ellipses and irregular curve:

9. Sketch an ellipse in rectangle No. 1 with a major axis of 4 inches and a minor axis of 2 1/2 inches, using the rectangular method. Do not erase your construction lines

but darken the finished ellipse.

10. In rectangle No. 2, sketch an ellipse with a major axis of 3 inches and a minor axis of 2 inches, using the trammel method. Darken the finished ellipse.

11. Sketch an ellipse in rectangle No. 3 with a major axis of 2 1/2 inches and a minor axis of 1 1/2 inches, using the free ellipse method. Your ellipse should be uniform on both ends. Darken the finished ellipse.

12. On scrap paper, draw an irregular curve similar to the one in Fig. 3-19. Make the curve a suitable size to fit rectangle No. 4 and position it over the rectangle. With a sharp pencil, press lightly to locate points along the curve on the drawing sheet approximately 1/2 inch apart. Sketch the irregular curve through these points. To obtain a smooth curve, make sure your strokes lead out of the previous curve into the next set of points. Sign your name and date the sketch.

SKETCHING OBJECTS ON PLAIN PAPER

13. Sketch on 8 1/2 x 11 inch plain paper (bond, drawing or tracing) those problems in Fig. 3-25 assigned by your instructor. Place only one problem on a sheet and do not dimension the object. Select a size for the object that presents a pleasing appearance on the sheet. Strive for good line quality and proportion. Sign your name and date the sketch.

SKETCHING OBJECTS ON CROSS SECTIONED PAPER

14. From among those problems in Fig. 3-26, sketch on 8 1/2 x 11 cross section paper (4, 5, 8 or 10 squares per inch) those problems assigned by your instructor. Place only one problem on a sheet and do not dimension the object. Select a size for the object that provides a suitable appearance. Strive for good line quality and proportion. Sign your name and date the sketch.

SKETCHING REAL OBJECTS

15. Sketch one view of those objects in the drafting room assigned by your instructor. Use plain or cross section paper and center the view on the sheet. Review the steps in sketching an object presented in this chapter. Your sketch should reflect your best sketching technique.

OUTSIDE CLASS ACTIVITY

16. Select some object at home, work or in the community and sketch one view that best describes object. Use plain or cross section paper and present your sketch in class.

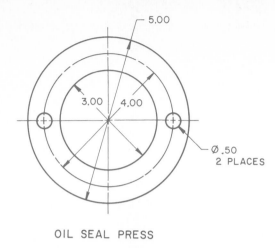

OIL SEAL PRESS

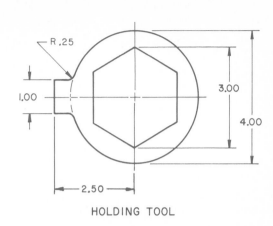

HOLDING TOOL

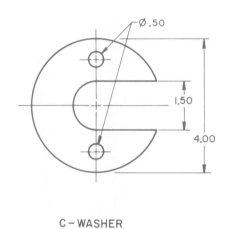

C – WASHER

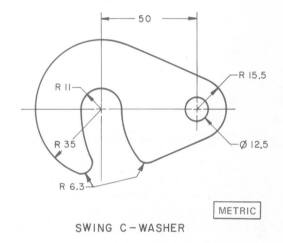

SWING C – WASHER

METRIC

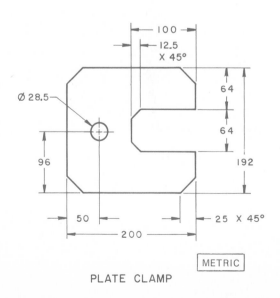

METRIC

PLATE CLAMP

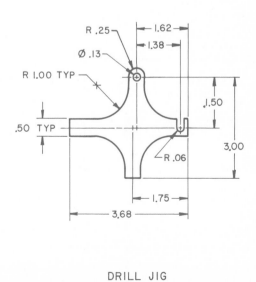

DRILL JIG

Fig. 3-25. Problems for sketching on plain paper.

Drafting for Industry

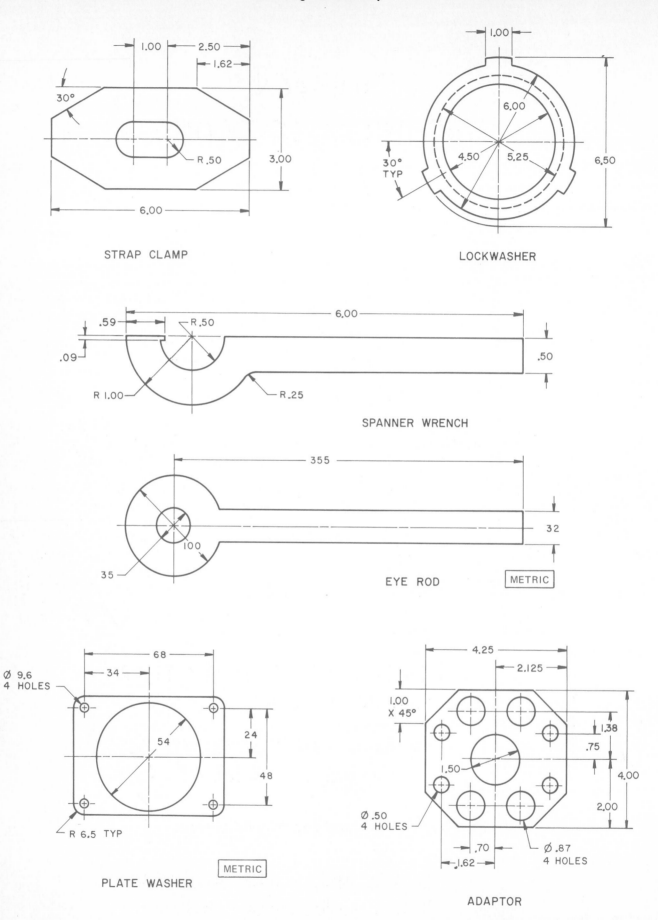

STRAP CLAMP

LOCKWASHER

SPANNER WRENCH

EYE ROD   METRIC

PLATE WASHER   METRIC

ADAPTOR

Fig. 3-26. Problems for sketching on cross section paper.

Chapter 4
TECHNICAL LETTERING

The purpose of lettering on a drawing is to further clarify the projections (views) by providing notes that specify materials and processes. Although more use is being made of lettering typewriters, transfer letters and other devices, most drafters are still required to letter some drawings freehand.

It is estimated that lettering consumes 20 percent of the drafter's time in industry. And it is generally agreed that lettering affects the appearance of a drawing more than any other single factor. Lettering that is difficult to read could contribute to costly errors in manufacturing and servicing of parts and machines. The mastery of lettering techniques becomes essential so that this phase of drafting may be done legibly and quickly.

A careful study of the techniques presented in this chapter and concentrated practice sessions will enable you to develop skill in freehand technical lettering.

PENCIL TECHNIQUE

Freehand lettering is accomplished with less pressure than when the pencil is guided by a drawing instrument. A softer lead (having less clay) is used to maintain equal density (darkness) with lines on the drawing. An HB, F or H pencil, sharpened to a conical point will provide pencil tracings of sufficient quality to reproduce good blueprints.

Your forearm should be fully supported on the table with your hand resting on its side. Your third and fourth fingers should rest on the board, Fig. 4-1, and your forefinger should be on top of and in line with the pencil.

Hold the pencil firmly, but not too tight. Should your fingers tire, pause and flex them a few times. It may help you improve your work. Rotate the pencil frequently to maintain a conical point and produce letters of uniform width.

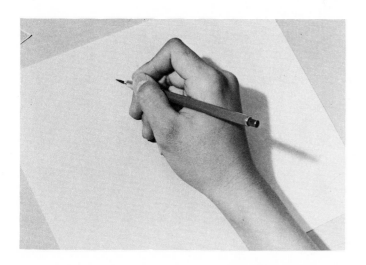

Fig. 4-1. Position of hand and arm for freehand lettering.

(a) ROMAN   BRAZIL, CHILE AND OTHER
It shows civilization at the time so

(b) ITALIC   *BRAZIL, CHILE AND MEXICO*
It shows civilization at a time so remote

(c) TEXT   Announcement Big Social Gathering

(d) GOTHIC   BRAZIL AND SOUTH AMERICAN
It shows civilization at such a remote

Fig. 4-2. Four main styles of lettering.

STYLES OF LETTERING

Letter styles may be classified in four groups: Roman, Italic, Text and Gothic. The Roman, Fig. 4-2 (a), is characterized by thick and thin lines with "accented" strokes. The Italic letters shown, (b), are similar to the Roman, but inclined. Text, (c), includes all styles of Old English Cloister, Church, Black and German Text. These two styles of letters have their place in printing, various sign applications, and on specialty drawings such as map work and technical illustrations.

For many years now, Gothic letters have become the standard style used in industry, (d). Because of its legibility and ease of execution, it is known as "single-stroke" Gothic. This refers to the width of various parts of letters being formed by a single stroke rather than a number of strokes.

Upper case (capital) letters are recommended for use on machine drawings. They may be either vertical or inclined, but never mixed on the same drawing. Lower case letters are used for notes on maps and other topographical drawings.

Architectural lettering is similar in style to vertical capital Gothic lettering, but less mechanical in appearance, Fig. 4-3. Lower case letters are seldom used in architectural work.

ARCHITECTURAL PLANS
NO. 12738

Fig. 4-3. A typical architectural style of lettering.

LEFT-HANDED DRAFTERS

Many drafters are left-handed and perform just as well as those who are right-handed. The lettering procedure given here is for right-handed persons, but it may be easily adapted by the left-hander.

The left-hander should follow a system of strokes that involve pulling the pencil or pen instead of pushing which tends to dig into the paper. The movement on horizontal strokes may best suit you from right to left and vertical strokes from the top down.

GUIDELINES

Two types of guidelines, Fig. 4-4, are used in freehand technical lettering to maintain uniformity in height and slope. These are very light lines drawn with a sharp pencil. A 4H or harder pencil is preferred, but your regular drawing pencil may be used. Take care, however, to draw lines that are light enough not to be seen at arms length from the drawing.

Horizontal guidelines may be spaced with dividers or with a scale, Fig. 4-5 (a). Spacing for the body of lower case letters is two-thirds the height of capitals. Small capitals when used with large capitals are two-thirds to four-fifths the height of large capitals, (b).

The spacing of lines for lettering appears best when the distance between lines is from one-half to one letter height, (b). Vertical or inclined guidelines are drawn at random with the triangle against the T-square or straightedge.

Several useful devices are available for drawing horizontal and vertical guidelines. The Ames Lettering Guide, Fig. 4-6, may be set for drawing guidelines for letters 1/16 to 2 inches in height.

To use the guide, place it in position along the T-square or straightedge, insert the point of a sharp pencil in the holes at the desired spacing and draw the horizontal guidelines. Vertical or inclined guidelines are drawn as shown in Fig. 4-4.

The Braddock-Rowe Triangle, Fig. 4-7, is also useful for drawing horizontal and slope guidelines. Numbers on the

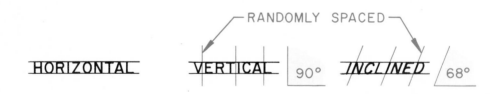

Fig. 4-4. Guidelines assist in keeping letters straight and at the correct angle.

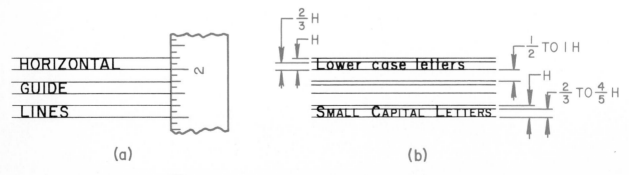

Fig. 4-5. Spacing of horizontal guidelines.

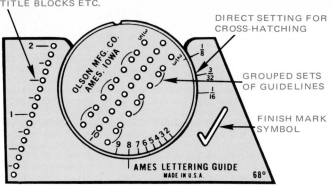

Fig. 4-6. Ames Lettering Guide illustrating a variety of uses.

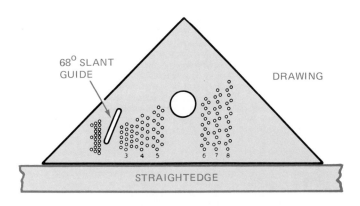

Fig. 4-7. Braddock-Rowe Triangle simplifies drawing lettering guidelines. (Teledyne Post)

triangle indicate the height for capital letters in thirty-seconds of an inch. Example: No. 8 is 8/32 or 1/4 inch from top to bottom lines between the group of three holes.

A third device for drawing guidelines is the Parallelograph, Fig. 4-8, which provides horizontal guideline spacing in 32nds of an inch and in millimetres. Slopes for 68 and 75 degree inclined letters are also provided.

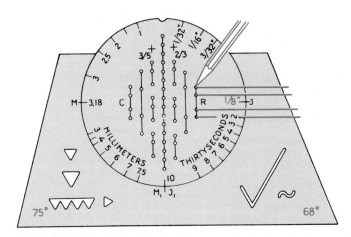

Fig. 4-8. Parallelograph lettering guide. (Gramercy)

SINGLE-STROKE GOTHIC LETTERS

Lettering strokes are the GUIDE POSTS to forming good letters. These strokes consist of straight-line stems, cross bars and well-proportioned ovals, carefully combined to produce a well balanced letter form.

The vertical Gothic alphabet and numerals shown in Fig. 4-9 are broken into groups of letters and numerals of similar strokes. Close study of this illustration will assist you in learning the order and direction of strokes used in forming each letter. The width of letters will vary from one space for the letter "I" to six spaces for the "W."

Fig. 4-9. Vertical capitals and numerals.

Draw vertical and inclined strokes with a movement of the fingers, Fig. 4-10. Form horizontal strokes by pivoting the hand at the wrist with a slight finger movement as needed to maintain a straight line, Fig. 4-11. Ovals are formed with a combination of hand and finger movement, Fig. 4-12. They are perfect ellipses with a major and minor axis.

Inclined Gothic capital letters and numerals are drawn in a manner similar to vertical letters and numerals, except the vertical axis is at an angle between 68 and 75 degrees, Fig. 4-13.

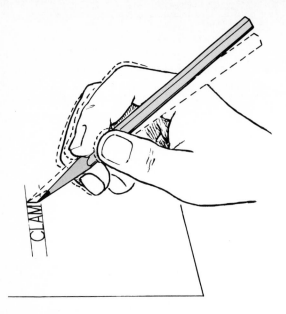

Fig. 4-10. Vertical and inclined strokes are made by finger movement only.

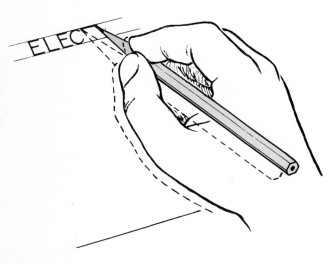

Fig. 4-11. Horizontal strokes are formed by a movement of the hand at the wrist, along with a slight finger movement.

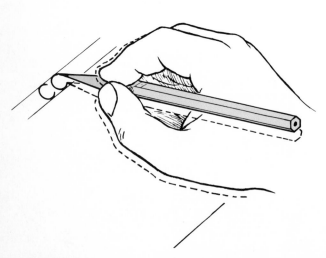

Fig. 4-12. Ovals are formed by movement of the hand and fingers in combination.

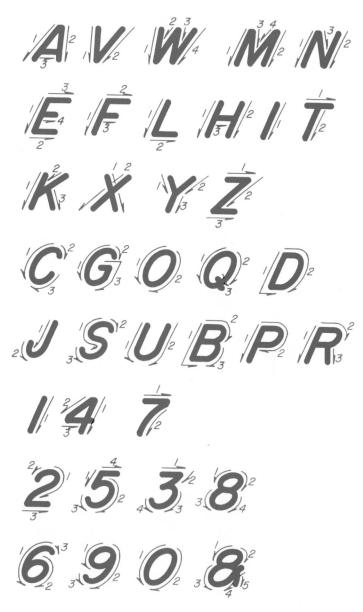

Fig. 4-13. Inclined Gothic capital letters and numerals.

Lower case vertical Gothic letters are formed as shown in Fig. 4-14; lower case inclined Gothic in Fig. 4-15. The body of the lower case letters is two-thirds the height of capital letters. Ascending or descending stems are equal in length to the height of the capitals.

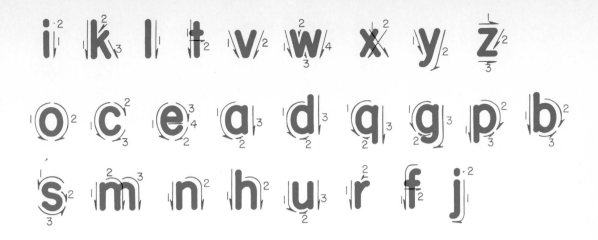

Fig. 4-14. Vertical Gothic lower case letters.

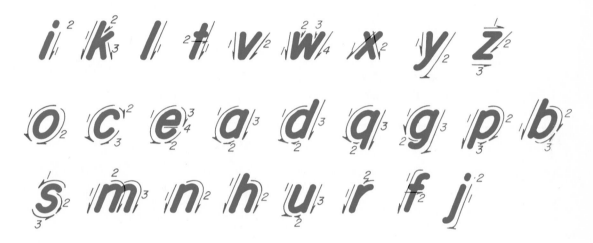

Fig. 4-15. Inclined Gothic lower case letters.

COMBINING LARGE AND SMALL CAPITALS

Some drafters use large and small capitals in combination for titles or notes on drawings. This combination is more easily read than lower case letters, Fig. 4-16. Height of small capitals should be two-thirds to four-fifths that of large capitals.

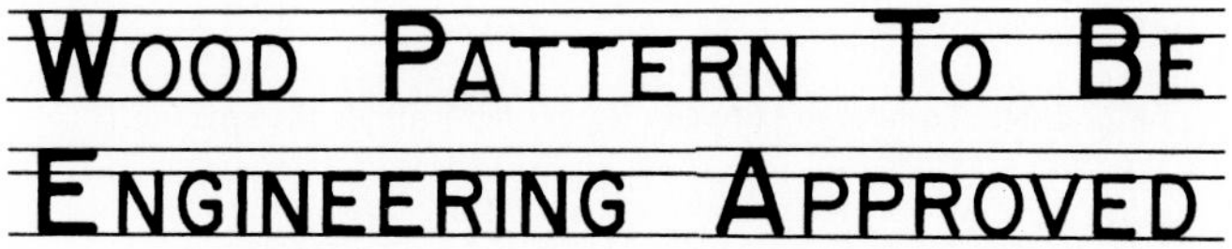

WOOD PATTERN TO BE
ENGINEERING APPROVED

Fig. 4-16. Large and small capitals used in combination.

Skill in lettering comes with careful study of the form of the letters — and by diligent practice. With practice, any student with a talent for drafting can learn to produce good letters.

PROPORTION IN LETTERS AND NUMERALS

Once the technique of forming the letters and numerals is understood, care must be given to proportioning each element.

Proportion is necessary to present a neat appearance where letters are formed into words and sentences, Fig. 4-17.

There may be times, however, when it is necessary or desirable to compress or expand letters or words. This need occurs when space is limited or if you want to attract attention, Fig. 4-18.

GOOD
PROPORTION

BAD
PROPORTION

Fig. 4-17. Proportion in forming individual letters is important.

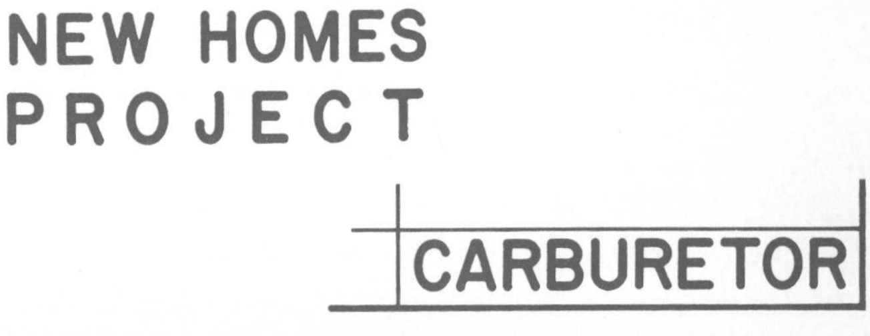

NEW HOMES
PROJECT

CARBURETOR

Fig. 4-18. Maintain good proportion in letters when compressed or expanded.

STABILITY IN LETTERS AND NUMERALS

Optical illusion is also a factor in lettering. For this reason, it is necessary to place the horizontal bar on letters such as B, E and H slightly above center. If not, they will appear top-heavy and unstable, Fig. 4-19. It is also necessary to draw the upper portion of letters such as B, K, R, S, X and Z and numerals 2, 3, 5 and 8 slightly smaller than the lower portion to show stability, Fig. 4-20.

B E H   **B E H**
STABLE   UNSTABLE

Fig. 4-19. Stability of letters.

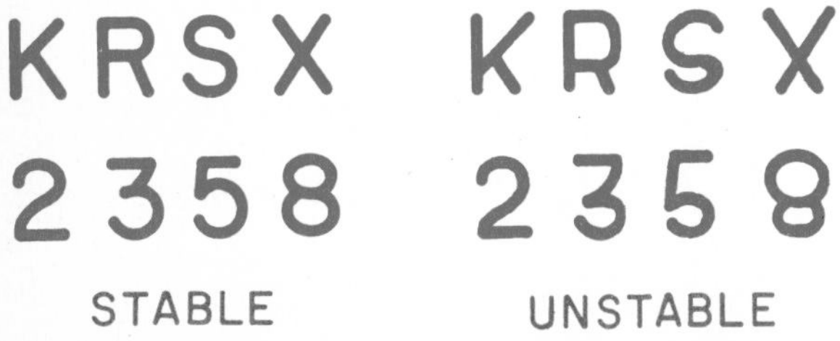

KRSX   **KRSX**
2358   **2358**
STABLE   UNSTABLE

Fig. 4-20. Stable letters and numerals are more pleasing in appearance.

QUALITY OF LETTERING

The appearance of lettering on a drawing is enhanced when uniformity is maintained in style, height, slope, spacing and line weight. The appearance of the drawing and skill of the drafters are reflected in the quality of lettering.

Care should be taken to form each letter correctly within the lightly drawn guidelines and maintain proper density. Lettering is a special technique which is different from writing, Fig. 4-21.

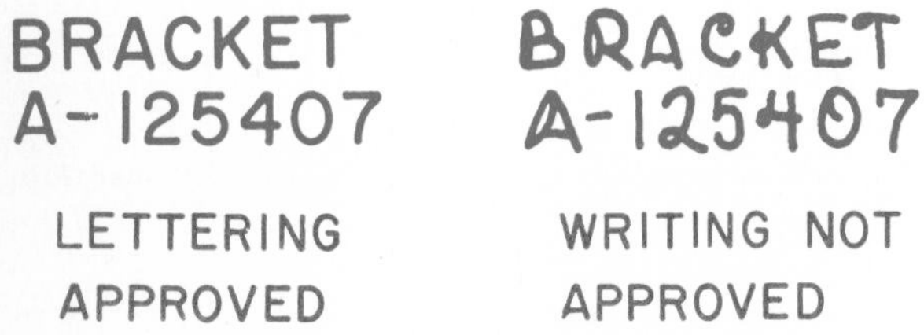

BRACKET   **BRACKET**
A-125407   **A-125407**

LETTERING   WRITING NOT
APPROVED   APPROVED

Fig. 4-21. Lettering requires a special technique that reflects the skill of the drafter.

COMPOSITION OF WORDS AND LINES

The manner in which letters are combined into words and sentences tends to reveal the drafter's lettering skill. Many beginning drafters tend to space the letters of words too

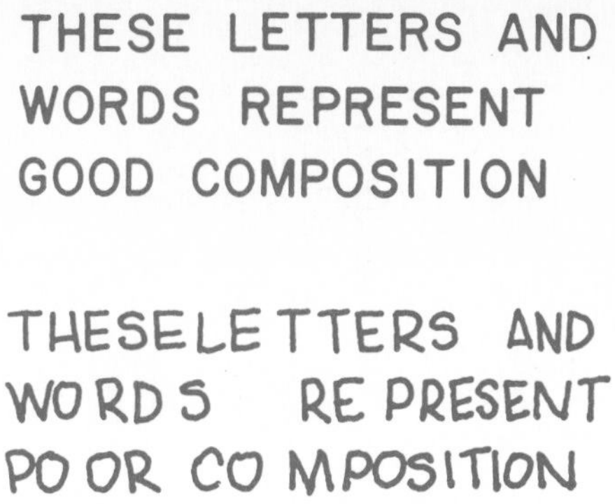

THESE LETTERS AND WORDS REPRESENT GOOD COMPOSITION

THESELETTERS AND WORDS REPRESENT POOR COMPOSITION

Fig. 4-22. Composition of words and lines.

widely and crowd the spacing between words. Lettering of this nature is difficult to read, Fig. 4-22.

Correct spacing of letters within words is based upon the total area between two letters, not just the distance between letters. When letters are spaced an equal distance apart, poor composition results, Fig. 4-23.

BORE TO OBTAIN
.0002 CLEARANCE

Fig. 4-23. Equal distant spacing of letters results in poor composition.

The shape of the letters themselves determines how much spacing should occur between any two letters. For example, when "A" and "M" are next to each other, less distance is required than for "I" and "M," Fig. 4-24. When "T" and "C" are next to each other they nearly overlap, as does "LT."

AM   IM   TC   BOLT

Fig. 4-24. Total area between letters determines the spacing.

Words should be separated by a space equivalent to the letter space "O," Fig. 4-25. Two letter spaces of "O" are allowed between the end of one sentence and the start of the next, Fig. 4-25.

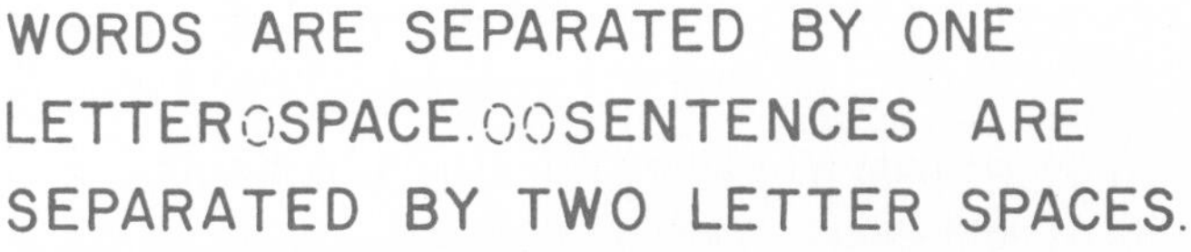

WORDS ARE SEPARATED BY ONE LETTER SPACE. SENTENCES ARE SEPARATED BY TWO LETTER SPACES.

Fig. 4-25. Spacing between words and sentences.

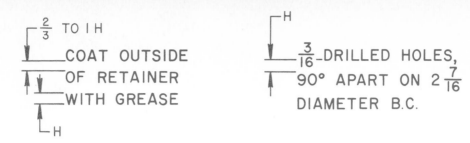

Fig. 4-26. Spacing between lines of letters.

The space between lines of letters should equal two-thirds letter height for readability and appearance, Fig. 4-26. Spacing may vary from one-half to one full letter height, depending on the space available, but it should be constant on a single drawing. Where fractions are involved, spacing between lines should be a full letter height.

SIZE OF LETTERS AND NUMERALS

Capital letters and whole numerals are usually a minimum of .12 inch in height for notes. Height for the drawing title and number is generally .25 inch.

The size of letters and numerals recommended in the Drawing Requirements Manual for the Departments of Defense and Commerce are as follows:

USE	SIZE (MIN.)
Drawing number	.38
Drawing Title, Code Identification No.	.20
Letters and Numerals	.156*
Tabulated and Section Letters	.40
The words "SECTION", "VIEW"	.20

*This applies to letters and numerals on Field of Drawing, in general notes, revision block and parts list. All tolerances are the same character size as the dimension. (Globe Engineering Documentation Services, Inc., Newport Beach, Calif.)

The IBM Corporation's Drafting Manual lists minimum letter sizes as shown in the table below.

The overall height of fractions is twice the height of whole numbers, Fig. 4-27 (a). Note that the numerator and denominator are smaller than the whole number, and they are separated by .08 inch so that they DO NOT touch the fraction bar. This avoids confusion in reading fractions on a drawing and provides legible copy from microfilm enlargements. Limit dimensions are shown with a space of .08 inch minimum between the upper and lower dimensions, Fig. 4-27 (b).

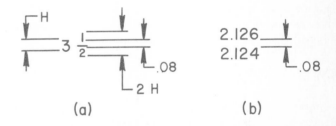

Fig. 4-27. Size of whole numbers, fractions and limit dimensions.

ITEM	SIZES							
	A		B		C		D and larger	
	inches	mm	inches	mm	inches	mm	inches	mm
DRAWING AND PART NUMBER	.250	6	.250	6	.312	8	.312	8
TITLE	.125	3	.156	4	.156	4	.188	5
LETTERS AND FIGURES FOR BODY OF DRAWING (INCLUDING DIMENSIONS AND NOTES)	.125	3	.156	4	.156	4	.188	5
TOLERANCES	.100	2.5	.125	3	.125	3	.156	4
DESIGNATION OF VIEWS, SECTIONS, DETAILS AND DATUM: "VIEWS", "SECTION" AND "DETAIL".	.156	4	.188	5	.188	5	.250	6
"A–A", "B", "X–X", etc.	.188	5	.250	6	.250	6	.312	8
SECURITY CLASSIFICATION, DRAWING STATUS, REFERENCE DRAWING, etc.	.250	6	.250	6	.250	6	.250	6
TYPED SCHEMATICS, DIMENSIONS, AND NOTES ON BODY OF DRAWING	.100	2.5	.125	3	.142	4	.142	4

MINIMUM LETTER SIZES

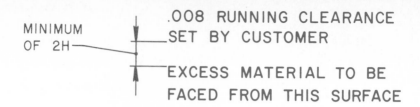

Fig. 4-28. Minimum spacing between separate notes.

NOTES ON DRAWINGS

Notes on drawings supplement the graphic presentations and are lettered on the drawing to be read horizontally. Capital letters at least .12 inch high are preferred.

Minimum spacing between lines within a note is 2/3 H (full letter height on drawings to be microfilmed) as shown in Fig. 4-26. Spacing between two separate notes should be at least two letter heights, Fig. 4-28.

BALANCING WORDS IN A SPACE

It is necessary at times to balance (space evenly) words within a limited space, such as a title block, Fig. 4-29. This may be done by first lettering the title or note on scrap paper using appropriate guidelines. Then, slip the copy under your tracing paper or vellum and adjust it to suit. In the event you are working on an opaque surface, measure the scrap copy and transfer the starting point to the drawing.

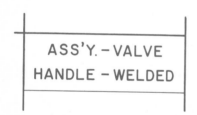

Fig. 4-29. Words balanced within a given space.

LETTERING TO BE MICROFILMED

Drawings to be microfilmed require special lettering. The letters must be large enough and dense enough to reproduce clearly after enlargement on a print one half its original size.

Some companies have modified certain letters and numerals used on drawings to be microfilmed in order to improve their clarity on blowbacks, Fig. 4-30.

LETTERING WITH PEN AND INK

Most drawings to be used for photographic reproductions in technical publications are inked. These usually are in the form of presentation drawings. Inked drawings are made if a number of copies of prints are to be made over a long period of time.

Inking of freehand letters on drawings requires a special technique and proper equipment. Several styles of pens are available, and standard lettering pen points come in a range of sizes from very fine to heavy.

Freehand inked letters should be done with a technical pen, Figs. 4-31 and 4-32. It has the advantage of enabling uniform stroke widths and does not vary, as do the more flexible pen points.

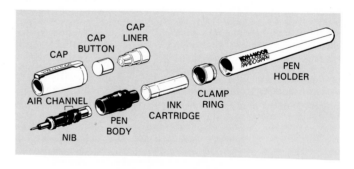

Fig. 4-31. Technical pen for freehand inking of letters.
(Koh-i-noor Rapidograph, Inc.)

Technical pens are made for use with specially formulated inks. They consist of a pen point, a needle running through the point to maintain the ink flow and an ink reservoir, Fig. 4-31.

Most technical pens are shaped to permit their use with

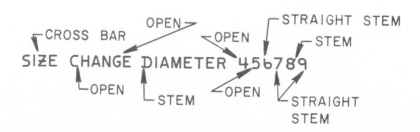

Fig. 4-30. Examples of modified letters and numerals used on drawings to be microfilmed.

Leroy lettering devices similar to the one shown in Fig. 4-33. These pens come in a range of sizes suitable for use with various size lettering templates. The technical pen also may be used for inking lines on a drawing (see page 50).

MECHANICAL LETTERING DEVICES

Since the inking of letters and numerals requires considerable skill to achieve satisfactory results, many drawings to be inked are lettered with mechanical devices, Fig. 4-33.

The lettering guide consists of a scriber, pen and template. The scriber is adjustable for making inclined as well as vertical lettering, and templates are available in a variety of letter styles and sizes. Templates are also available for various graphic symbols such as electronic and architectural, Fig. 4-34.

To use the Leroy lettering guide:

1. Select template size and style desired and lay it along straightedge.

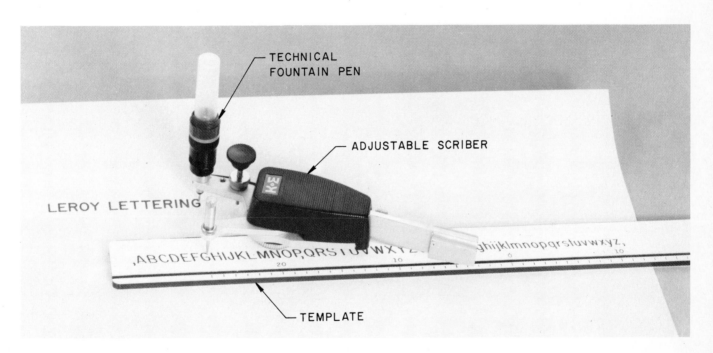

Fig. 4-33. Leroy lettering guide. (Keuffel & Esser Co.)

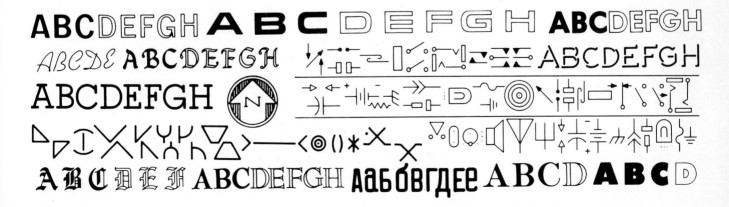

Fig. 4-34. A wide variety of letter and symbol templates are available for use with the Leroy lettering guide. (Keuffel & Esser Co.)

2. Place tail pin of scriber in straight groove of template and tracer pin in letter to be traced.
3. Adjust straightedge and template to correct position and trace letter.
4. Letters are spaced by eyeing space between letter being positioned and previous letter.
5. Words are spaced by positioning letter on template that just precedes letter of next word. For example, in note "SEAM WELD," Fig. 4-35, the "V" is positioned as if it were the next letter after the "M" in SEAM. The "W" is then correctly spaced to start the next word. In the event the letter preceding the next word is narrower (I) or wider (W) than a normal letter, make the adjustment by eye.

Fig. 4-35. Spacing words with the Leroy lettering guide.

6. Sentences are spaced in a similar manner, except two letter spaces precede the first letter of the next sentence, Fig. 4-36.

Fig. 4-36. Spacing between sentences with the Leroy lettering guide.

7. To estimate space required for a word or group of words, count number of letters and spaces. Then use scale on lower edge of each template to lay off space requirements.

Another type of lettering device is the Wrico lettering guide, Fig. 4-37. This device makes use of a template and a technical fountain pen. A variety of templates are also available for use with the Wrico lettering guide.

Fig. 4-37. Wrico lettering guide.   (Wood-Regan Instrument Co.)

A third type of mechanical lettering device is the height and slant control scriber, Fig. 4-38. This device operates from a template like the Leroy, but it is constructed in a way that permits the expansion or compression of letters by varying the height of the letter to the width.

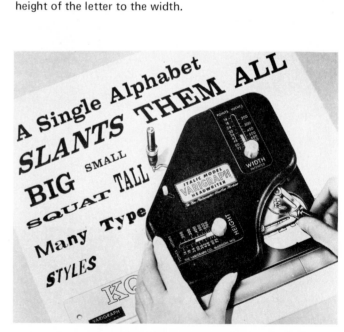

Fig. 4-38. Operation of the Varigraph lettering instrument.

NOTES AND SYMBOLS IN COMPUTER SYSTEMS

The computer is used by many industries today to prepare notes and symbols for drawings, Fig. 4-39. Computer-aided

Fig. 4-39. Direct entry of text material into the computer.

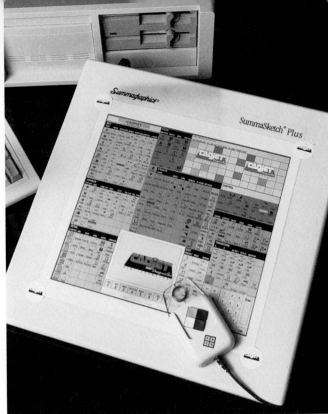

Fig. 4-40. The drafter selects the desired symbol and inputs it to the computer. (Summagraphics)

drafting speeds the lettering process and provides uniform, legible notes and symbols.

The notes and symbols are input to the computer and are output on the drawing by a plotter or printer. When the entire drawing is developed by computer graphics, the notes may be input through the keyboard and the symbols selected from the menu tablet, Fig. 4-40. The notes and symbols appear on the drawing in the proper position and are printed along with the

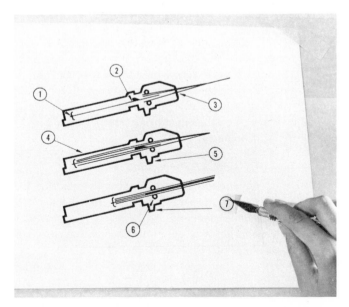

Fig. 4-41. Pressure sensitive symbol is applied to a drawing. (Formatt)

graphical solution.

PRESSURE SENSITIVE LETTERS AND SYMBOLS

Special printed sheets and tapes of pressure sensitive letters and symbols are available for use in lettering, dimensioning or applying certain symbols to drawings. These materials save industry considerable time in drafting. They are particularly useful where standard notes or symbols are frequently used. Fig. 4-41 shows a pressure sensitive symbol being transferred to a drawing.

PROBLEMS AND ACTIVITIES

FORMING SINGLE-STROKE GOTHIC LETTERS

Prepare a lettering layout sheet. Use one of the guideline devices previously described to lay out horizontal and slope guidelines as shown in Fig. 4-42, or use a commercial lettering sheet as directed by your instructor.

When all materials are ready, do the following Gothic lettering activities.

1. Draw vertical capitals as shown in Fig. 4-9. Letter alphabet and numerals as many times as space permits. Spacing for letter height is 3/8 inch (10 mm) and space between lines of letters is 3/16 inch (5 mm). Use vertical guidelines to keep letters uniform.

2. Draw inclined capitals as shown in Fig. 4-13. Letter alphabet and numerals as many times as space permits. Spacing for letter height is 1/4 inch (6 mm) and 1/8 inch (3 mm) between lines. Use 68 degree inclined guidelines.

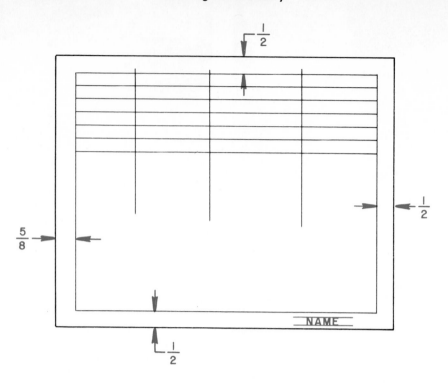

Fig. 4-42. Lettering activity sheet layout.

3. Draw vertical lower case letters as shown in Fig. 4-14. Repeat alphabet as many times as space permits. Spacing for letter height is 3/8 inch (10 mm) and 3/16 inch (5 mm) between lines. Use vertical guidelines.

4. Draw inclined lower case letters as shown in Fig. 4-15. Repeat alphabet as many times as space permits. Spacing for letter height is 1/4 inch (6 mm) and 1/8 inch (3 mm) between lines. Use inclined guidelines.

LETTERING DRAWING NOTES — INCH, (METRIC)

5. Letter the following drawing note in 1/8 inch (3 mm) vertical capitals on a drawing sheet. Decide on a pleasing spacing between lines and a line length not to exceed four inches. Center the note in the upper left-hand quadrant (quarter) of the sheet, starting one inch from the top border. Do not include the quotation marks in your note.

"Dimensions throughout are to 90 degree bend as assembled on product. Part should be overbent to 91 degrees for additional tension."

6. Use 3/16 inch (5 mm) inclined capitals to letter the following notes in the upper right-hand quadrant of the same sheet.

"Slot must have no external burrs"

".203 (5.2) Dia hole — pierce from both sides — four places"

7. Use 1/8 inch (3 mm) vertical capitals and lower case letters in lettering the following note in the lower left-hand quadrant of the sheet above.

"Note: An easement of four feet (1.3 m) on either side of this line is reserved for utility line thru properties."

8. Use 3/16 inch (5 mm) inclined initial capitals and small capitals to letter the following note in the lower right-hand quadrant of the sheet used for problems No. 5, 6 and 7.

"Emboss 5/16 (7.9) Dia x 1/16 (1.5) deep — inside only — four places."

9. Review several blueprints from industry and evaluate the lettering and dimension numerals as to style, size, uniformity and quality of reproduction. Be prepared to discuss your findings in class.

10. Select a note from an industrial blueprint and letter the note in the same size and style lettering. Compare your work with that done by the industrial drafter.

11. Select a title block from an industrial blueprint. Lay out on a sheet and letter the content in the same size and style. Compare your work with that on the blueprint.

12. Lay out a lettering sheet as shown in Fig. 4-42 for 3/16 inch (5 mm) letters. Letter the alphabet and numerals lightly in pencil on the upper half of this sheet and then ink the copy freehand.

13. With one of the mechanical lettering devices, letter in ink the following note in .175 inch (4.5 mm) high capital letters on the lower half of the sheet used in No. 12. Space allowed for the note is 2 1/2 inches (64 mm) in length and depth as required.

"Rivets must not be curled too tightly since they are at moving joints. Rivet heads should be outside."

14. Using pressure sensitive letters, prepare the following title.

"FORK BRACKET, SPINDLE RAM ASSY"

Chapter 5
GEOMETRY IN TECHNICAL DRAFTING BASIC

Engineers, architects, designers and drafters regularly apply the principles of geometry to the solution of technical problems. This is clearly evident in highway interchanges, architectural structures and complex machine parts, Fig. 5-1.

Since much of the work done in the drafting room is in geometric constructions, every person engaged in technical work should be familiar with solutions to common problems in this area. Methods of problem solving presented in this chapter are based on the principles of plane geometry.

However, constructions are modified and many short cuts are introduced to take advantage of the instruments available in modern drafting rooms.

Accuracy is very important in drawing geometric constructions. A slight error in laying out a problem could result in costly errors in the final solution. Use a sharp 2H to 4H lead in your pencil and compass. Make construction lines sharp and very light. They should not be noticeable when the drawing is viewed from an arm's length.

Fig. 5-1. Geometric construction is used in building highways and architectural structures and in manufacturing machine parts.
(The Gillette Company)

79

STRAIGHT LINES

There are a number of geometric constructions the drafter needs to perform using straight lines. Some examples are presented in the following paragraphs.

BISECT A LINE BY THE TRIANGLE AND T-SQUARE METHOD

1. Given line AB in Fig. 5-2 (a).
2. Draw 45 degree or 60 degree lines AC and BC.
3. Line CD, drawn through their intersection and perpendicular to line AB, bisects line AB.

BISECT A LINE BY THE COMPASS METHOD

1. Given line EF in Fig. 5-2 (b).
2. Using E and F as centers, strike radius arcs greater than one-half EF, scribing points G and H.
3. Draw line GH as a perpendicular bisector of line EF.

LINE DRAWING TECHNIQUE

DRAW A LINE THROUGH A POINT AND PARALLEL TO ANOTHER LINE, USING TRIANGLE AND T-SQUARE

1. Given the line AB and the point P through which a line parallel to AB is to be drawn, Fig. 5-3 (a).
2. Position the T-square or straightedge so the edge of the triangle lines up with line AB.
3. Without moving the T-square, slide the triangle until its edge is in line with point P.
4. Draw the parallel line CD.

DRAW A LINE THROUGH A POINT AND PARALLEL TO ANOTHER LINE, USING A COMPASS

1. Given the line EF and the point P through which a line parallel to EF is to be drawn, Fig. 5-3 (b).
2. With point P as the center, strike arc GK. With the same radius and K as the center, strike arc PJ.
3. With PJ as the radius and K as the center, strike an arc to locate G.
4. Draw line GP parallel to EF.

DRAW A LINE THROUGH A POINT AND PERPENDICULAR TO A GIVEN LINE, USING A TRIANGLE AND T-SQUARE

1. Given the line AB and the point P through which a line perpendicular to AB is to be drawn, Fig. 5-4 (a).

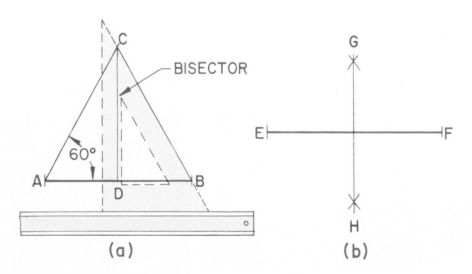

(a)   (b)

Fig. 5-2. Bisecting a line.

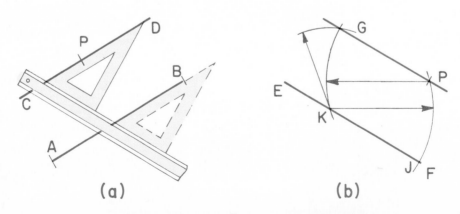

(a)   (b)

Fig. 5-3. Drawing a line through a point and parallel to another line.

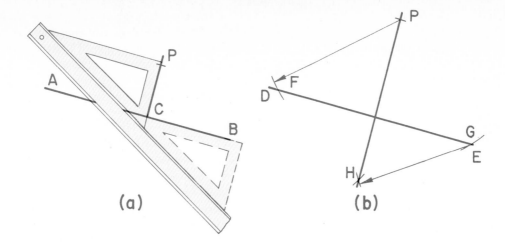

Fig. 5-4. Drawing a line through a point and perpendicular
to another line.

2. Position the T-square and triangle so an edge of the triangle adjacent to the right angle lines up with the line AB.
3. Without moving the T-square, slide the triangle so its other edge is in line with point P.
4. Draw the required perpendicular line PC.

DRAW A LINE THROUGH A POINT AND PERPENDICULAR TO A GIVEN LINE,

USING A COMPASS

1. Given the line DE and the point P through which a line perpendicular to DE is to be drawn, Fig. 5-4 (b).
2. With P as the center and using any convenient radius, strike arc FG.
3. Using the intersections of arc FG with line DE as centers, strike intersecting arcs at H.
4. Draw line PH perpendicular to DE.

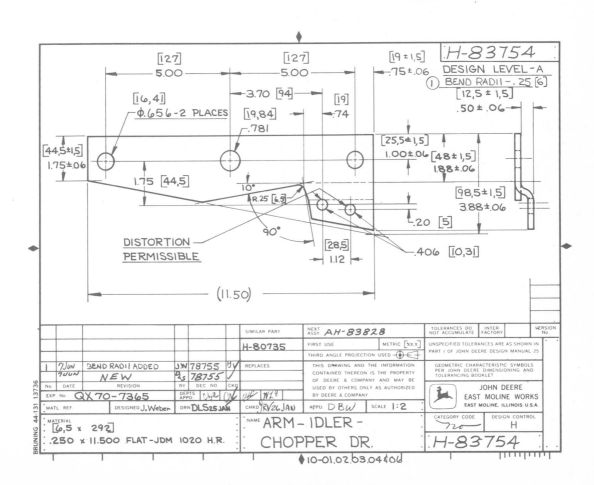

Fig. 5-5. A machine part requiring application of geometry in its layout.

DIVIDING A LINE

DIVIDE A LINE INTO A GIVEN NUMBER OF EQUAL
PARTS BY THE VERTICAL LINE METHOD

1. Given the line AB to be divided into eleven equal parts, Fig. 5-6 (a).
2. With the T-square and triangle, draw a vertical line at B.
3. Locate the scale with one point at A and adjust so that a multiple of equal divisions (in this case eleven 1/2 inch divisions) lay between A and the vertical line BC.
4. Mark vertical points at each of the eleven divisions and project a vertical line parallel to BC from these divisions to line AB. These verticals divide line AB into eleven equal parts.

DIVIDE A LINE INTO A GIVEN NUMBER OF EQUAL
PARTS BY THE INCLINED LINE METHOD

1. Given the line DE to be divided into six equal parts as in Fig. 5-6 (b).
2. Draw line from D at any convenient angle, with scale or dividers, lay off a multiple of equal divisions (six in this case).
3. Draw a line between F (the last division) and E.
4. With lines parallel to FE, project the divisions to line DE to divide the line into six equal parts.

DIVIDE A LINE INTO PROPORTIONAL PARTS
BY THE VERTICAL LINE METHOD

1. Given line AB to be divided into proportional parts of 1, 3 and 5, Fig. 5-7.
2. With T-square and triangle, draw a vertical line at B.
3. Locate the scale with one point on A and adjust so that a multiple of equal units (in this case 1 + 3 + 5 = 9 units) lays between A and the vertical line BC.
4. Vertically mark each of the proportions and project a vertical line from these divisions to line AB. These verticals divide line AB into proportional parts of 1, 3 and 5 units.

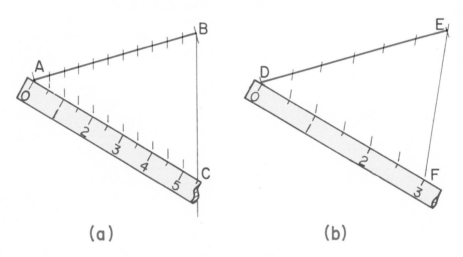

Fig. 5-6. Dividing a line into a number of equal parts.

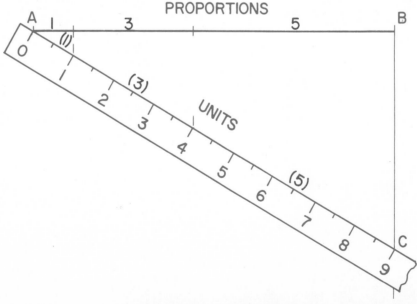

Fig. 5-7. Dividing a line proportionally.

ANGLES

Another geometric element in drafting is angle construction. The common terms applied to angles in drafting are illustrated in Fig. 5-8. The following symbols are used to represent angles and angular constructions: $\angle$ = angle; $\triangle$ = triangle; $\perp$ = $\llcorner$ = perpendicular. Techniques for drawing angular constructions are discussed in this section.

TRANSFER AN ANGLE
1. Given angle BAC to be transferred to a new position at A′ B′, Fig. 5-10 (a).
2. Strike arc D at any convenient radius at centers A and A′.
3. Adjust compass for arc between E and F and strike arc of same radius at center E′, (c).
4. Draw line A′ C′ thru F′ for new angle B′ A′ C′.

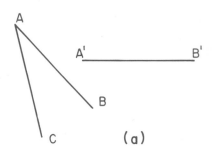

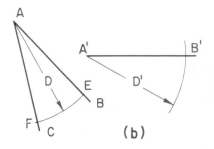

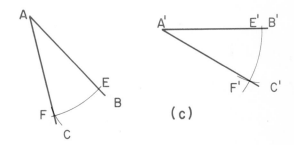

Fig. 5-10. Transferring an angle.

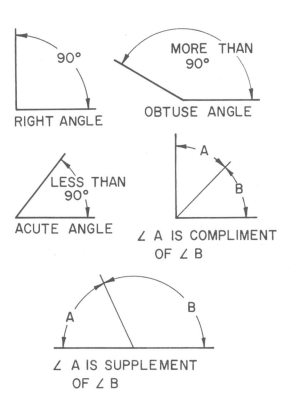

Fig. 5-8. Angle terminology.

BISECT AN ANGLE
1. Given angle BAC to be bisected, Fig. 5-9 (a).
2. Strike arc D at any convenient radius, (b).
3. Strike equal arcs and with radius slightly greater than one-half BC to intersect at E, (c). The line AE bisects angle BAC, (d).

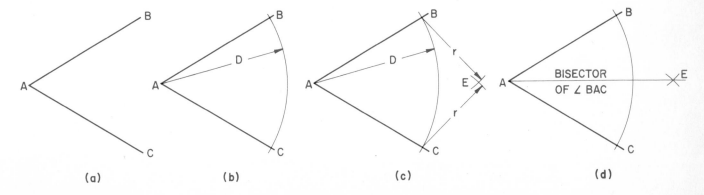

Fig. 5-9. Steps in bisecting an angle.

DRAW A PERPENDICULAR (90 degrees) TO
ANOTHER LINE BY THE 3, 4, 5 METHOD

1. Given line AB. Draw line BC perpendicular to line AB, Fig. 5-11.
2. Select a unit of any convenient length (1/4, 1/2, 1 inch) and lay off 3 units on line AB, (a).
3. With a radius of 4 units, strike arc C with center at B, (a).
4. With a radius of 5 units, strike arc D with center at A to intersect with arc C at F, (b). BF is perpendicular to line AB.

LAY OUT ANY GIVEN ANGLE
WITH THE PROTRACTOR

1. Given line AB, draw an angle of 35 degrees, counterclockwise at C, Fig. 5-12.
2. Locate the protractor accurately along line AB with the center at C.
3. Mark a short line in line with 35 degrees at D.
4. The angle DCB equals 35 degrees.

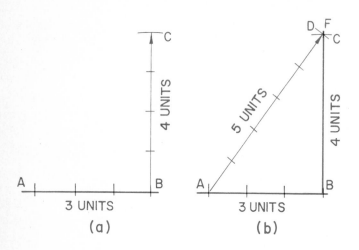

Fig. 5-11. Drawing a perpendicular line by the 3, 4, 5 method.

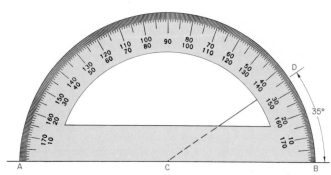

Fig. 5-12. Laying out angles with the protractor.

POLYGONS

A geometric figure enclosed with straight lines is called a polygon. Such figures are called regular polygons when their sides and interior angles are equal. Fig. 5-13 illustrates some of the more common polygons.

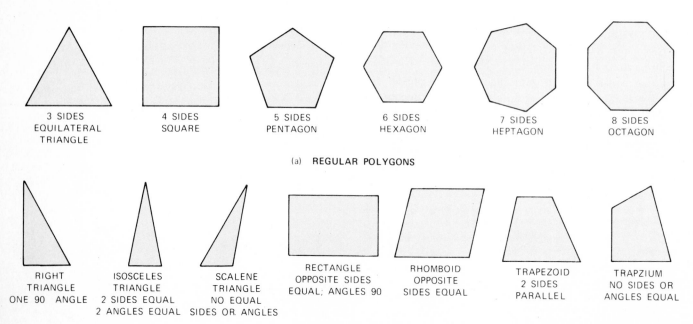

Fig. 5-13. Some common polygons.

Geometry in Technical Drafting — Basic

CONSTRUCTING TRIANGLES

CONSTRUCT A TRIANGLE WITH THREE SIDES GIVEN
1. Given the sides A, B, C, Fig. 5-14 (a).
2. Draw a base line and lay off side C, (b).
3. With side A as radius, lay off arc A.
4. With side B as radius, lay off arc B to intersect with arc A, (c).
5. Draw sides A and B to form required triangle, (d).

CONSTRUCT A TRIANGLE WITH TWO SIDES AND INCLUDED ANGLE GIVEN
1. Given the sides A and B and the included angle, as shown in Fig. 5-15 (a).
2. Draw a base line and lay off side B, (b).
3. Construct the angle at one end of line B and lay off side A.
4. Join the end points of the two given lines to form the required triangle, (c).

CONSTRUCT A TRIANGLE WITH TWO ANGLES AND INCLUDED SIDE GIVEN
1. Given angles A and B and included side AB, Fig. 5-16 (a).
2. Draw a base line and lay off side AB, (b).
3. Construct angles A and B.
4. Extend the sides of angles A and B until they intersect at C, (c). Triangle ABC is the required triangle.

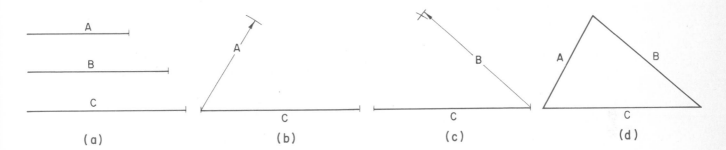

Fig. 5-14. Constructing a triangle with three sides given.

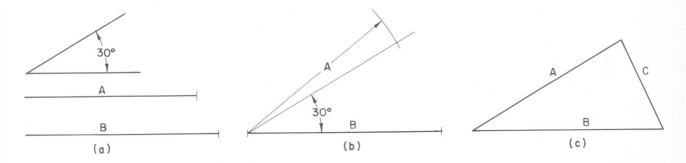

Fig. 5-15. Constructing a triangle with two sides and included angle given.

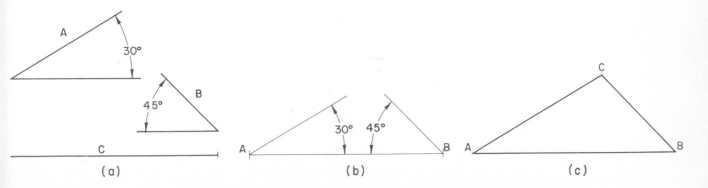

Fig. 5-16. Constructing a triangle with two angles and included side given.

CONSTRUCT AN EQUILATERAL TRIANGLE

An equilateral triangle has three equal sides and three equal (60 degree) angles.

1. Given side AB, Fig. 5-17 (a).
2. Draw a base line and lay off side AB, (b).
3. The T-square and triangle method is shown in Fig. 5-17 (c), and the compass method is shown in Fig. 5-17 (d). Triangle ABC is the required triangle, Fig. 5-17 (c and d).

CONSTRUCT AN ISOSCELES TRIANGLE

An isosceles triangle has two sides and two equal angles, Fig. 5-18.

1. If the length of the two equal sides and base are given, construct the triangle as shown in Fig. 5-14.
2. If the two equal angles and base are given, construct as shown in Fig. 5-16.

CONSTRUCT A RIGHT TRIANGLE

A right triangle has one 90 degree angle, Fig. 5-19 (a).

1. If the length of the two sides is known, construct a perpendicular as shown in Fig. 5-11. Lay off the lengths of the sides and join the ends to complete the triangle.
2. If the length of the hypotenuse (side opposite the right angle) and the length of one side are known, draw a semicircle with a radius (R_1) one-half the length of the hypotenuse AB, Fig. 5-19 (b).
3. With a compass, scribe arc AC equal to the length of the side (R_2).
4. The triangle ACB is the required right triangle and the 90 degree angle is at C.

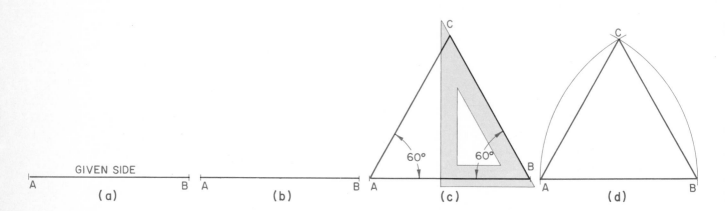

Fig. 5-17. Constructing an equilateral triangle by triangle method (c) and compass method (d).

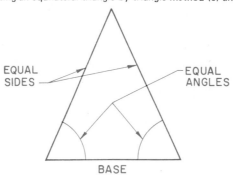

Fig. 5-18. Isosceles triangle.

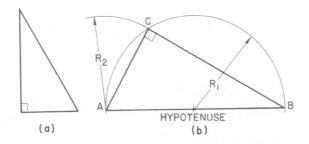

Fig. 5-19. Right triangles.

CONSTRUCTING SQUARES

CONSTRUCT A SQUARE WITH THE LENGTH OF THE SIDE GIVEN, USING THE TRIANGLE AND T-SQUARE METHOD

A square is a polygon with four equal sides and four right angles.

1. Given the length of side AB, lay off AB as the base line, Fig. 5-20 (a).
2. With triangle and T-square, project line BC and measure to length.
3. Draw horizontal line CD and vertical line AD to complete the required square ABCD.

CONSTRUCT A SQUARE WITH THE LENGTH OF THE SIDE GIVEN, USING THE COMPASS METHOD

1. Given the length of side EF, Fig. 5-20 (b), lay off EF as the base line.
2. Construct a perpendicular to line EF at E by extending base line EF to the left and striking equal arcs E_1 and E_2 of any convenient length on the base line.
3. From these two points, strike equal arcs to intersect at J.
4. Draw a line from E through J equal in length to EF to form EH.
5. From points F and H, using a compass arc equal to EF, lay off intersecting arcs FG and HG.
6. Lines drawn to this intersection complete the required square EFGH.

CONSTRUCT A SQUARE WITH THE LENGTH OF THE DIAGONALS GIVEN

1. Given the length of the diagonal AB, Fig. 5-21 (a).
2. With a radius (R_1) of one-half the diagonal, construct a circle, (b).
3. Draw 45 degree diagonals AC and DB.
4. Draw lines joining the points where the diagonals touch the circle, forming the required square ABCD, (c).

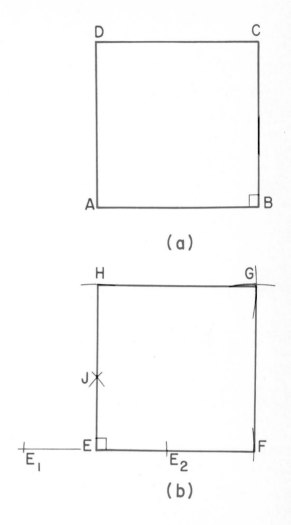

(a)

(b)

Fig. 5-20. Constructing the square.

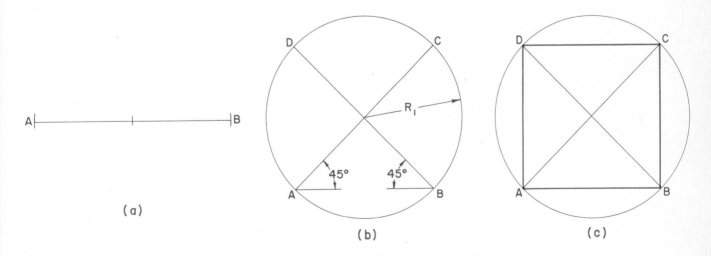

(a)   (b)   (c)

Fig. 5-21. Constructing a square with the diagonals given.

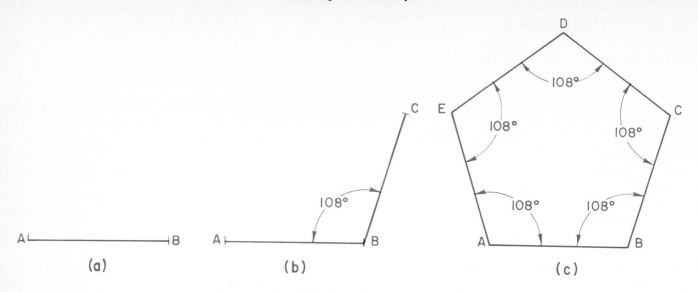

Fig. 5-22. Constructing a pentagon with length of side given.

CONSTRUCTING OTHER POLYGONS

CONSTRUCT A PENTAGON WITH THE LENGTH OF SIDE GIVEN, USING THE PROTRACTOR METHOD
1. Given the length of side AB, Fig. 5-22 (a).
2. With protractor, lay off side BC equal in length to AB and at an angle of 108 degrees with center at B, (b).
3. Lay off side AE in a similar manner and continue until the required regular pentagon ABCDE is formed, (c).

CONSTRUCT A PENTAGON WITH THE CIRCUMSCRIBED CIRCLE GIVEN, USING THE COMPASS METHOD
1. Given the radius FG of the circumscribed circle, as shown in Fig. 5-23 (a).
2. Bisect the radius FG, (a).
3. With H as the center, scribe arc HD to intersect with the diameter of the circle at J, (b).
4. With D as the center, scribe arc DJ to intersect with the circle at E.
5. From the same center at D, scribe an equal arc to intersect with the circle at C.
6. With E and C as centers, scribe arcs EA and CB and draw lines between the points of intersection to form the required regular pentagon ABCDE, (c).

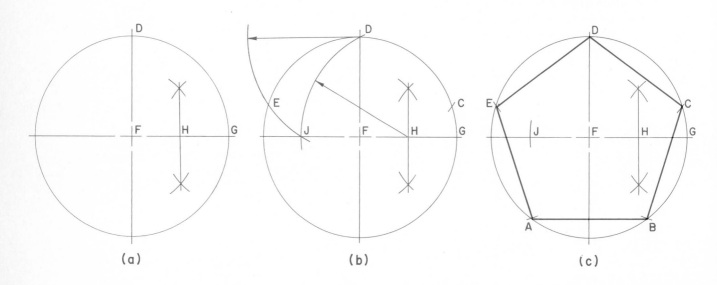

Fig. 5-23. Constructing a pentagon with circumscribed circle given.

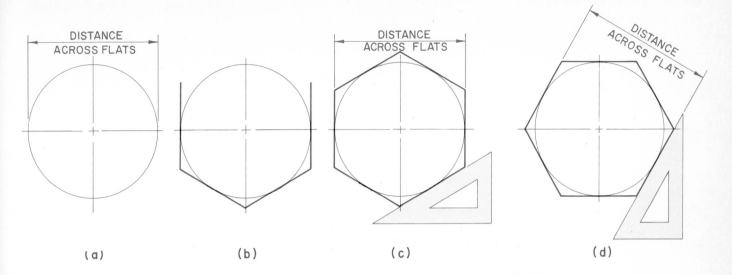

Fig. 5-24. Constructing a hexagon with the distance across the flats given.

CONSTRUCT A HEXAGON WITH THE DISTANCE ACROSS THE FLATS GIVEN

1. Draw a circle equal in diameter to the distance across the flats, Fig. 5-24 (a).
2. With T-square and 30-60 degree triangle, draw lines tangent to the circle, (b).
3. Draw the remaining sides to form the required hexagon, as shown in (c).
4. An alternate position for the hexagon with distance across the flats is shown in Fig. 5-24 (d).

CONSTRUCT A HEXAGON WITH THE DISTANCE ACROSS CORNERS GIVEN

1. Draw a circle equal in diameter to the distance across the corners, Fig. 5-25 (a).
2. With T-square and 30-60 degree triangle, draw 60 degree diagonal lines across the center of the circle, (b).
3. With the T-square and triangle, join the points of intersection of diagonal lines with the circle. This forms required hexagon with distance across corners given, (c). An alternate position is shown in Fig. 5-25 (d).

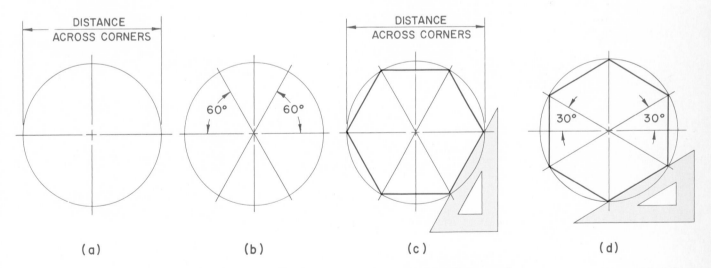

Fig. 5-25. Constructing a hexagon with the distance across the corners given.

CONSTRUCT AN OCTAGON WITH THE DISTANCE
ACROSS THE FLATS GIVEN, USING THE
CIRCLE METHOD

1. Draw a circle equal in diameter to the distance across the flats, Fig. 5-26 (a).
2. With T-square and 45 degree triangle, draw the eight sides tangent to the circle, (b).

CONSTRUCT AN OCTAGON WITH THE DISTANCE
ACROSS THE CORNERS GIVEN, USING THE
CIRCLE METHOD

1. Draw a circle equal in diameter to the distance across the corners, Fig. 5-26 (c).
2. With T-square and 45 degree triangle, lay off 45 degree diagonals with the horizontal and vertical diagonals (diameters).
3. With triangle, draw the eight sides between the points where the four diagonals intersect with the circle, (d).

CONSTRUCT AN OCTAGON WITH THE DISTANCE
ACROSS THE FLATS GIVEN, USING THE
SQUARE METHOD

1. Draw a square with the sides equal to the distance across the flats of the required octagon, Fig. 5-27 (a).
2. Draw diagonals with a radius equal to one-half the diagonal and, using corners of square as centers, scribe arcs, (b).
3. With 45 degree triangle and T-square, draw the eight sides to complete the required octagon, (c).

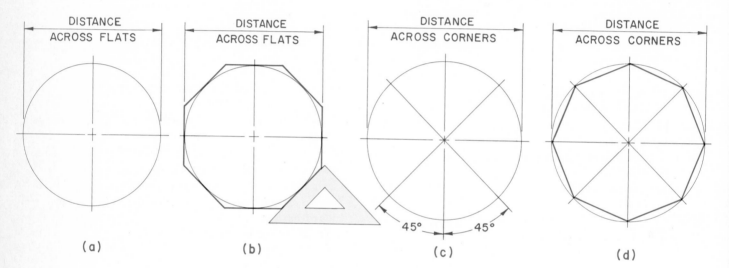

Fig. 5-26. Constructing an octagon by the circle method.

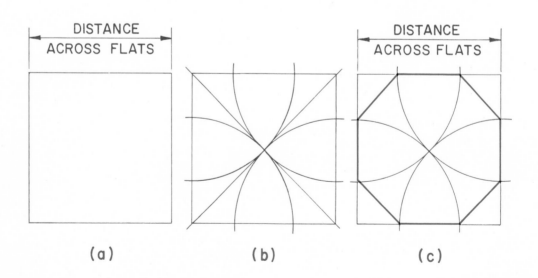

Fig. 5-27. Constructing an octagon by the square method.

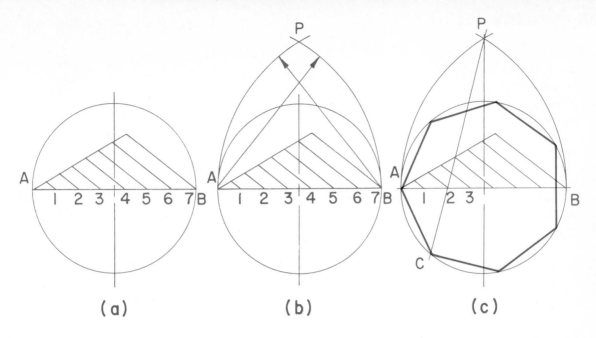

Fig. 5-28. Constructing a regular polygon having any number of sides and the diameter of a circumscribed circle given.

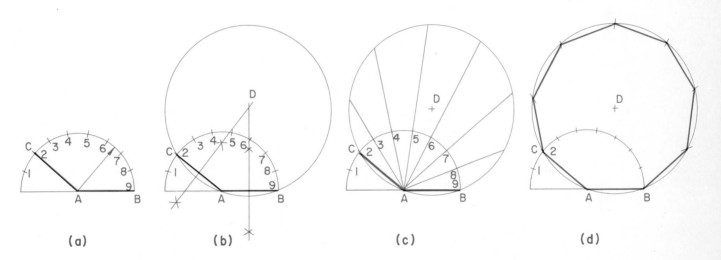

Fig. 5-29. Constructing a regular polygon having any number of given sides and length of side given.

CONSTRUCT A REGULAR POLYGON HAVING ANY NUMBER OF SIDES AND THE DIAMETER OF A CIRCUMSCRIBED CIRCLE GIVEN

1. Draw the circle and divide its diameter into the required number of equal parts (seven in the example), as shown in Fig. 5-28 (a). Use the Inclined Line Method illustrated in Fig. 5-6.
2. With a radius equal to the diameter and with centers at the diameter ends A and B, draw arcs intersecting at P, as shown in (b).
3. Draw a line from point P through the second division point of diameter AB until it intersects with the circle at C, (c). The chord AC is one side of the polygon.
4. Lay off the distance AC around the circle to complete the regular polygon with the required number of sides, as shown in (c).

CONSTRUCT A REGULAR POLYGON HAVING ANY NUMBER OF SIDES AND THE LENGTH OF SIDE GIVEN

1. Draw side AB equal to the given side, Fig. 5-29 (a).
2. Extend AB to the left and draw a semicircle with the center at A and a radius equal to AB.
3. With the dividers, divide the semicircle into the required number of equal parts (nine in the example).
4. From point A to the second division point, draw line AC.
5. Construct perpendicular bisectors of lines AB and AC and extend the bisectors to meet at D, (b).
6. Using D as the center and a radius equal to DB, construct a full circle to pass through B, A and C.
7. From point A, draw lines through remaining division points on the semicircle to intersect with larger circle, (c).
8. Join these points on the larger circle to form the regular polygon with the length and number of sides required, (d).

TRANSFERRING PLANE FIGURES

TRANSFER A PLANE FIGURE BY THE TRIANGLE METHOD

Plane figures with straight lines may be transferred or duplicated by the triangle method.

1. Given the polygon ABCDE, Fig. 5-30 (a).
2. At corner A, draw straight lines AC and AD to form triangles, (b).
3. Using A as the center, draw an arc outside the polygon cutting across the extended lines.
4. Draw the line A'B' in the transferred position desired, as shown in (c).
5. Draw new arc with radius R'.
6. Transfer arcs 1', 2' and 3' to locate triangle lines.
7. Using A as center, set off distances A'C', A'D' and A'E' equal to AC, AD and AE.
8. Draw straight lines to form transferred polygon, A'B'C'D'E'.

TRANSFER A PLANE FIGURE INVOLVING IRREGULAR CURVES

Plane figures involving irregular curves may be transferred or duplicated by the squares method.

1. Given the plane figure shown in Fig. 5-31 (a).
2. Draw rectangle to enclose the plane figure, (b).
3. Select points on the irregular curve that locate strategic points or change of direction and draw coordinates as shown.
4. Reproduce the rectangle and coordinate points in the new position and draw the straight lines and irregular curve as shown in (c).

Plane figures may also be duplicated, enlarged or reduced in size by the squares method shown in Fig. 5-32. This method is similar to the coordinate lines method except squares are drawn, and the irregular curve plotted on the squares.

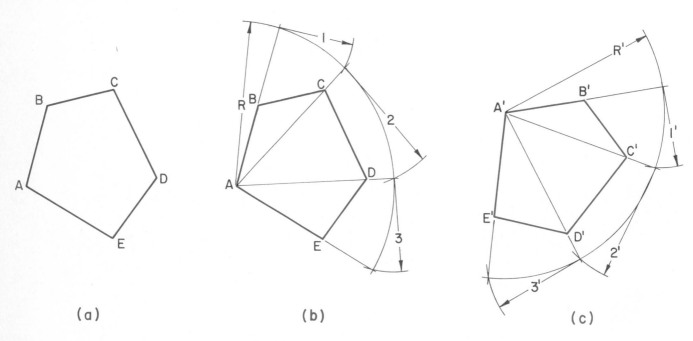

(a)   (b)   (c)

Fig. 5-30. Transferring a given plane figure.

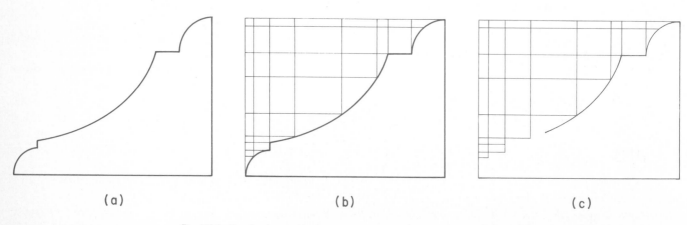

(a)   (b)   (c)

Fig. 5-31. Duplicating an irregular curve by the coordinate lines method.

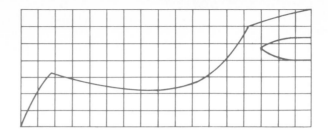

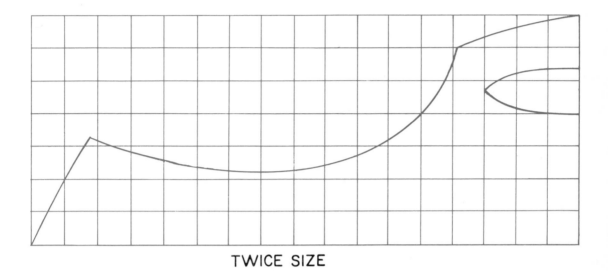

TWICE SIZE

Fig. 5-32. Enlarging an irregular curve by the squares method.

Industry photo. Drawings needed for the design and construction of this building required much use of geometry.

CIRCLES AND ARCS

A circle is a closed plane curve having all points on the curve equally distant from a point within the circle called the center. An arc is any part of a circle or other curved line, Fig. 5-33.

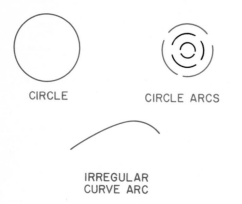

CIRCLE   CIRCLE ARCS

IRREGULAR
CURVE ARC

Fig. 5-33. Circle and arcs.

CONSTRUCTING CIRCLES

CONSTRUCT A CIRCLE THROUGH THREE GIVEN POINTS

1. Given the three points A, B and C, draw connecting lines between the points, Fig. 5-34 (a).
2. Construct the perpendicular bisector of each line, (b).
3. Point of intersection of bisectors is the center of the circle which passes through all three points, (c).

LOCATE CENTER OF CIRCLE OR ARC

1. Draw two nonparallel chords, AB and BC, Fig. 5-34 (c).
2. Construct perpendicular bisectors of chords.
3. Point of intersection of bisectors is the center of the circle or arc.

CONSTRUCT A CIRCLE WITHIN A SQUARE

1. Given the square ABCD, draw diagonals AC and BD to locate center of circle, Fig. 5-35 (a).
2. Locate the center of one side, E of side CD, (b).
3. Distance OE is the radius of the circle inscribed within the square, (c).

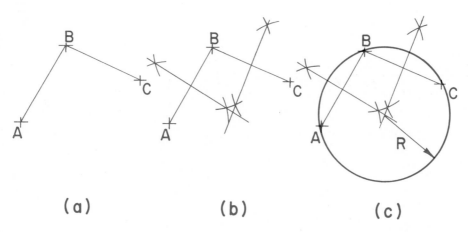

(a)   (b)   (c)

Fig. 5-34. Constructing a circle through three points.

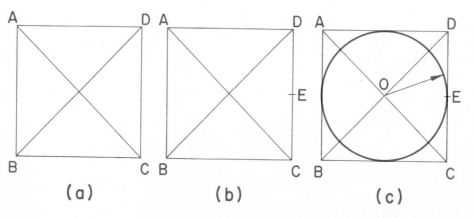

(a)   (b)   (c)

Fig. 5-35. Constructing a circle within a square.

DRAWING TANGENTS

A tangent is a line or curve that touches the surface of an arc or circle at only one point.

CONSTRUCT A LINE TANGENT TO A CIRCLE OR ARC BY THE TRIANGLE METHOD

1. With triangle and straightedge, adjust so that one edge of the right angle coincides with line joining the circle center and point of tangency P, Fig. 5-36 (a).
2. Slide the triangle along the straightedge until the other edge of the right angle passes through P, forming the tangent line.

CONSTRUCT A LINE TANGENT TO A CIRCLE OR ARC, USING THE COMPASS METHOD

1. With point of tangency P as the center, and the circle radius PC, construct an arc through C on through the circle at A, Fig. 5-36 (b).
2. With A as the center and the same radius PC, construct a semicircular arc to pass through P.
3. Extend line CA until it intersects with semicircle at B.
4. Line BP is perpendicular to PC and is the required tangent line.

CONSTRUCT A CIRCLE OR ARC TANGENT TO A STRAIGHT LINE AT A GIVEN POINT

1. Draw a perpendicular to line AB at point P, as shown in Fig. 5-37 (a).
2. Lay off the radius CP of the required circle on the perpendicular, (b).
3. Draw circle, (c).

An application of this construction to a drawing problem is shown in Fig. 5-37 (d).

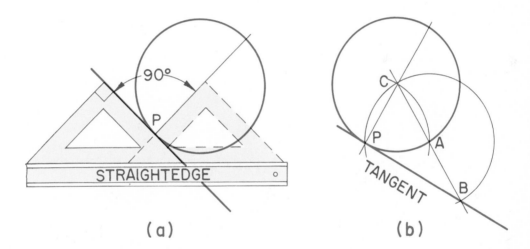

(a) (b)

Fig. 5-36. Constructing a line tangent to a circle or arc.

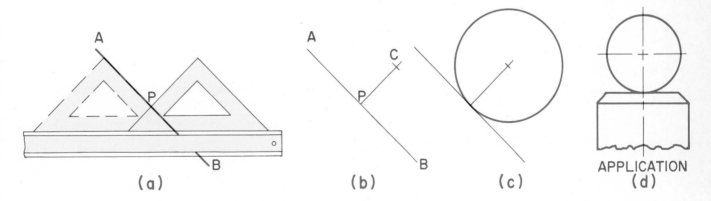

(a) (b) (c) APPLICATION (d)

Fig. 5-37. Constructing a circle tangent to a line at a given point.

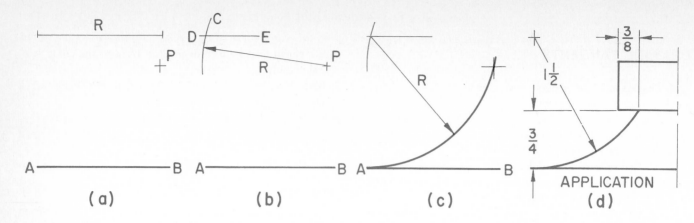

Fig. 5-38. Constructing an arc through a given point and tangent to a straight line.

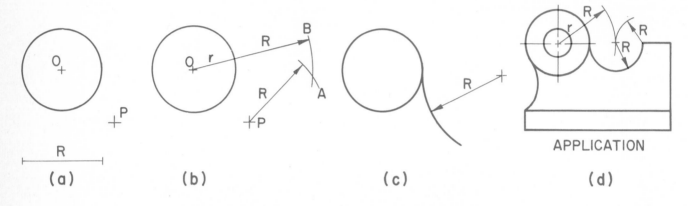

Fig. 5-39. Constructing an arc through a given point and tangent to a circle.

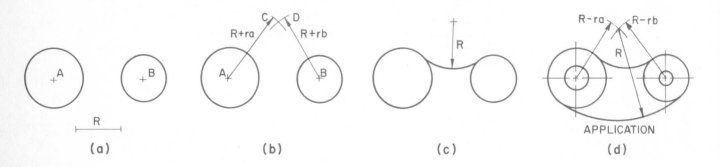

Fig. 5-40. Constructing an arc tangent to two circles.

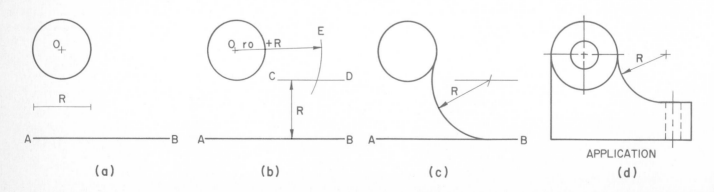

Fig. 5-41. Constructing an arc tangent to a straight line and a circle.

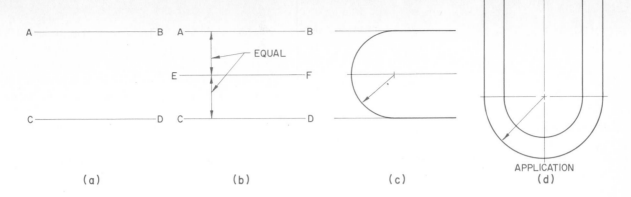

(a)   (b)   (c)   APPLICATION (d)

Fig. 5-42. Above. Constructing an arc tangent to two parallel lines.   Fig. 5-43. Below. Constructing an arc tangent to two nonparallel lines.

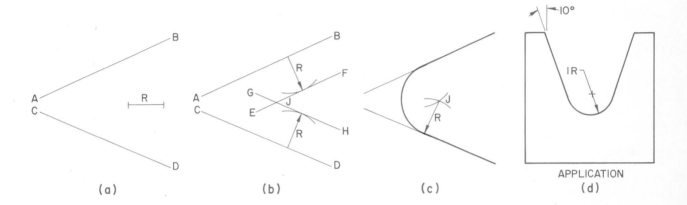

(a)   (b)   (c)   APPLICATION (d)

CONSTRUCT A CIRCLE OR ARC THROUGH A GIVEN POINT AND TANGENT TO A STRAIGHT LINE

1. Given point P, line AB and radius R, Fig. 5-38 (a).
2. Strike arc C with radius R using P as the center, (b).
3. Draw line DE parallel to line AB at distance R from AB.
4. The intersection of arc C and line DE is the center of the tangent circle or arc, (c).

An application is shown in Fig. 5-38 (d).

CONSTRUCT A CIRCLE OR ARC THROUGH A GIVEN POINT AND TANGENT TO A CIRCLE

1. Given point P, radius R and circle O, Fig. 5-39 (a).
2. Strike arc A with radius R using P as the center, (b).
3. Strike arc B using the center of the circle as the arc center and a radius of R + r.
4. The intersection of arcs A and B is the center of the required arc through a point and tangent to a circle, (c).

An application is shown in Fig. 5-39 (d).

CONSTRUCT A CIRCLE OR ARC TANGENT TO TWO CIRCLES

1. Given circles A and B, and radius R, Fig. 5-40 (a).
2. Using A as a center and a radius of R + ra, strike arc C, as shown in (b).
3. Using B as a center and a radius of R + rb, strike arc D.
4. The intersection of arcs C and D is the center of the required arc tangent to two circles, (c).

  An application is shown in Fig. 5-40 (d).

  NOTE: In order to locate the center of an arc tangent to two enclosed circles, subtract the circle radii from the arc radius, (d).

CONSTRUCT A CIRCLE OR ARC TANGENT TO A STRAIGHT LINE AND A CIRCLE

1. Given line AB, circle O and radius R, Fig. 5-41 (a).
2. Draw line CD parallel to AB at a distance R from AB, (b).
3. Strike arc E, to intersect line CD, using O as the center and a radius of R + ro.
4. The intersection of arc E and line CD is the center of the required arc tangent to line AB and circle O, (c).

  An application of this construction to a drawing problem is shown in Fig. 5-41 (d).

CONSTRUCT A CIRCLE OR ARC TANGENT TO TWO PARALLEL LINES

1. Given parallel lines AB and CD, Fig. 5-42 (a).
2. Construct parallel line EF equidistant between lines AB and CD, (b).
3. Set compass on line EF and adjust until it just touches lines AB and CD, and strike required arc, (c).

  An application of this construction to a drawing problem is shown in Fig. 5-42 (d).

CONSTRUCT A CIRCLE OR ARC TANGENT TO TWO NONPARALLEL LINES

1. Given nonparallel lines AB and CD and the radius R, Fig. 5-43 (a).
2. Construct lines EF and GH parallel to AB and CD at a distance R from AB and CD, (b).
3. The intersection of the lines at J is the center of the required arc tangent to lines AB and CD, (c).

  An application of this construction to a drawing problem is shown in Fig. 5-43 (d).

CONNECT TWO PARALLEL LINES WITH REVERSING ARCS OF EQUAL RADIUS

1. Given lines AB and CD, Fig. 5-44 (a).
2. Draw a line between B and C and divide line BC into two equal parts BE and EC, (b).
3. Construct perpendicular bisectors of lines BE and EC, (c).
4. Draw perpendiculars to lines AB and CD at B and C.
5. The points of intersection of the bisectors and perpendiculars at F and G are the centers for drawing the equal arcs FB and GC. These form the required arcs which are tangent to each other and to parallel lines AB and CD.

An application of this construction in a highway drawing is shown in Fig. 5-44 (d).

CONNECT TWO PARALLEL LINES WITH REVERSING ARCS OF UNEQUAL RADIUS

1. Given lines AB and CD, and radius of one arc R1, Fig. 5-45 (a).
2. Draw the line BC, (b).
3. Draw a line perpendicular to AB at B and lay off radius R1 at E. With E as center, strike arc R1 from B to intersect line BC at F, (c).
4. Draw a line perpendicular to CD at C and extend line EF to intersect this perpendicular at G.
5. Using G as the center and GC as the radius, draw arc GC from C to F. These form the required arcs which are tangent to each other and to parallel lines AB and CD.

CONNECT TWO NONPARALLEL LINES WITH REVERSING ARCS

1. Given nonparallel lines AB and CD, and the radius R1, Fig. 5-46 (a).

2. Draw a line perpendicular to CD at C, lay off radius R1 at E. With E as the center, strike arc R1 from C, (b).
3. Draw a line perpendicular to AB at B. Lay off BF equal to CE.
4. Draw line FE and bisect with perpendicular bisector GH.
5. Extend lines FB and GH until they intersect at J. Draw line JE, (c).
6. Using J as a center and JB as the radius, draw an arc to connect line AB to the other arc at K. These are the required arcs and are tangent to each other and to lines AB and CD.

LAY OFF THE LENGTH OF THE CIRCUMFERENCE OF A CIRCLE (ALSO REFERRED TO AS THE RECTIFIED LENGTH), USING THE CONSTRUCTION METHOD

1. Given the circle O, draw line AB tangent to the vertical center line the length equal to three times the diameter of the circle, Fig. 5-47 (a).
2. From point C on the horizontal center line, using a radius equal to the radius of the circle, strike an arc to intersect with the circle at point D.
3. From point D, draw line ED perpendicular to the vertical center line.
4. Line EB is the approximate length of the circumference of circle O (error equal to less than one inch in 20,000 inches, or well within the range of accuracy of draftsmen using mechanical instruments).

LAY OFF THE LENGTH OF THE CIRCUMFERENCE OF A CIRCLE, USING THE EQUAL CHORD METHOD

1. Given the circle X, draw line YZ, Fig. 5-47 (b).
2. Using the dividers, divide one quarter of the circle into an equal number of chord lengths (accuracy is increased when a greater number of chords are used).

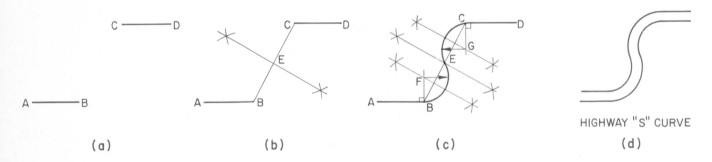

(a) (b) (c) HIGHWAY "S" CURVE (d)

Fig. 5-44. Connecting two parallel lines with reversing arcs of equal radius.

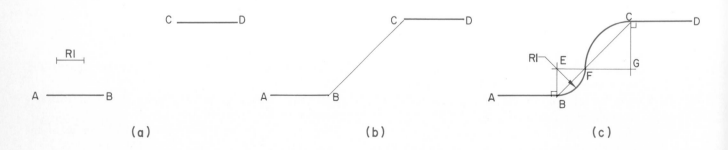

(a) (b) (c)

Fig. 5-45. Connecting two parallel lines with reversing arcs of unequal radius.

3. Lay off on line YZ four times the number of chord lengths in the quarter circle. The length YZ is the required approximate length of circle X.

LAY OFF THE LENGTH OF THE CIRCUMFERENCE OF A CIRCLE, USING THE MATHEMATICAL METHOD

The circumference of a circle may also be calculated very accurately by multiplying the diameter times 3.1416 and laying off this length on a straight line.

Example: D = 3 inches
$$\begin{array}{r} 3.1416 \\ \underline{\times\ 3} \\ 9.4248 \text{ inches} \end{array}$$

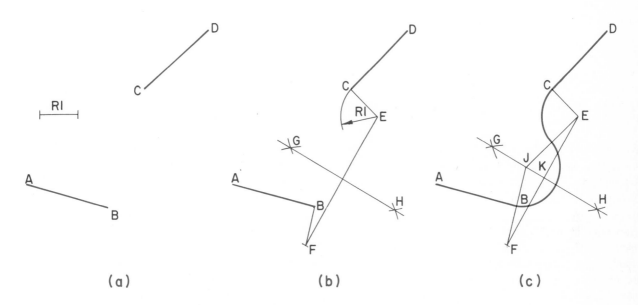

(a)   (b)   (c)

Fig. 5-46. Connecting two nonparallel lines with reversing arcs.

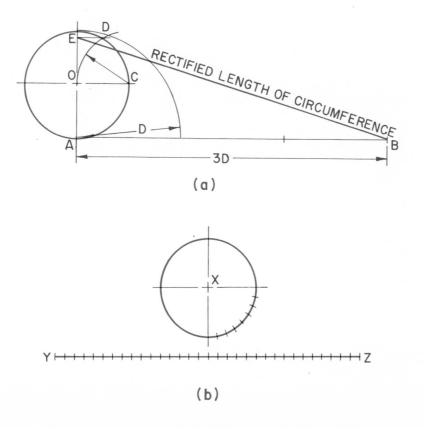

(a)

(b)

Fig. 5-47. Laying off approximate length of circumference of a circle.

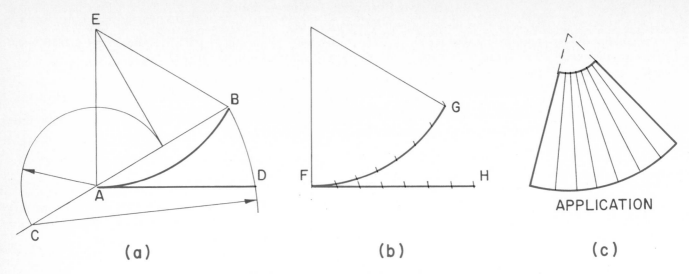

Fig. 5-48. Laying off the length of a circular arc.

LAY OFF THE LENGTH OF A CIRCLE ARC, USING THE CONSTRUCTION METHOD

1. Given circle arc AB, Fig. 5-48 (a).
2. Draw chord AB and extend it to C so that CA is equal to one-half AB.
3. Draw tangent line AD perpendicular to EA at A.
4. Using C as the center and CB as the radius, strike an arc to intersect AD. The length AD is the approximate length of arc AB (error is less than one inch in 1000 inches for angles up to 60 degrees).

LAY OFF THE LENGTH OF A CIRCLE ARC, USING THE EQUAL CHORD METHOD

1. Given circle arc FG, Fig. 5-48 (b).
2. Draw a tangent line FH.
3. Using the dividers, divide arc FG into an equal number of chord lengths (accuracy is increased when a greater number of chords are used).
4. Lay off the same number of chord lengths along line FH. The length FH is the required approximate length of arc FG.

An application of this construction would be in ascertaining the "stretchout" length of a sheet metal part (c).

LAY OFF A CIRCLE ARC, USING THE CONSTRUCTION METHOD

1. Given the length of tangent line AB and the circle arc AD, Fig. 5-49 (a).
2. Divide line AB into four equal parts.
3. Using the first division point C as the center, strike arc CB to intersect the circle arc at D.
4. Arc AD is the required arc and is equal in length to line AB within six parts in a thousand for angles less than 90 degrees.

LAY OFF A GIVEN LENGTH ON A CIRCLE ARC, USING THE EQUAL CHORD METHOD

1. Given the length of tangent line EF and the circle arc EG, Fig. 5-49 (b).
2. Using the dividers, divide tangent line EF into an equal number of parts (accuracy is increased when a greater number of divisions are used).
3. Lay off the same number of chord lengths along circle arc EG. The length of circle arc EG is the required approximate length of line EF.

This type of construction is used in ascertaining the true length of a formed metal part, Fig. 5-49 (c).

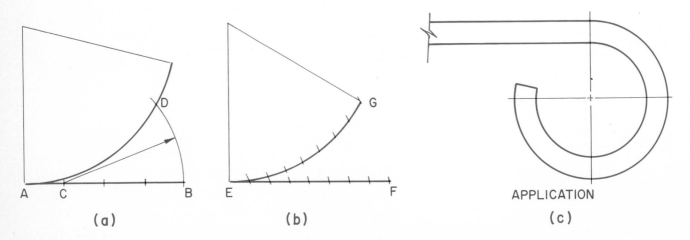

Fig. 5-49. Laying off a given length on a circle arc.

Geometry in Technical Drafting — Basic

PROBLEMS AND ACTIVITIES

The following problems have been designed to give you experience in performing simple geometric constructions used in drafting. Practical applications of geometry applied to drafting are included to acquaint you with typical problems the drafter, designer or engineer must solve.

All problems are to be drawn using Layout I, in the Reference Section. Place drawing paper horizontally on drawing board or table. Use the title block shown in Layout I and lay out your problems carefully, making the best use of space available. A freehand sketch of the problem and its solution will aid in this regard.

Accuracy is extremely important in drawing geometrical constructions. Use a hard lead (2H to 4H) sharpened to a fine conical point and draw light lines. When the construction is complete, darken the required lines and leave all construction lines as drawn to show your work.

Problems which call for "any convenient length" or "any convenient angle" should not be measured with the scale nor laid out with the T-square and triangles. Do those problems assigned by your instructor.

A suggested layout for the first four problems is shown in Fig. 5-50.

1. Draw a horizontal line three inches long (mark the ends with a vertical mark) and bisect it by the triangular method.
2. Draw a horizontal line of any convenient length and bisect it by the compass method.
3. Draw a line AB at any convenient angle and construct another line one inch from it and parallel to it, using the triangle-straightedge method.
4. Using the trangle-straightedge method, draw a line at any convenient angle and construct a perpendicular to it from a point off the line.
5. Draw a line of any convenient length and angle and divide it geometrically into seven equal parts, using the vertical line method.
6. Lay out an angle of any convenient size and bisect it.
7. Transfer the angle in problem No. 6 to the next section on your sheet and show its new location revolved approximately 90 degrees.
8. Draw a horizontal line and construct a perpendicular to it by the 3, 4, 5 method.
9. Construct a triangle given the following: side AB = 3 1/4 inches, ∠ A = 37 degrees, ∠ B = 70 degrees.
10. Construct a triangle given the following: sides AB = 3, BC = 1 1/4, CA = 2 1/8 inches. Measure each angle with the protractor and add the three together. If your answer is 180 degrees, you have measured accurately. If not, check your measurements again.
11. Construct a triangle given the following: sides AB = 1 1/2, AC = 2 inches, ∠ A = 30 degrees.
12. Construct an equilateral triangle given side AB = 2 3/4 inches.
13. Draw a 1 1/2 inch line inclined slightly (approximately 10 degrees, but do not measure) from the horizontal. Using this line as the first side, construct a square by any

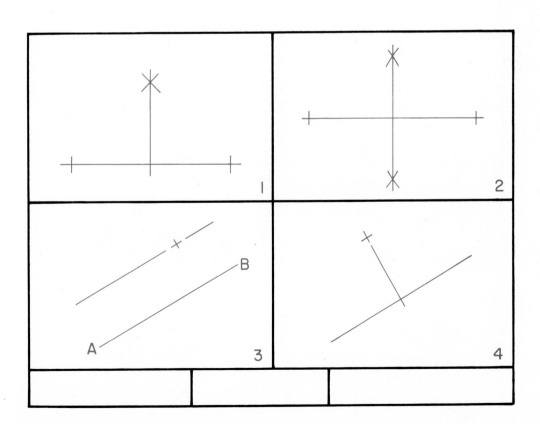

Fig. 5-50. Suggested layout for geometrical problem solutions.

101

method described in this chapter.

14. Draw a circle 2 1/2 inches in diameter in the center of one of the four sections of the sheet. Inscribe the largest square possible within this circle.

15. Construct a pentagon within a 2 1/2 inch circle, using the compass method.

16. Construct a hexagon measuring three inches across the corners.

17. Construct a hexagon with a distance of 2 1/2 inches across the flats when measured horizontally.

18. Construct an octagon with a distance across the flats of 2 1/2 inches, using the square method.

19. Construct a seven sided regular polygon using the method shown in Fig. 5-29, given the length of one side as 1 1/4 inches.

20. Draw an irregular shaped pentagon approximately two inches across corners in the upper left-hand portion of one section of a sheet. Transfer the polygon to the lower right-hand corner in a 180 degree revolved position, using the triangle method.

21. Without measuring, place three points approximately 1 1/2 inches from the assumed center of the sheet section at the ten, three and six o'clock positions. Construct a circle to pass through all three points.

22. Using a circle template, draw a semicircular arc approximately 1 1/2 inches in radius. Then locate the center of the arc.

23. Draw a circle 2 1/2 inches in diameter and construct a tangent at the approximate two o'clock position.

24. Draw a two inch diameter circle in the lower left-hand portion of a sheet section. Draw a line approximately one inch above and inclined toward the far corner. Construct a one inch circle arc tangent to the circle and straight line.

25. Draw two nonparallel lines and construct a circular arc of any convenient radius tangent to the two lines.

26. Draw a circle whose diameter is 1 1/2 inches. Obtain the length of the circumference of the circle by the three different methods indicated and compare your findings. Lay off the distance as follows:

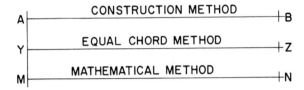

27. Draw a 2 1/4 inch diameter circle and find the length of the arc from the six o'clock to the approximate four o'clock position, using the construction method.

28. Draw a 2 1/2 inch diameter circle and lay off one inch along an arc. Start at the six o'clock position and use the construction method.

A high degree of technical skill is required for complex designs, such as aircraft.   (Beech Aircraft Corp.)

Chapter 6
GEOMETRY IN TECHNICAL DRAFTING ADVANCED

The more complex and advanced a geometric construction becomes, the more difficult it is to solve the drafting problems it presents. Again, however, each problem at hand can be worked out by applying the principles of plane geometry and utilizing one or more of your drafting instruments.

The preceding chapter was concerned with basic geometric problems. This chapter is designed to provide specific definitions and details on more advanced forms of geometric construction such as: conic sections, the Spiral of Archimedes, the helix, cycloids and the involute.

CONIC SECTIONS

Conic sections are curved shapes produced by passing planes through a right circular cone. A right circular cone, Fig. 6-1, has a circular base and an axis perpendicular to the base at its center. An element is a straight line drawn from any point on the base to the peak of the cone.

Four types of curves result from cutting planes at different angles: circle, ellipse, parabola and hyperbola, Fig. 6-2. The circle is a special, often used conic section discussed at length in Chapter 5.

THE ELLIPSE

An ellipse is formed when a plane is passed through a right circular cone, making an angle with the axis greater than the elements, Fig. 6-2 (c).

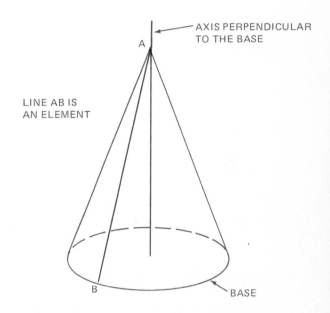

Fig. 6-1. A right circular cone has a circular base and an axis perpendicular to the base at its center.

An ellipse is also seen when a circle is viewed at an angle. It may be defined as a curve formed by a point moving in a plane in such a way that the sum of its distances from two fixed points is a constant and is equal to the major axis (longest diameter). The two fixed points, called foci, are often used in constructing an ellipse.

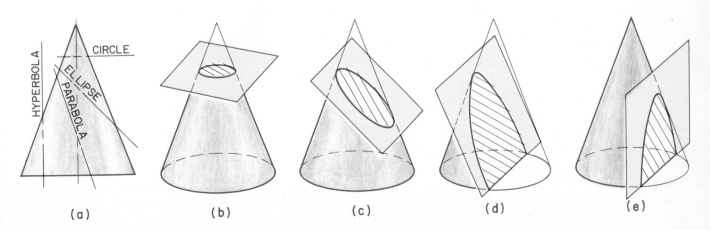

Fig. 6-2. Conic sections of a right circular cone (a) include: circle (b), ellipse (c), parabola (d), hyperbola (e).

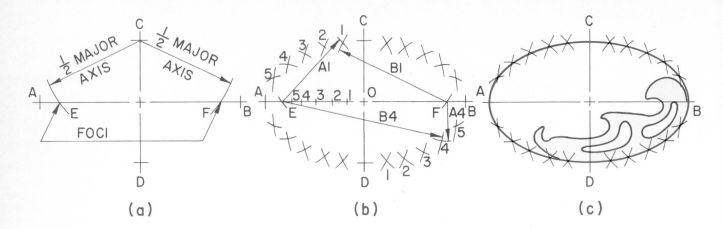

Fig. 6-3. Constructing an ellipse by the foci method.

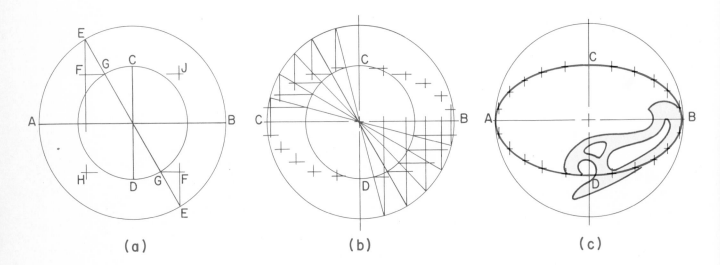

Fig. 6-4. Constructing an ellipse, using the concentric circle method.

CONSTRUCT AN ELLIPSE, USING THE FOCI METHOD

1. Given AB, the major axis and CD the minor axis of the ellipse, Fig. 6-3 (a).
2. Locate foci E and F on the major axis by striking arcs CE and CF with radius equal to one-half the major axis.
3. On the major axis between E and O, (b), mark points at random equal to the number of points desired in each quadrant of the ellipse. To insure a smooth curve, space points closely near E.
4. Begin construction with a point in the upper left-hand quadrant of the ellipse. Using E and F as centers, and radii equal to A1 and B1, strike intersecting arcs at point 1.
5. Use A2 and B2 for intersecting arcs at point 2 and continue until all points are plotted.
6. The three remaining quadrants may be plotted, using the same compass settings for the lower left quadrant and reversing the centers for the radii for the plotting of points in the two right-hand quadrants.
7. Sketch a light line through the points. Then, with the aid of an irregular curve, darken the final ellipse, (c).

CONSTRUCT AN ELLIPSE, USING THE CONCENTRIC CIRCLE METHOD

1. Given the major (AB) and minor (CD) axes, draw circles of these diameters, Fig. 6-4 (a).
2. Draw diagonal EE at any point.
3. At points where the diagonal intersects with the major axes circle, draw lines EF parallel to the minor axis, as shown in Fig. 6-4 (a).
4. At points where the diagonal intersects with the minor axes circle, draw lines FG parallel to the major axis. The intersections at F are points on the ellipse curve.
5. Two additional points, H and J, may be located in the other quadrants by extending lines EF and FG, as shown in Fig. 6-4 (a).
6. Draw as many additional diagonals as needed to produce a smooth ellipse curve and project their points of intersection, (b).
7. Sketch a light line through the points. Then, with the aid of an irregular curve, darken the final ellipse, as shown in Fig. 6-4 (c).

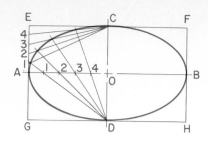

MAJOR AND MINOR
AXES GIVEN

(a)

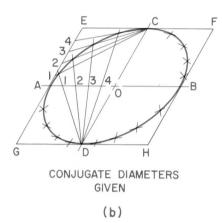

CONJUGATE DIAMETERS
GIVEN

(b)

Fig. 6-5. Constructing the parallelogram ellipse.

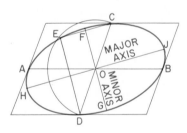

Fig. 6-6. Locating the major and minor axes of an ellipse, given the
conjugate diameters.

CONSTRUCTING AN ELLIPSE, USING THE PARALLELOGRAM METHOD

1. Given the major and minor axes, AB and CD, Fig. 6-5 (a), or the conjugate diameters, (b).
2. Construct circumscribing rectangle or parallelogram using the major and minor axes as center lines.
3. Divide AO and AE into the same number of equal parts.
4. Draw D1 to intersect with C1, D2 with C2, and so on. These points of intersection are plotting points for the ellipse.
5. Locate points in the remaining quadrants in a like manner.
6. Sketch a light line through the points and then, with the aid of an irregular curve, darken the final ellipse.

To locate the major and minor axes for an ellipse with conjugate diameters given, draw a semicircle using center O and a diameter of CD, Fig. 6-6. intersecting the ellipse at E, Fig. 6-6. The minor axis, FG, is parallel to ED and the major axis HJ is parallel to EC and perpendicular to minor axis FG.

CONSTRUCT AN ELLIPSE, USING THE FOUR-CENTER APPROXIMATION METHOD

1. Given the major and minor axes AB and CD, Fig. 6-7 (a).
2. Draw the line CB. Using the center O and the radius OC, strike an arc, intersecting OB at E.
3. With the radius EB, and using C as the center, strike an arc intersecting CB at F.
4. Construct a perpendicular bisector of FB and extend to intersect with the major and minor axes at G and H, (b).
5. Points G and H are the centers for two of the four arcs of the ellipse. With a compass and using O as the center, locate J and K symmetrically with G and H.
6. Draw a line from H extending through J and lines from K through J and G.
7. Using centers J and G, strike the arcs JA and GB from P1 to P2 and from P3 to P4.
8. With H and K as centers, strike arcs HC and KD from P2 to P3 and P4 to P1.

These four arcs will be tangent to each other, forming the four-center approximate ellipse.

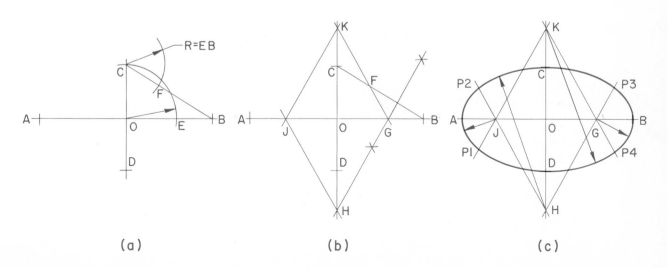

(a)   (b)   (c)

Fig. 6-7. Constructing the four-center approximate ellipse.

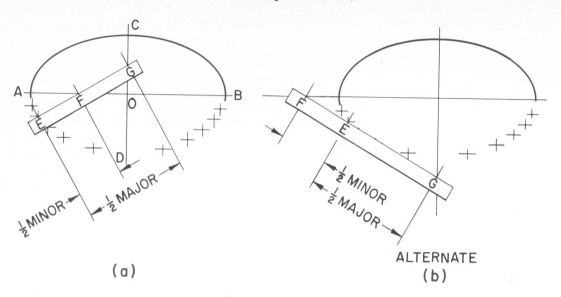

Fig. 6-8. Constructing the trammel ellipse.

CONSTRUCT AN ELLIPSE, USING THE TRAMMEL METHOD

1. Given the major and minor axes AB and CD, Fig. 6-8 (a).
2. On a straightedge, such as a piece of paper, lay off points E, F and G so that EF is equal to one-half the minor diameter (OC) and EG is equal to one-half the major diameter (OA). This marked straightedge serves as a trammel.
3. Place the trammel so that G is on the minor axis and F is on the major axis.

4. Point O will fall on the ellipse curve. Mark at least five points (more on large ellipses) on each quadrant.
5. Sketch a light line through the points, then, with the aid of an irregular curve, darken the final ellipse.

An alternate method of marking and using the trammel is shown in (b).

This trammel method is one of the most accurate means of constructing an ellipse. An ellipsograph, which draws ellipses mechanically, is based on the trammel principle, Fig. 6-9.

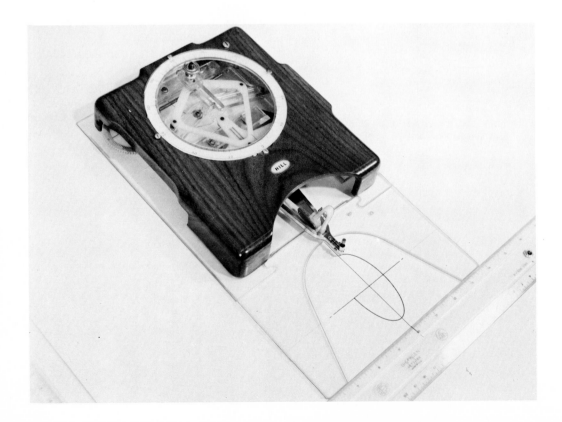

Fig. 6-9. The ellipsograph is based on the trammel principle of constructing ellipses.
(Hill Products Co.)

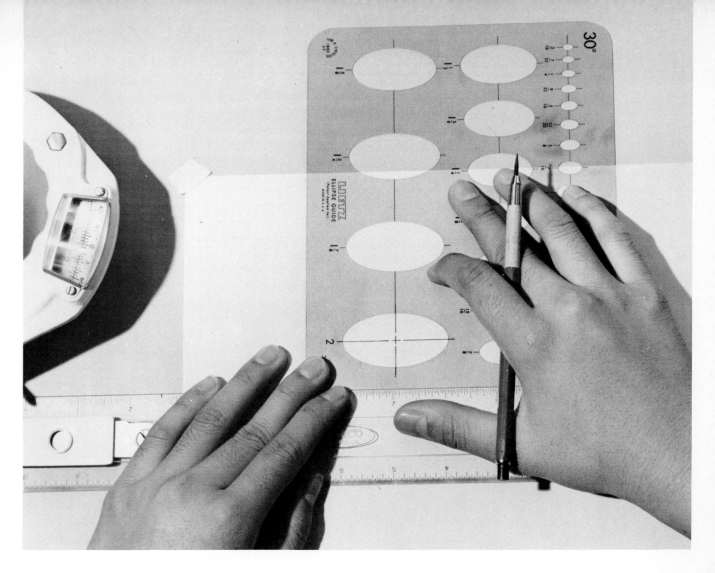

Fig. 6-10. Drawing an ellipse with the aid of an ellipse template.

DRAW AN ELLIPSE, USING THE TEMPLATE METHOD

Considerable time can be saved in ellipse construction by using an ellipse template, as shown in Fig. 6-10. Templates come in a variety of sizes, and are usually designated by the angle at which a circle of that size is viewed, see Fig. 6-11. The selection of proper ellipse templates is discussed in Chapter 11.

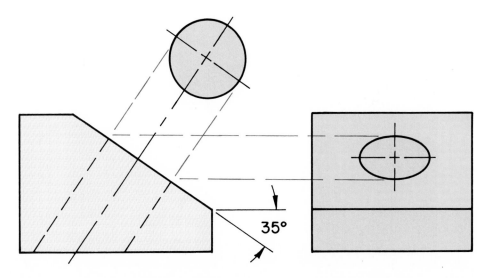

Fig. 6-11. A 35 degree ellipse template is used in drawing this ellipse.

THE PARABOLA

The parabola is formed when a plane cuts a right circular cone, making the same angle with the axis as the elements, Fig. 6-2 (d). The parabolic curve is used in engineering and construction on vertical curves (overpasses) on highways and dams, Fig. 6-12. It is also used in designing bridge arches and in forming the shape of reflectors for sound and light.

The parabola may be defined mathematically as a curve generated by a point moving so that its distance from a fixed point (the focus) is always equal to its distance from a fixed line (the directrix), Fig. 6-13.

CONSTRUCT A PARABOLA, USING FOCUS METHOD
1. Given the focus F and the directrix AB, Fig. 6-14 (a).
2. Draw line CD parallel to directrix at any distance. See EG.
3. With a radius of EG, and using F as the center, strike an arc to intersect line CD at H and J. These points of intersection are points on the parabola.
4. In a like manner, locate as many points as necessary to draw the parabola.
5. Sketch a light line through the points and use the irregular curve to darken the line.
6. The vertex (V) of the parabola is located half way between the origin (E) and the focus (F).

CONSTRUCT A TANGENT TO A PARABOLA
1. Given the parabola AB, its axis CD, focus F and point of tangency P, Fig. 6-14 (b).
2. Draw line PO parallel to the axes and line PF through the focus, and bisect the angle OPF. The bisector PQ is tangent to the parabola at point P.

CONSTRUCT A PARABOLA, USING THE TANGENT METHOD
1. Given the points A and B and the distance CD from line AB to the vertex, Fig. 6-15.
2. Extend CD to E so that DE is equal to CD.
3. Draw lines AE and BE which are tangent to the parabola at points A and B.
4. Divide lines AE and BE into the same number of equal parts (accuracy increases with number of divisions). Number the points from opposite ends.
5. Draw lines between the corresponding points: 1 and 1, 2 and 2, etc.
6. These lines are tangent to the required parabola.
7. Sketch a light line tangent to these lines, and use the irregular curve to darken the lines.

CONSTRUCT A PARABOLIC CURVE THROUGH TWO GIVEN POINTS
1. Given points A and B, Fig. 6-16 (a).
2. Assume any point C and draw tangents CA and CB.
3. Construct the parabolic curve, using the tangent method shown in Fig. 6-15.

The parabolic curves shown in Fig. 6-16 are three of many possibilities. These curves are frequently used in industrial and product design because of their pleasing appearance. Note that the distances AO and OB are not necessarily equal. When they are, as in Fig. 6-16 (a), the bisector of angle AOB is also the axis of the parabola.

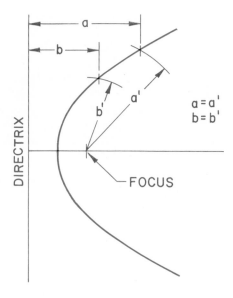

Fig. 6-13. The parabolic curve.

Fig. 6-12. Parabolic curves were used in the design of each lock of this dam.

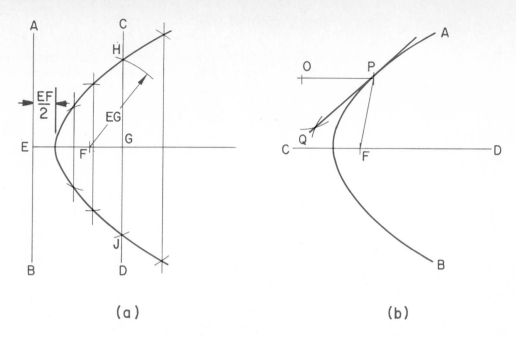

(a)   (b)

Fig. 6-14. Generating a parabola (a) and constructing a tangent to a parabola (b).

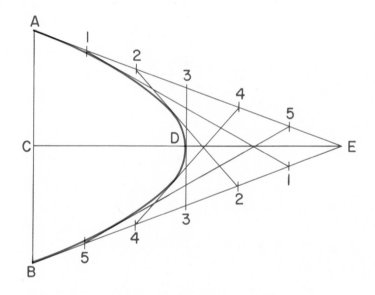

Fig. 6-15. Constructing a parabola by the tangent method.

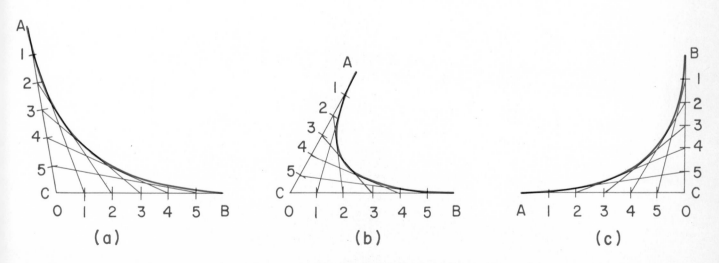

(a)   (b)   (c)

Fig. 6-16. Constructing parabolic curves through two given points.

THE HYPERBOLA

The hyperbola is the curve formed when a plane cuts a right circular cone making an angle with the axis smaller than that made by the elements, Fig. 6-17.

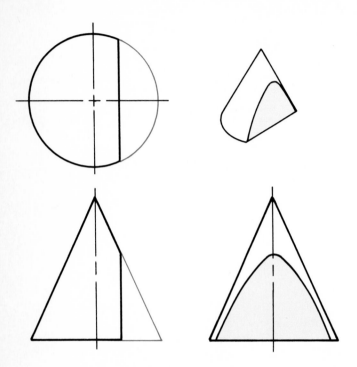

Fig. 6-17. The hyperbola.

Mathematically, the hyperbola may be described as a plane curve traced by a point moving so that the difference of its distance from the two fixed points (foci) is a constant equal to the transverse axis.

Hyperbolic curves are used in space probes, while the equilateral hyperbola can be used to indicate varying pressure of gas as the volume varies. Gas pressure varies inversely as the volume changes.

CONSTRUCT A HYPERBOLA, USING THE FOCI METHOD

1. Given the foci F1 and F2 and the transverse axis (constant difference) AB, Fig. 6-18 (a).
2. Lay off a convenient number of points to the right of F2.
3. With F1 and F2 as centers and A4 (in the example) as radius, draw arcs C, D, E and G.
4. With F1 and F2 as centers and B4 as radius, draw intersecting arcs with C, D, E and G. These points of intersection are points on the hyperbola.
5. Continue to lay off intersecting arcs, using A1 and B1, A2 and B2, etc., as radii.
6. Sketch a light line through the points and use an irregular curve to darken the final curve.

Asymptotes ML and JK, Fig. 6-18 (b), are two straight lines which the hyperbolic curves approach as they extend toward infinity. They are located by:

1. Drawing a circle which has its center (O) midway on the transverse axis and its circumference through the foci.
2. Erecting perpendiculars to the transverse axis at points A and B.
3. The asymptotes extend through the points of intersection CD and EG.

CONSTRUCT A TANGENT TO A HYPERBOLA

1. Given the hyperbola LBK and the point P shown in Fig. 6-18 (b).
2. Draw lines from P to foci F1 and F2.
3. Bisect angle F1 P F2. The bisector HP is the required tangent.

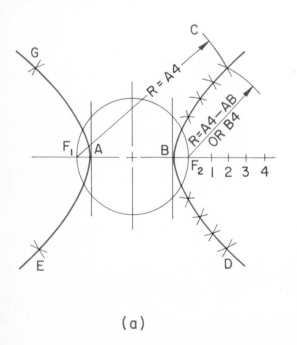

(a)

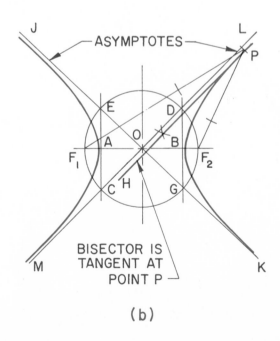

(b)

Fig. 6-18. Constructing the hyperbolic curve when the foci and the transverse axis are given.

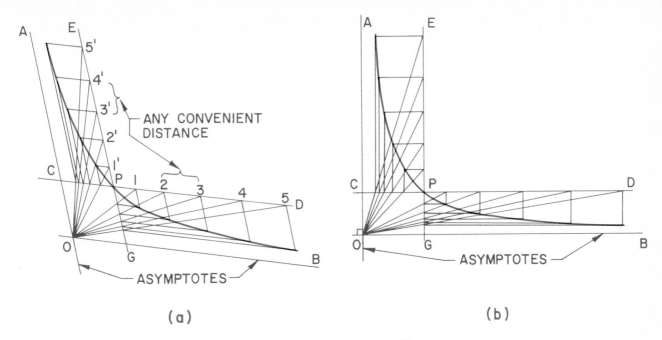

Fig. 6-19. Constructing a hyperbola through a point with the asymptotes given.

CONSTRUCT A HYPERBOLA WITH THE ASYMPTOTES AND ONE POINT ON THE CURVE GIVEN

1. Given asymptotes OA and OB and point P on the curve, Fig. 6-19 (a).
2. Through point P, draw lines CD and EG parallel to the asymptotes.
3. From O, the origin, draw a number of radial lines intersecting CD at points 1, 2, 3, 4, 5, etc., and EG at points 1', 2', 3', 4', 5', etc.
4. Draw lines parallel to the asymptotes at points 1 and 1', 2 and 2', etc.
5. The intersections of these lines are points on the hyperbola. Continue until a sufficient number of points have been located to produce a smooth and accurate curve.
6. Sketch a light line through the points and finish with the aid of an irregular curve.

Asymptotes which are at right angles to each other, Fig. 6-19 (b), produce a hyperbola that is called rectangular or equilateral. When a point on the hyperbolic curve is given, the construction is the same as above.

Geometry in technical drafting played an important part in the drawings used to manufacture these photocopiers. (Canon U.S.A. Inc.)

OTHER CURVES

Other plane curves commonly used in engineering, design and drafting are the Spiral of Archimedes, the helix, the cycloids and the involute. The construction and application of these curves are discussed in this section.

CONSTRUCT A SPIRAL OF ARCHIMEDES

The Spiral of Archimedes is formed by a point moving uniformly around and away from a fixed point, Fig. 6-20.

1. Given the rise OB to move uniformly 1 1/2 inches away from O in one revolution around a fixed point, Fig. 6-20.
2. Draw horizontal line AB through O and lay off a convenient number of equal parts totaling 1 1/2 inches (for example, 12 parts of 1/8 inch each).
3. Using O as the center and O-12 as the radius, draw a circle.
4. Divide circle into 12 equal parts (30 degrees each) and number each line, starting with first line after OB. The number of radial divisions and divisions along line OB must be equal.
5. Using the center O, draw concentric arcs starting on line OB with equal part number 1 and joining line 1.
6. Continue with concentric arcs from each of the equal parts (uniform rise) to the corresponding numbered line.
7. The points of intersection of the concentric arcs and radial lines are points on the spiral curve.
8. Sketch light line through these points. Finish with irregular curve.

Spiral of Archimedes curve is used in design of cams to change uniform rotary motion into uniform reciprocal motion.

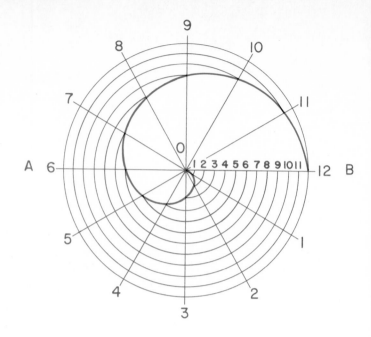

Fig. 6-20. Spiral of Archimedes.

CONSTRUCT A HELIX

The helix is a space or three-dimensional curve rather than a plane curve as those previously discussed. The path of a point on the curve can best be described as moving around a cylinder at a uniform angular rate and also moving parallel to the axis at a uniform linear rate.

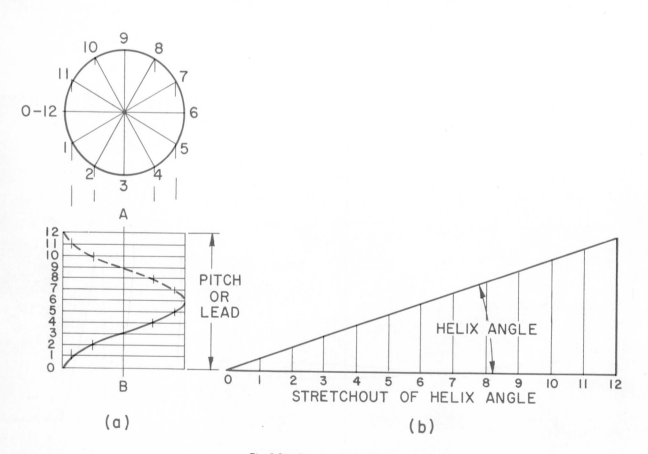

(a)   (b)

Fig. 6-21. Constructing the helix.

1. Given the diameter of the cylinder and the pitch or lead (distance parallel to the axis AB), as illustrated in Fig. 6-21 (a).
2. Draw the top view as a circle equal to the diameter of the cylinder and divide into any number of equal parts, for example, 12 at 30 degrees each. Number the divisions, as in Fig. 6-21 (a).

CYCLOIDS

Curves in the cycloid group are those formed by the path of a fixed point on the circumference of a rolling circle and are useful in the design of cycloidal gear teeth. When the circle rolls along a straight line, the path of the fixed point forms a cycloid, Fig. 6-22.

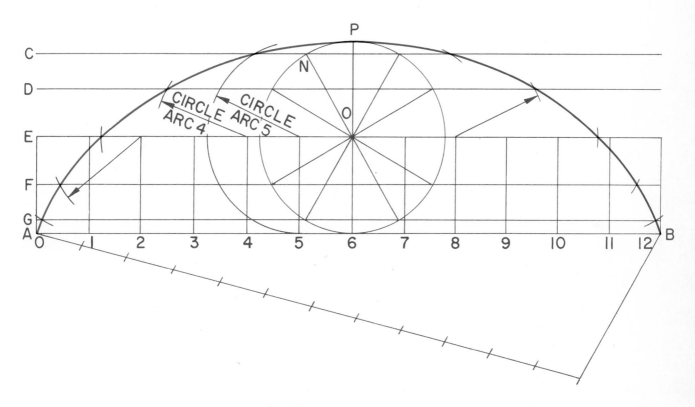

Fig. 6-22. Constructing a cycloid.

3. Draw the front view of the cylinder whose length is equal to the pitch or lead.
4. Divide the front view along the axis into the same number of equal parts (12) as for the top view. Number the divisions.
5. Project the points of intersection of the radial lines with the circumference in the top circle to their corresponding linear division in the front view.
6. These points of intersection in the front view are points on the helix curve.
7. Sketch a light line through the points and finish the curve with the aid of an irregular curve.

A stretchout of the development of the helix angle is shown in Fig. 6-21 (b).

Typical uses of the helix are bolt and screw threads, auger bits used in boring wood, flutes on a drill and a helical gear. The helix shown in Fig. 6-21 is a right-hand helix and advances into the work or mating part when turned clockwise.

On a left-hand helix, the path moves from right to left and advances into the work or mating part when turned counterclockwise.

1. Given the generating circle and tangent line AB equal in length to the circumference of the circle.
2. Divide the circle and line AB into the same number of equal parts, (12).
3. Draw lines C, D, E, F and G through the division points on the circle and parallel to line AB.
4. Project the division points on line AB to line E by drawing perpendiculars.
5. Using the intersections of the perpendiculars with line E as centers, draw circle arcs of radius OP representing the various positions of the rolling circle as it moves to the left.
6. Assume the fixed point P is at its highest point on the curve at division line 6. When it moves directly above line 5, P will have moved to the level formerly occupied by N on line C. The intersection of circle arc 5 and line C is the next point on the cycloid curve.
7. Similarly, the next point on the curve to the left is the intersection of circle arc 4 and line D. Locate the remaining points on the curve by intersecting circle arcs.
8. Sketch a light line through the points of intersection and finish the cycloid curve using the irregular curve.

CONSTRUCT AN EPICYCLOID

The epicycloid is formed by the generating circle with its fixed point rolling on the outside of another circle. The construction is similar to that of the cycloid except line AB and other horizontal lines are concentric circle arcs, Fig. 6-23.

CONSTRUCT A HYPOCYCLOID

The hypocycloid is formed by a fixed point on the generating circle rolling on the inside of another circle. The construction of this curve is similar to that of the cycloid and epicycloid. The hypocycloid is shown in Fig. 6-24.

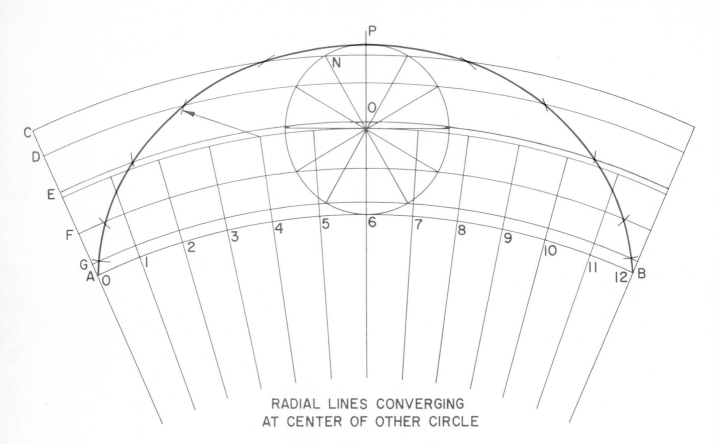

Fig. 6-23. Constructing an epicycloid.

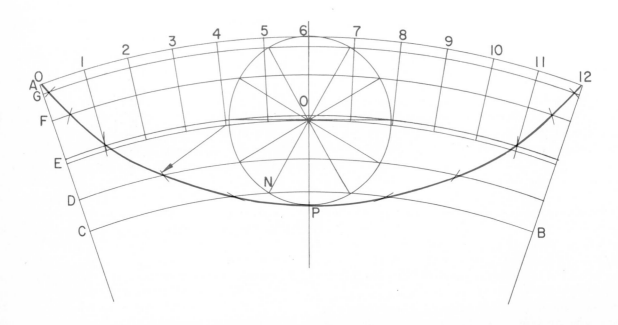

Fig. 6-24. Constructing a hypocycloid.

INVOLUTES

The involute is the curve formed when a tightly drawn chord unwinds from around a circle or a polygon such as a triangle or square, Fig. 6-25. Involute curves may start on the surface of the circle or polygon or they may begin a distance away from the geometric form as in (b).

CONSTRUCT AN INVOLUTE OF AN EQUILATERAL TRIANGLE

1. Given the triangle ABC, extend side CA to D, Fig. 6-25 (a).
2. With A as the center and AB as the radius, strike arc BD.
3. Extend side BC to E, and using C as the center and CD as the radius, strike arc DE.
4. In a similar manner strike arc EF.
5. Continue process until curve of desired size is completed.

CONSTRUCT AN INVOLUTE OF A SQUARE

1. Given the square ABCD and a starting point P off the surface of the square, Fig. 6-25 (b).
2. Extend side AB to P and side DA to E, and using A as center and radius AP, strike arc PE.
3. Extend side CD to F and using D as the center and DE as radius, strike arc EF.
4. Continue until an involute of the desired size is achieved.

CONSTRUCT AN INVOLUTE OF A CIRCLE

1. Given circle O and starting point P on the surface of the circle, Fig. 6-25 (c).
2. Divide the circle into a number of equal parts (12) and draw tangents at the division points.
3. Beginning at first division point after the starting point P, lay off on tangent A a distance equal to length of arc AP.

Fig. 6-26. Gears with involute curve tooth design. (Cincinnati Gear Co.)

4. On tangent B, lay off a distance equal to the length of arc AP + BA (the length of two circle arcs).
5. Continue with tangent C with a distance equal to three circle arcs, and so on until the distance on the final tangent has been set off.
6. Sketch a light line through these points and finish the involute of a circle with the irregular curve.

The involute of the circle is the curve form used in the design of involute gear teeth, Fig. 6-26.

Fig. 6-25. Constructing involutes of polygons and circles.

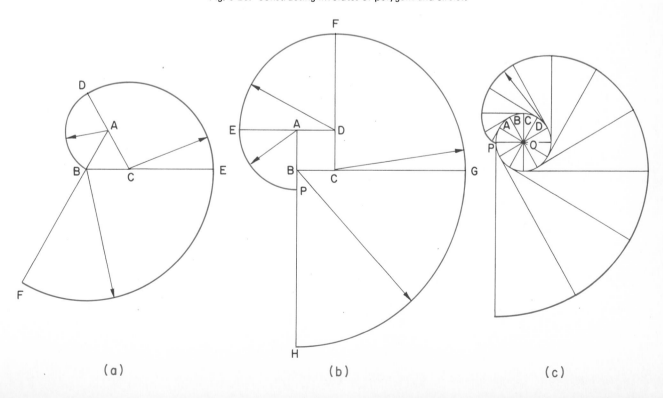

(a)   (b)   (c)

PROBLEMS AND ACTIVITIES

The following problems involve more complicated geometric constructions. Again, practical applications are included to acquaint you with typical geometric problems the drafter, designer or engineer must solve.

All problems are to be drawn on drawing sheets of suitable size and in accordance with layout shown in the Reference Section. Place drawing sheets horizontally on the drawing board or table. Use the title block shown in Layout I, in the Reference Section, and lay out your problems carefully to make the best use of space available. A freehand sketch will help.

Accuracy is extremely important. Use a hard lead (2H to 4H), sharpened to a fine, conical point. Draw light lines, darkening in the required lines when the construction is complete. Leave all construction lines as drawn to show your work.

1. Using a scale of any convenient size, draw the outline of an elliptical swimming pool with a major diameter of 20 feet and a minor diameter of 12 feet. Use the foci method.
2. The six spokes in a gear wheel have an elliptical cross section with a major diameter of 2.50 inches and a minor diameter of 1.50 inches. Construct the ellipse twice size, using the concentric circle method.
3. The design for a bridge support arch is elliptical in shape. It has a span of 36 feet and the rise at the center of the ellipse is 12 feet above the major diameter. Construct the half ellipse representing the arch, using the trammel method.
4. Using an ellipse template, draw an ellipse and letter in the size and degree of the ellipse.
5. A parabolic reflector for a flood light has the following characteristics: focus is five inches from the directrix; the rise is four inches. Draw the parabolic form of the reflector.
6. A highway vertical curve (overpass) has a horizontal span of 200 feet and a rise of 25 feet. The curve form is parabolic. Draw the form of this curve. Hint: The apex of the two tangent lines from the end points of the span must be 50 feet above the end points.
7. Construct the hyperbola and its asymptotes, using the foci method and given the following characteristics: horizontal transverse axis 3/4 inch. This axis and the foci are 1 1/4 inches apart.
8. Construct the equilateral hyperbola given the following characteristics: asymptotes located near and parallel to the lower and left-hand borders of your sheet section; a point 1/4 inch to the right of the vertical asymptote and 3/4 inch above the horizontal asymptote through which the hyperbola is to pass.
9. Starting in the center of a sheet section, draw one revolution of a Spiral of Archimedes with the generating point moving uniformly in a counterclockwise direction and away from the center at the rate of 1 1/4 inches per revolution.
10. Construct a section of a horizontal right-hand helix given the following characteristics: diameter 2 inches, length 1 1/2 inches and a lead of 1 inch.
11. Construct a cycloid generated by a one inch diameter circle rolling along a horizontal line.
12. Construct the involute of a point starting at the apex of a 1/2 inch equilateral triangle for one revolution clockwise.

Chapter 7
MULTIVIEW DRAWING

The multiview drawing is the major type of drawing used in industry. It is a projection drawing that incorporates several views of a part or assembly on one drawing, Fig. 7-1.

The various views of the object are carefully selected to show every detail of size and shape, as well as the processes to be performed. Usually, three views are drawn. However, drawings may vary from one or two views for a simple part to four or more views for a complicated part or assembly.

In addition, the views are arranged in a manner that is standard throughout industry. The top view always appears above the front view and the right-side view normally appears to the right of the front view. When used, left-side view usually is placed directly to the left and in line with front view.

This system of drawing is known as orthographic projection, and the terms "multiview drawings" and "orthographic projection drawings" are used interchangeably.

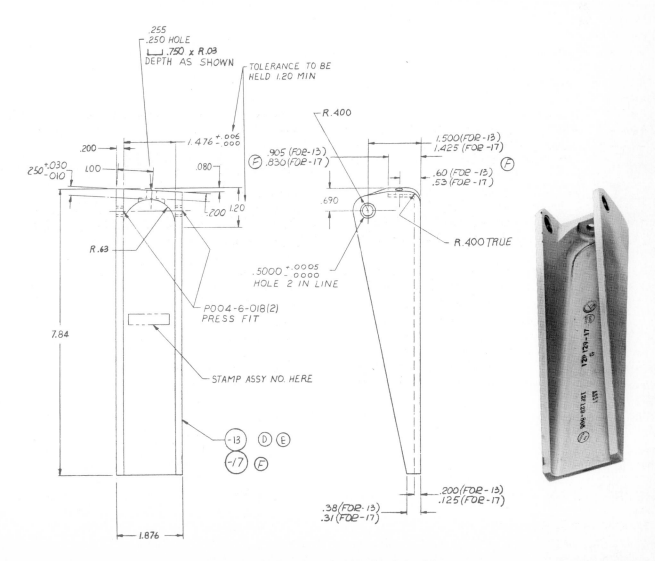

Fig. 7-1. A typical multiview drawing of an industrial part.

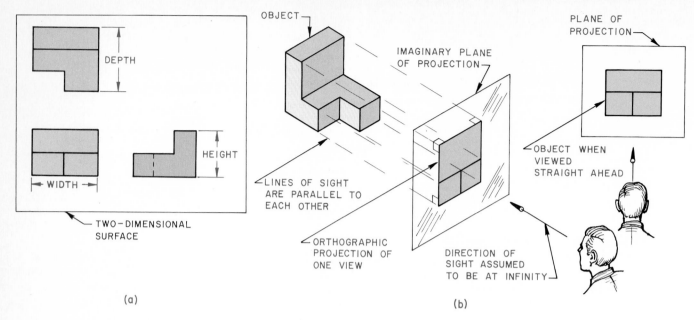

(a)

(b)

Fig. 7-2. Visualizing one view of an orthographic projection.

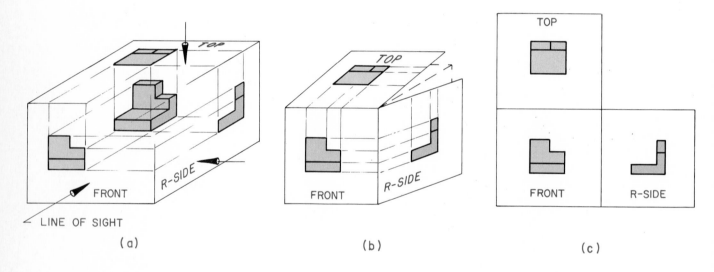

(a)

(b)

(c)

Fig. 7-3. The "glass box" illustrates the visualization system of orthographic projection.

ORTHOGRAPHIC PROJECTION

An orthographic projection drawing is a representation of the separate views of an object on a two-dimensional surface. It reveals the width, depth and height of the object, Fig. 7-2 (a).

The projection is achieved by viewing the object from a point assumed to be at infinity (an indefinitely great distance away). The lines of sight (or projectors) are parallel to each other and perpendicular to the plane of projection, (b).

THE PROJECTION TECHNIQUE

Persons experienced in drafting are readily able to "picture" different views of an object in their minds. This mental process is known as "visualizing the views" by looking at

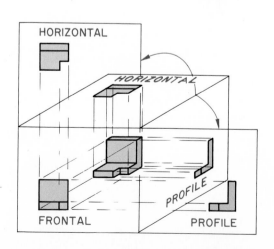

Fig. 7-4. Principal or coordinate planes of projection.

Multiview Drawing

the actual object or a three-dimensional picture of the object. This is one of the most important aspects of drafting technique that you can learn.

The "glass box," Fig. 7-3, is helpful in developing skill in visualizing the view. Each face of the object is viewed from a position that is 90 degrees, or perpendicular, to the projection plane for that view, Fig. 7-3 (a). The views are obtained by projecting the lines of sight to each plane of the glass box.

side or end elevation. These three planes are at right angles to each other when in their natural position with the glass box closed, Fig. 7-5. The frontal plane is considered to be lying in the plane of the drawing paper. The horizontal and profile planes are revolved into position on the drawing in the same plane as the drawing paper.

These three planes are referred to as principal planes because they are the views shown on most industrial drawings.

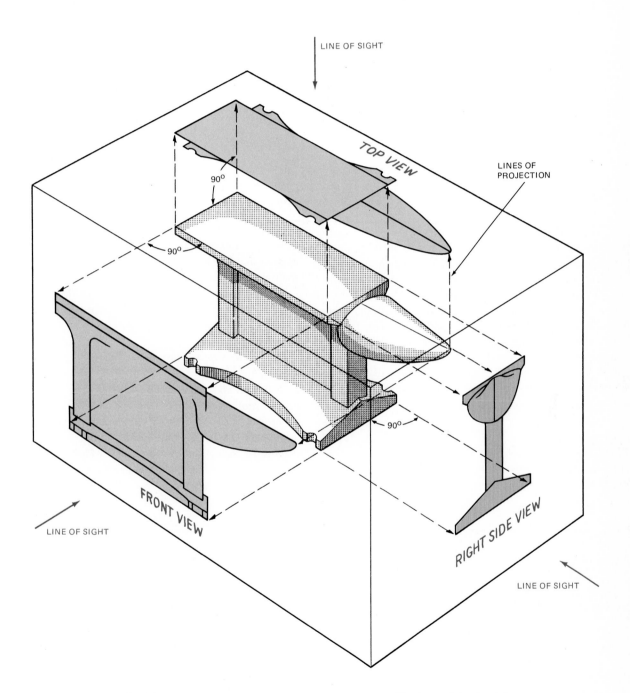

Fig. 7-5. Views projected on the frontal, horizontal and profile planes of the glass box.

The projection shown in the FRONTAL PLANE, Fig. 7-4, is called the front view or front elevation. On the HORIZONTAL PLANE, it is called the top view or plan view. If it is on the PROFILE PLANE, it is called the side or end view, or

They are also called coordinate planes because of their right-angle relationship in the folded box. When they unfold, they establish a definite coordinate relationship between all views of an orthographic projection.

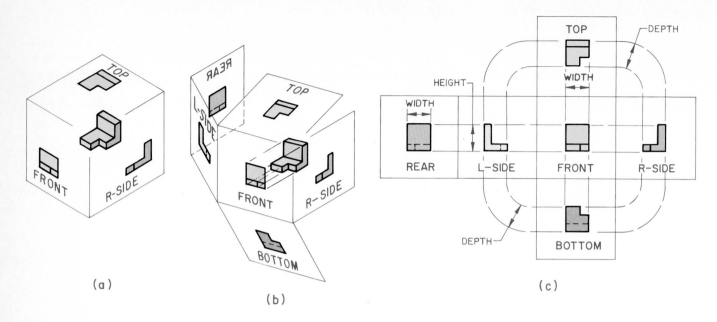

Fig. 7-6. Unfolding the glass box to show the six coordinate planes.

However, six views are possible from the six sides or planes of the glass box, Fig. 7-6. Note the manner in which the box is unfolded as shown in Fig. 7-6 (b and c). This establishes the coordinate relationship of the three additional views.

Also note the lines of projection from one view to the next. This insures that the height dimension will be the same for the rear, left-side, front and right-side views, and they are all aligned with coordinates, (c). The top, front and bottom views are all aligned and have the common dimension of width as does the rear view. The top, right-side, bottom and left-side views have depth dimensions that are common.

ALTERNATE LOCATION OF VIEWS

There are occasions when the location of views as shown in Fig. 7-7 (a) is not feasible due to space limitations. For example, an expanded "List of Materials" on a drawing, which appears above the title block, may crowd the usual location of the right side view, Fig. 7-7 (a).

To compensate for this lack of space, it is permissible to project the right side view directly across from the top view as if the profile plane were hinged to the horizontal plane, as shown in (b).

The projection would appear on the drawing as shown in Fig. 7-7 (c).

Although the usual practice is to locate required views in normal projected positions, the side view or profile planes may be projected off the top or bottom views. Likewise, the rear view may be projected upward from the top view or downward from the bottom view. Each is revolved into position in the same plane as the front view when alternate locations are necessary.

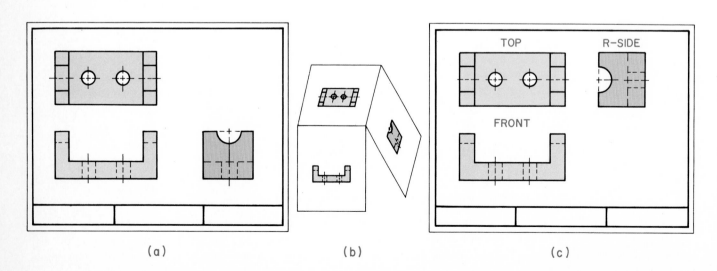

Fig. 7-7. Alternate location of right-side view.

PROJECTION OF ELEMENTS

Drawing the several views of an object is done by making measurements of certain points, lines and surfaces in one view and projecting these to the other views. Features such as circular holes or arcs should be located initially in the view where they appear as circles or arcs, then projected to the remaining views, Fig. 7-8.

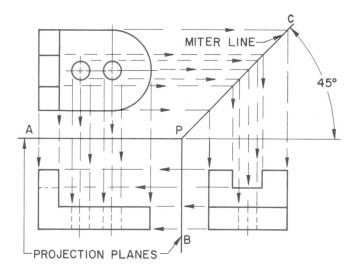

Fig. 7-8. Projecting elements between views.

Projection of the elements provides for greater accuracy in the alignment of views, and it is faster than measuring each view separately with the scale or dividers. A single, 45 degree miter line, PC in Fig. 7-8, is drawn to project point, line and surface measurements from one view to another. This miter line meets the projection planes AP and PB at point P equidistant between the views.

PROJECTION OF POINTS

A point is defined as something having position, but not extension. It has location in space, but has no length, depth or height. A point on a drawing should be indicated by a small cross (+) mark and never by a dot. It may be the intersection of two lines, the end point of a line or the corner of an object.

Point P in space is located by measuring three directions (length, depth and height) from the planes of projection. In Fig. 7-9 (a), these measurements are made from the frontal (P_F), horizontal (P_H) and profile (P_P) planes. Point P in the orthographic projection, Fig. 7-9 (b), is located by making each measurement once in the appropriate view and projecting measurements with the triangle and T-square to the other views.

Measurements need not be made from planes of projection on a drawing. Usually, the first point representing a corner, center line or some other feature of the object is properly located. Then space between views is allotted to produce a balanced drawing. Other points, lines or surfaces are located from this first point, Fig. 7-10.

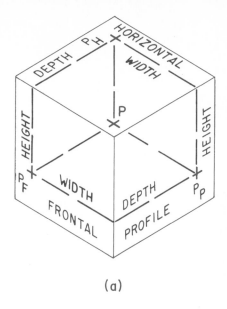

(a)

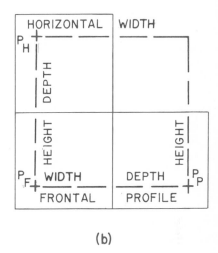

(b)

Fig. 7-9. Locating points in space on the projection planes.

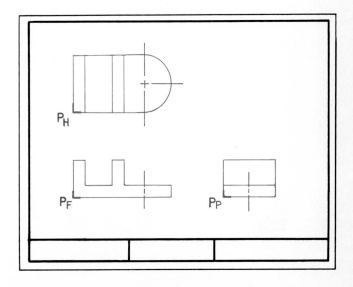

Fig. 7-10. Locating a starting point on a drawing.

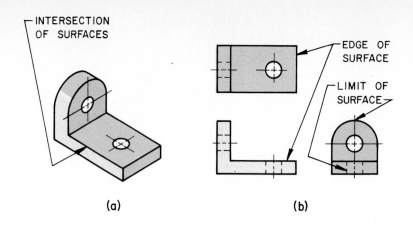

Fig. 7-11. Lines represent intersection of surfaces, edge of surfaces or limits of surfaces.

PROJECTION OF LINES

A straight line is defined as the shortest distance between two points. A curved line is a line following any of a variety of arcs or curved forms.

In a drawing, lines may represent the intersection of two surfaces, Fig. 7-11 (a); the edge view of a surface, (b); or, the limits of a surface, (b).

There are four kinds of straight lines found on objects in drawings: horizontal, vertical, inclined and oblique. Each line is projected by locating its end point, as covered in the preceding section.

Horizontal Lines are parallel to the horizontal plane of projection and one of the other planes, while being perpendicular to the third plane, AB in Fig. 7-12. A horizontal line appears true length in two of the planes and as a point in the third.

Vertical Lines are parallel to both the frontal and profile planes and perpendicular to the horizontal plane, EG in Fig. 7-12. A vertical line appears true length in the frontal and profile planes and as a point in the horizontal plane.

Inclined Lines are parallel to one plane of projection and inclined in the other two planes, CE in Fig. 7-12. An inclined line appears true length in one of the planes and foreshortened (not as long) in the other two.

Oblique Lines are neither parallel nor perpendicular to any of the planes of projection, line DF in Fig. 7-12. An oblique line appears foreshortened in all three planes of projection.

Curved Lines may be circular, elliptical, parabolic, hyper-

bolic or some other geometric curve form. They may also be irregular curves. For curves other than circles and arcs of circles, a number of points must be located on the curve and projected to the view concerned.

PROJECTION OF SURFACES

Plane and curved surfaces represent most of the surface features found on machine parts. Examples of plane surfaces are found on cubes and pyramids; circular curved surfaces are found on cylinders and cones.

Like the lines they outline, surfaces may be horizontal, vertical, inclined, oblique or curved. These surfaces are drawn by locating the end points of the lines that outline their shapes.

Horizontal Surfaces are parallel to the horizontal projection plane and appear in their true size and shape in the top view, A and B in Fig. 7-13. Horizontal surfaces (A and B) appear as lines (4-13) in the frontal and profile planes of projection.

Vertical Surfaces are parallel to one or the other of the frontal or profile planes. They appear in their true size and shape in this plane, G and F in Fig. 7-13. They are perpendicular to the other two planes and appear as lines in these planes (8-9 in the frontal plane and 6-8 in the profile plane), Fig. 7-13.

Inclined Surfaces are neither horizontal nor vertical, D in Fig. 7-13. They are perpendicular to one of the projection planes and appear as a true length line in this view (1-5 in side view). In the other two planes or views, top and front in Fig.

Fig. 7-12. Kinds of straight lines used on drawings.

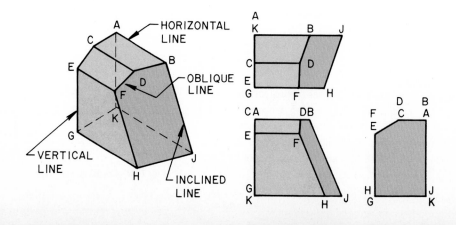

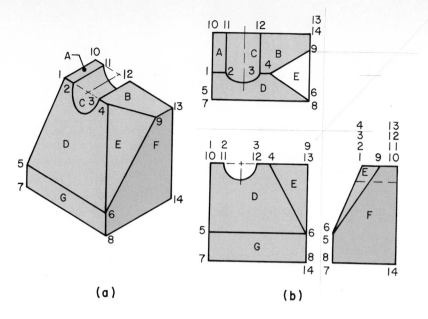

(a) (b)

Fig. 7-13. Projecting plane and circular curved surfaces.

7-13 (b), inclined surfaces appear foreshortened.

Oblique Surfaces are neither parallel nor perpendicular to any of the planes of projection, E in Fig. 7-13. They appear as a surface in all views but not in their true size and shape.

Curved Surfaces may be a single curved surface (cone or cylinder), double curved surface (sphere, spheroid or torus) or warped surface (machine screw thread or helix), Fig. 7-14.

Curved surfaces of the circular curve forms (cylinder) appear as circles in one view and as rectangles in the other views, Fig. 7-14. Three views of curved surface objects are shown here, but two are usually sufficient since two of the views are identical. Methods of obtaining the true size and shape of various lines and surfaces is covered in detail in Chapter 12.

Fig. 7-14. Examples of curved surface objects.

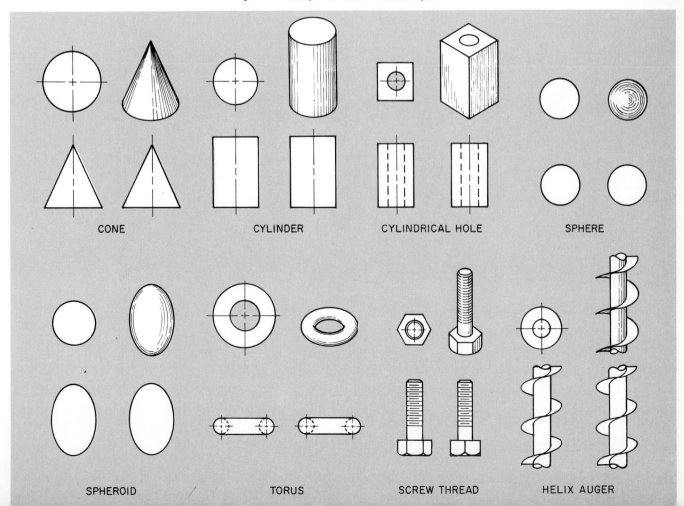

CONE   CYLINDER   CYLINDRICAL HOLE   SPHERE

SPHEROID   TORUS   SCREW THREAD   HELIX AUGER

PROJECTION OF ANGLES IN MULTIVIEW DRAWINGS

Angles that lie in a plane parallel to one of the projection planes will project their true size on that plane, A in Fig. 7-15. Angles that lie in a plane inclined to the projection plane will project smaller or larger than true size, depending on their location, B in Fig. 7-15.

Also, when a 90 degree angle on an inclined plane has one of its legs parallel to two projection planes, it will project true size in the other plane, as shown in C in Fig. 7-15. Angles on an oblique plane will always project smaller or larger than true size depending on their location, as shown in D in Fig. 7-15.

To project angles lying in an inclined or oblique plane, locate their end points by projection and draw lines between these points.

SELECTION OF VIEWS

First considerations in making multiview drawings are the selection and arrangement of views to be drawn. Select views which clearly describe the details of the part or assembly. Give prime consideration to the front view in order to select the view that best characterizes the shape of the object.

The number of views to be drawn depends upon the shape and complexity of the part. Often, two views will provide all the details necessary to construct or assemble the object. This is particularly true of cylindrical or round objects, as shown in Fig. 7-16 (a).

Flat objects made from relatively thin sheet stock may be adequately represented with only one view by noting the stock thickness on the drawing, Fig. 7-16 (b). The views for the object at (c) were not well chosen, resulting in a poor arrangement.

The views at Fig. 7-16 (d) have been well selected to best describe the part. They reduce the number of hidden lines in all views and provide for a balanced arrangement of the three required views.

SUMMARY OF FACTORS INVOLVED

Four factors serve as guidelines in the selection of views. No single factor should determine the selection. Rather, a composite consideration of all factors is most likely to result in the best selection of views.

Give prime consideration to selection of front view:
1. Select a view that is most representative of contour or shape of object.
2. Consider natural or functioning position of object.
3. Place principal surface area parallel or perpendicular to one or more planes of projection.
4. Consider orientation of view which produces least hidden lines in all views.
   Consider space requirements of entire drawing.
1. Long and narrow objects may suggest top and front view.
2. Short and broad objects may suggest front and side view.
3. When "Title Block" or "List of Materials" tends to crowd right side view, consider an alternate position of view or select another view.
   Choose between two equally important views unless space requirements or other factors prohibit.
1. Right-side view is preferred over left-side when a choice is available.
2. Top view is preferred over bottom view when a choice is available.
   Consider the number of views to be drawn.
1. Use only number of views necessary to present a clear understanding of object.
2. One or two views may be sufficient for a relatively simple object; three or more views may be required for more complex objects.

SPACE ALLOTMENT FOR MULTIVIEW DRAWING

Crowding the views in multiview drawings detracts from their appearance. It makes reading and understanding of the

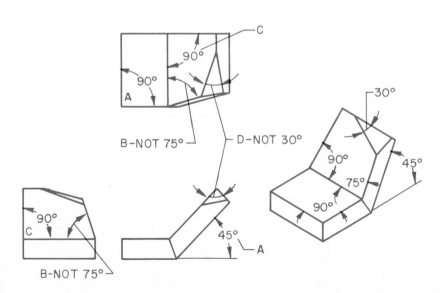

Fig. 7-15. True size and projected size of angles in orthographic projection.

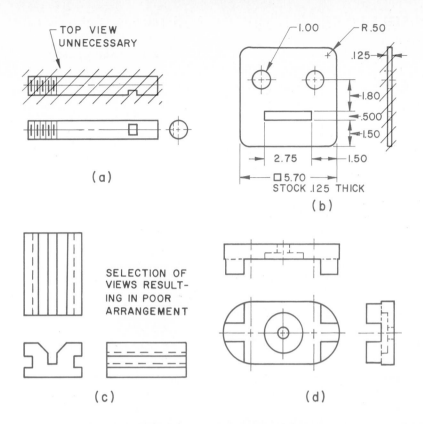

Fig. 7-16. Factors in the selection of views.

drawings more difficult.

The practice in industrial drafting rooms is to provide ample space between the views of a drawing for dimensions, callouts and notes. This serves to make the views more distinct and provides a neat appearance. Anticipate the number of dimension lines or notes to be used and allow sufficient space.

Once the scale of the drawing has been determined, figuring space allotment is rather easy. Add the combined width and depth of the front and side views, Fig. 7-17. Lay off this total (6.5 inches, for example) to scale along the lower border. Allow approximately 1 inch between views and lay off this distance beyond the first measurement. Divide the amount remaining between the two end spaces, which allow about 1.50 inches for each.

Vertical spacing is figured in a like manner by laying off the distances along a vertical border line, Fig. 7-17.

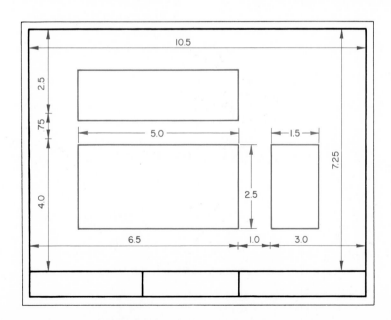

Fig. 7-17. Spacing views on a drawing sheet.

PROJECTION OF INVISIBLE LINES AND SURFACES

Surfaces and intersections that are hidden behind a portion of the object in a particular view are usually represented by "hidden" or "invisible lines." Obviously, these terms are used to refer to the surface or intersection that the invisible line represents, rather than the line itself being invisible.

Invisible lines and arcs begin with a dash, unless the hidden line or arc is a continuation of a visible line or curve. If it is a continuation, a space is left to show exactly where the invisible line or arc begins, Fig. 7-18 (1). When invisible surfaces actually intersect on the object itself, the dashes join at the intersection with a "+" or "T", Fig. 7-18 (2).

Invisible lines that cross but do not intersect each other on the object are drawn with a gap at the crossover on the drawing (3). Angular lines that come to a point are shown with the dashes joined at the point, such as the bottom of a drilled hole, Fig. 7-18 (4). Invisible lines meet at the corners with an "L" (5) unless the corner is joined by a visible line, then a gap is used (6). Parallel invisible lines have their dashes staggered (7).

Invisible lines are usually omitted in sectioned views unless absolutely necessary for clarity, Fig. 7-18 (8). Industrial draftsmen frequently omit some invisible lines from views when the drawing is clear without them. This avoids a cluttered appearance. However, it is good practice for the beginning draftsman to include all invisible lines in regular views until he has gained a fuller understanding of the problem of clarity in a drawing.

PRECEDENCE OF LINES

Occasionally you will find that certain lines coincide in the projection of views in multiview drawings. Should this occur, visible lines take precedence over all others. The following priority of line importance governs precedence of lines:
1. Visible lines.
2. Invisible lines.
3. Cutting-plane line.
4. Center line.

5. Break line.
6. Dimension and extension lines.
7. Section line (crosshatching).

REMOVED VIEWS

Sometimes it is desirable to show a complete or partial view on an enlarged scale to clarify the detail of the part, Fig. 7-19. This particular view is removed to a nearby area of the drawing, yet the same orientation of the view is maintained. The removed view is appropriately identified and referred to on the regular view.

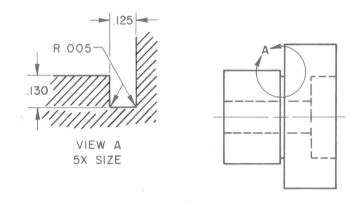

Fig. 7-19. A removed partial view enlarged to clarify a machining detail.

PARTIAL VIEWS

Because of their shape, some objects may not require all views to be full view. Objects which are symmetrical in one view may require only a half view in that plane, Fig. 7-20 (a). A partial view may be broken on the center line (a), or at another place with a broken line (b).

Objects with different side views should have two partial side views drawn. Each of these views should include lines for that view only, thus avoiding a confusion of lines from the other view, Fig. 7-20 (c).

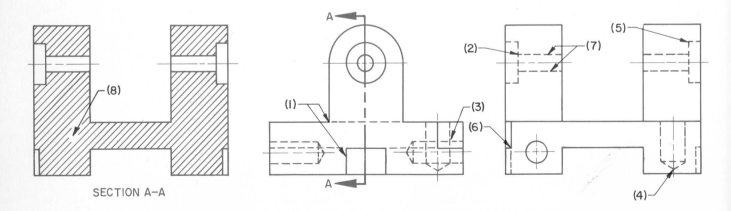

Fig. 7-18. Projection of invisible lines and surfaces.

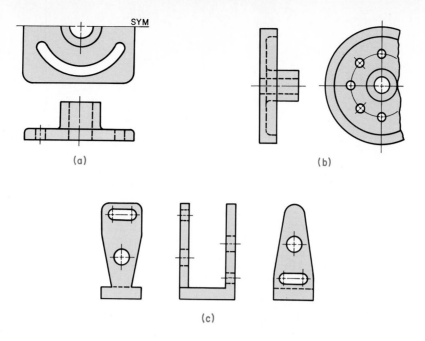

Fig. 7-20. Partial views of symmetrical and nonsymmetrical objects.

CONVENTIONAL DRAFTING PRACTICES

A number of conventional drafting practices are used in American industry to reduce costs, speed the drafting process and clarify drawings. Some of these practices are presented here; others appear in chapters related to the specific practice being described.

FILLETS AND ROUNDS

When making metal castings, it is necessary to avoid sharp interior corners to prevent fractures of the metal. Also, sharp exterior corners are difficult to form in the mold. To eliminate these problems, patterns for the castings are made with rounded corners. The small, rounded internal corner is known as a "fillet."

The small, rounded external corner is known as a "round," Fig. 7-21 (a). Since there is no sharp line intersecting the two surfaces, an assumed line of intersection of the two surfaces is drawn, (b). The view at (c) shows fillets and rounds represented on plane surfaces.

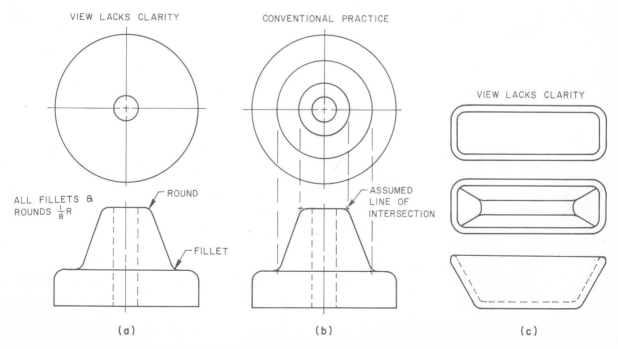

Fig. 7-21. Conventional representations of fillets and rounds.

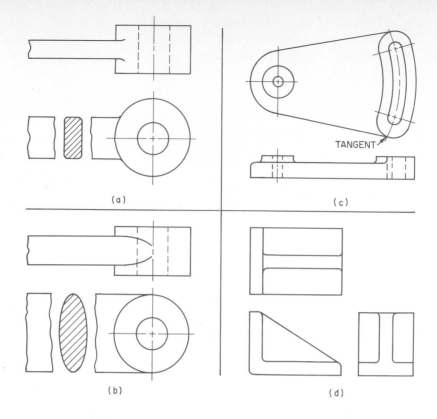

Fig. 7-22. Representation of runouts on drawings.

RUNOUTS

A runout is the intersection of a fillet or round with another surface. The shape of an arm, spoke or web also affects the shape of the runout, Fig. 7-22. The arc of the runout should be the same radius as the fillet or round. It may be drawn freehand, with irregular curve or compass.

RIGHT AND LEFT-HAND PARTS

Whenever possible, industry will make opposite parts identical to reduce the number of different parts required for an assembly. Examples of identical parts used on opposite sides are automobile wheels and tires, and doors for kitchen cabinets which may be hung as right or left-hand doors. In some cases, opposite parts are not identical (such as automobile door handles and cabinet drawer slides), Fig. 7-23.

When opposite parts cannot be made interchangeable, the conventional practice is to draw one part and to note RH PART SHOWN, LH PART OPPOSITE, Fig. 7-24. This works quite well for most simple parts and saves considerable drafting time. Where there is a chance for confusion in details of the opposite part, both should be drawn.

CYLINDER INTERSECTIONS

The conventional representation for small cylinder intersections with plane or cylindrical surfaces is shown in Fig. 7-25 (a). The same is true for cuts in small cylinders, (b). For keyseats and small drilled holes, the intersection is so unimportant that clarity on the drawing is served by treating these intersections conventionally, (c). Intersection of larger cylinders should be plotted as discussed in Chapter 14.

Fig. 7-23. These automobile door handles show that opposite parts are not always indentical. (Fisher Body Div., GMC)

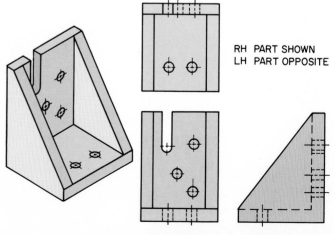

RH PART SHOWN
LH PART OPPOSITE

Fig. 7-24. A drawing calling for opposite parts.

128

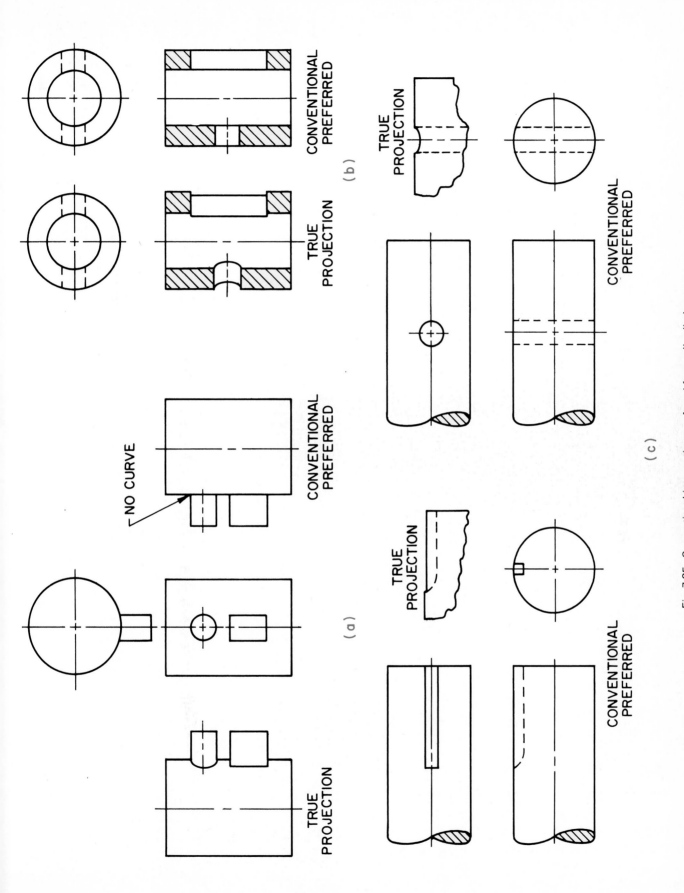

Fig. 7-25. Conventional intersections preferred for small cylinders.

REVOLVING RADIAL FEATURES

Some objects that have radially arranged features appear confusing when true orthographic projection techniques are followed. For example, the lugs on a flange appear awkward and out of position with the flange in true projection, Fig. 7-26 (a). When revolved to a position on the center line, the ribs appear to be symmetrical, (b).

A number of small holes arranged radially in a plate are also confusing in the side view when true projection techniques are followed, Fig. 7-26 (d). Conventional practices call for the holes to be revolved to positions of symmetry, (e). Ribs on a hub are another example of radial features that appear more clearly when revolved to a position of symmetry, (h).

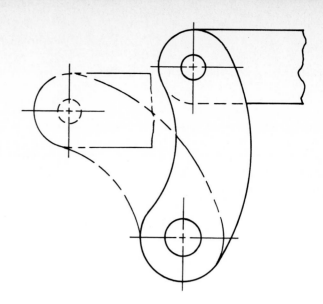

Fig. 7-27. Alternate position shown in phantom lines.

REPEATED DETAIL

Drawings of coil springs, radial flutes and other repeated details would require considerable drafting time if these were drawn in full views. Conventional drafting practices require the drawing of one or two of the individual details, with the remainder represented by phantom lines, Fig. 7-28.

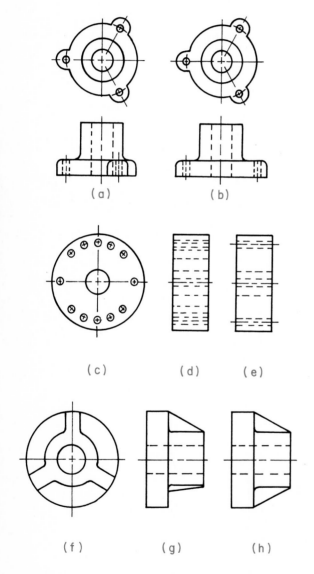

(a)   (b)

(c)   (d)   (e)

(f)   (g)   (h)

Fig. 7-26. Revolve radial features to achieve symmetry and a more understandable drawing.

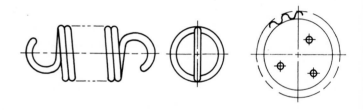

Fig. 7-28. Phantom lines used to represent repeated detail.

Fig. 7-29. Principal coordinate planes of projection and quadrants.

PARTS IN ALTERNATE POSITION

At times it is necessary to draw the alternate position of parts to show limits and necessary clearance during operation of the part. The part is drawn in its alternate position by using a phantom line, Fig. 7-27.

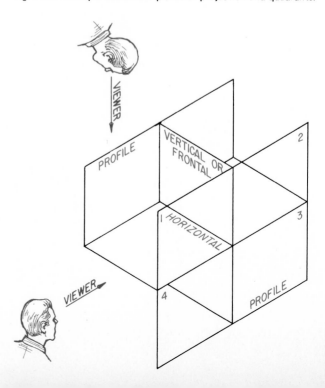

FIRST AND THIRD-ANGLE PROJECTIONS

In orthographic projection, drawings are referred to as "first-angle" or "third-angle" projections. These two projections are derived from a theoretical division of all space into four quadrants by a vertical plane and a horizontal plane, Fig. 7-29. The quadrants are numbered 1 through 4, starting in the upper front quadrant and continuing clockwise when viewed from the right side. The viewer of the four quadrants is considered to be in front of the vertical or frontal plane, and above the horizontal plane. The position of the profile plane is not affected by the quadrants. It is considered to be either to the right or left of the object as desired.

Third-angle projection is used in the United States and Canada. Most European countries use first-angle projection.

The main difference between the two is how the object is projected and the positions of the views on the drawing.

In third-angle projection, the projection plane is considered to be between the viewer and the object, and the views are projected forward to that plane, Fig. 7-30 (a). The views appear in their natural positions when the views are revolved into the same plane as the frontal plane, (b). The top view appears above the front view, the right-side view is to the right of the front view, the left view to the left of the front view, and so on.

In first-angle projection, the projection plane is on the far side of the object from the viewer, Fig. 7-31. The views of the object are projected to the rear and onto the projection plane instead of being projected forward.

The individual views are the same as those obtained in

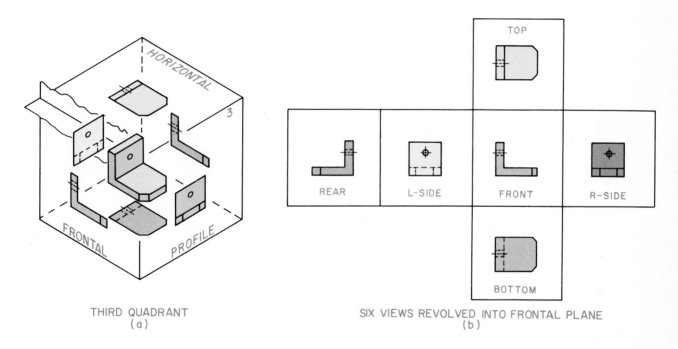

THIRD QUADRANT
(a)

SIX VIEWS REVOLVED INTO FRONTAL PLANE
(b)

Fig. 7-30. Above. Views are projected forward in third-angle projection.
Fig. 7-31. Below. Views are projected rearward in first-angle projection.

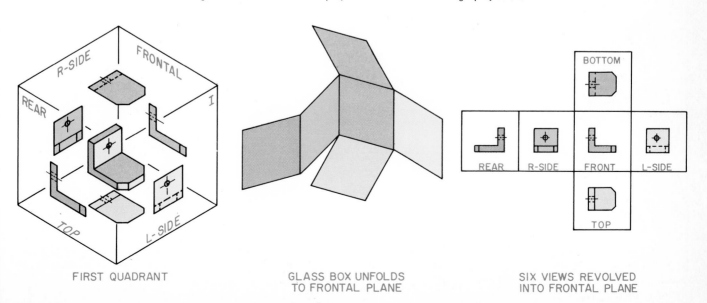

FIRST QUADRANT

GLASS BOX UNFOLDS TO FRONTAL PLANE

SIX VIEWS REVOLVED INTO FRONTAL PLANE

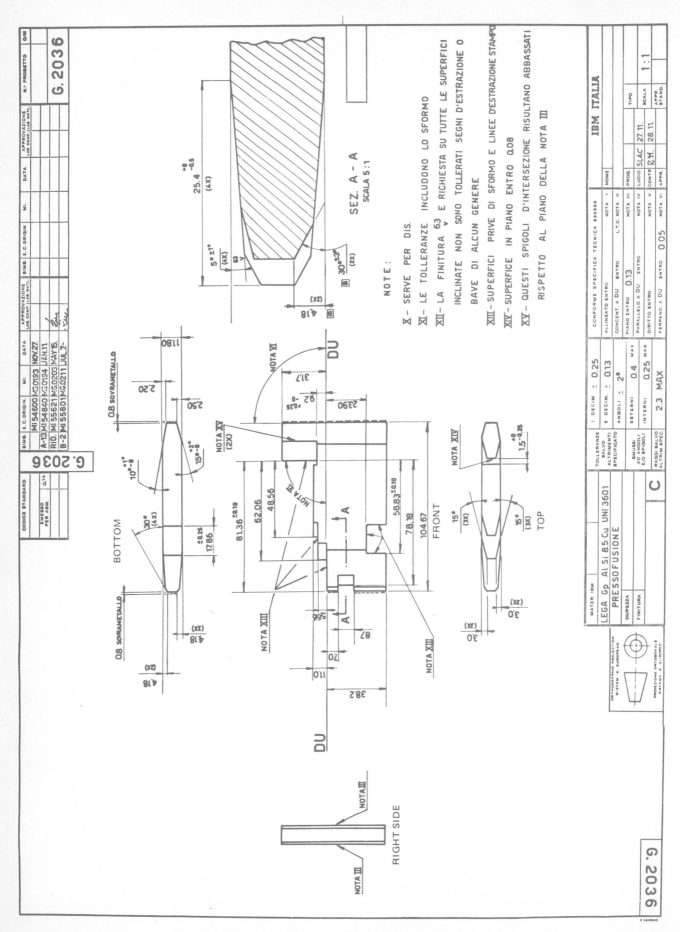

Fig. 7-32. First-angle projection is used in most European countries. (IBM Corp.)

third-angle projection, but their arrangement on the drawing is different when revolved into the frontal plane, Fig. 7-32. The glass box is still hinged to the frontal plane, but the frontal plane is behind the object. The top view appears below the front view; the right side appears to the left of the front view; the left side appears to the right of the front view.

It is possible to place an object in any of the four quadrants, but the second and fourth are not practical. The third-angle of projection is followed entirely in this text. However, it is well for you to understand first-angle projection so you will be able to interpret a drawing prepared in another country.

Industries who serve customers in the international market sometimes mark their drawings to indicate first-angle or third-angle projection, Fig. 7-33.

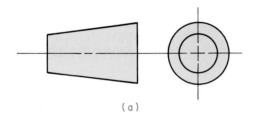

(a)

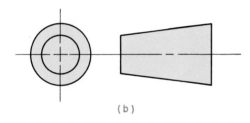

(b)

Fig. 7-33. Symbols for indicating (a) first-angle and (b) third-angle projections.

VISUALIZING AN OBJECT FROM A MULTIVIEW DRAWING

Most students in beginning drafting have difficulty visualizing the machine part or assembly from a multiview drawing. Mastery of a few simple techniques will enable you to solve the "mystery." Follow the steps listed below:

1. Break each view down to a basic geometric shape (rectangle, circle, cone, triangle, etc.), Fig. 7-34.
2. Consider possibilities for each shape. Is the circle a hole or a protruding shaft? Is rectangle a base plate or web? Is triangle a brace or support?
3. Check basic geometric shape in one view against its shape in another view. What do hidden lines and center lines represent?
4. Check length, depth or height dimensions in two or more views and compare with geometric shapes.
5. Put various geometric components together mentally and you should begin to visualize the object.
6. Make a freehand sketch of object to clear up any uncertain details.

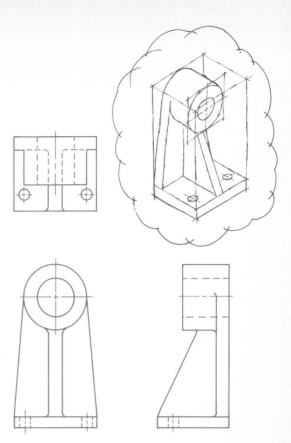

Fig. 7-34. Visualizing an object from a multiview drawing.

LAYING OUT AND FINISHING A DRAWING

It is good practice to draw light layout and construction lines until you have solved the essential problems in each view. All lines should then be darkened to complete the drawing. If changes are necessary, they are relatively simple to make when lines are lightly drawn.

Observe the following steps until they become part of your drafting practice:

1. Draw light lines at first (heavy lines tend to "ghost," making them difficult to erase).
2. Select a sheet size and drawing scale that will avoid crowding of views, dimensions and notes.
3. Check your measurements carefully in "blocking out" required views.
4. Locate and lay out arcs and circles first, then straight lines.
5. During layout, do not include hidden lines, center lines or dimension lines. A short mark for a line, or a dimension figure lightly noted near its location, will serve as a reminder it is to be included.
6. Check your layout carefully for missing lines, dimensions, notes or special features required in problem assignment.
7. Remove unnecessary construction lines and give drawing a general cleaning.
8. Darken the lines. Start with arcs and circles, then do the lines from top down, and from left to right (unless you are left handed).
9. Letter the notes and title block.
10. Check finished drawing carefully for spelling, line weight and general appearance.

PROBLEMS AND ACTIVITIES

Problems with missing lines and missing views are given to provide you with experience in multiview projection techniques. Further problems in multiview drawing follow.

MISSING LINES

Study the views. Use drawing sheet Layout II, in the Reference Section. Sketch in the views of the problems presented in Fig. 7-35 and add the missing lines.

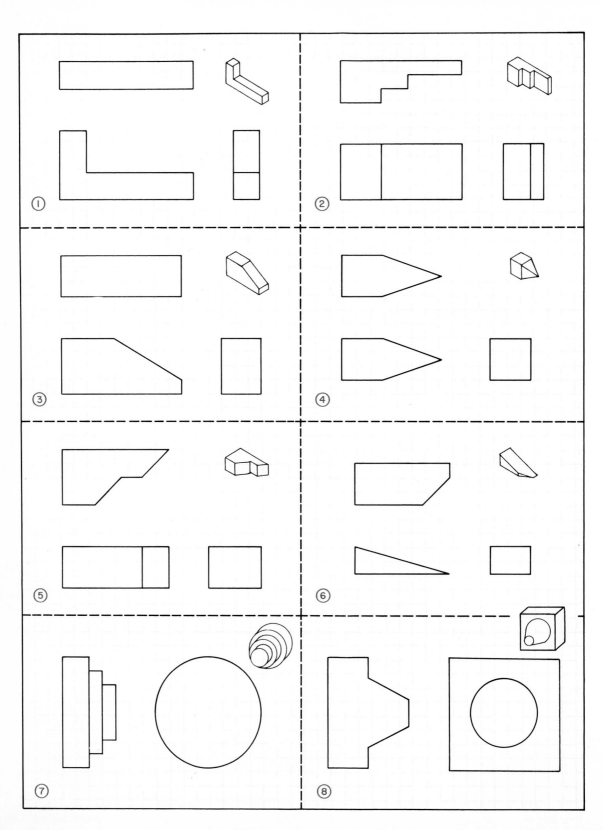

Fig. 7-35. Missing line multiview problems.

Multiview Drawing

MISSING VIEWS

Study the views given in Fig. 7-36. Make a sketch of the two views given and the missing view. Have your sketch approved by your instructor. Then make drawings, using drawing sheet Layout II, found in the Reference Section.

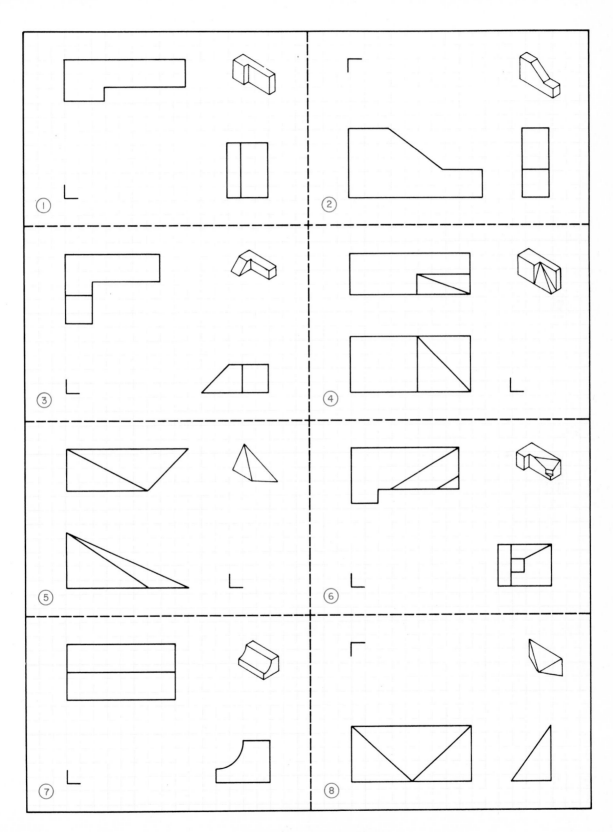

Fig. 7-36. Missing view problems.

IDENTIFICATION OF POINTS, LINES AND SURFACES

This problem calls for identification of points, lines and surfaces in four multiview drawings to be listed in chart form, Fig. 7-37. A sample entry has been completed in the first row for problem A shown in Fig. 7-38.

Check the views in problem A for the projection of points, lines and surfaces listed in the first row of the chart in Fig. 7-37. Given: line 35-36 in the right-side view. Identify its elements in the two remaining views of the multiview drawing and in the pictorial drawing.

A study of the views reveals that line 35-36 appears as line 9-8 in the pictorial view. It also is in the same plane as line 7-6, but only the nearest point or line in the line of projection is listed, and as surface D; it appears as point 30 in the front view; and as line 23-26 and surface V in the top view.

Prepare a chart with nineteen rows for problems and identification items, using Layout II, Reference Section, in the manner illustrated in Fig. 7-37. Letter the headings and the information required. Use 1/8 inch capital letters and numerals.

Continue identification of the remaining items for problem A and the following for problems B, C and D:

Problem B
1. Line 3-6, front view.
2. Surface Z, front view.
3. Line 6-10, front view.
4. Surface B, pictorial view.
5. Line 1-2, top view.

Problem C
1. Line 28-30, front view.
2. Surface V, top view.
3. Line 4-7, pictorial view.
4. Surface C, pictorial view.
5. Line 22-27, top view.

Problem D
1. Line 26-27, front view.
2. Surface Z, right-side view.
3. Line 23-25, top view.
4. Surface B, pictorial view.

READING A DRAWING

Reading an industrial drawing will aid you in visualizing the views. The following questions will guide your reading of the drawing in Fig. 7-39. Answer these questions on a separate sheet of paper.

1. What views are shown?
2. What is the name of the part?
3. What is the number of the part?
4. From what size stock is the part to be made?
5. Starting with the circled letters, match the lines and surfaces in the two views. Note there are two extra numbers for which no letter matches.
6. Why were the two views chosen by the drafter? Would a top view make the drawing any clearer?
7. Is the circle at number 7 a recessed or protruding cylinder? What confirms this?
8. What size hole is drilled through at number 1?
9. Give the pilot drill size for the threaded hole at number 8.
10. How thick is the piece at S?

MULTIVIEW PROBLEMS

Study the problems shown in the pictorial views in Fig. 7-40. Select and sketch the necessary views on sketch paper for each problem assigned. Have these checked by your instructor. Prepare a multiview drawing, using Layout I, Reference Section. Do not dimension the drawing.

	PICTORIAL VIEW			FRONT VIEW			SIDE VIEW			TOP VIEW		
	POINT	LINE	SURFACE	POINT	LINE	SURFACE	POINT	LINE	SURFACE	POINT	LINE	SURFACE
PROBLEM A												
I.		9-8	D	30				35-36			23-26	V
2.			C									
3.											24-26	
4.											23-26	
5.					27-28							
PROBLEM B												
I.												

Fig. 7-37. Chart for recording information on multiview drawings.

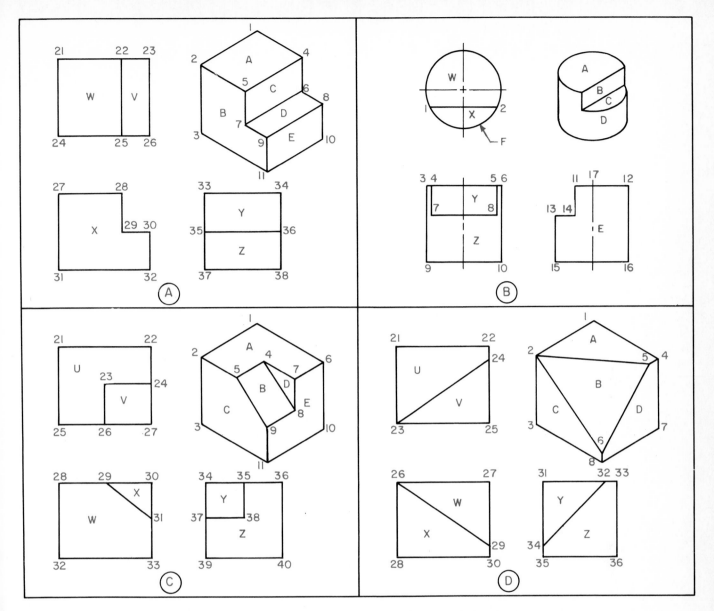

Fig. 7-38. Identification of points, lines and surfaces in multiview drawings.

MULTIVIEW DRAWINGS WITH REMOVED VIEWS

Study the parts shown in Fig. 7-41. Then select and draw the necessary views, including a removed view of the feature indicated. Unless otherwise indicated, the removed view is to be drawn two times size. A freehand sketch is not required. However, it is a good practice to first sketch all drawn views until you have become proficient in visualizing and spacing multiview drawings on a sheet. Use Layout I, and select the appropriate scale for the drawing.

MULTIVIEW DRAWINGS WITH PARTIAL VIEWS

When a machine part is symmetrical in one view, a partial view may be drawn. In others, because of dissimilarity in side views, two partial-side views should be drawn. Select views carefully and draw necessary partial views for objects in Fig. 7-42. Use Layout I, select appropriate scale.

MULTIVIEW PROBLEMS INVOLVING CONVENTIONAL DRAFTING PRACTICES

Select and draw the necessary views for the parts shown in Fig. 7-43. Use standard conventional drafting practices where applicable. Prepare Layout I and select the appropriate scale for the drawing.

FIRST-ANGLE PROJECTION

Although most of your work in drafting will be done in third-angle projection, making a drawing in first-angle projection is the best way to understand the system used in most European countries. Make the necessary first-angle projection views of the BEARING SUPPORT in Fig. 7-40 or the ACTUATING LEVER in Fig. 7-43. Use Layout I.

PROBLEMS IN MACHINE PARTS

Study the machine parts in Fig. 7-44 and draw the necessary views of each part to adequately describe it.

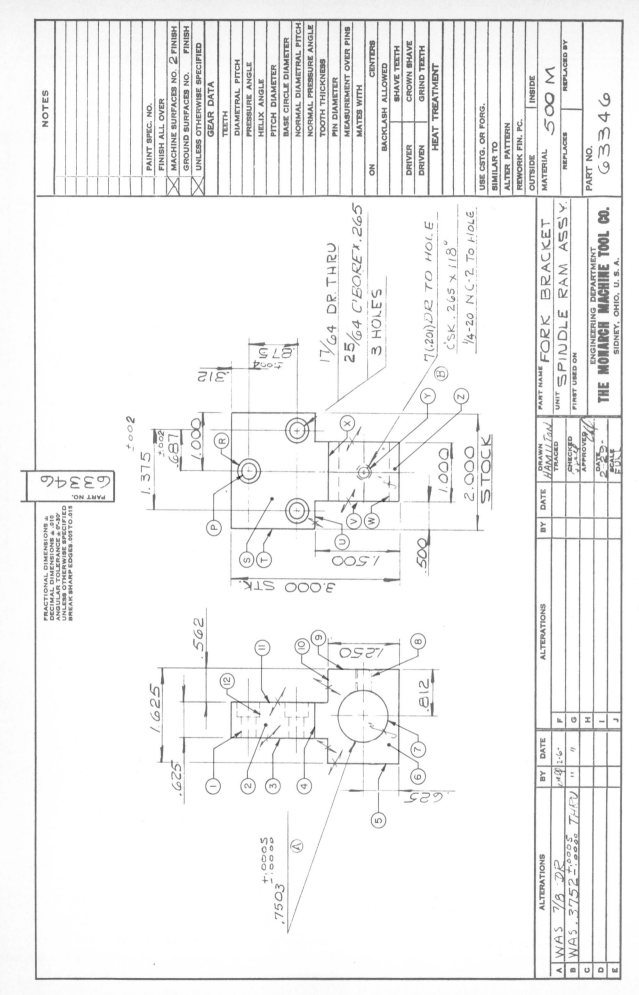

Fig. 7-39. Problem in reading a drawing.

138

Multiview Drawing

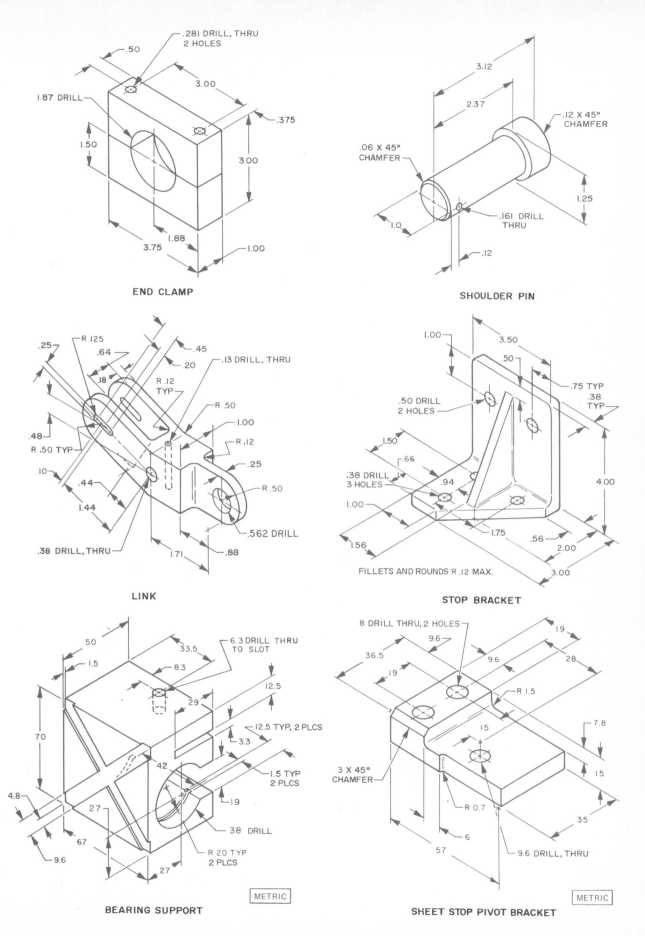

END CLAMP

.281 DRILL, THRU
2 HOLES
.50
3.00
.375
1.87 DRILL
1.50
3.00
1.88
3.75
1.00

SHOULDER PIN

3.12
2.37
.12 X 45°
CHAMFER
.06 X 45°
CHAMFER
1.25
1.0
.161 DRILL
THRU
.12

LINK

R .125
.25
.64
.45
.20
.18
.13 DRILL, THRU
R .12
TYP
R .50
1.00
.48
R .12
R .50 TYP
.10
R .50
.44
1.44
.562 DRILL
.38 DRILL, THRU
1.71
.88
.25

STOP BRACKET

1.00
3.50
.50
.50 DRILL
2 HOLES
.75 TYP
.38
TYP
1.50
.66
.38 DRILL
3 HOLES
.94
1.00
4.00
1.56
1.75
.56
2.00
3.00
FILLETS AND ROUNDS R .12 MAX.

BEARING SUPPORT

50
33.5
6.3 DRILL THRU
TO SLOT
1.5
8.3
12.5
29
70
12.5 TYP, 2 PLCS
3.3
42
1.5 TYP
2 PLCS
4.8
19
27
38 DRILL
67
R 20 TYP
2 PLCS
9.6
27

METRIC

SHEET STOP PIVOT BRACKET

8 DRILL THRU, 2 HOLES
9.6
19
36.5
9.6
28
19
R 1.5
15
7.8
3 X 45°
CHAMFER
15
R 0.7
35
6
9.6 DRILL, THRU
57

METRIC

Fig. 7-40. Multiview problems.

139

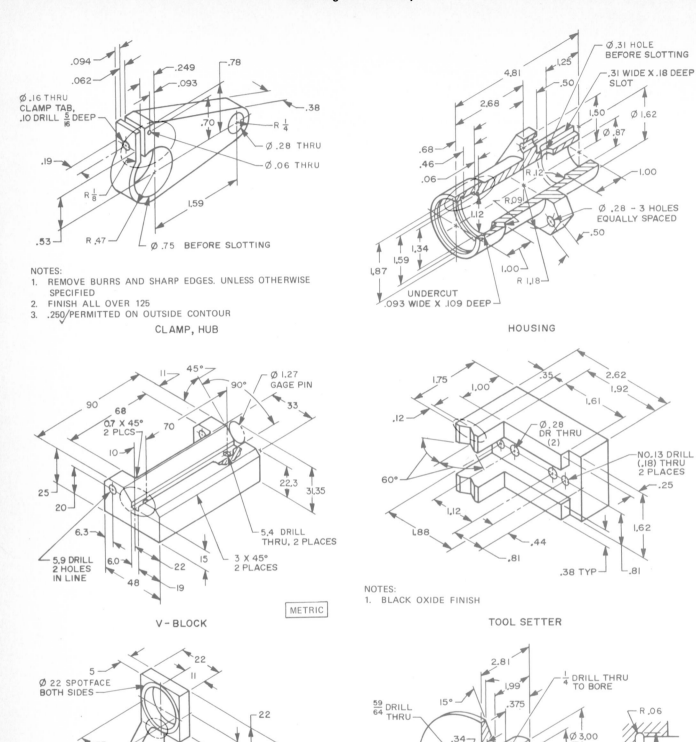

CLAMP, HUB

.094
.062
Ø .16 THRU
CLAMP TAB,
.10 DRILL 5/16 DEEP
.249
.093
.78
.38
.70
R 1/4
Ø .28 THRU
Ø .06 THRU
.19
R 1/8
1.59
.53
R .47
Ø .75 BEFORE SLOTTING

NOTES:
1. REMOVE BURRS AND SHARP EDGES. UNLESS OTHERWISE SPECIFIED
2. FINISH ALL OVER 125
3. .250/ PERMITTED ON OUTSIDE CONTOUR

HOUSING

Ø .31 HOLE BEFORE SLOTTING
.31 WIDE X .18 DEEP SLOT
4.81
1.25
.50
2.68
1.50
Ø 1.62
Ø .87
.68
.46
.06
R .12
R .09
1.00
1.12
Ø .28 – 3 HOLES EQUALLY SPACED
.50
1.87
1.59
1.34
1.00
R 1.18
UNDERCUT
.093 WIDE X .109 DEEP

V – BLOCK

11
45°
90°
Ø 1.27 GAGE PIN
90
68
0.7 X 45° 2 PLCS
70
33
10
25
20
22.3
31.35
6.3
5.4 DRILL THRU, 2 PLACES
3 X 45° 2 PLACES
5.9 DRILL 2 HOLES IN LINE
6.0
22
15
48
19
METRIC

TOOL SETTER

1.75
1.00
.35
2.62
1.92
1.61
.12
Ø .28 DR THRU (2)
NO.13 DRILL (.18) THRU 2 PLACES
.25
60°
1.12
1.62
1.88
.44
.81
.38 TYP
.81

NOTES:
1. BLACK OXIDE FINISH

HANGER, SINGLE BEARING

5
22
11
Ø 22 SPOTFACE BOTH SIDES
22
15.7
11
5
R 1.5 2 PLACES
R
37
6.3
48
5.9
7
7
6.3
25
6.3
12.5
25
METRIC

STOP, STOCK

2.81
1/4 DRILL THRU TO BORE
1.99
15°
.375
59/64 DRILL THRU
.34
Ø 3.00
Ø 2.00
A
7.50
1.68
.31

R .06
Ø 2.04
VIEW AT A

Fig. 7-41. Multiview problems requiring removed views.

Multiview Drawing

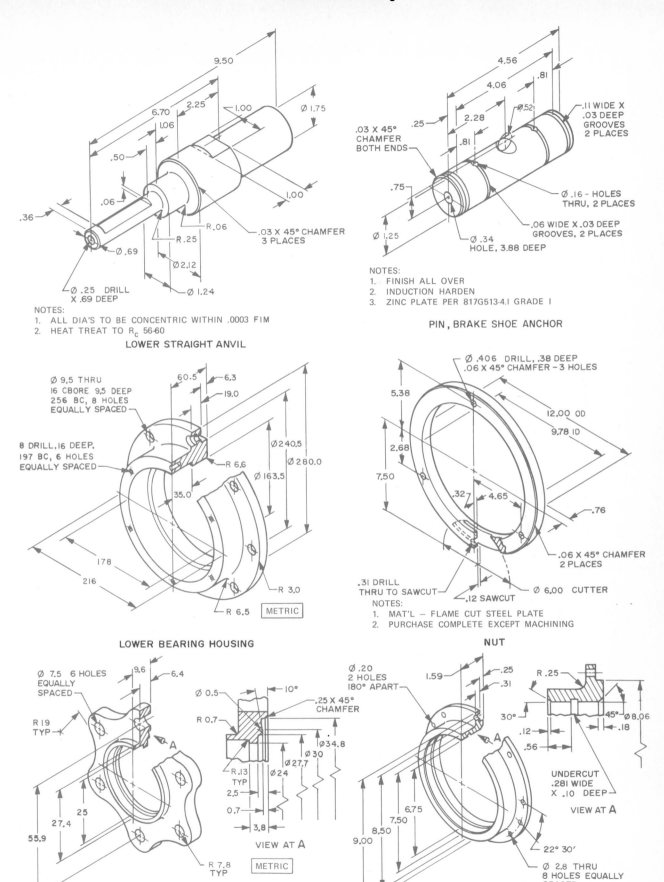

9.50

2.25

6.70

1.06

1.00

Ø 1.75

.50

.06

.03 X 45° CHAMFER
3 PLACES

R.06

R.25

.36

Ø .69

Ø2.12

1.00

Ø 1.24

Ø .25 DRILL
X .69 DEEP

NOTES:
1. ALL DIA'S TO BE CONCENTRIC WITHIN .0003 FIM
2. HEAT TREAT TO R$_c$ 56-60

LOWER STRAIGHT ANVIL

4.56

4.06

.81

.25

2.28

.81

Ø.52

.11 WIDE X
.03 DEEP
GROOVES
2 PLACES

.03 X 45°
CHAMFER
BOTH ENDS

.75

Ø 1.25

Ø .34
HOLE, 3.88 DEEP

Ø .16 - HOLES
THRU, 2 PLACES

.06 WIDE X .03 DEEP
GROOVES, 2 PLACES

NOTES:
1. FINISH ALL OVER
2. INDUCTION HARDEN
3. ZINC PLATE PER 817G513-4.1 GRADE I

PIN, BRAKE SHOE ANCHOR

Ø 9.5 THRU
16 CBORE 9.5 DEEP
256 BC, 8 HOLES
EQUALLY SPACED

60.5

6.3

19.0

8 DRILL, 16 DEEP,
197 BC, 6 HOLES
EQUALLY SPACED

Ø240.5

Ø 280.0

R 6.6

Ø 163.5

35.0

178

216

R 3.0

R 6.5   METRIC

LOWER BEARING HOUSING

Ø .406 DRILL, .38 DEEP
.06 X 45° CHAMFER – 3 HOLES

5.38

12.00 OD

9.78 ID

2.68

7.50

.32

4.65

.76

.31 DRILL
THRU TO SAWCUT

.12 SAWCUT

Ø 6.00 CUTTER

.06 X 45° CHAMFER
2 PLACES

NOTES:
1. MAT'L – FLAME CUT STEEL PLATE
2. PURCHASE COMPLETE EXCEPT MACHINING

NUT

Ø 7.5 6 HOLES
EQUALLY
SPACED

9.6

6.4

R 19
TYP

Ø 0.5

R 0.7

10°

.25 X 45°
CHAMFER

Ø34.8

Ø 30

Ø27.7

Ø 24

R.13
TYP

2.5

0.7

3.8

25

27.4

55.9

R 7.8
TYP

VIEW AT A

METRIC

FLANGE

Ø .20
2 HOLES
180° APART

1.59

.25

.31

R .25

30°

45°

Ø 8.06

.18

.12

.56

6.75

7.50

8.50

9.00

UNDERCUT
.281 WIDE
X .10 DEEP

VIEW AT A

22° 30'

Ø 2.8 THRU
8 HOLES EQUALLY
SPACED

DYNO PILOT CAP

Fig. 7-42. Multiview problems requiring partial views.

Drafting for Industry

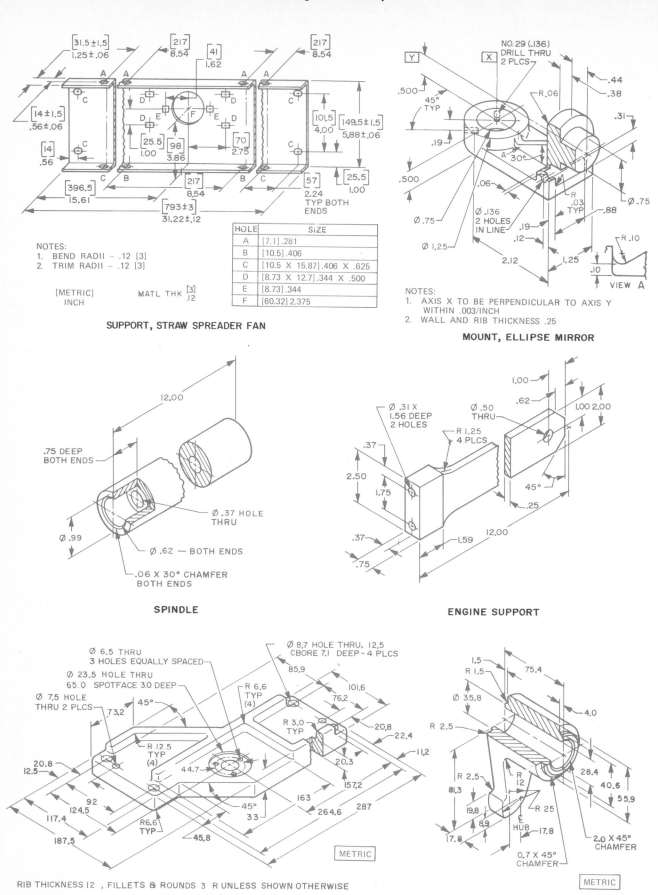

Fig. 7-43. Multiview problems requiring standard drafting conventions.

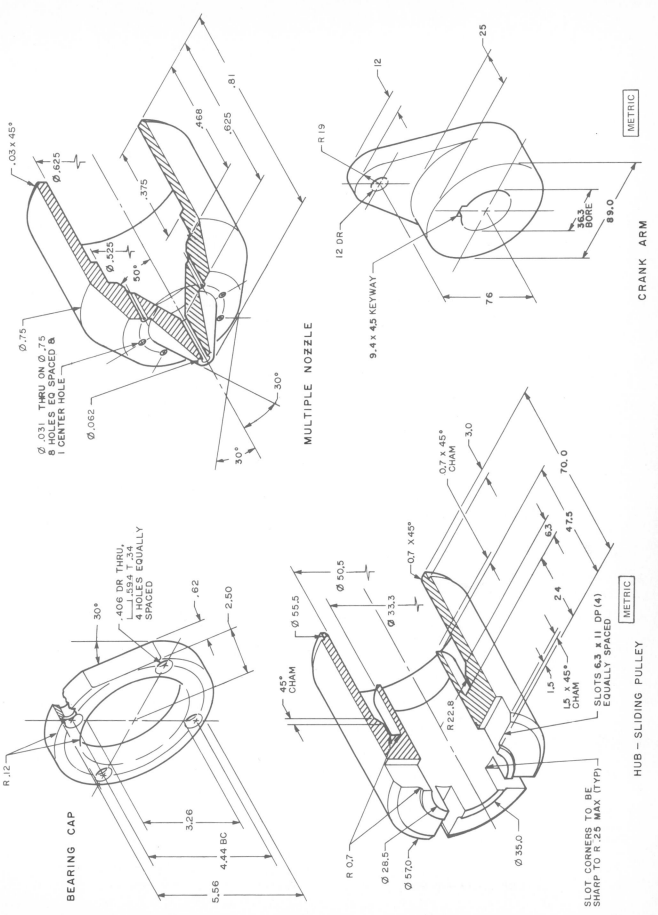

MULTIPLE NOZZLE

Ø.03 x 45°

Ø.625

Ø.525

50°

Ø.75

Ø.031 THRU ON Ø.75
8 HOLES EQ SPACED &
1 CENTER HOLE

Ø.062

.468

.625

.81

.375

30°

30°

R 19

12 DR

9.4 x 4.5 KEYWAY

12

25

36.3 BORE

89.0

76

CRANK ARM

METRIC

BEARING CAP

30°

.406 DR THRU,
⌴.594 T.34,
4 HOLES EQUALLY
SPACED

R.12

.62

2.50

3.26

4.44 BC

5.56

HUB – SLIDING PULLEY

45° CHAM

Ø 55.5

Ø 50,5

Ø 33,3

50,5

R 22.8

R 0,7

Ø 28.5

Ø 57,0

Ø 35.0

0.7 x 45° CHAM

0.7 X 45°

3.0

6.3

47.5

70.0

24

1.5

1.5 x 45° CHAM

SLOTS 6,3 x 11 DP(4)
EQUALLY SPACED

SLOT CORNERS TO BE
SHARP TO R .25 MAX (TYP)

METRIC

Fig. 7-44. Machine parts for which multiview drawings are to be prepared.

Chapter 8
DIMENSIONING FUNDAMENTALS

Dimensioning is the process of defining the size, form and location of geometric components on engineering or architectural drawings. It is one of the most important operations in producing a detail drawing and should be given very careful attention.

Two general types of dimensions are used on drawings, size dimensions and location dimensions, Fig. 8-1. Size dimensions define the size of geometric components of a part (diameter of a cylinder or width of a slot). Location dimensions define the location of these geometric components in relation to each other (distance from edge of part to center of hole).

However, broken and full dimension lines should not be mixed on a single drawing. The first dimension line is spaced .40 inch (3/8 to 1.0 inch or 10 to 25mm from the view, depending on space available on the drawing) as shown in Fig. 8-2 (a).

The practice in industry is to keep dimension lines away from the view for greater clarity. When the minimum distance of .40 inch is used, adjacent dimension lines should be spaced at least .25 inch apart. Dimensions spaced one inch or more from the view may have subsequent lines spaced less than one inch, depending on the size of the drawing.

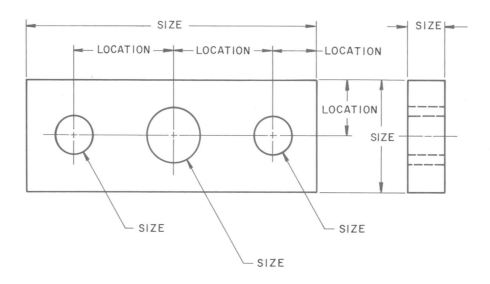

Fig. 8-1. Dimensions provide size and location on drawings.

ELEMENTS IN DIMENSIONING

A standard set of lines and notes are recommended for use on industrial drawings. All lines used in dimensioning are drawn as light, thin lines, using a 2H or 4H pencil, Fig. 8-2.

DIMENSION LINES

A dimension line is a line with arrowheads at each end to indicate the direction and extent of a dimension, Fig. 8-2 (a). The dimension line may be broken and the dimension numeral inserted, (b). Or the line may be a full unbroken line with the dimension numeral located above or below, (a).

EXTENSION LINES

Extension lines are used to indicate the termination of a dimension, Fig. 8-2 (a). They are usually drawn perpendicular to the dimension line with a visible gap of approximately .06 inch (1.5mm) from the object. Extension lines extend approximately .12 (1/8 inch or 3mm) beyond the dimension line. When extension lines are used to locate a point, they must pass through the point as in (c).

Crossing of dimension or extension lines should be avoided by placing the shortest dimensions nearest the object and progressing outward according to size. Dimension lines should be located so they are not crossed by any line. When it is

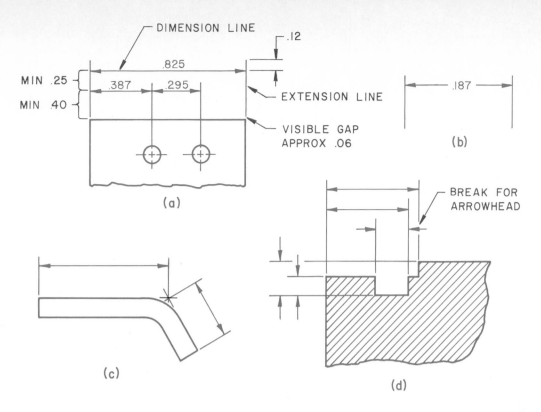

Fig. 8-2. Dimension and extension lines.

necessary to cross a dimension line with an extension line, the extension line is broken, Fig. 8-2 (d).

LEADERS

Leaders are thin, straight lines leading from a note or dimension to a feature on the drawing. Leaders terminate with an arrowhead or dot, Fig. 8-3. Those which terminate on an edge or at a specific point should end with an arrowhead. Dots are used with leaders that terminate inside the outline of an object, such as a flat surface.

The dimension or note end of the leader contains a horizontal bar approximately .12 (1/8) inch in length. Preferably, the leader angle should be 45 to 60 degrees. Leaders should never be drawn parallel to extension or dimension lines. Leaders drawn to a circle or circular arc should be in line with center of the particular feature.

FEATURES INDICATED BY "X"

Some industries indicate the number of repetitive features (holes, slots, etc.) by "X," such as: 2 X ⌀.375 for two holes, Fig. 8-3. This provides for clarity and speed in drafting.

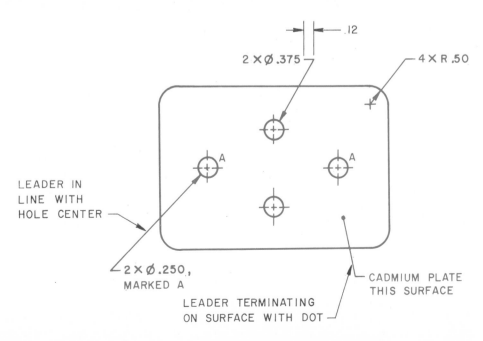

Fig. 8-3. Leaders used in dimensioning.

DIMENSIONAL NOTES

Dimensional notes are used to describe size or form, such as specifying holes, chamfers and threads, Fig. 8-4. They serve the same purpose as dimensions. Notes are used with dimension and extension lines to provide specific information about details on the drawing. Dimensional notes always appear horizontally and parallel to the bottom of the drawing.

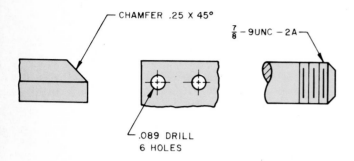

Fig. 8-4. Dimensional notes with leaders used to describe size and form.

ARROWHEADS

Arrowheads are drawn at the termination of dimension lines and leaders. They can be drawn freehand, but must be distinct and accurate. Arrowheads for all dimension lines and leaders on same drawing should be approximately same size as height of whole numerals, usually .12 (1/8) inch, Fig. 8-5 (a).

The width of the base of the arrowhead should be one-third of its length. It is drawn with a single stroke forming each side, either toward the point or away from it depending on the position of the arrowhead and the preference of the draftsman, Fig. 8-5 (b). A third stroke forms the curved base, (c), and the arrowhead is filled in for a distinctive appearance, (d).

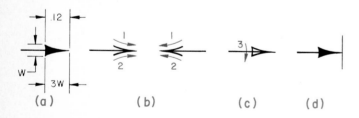

Fig. 8-5. Drawing the filled-in-head arrowhead.

DIMENSION FIGURES

Dimension figures should be clearly formed to prevent any possibility of being misread. Some industries are using a style of numerals which provide positive recognition even when a portion of the numeral is lost in reproduction or is not clear on the drawing (see Fig. 4-30, page 74).

The height of dimension figures is the same as the letter height on the drawing, usually .12 (1/8) inch. Fractions are twice the height of whole numbers.

PLACEMENT OF DIMENSION FIGURES

Common fractions are centered in a break in the dimension line, Fig. 8-6 (a). Decimal dimensions may be centered in a break in the line, (b), or placed above the line and metric dimensions below the line, (c).

Some industries practicing dual dimensioning (decimal and metric) will show the decimal dimension above the line and the metric following in parentheses, Fig. 8-6 (d). A comma is used in place of a decimal point in metric dimensioning in many European countries. However, it has not yet become standardized for worldwide practice.

STAGGERED DIMENSIONS

Where a number of dimensions are grouped in parallel, the dimension figures are staggered to save space, Fig. 8-7.

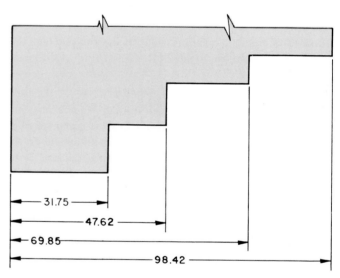

Fig. 8-7. Methods of placing staggered dimensions on a drawing.

UNIDIRECTIONAL AND ALIGNED SYSTEMS

In the unidirectional system, all dimension figures are placed to be read from the bottom of the drawing, Fig. 8-8 (a). This is the recommended industry standard. Aligned dimensions are placed parallel to their dimension lines and are read from the bottom or right side of the drawing, (b).

There are zones that should be avoided in the aligned

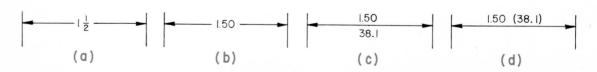

Fig. 8-6. Placement of dimension figures in fractions, decimals and metric units.

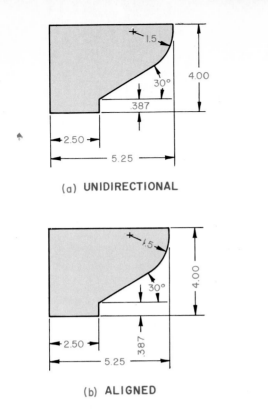

(a) **UNIDIRECTIONAL**

(b) **ALIGNED**

Fig. 8-8. Unidirectional and aligned systems of dimensioning.

system due to the awkward angle of reading, Fig. 8-9. The unidirectional system of dimensioning has been widely adopted and is gaining favor in industry.

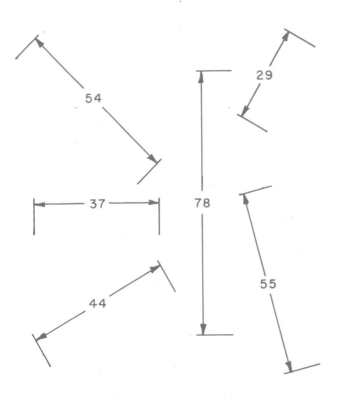

Fig. 8-9. Unidirectional dimensions placed in awkward angles are easier to read than aligned dimensions.

DIMENSIONING WITHIN THE OUTLINE OF AN OBJECT

Dimensions should be kept outside the views of an object whenever possible. Exceptions are permissible where directness of application makes it necessary, Fig. 8-10 (a). When it is necessary to dimension within the sectioned part of a sectional view, crosshatch lines are omitted from the dimension area, (b).

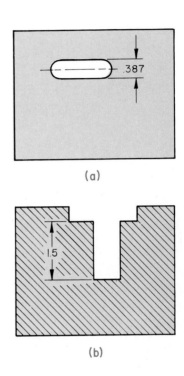

(a)

(b)

Fig. 8-10. Permissible dimensions within the outline of an object.

MEASURING SYSTEMS

Linear dimensions on a drawing are expressed in decimals or common fractions of an inch and as millimetres in the metric system. When linear dimensions exceed a certain length (144 to 192 inches, depending on the particular industry), the dimension value is given in feet and inches. In such cases, abbreviations for feet (ft. or ') and inches (in. or '') are usually shown after the values. Linear dimensions in excess of 10 000 millimetres (mm) are expressed in metres (m) or metres and decimal portions of metres.

DECIMAL INCH DIMENSIONING

Decimal dimensioning is preferred in most manufacturing industries today because decimals are easier to add, subtract, multiply and divide. Preferably, decimal dimensioning should employ a two-place increment of .02 inch (fiftieths of an inch) such as .04, .06, 3.12 and so on.

When decimals of .02 inch increments are divided by two, the quotient is a two-place decimal as well. Decimal dimensions in sizes other than .02 inches should be used where more exacting requirements must be met.

147

DECIMAL DIMENSIONING RULES

1. Omit zeros before the decimal point for values of less than one, Fig. 8-11 (a).

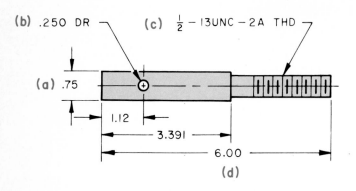

(b) .250 DR

(c) $\frac{1}{2}$ – 13UNC – 2A THD

(a) .75

1.12

3.391

6.00

(d)

Fig. 8-11. Applying the rules of decimal dimensioning to a drawing.

2. Common fraction decimal equivalents of .250, .500 and .750 may be shown as two-place decimals unless they are used to designate a drilled hole size, a material thickness or a thread size, (b).

3. Standard nominal sizes of materials, threads and other features produced by tools that are designated by common fractions may be shown as common fractions. Examples: $\frac{3}{4}$ – 10 UNC – 2A THD; .250 DIA ($\frac{1}{4}$); or $\frac{3}{8}$ HEX (c).

4. Decimal points must be definite, uniform and large enough to be visible on reduced size drawings. The decimal point should be in line with the bottom edge of the numerals and letters to which it relates, (d).

RULES FOR ROUNDING OFF DECIMALS

When it is necessary to round off decimals to a lesser number of places, the following rules apply.

1. When the next figure beyond the last digit to be retained is less than 5, use the shortened form unchanged.

   Example:
   Shorten to two decimal places. 2.62385
   Next figure beyond is less than five (5). 2.62385
   Decimal is rounded off. 2.62

2. When the next figure beyond the last digit to be retained is greater than 5 (this includes 5 followed by one or more quantitative digits), increase the digit by one (1).

Example:
Shorten to three decimal places. 2.62385
Next figure beyond is greater than five (5). 2.62385
Decimal is rounded off. 2.624

3. When the next figure beyond the last place to be retained is 5, and:

   a. The last digit of the shortened form is an odd number, increase the last digit by one.
   Example:
   Shorten to four decimal places. 2.62375
   Last digit of the shortened form is an odd digit. 2.62375
   The last digit of shortened form is increased by one. 2.6238

   b. If the last digit of the shortened form is an even digit, the shortened form remains unchanged.
   Example:
   Shorten to four decimal places. 2.62385
   Last digit of the shortened form is an even digit. 2.62385
   The last digit of shortened form is left unchanged. 2.6238

FRACTIONAL DIMENSIONING

Common fraction dimensioning is used on drawings in architectural and structural fields where close tolerances are not important. The horizontal fraction bar is used with all fractions and is located at mid-point on the vertical height of numerals and capital letters. Where older drawings have been dimensioned with common fractions, some manufacturing industries change these fractions to decimals by referring to a conversion chart. A typical conversion chart is shown in the Appendix on page 549.

METRIC DIMENSIONING

Metric dimensioning, like decimal inch dimensioning, employs the base-ten number system which makes it easy to move from one multiple or submultiple to another by shifting the decimal point. The unit of linear measure in the metric system is the meter. However, the millimeter is used on most drawings dimensioned in the metric system where the linear dimension is less than 10000 millimeters.

The letter abbreviation for millimeters (mm) following the dimension figure is omitted when all dimensions are in millimeters.

The metric system has, since 1960, been referred to

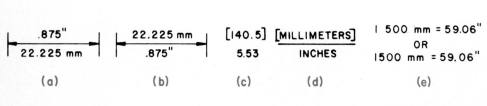

.875"
22.225 mm

(a)

22.225 mm
.875"

(b)

[140.5]
5.53

(c)

[MILLIMETERS]
INCHES

(d)

1 500 mm = 59.06"
OR
1500 mm = 59.06"

(e)

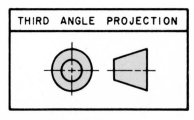

THIRD ANGLE PROJECTION

(f)

Fig. 8-12. Dual dimensioning with dimensions shown in the decimal inch and millimeters.

Dimensioning Fundamentals

internationally as ''Systeme International d' Unites'' or the International System of Units. The universal abbreviation SI indicates this system.

METRIC DIMENSIONING RULES

1. A period is used for the decimal point in English speaking countries; most others use a comma.
2. Whenever a numerical value is less than one millimeter, a zero should precede the decimal point.
3. Digits in metric dimensions shall not be separated into groups by use of commas nor spaces.
4. Use multiple and submultiple prefixes of 1000, such as kilometer (km), meter (m) and millimeter (mm) whenever possible. Avoid the use of centimeter (cm).
5. Do not mix SI units with units from a different system.

DUAL DIMENSIONING

A system of dimensioning, referred to as ''dual dimensioning,'' employs the English inch and metric (SI) dimensions on the same drawing. In dual dimensioning, the English measurement is usually given in decimal inches and the metric in millimeters. If the drawing is intended primarily for use in the United States, usually the decimal inch dimension appears above the line and millimeters below, Fig. 8-12 (a). In metric countries the millimeter dimension is shown above the line and the decimal inch below, (b).

Some industries using dual dimensioning place the metric dimension in brackets above the decimal dimension, (c). It is recommended that a note be used adjacent to or within the title to show how the inch and millimeter dimensions are identified, as shown in (d). Whole numbers in the metric system may be separated according to groups by a letter space or simply run together, (e). It is recommended that all dual dimensioned drawings indicate the angle of projection used, to eliminate any confusion when used in different countries, (f).

Dual dimensioning is being replaced by all metric dimensioned drawings. Where drawings are dimensioned in a single system, millimeters or decimal inches, individual identification of linear units is not required. However, a note on the drawing will state ''UNLESS OTHERWISE SPECIFIED, ALL DIMENSIONS ARE IN MILLIMETERS'' (or INCHES). Where some inch dimensions are shown on a millimeter-dimensioned drawing, the abbreviation IN. shall follow the inch value (on decimal-inch drawings, the symbol mm shall follow millimeter values).

Drafting is an important means of communications in industry.

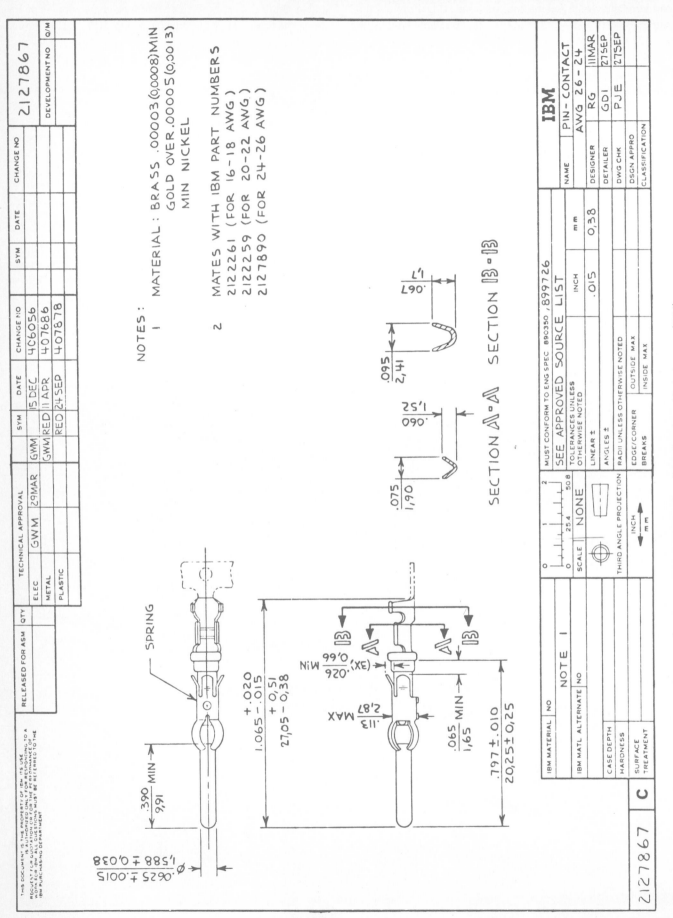

Fig. 8-13. A dual dimensioned drawing as used in industry. (IBM)

Dimensioning Fundamentals

An example of a drawing employing dual dimensioning is shown in Fig. 8-13. Tables are provided in the Reference Section for conversion of common fractions and decimals to millimeters and vice versa.

DIMENSIONING FEATURES FOR SIZE

A drawing of the features of a part or assembly consists of geometric shapes such as cylinders, cones, pyramids and spheres. Dimensioning consists of describing the size and position of each feature.

CYLINDERS

Cylinders may be solids (as in a projecting stem) or negative volumes (as in a hole), Fig. 8-14 (a). Cylindrical features are dimensioned for diameter and length. In single view drawings where the circularity of the feature or part is not shown, the dimension should be preceded by the symbol for diameter, (b), (the international symbol ϕ).

The diameter symbol is not necessary for cylinders on the side view where a circular view is shown, Fig. 8-14 (c), or where the dimension is on the circular view itself, (d). Where a leader is used to indicate the diameter, the value of the diameter should be preceded by the symbol for diameter, (c).

CIRCULAR ARCS

Circular arcs are dimensioned by indicating their radius with a dimension line, Fig. 8-15. The dimension line is drawn from the radius center and ends with an arrowhead at the arc. The dimension is inserted in the line preceded by "R", (a). Dimensions of radii, where space is limited, may be indicated outside the arc as at (b) or with a leader, (c). A small cross should be used to indicate a radius located by a dimension as at (d).

FORESHORTENED RADIUS

Sometimes the center of an arc radius exists outside the drawing itself or interferes with another view. In this case, the radius dimension line should be shown foreshortened and the arc center located with coordinate dimensions, Fig. 8-16. That portion of the dimension line next to the arrow head and arc is shown radially (in line with the arc center).

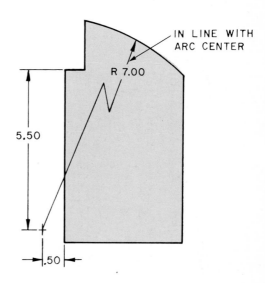

Fig. 8-16. Technique for dimensioning a foreshortened radius.

TRUE RADIUS INDICATION

A true arc on an inclined surface, when dimensioned as shown in Fig. 8-17 (a), may be misleading to the worker. This dimension is clarified by the addition of "TRUE R" before the radius dimension, (b), which means "true radius on the surface."

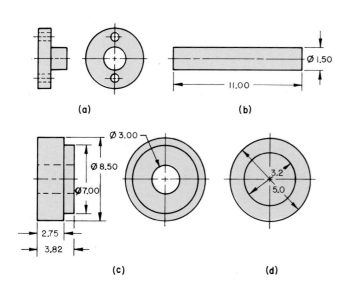

Fig. 8-14. Proper methods for dimensioning cylinders.

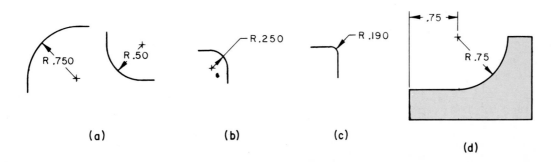

Fig. 8-15. A number of ways that arcs may be correctly dimensioned.

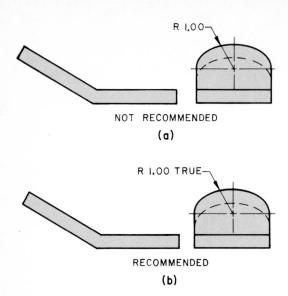

Fig. 8-17. Dimensioning the true radius indiciation for an arc.

FILLETS AND CORNER RADII

Fillets and corners may be dimensioned by a leader as shown in Fig. 8-18 (a). Where there are a large number of fillets or rounded edges of the same size on a part, the preferred method is to specify these with a note rather than to show each radius, (b).

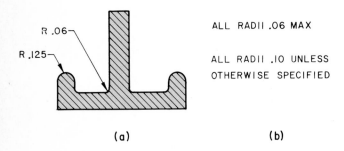

ALL RADII .06 MAX

ALL RADII .10 UNLESS OTHERWISE SPECIFIED

(a)   (b)

Fig. 8-18. Specifying radii for fillets and corners.

ROUND HOLES

Holes are dimensioned on the view in which they appear as circles. Small holes are dimensioned with a leader and larger holes by a dimension at an angle (usually 30, 45, or 60 degrees) across the diameter, Fig. 8-19. Holes that are to be drilled reamed, punched, etc., are specified by a note. The depth of holes may be specified by a note or dimensioned in a sectioned view. Additional information on the dimensioning of holes is discussed in Chapter 18.

HIDDEN FEATURES

Dimensions should be drawn to visible lines whenever possible. Where hidden features exist and are not visible for dimensioning in another view, a section view should be used, Fig. 8-20. An exception to this general rule is a diameter dimension on a partial section where no other view is used to show the circular diameter, Fig. 8-21.

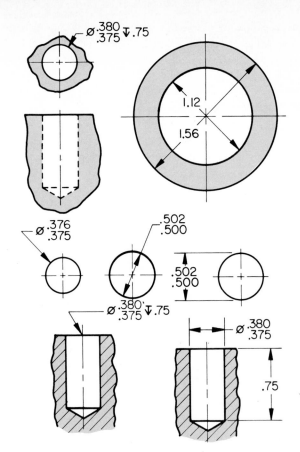

Fig. 8-19. Dimensioning round holes.   (ANSI)

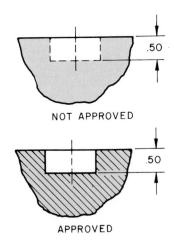

NOT APPROVED

APPROVED

Fig. 8-20. Correct and incorrect methods of dimensioning hidden features.

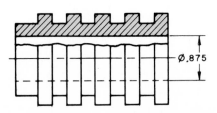

Fig. 8-21. Partial section is an exception to general rule of avoiding dimensions to hidden lines.

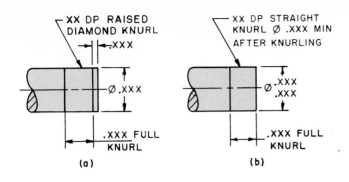

Fig. 8-22. Specification of knurls.

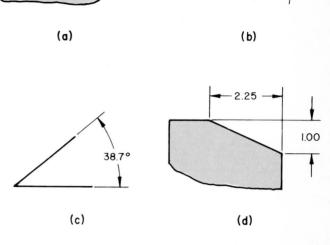

Fig. 8-23. Dimensioning angles according to specifications required.

KNURLS

Knurls are specified by diameter, and type and pitch of the knurl, Fig. 8-22 (a). When control of the diameter of a knurl is required for an interference fit between parts, this is also specified, Fig. 8-22 (b). The length along the axis may be specified if required.

ANGLES

Angular dimensions may be specified in degrees, minutes and seconds, Fig. 8-23 (a, b and c). The symbols for these are: degrees °, minutes ', and seconds ''. Angles specified in degrees alone have the numercial value followed by the sumbol ° or by the abbreviation DEG, (a). When an angle is expressed in minutes alone, the number of minutes is preceded by 0 °, (b). Angles may be specified in degrees and decimal parts of a degree, (c).

Angles may also be specified by coordinate dimension, (d).

CHAMFERS

Chamfers of 45 degrees may be dimensioned by a note as in Fig. 8-24 (a). Use of the word CHAMFER is optional in the note. All chamfers, other than 45 degrees, are dimensioned by giving the angle and the measurement along the length of the part, (b). Chamfers should never be dimensioned along their angular surface.

Internal chamfers are dimensioned in the same manner except in cases where the diameter requires control, (c).

COUNTERBORE OR SPOTFACE SYMBOL

A counterbore or spotface is indicated by the symbol shown in Fig. 8-25 (a). The symbol precedes the dimension.

COUNTERSINK SYMBOL

A countersink is indicated by the symbol shown in Fig. 8-25 (b). The symbol precedes the dimension.

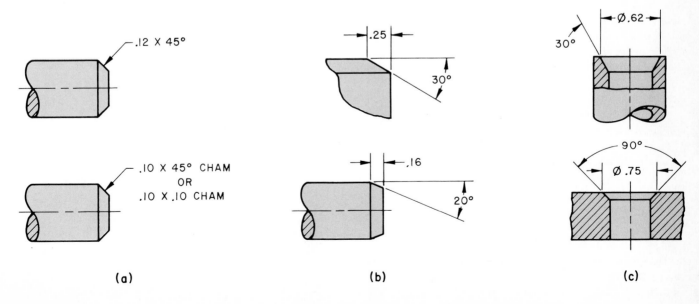

Fig. 8-24. Chamfers may be dimensioned in a variety of ways according to their location.

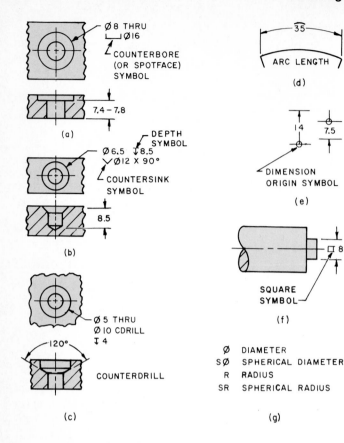

Fig. 8-25. Standard symbols used in dimensioning.

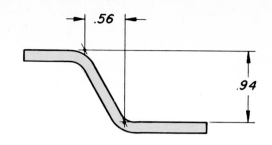

Fig. 8-26. Dimensioning offsets from points of intersection. (General Motors Corp.)

OFFSETS

Offsets on a part should be dimensioned from the points of intersection of the tangents along one side of the part as shown in Fig. 8-26.

KEYSEATS

Keyseats (recess in the shaft or hub) are dimensioned as shown in Fig. 8-27 (a). Woodruff keyseats are dimensioned as shown in Fig. 8-27 (b). For additional information see keys reference section pages 587-590.

NARROW SPACES AND UNDERCUTS

Dimension figures should not be crowded into narrow spaces. One of the methods shown in Fig. 8-28 will help maintain clarity on the drawing. Note the breaks in extension lines near the arrowheads in the lower illustration.

COUNTERDRILL

Counter drilled holes are dimensioned by specifying the diameter of the hole and the diameter, depth and included angle of the counterdrill, Fig. 8-25 (c).

DEPTH SYMBOL

The depth of a feature is indicated by the symbol shown in Fig. 8-25 (b). The symbol precedes the depth dimension.

DIMENSION ORIGIN SYMBOL

The origin of a toleranced dimension between two features is indicated as shown in Fig. 8-25 (e).

SQUARE SYMBOL

A square shaped feature is indicated by a single dimension preceded by the symbol shown in Fig. 8-25 (f).

ARC LENGTH SYMBOL

Arc length measured on a curved outline is indicated by the symbol shown in Fig. 8-25 (d), which is placed above the dimension.

DIAMETER AND RADIUS SYMBOLS

Diameter, spherical diameter, radius and spherical radius are indicated by the symbols shown in Fig. 8-25 (g). These symbols precede the value of a dimension or tolerance given for a diameter or radius.

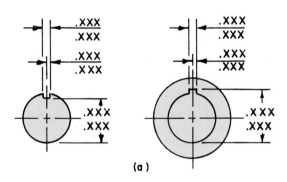

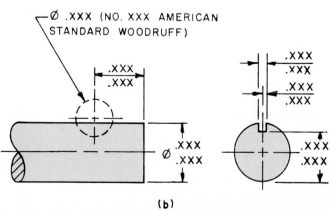

Fig. 8-27. Dimensioning keyseats.

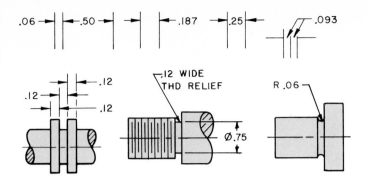

Fig. 8-28. Techniques generally used for dimensioning narrow spaces.

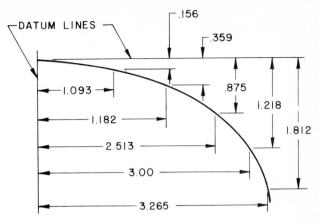

Fig. 8-30. Dimensioning an irregular curve using datum lines.

TAPERS—CONICAL AND FLAT

Conical tapers are dimensioned by specifying one of the following:

1. A basic taper and a basic diameter, Fig. 8-29 (a).
2. A size tolerance combined with a profile of a surface tolerance applied to the taper.
3. A toleranced diameter at both ends of a taper and a toleranced length, (b).

Flat tapers are dimensioned by specifying a toleranced slope and a toleranced height at one end, (c).

Taper and slope for conical and flat tapers are depicted by the symbols shown. The vertical leg of the symbol is always to the left.

IRREGULAR CURVES

Irregular curves may be dimensioned by the coordinate, or offset, method as shown in Fig. 8-30. Each dimension line is extended to a datum line.

SYMMETRICAL CURVES

Curves which are symmetrical may be dimensioned on one side of the axis of symmetry only, Fig. 8-31 (a). When only one half of a symmetrical part is shown, it is dimensioned as shown in (b).

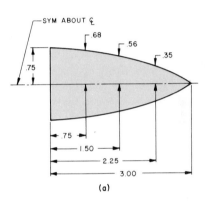

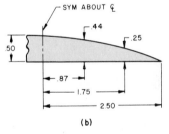

Fig. 8-31. Dimensioning curves symmetrical to a center line.

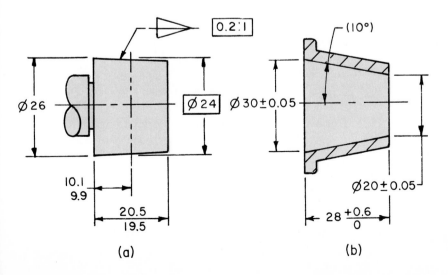

Fig. 8-29. Dimensioning tapers.

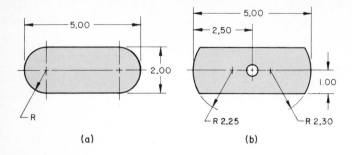

Fig. 8-32. Two methods used to dimension rounded ends on a drawing.

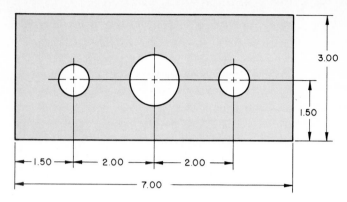

Fig. 8-34. Dimensioning features for position using the point-to-point method.

ROUNDED ENDS

For parts having fully rounded ends, overall dimensions should be given for the part and the radius of the end indicated but not dimensioned, Fig. 8-32 (a). Parts having partially rounded ends should have the radii dimensioned, (b).

SLOTTED HOLES

Slotted holes are treated as two partial holes separated by space. A slot of regular shape is dimensioned for size by length and width dimensions, Fig. 8-33. The slot is located on the part by a dimension to its longitudinal center and either one end or a center line.

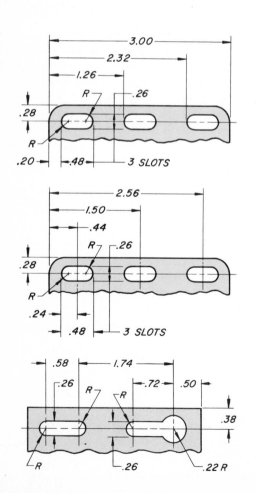

Fig. 8-33. Standard methods for dimensioning slotted holes. (General Motors Corp.)

DIMENSIONING FEATURES FOR POSITION

Position dimensions specify the location or distance relationship of one feature of a part with respect to another feature or datum. Features may be located with respect to one another by either linear or angular expressions.

POINT-TO-POINT DIMENSIONING

Point-to-point dimensions are usually adequate for simple parts, Fig. 8-34, but parts which contain features mating with another part should be dimensioned from a datum. (See Chapter 18.) In point-to-point dimensioning, one dimension is omitted to avoid locating a feature from more than one point and possibly causing unsatisfactory mating of parts.

COORDINATE DIMENSIONING SYSTEMS

Coordinate dimensioning is a type of dimensioning useful in locating holes and other features on parts. Basically, two different systems are employed: rectangular coordinate and polar coordinate dimensioning.

Rectangular Coordinate Dimensioning is useful in locating holes and other features that lie in a rectangular or noncircular pattern, Fig. 8-35 (a). These dimensions are at right angles to each other and from a datum plane.

With the assistance of a computer, holes distributed about a bolt circle can be located accurately for dimensioning, using the rectangular coordinate system as in Fig. 8-35 (b). Rectangular coordinate dimensioning is used on drawings of parts that are to be numerically machined.

Polar Coordinate Dimensioning should be used when holes or other features to be located lie in a circular or radial pattern, Fig. 8-36. A radial dimension is given from the center of the pattern and an angular dimension from a datum plane.

When it is necessary to hole closer tolerances for mating features, true-positioning dimensioning should be used as discussed in Chapter 18.

TABULAR DIMENSIONING

Tabular dimensioning is a form of rectangular coordinate dimensioning. The location of dimensions for features are given from datum planes and listed in a table on the drawing sheet, Fig. 8-37. Dimensions are not all applied directly to

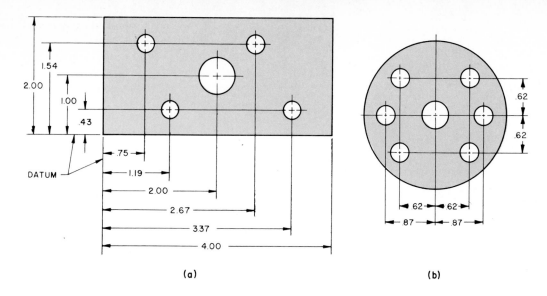

Fig. 8-35. Rectangular coordinate dimensioning using datum lines and center lines.

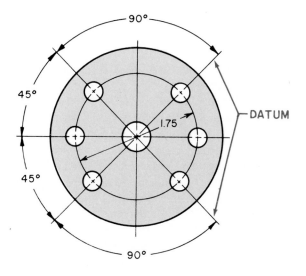

Fig. 8-36. Both radial dimensions and datum planes used in polar coordinate dimensioning.

the views. This method of dimensioning is useful where a large number of similar features are to be located.

ORDINATE DIMENSIONING

Ordinate dimensioning is very similar to the rectangular coordinate system. It makes use of datum dimensioning in which all dimensions are from two or three mutually perpendicular datum planes.

Ordinate dimensioning differs from the coordinate system in that the datum planes are indicated as zero coordinates. Dimensions from these planes are shown on extension lines without the use of dimension lines or arrowheads, Fig. 8-38.

This system of dimensioning, which is sometimes referred to as "arrowless" dimensioning, lends itself to work which is to be programmed for numerical machining.

UNNECESSARY DIMENSIONS

Any dimension not needed in the manufacture or assembly of an item is an unnecessary dimension. This includes

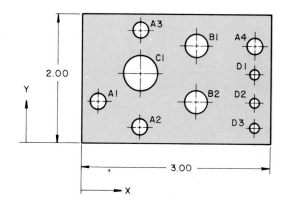

	REQD	4	2	1	3
	HOLE DIA	.250	.312	.500	.125
POSITION		HOLE SYMBOL			
X →	Y ↑	A	B	C	D
.250	.625	A1			
1.000	.250	A2			
1.000	1.750	A3			
2.750	1.500	A4			
1.750	1.500		B1		
1.750	.625		B2		
1.000	1.000			C1	
2.750	1.000				D1
2.750	.625				D2
2.750	.250				D3

Fig. 8-37. Tabular dimensioning includes the drawing and table listing.

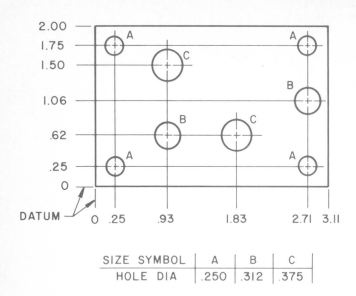

SIZE SYMBOL	A	B	C
HOLE DIA	.250	.312	.375

Fig. 8-38. Ordinate dimensioning using distances measured from zero coordinates.

dimensions repeated on the same or on another view, Fig. 8-39 (a). Also, it is not necessary to include all "chain" dimensions when the overall dimension is given.

When an entire series of chain dimensions is given, difficulties may arise in manufacturing where there is an accumulation of tolerances (variations permitted in measurements). The exception to providing all of the individual dimensions in a chain is in the architectural and structural industries where the interchangeability of parts and close tolerances are of no great concern, Fig. 8-39 (b).

NOTES

Notes are used on drawings to supplement graphic information, dimensions and, in some cases, to eliminate the necessi-

ty of repetitive dimensions. All notes should be brief and clearly stated. Only one interpretation must be apparent.

Standard abbreviations and symbols may be used in notes where feasible. Abbreviations do not require periods unless the resultant abbreviated letters spell a word or are subject to misinterpretation.

SIZE, SPACING AND ALIGNMENT OF NOTES

Lettering in notes is the same size as dimensions on the drawing, usually .12 (1/8) inch in height. Spacing between lines within a note should be from one-half to one full letter height. At least two letter heights should be allowed between separate notes.

All notes should be placed on drawing parallel to bottom of the drawing. Lines of a single note and successive notes in a list of general notes should be aligned on the left side.

GENERAL NOTES

Notes which convey information that apply to the entire drawing are called general notes. Some examples are:
1. CORNER RADII .12 ± .06
2. REMOVE ALL BURRS AND BREAK ALL SHARP EDGES .010R MAX
3. ALL MARKINGS TO BE PER MIL—STD—130B FOR LIQUID OXYGEN SERVICE
4. MAGNETIC INSPECT PER MIL—STD—I-6868

No period is required at the end of the note, unless more than one statement is included within any one note.

General notes are usually placed on the right-hand side of the drawing above the title block or to the left of the title block, Fig. 8-40 (a). They are numbered from the top down or from the bottom up, depending on the industry policy.

Notes are sometimes included in the title block of the drawing, such as general tolerances, material specification and heat treatment specification, Fig. 8-40 (b). Notes shown at (c) are also general notes since they apply to entire drawing.

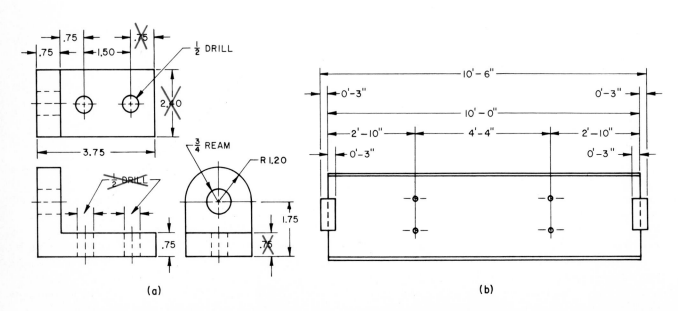

(a)   (b)

Fig. 8-39. Duplicate dimensions in machine and structural drawings.

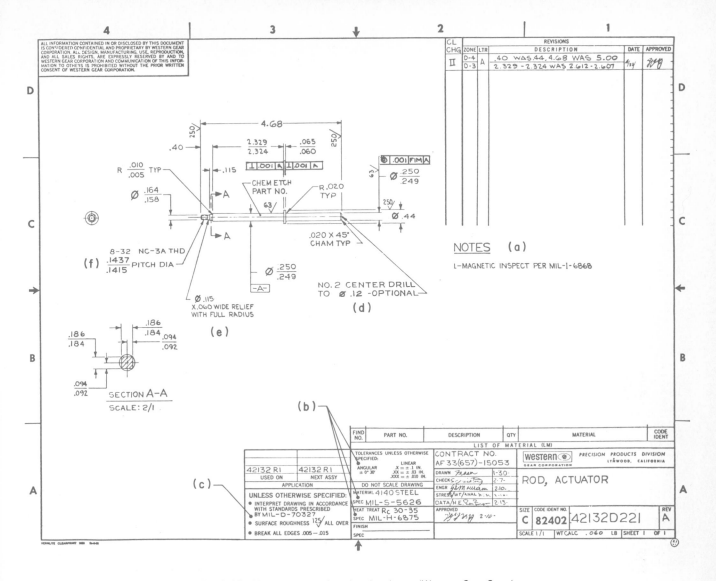

Fig. 8-40. Notes on an engineering drawing. (Western Gear Corp.)

LOCAL NOTES

Local notes provide specific information, such as designating a machine process, Fig. 8-40 (d and f), and actual dimension, (e), or a standard part. (For example, a standard size hexagonal cap screw: No. 6-32UNC-2B HEX CAP SCR.)

Some examples of local notes are:

1. .344 ± φ .002, 36 HOLES
2. R.06 — 2 PLACES
3. PAINTED AREA 1.75 SQUARE AS INDICATED, ONE SIDE ONLY

Local notes are usually located close to a specific feature by a leader extending from the beginning or final word of the note, Fig. 8-41.

Local notes may be included with the list of general notes. They refer to a specific feature or area on the drawing. Enclose the number in a square or triangle (called a "flag") and place it on the field near the feature, directed by a leader, Fig. 8-41.

To avoid a crowded appearance, or the necessity to relocate a note, notes should not be placed on the drawing until after the dimensions have been added.

RULES FOR GOOD DIMENSIONING

The following rules should serve as guides to good dimensioning practices:

1. Take time to plan the location of dimension lines. Avoid crowding by providing adequate room for spacing (at least .40 inch (10 mm) for first and .25 inch (6 mm) for successive lines).

2. Dimension lines should be thin and contrast noticeably with visible lines of the drawing.

3. Dimension each feature in the view which most completely shows the characteristic contour of that feature.

4. Dimensions should be placed between the views to which they relate and outside the outline of the part.

5. Extension lines are gapped away from the object approximately .06 inch (1.5 mm) and extend beyond the dimension line approximately .12 inch (3 mm). Extension lines may cross other extension lines or object lines when necessary. Avoid crossing dimension lines with extension lines or leaders whenever possible.

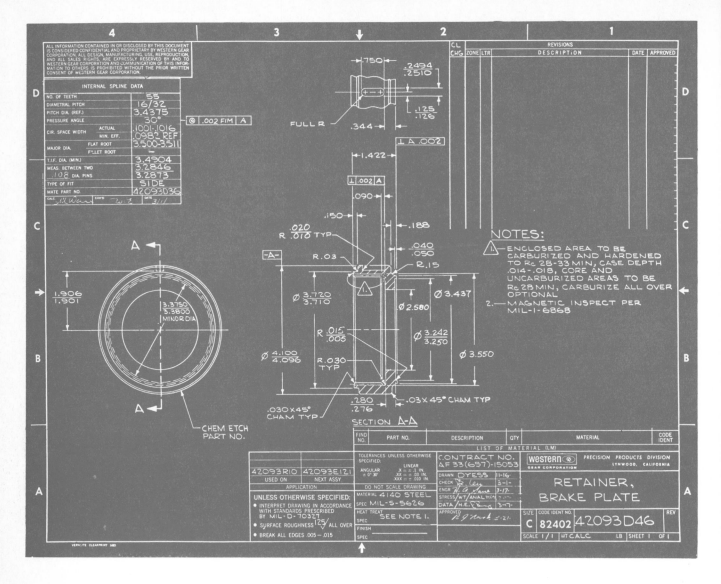

Fig. 8-41. Local note directed to a feature on the drawing by a reference number and "flag." (Western Gear Corp.)

6. Show dimensions between points, lines or surfaces which have a necessary and specific relation to each other.

7. Dimensions should be placed on visible outlines rather than hidden lines.

8. State each dimension clearly so that the engineering intent can be interpreted in only one way.

9. Dimensions must be sufficiently complete for size, form and location of features so that no scaling of the drawing, calculating nor assuming of distances is necessary.

10. Avoid duplication of dimensions. Only those dimensions that provide essential information should be shown.

11. One dimension in a chain dimension should be omitted (architectural and structural drawing excepted) to avoid location of a feature from more than one point.

PROBLEMS AND ACTIVITIES

The following dimensioning problems are designed for A or B size drawing sheets. Use a separate sheet for each problem assigned by the instructor. Dimensions should not be crowded. Provide ample space between views by the selection of the proper scale for the size of drawing sheet being used.

1. Locate and construct by freehand sketching techniques the required extension lines, dimension lines and leaders to correctly dimension the parts shown in Fig. 8-42. No dimension figures are to be included. Letter the title of each part in the title block of the sketch.

2. Draw and dimension the parts shown in Fig. 8-43. Check each drawing against the rules for good dimensioning.

3. Draw the necessary views and dimension the parts shown in Fig. 8-44. Check each drawing against the rules for good dimensioning.

SELECTED ADDITIONAL READING

1. DIMENSIONING AND TOLERANCING FOR ENGINEERING DRAWINGS, ANSI Y14.5M-1982, American National Standards Institute, 1430 Broadway, New York, NY 10018.

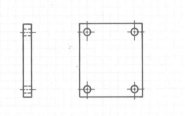

PRESSURE REGULATOR PLATE

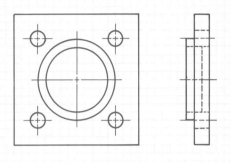

BEARING HOUSING

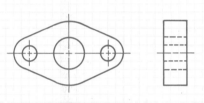

FLANGE, WATER INLET

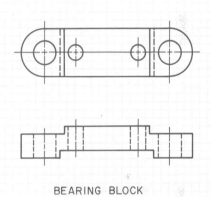

BEARING BLOCK

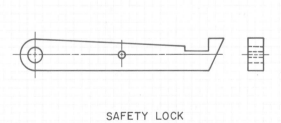

SAFETY LOCK

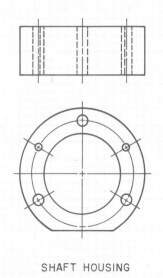

SHAFT HOUSING

Fig. 8-42. Sketching problems for dimensioning line layout.

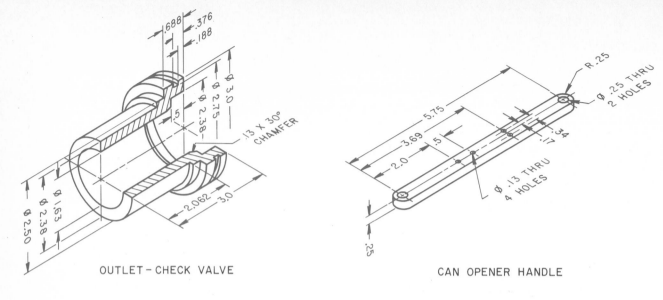

OUTLET-CHECK VALVE

CAN OPENER HANDLE

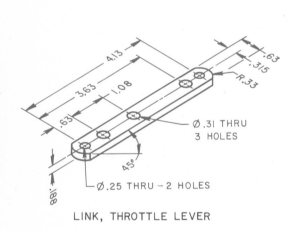

LINK, THROTTLE LEVER

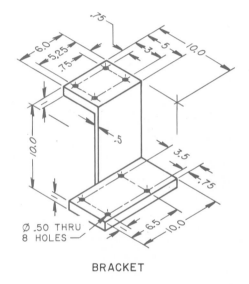

BRACKET

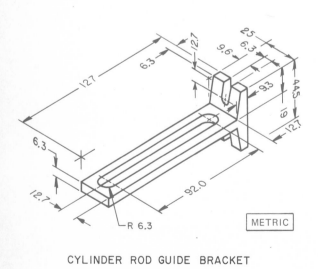

METRIC

CYLINDER ROD GUIDE BRACKET

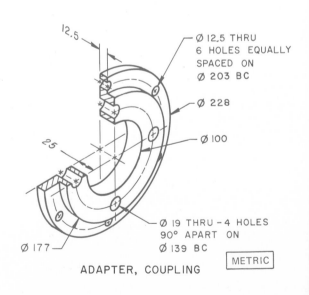

METRIC

ADAPTER, COUPLING

Fig. 8-43. Fundamental dimensioning problems.

Dimensioning Fundamentals

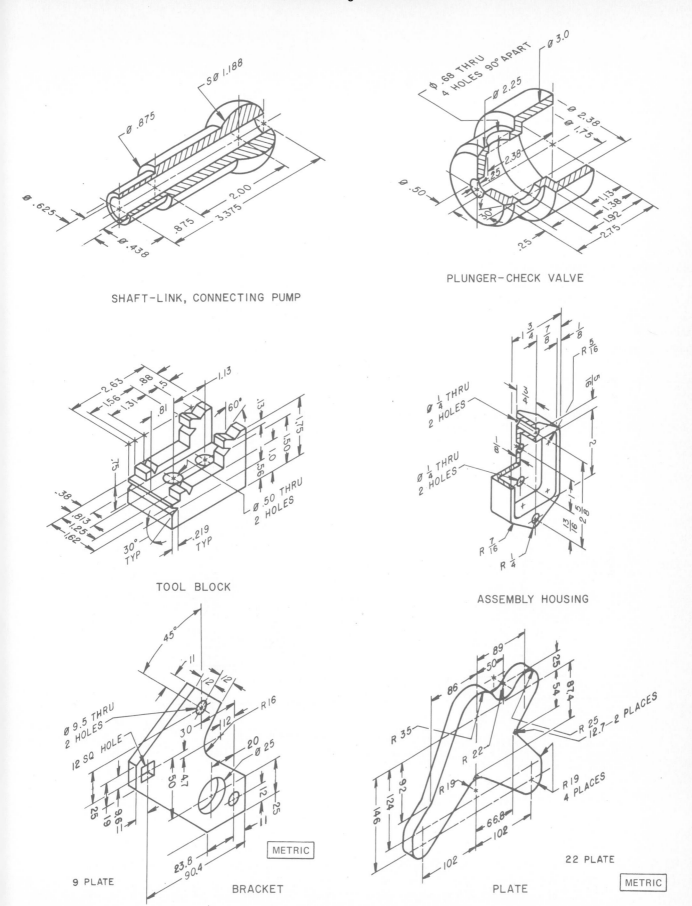

SHAFT-LINK, CONNECTING PUMP

PLUNGER-CHECK VALVE

TOOL BLOCK

ASSEMBLY HOUSING

BRACKET

PLATE

Fig. 8-44. Additional dimensioning problems.

Chapter 9
THE DESIGN METHOD

Engineers, industrial designers and drafters make extensive use of the design method in arriving at a final solution to a design problem. Examples of results of the design method used in one industry are a MECHANICAL COMPONENT ASSEMBLY, Fig. 9-1, and a TORQUE-RESOLVER CAPSULE, Fig. 9-2. From basic sketches like these, by technical design personnel in industry, come the machines and products of our technological age.

Skill in the use of the design method is a valuable aid to the drafter, designer or engineer in industry.

WHAT IS THE DESIGN METHOD?

Many of the problems you will solve as a student of design drafting will be presented to you in a clear-cut fashion and the solutions indicated. For example, supplying the third view

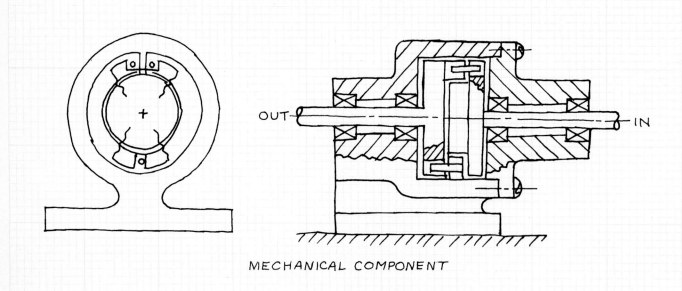

COMPONENT, LOW COST, BI-DIRECTIONAL, MANUALLY
OVERRIDABLE. NO-BACK DEVICE

OUT

IN

MECHANICAL COMPONENT

Fig. 9-1. A design engineer's preliminary sketch of a machine part.
(Sperry Flight Systems Div.)

The lunar vehicle "Rover" was first created by designers, using the design method. Its various components were developed through the use of sketches, then a prototype was constructed and tested on earth, Fig. 9-3. In Fig. 9-4, Rover is shown on the moon's surface during our astronauts' third lunar mission.

when two views are given. All that is left for you to do is solve for the missing view.

While these experiences are necessary in learning the basics of drafting and other subjects, there are other problems you will need to solve that are not clearly identified. See Fig. 9-5. For example, the design for a means of carrying the family-

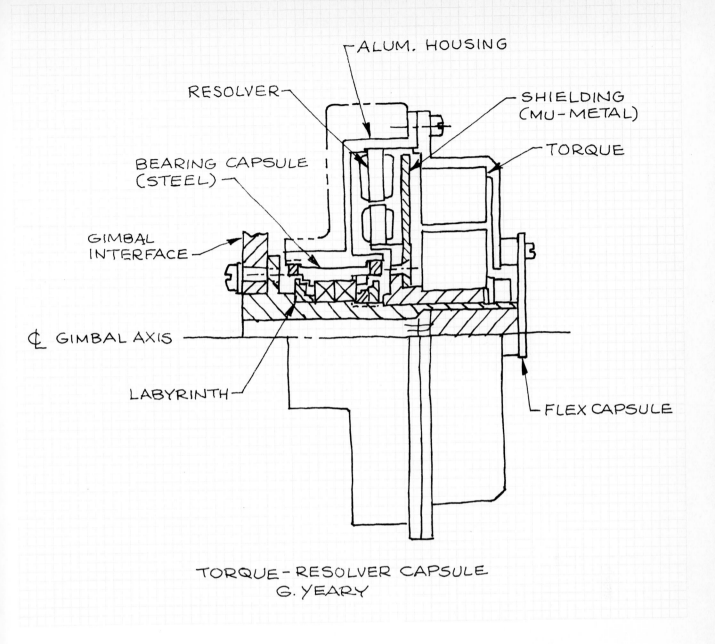

ALUM. HOUSING

RESOLVER

SHIELDING
(MU-METAL)

BEARING CAPSULE
(STEEL)

TORQUE

GIMBAL
INTERFACE

$\mathbb{C}$ GIMBAL AXIS

LABYRINTH

FLEX CAPSULE

TORQUE-RESOLVER CAPSULE
G. YEARY

Fig. 9-2. Sketching is one of the most useful techniques to a designer in "thinking through" a solution to a problem.

Fig. 9-3. "Rover" trainer model is used in training astronauts on earth. It is an operational prototype design of actual lunar vehicle. (NASA)

Fig. 9-4. The design method was used extensively in the development of "Rover" shown here on the moon's surface. (NASA)

Fig. 9-5. This device makes is possible for an automotive technician to "drive" a car in place. The technician can check for noises that would occur at normal driving speed. (Dyna)

Fig. 9-6. A bicycle locking mechanism which solved the problem of its designers. It secures a bicycle while parked and eliminates the necessity of carrying a chain and padlock.
(Ron Anderson, James Shaw, Clay Willits and Lyle Zeigler)

camping equipment for an overnight or extended camping trip. This is known as "creative problem solving." See Fig. 9-7.

There is a method of approaching the solution to problems like this that aid in their solution, it is known as the "design method." This method is discussed here to assist you in developing the technique of solving problems which face the design drafter.

DEFINITION OF TERMS

Some of the terms used in design and problem solving activities are as follows:

PROBLEM
A situation, question or matter requiring choices and action for solution.

CREATIVITY
The ability to develop a number of original or unique ideas which could solve or contribute to the solution of a problem.

DESIGN
As used in this text, design is the result of creative imagination that forms ideas as preliminary solutions to problems. Evaluation of these ideas provides functional solutions to problems. It is not necessarily void of art and beauty, but its emphasis is inventive and ingenious ways of solving technical problems.

Another example of design problem solving is the bicycle cable lock designed by a group of students, Fig. 9-6. The hardened flexible cable, stored on a spring-loaded reel inside the housing, is pulled out and around the bicycle rack, pole or tree. The end snaps back into a lock in the housing.

PROBLEM SOLVING METHOD
The procedure of systematically approaching a problem and arriving at its solution.

STEPS IN DESIGN PROBLEM SOLVING

In order to systematically approach the solution to a problem, it is necessary to break the solution into a number of steps. These steps are listed and explained here to provide a clear understanding of their nature and use.

STEP 1: PROBLEM DEFINITION

Before much progress can be made toward the solution of a problem, the problem must be clearly defined. This step is divided into four parts:

STATEMENT
The statement of the problem should be concise and it should describe the need that is the basis of the problem:

"Design a device for carrying the family camping equipment." Note that there are no stated requirements or limitations such as how much or what size.

REQUIREMENTS

The requirements of the problem should be listed (required features and desired features):

"The space provided must be outside the passenger seating and trunk space of the car. Space is needed for the equipment of five persons — bed rolls, cooking utensils, dishes, food supply, hiking and recreational equipment."

LIMITATIONS OR RESTRICTIONS

The limitations or restrictions on the design should be specified: (1) limiting physical factors (size, weight and materials) and (2) limiting monetary factors.

"The space provided should have easy access, contents protected from weather and road dust, and cause only minimal drag on the car while driving. Total cost is a factor."

RESEARCH

The research of the problem should further aid in defining the problem. Information must be gathered relative to the number and size of items to be included and what weather protection is necessary. Questions need to be asked that will reveal the various requirements of the problem statement:

"How many bed rolls and what space is required? What cooking utensils, equipment and dishes are needed and what space is required? How much space will be required for food supplies? What recreational equipment is to be taken and what space requirements are needed?"

Answers for these questions are obtained from previous experience, actual measurements, research in technical publications and by discussions with knowledgeable persons.

"Five bed rolls, 15 inches diameter by 20 inches in length; cooking utensils, equipment and dishes approximately 8 cubic feet; food supply approximately 4 cubic feet; and, recreational equipment approximately 3 cubic feet with a length of at least 36 inches."

STEP 2: PRELIMINARY SOLUTIONS

The second step in the design problem solving process is the most creative and will be facilitated by the background (experiences and personal knowledge) of the designer. If the designer has had considerable experience with similar problems, and if the problem has been carefully researched, she or he will be able to develop more creative and useful solutions.

There are two methods by which this step may be developed — individual methods and group methods. An individual may work alone and list all of the solutions that come to mind, no matter how "far-out" they may seem at the time. The designer should try to think of unique uses of existing items for possible solutions as well as standard or customary ways of solving the problem.

Small groups (up to 10 or 12 individuals) can also work effectively in the above manner. "Brainstorming" is a means of people working together for creative solutions to problems. Each person makes suggestions as they come to mind. New and unusual solutions often come forward as various individuals are stimulated by the suggestions of others.

Some preliminary solutions which may be offered for the

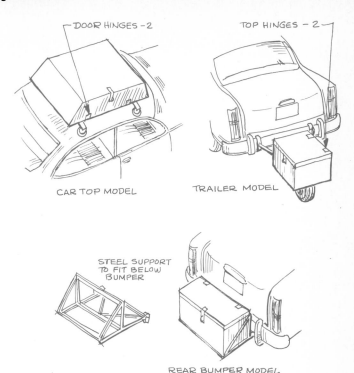

Fig. 9-7. Rough sketches of three preliminary solutions to family camping carrier.

family camping equipment carrier are:
1. Trailer.
2. Waterproof box.
3. Plastic pouch.
4. Device mounted on the trunk lid.
5. Car-top carrier.
6. Carrier fastened to front bumper (for rear engine cars).
7. Pouches mounted on the sides of the car.
8. Carrier on each front fender.
9. Device fastened to rear bumper and free of trunk lid.

No attempt should be made during the listing of preliminary ideas to evaluate the various solutions suggested. This takes place in the next step and it is important to get as many preliminary solutions as possible.

STEP 3: PRELIMINARY SOLUTION REFINEMENT

Combine or resolve all preliminary ideas into as few solutions as possible. Review and evaluate the better preliminary solutions in terms of the problem definition. Make rough sketches to further analyze each solution. Eliminate those that do not show promise. Refine and analyze the remaining preliminary solutions until you have only three or four. Evaluate these in terms of the general problem statement.

"Design a device for carrying the family camping equipment. The device must be located outside the car, have easy access, protect contents against the weather, provide minimal drag on car and cost is a factor."

When all factors were considered, three preliminary solutions were selected for final consideration: (1) a trailer, (2) a car-top carrier and (3) a device fastened to the rear bumper free of the trunk lid, Fig. 9-7.

DECISION CHART			
FACTORS OF COMPARISON	RATINGS OF BEST PRELIMINARY SOLUTIONS		
	NO. 1 TRAILER	NO. 2 CAR-TOP	NO. 3 REAR BUMPER
Located Outside Car	1	1	1
Easy Access	1	2	1
Weather Protect	2	1	2
Drag on Car	3	1	2
Cost	3	1	2
Difficulty of Construction	2	1	3
RATING TOTALS	12	7	11

Fig. 9-8. A decision chart used in rating preliminary solutions to problems.

STEP 4: DECISION AND IMPLEMENTATION

When the best preliminary solutions have been selected, a decision chart is prepared showing comparison of the solutions on the main requirements and limitations of the problem, Fig. 9-8.

A rating system of weighing each factor for comparison on a scale of 1 to 3 may be used, with 1 being the best rating. Where two or more preliminary solutions seem equally desirable on a particular factor of comparison, an equal rating of 1 or 2 should be given. The car-top carrier solution was given the best (lowest score) rating of the three preliminary solutions considered.

This solution is implemented by preparing a working drawing, Fig. 9-9, and building a model and/or prototype of the design solution. Implementation also includes going into production with the final design solution should the problem call for more than one item.

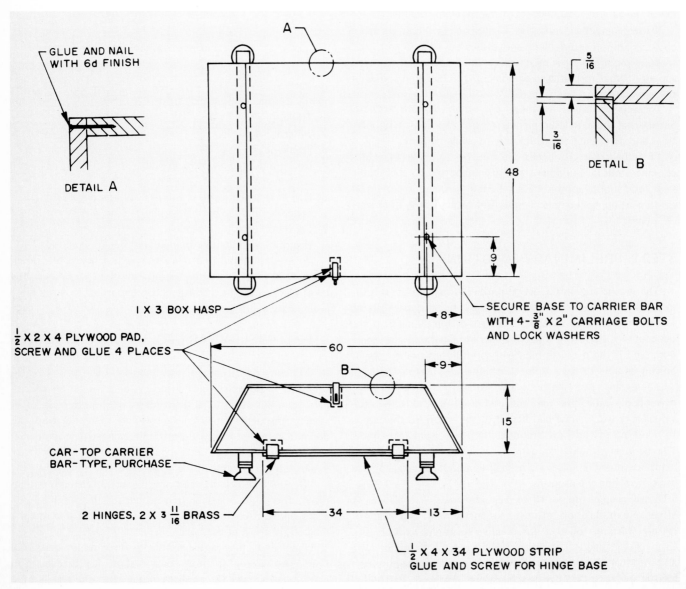

Fig. 9-9. A working drawing of the car-top carrier.

MODELS

The use of models in the design process is increasing in industry, Fig. 9-10. Two important advantages in using models, Fig. 9-11, are improved communication between technical personnel and greater visualization of problems and their solution by nontechnical and management personnel.

Models are also used extensively in the presentation and promotion of a product or design solution. In addition, they are useful in training programs for personnel who will use the equipment.

The term model is used to refer to three-dimensional scale models, mock-ups and prototypes. The characteristics and uses of each type are discussed here as a function of the design method.

SCALED MODELS

A scaled model is a replica of the actual or proposed object. It is made smaller or larger to show proportion, relative size of parts and general overall appearance. Some scaled models are

Fig. 9-10. A model maker at work on a clay model for a new automobile design. (General Motors Corp.)

Fig. 9-11. A design model. (Dow Chemical U.S.A.)

Fig. 9-12. A working model of the Glen Canyon Dam water spillway outlets.
(U. S. Dept. Interior, Bureau of Reclamation)

working models used to aid engineers and designers in their analysis of the function and value of certain design features, Fig. 9-12. The scale of models may vary depending on the size of the actual object and the purpose of the model.

Construction of models in industry is the work of professional model makers, Fig. 9-13. Materials used for model making include all easily worked media such as balsa wood, clay, plaster and aluminum. Standard parts such as wheels, scaled furniture and machine equipment, decals and simulated construction materials are available from model supply dealers.

MOCK-UPS

A mock-up is a full-size model which simulates an actual machine or part, Fig. 9-14. The full-size mock-up presents a more realistic appearance than a scale model and aids in checking design appearance and function. The mock-up may be used as a simulator for training purposes, but it is not an operational model.

PROTOTYPE

The prototype is a full-size operating model of the actual object, Fig. 9-15. It is usually the original full-size working model that has been constructed by craftsmen making each part individually. Its purpose is to correct design and operational flaws before starting mass production of the object.

Fig. 9-13. Professional model makers constructing a full-scale model of a future car.   (General Motors)

Fig. 9-14. A research pilot checks the instrument panels in a mock-up
of the crew station for a future space shuttle vehicle. (NASA)

Fig. 9-15. A coffee maker prototype, designed on computer, being put through its test run.
(The Gillette Company)

PROBLEMS AND ACTIVITIES

1. List as many useful items as you can which could be made from each of the following items. Let your imagination be your guide. Include all items regardless of practicality for production:
   a. Bale of straw.
   b. One square yard of heavy canvas.
   c. Piece of white pine lumber 1″ x 10″ x 12′.
   d. Sheet of vinyl plastic .005 inch in thickness and 10 feet square.
   e. Unfinished interior flush panel door 2′− 6″ x 6′− 8″.
   f. Empty five-gallon paint can.
   g. Discarded but undamaged automobile radiator core.
   h. Sheet of window glass 30″ x 40″.
   i. Rubber garden hose 50 feet long.
   j. A number of broken and unbroken glass bottles.

2. Creative thinking is a matter of developing the mind to be resourceful and imaginative. Try your creativeness by listing as many different ways you can think of to solve the following:
   a. Paint a board 2″ x 10″ x 10″ on all surfaces.
   b. Transfer a gallon of water to another location 10 feet away and 45 degrees above its present location.
   c. Make a 1 inch square hole in the side of a tin can.
   d. Design a clamping device for gluing wood edge-to-edge without the use of a threaded piece.

3. Make a list of six or more items, not currently available on the market, that in your judgment have sales potential. Be sure the items could be produced by your class, using materials available to you commercially and using equipment at school or at home.

4. Select one item from Problem 3 and try to interest 3 to 5 members of your class in the design and production of the item. With the approval of your instructor, follow the project through the design stage and development of the prototype.

   Follow the steps outlined in this chapter on the design method and evaluate the design and production problems. Establish a selling price for which you could manufacture the item and make a reasonable profit.

5. Select any one of the following problems and develop a solution for it, making use of the design method of problem solving outlined in this chapter. Report your work and activities in writing through each of the steps. Develop sketches for at least two of the best possible solutions and a dimensioned instrument drawing for the proposed final solution.

   a. Portable shoeshine equipment with folding seat and storage space for tools and supplies.
   b. A system for screening gravel (rock) into various sizes such as 1/2, 1, 1 1/2 and 2 inches.
   c. A study area that provides space and lighting for writing, reading and storage for materials commonly used in the area.
   d. Means of utilizing for the good of mankind common types of plastic containers collected in the home through the purchase and consumption of food and household products.

6. One of the most difficult types of problems to solve is one that calls for new uses or unique applications of commonly used items in ways they have not been used before. For example, a small gasoline engine is commonly used as a power unit for such items as lawn mowers, motor bikes, snow blowers and tree saws. How many ideas can you suggest for the small gasoline engine that would be useful and yet unique or never applied before?

   Try your creative imagination and list some possible uses. You may add other pieces of equipment or material as needed.

7. Try the same approach used in 6 with the following:
   a. An electric motor from a cordless toothbrush, shaver or hedge trimmer.
   b. A spare tire and wheel from an automobile and tire pump.
   c. Bicycle wheels without tires.
   d. The parts from several radios.

8. Ask your instructor, a member of your family or an acquaintance to help you get in touch with an engineer or industrial designer. Make an appointment for an interview with him or her to find out all you can about the nature of the work and the design method used to solve problems. Ask to be shown sketches used in the solution of problems. Report your findings to your class.

9. Select one or more famous creative inventors or designers and read their biography. See if you can identify things that caused them to develop their creativeness. Try to sum up these characteristics in two or three statements that could serve as a guide to others who want to develop their creative abilities. Report your findings to your class.

10. Review the local newspaper for news articles relating to some school or community need that could be solved using the design method. Take one such problem. Develop solution by applying the design method. Prepare written report of each step in development of solution.

Chapter 10
SECTIONAL VIEWS

A sectional view is that view "seen" beyond an imaginary cutting plane passing through an object at right angles to the direction of sight, Fig. 10-1 (a). Sectional views are used to show the interior construction or details of hidden features that cannot be shown clearly by outside views and hidden lines. Sectional views, (b), are also used to show the shape of exterior features such as automobile body components and airplane wing and fuselage sections.

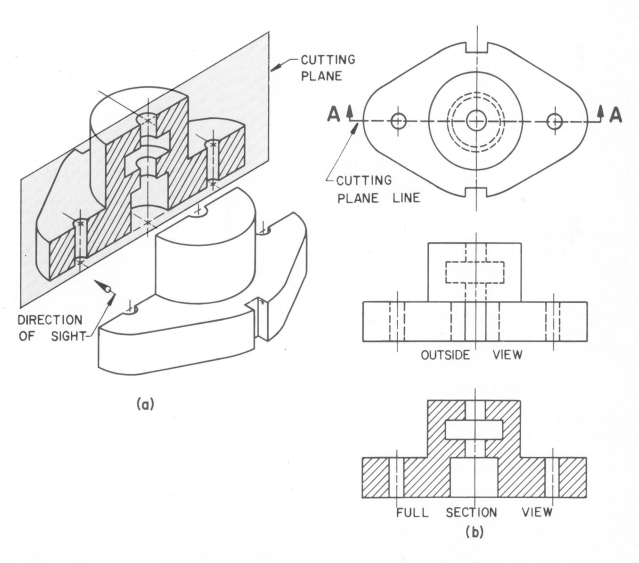

Fig. 10-1. Sectional views show the construction details and shape of parts.

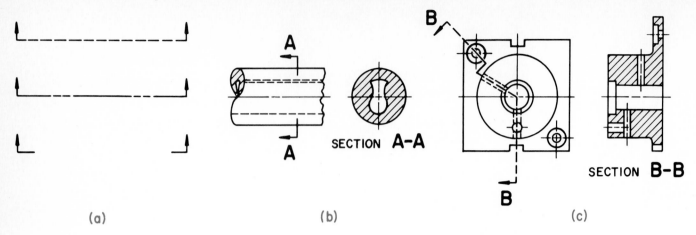

Fig. 10-2. Section cutting plane lines and their proper use on drawn parts.

SECTION CUTTING PLANE LINES

The cutting plane on which a section has been taken is indicated by a heavy dash line about 1/4 inch long, Fig. 10-2 (a). Or, a heavy line of alternating long dashes 3/4 to 1 1/2 inches (or more) with a pair of short dashes 1/8 inch long spaced 1/16 inch apart, may be used. Some industries favor a simplified representation of the cutting plane line that includes only the ends, (b).

On objects having one major center line, the cutting plane line may be omitted if it is clear that the section is taken along the center line. Arrowheads at the ends of the cutting plane lines are used to indicate the direction in which the sections are viewed, Fig. 10-2 (b). The cutting plane may be bent or offset to show details of hidden features to better advantage, (c).

PROJECTION AND PLACEMENT OF SECTION VIEWS

Whenever possible, a sectional view should be projected from, and perpendicular to, the cutting plane. It should be placed behind the arrows, Fig. 10-3. This arrangement should be maintained whether the section is adjacent to or removed some distance from the cutting plane.

SECTION LINING

The exposed (cut) surface of the sectional view is indicated by section lines, sometimes called crosshatching. These lines emphasize the shape of a detail part or differentiate one part from another on an assembly drawing.

Section lines are thin, parallel lines drawn with a sharp pencil or No. 0 (0.3 mm) pen. Where it is necessary to locate dimensions within a sectional view, section lines should be voided around these dimensions to provide clarity to the drawing.

DIRECTION OF SECTION LINES

Section lines should be drawn at an angle of 45 degrees to the main outline of the view, Fig. 10-4 (a). On adjacent parts,

section lines should be 45 degrees in the opposite direction, (b). For a third part, adjacent to the other two parts, section lines should be drawn at an angle of 30 or 60 degrees, (c).

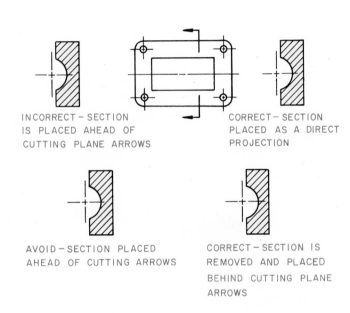

Fig. 10-3. Proper and improper placement of sectional views, in relation to the cutting plane.

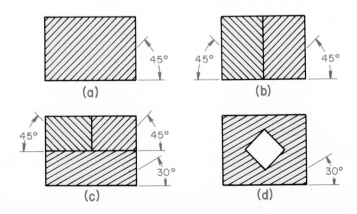

Fig. 10-4. Direction of section lines for single and adjacent parts.

Where the angle of section lining is parallel, or nearly parallel with the outline of the part, another angle should be chosen, (d). Section lines should not be intentionally drawn to meet at common boundaries.

SPACING OF SECTION LINES

Section lines should be uniformly spaced throughout the section, Fig. 10-5. However, spacing may be varied according to the size of the drawing. Section lines should be spaced approximately .12 (1/8) to .18 (3/16) inch apart. Consult the chapter on reproduction methods for spacing of section lines on drawings to be microfilmed.

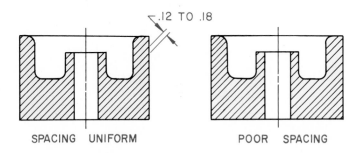

Fig. 10-5. Uniform spacing of section lines enhances the appearance of a drawing.

In general, spacing should be as generous as possible to save time and to improve the appearance of the drawing. Section lines should end at the visible outline of the part without gaps or overlaps.

HIDDEN LINES BEHIND THE CUTTING PLANE

Hidden lines behind the cutting plane, Fig. 10-6 (a), should be omitted unless needed for clarity. In half sections, hidden lines are shown on the unsectioned half ONLY if needed for dimensioning or clarity on the drawing, (b).

VISIBLE LINES BEHIND THE CUTTING PLANE

In general, visible lines behind the cutting plane are shown. They may, however, be omitted where clarity in the sectioned view is assured and the omission represents an appreciable saving in drafting time, Fig. 10-7.

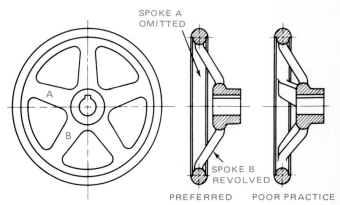

Fig. 10-7. Omission of visible lines in sectional view. (ANSI)

TYPES OF SECTIONS

A variety of types of sections have been adopted as standard sectioning procedure. Each type has a unique function in drafting. Types of sections and their uses are discussed and illustrated in the following:

FULL SECTIONS

A full sectional view is one in which the cutting plane passes entirely through an object, and the cross section behind the cutting plane is exposed to view, Fig. 10-8. The cutting plane line and section title may be omitted if the sectional view is in orthographic projection position.

A full sectional view usually replaces an exterior view in order to show some interior feature.

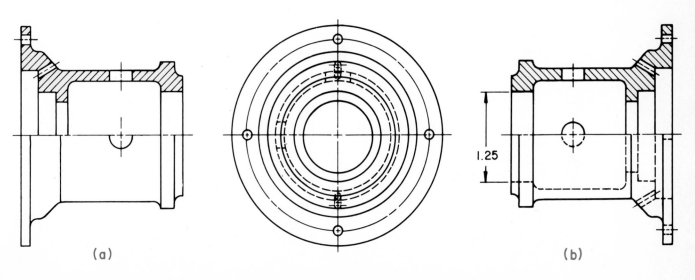

Fig. 10-6. Use of hidden lines in sectional views.

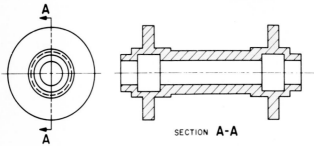

Fig. 10-8. A full sectional view of the corresponding machined shaft.

REVOLVED SECTIONS

A revolved section is obtained by passing a cutting plane perpendicularly through the center line or axis of the part to be sectioned. The resulting section is revolved 90 degrees in place, Fig. 10-10. The cutting plane line is omitted for symmetrical sections. Visible lines may be removed on each side of the section and break lines used for clarity.

A revolved section is used to show the true shape of the cross section of an elongated object such as a bar, or some feature of a part such as a rib, arm or linkage.

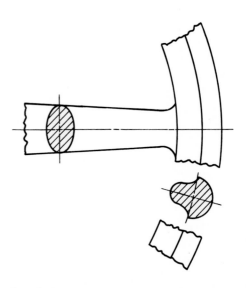

Fig. 10-10. Typical layout for drawing a revolved section.

HALF SECTIONS

A half section of a symmetrical object shows the internal and external features in the same view, Fig. 10-9. Two cutting planes are passed at right angles to each other along the center lines or symmetrical axes, (a). One-quarter of the object is considered removed and a half-sectional view is exposed. The cutting plane lines and section titles are omitted.

A half section is used when it is desired to show both the interior and exterior features of a symmetrical object on a single view, Fig. 10-9 (b).

REMOVED SECTIONS

A removed section, Fig. 10-11, is one that has been moved out of its normal projected position in the standard arrangement of views. The removed section should be labeled SECTION A-A, and placed in a convenient location on the same sheet. If it becomes necessary to locate the removed section on another sheet of a multiple-sheet drawing, appro-

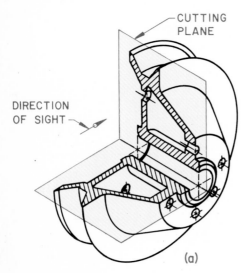

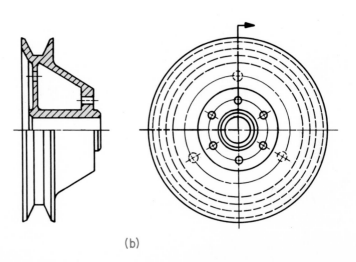

Fig. 10-9. A half section shows both interior and exterior features of an object.

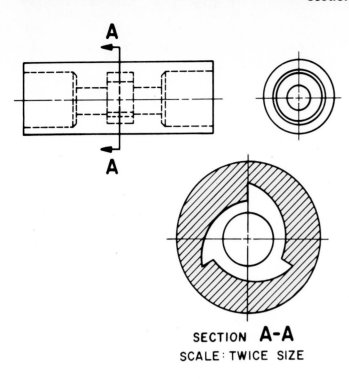

SECTION A-A
SCALE: TWICE SIZE

Fig. 10-11. A removed section, enlarged and located near the actual sectioning.

one plane. Offsets in the cutting plane are not shown in the sectional view.

Offset sections are useful when it is desired to obtain a sectional view with features in more than one cutting plane.

BROKEN-OUT SECTIONS

A broken-out section appears in place on the regular view and the partial section is limited by a break line, Fig. 10-13.

Broken-out sections are used to show interior detail of objects where less than a half section is required to convey the necessary information.

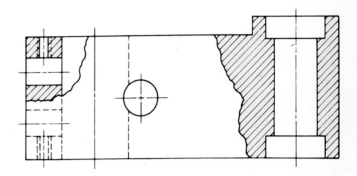

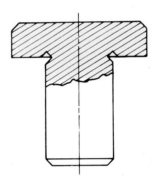

Fig. 10-13. Examples of broken-out sections to show interior details. Note the break lines.

priate identification and zoning references should be placed on related sheets.

A removed section is similar to a revolved section except that the cross section is removed from the actual view of the part.

OFFSET SECTIONS

An offset section is one in which the cutting plane is not one continuous plane. It is stepped, or offset, to pass through features which lie in more than one cutting plane, Fig. 10-12. The path of cutting plane is shown on the view to be sectioned and the features drawn in the sectional view as if they were in

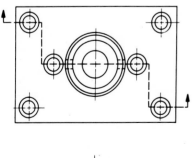

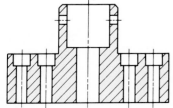

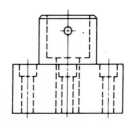

Fig. 10-12. Left. Drawing of an offset section showing the section line. Right. The actual machined part.

ALIGNED SECTIONS

Certain objects may be misleading when a true projection is made. In an aligned section, features such as spokes, Fig. 10-14, holes and ribs are drawn as if rotated into, or out of, the cutting plane.

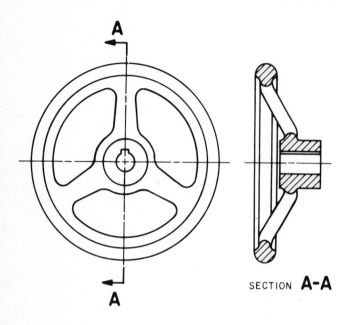

Fig. 10-14. An aligned section with one spoke rotated out of the cutting plane.

Aligned sectional views are used when actual or true projection would be confusing. Note that the offset features have been rotated to align with the center line and projected to the sectional view for clarity. If these features were projected directly, their lengths and positioned locations would be distorted. Also see Fig. 10-2 (c).

THIN SECTIONS

Structural shapes, sheet metal, packing gaskets, etc., are often too thin for section lining and may be shown solid, Fig. 10-15. Where two or more thicknesses are shown, a space should be left between them which is compatible with the microfilm process.

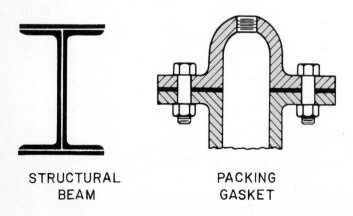

STRUCTURAL BEAM

PACKING GASKET

Fig. 10-15. Structural beam and packing gasket shown solid in section.

AUXILIARY SECTIONS

An auxiliary section is a removed section of an auxiliary view, Section A-A, Fig. 10-16. The section should be shown in its normal auxiliary position and, if necessary, should be identified with a cutting plane line and letters. For information on the construction of auxiliary views, see Chapter 12.

The auxiliary section is used to add clarity to critical areas of a drawing.

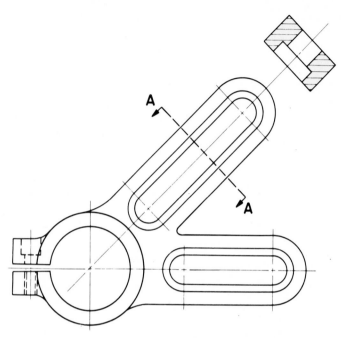

Fig. 10-16. An auxiliary section and rotated features.

PARTIAL SECTIONS

Partial sections are used to show details of objects without the necessity of drawing complete conventional views, Fig. 10-17.

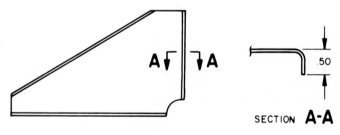

Fig. 10-17. Partial sections save drafting time. (General Dynamics, Engineering Dept.)

PHANTOM SECTIONS

The phantom (hidden) section, although not used extensively in industry, is used to show interior construction while retaining the exterior detail of a part. Fig. 10-18 (a) illustrates a phantom section through a piston. This type section is also used to show the positional relationship of an adjacent part, as in the construction industry, (b). The symbol used for section lining is a series of short dashes.

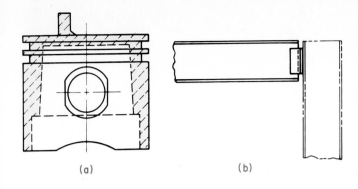

Fig. 10-18. Applications of the phantom section.

UNLINED SECTIONS

For clarity in sections of assembly drawings, standard parts such as bolts, nuts, rods, shafts, bearings, rivets, keys, pins and similar objects whose axes lie in the cutting plane, should not be sectioned. When the axes of these parts lie at right angles to the cutting plane, the parts should be sectioned. See Figs. 10-19 and 10-20.

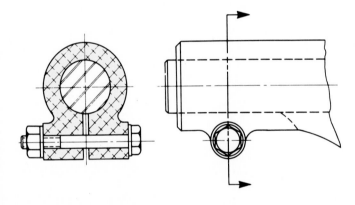

Fig. 10-19. Sectional view of a shaft lying at right angles to the cutting plane.

CONVENTIONAL PRACTICES

Certain practices in regard to sectional views have become standard procedure. Those not covered under types of sections are discussed here.

RECTANGULAR AND CYLINDRICAL BAR AND TUBING BREAKS

In drawing a long bar or tubing with uniform cross section, it is usually not necessary to draw its full length. Often, the piece is drawn to a larger scale, and a break made and sectioned as shown in Fig. 10-21. The true length of the piece is indicated by a dimension, Fig. 10-22.

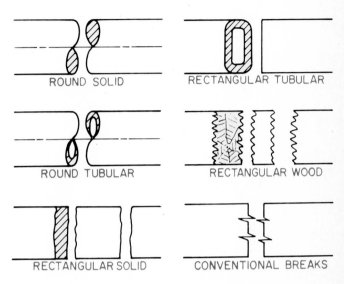

Fig. 10-21. Symbols used when sectioning bar and tubing.

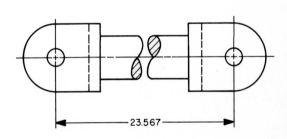

Fig. 10-22. Conventional breaks permit long, uniformly shaped objects to be drawn to a scale large enough to present its details clearly.

The conventional breaks for cylindrical bars and tubing are known as "S" breaks and may be drawn with a template or constructed as follows:

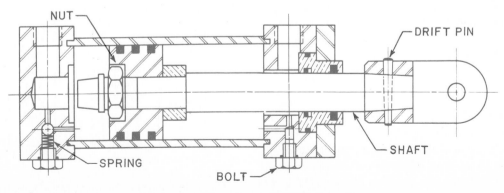

Fig. 10-20. Shafts, bolts, nuts and pins in section.

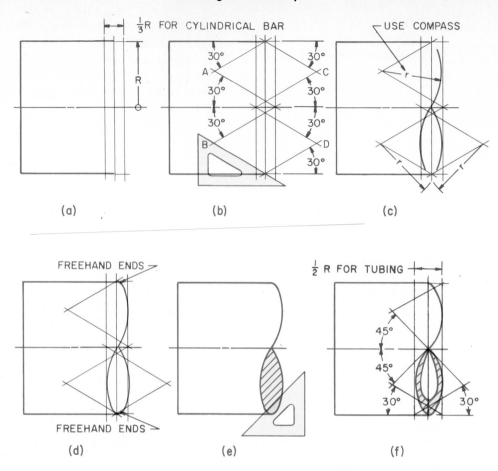

Fig. 10-23. Constructing the "S" break for tubing and cylindrical bar.

1. Draw rectangular view for bar or tube, Fig. 10-23 (a).
2. Lay off fractional radius widths on end to be sectioned.
3. Scribe 30 degree construction lines to locate radii centers A, B, C and D, (b). (For a wider sectional face, use 45 degree projection on four angles adjacent to center line.)
4. Set compass on radius center and adjust so arc passes through center point (P), stopping short of outside edge, (c). Only three radii are used, depending on placement of sectional face.
5. Complete ends of "S" curve freehand, (d).
6. Draw inside curve freehand when tubing is being represented.
7. Add section lining to visible sectioned part, (e and f).

NOTE: When an "S" break is shown with stock continuing on both sides of the break, as in Fig. 10-21, the sectional faces are diagonally opposite. The "S" break can also be drawn freehand by estimating the width of the "S" and crossing at the center line.

INTERSECTIONS IN SECTIONS

When a section is drawn through an intersection, and the offset or curve of the true projection is small, the intersection may be drawn conventionally without offset or curve, Fig. 10-24 (a and c). Intersections of a larger configuration may be projected as shown at (b), or approximated by circular arcs as at (d).

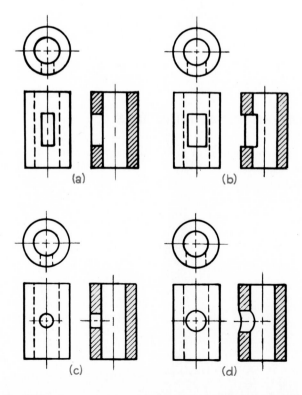

Fig. 10-24. Conventional representation of intersections in sections. (ANSI)

RIBS, WEBS, LUGS AND GEAR TEETH IN SECTION

Ribs and webs are used to strengthen machine parts, Fig. 10-25 (a). When the cutting plane extends along the length of a rib, web, lug, gear tooth or similar flat element, the element is not sectioned in order to avoid a false impression of thickness or mass.

An alternate method of section lining is shown at Fig. 10-25 (b). Spacing is twice that of the regular section and is used where the actual presence of the flat element is not sufficiently clear without section lining.

When the cutting plane cuts an element crossways, the element is sectioned in the usual manner, (c). Gear teeth are NOT sectioned when the cutting plane extends through the LENGTH of the tooth, (d). Gear teeth ARE sectioned when the cutting plane cuts ACROSS the teeth such as in the profile view of the worm.

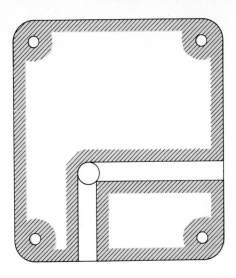

Fig. 10-26. Using the outline sectioning method for a large object.

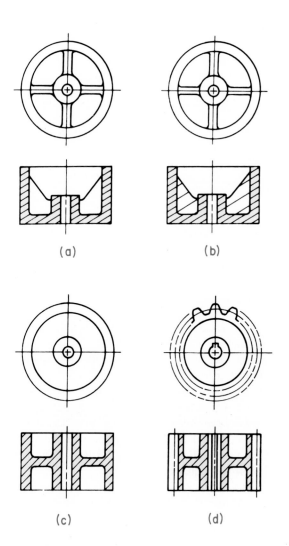

Fig. 10-25. Ribs, webs and gear teeth in section.

OUTLINE SECTIONING

When clarity in the drawing is assured, section lines should be shown only along the borders of the part, Fig. 10-26. This convention is particularly useful on large parts where considerable time would be required in sectioning.

SECTION TITLES

The letters which identify the cutting plane are included in the title of the section, such as SECTION A-A, SECTION B-B, etc. When the single alphabet is exhausted, multiples of letters may be used such as SECTION AA-AA, SECTION BB-BB. The section title always appears directly under the section view.

When a sectional view is located on a sheet other than the one containing the cutting plane indication, the number of the sheet and zone on which the cutting plane indication appears should be referenced for easy location. A similar cross reference should be located on the view containing the cutting plane line.

SCALE OF SECTIONS

Preferably, sections should be drawn to the same scale as the outside view from which they are taken. When it is desirable or necessary to draw the section to a different scale, the scale should be specified directly below the section title, such as:

SECTION A-A
SCALE: 2/1   or   SECTION B-B
SCALE: 4 X SIZE

MATERIAL SYMBOLS IN SECTION

Symbolic cross sectioning for the graphic indication of various materials is shown in Fig. 10-27. On detail drawings it is recommended that general purpose (cast-iron) section lining be used for all materials, except parts made of wood.

Detail drawings and assembly drawings of multimaterial parts should make use of the symbols for various materials for clarity on the drawing. This calls attention to the general types of material of which the various parts are made. See Figs. 10-19 and 10-20.

Symbolic sectioning, however, serves no practical purpose on detail drawings. The general purpose symbol requires less time to construct, and the materials, processes and protective treatment necessary to meet the design requirements of a part are normally indicated on the drawing or parts list.

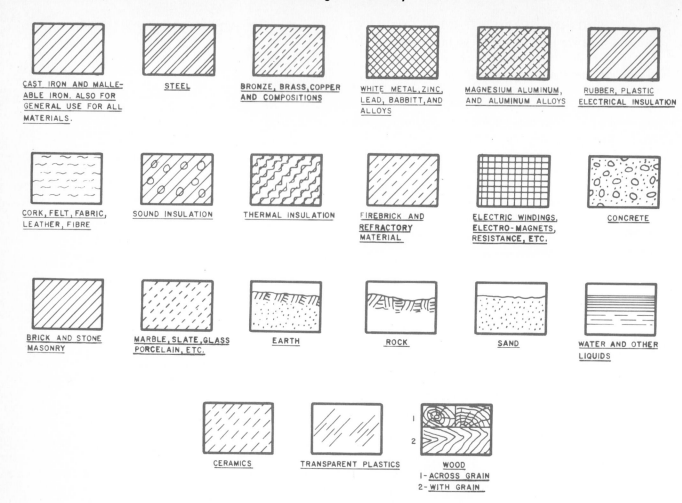

Fig. 10-27. Material symbols in section.
(Deere and Co.)

PROBLEMS IN SECTIONING

The problems which follow have been selected to provide experience in drawing various types of sections and to assist in developing an understanding of their applications. Use a CAD system, or select A or B size sheets and an appropriate scale for producing a well balanced drawing.

1. Refer to Fig. 10-28 and draw the necessary views, including a full section of the parts shown. On those parts where the cutting plane line has not been given, select the position of the sectional view which will best aid in clarifying interior detail.

2. Make a half section drawing of the parts shown in Fig. 10-29.

3. Draw the necessary views of the objects shown in Fig. 10-30 and show revolved sections of the appropriate features.

4. Draw the necessary views of the objects shown in Fig. 10-31, including a removed sectional view of the features indicated. Use sectional views for the regular views where clarity will be improved.

5. Draw the necessary views, including offset sections as indicated of the objects shown in Fig. 10-32.

6. Draw the necessary views of the CLUTCH PISTON and SHEAVE, Fig. 10-33, including an aligned section.

7. Make a two-view drawing of the PIVOT PIN, including a broken-out section, Fig. 10-33.

8. Draw the necessary views to clearly describe the objects shown in Fig. 10-34. Use drafting conventions discussed in this chapter to represent conventional breaks and intersections.

9. Draw the necessary views of the FAN BRACKET and ANGLE BRACKET, Fig. 10-35. Either method of showing webs in section may be used.

Sectional Views

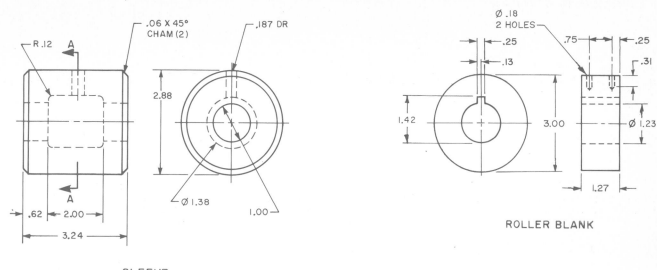

SLEEVE

ROLLER BLANK

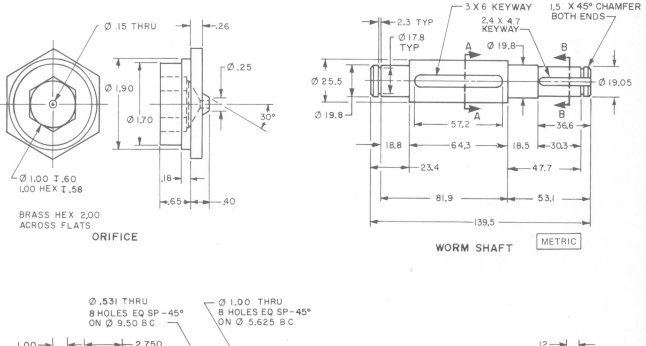

BRASS HEX 2.00
ACROSS FLATS
ORIFICE

WORM SHAFT METRIC

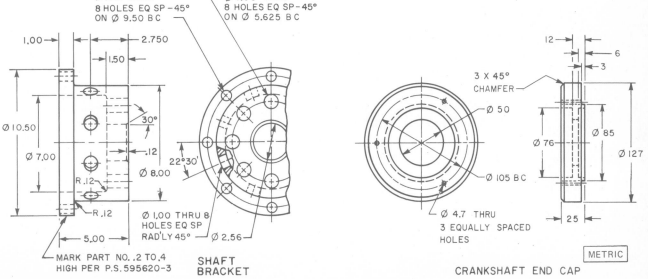

SHAFT
BRACKET

CRANKSHAFT END CAP

Fig. 10-28. Parts for which full sectional views are to be drawn.

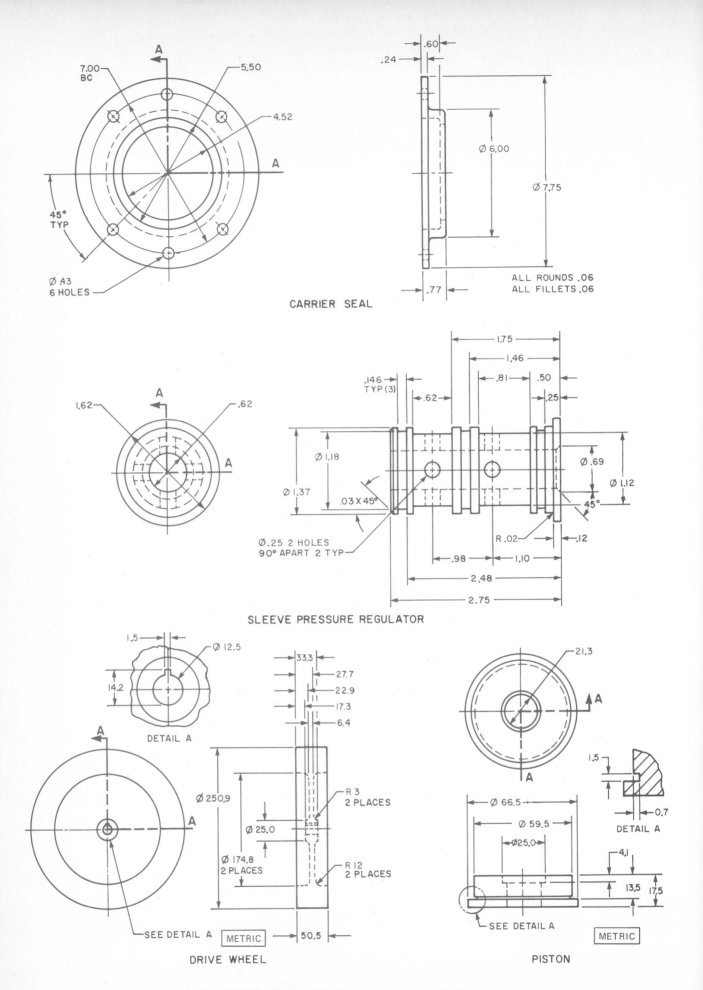

Fig. 10-29. Parts for which half sectional views are to be drawn.

Sectional Views

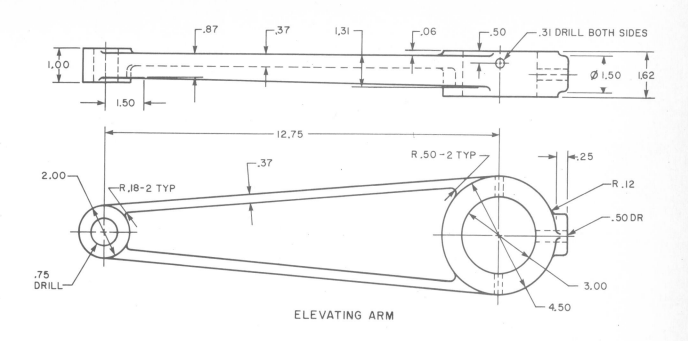

ELEVATING ARM

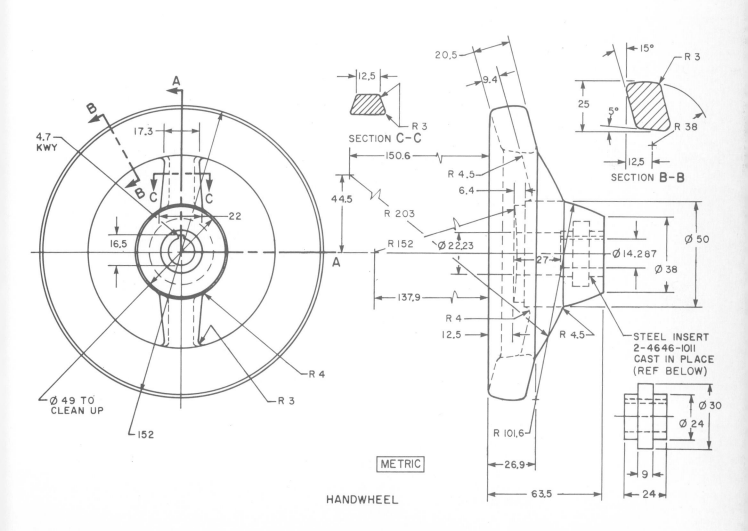

SECTION C-C

SECTION B-B

HANDWHEEL

METRIC

Fig. 10-30. Revolved sectional view problems.

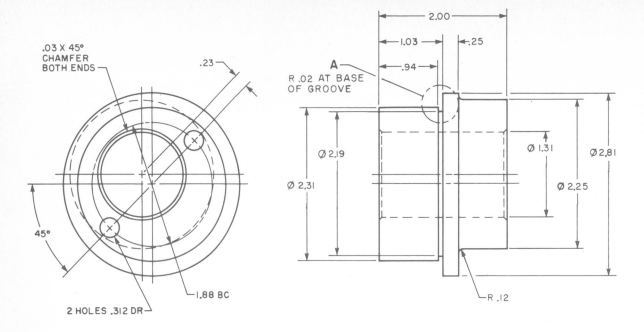

LOWER ECCENTRIC

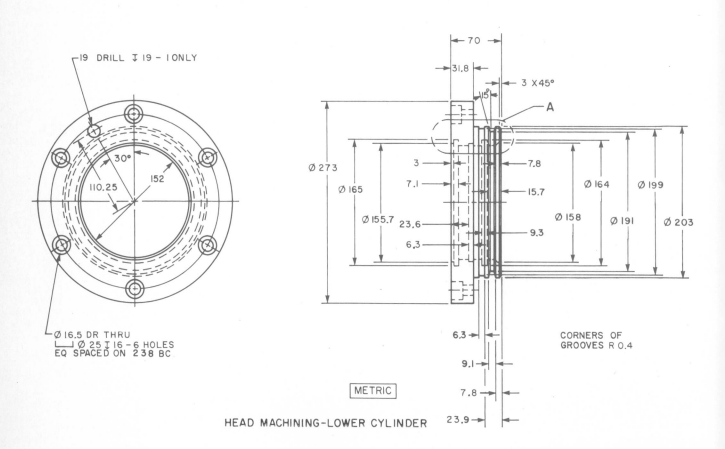

METRIC

HEAD MACHINING-LOWER CYLINDER

Fig. 10-31. Removed sectional view problems.

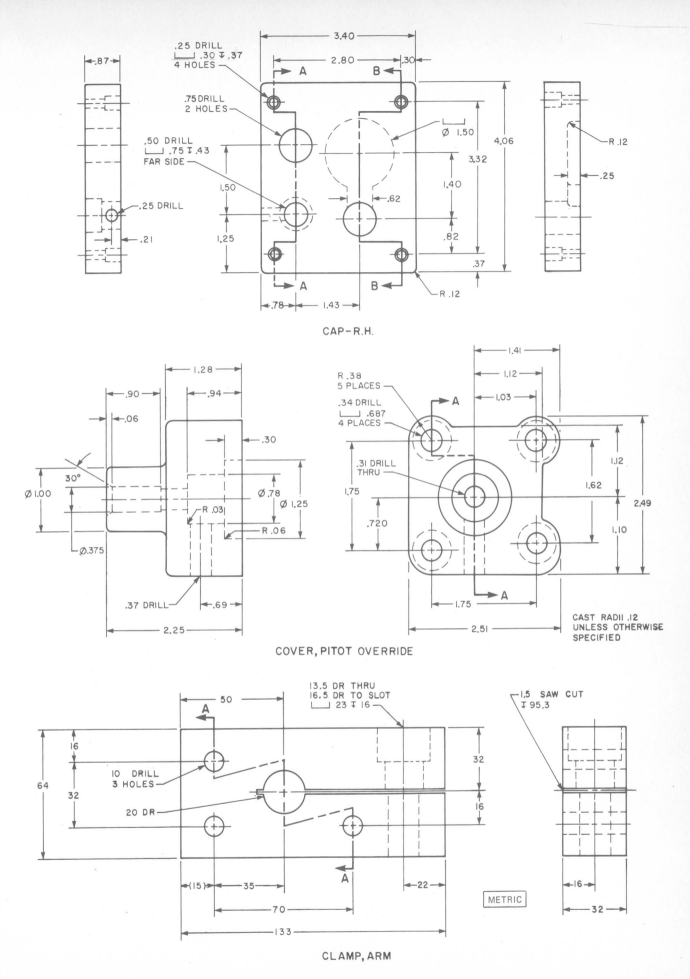

CAP–R.H.

COVER, PITOT OVERRIDE

CLAMP, ARM

Fig. 10-32. Parts for which offset sectional views are to be drawn.

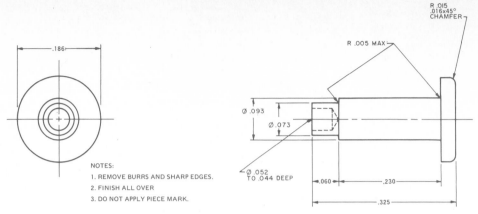

NOTES:
1. REMOVE BURRS AND SHARP EDGES.
2. FINISH ALL OVER
3. DO NOT APPLY PIECE MARK.

PIN, PIVOT

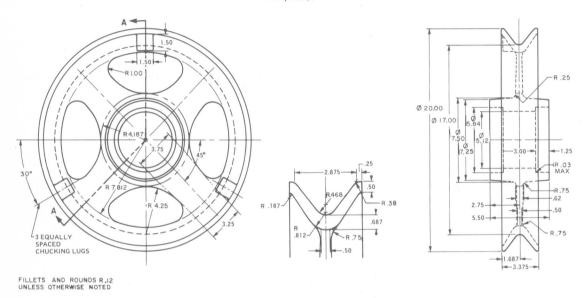

FILLETS AND ROUNDS R .12
UNLESS OTHERWISE NOTED

SHEAVE

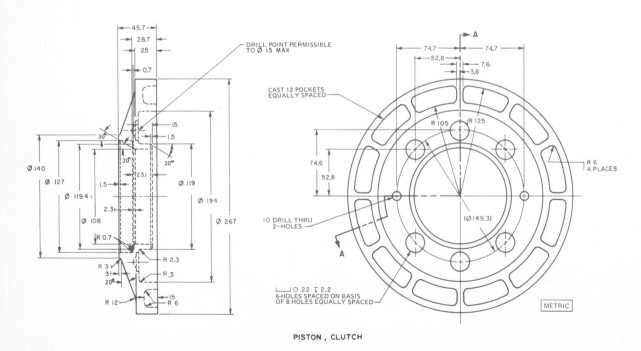

PISTON, CLUTCH

Fig. 10-33. Aligned and broken-out sectional view problems.

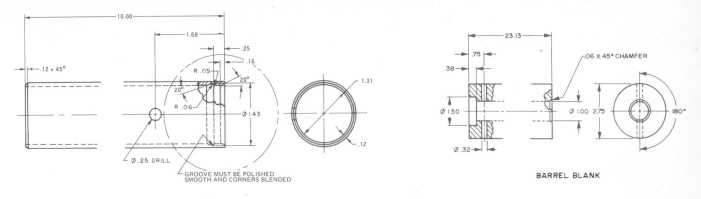

BARREL

BARREL BLANK

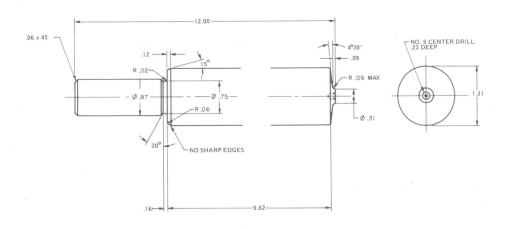

ROD

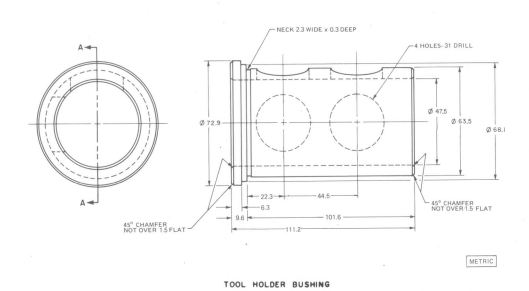

METRIC

TOOL HOLDER BUSHING

Fig. 10-34. Sectional view problems featuring conventional drafting practices.

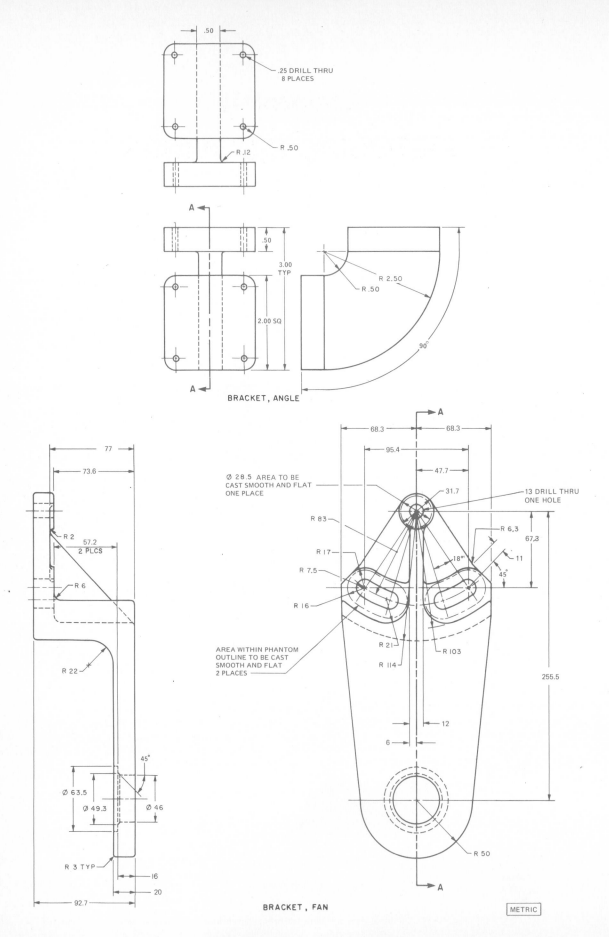

BRACKET, ANGLE

BRACKET, FAN

METRIC

Fig. 10-35. Sectional view problems with webs.

Chapter 11
PICTORIAL DRAWINGS

Pictorial drawings are more representative of a machine part or structure than multiview-orthographic drawings. This is particularly true for the nontechnical person who needs information from a drawing.

Pictorial drawings are used to supplement multiview drawings when the pictorial clarifies information contained in the multiview. Pictorials may even be used as a substitute for multiview drawings when the task to be performed is not highly complex, as in the assembly of machine parts, Fig. 11-1.

Pictorial drawings are widely used for assembly drawings, piping diagrams, service and repair manuals, sales catalogs and technical training manuals. Pictorials are also used by the general public in the assembly of prefabricated furniture, swing sets and do-it-yourself kits.

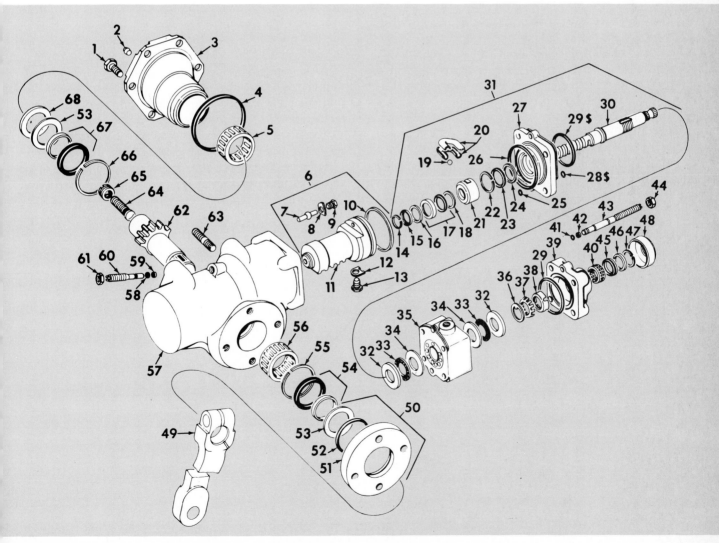

Fig. 11-1. An exploded pictorial drawing clarifies the assembly of parts.
(International Harvester Co.)

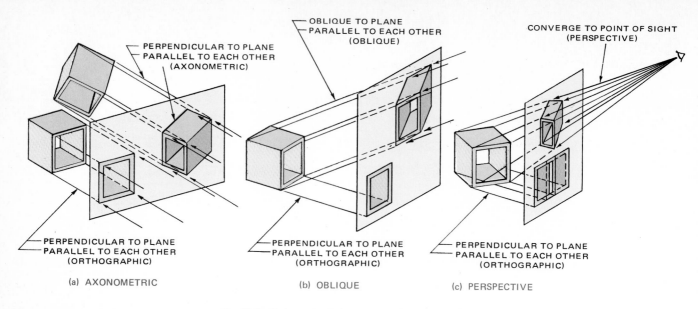

Fig. 11-2. Basic types of pictorial projections.   (ANSI)

TYPES OF PICTORIAL PROJECTIONS

There are three basic types of pictorial projections used in industry: axonometric, oblique and perspective, Fig. 11-2. Under each of these three groupings are several subtypes. This chapter presents the various types of pictorials and the techniques used in drawing them. Recommended methods of dimensioning and sectioning are also covered.

AXONOMETRIC PROJECTION

In axonometric projection, the lines of sight are perpendicular to the plane of projection. In this sense, axonometric projection is a form of orthographic projection, Fig. 11-3. It should be noted, however, that while the lines of sight of the axonometric projection are perpendicular, the three faces of the object are all inclined to the plane of projection, giving a three-dimensional pictorial effect.

Axonometric projection is distinguished from orthographic projection in that the object is inclined to the plane of projection and the principal axes may form any angle except 90 degrees.

There are three types of axonometric projections: isometric, dimetric and trimetric, Fig. 11-4. The differences between the three types of axonometric projections are in the angles the three principal faces and the three principal axes

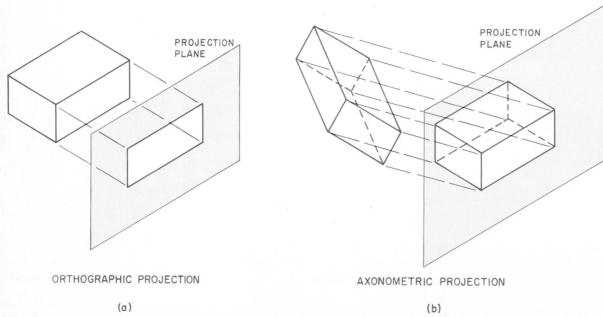

Fig. 11-3. Axonometric projection compared to orthographic projection.

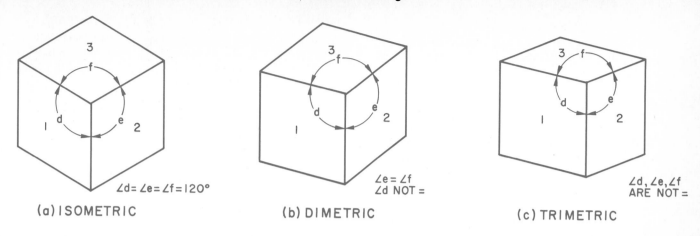

Fig. 11-4. Relationship of the three types of axonometric projections.

make with the plane of projection.

In an ISOMETRIC PROJECTION, Fig. 11-4 (a), the three principal faces (1, 2, 3) of a rectangular object are equally inclined to the plane of projection and the three axes make equal angles (d, e, f) with each other.

A DIMETRIC PROJECTION, Fig. 11-4 (b), has two faces equally inclined (1 and 3 in our example). Two axes make equal angles (e and f) with the plane of projection, while the third face and angle are different.

In a TRIMETRIC PROJECTION, Fig. 11-4 (c), all three faces and axes make different angles with the plane of projection. The axes of axonometric projections, so long as they retain their relationship, may be placed in a variety of positions in addition to those shown in Fig. 11-4.

ISOMETRIC PROJECTION

Isometric means "equal measure." The three principal faces and three principal edges make equal angles with the plane of projection. An isometric projection is a true orthographic projection (parallel projection lines) of an object on the projection plane. It may be produced by revolving the object in the multiview 45 degrees about the vertical axis XC, Fig. 11-5 (b), and tilting the cube forward until the body diagonal XY is perpendicular to the plane of projection, (c).

The vertical axis, XC, is at an angle of 35° 16' with the plane of projection and appears vertical on that plane. Principal axes AX and XB appear on the plane of projection at 30° with the horizontal. The three front edges AX, XB and XC, called isometric axes, are separated by equal angles of 120° in the isometric projection, (c).

Angles of 90° in the orthographic view appear as large as 120° or as small as 60° in the isometric view depending on the viewing point.

Lines along, or parallel to, the isometric axes are called ISOMETRIC LINES and are foreshortened in an isometric projection to approximately 81 percent of their true lengths. See (a, c and d) in Fig. 11-5.

Lines which are not parallel to the isometric axes are called NON-ISOMETRIC LINES, Fig. 11-5 (d). The faces of the cube shown in Fig. 11-5 (d) are called ISOMETRIC PLANES and

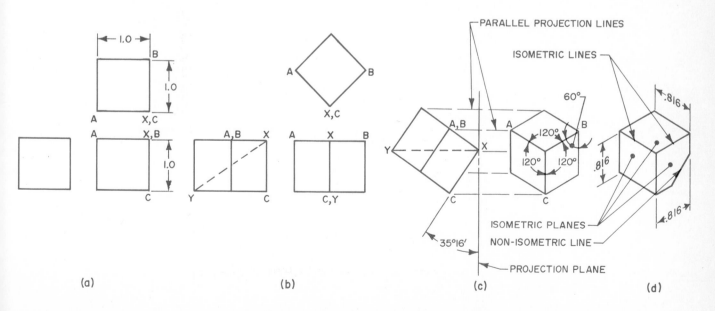

Fig. 11-5. Constructing an isometric projection by revolution.

include all planes parallel to these planes.

An isometric projection, Fig. 11-6, can also be obtained by means of an auxiliary projection.

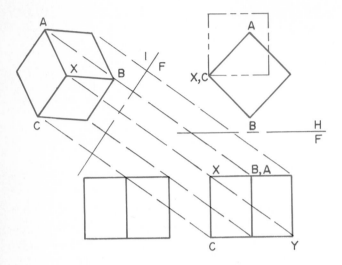

Fig. 11-6. Constructing an isometric projection by the auxiliary view method.

Isometric projections are true projections. However, for objects more complicated than the cube (see Fig. 11-5), the process requires that the object first be drawn in orthographic projection. Then the isometric projection is constructed by revolution, auxiliary projection or by the use of a special scale similar to the one shown in Fig. 11-7. Since direct measurements can then be made on the isometric axes, it is common practice to make an isometric drawing instead of an isometric projection.

ISOMETRIC DRAWING AND PROJECTION COMPARED

Basically, an isometric drawing can be constructed without first making a multiview drawing. This simplified process is possible because measurements may be made with a regular scale on the isometric axes of the drawing.

The chief difference between isometric projection and isometric drawing is that the latter is somewhat larger, Fig. 11-8. Actual measurement of full lengths are used in the drawing, while foreshortened lengths are projected in the isometric projection.

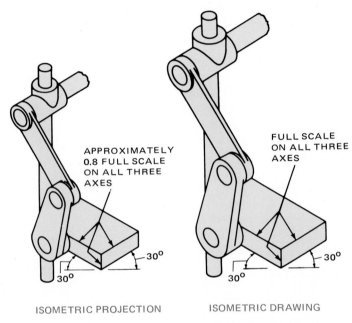

ISOMETRIC PROJECTION   ISOMETRIC DRAWING

Fig. 11-8. An isometric drawing of an object is similar to an isometric projection except somewhat larger.   (ANSI)

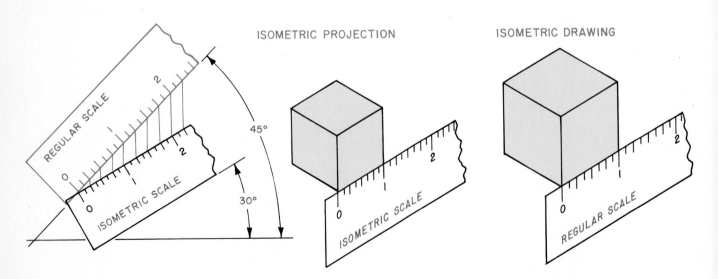

Fig. 11-7. Construction of an isometric scale.

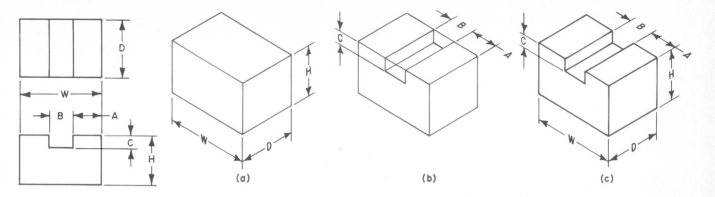

Fig. 11-9. Constructing an isometric drawing of an object having only normal surfaces.

ISOMETRIC DRAWING OF AN OBJECT WITH NORMAL SURFACES

Normal surfaces are those surfaces parallel to the principal planes of projection, not including inclined or oblique surfaces. It is easier to understand how to construct an isometric drawing if normal surfaces involving only isometric lines are considered at first.

The block shown in Fig. 11-9 is constructed as follows:

1. Draw an isometric block equal to width, depth and height of object shown in multiview, Fig. 11-9 (a).
2. Lay off dimensions along isometric lines for cut through top and draw isometric lines, (b).
3. Erase unnecessary construction lines and darken pictorial, (c).

Note: Hidden lines are omitted from isometric drawings unless needed for clarity.

AN ISOMETRIC DRAWING INVOLVING NON-ISOMETRIC LINES

Isometric drawings of objects with inclined or oblique surfaces involve non-isometric lines and cannot be measured directly on the drawing. These surfaces are drawn by locating and joining the end points of the lines enclosing the surfaces. All measurements on isometric drawings must be made parallel to the isometric axes, since non-isometric lines are not shown true length in the pictorial.

Construct an object involving inclined and oblique surfaces, using the following procedure:

1. Draw an isometric block equal to width, depth and height of object shown in multiview, Fig. 11-10 (a).
2. Lay off, along isometric lines, end points of non-isometric lines enclosing inclined or oblique surfaces, (b).
3. Join these points, using a straightedge, to form surfaces, (c).
4. Erase unnecessary construction lines and darken pictorial.

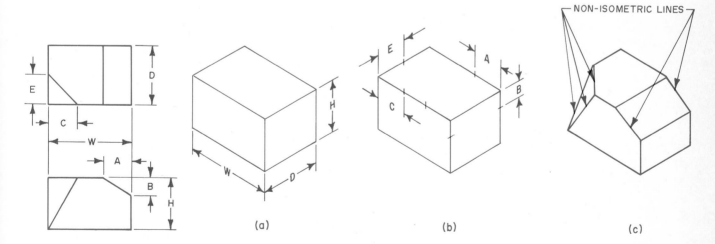

Fig. 11-10. Construction of an isometric drawing of an object involving non-isometric lines.

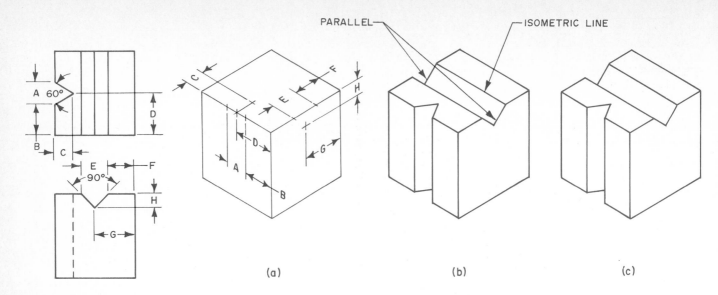

Fig. 11-11. Constructing angles in isometric drawings.

CONSTRUCTING ANGLES IN ISOMETRIC DRAWINGS

Angles do not appear in their true size in isometric drawings and cannot be drawn true value. An angle is drawn by locating the end points of its sides and connecting these points to form the required angle.

The following procedure is used in drawing angles in isometric:

1. Draw isometric block to encompass object, Fig. 11-11 (a).
2. Lay off, along isometric lines, distances necessary to locate end points of lines forming angle, (a).
3. Join end points, using a straightedge, to form angles, (b). Angles may be projected to opposite side of block, using isometric lines, since these are normal lines in orthographic views. The angle cut on opposite end should be drawn parallel to first cut.
4. Erase unnecessary construction lines and darken pictorial, (c).

COORDINATE METHOD OF LAYING OUT AN ISOMETRIC DRAWING

In the preceding sections, the "block" method of laying out isometric drawings has been employed. Isometric drawings may also be laid out by use of the coordinate method shown in Fig. 11-12. For certain objects which are not basically cubic (such as truncated pyramid), the coordinate method is faster.

Use the following procedure to construct an isometric drawing by the coordinate method:

1. Locate a starting point such as lower front corner and draw two horizontal (30°) axes, Fig. 11-12 (a). (Note: Technique of centering an isometric drawing in a space is covered later in this chapter.)
2. Lay off lengths of two base edges, (a).
3. Construct base line FG of assumed isometric plane which passes through point E, by measuring offset of distance X, (a).
4. Locate distance Y from corner A of pyramid and project to intersect line FG and Y', (a).
5. Construct a vertical line at Y' and lay off height of point E, (b). Draw line EC.
6. Continue to locate remaining points of truncated cut in a similar manner.
7. Erase unnecessary lines and darken finished drawing, (c).

Fig. 11-12. Constructing an isometric drawing, using coordinate or offset method.

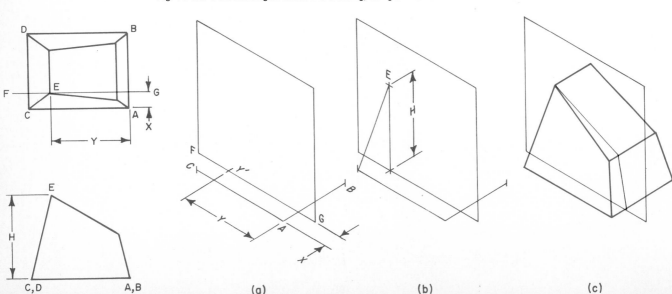

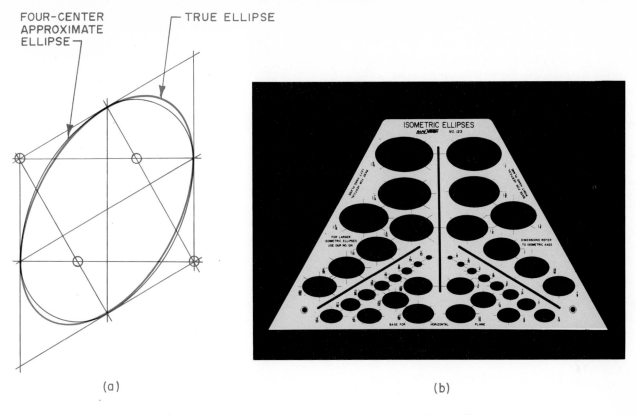

Fig. 11-13. Four-center approximate ellipse compared with true ellipse.

CIRCLES AND ARCS IN ISOMETRIC — FOUR-CENTER APPROXIMATE METHOD

Circles and arcs will appear as ellipses or partial ellipses in isometric drawings. One method of constructing these is the four-center approximate method. It is an approximate of the true ellipse, Fig. 11-13 (a). True ellipses may be drawn with the isometric ellipse template, (b), or by the coordinate method explained later in this chapter.

The four-center approximate method is fast and effective for most isometric drawings where tangent circles are not involved. The procedure for drawing the four-center approximate ellipse is as follows:

1. Locate center of circle and draw isometric center lines, as shown in Fig. 11-14 (a).

2. With O as center and a radius equal to radius of actual circle, strike arcs A, B, C and D to intersect isometric center lines, (b).

3. Through each of these points of intersection, draw a line perpendicular to other center line, (c).

4. Four intersecting points, E, F, G and H, of perpendiculars are radius centers of approximate ellipse. Radii of arcs are distances from center along perpendiculars to intersection of isometric center lines, that is EA, FB, GD and HA, as shown in (c).

5. Draw four arcs CA, BD, AB and CD to complete ellipse, as shown in (d).

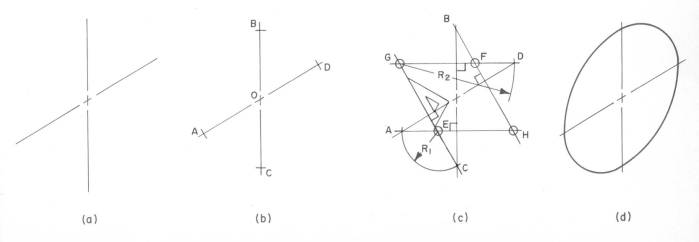

Fig. 11-14. Four-center approximate method of constructing an isometric ellipse.

An isometric arc is constructed using the four-center approximate method by selecting the required portion of the full circle, Fig. 11-15, and drawing that segment.

Isometric circles constructed by the four-center approximate method are shown in the three principal faces or planes of the isometric drawing, Fig. 11-16. The same procedure is used in drawing isometric ellipses by this method regardless of position.

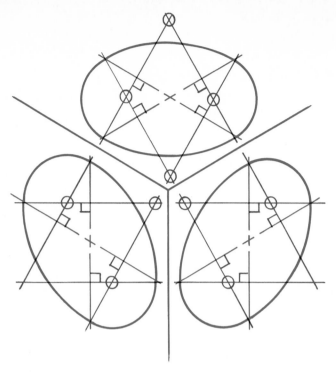

Fig. 11-16. Four-center approximate ellipses in three principal isometric planes.

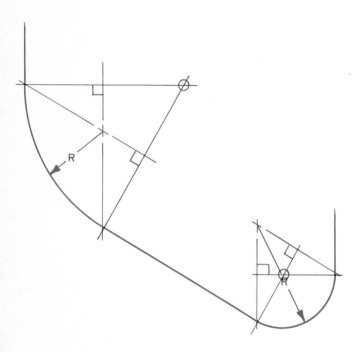

Fig. 11-15. Drawing arc by four-center approximate method.

COORDINATE METHOD OF DRAWING ISOMETRIC CIRCLES AND ARCS

The coordinate method of drawing an isometric circle is a process of plotting coordinate points on a true circle. These

points are then transferred to an isometric square the same size as the circle to be drawn, Fig. 11-17. This method results in a true projection of the isometric ellipse.

The procedure used in drawing an isometric circle by coordinate method is as follows:

1. Locate center of isometric circle and draw center lines, Fig. 11-17 (a).
2. Strike arcs equal to radius of required circle and construct an isometric square, (b).
3. Draw a semicircle adjacent to one side of isometric square, (b).

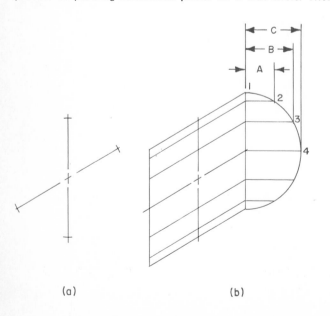

(a)   (b)

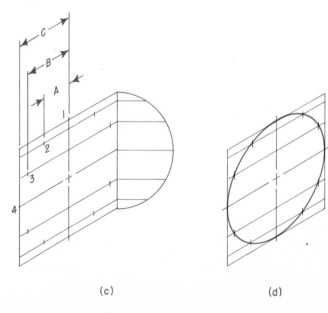

(c)   (d)

Fig. 11-17. Constructing an isometric ellipse, using coordinate method.

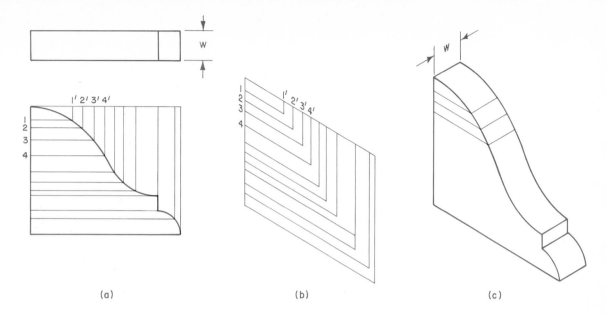

Fig. 11-18. Transferring an irregular curve to an isometric drawing, using coordinate method.

4. Divide semicircle into an even number of equal parts and project divisions to side of square, (b).

5. From these points, draw lines across isometric square parallel to center line, (b).

6. Transfer points 1, 2, 3 and 4 from semicircle to upper left quarter of isometric square by setting off appropriate distances, (c). Repeat for upper right quarter and project these intersections to two lower quarters.

7. Draw a smooth curve through points to form isometric ellipse, (d).

Arcs are constructed by the coordinate method in the same manner prescribed for circles. The required arc is divided into a number of equal parts and these points are projected to the isometric view. A smooth arc is then drawn through the projected coordinate points.

CONSTRUCTING IRREGULAR CURVES IN ISOMETRIC DRAWINGS

Irregular curves can be constructed in isometric drawings by using the coordinate method as shown in Fig. 11-18. The procedure is as follows:

1. Select a sufficient number of points on irregular curve in orthographic view to produce an accurate representation when transferred to isometric view, Fig. 11-18 (a). (Be sure to locate a point at each sharp break or turn.)

2. Draw coordinates through each point and parallel to two principal axes of orthographic view, (a).

3. Draw an isometric rectangle in corresponding plane of isometric view equal in size to one containing irregular curve in orthographic view, (b).

4. Draw isometric coordinate lines with spacing equal to coordinates in orthographic view, (b).

5. Draw a smooth curve through points of intersecting coordinates to form required irregular curve in isometric drawing, (c).

6. Project coordinate points to form thickness of object, (c).

SECTIONAL VIEWS IN ISOMETRIC

Sectional views in isometric are an effective means of graphically describing the interior of complex machine parts or assemblies. Half and full sections are frequently used, Fig. 11-19. When an isometric half section is used, the correct view

HALF SECTION

(a)

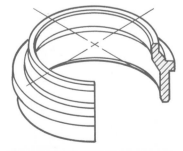

INCORRECT HALF SECTION

(b)

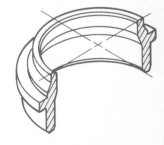

FULL SECTION

(c)

Fig. 11-19. Isometric half and full sectional views.

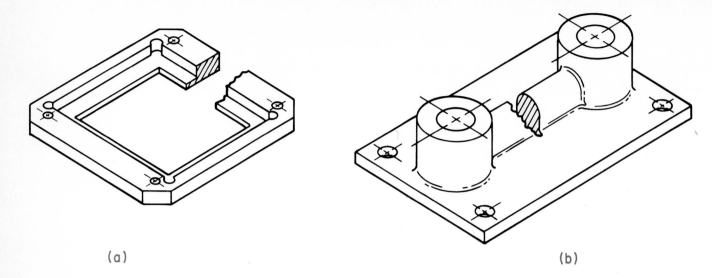

(a)

(b)

Fig. 11-20. Broken-out sections in isometric views.

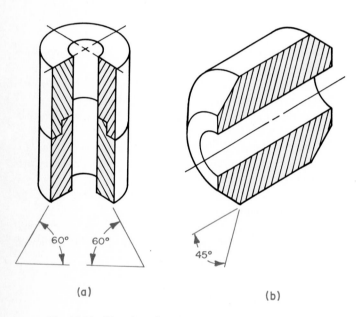

(a)

(b)

Fig. 11-21. Direction of section lining in isometric views.

is to position the part where both sides of the removed section are visible, (a). Occasionally a broken-out section is useful in showing a particular feature in an isometric view, Fig. 11-20.

Section lines are normally drawn at an angle of 60°, Fig. 11-21 (a). This most nearly resembles the 45° crosshatching in multiviews. The 60° angle should be changed if this causes section lining to be parallel or perpendicular to the visible outline of the object, (b).

The direction of section lining in a full section remains the same on all portions "cut" by the cutting plane, Fig. 11-21 (b), unless the view is of an assembly with several parts. Section lining in an isometric half section should be drawn to coincide if the two planes of the section were revolved together, Fig. 11-21 (a).

The steps in constructing an isometric sectional view are as follows:

1. Draw an isometric "box" the size of three overall dimensions of object. Lines should be very light construction weight, Fig. 11-22 (a).
2. Draw outline along cutting plane, (a).

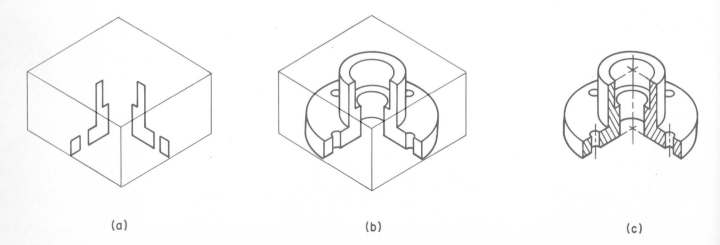

(a)

(b)

(c)

Fig. 11-22. Steps in constructing an isometric sectional view.

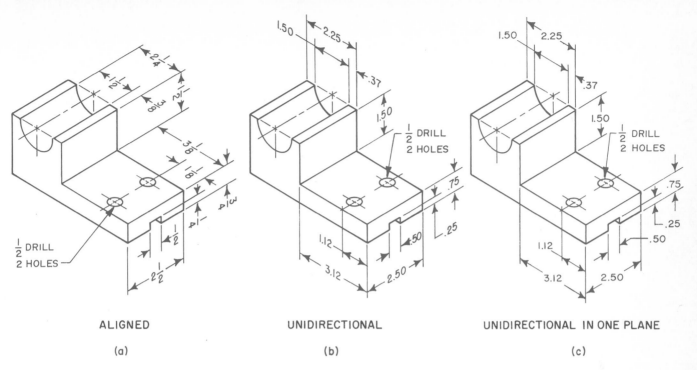

Fig. 11-23. Dimensioning isometric drawings.

3. Add remaining details, (b).
4. Erase construction lines and add crosshatching, (c).

ISOMETRIC DIMENSIONING

The general rules for dimensioning multiview drawings also apply to isometric drawings. The aligned or isometric plane, Fig. 11-23 (a), and the unidirectional system, (b and c), are the approved systems. The unidirectional system is gaining favor in industry because all dimensions are easily read from the bottom of the drawing. In either system, the dimension lines, extension lines and dimension figures should lie in the correct plane for the feature dimensioned. The dimension figures in the unidirectional system may all be shown in one plane for convenience and speed in drawing, (c). However, the dimension and extension lines must be properly aligned.

Notes in the aligned system lie in one of the principal planes, Fig. 11-23 (a). Notes in the unidirectional system are placed horizontally so that they lie in or are parallel to the picture plane, (b and c).

Some incorrect practices in isometric dimensioning are shown in Fig. 11-24 (a). The correct way to indicate these dimensions is shown in (b).

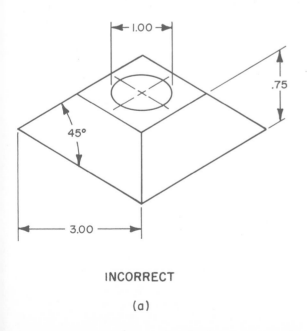

INCORRECT

(a)

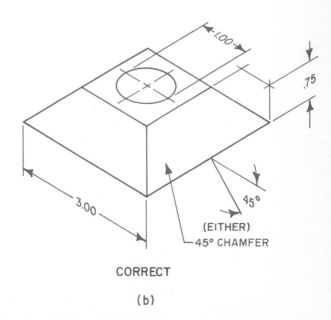

CORRECT

(b)

Fig. 11-24. Incorrect and correct methods of locating isometric dimensions.

REPRESENTING SCREW THREADS IN ISOMETRIC

The time consumed and clarity gained in the actual representation of screw threads in isometric does not justify the use of this procedure and is seldom used. The practice is to represent crest lines of threads with a series of isometric circles (ellipses) uniformly spaced, Fig. 11-25 (a). It is not necessary to duplicate the actual pitch of the thread. The dimension will identify thread characteristics. Shading may be used to increase effectiveness of the thread representation, (b).

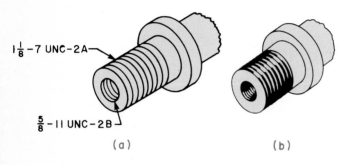

$1\frac{1}{8} - 7$ UNC-2A

$\frac{5}{8} - 11$ UNC-2B

(a)   (b)

Fig. 11-25. Representation of screw threads in isometric.   (ANSI)

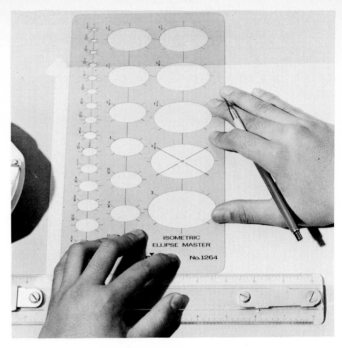

Fig. 11-26. Aligning ellipse template along isometric center lines to draw an isometric circle.

ISOMETRIC ELLIPSE TEMPLATES

As you have observed, the four-center approximate and coordinate methods of constructing isometric circles (ellipses) are very time-consuming. But it is important to know how to use these methods when isometric ellipse templates are not available.

When ellipse templates are available in the correct size, they should be used to speed the process of drafting and improve the appearance of the finished drawing. These templates are available in a variety of sizes. Their use simply requires alignment of the template along the isometric center lines of the circular feature, Fig. 11-26.

ALTERNATE POSITIONS OF ISOMETRIC AXES

It may be desirable to draw an object in isometric with the axis in a position other than normal. For example, the object could be viewed from below (called reverse position), Fig. 11-27 (a). Or, for long objects, it could be shown horizontally,

(c). The isometric axis may be located in any of a number of positions as long as equal spacing of 120° is maintained between the three axes.

CENTERING AN ISOMETRIC DRAWING

The technique of locating an isometric drawing in the center of a sheet or at any other location, is a matter of finding the center of the object and positioning this point in the desired location, Fig. 11-28. Match this center point with the desired center location on the sheet and the entire isometric drawing will be correctly positioned.

ADVANTAGES AND DISADVANTAGES OF ISOMETRIC DRAWINGS

The isometric drawing is one of the easiest to construct since the same scale is used on all axes. It has an advantange over orthographic projection in that three sides of the object may be shown in one view presenting a more realistic

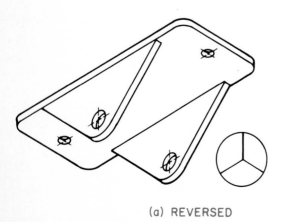

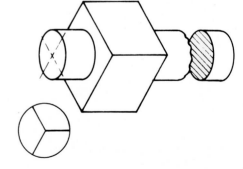

(a) REVERSED   (b) NORMAL POSITION   (c) HORIZONTAL

Fig. 11-27. Alternate positioning of isometric axes to reveal certain features of an object or to view object in its normal position.

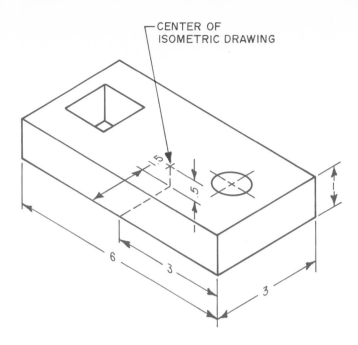

Fig. 11-28. Centering an isometric drawing in a desired location on a sheet.

representation of the object. Circles are not greatly distorted as in the receding views of an oblique drawing (discussed later in this chapter).

There are certain disadvantages inherent in isometric drawings. One is the tendency for long objects to appear distorted, because parallel lines on an object remain parallel rather than coverging toward a point as is the case when the actual object is viewed or seen in a perspective drawing. Also, the symmetry of an isometric drawing causes some lines to meet or overlap, confusing the reader of the drawing.

DIMETRIC PROJECTION AND DIMETRIC DRAWING

Dimetric projection is a type of axonometric projection. Dimetric differs from isometric in that only two of the axes make equal angles (any angle larger than 90° and less than 180° and not 120° — the isometric axes) with the plane of projection. The third axis makes either a larger or smaller angle.

The two axes making equal angles with the projection plane, or lines parallel to these, are foreshortened equally. The third axis and lines parallel to it are further foreshortened or enlarged depending on how the object is viewed. Like isometric projection, dimetric projection can be constructed by the revolution method or by the auxiliary view method.

CONSTRUCTING A DIMETRIC PROJECTION

The only difference between dimetric and isometric projection is in the angle (line of sight) the object is viewed. In Fig. 11-29, a cube is revolved in two views to the required line of sight for the two equal angle axes and projected from the orthographic to the dimetric projection. This produces a true measure of the cube as well as the angle of the dimetric axes.

Careful projection techniques will produce angles and

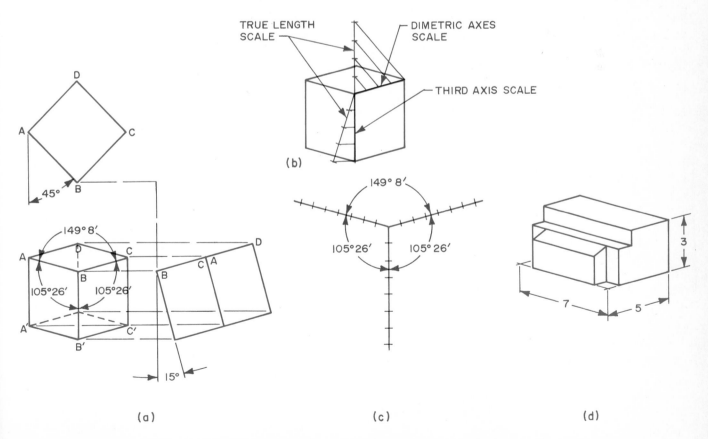

Fig. 11-29. Steps of procedure in constructing a dimetric projection.

Drafting for Industry

lengths accurate enough for most pictorial drafting requirements. Angles and measurements requiring mathematical accuracy should be calculated by trigonometry.

Given the cube and the angle of rotation of the dimetric axes, use the following procedure to find the angle of the dimetric axes and the true measure on all three axes in the dimetric projection:

1. Draw two views of cube (top and right side in our example) which are equally inclined to plane of projection, Fig. 11-29 (a).
2. Project points A, B, C and D to front view to form dimetric projection of cube.
3. Lay off actual scale of orthographic cube on a line at an angle with one of dimetric axes and a second line at an angle along third axis, (b). Divide each actual measurement line into a number of four equal parts, for example, and project these geometrically to divide each axis into equal parts proportionally.

The two dimetric axes will be foreshortened, but will have the same measure. The third axes will be further

foreshortened due to the line of sight for this particular cube. (In some cases, third axis may appear longer than dimetric axes, depending on line of sight.)

4. Transfer dimetric units of measure to dimetric scale, (c). This scale can be extended with like units for measuring greater lengths, and scale can become a permanent scale for measuring any dimetric projection drawn at this line of sight. An example of a dimetric projection of an object drawn at same angle as cube and using the dimetric scale just produced is shown at (d).

The number of variations in the angle of sight of a dimetric projection is infinite. This permits a view of an object which best portrays its features. It must be remembered that one of the angles of sight must be 45° (direction in which it is desired leg of "Y" axes should lie). The other angle of sight may be any angle other than 35° 16'. A dimetric projection of a gyro instrument is shown in Fig. 11-30.

A dimetric projection can also be constructed by the successive auxiliary view method as shown in Fig. 11-31. This method is discussed in Chapter 12.

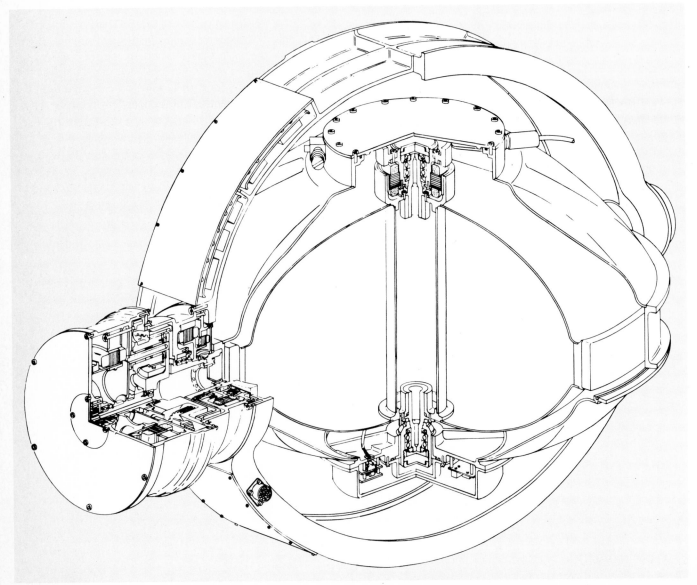

Fig. 11-30. A dimetric projection with a cutaway section.   (Sperry Flight Systems Div.)

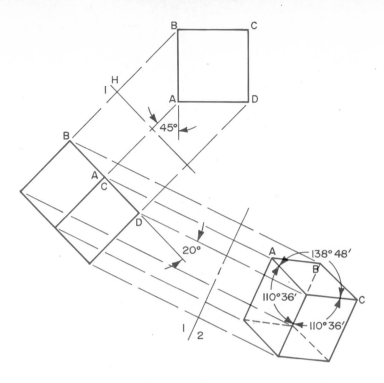

Fig. 11-31. Constructing a dimetric projection using the successive auxiliary view method.

Many industries modify their dimetric projection drawings in the interest of time. They construct an approximate dimetric drawing where regular scales and angles common to the triangles or compass are used. The full-size scale and the three-quarter or half scale are frequently selected for the dimetric axes or the third axis, depending on which is to be reduced. Construction of an approximate dimetric drawing is discussed in the next section.

CONSTRUCTING AN APPROXIMATE DIMETRIC DRAWING

An approximate dimetric drawing (not a true projection) can be drawn by substituting angles closely approximating those of a true projection. These angles can easily be achieved with the protractor or adjustable triangle.

The regular full-size scale or a foreshortened scale can be used on the different axes depending on their positions in the dimetric view. Some common axes for approximate dimetric drawings are shown in Fig. 11-32 along with suggested scales for the axes.

Approximate dimetric drawing should not be attempted until the student is familiar with the principles of dimetric projection. One must have a "feel" for the angle of sight and the proportioning of measurements on the axes.

TRIMETRIC PROJECTION AND TRIMETRIC DRAWINGS

In trimetric projection all three axes make different angles with the plane of projection. Angles must be selected which best portray the features of the object in the line of sight desired. Construction of a trimetric projection is essentially the same as a dimetric projection, except no two axes or surfaces are viewed alike on the plane of projection.

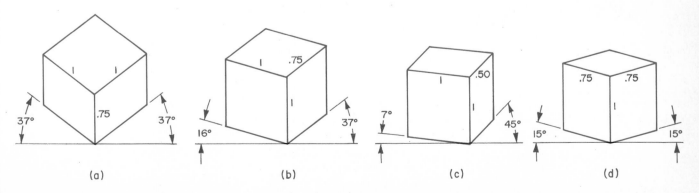

Fig. 11-32. Some common axes and scales for constructing approximate dimetric drawings.

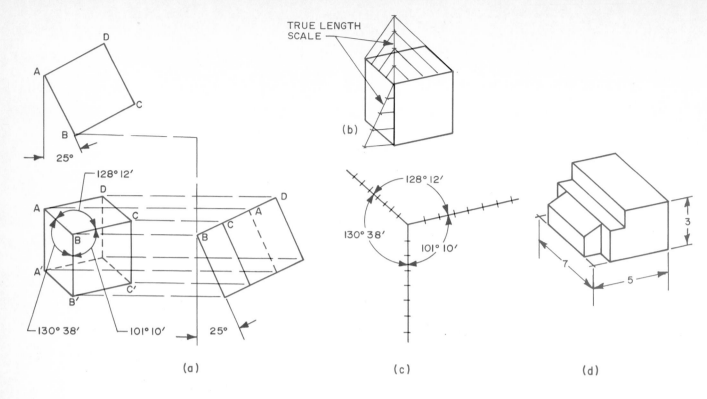

Fig. 11-33. Steps of procedure in constructing a trimetric projection.

CONSTRUCTING A TRIMETRIC PROJECTION

The construction of a trimetric projection is shown in Fig. 11-33. The same steps of procedure may be used except the cube has now been inclined at unequal angles on the viewing plane, (a). The trimetric scale is constructed in the same manner as the dimetric, except a scale must be developed for each axis, (c). The object shown in Fig. 11-29 (d) has been constructed as a trimetric projection in Fig. 11-33 (d).

CONSTRUCTING AN APPROXIMATE TRIMETRIC DRAWING

The construction of an approximate trimetric drawing is similar to that of an approximate dimetric drawing. It should not be attempted, however, until the student becomes thoroughly familiar with true trimetric projection. Fig. 11-34 illustrates some common axes for approximate trimetric drawings.

ELLIPSE GUIDE ANGLES FOR DIMETRIC AND TRIMETRIC DRAWINGS

The four-center approximate method for constructing ellipses in isometric is satisfactory for the dimetric axis in which the scale is the same on both edges. Ellipses may be drawn by the coordinate plotting method on all axes.

However, the most satisfactory method of drawing ellipses in dimetric and trimetric drawings is to find the angle from which a circle is viewed in a particular plane; then select the appropriate ellipse template to use in the drawing. This is accomplished by studying auxiliary views (see Chapter 12). Simply find the angle between the edge view of the principal planes in the dimetric or trimetric projection and the line of sight, as shown in Fig. 11-35.

The ellipse angle for a dimetric drawing is identical for the two dimetric axes and different for the third. The ellipse angle differs for each of the three trimetric axes.

Given a trimetric projection of the cube in Fig. 11-33, you can find the ellipse guide angle and direction of the major diameter of the ellipse on each principal plane of the trimetric cube by proceeding as follows:

1. Line OE is perpendicular to plane ABCO, OC is perpendicular to AOEF and OA to OCDE (planes of a cube are perpendicular to each other), Fig. 11-35 (a).
2. Construct true length lines GH, HJ and JG on each plane by

Fig. 11-34. Some common axes and scales for constructing approximate trimetric drawings.

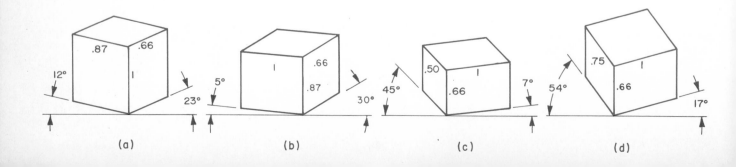

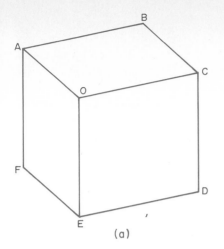

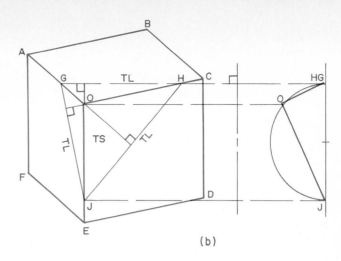

(a)

(b)

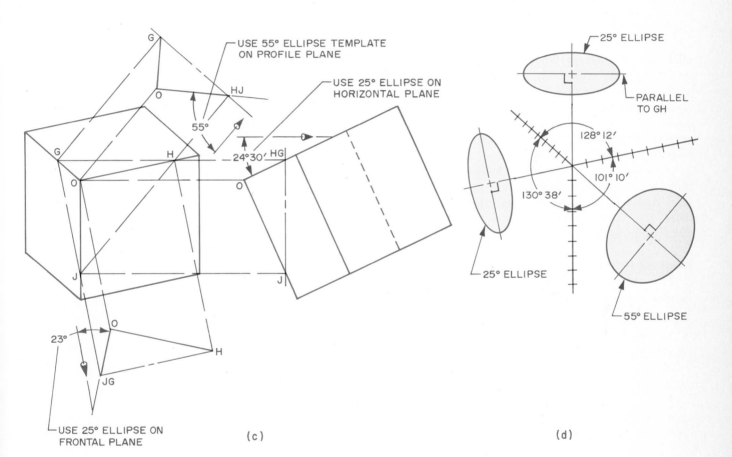

USE 55° ELLIPSE TEMPLATE
ON PROFILE PLANE

USE 25° ELLIPSE ON
HORIZONTAL PLANE

55°

24°30'

23°

USE 25° ELLIPSE ON
FRONTAL PLANE

(c)

25° ELLIPSE

PARALLEL
TO GH

128° 12'

101° 10'

130° 38'

25° ELLIPSE

55° ELLIPSE

(d)

Fig. 11-35. Determining ellipse guide angles for dimetric and trimetric drawings.

drawing them perpendicular to extensions of lines OC, OA and OE. When two lines known to be perpendicular appear on the drawing as perpendicular, one line, or both, is true length, (b).

3. Intersection of true length lines form true size plane GHJ. (A plane whose edges appear true length will appear true size.)

4. Plane GHJ is a frontal plane. This plane will project as a vertical edge in side view, (b).

5. To find position of 90° angle of cube, construct a semicircle with a diameter equal in length to edge view of plane GHJ. Then project point O to semicircle and inscribe position of 90° corner of cube. (Lines from end points of

diameter of a semicircle, joined at a point on semicircle, form a right angle.)

6. Draw auxiliary views showing edge view of true size plane by drawing point views of lines GH, HJ and JG, (c).

7. Angle between edge view and line of sight for a particular view is ellipse angle.

8. Position ellipse guide so major diameter is parallel to true length line on that particular plane, (d).

Circles will appear as ellipses in dimetric and trimetric drawings and the major axis of the ellipse will appear as a true length line. The direction of true length lines on the principal planes of the trimetric drawing in Fig. 11-35 may be found by locating a true size plane in the frontal view, (b). A plane which appears true size in the frontal view will appear as an edge in the profile view. Draw this edge view of the true size plane parallel to the vertical edges in the frontal view.

ADVANTAGES AND DISADVANTAGES OF DIMETRIC AND TRIMETRIC DRAWINGS

Dimetric and trimetric drawings permit an infinite number of positions for a machine part or other product to be viewed pictorially. This is of particular advantage in showing certain features of the object.

The greatest disadvantage of dimetric and trimetric drawings is that they are considerably more difficult to construct. However, once a desired position has been decided upon and a scale developed, including the angle and location of ellipses, the dimetric and trimetric drawings can be produced quite easily.

OBLIQUE PROJECTIONS AND DRAWINGS

An oblique projection or drawing is a type of pictorial drawing somewhat like the axonometric drawing (isometric, dimetric and trimetric). Only one plane of projection is used and, while the lines of sight are parallel to each other, they meet the plane of projection at an oblique angle, Fig. 11-36.

The object is positioned with its front view parallel to the plane of projection so this view, and surfaces parallel to it (axes X and Z), will project in their true size (to scale) and shape. The top and side views are viewed at an oblique angle and are distorted along the Y axes.

OBLIQUE PROJECTION

True oblique projection is obtained by the projection techniques shown in Fig. 11-36. Lines of sight are selected at an angle suitable to show the desired features of the object. The front view should be the one most characteristic of the object.

At least two orthographic views are necessary to construct an oblique projection. The top and right side view have been used in Fig. 11-36. All points are projected to the projection plane by oblique projectors parallel to the line of sight. From the plane of projection, the points are projected horizontally or vertically to intersect with their corresponding point. These points are connected to form the required oblique projection.

Lines along the X and Z axes will intersect at right angles and the angle formed by the Y axis and the horizontal is governed by the lines of sight selected in the two orthographic views. The true size of this angle can be found by methods explained in Chapter 12.

Oblique projection of a plain block is a relatively easy procedure. However, with more complex machine parts or industrial products, the oblique projection is a time-consuming procedure and little use is made of this method in industry.

OBLIQUE DRAWING

The oblique drawing is based on the principles of oblique projection, although its construction is much faster and the results often more satisfactory than oblique projection itself.

There are three types of oblique drawings: cavalier, cabinet and general. The three differ only in the ratio of the scales used on the front axes (X and Z) and the receding axis (Y).

The CAVALIER oblique is a drawing based on an oblique projection in which the lines of sight make an angle of 45° with the plane of projection. In a true cavalier projection, the receding lines along the Y axis project in their true length. Therefore, the cavalier oblique drawing is usually drawn with a receding axis of 45° (approximating a line of sight of 45°) and the same scale is used on all three axes, Fig. 11-37.

The equal scale on all axes is the principal advantage of the cavalier oblique drawing over other types. However, it presents a distorted appearance for objects whose depth approaches or exceeds the width. Unless otherwise indicated, an oblique

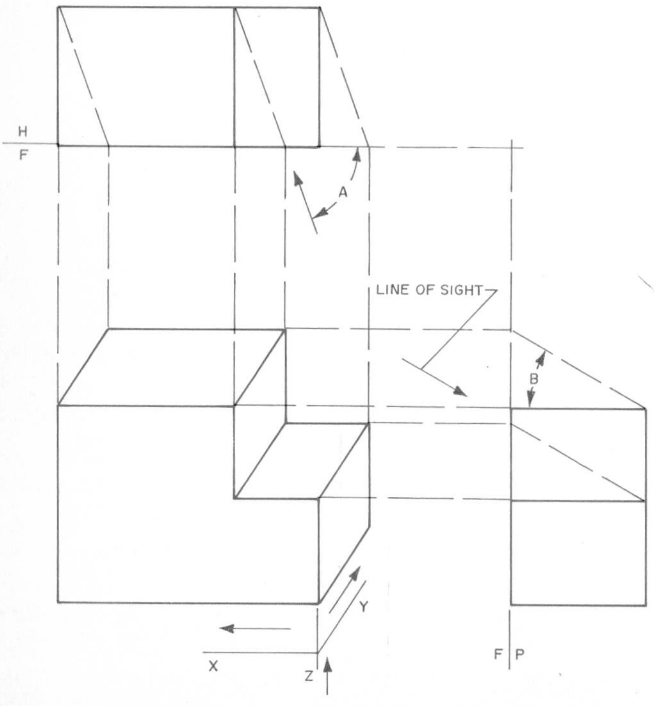

Fig. 11-36. Construction of an oblique projection.

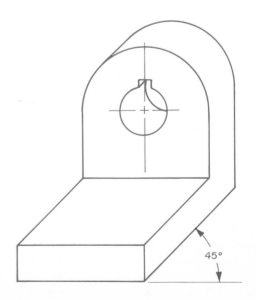

Fig. 11-37. A cavalier oblique drawing.

drawing refers to a cavalier oblique.

The CABINET oblique drawing is based on an oblique projection in which the lines of sight make an angle of 63° 26'. In this type of projection the receding lines project one-half their true length. Therefore, the scale on the receding axis of the cabinet oblique drawing is one-half that for the other axes, Fig. 11-38.

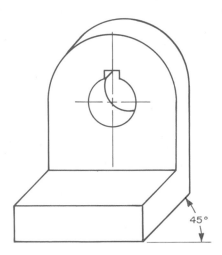

Fig. 11-38. A cabinet oblique drawing.

The GENERAL oblique drawing is based on a type of oblique projection in which the line of sight is other than 45° or 63° 26', and the scale on the receding axis is other than one-half or full size. The general oblique drawing may be drawn at any angle between 0° and 90° and a scale between one-half and full used on the receding axis, Fig. 11-39.

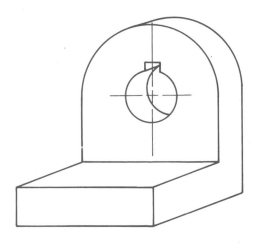

Fig. 11-39. A general oblique drawing.

While oblique drawings may be drawn with a receding axis at any angle between 0° and 90°, the most common angles are 45° and 30° with the horizontal since these can readily be drawn with triangles. The three types differ mainly in the ratio of the scale used on the receding (Y) axis compared to the other axes, Fig. 11-40.

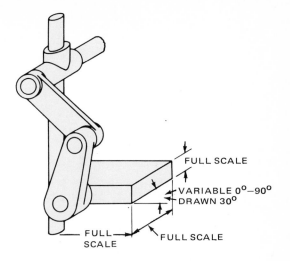

OBLIQUE PROJECTION — CAVALIER

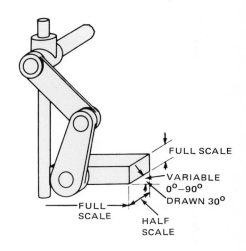

OBLIQUE PROJECTION — CABINET

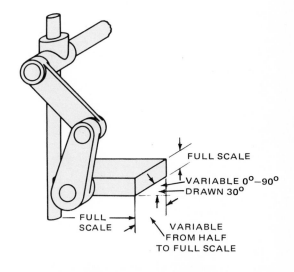

OBLIQUE PROJECTION — GENERAL

Fig. 11-40. A comparison of types of oblique drawings. (ANSI)

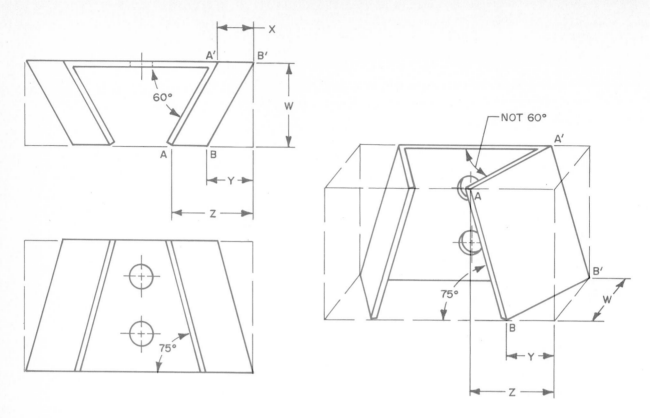

Fig. 11-41. Laying off angles in oblique drawings.

ANGLES IN OBLIQUE DRAWINGS

Angles which are shown true size in the frontal orthographic view will appear true size in the frontal plane of the oblique. Angles which lie in planes other than the frontal plane must be located by finding the end points of the lines forming the angle, Fig. 11-41. The top and front views of the object are enclosed in rectangles which outline the overall size of the object and are used in laying out the oblique view.

The angles, lines AA' and BB' made with the back of the object, are found by setting off the appropriate measurements from the orthographic views on the oblique cube. All measurements must be made along the oblique axes or lines parallel to these axes and foreshortened on the Y axes for the cabinet and general oblique drawings.

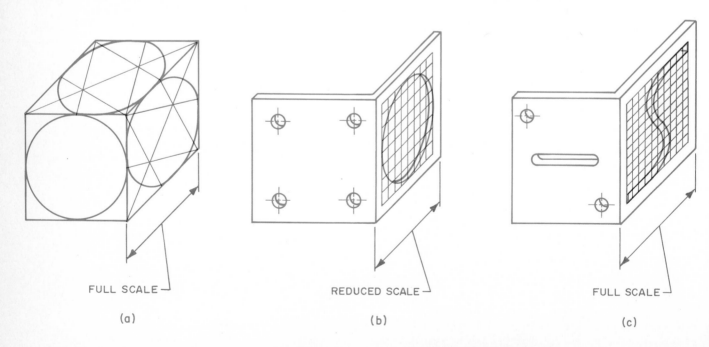

FULL SCALE

(a)

REDUCED SCALE

(b)

FULL SCALE

(c)

Fig. 11-42. Constructing arcs, circles and irregular curves in oblique drawings.

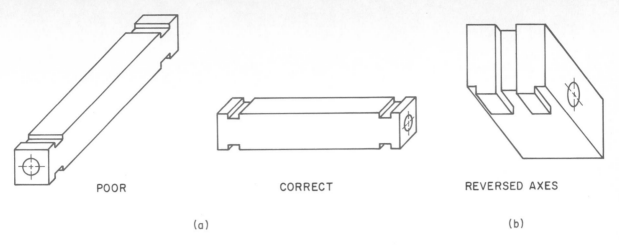

POOR   CORRECT   REVERSED AXES

(a)   (b)

Fig. 11-43. Selecting positions for oblique drawings.

ARCS, CIRCLES AND IRREGULAR CURVES IN OBLIQUE DRAWINGS

Arcs and circles located in the frontal plane of an oblique drawing will appear in their true shape. When arcs and circles are located in the principal cavalier oblique planes other than the frontal, these may be drawn using the four-center approximate method, Fig. 11-42 (a).

Arcs and circles for cabinet and general oblique drawings are drawn in their true shape in the frontal plane. They must then be transferred to the other oblique planes using the coordinated method and oblique "squares," due to the foreshortening of the Y axes, (b). Irregular curves are also transferred to oblique drawings by means of the coordinate method, (c).

SELECTION OF POSITION FOR OBLIQUE DRAWINGS

One of the advantages of oblique drawings is that arcs and circles are drawn in their true shape when they are located in the frontal plane. When possible, consideration should be given to selecting the view containing arcs and circles as the front view.

There may be other reasons, however, why this choice of front view may not be the best. For example, elongated objects should be located parallel to the frontal view to minimize distortion, Fig. 11-43 (a). It may also be desirable, because of certain features, to view the object from below. This can be done by drawing the A and Z axes in their normal positions, but drawing the receding axis Y down at the oblique angle instead of up, (b).

SECTIONAL VIEWS IN OBLIQUE

Sectional views may be used in an oblique drawing to provide a better view of interior detail, Fig. 11-44. The half section is used more often than the full section because the latter usually does not show sufficient exterior detail of the part. Correct positioning of the part is as important for oblique sections as it is for exterior oblique drawings.

OBLIQUE DIMENSIONING

Dimensioning of oblique drawings is similar to the dimensioning of isometric drawings. Approved practice calls for the dimension and extension lines to lie in the correct plane for the feature dimensioned, Fig. 11-45. The aligned or unidirectional system may be used.

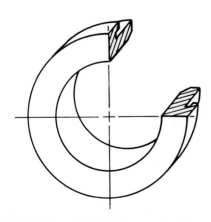

Fig. 11-44. An oblique half section of a ring seal.

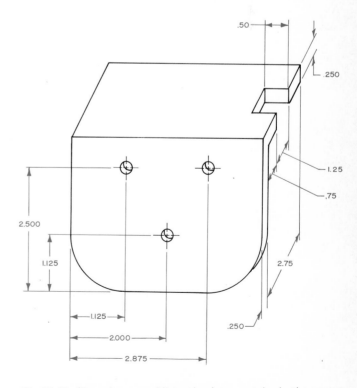

Fig. 11-45. Dimensions on oblique drawings must be in the correct plane for the feature dimensioned.

211

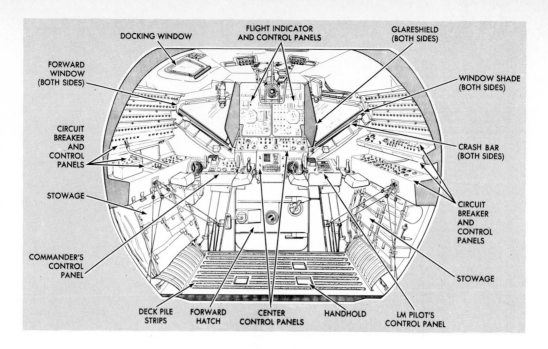

Fig. 11-46. Perspective drawing of crew compartment interior of a space vehicle ascent stage. (NASA)

ADVANTAGES AND DISADVANTAGES OF OBLIQUE DRAWINGS

The oblique drawing has the advantage of showing an object in its true shape in the frontal plane. Therefore, it may be a faster method of pictorial presentation when arcs and circles are located in this plane. While the cavalier oblique is somewhat distorted on the receding planes, this is compensated for in the cabinet and general oblique drawings with reduced scales on the Y axis.

A definite limitation of the cabinet and general oblique drawing is that circular and curved features on the receding planes must be drawn by the time-consuming coordinate method. Long objects appear distorted when it becomes necessary to draw these features on the Y axis, particularly in the cavalier oblique. When cabinet and general oblique drawings are used, a second scale must be employed on the Y axis.

PERSPECTIVE

Perspective drawings are the type of pictorial drawings that most nearly represent what is seen by the eye or camera. In pictorial drawings studied earlier, parallel lines remained parallel. In perspective drawings, however, parallel lines tend to converge as they recede from the person's view, Fig. 11-46.

The three basic types of perspective drawings are the one-point (parallel), two-point and three-point methods, so named for the number of vanishing points required in their construction, Fig. 11-47. The construction of the first two of these basic types is discussed in this section.

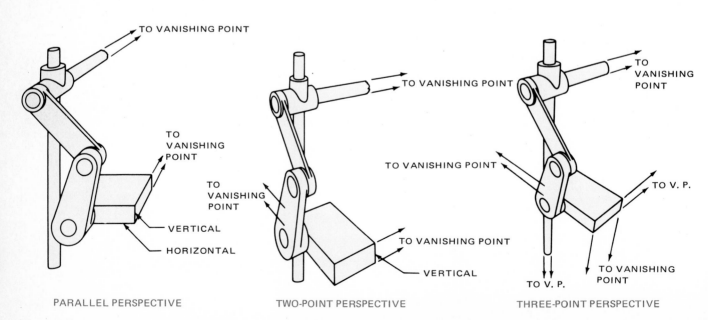

Fig. 11-47. A comparison of one-point, two-point and three-point perspectives. (ANSI)

The principles of perspective are illustrated in this photo of train tracks. The rails appear
to get closer in the distance.

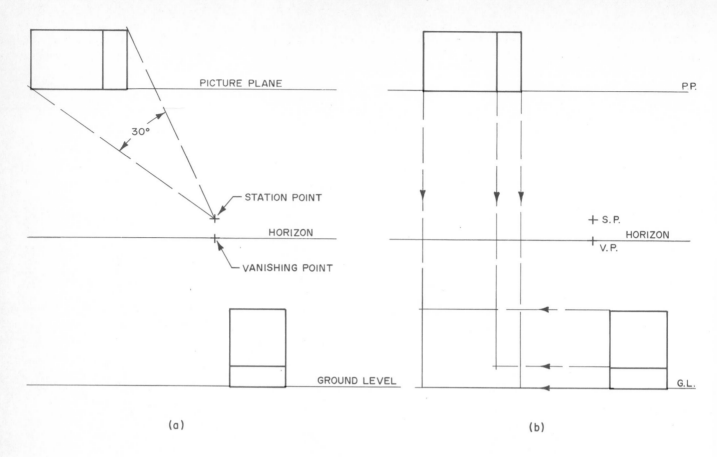

PICTURE PLANE

30°

STATION POINT

HORIZON

VANISHING POINT

GROUND LEVEL

(a)

P.P.

+ S.P.

HORIZON

V.P.

G.L.

(b)

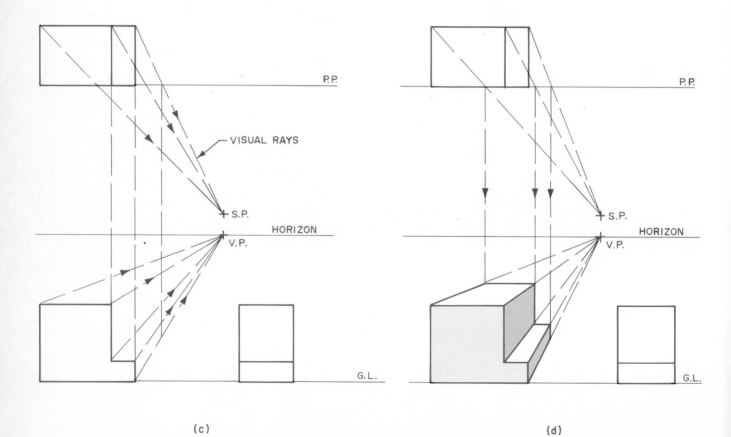

P.P.

VISUAL RAYS

+ S.P.

HORIZON

V.P.

G.L.

(c)

P.P.

+ S.P.

HORIZON

V.P.

G.L.

(d)

Fig. 11-48. Construction of a one-point, or parallel, perspective drawing.

Pictorial Drawings

TERMINOLOGY IN PERSPECTIVE DRAWINGS

There are certain terms that must be defined before a discussion of perspective drawings can proceed. Those terms commonly used are presented here and illustrated for further clarification in Fig. 11-48.

1. STATION POINT is an assumed point representing the position of the observer's eye, sometimes called the "point of sight." The location of this point greatly affects the perspective produced. To select the best view, locate the station point (SP) at a distance from the object to form a viewing angle of the entire object of approximately 30°. See Fig. 11-48 (a). Move the station point to the right or left depending on the particular view of the object to be emphasized. The elevation of the station point is on the horizon line and determines whether the object is viewed from above, on center or from below.

2. VANISHING POINTS are points in space where all parallel lines, which are not parallel to the picture plane, meet. For horizontal parallel lines, the vanishing points (VP) are always located on the horizon line, as shown in the drawing

Fig. 11-48 (c).

3. VISUAL RAYS are lines of sight from the object to the station point. They represent the light rays that produce an image in the eye, (c).

4. PICTURE PLANE is the projection plane on which the perspective is viewed. It may be aligned with the object or be in front of it. This is a vertical plane for most perspectives, and appears as a line in the top view and as a plane in the front view. It could appear, however, in any position. For example, for a "bird's eye" perspective, the plane would be horizontal in the top view and appear as an edge in the front view.

5. HORIZON LINE is a horizontal line on which the vanishing points are located and where receding lines tend to converge. It may appear at any level; above, behind or below to produce the perspective view desired. Fig. 11-49 illustrates the effect of various levels of the horizon line and positions of the station point.

6. GROUND LINE is the base line or position of rest for the object.

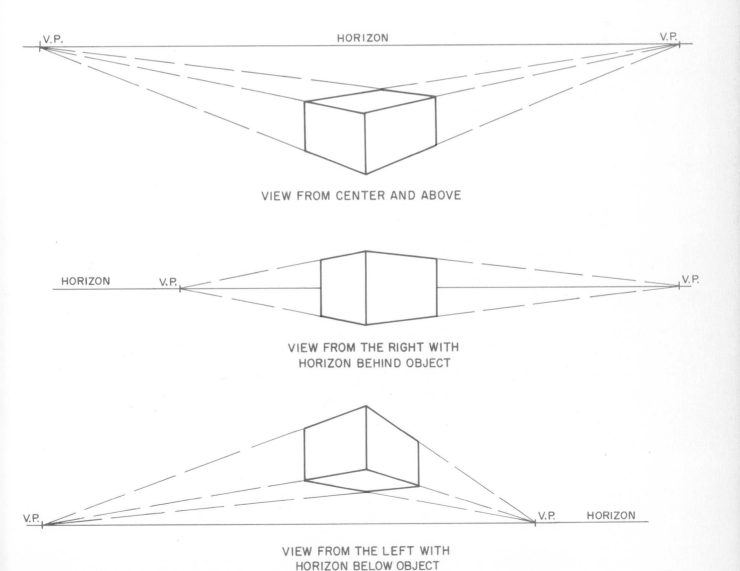

VIEW FROM CENTER AND ABOVE

VIEW FROM THE RIGHT WITH
HORIZON BEHIND OBJECT

VIEW FROM THE LEFT WITH
HORIZON BELOW OBJECT

Fig. 11-49. Effect of varying positions of horizon line and station point on perspective drawings.

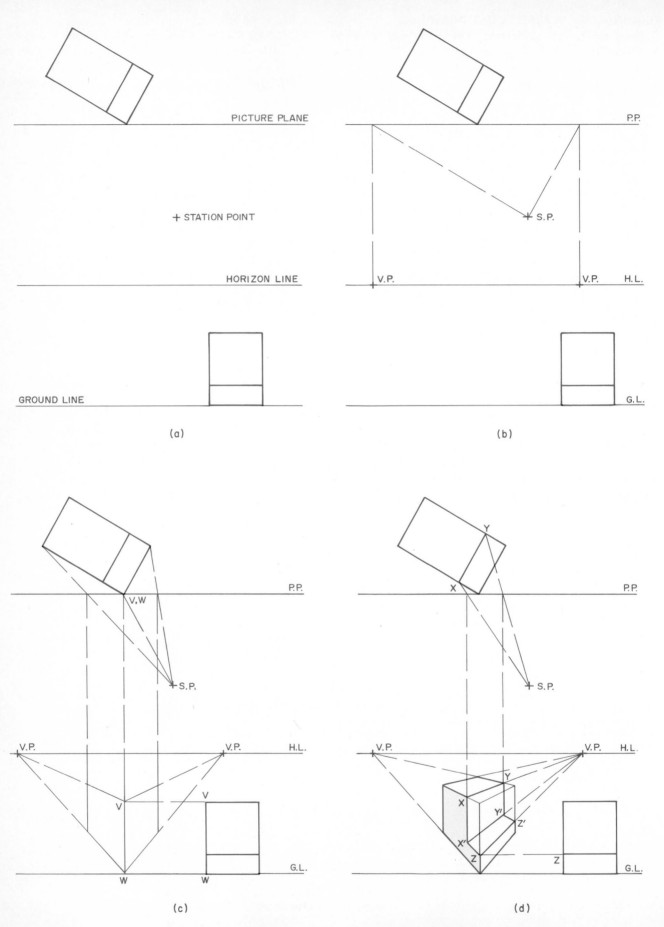

PICTURE PLANE

+ STATION POINT

HORIZON LINE

GROUND LINE

P.P.

+ S.P.

V.P.　　　　　　V.P.　　H.L.

G.L.

(a)　　　　　　　　　　　　(b)

P.P.

V,W

+ S.P.

V.P.　　　　　　V.P.　　H.L.

V　　　　　V

V

G.L.

W　　　W

(c)

Y

X

P.P.

+ S.P.

V.P.　　　　　　　V.P.　　H.L.

Y

X

Y'

Z'

X'

Z

Z

G.L.

(d)

Fig. 11-50. Construction of a two-point or angular perspective drawing.

ONE-POINT OR PARALLEL PERSPECTIVE

The one-point perspective has only one vanishing point, and the frontal plane of the object is parallel to the picture plane. Because of the latter, this perspective is also known as a parallel perspective.

Given the top view parallel to the picture plane and side view of the object, the station point, horizon line and ground line, the construction procedure for one-point perspective is as follows:

1. Locate vanishing point on horizon line directly below station point, Fig. 11-48 (a).
2. Construct frontal plane of object by projections from top and side views. This will be a true size (scale) projection since surface lies in picture plane, (b).
3. Draw projectors from rear points of object (top view) to station point, (c).
4. Draw projectors from points on front view to vanishing point, (c).
5. Construct depth of perspective view by drawing vertical projectors from points where projectors from top view to station point cross picture plane, (d). These vertical projectors intersect projectors to vanishing points to form one-point or parallel perspective.

The one-point perspective is particularly useful in illustrating the interior of a room or structure. See Fig. 11-46.

TWO-POINT OR ANGULAR PERSPECTIVE

In the two-point or angular perspective (called angular because of angle object makes with picture plane), two sets of principal planes of the object are inclined to the picture plane. The third set (usually top view) remains parallel to its plane of projection. Parallel lines of the two sets of inclined planes converge at vanishing points on the horizon line, Fig. 11-50.

Given the top and side views and the station point, the construction procedure for the two-point perspective is as follows:

1. Draw picture plane, horizon line and ground line, Fig. 11-50 (a).
2. Draw projectors from station point parallel to forward edges of object to intersect picture plane, (b).
3. Project these points vertically down to horizon line to establish two vanishing points.
4. From where it appears at picture plane, project line VW to ground line, (c).
5. Line VW is parallel to and lies on picture plane and, therefore, appears true length in perspective when projected from side view.

6. Project end points of line VW to left and right vanishing points to establish two of perspective planes.
7. Draw projectors from exterior corners of object in top view to station point.
8. Project intersections of these projectors with picture plane to perspective view to establish limits of object.
9. Draw projectors from points X and Y to station point. Where these projectors cross picture plane, drop vertical projectors to locate these lines in perspective view, (d).
10. Project point Z to perspective true length corner line and from there toward vanishing points to complete shoulder cut in object.
11. Complete remaining lines of perspective as shown; erase construction lines.

The two-point perspective drawing is most useful in drawing perspectives of large structures such as buildings in architectural work or engineering projects such as bridges or petroleum plant piping installations.

PERSPECTIVE OF OBJECTS LYING BEHIND PICTURE PLANE

When objects lie behind and do not touch the picture plane, the lines of the object must be extended to the picture plane to establish their piercing points, Fig. 11-51. From these piercing points, verticals are dropped to the ground line to establish true length lines from which receding lines are drawn to the vanishing points. On these receding lines, lengths are projected from the intersection of visual rays with the picture plane.

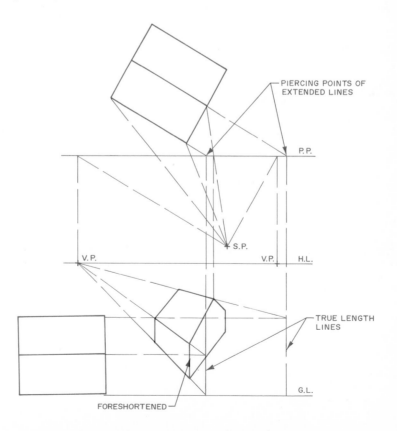

Fig. 11-51. Lines on objects which lie behind picture plane are foreshortened.

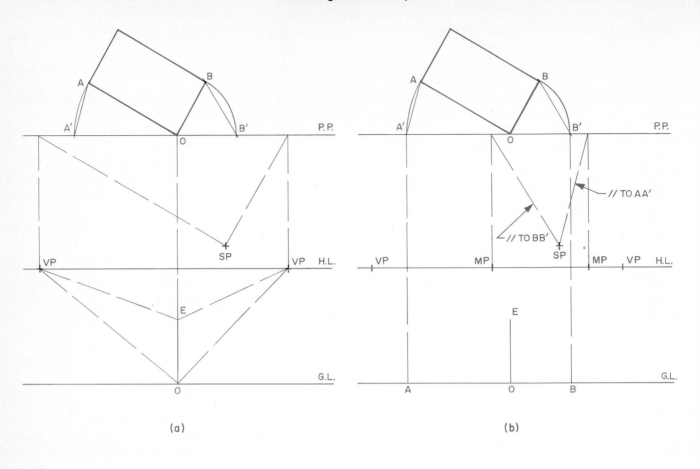

(a)

(b)

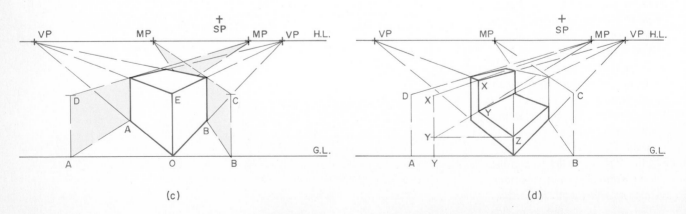

(c)

(d)

Fig. 11-52. Establishing measuring points in construction of a perspective drawing.

Pictorial Drawings

MEASURING-POINT SYSTEM IN PERSPECTIVES

The measuring-point system is a means of making accurate measurements in perspective drawings without having to retain the top view once the vanishing points and measuring points have been established, Fig. 11-52 (c). That is, actual measurements may be made on the ground line rather than projecting these from the top view. The process of locating measurements on the perspective view is more direct; large top views that require considerable drawing table space (architectural plan, for example) may be removed.

Given top view and its position, station point, horizon line and ground line, follow this procedure for drawing two-point perspective using measuring-point system as follows:

1. Locate vanishing points as in a regular two-point perspective. See Fig. 11-50 (b).
2. Draw front corner line OE true length from known height measurement in side view, and project its planes to their respective vanishing points, Fig. 11-52 (a).
3. Using point O in top view (corner of block on picture plane) as a center of radius, revolve edges OA and OB into picture plane.
4. Draw lines AA' and BB'.
5. Draw lines parallel to AA' and BB' from picture plane and passing through station point, (b).
6. Project points of intersection with picture plane in step 5

to horizon line to establish measuring points.
7. Project lines A'O and OB' to ground line to form true length lines AO and OB. (These lines have been revolved into picture plane where they appear true length. They could have been laid off true length on ground line by direct transfer of measurements from top view.)
8. Project true length height of AD and BC from OE, (c).
9. Extend planes AD and BC to their respective measuring points. Intersections of these planes with receding planes from corner OE establish limits of latter.
10. Complete perspective of block by drawing remaining vertical or receding lines.
11. Measurements for other features may be laid off true length on ground line from known measurements in top, as an example, shoulder cut X-Y-Z in (d).

CIRCLES IN PERSPECTIVE

Circles which are parallel to the picture plane will appear as circles in perspective drawings, Fig. 11-53. If the circle lays on the picture plane, it will be true size. If the circle is on a plane behind the picture plane, it will appear in a reduced size but as true circle (rear surface of block), Fig. 11-53. These circles may be drawn with a compass or circle template by locating their centers in the perspective view and determining their diameters by projection.

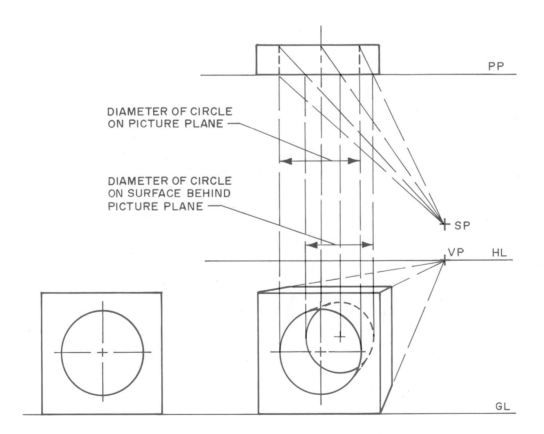

Fig. 11-53. Constructing circles on surfaces parallel to picture plane in perspective drawings.

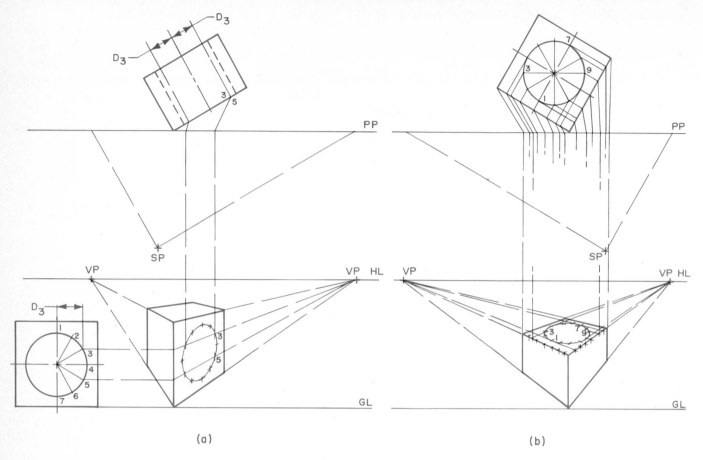

Fig. 11-54. Constructing circles on surfaces inclined to picture plane in perspective drawings.

(a)

(b)

Circles that lay on a plane inclined to the picture plane will appear elliptical in the perspective view and may be located by means of the enclosing square or coordinate method, Fig. 11-54.

In Fig. 11-54 (a), the technique for drawing a perspective representation of a circle in a vertical plane inclined to the picture plane is shown. Given the usual information, the procedure is as follows:

1. Divide orthographic front view into a number of parts (30°–60° divisions are suggested).
2. Transfer these division points to their respective locations in top view.
3. Project points from front view to front corner of block containing circle in perspective view and from there to vanishing point.
4. Project division points from top view to picture plane by visual rays to station point and from picture plane, vertically, to perspective view to intersect with their corresponding lines.
5. Draw perspective ellipse with use of an irregular curve or an ellipse template.

The construction of circles on horizontal surfaces in perspective drawings is shown in Fig. 11-54 (b). Given the usual information, the procedure is as follows:

1. Divide circle in top view into a number of parts (30°–60° divisions are suggested).
2. Project these division points to front two edges of block on

lines parallel to adjacent sides.

3. Establish height of block on picture plane corner in perspective (true length since corner is on picture plane) and draw receding lines to vanishing points.
4. Project division points from top view to picture plane by visual rays to station point and from picture plane, vertically, to perspective view to intersect with their corresponding front edge lines.
5. From these points of intersection on frontal edges, draw receding lines to two vanishing points. Points of intersection of corresponding lines are points on ellipse.
6. Draw perspective ellipse using an irregular curve or an ellipse template.

IRREGULAR CURVES IN PERSPECTIVE

Irregular curves may be drawn in perspective by means of the coordinate method, Fig. 11-55. Points are located along the curve in the orthographic views and projected to the picture plane and vertical true length line in the perspective.

PERSPECTIVE GRID

A perspective grid is a graph-oriented method of making accurate perspective drawings without having to establish (and project from) vanishing, measuring and sighting points, Fig. 11-56. There are many variations, such as the one-point method for interior views, the three-point oblique, the cylindrical grid for representing aircraft fuselages and others.

220

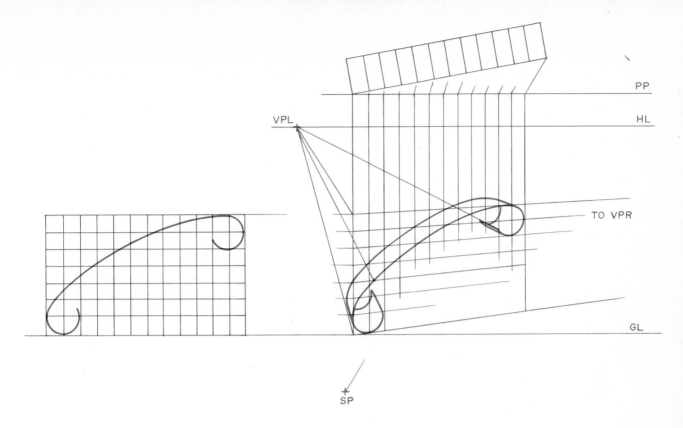

Fig. 11-55. Constructing irregular curves in perspective.

However, the cube grid is the most widely used for general purpose illustration, Fig. 11-57.

Grids include a scale from which measurements may be projected on the grid to maintain accurate representation of proportion in the object being drawn. Instead of actually drawing on the grid, an overlay sheet of tracing paper or vellum is used to preserve the grid for further use as well as to eliminate the grid pattern on the perspective drawing.

Grids save time in the preparation of perspective drawings and permit several draftsmen-illustrators to draw related

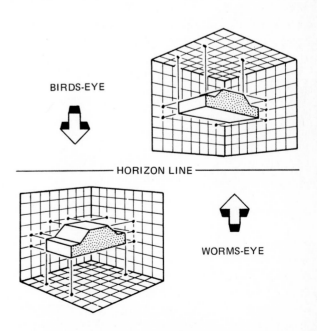

Fig. 11-56. A perspective drawing produced by means of a grid.

Fig. 11-57. A cube-type grid showing projection of an object. (General Motors Drafting Standards)

components independent of each other and be sure they will match with respect to size, angle and perspective. Perspective grids also permit additions or revisions to drawings at a future date with speed and accuracy. Perspective grids may be purchased at local drafting and art supply stores.

PERSPECTIVE DRAWING BOARDS

Special perspective drawing boards are available that also speed the process of constructing perspective drawings, Fig. 11-58. These boards are fairly simple to use and come with a variety of scales, permitting direct reading for layout of perspectives. Perspective boards are most valuable to the draftsman-illustrator who makes frequent use of perspective drawings in his work.

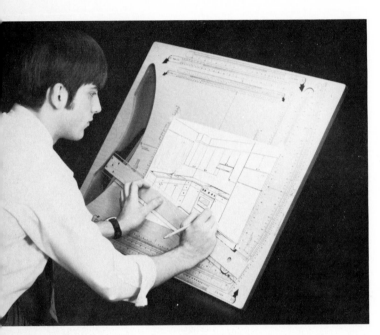

Fig. 11-58. Using a Klok perspective drawing board.
(Modulux Div.)

SKETCHING IN PERSPECTIVE

The technique of sketching in perspective is based upon an understanding of the principles of sketching discussed in Chapter 3. Also, the mechanical projection of perspective drawings discussed earlier in this chapter.

Developing a technique for sketching in perspective is dependent on good proportion of the objects being represented. The steps in sketching a two-point or angular perspective are as follows:

1. Establish two vanishing points on horizon line as far apart as desired, Fig. 11-59 (a).
2. Sketch a true length (scale) vertical line to represent front corner of object to be sketched. For illustration (a), this has been centered between two vanishing points and located below horizon line. This positions front faces of object at 45° with picture plane and a view from above object.
3. Sketch lines from ends of front vertical line to two vanishing points.

4. Sketch vertical lines at a distance one-half true distance (scale) from front vertical line to establish length of two frontal planes.
5. From upper-rear corners of these planes, sketch receding lines to opposite vanishing points to form top surface of object.

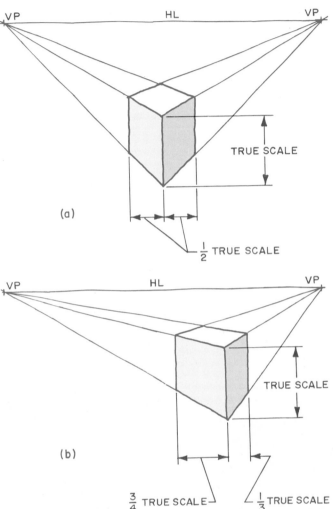

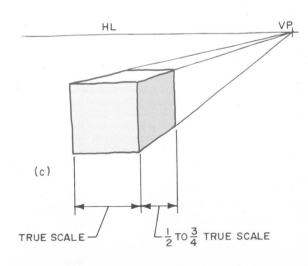

Fig. 11-59. Variations of perspective sketches.

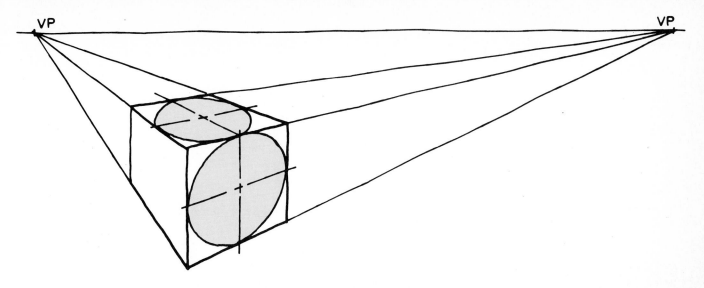

Fig. 11-60. Sketching circles in perspective.

Other features may be added to the sketch by measuring on the front corner and projecting to their proper location of depth on a half-scale basis. The objects may be positioned so that its faces make angles of 30° and 60° with the picture plane by locating the front vertical line as shown in Fig. 11-59 (b). A one-point or parallel perspective would be sketched as shown in Fig. 11-59 (c).

Circles are sketched in perspective by drawing a square, locating the center points of the sides of the square, and sketching the enclosed circle (ellipse), Fig. 11-60.

CAD GENERATED PICTORIALS AND MODELS

The use of computer-aided design brings new methods to create and view three-dimensional drawings. Although you can still create pictorial drawings using the techniques discussed in this chapter, you can also use the 3D drawing functions of CAD to create true 3D models. The computer thinks of the design as having material and volume. Fig. 11-61 shows the 3D model of a juice container. The designer can view the model from any angle, and with shading.

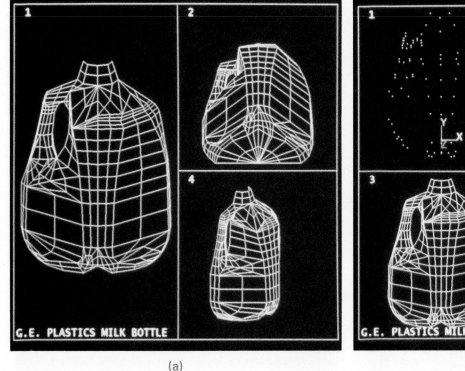

(a)

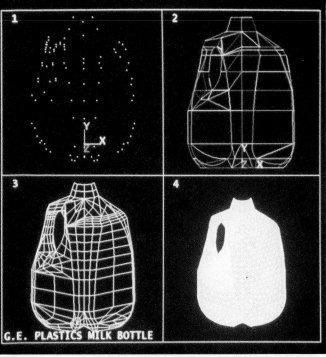

(b)

Fig. 11-61. A 3D modeling program was used to create this design. A—Viewing the container from various angles. B—Adding shading and performing strength testing on the product. (GE Plastics, Inc.)

PROBLEMS IN PICTORIAL DRAWING

The following problems will provide you with the opportunity to apply the techniques of pictorial drawing. The problems may be placed on A or B size sheets.

1. Make isometric drawings of any two of the parts in Fig. 11-62. Dimension the drawings.
2. Prepare a dimetric scale and use this scale in drawing an approximate dimetric drawing of any of the parts in Fig. 11-62. Dimension the drawing.
3. Make an approximate trimetric drawing of the BEARING MOUNT in Fig. 11-63. Select an axis from those shown in Fig. 11-34 which best displays the features of the object. Find the ellipse guide angle and use an ellipse template in drawing the circular features.
4. Make an oblique drawing of the WEAR PLATE in Fig. 11-63. Dimension the drawing.
5. Construct a one-point perspective drawing of the MOUNTING BRACKET in Fig. 11-63. Do not dimension.
6. Make pictorial drawings of the remaining objects in Figs. 11-63 and 11-64. You may select the type of pictorial drawing to be used.
7. Select an object at home, school or work and prepare a pictorial drawing of the object. Use any type of pictorial drawing you prefer.
8. Design some object which you can use around home or work. Prepare a dimensioned pictorial drawing of the object.

Industry photo. Perspective lines take on a realistic meaning in long range photography.

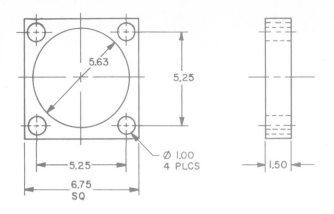

FLANGE, MOUNTING

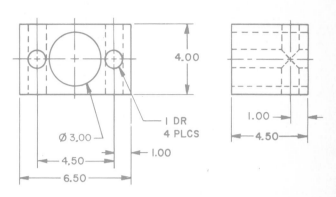

BLOCK, MOUNTING

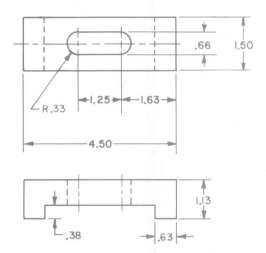

DOUBLE END STRAP

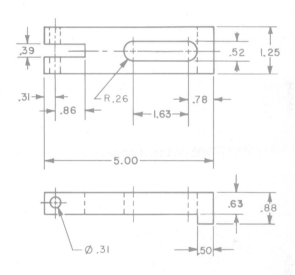

CAM STRAP

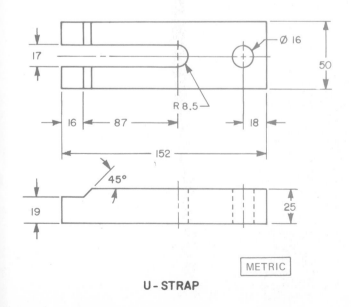

METRIC

U-STRAP

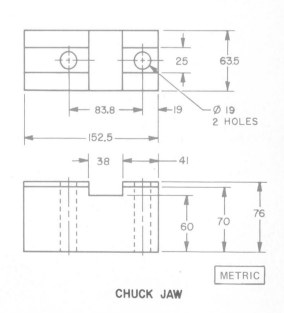

METRIC

CHUCK JAW

Fig. 11-62. Problems for isometric and dimetric drawings.

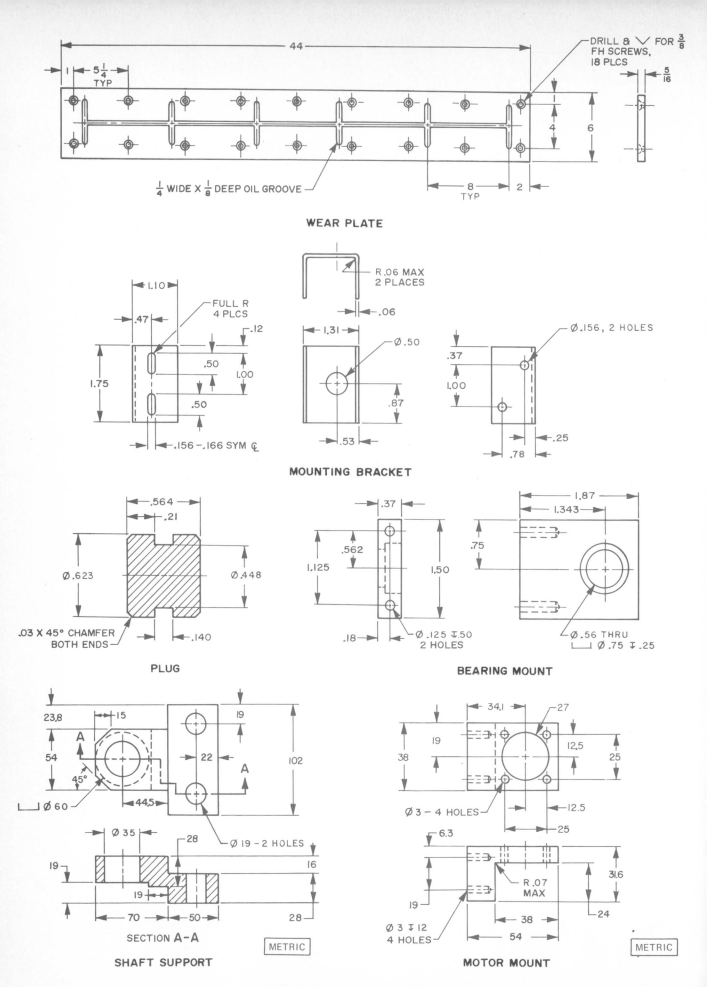

Fig. 11-63. Pictorial drawing problems.

226

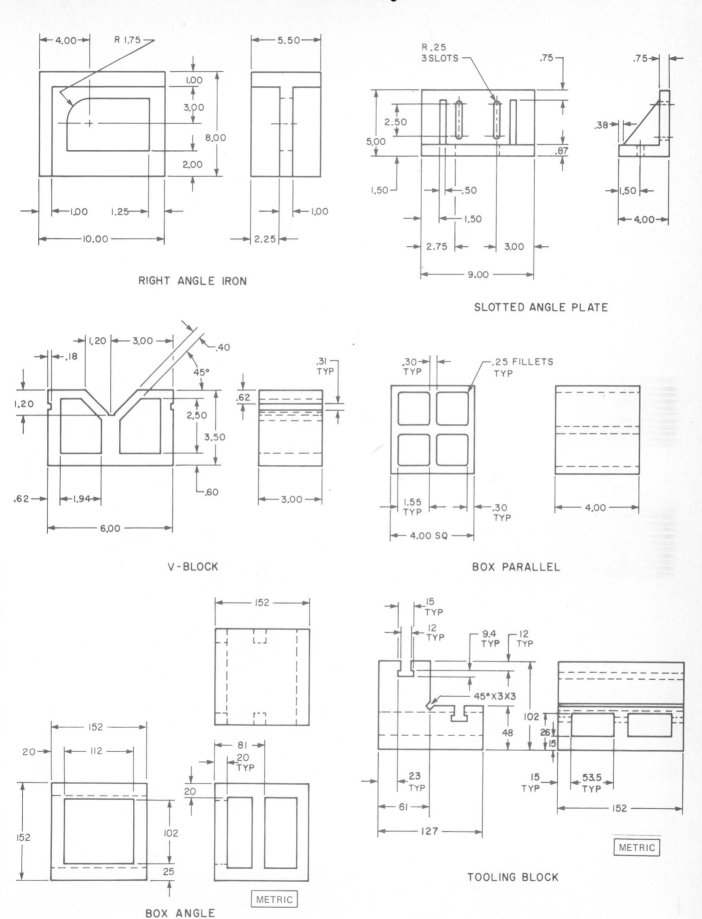

RIGHT ANGLE IRON

SLOTTED ANGLE PLATE

V-BLOCK

BOX PARALLEL

BOX ANGLE

TOOLING BLOCK

Fig. 11-64. Additional pictorial drawing problems.

PART II
DESCRIPTIVE GEOMETRY

Descriptive geometry is that phase of drafting dealing with spatial relationships and graphical analysis of problems. An introduction was presented in Chapter 7, Multiview Drawings, on the fundamentals of projection drawing.

In this part of the text, the principles and applications of descriptive geometry useful in problem solving are presented under the headings of AUXILIARY VIEWS, REVOLUTIONS, INTERSECTIONS AND DEVELOPMENTS.

An example of a massive machine part for which descriptive geometry was used in determining the lines of intersection, true sizes of angles and planes.   (Ex-Cell-O-Corp.)

Chapter 12
AUXILIARY VIEWS

The principal views of multiview drawings (top, front and side views) are normally adequate for describing the shape and size of most objects. However, objects having features inclined or oblique to the principal projection planes require special projection procedures, Fig. 12-1, which are possible through auxiliary views.

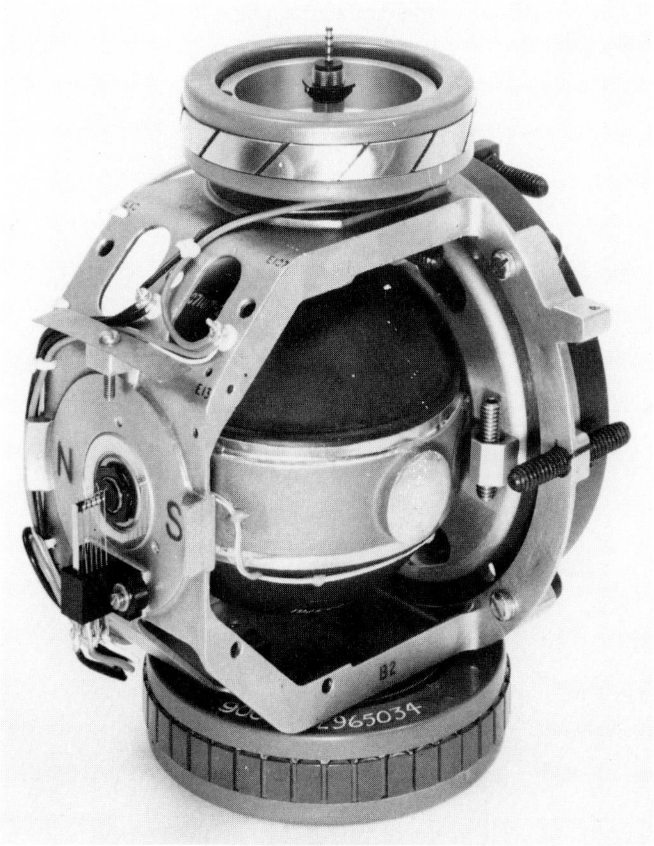

Auxiliary views provide the following basic information for features appearing on other than principal surfaces: (1) true length of a line, (2) identifying the point view of a line, (3) finding the edge view of a plane, (4) determining the true size of a plane, (5) finding the true angle between planes.

Auxiliary (supplementary) projections are necessary to sufficiently analyze and describe these features for production purposes. Two kinds of auxiliary views are used in describing and refining the design of industrial products, primary and secondary auxiliary views.

Primary auxiliary views are projected from orthographic views—horizontal, frontal or profile. Secondary auxiliary views are projected from a primary auxiliary and a principal view. Views projected after a secondary auxiliary are known as successive auxiliary views.

Fig. 12-1. Gimbal ring for the directional gyro required auxiliary view development of its inclined surfaces. (Sperry Flight Systems Div.)

Other uses of auxiliary views are in the sheet metal and packaging industries. In many cases, patterns must be developed for various-shaped surfaces at an angle with other surfaces, Fig. 12-2. The projection of these features in normal views always results in foreshortened and distorted views and surfaces which are not true to shape and size.

Fig. 12-2. Auxiliary views were used to develop the inclined surfaces of these satellite dishes. (Winnebago)

PRIMARY AUXILIARY VIEWS

A primary auxiliary view is a first, or direct, projection of an inclined surface. It is projected perpendicularly to one of the principal orthographic views: horizontal, frontal or profile, Fig. 12-3. A primary auxiliary view is perpendicular to only one of the principal views. A view that is perpendicular to two of the principal views is one of the normal views (discussed in Chapter 7) and not an auxiliary.

TRUE SIZE AND SHAPE OF INCLINED SURFACES

The primary auxiliary view is useful in determining the true size and shape of a surface that is inclined and not true size in any of the principal views of projection, Fig. 12-3. The auxiliary view is projected from the principal view in which the inclined surface appears as an edge. In the example in Fig. 12-4, the auxiliary view appears as an edge in the frontal plane. Therefore, the auxiliary projection is called a frontal projection.

Primary auxiliary planes of projection are always numbered

1 and preceded by the letter indicating a frontal, horizontal or profile plane. In the example (Fig. 12-4), reference plane F-1 is drawn at any convenient location and parallel to the line in the front view that represents the edge of the inclined surface, (b). It could have been drawn to coincide with the front edge of the auxiliary view or the center of the view in the event these locations would have been more convenient.

Note in Fig. 12-4 (a) that the primary auxiliary plane, F-1, is perpendicular to the frontal plane. Also, the line of sight is perpendicular to the auxiliary plane, just as any of the normal projection planes are viewed. Distances along the plane of projection are projected perpendicularly in their true length, (b). Measurements of depth are transferred with a scale or dividers in their true length from the top or side view to the auxiliary view.

This same object could have been arranged so that the inclined surface appears as an edge in the top view, Fig. 12-5. The inclined surface is projected in the same manner as for a frontal auxiliary projection. However, the plane of projection, H-1, is a horizontal (top view) auxiliary, perpendicular to the horizontal plane. The line of sight is perpendicular to the inclined surface in the horizontal plane, and the result is a horizontal auxiliary.

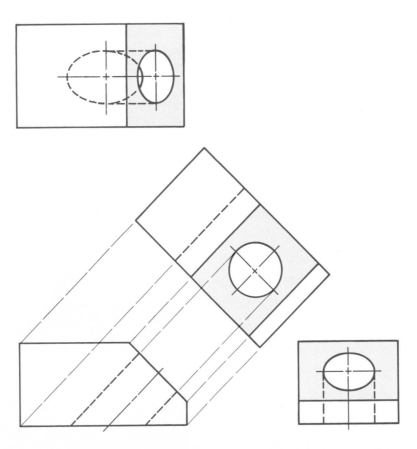

Fig. 12-3. Primary auxiliary view of an inclined surface.

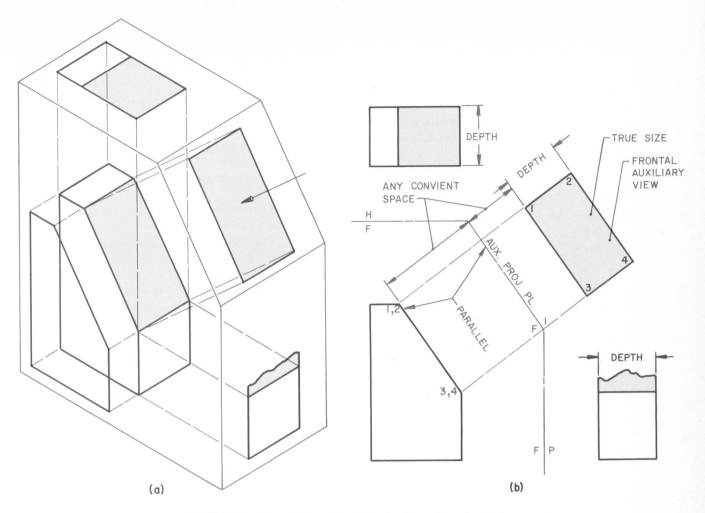

Fig. 12-4. True size and shape of an inclined surface — frontal projection.

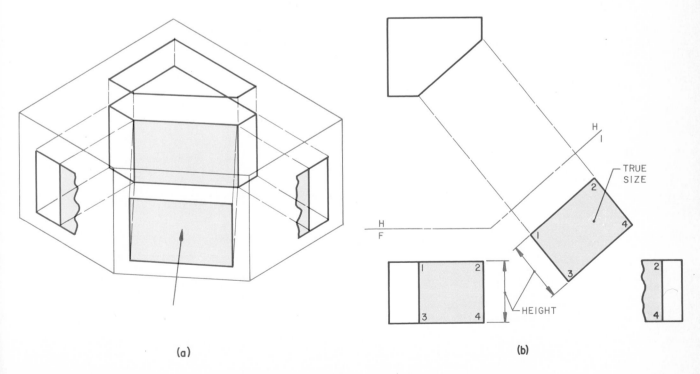

Fig. 12-5. True size and shape of an inclined surface in horizontal projection.

A profile auxiliary view would be projected in a like manner from the profile or side view when the inclined surface appears as an edge in this view, Fig. 12-6.

LOCATION OF A POINT IN AN AUXILIARY VIEW

The location of points in an auxiliary view is the first important step in understanding the projection and development of auxiliary views. Point "A" is shown in the "glass box" and it has been projected to the frontal, horizontal and auxiliary planes, Fig. 12-7 (a). Since the inclined surface is perpendicular to the frontal plane, a frontal auxiliary is projected. Point A is located in the auxiliary view by projecting it perpendicularly to the plane of projection and setting off the depth from the top view, (b).

LOCATION OF A TRUE LENGTH LINE
IN AN AUXILIARY VIEW

The auxiliary view method is useful in determining the location and true length of a line which is inclined to the principal views in orthographic projection. Location and determination of the line's true length consists of locating its two end points in the auxiliary view (discussed in LOCATION OF A POINT) and connecting these points to form the required line.

Given the top and front views, the procedure is as follows:

1. Draw an auxiliary viewing plane F-1 parallel to line AB in front view, Fig. 12-8 (b). (Viewing plane could have just as well been established for top view.)
2. Line of sight is perpendicular to viewing plane F-1 and line AB. Any line viewed perpendicularly will be seen in its true length.
3. Measure distance line AB lies away from frontal plane as viewed in horizontal plane. Plot these distances perpendicularly away from auxiliary viewing plane F-1 to locate points A and B, (c).
4. Connect end points for true length line AB, (d). Any given line whose location is plotted on a viewing plane parallel to given line will appear in its true length when viewed perpendicularly to plane.

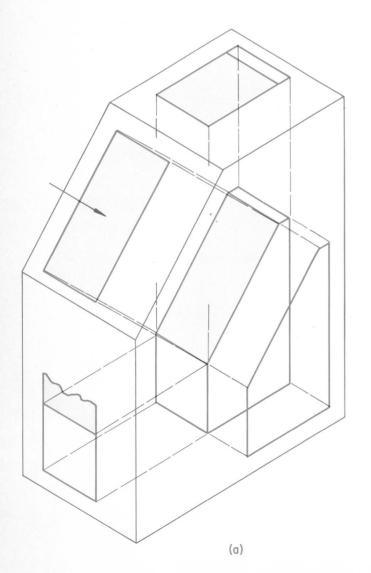

(a)

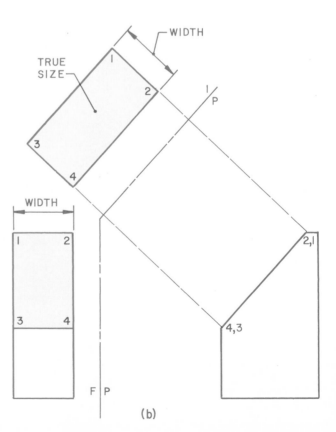

(b)

Fig. 12-6. True size and shape of an inclined surface in profile projection.

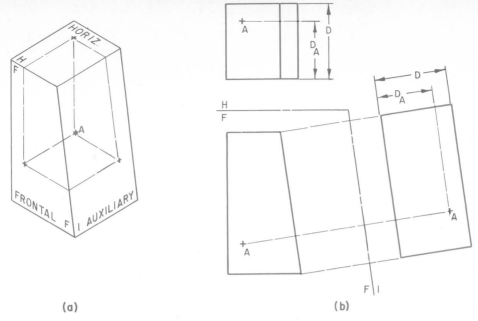

(a)

(b)

Fig. 12-7. Location of a point in an auxiliary view is the first important step in understanding auxiliary projection of surfaces and features for accurate size and shape description.

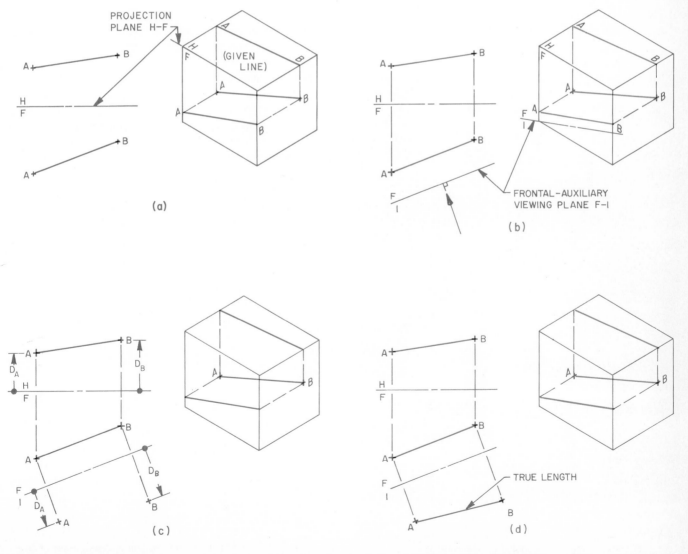

Fig. 12-8. Steps in determining the location and true length of a line, using the auxiliary view projection method.

233

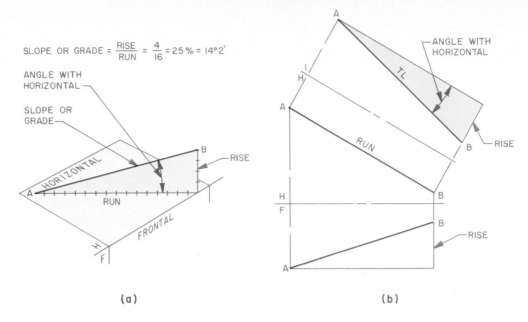

SLOPE OR GRADE = $\frac{RISE}{RUN}$ = $\frac{4}{16}$ = 25 % = 14°2'

Fig. 12-9. Finding the slope of a line.

(a)

(b)

SLOPE OF A LINE

The slope of a line is the angle the line makes with the horizontal plane. The extent of slope may be expressed in percent or degrees of grade, Fig. 12-9 (a). When the line of slope is oblique to principal planes, an auxiliary view is required to find the true length of the line. Since slope is the angle the line makes with the horizontal, a horizontal auxiliary view is used, (b). The slope line, AB, appears true length and the angle it makes with the horizontal is the required angle.

Pipeline systems that depend on gravity flow must be planned carefully in advance and the slope specified on a drawing for field use, Fig. 12-10. The slope of a line may be designated as positive or negative. If positive, the slope is upward from the designated point; downward if negative.

POINT METHOD APPLIED TO THE DEVELOPMENT OF AN INCLINED SURFACE

The application of the point method to the development of an inclined surface in a primary auxiliary view is shown in Fig. 12-11.

Fig. 12-10. Slope for this pipeline had to be determined before field drawings could be prepared for its construction. (Exxon Co. USA)

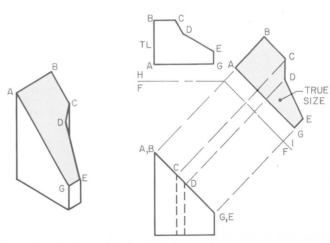

Fig. 12-11. Application of the point method in development of an inclined surface in a primary auxiliary view.

234

TRUE ANGLE BETWEEN A LINE AND A PRINCIPAL PLANE

The true angle (θ) formed between a line in a primary auxiliary view and a principal plane may be determined in a view where the principal plane appears as an edge and the line in its true length (TL). In Fig. 12-12, the angle between the horizontal plane and the true length line AB is a true angle. Note that the reference plane has also been drawn through point A to show the true angle.

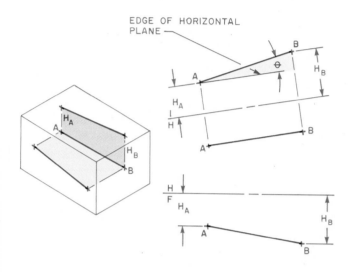

Fig. 12-12. Determining the true angle between a line and a principal plane.

The true angle could also be developed for the frontal auxiliary plane, Fig. 12-13. The line of sight is parallel to the frontal plane and perpendicular to line AB. A reference plane has been drawn through point A and the true angle projected away from the frontal plane, as shown in the pictorial view in Fig. 12-13.

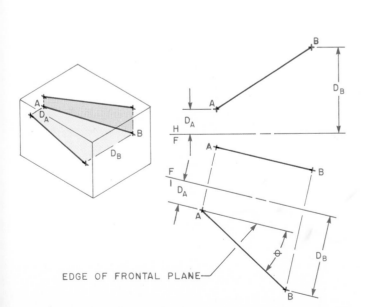

Fig. 12-13. True angle between a line and the frontal plane.

DETERMINATION OF TRUE ANGLE BETWEEN TWO PLANES

Frequently in the design and manufacture of parts, it is necessary to determine the angle between two planes in order to correctly specify the design of the parts, Fig. 12-14. The true angle between two planes, called a "dihedral angle," can be found in a primary auxiliary view. This occurs when their line of intersection is true length in one of the principal views and their line of intersection appears as a point in the auxiliary view.

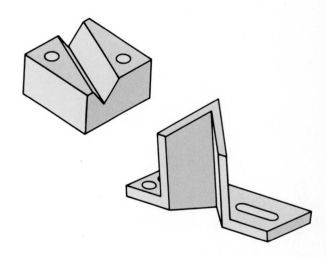

Fig. 12-14. Typical applications requiring angle determination before parts can be specified.

In Fig. 12-15, the line of intersection AB is seen in its true length in the top view since it is parallel to the horizontal plane. A horizontal auxiliary is drawn perpendicular to line AB, providing a point view of line AB. The angle between the two planes is a true angle.

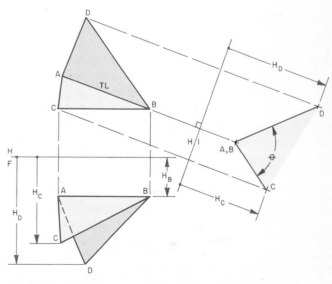

Fig. 12-15. True angle between two planes as developed in a primary auxiliary view.

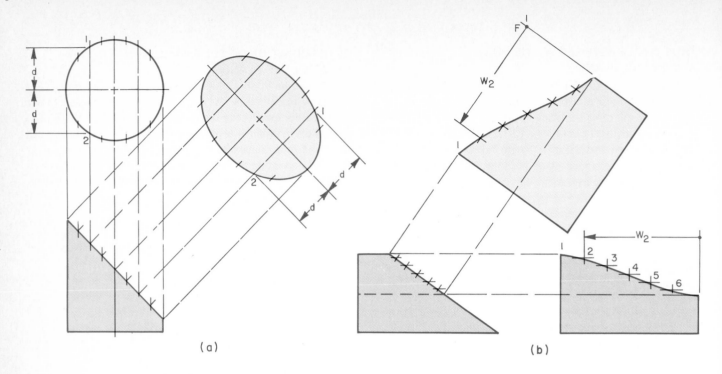

Fig. 12-16. Primary auxiliary view of a circle and an irregular curve on inclined surfaces.

PROJECTING CIRCLES AND IRREGULAR CURVES IN AUXILIARY VIEWS

Circles and irregular curved lines may be projected in auxiliary views by the identification of a sufficient number of points in the primary views. These points are projected to the auxiliary view and are plotted to provide a smooth curve, Fig. 12-16 (a). The steps involved in the projection of circles and irregular curved lines in a primary auxiliary view are as follows:

1. Draw two principal views showing circle or irregular curve and angle of inclined surface, Fig. 12-16 (b).
2. Locate points on curve in principal view. Any desired number and location of points may be selected as long as these are a sufficient number to adequately project charac-ter of curve.
3. Project points on curve to line in principal view represent-ing edge of inclined surface.
4. Draw reference plane F-1 parallel to edge view of inclined surface, (b). In irregular curve example, reference plane has been drawn in line with far side of object. It could have been drawn to left of auxiliary view, or in center and measured accordingly in side view and results would have been the same.
5. Project points marked on inclined surface perpendicularly to auxiliary view and measure off location of points as shown.
6. Sketch a light line through these points and finish with an irregular-curve instrument.

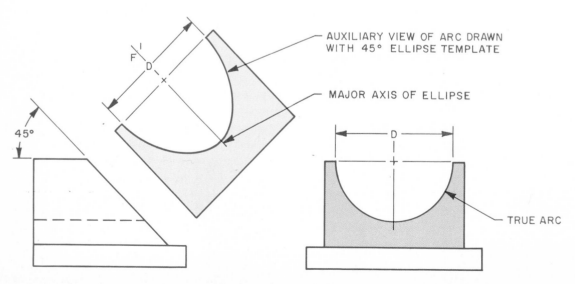

Fig. 12-17. Primary auxiliary view of circular features drawn with the aid of an ellipse template.

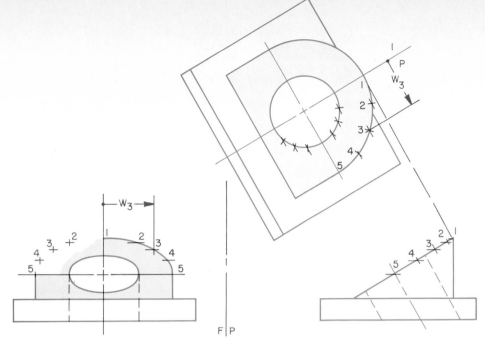

Fig. 12-18. Use of an auxiliary view in the construction of a
principal view.

DRAWING CIRCLES AND CIRCLE ARCS IN AUXILIARY VIEWS BY ELLIPSE TEMPLATE METHOD

Circular shapes may be drawn in auxiliary views by the projection method. However, the use of ellipse templates speeds up the process and produces much better results. Take the following steps in drawing regular circular features appearing as ellipses in auxiliary views:

1. Locate reference plane for auxiliary view (in center of view), Fig. 12-17.
2. Project, to auxiliary view, points and lines locating center line, major and minor axes of circular feature.
3. Identify angle inclined surface makes with principal plane.
4. Select an ellipse template of correct angle and major diameter. Draw required ellipse or partial ellipse.

CONSTRUCTION OF A PRINCIPAL VIEW WITH THE AID OF AN AUXILIARY VIEW

Sometimes the true size and shape of a feature exists on the inclined surface of a part, such as a circular curve or hole. When this occurs, it is necessary to construct the auxiliary view, then reverse the projection procedure in constructing the principal view where the feature is shown foreshortened, Fig. 12-18. Take the following steps:

1. Construct view in which inclined surface appears as an edge (profile view in our example). It may be advantageous to partially complete view until features concerned have been drawn in auxiliary view.
2. Construct auxiliary view perpendicular to a reference plane which is parallel to inclined surface. (Line P-1, in our example, which is located in center of feature.)
3. Locate a sufficient number of points on outline of feature to assure an accurate representation.
4. Project points from auxiliary view to its principal view, then to other principal view in which feature appears in a foreshortened plane. Measurements are taken from reference plane P-1 in the auxiliary view and transferred to the

principal view.
5. Complete view by joining points. (In our example, points were joined into a smooth curve with an irregular-curve instrument. Center lines could have been located and an ellipse template of correct size used.)

Note: Auxiliary view in Fig. 12-18 is a complete rather than a partial auxiliary, since entire part is shown, not just inclined surface. Hidden lines are normally omitted in auxiliary views unless required for clarity.

SECONDARY AUXILIARY VIEWS

Oblique lines and surfaces are neither parallel nor perpendicular to any of the principal planes of projection. As such, a secondary auxiliary view is required to describe their true size and shape.

In structures having many oblique surfaces, a number of auxiliary views are required, Fig. 12-19. All auxiliary views beyond the secondary auxiliary view are called "successive auxiliary views" and are projected from secondary or prior successive auxiliary views.

Fig. 12-19. Telstar satellite has a number of auxiliary surfaces which require secondary auxiliary projection techniques in their definition. (Bell Laboratories)

SECONDARY AUXILIARY VIEW OF AN OBJECT

Primary auxiliary views require two principal views for their projection. Secondary auxiliary views are projected from a primary auxiliary view and one of the principal views, Fig. 12-20.

The line of sight at which the object is viewed is indicated in the top and front views, Fig. 12-20 (a). The line of sight is shown in its true length in the primary auxiliary view, (b).

An auxiliary must be viewed in the direction of its line of sight. To do this, the secondary auxiliary viewing plane 1-2 is constructed at right angles to the true length line of sight, giving a point view of the line of sight, (c).

Two plane surfaces are projected to the secondary and given letters to aid in identifying the formation of the object. In section (d), the object is completed, showing visible surfaces and lines, plus those edges which are hidden from view.

The steps in the construction of a secondary auxiliary view

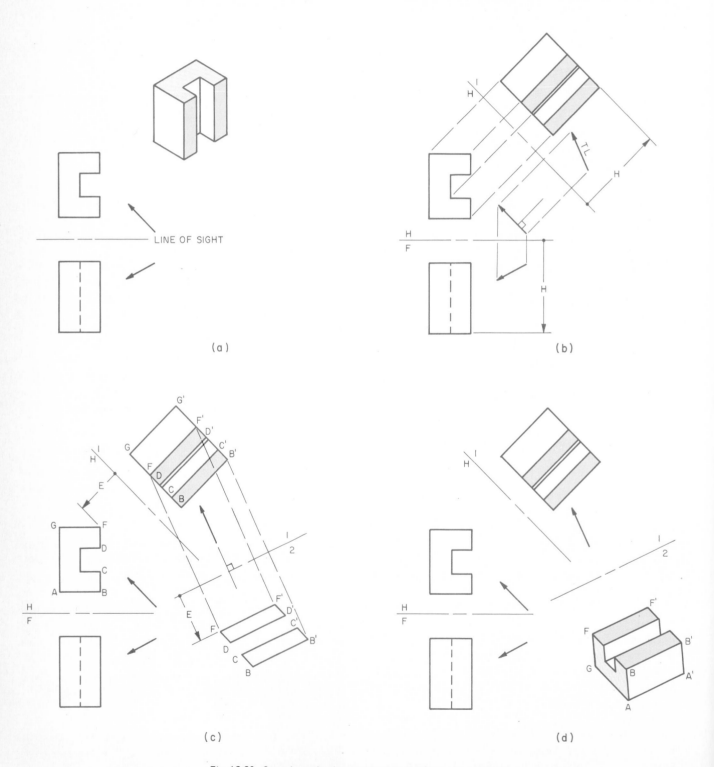

Fig. 12-20. Steps in projecting a secondary auxiliary of an object.

are shown in Fig. 12-20. No oblique surfaces were involved in the projection of this object. (How to obtain point-views of lines and true sizes and shapes of oblique surfaces will be covered later in this chapter.)

Given the top and front views, follow this procedure for drawing a secondary auxiliary view of an object to find the

POINT VIEW OF A LINE IN A SECONDARY AUXILIARY

Finding the point view of a line is basic to the projection of a secondary auxiliary view. Given the top and front view of a line, Fig. 12-21 (a), proceed as follows:

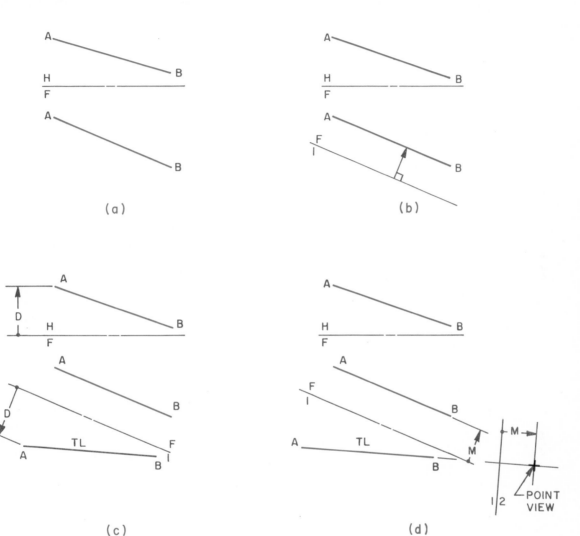

Fig. 12-21. Steps in finding point view of a line in a secondary auxiliary.

view of the object when viewed in the line of sight:
1. Project a primary auxiliary view of object off of a principal view so that it is perpendicular to chosen line of sight, Fig. 12-20 (b).
2. Construct true length line of sight.
3. Draw secondary auxiliary projection plane 1-2 perpendicular to true length line of sight in primary auxiliary, (c). Projection of this line will give a point view in secondary auxiliary.
4. Project two planes by locating their points of intersections and dimensions from H-1 plane as shown in (c).
5. Locate remaining points and draw connecting lines to complete object, (d). Line of sight will indicate which surfaces and edges are visible and which are hidden.

1. Construct a primary auxiliary view of line AB by drawing reference plane F-1 parallel to the front view of the line, as shown in (b). (The top view could also have been chosen in this example.)
2. Find the true length of line AB by projecting it perpendicular to the reference plane. Locate the distance the end points are from the H-F reference plane in the top view, as shown in (c).
3. Draw reference plane 1-2 for secondary auxiliary perpendicular to line AB and measure off distance M to locate view point for line AB, (d).
4. Distance M is perpendicular to reference planes F-1 and 1-2, both of which appear as an edge in the front and secondary auxiliary views.

DETERMINATION OF TRUE ANGLE BETWEEN TWO PLANES IN A SECONDARY AUXILIARY

It has been illustrated that the true angle between two planes could be determined when the line of intersection was viewed as a point, and the planes appeared as edges. However, when the line of intersection of an angle between two planes is an oblique line, a secondary auxiliary is required to determine the true angle.

Two planes, shown in Fig. 12-22, intersect in an oblique line to form an angle. Given the top and front views of the two planes, proceed as follows to find the true size of an angle between two planes:

1. Construct a primary auxiliary view to develop true length of line of intersection AC of angle, (b).
2. Construct a secondary auxiliary view to develop point view of line of intersection, (c). (Refer to Fig. 12-21.) Plane of angle is perpendicular to true length line of intersection in

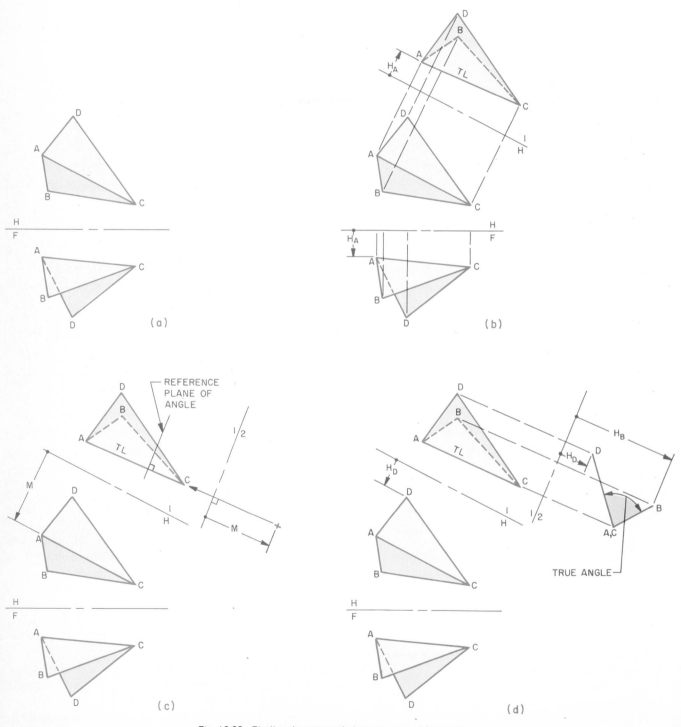

Fig. 12-22. Finding the true angle between two oblique planes.

primary auxiliary.

3. True angle is formed by locating points B and D, which complete edge view of two planes enclosing angle, as shown in (d).

4. When line of intersection of two planes appears as a point and planes appear as edges, true size of angle can be measured.

TRUE SIZE OF AN OBLIQUE PLANE

Frequently it is necessary for the draftsman or engineer to specify the exact size and shape of an oblique surface of an industrial product. Since some of these surfaces (Fig. 12-23) must function within a very close tolerance range, they must be located and specified on the drawing in a precise manner.

Fig. 12-23. Auxiliary projection was used to identify true size of oblique planes on members of these structures.   (DuPont)

The true size and shape of inclined surfaces may be determined by the use of a primary auxiliary. Except, however, for oblique surfaces which lie in a plane that is not perpendicular to any of the principal planes of orthographic projection. These surfaces require the use of a secondary auxiliary to identify their true size and shape.

Given the top and front views of an oblique plane, Fig. 12-24 (a), proceed to develop its true size as follows:

1. Draw horizontal line AA' in top view, (b). (Front view could also have been used.)
2. Project line to front view where it appears true length. (A line which is parallel to one of principal planes will appear in its true length in an adjacent view.)
3. Draw primary auxiliary plane F-1 perpendicular to line AA' and line of sight. Project a point view of line AA' in primary auxiliary view, (c).
4. Project points B and C to this view where plane will appear as an edge.
5. Draw secondary auxiliary view plane 1-2 parallel to edge view of plane ABC, (d).

6. Construct projectors for all three points of plane ABC. Locate points on these projectors by taking measurements perpendicular from primary auxiliary plane edge view F-1. This produces plane ABC in its true size.

APPLICATION OF SECONDARY AUXILIARY METHOD IN DEVELOPING TRUE SIZE OF AN OBLIQUE PLANE

Assume that the object shown in Fig. 12-25 is a sheet metal cover for a special machine. One problem that must be resolved in making a drawing of the object is to find the true size and shape of the surface of the oblique plane.

Given the front and top views of the sheet metal cover with an oblique surface, take the following steps in your definition of surface ABCD:

1. Draw reference plane H-F between top and front view at any desired location, Fig. 12-25.
2. Draw horizontal line CC' in front view and project it perpendicular to top view to establish a true length line on surface ABCD.
3. Draw reference plane H-1 perpendicular to line of sight to

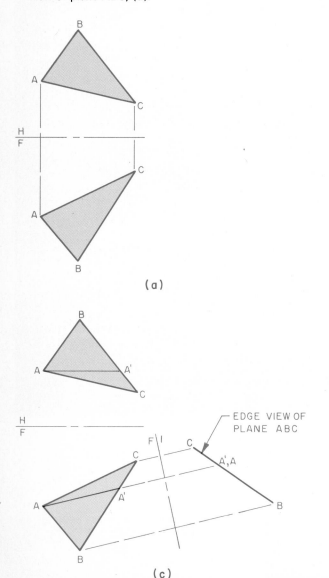

(a)

(c)

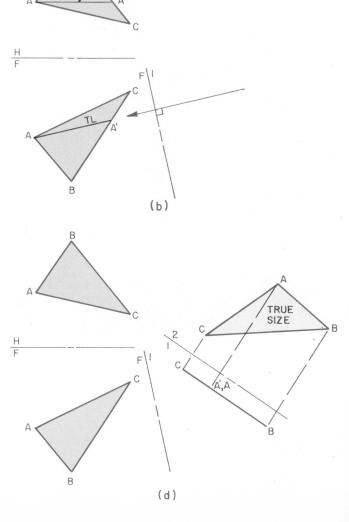

(b)

(d)

Fig. 12-24. Steps in finding the true size and shape of an oblique plane.

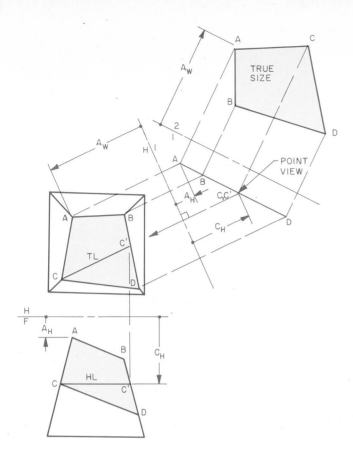

Fig. 12-25. Secondary auxiliary method applied in the development of true size and shape of an oblique plane.

produce a point view of line CC' in primary auxiliary view.

4. Project points A, B and D and measure their location from reference plane H-F in front view to form edge view of ABCD. (All should lie in a straight line.)

5. Draw reference plane 1-2 perpendicular to line of sight for edge view of plane ABCD.

6. Project points A, B, C and D and measure their location from reference plane H-1 in top view.

7. Connect points to complete true size of plane ABCD.

Note: Secondary auxiliary view shown in Fig. 12-25 is a partial auxiliary view. It only shows oblique surface. Remainder of object could have been projected, but it is

not needed in this view and would have been out of true size and shape.

PROBLEMS IN AUXILIARY VIEWS

The following have been planned to provide experience in primary and secondary auxiliary projection in the solution of problems in drafting and design.

Size A drafting sheets should be used in solving the problems. The problems are shown on 1/4 inch section paper to facilitate location on the drawing sheet. Label the points, lines and planes and show your construction procedure.

PRIMARY AUXILIARY PROJECTION
1. Refer to Fig. 12-26 (a and b) and construct an auxiliary view of the inclined surface "X" of each object. Use a frontal auxiliary for (a) and a horizontal auxiliary for (b).
2. Refer to Fig. 12-26 and solve each section as follows: In (c), find the true length of the line utilizing a frontal auxiliary. In (d), find the slope of the line.
3. Refer to Fig. 12-27 and develop the true angle between the planes for (a and b).
4. Refer to Fig. 12-27 and develop the true size and shape of the features in the auxiliary view for the objects in (c and d).
5. Refer to Fig. 12-28 (a and b) and draw the elliptical features in the auxiliary view, using ellipse templates.

SECONDARY AUXILIARY PROJECTION
6. Refer to Fig. 12-28 (c and d) and draw a secondary auxiliary view of the object as indicated by the lines of sight.
7. Refer to Fig. 12-29 (a) and find the point view of the line in the secondary auxiliary view.
8. Refer to Fig. 12-29 (b) and find the true angle between the oblique planes.
9. Refer to Fig. 12-29 (c and d) and construct the true size and shape of the oblique surfaces by means of secondary auxiliary views.
10. Refer to Fig. 12-30 and draw the necessary views, including auxiliary views, of the parts shown. Dimension your drawings.

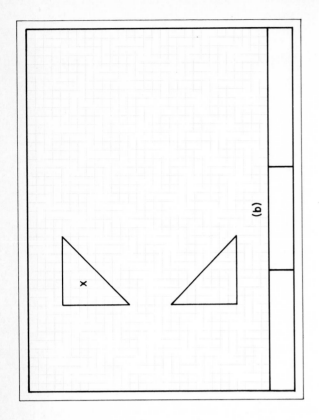

(b)

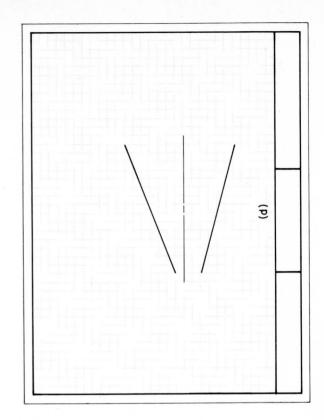

(d)

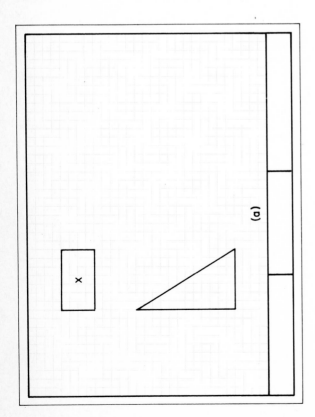

(a)

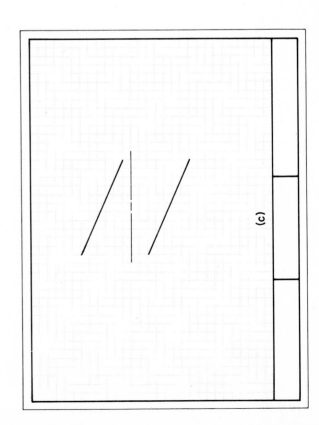

(c)

Fig. 12-26. Problems in primary auxiliary projection.

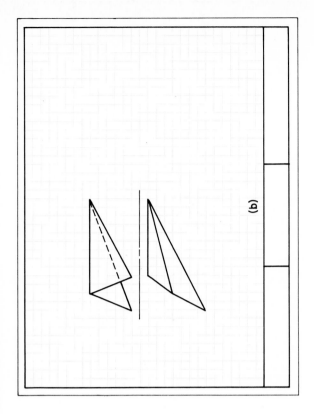

(a)

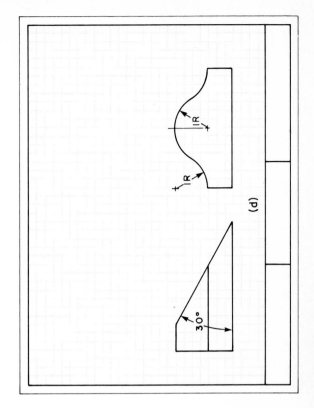

(b)

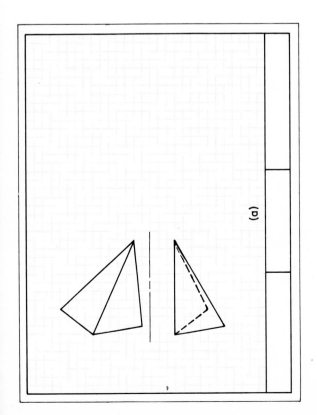

(c)

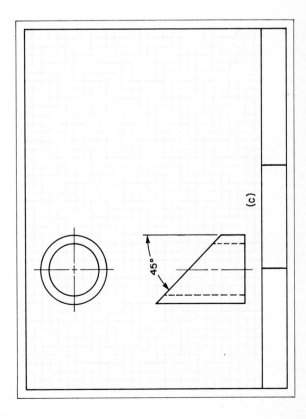

(d)

Fig. 12-27. Additonal problems in primary auxiliary projection.

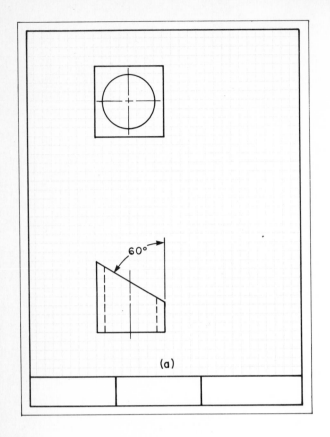

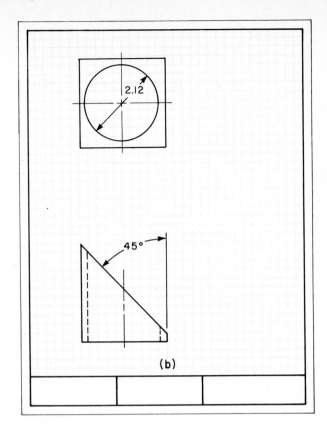

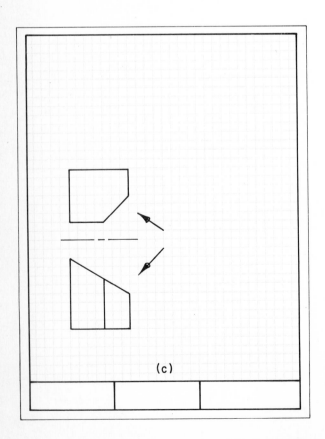

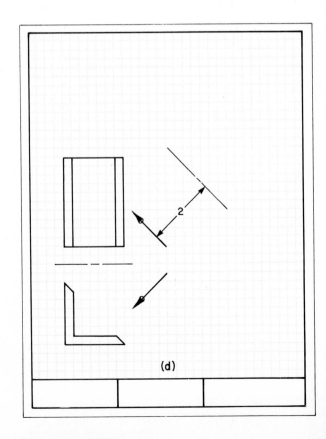

Fig. 12-28. Problems in primary and secondary auxiliary projections.

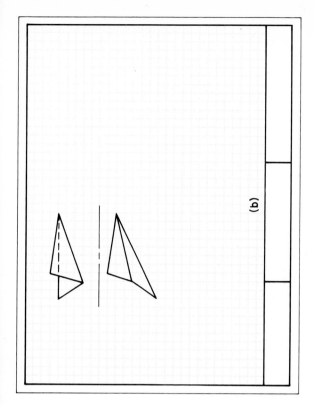

(a)

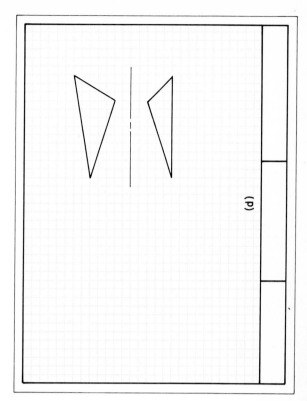

(b)

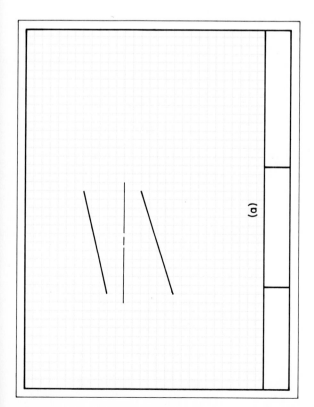

(c)

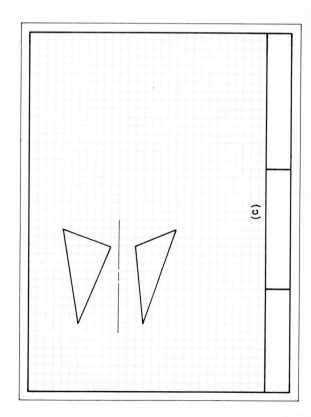

(d)

Fig. 12-29. Secondary auxiliary view problems.

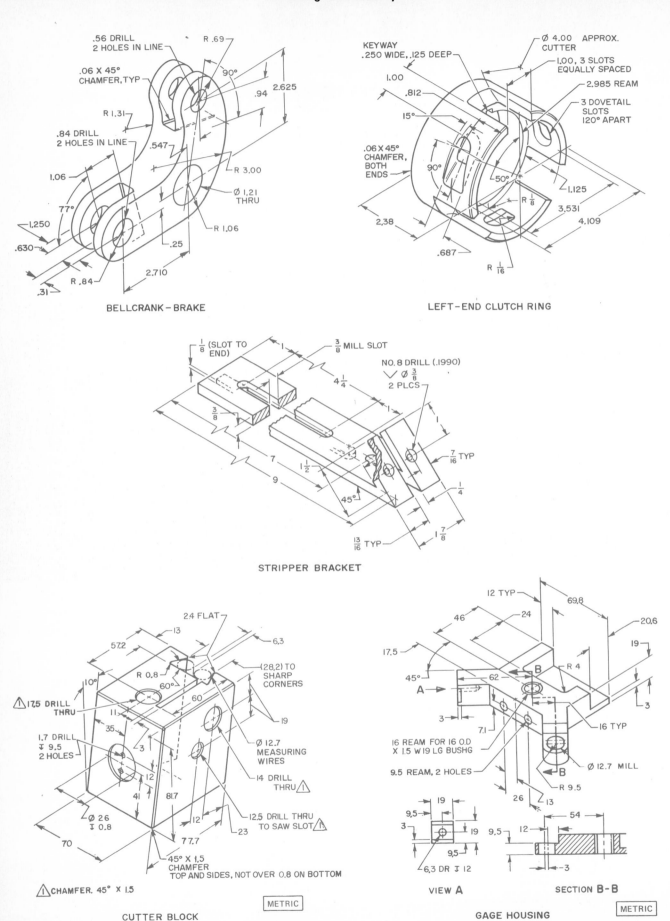

Fig. 12-30. Machine parts requiring auxiliary views in their description.

Chapter 13
REVOLUTIONS

Revolution is another method available to the drafter-designer in defining spatial relationships of rotating or revolving parts, Fig. 13-1. As the principles of revolution are learned and applied, the similarities between this method and the auxiliary view method will be apparent. Each method tends to enhance one's understanding of the other.

REVOLUTIONS AND AUXILIARY VIEWS COMPARED

To obtain an auxiliary view of a surface or object, the observer shifts position to a point where he or she may view the object perpendicular to the inclined surface, Fig. 13-2 (a). In a revolution of the object, the observer is assumed to remain in the original position while the object is revolved, Fig. 13-2 (b). The view shown as an auxiliary or as a revolved view is exactly the same.

Fig. 13-1. Machine parts such as handles and levers may be checked for clearances by the revolution method.
(Motorola, Inc.)

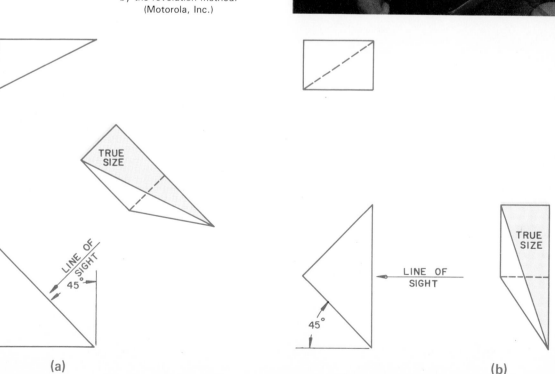

(a)   (b)

Fig. 13-2. Viewing an object (a) as an auxiliary view and (b) as a revolution.

249

The method of revolution may be further shown as an imaginary plane section of a cone. When it is inclined to the observer, it appears foreshortened and not in its true size, Fig. 13-3 (a). If we imagine the cone being revolved so that the plane AB'C' is perpendicular to the line of sight, the plane appears in its true size, (b). Line AC in view (a) is not shown in its true length because it is not on a plane perpendicular to the line of sight. Line AC' is a true length line in view (b).

SPATIAL RELATIONSHIPS IN REVOLUTION

Some of the more common spatial relationships obtainable through the use of revolution are presented here. Knowledge of the principles and techniques of revolution should enable you to select this method when it is advantageous over other methods.

REVOLUTION OF A LINE TO FIND ITS TRUE LENGTH

The revolution method may be employed to find the true length of a line in space. Fig. 13-4 (a) shows a frontal line being revolved as an element of a cone to find its true length. When the line is thought of as an element of a cone, it is easier to visualize the revolution process.

Given the top and front views of an oblique line AB in space, proceed as follows to obtain its true length:

1. Line is drawn conventionally in horizontal and frontal views, Fig. 13-4 (b).
2. Assume line appears as an element of a cone section in top and front views, (c).
3. Revolve line in top view, using point A as center of radius, to a position parallel to horizontal plane, (d).
4. Any line parallel to the horizontal plane in the top view will be seen true length in the adjacent (front) view. Therefore, line AB' is shown true length in the front view, (d).

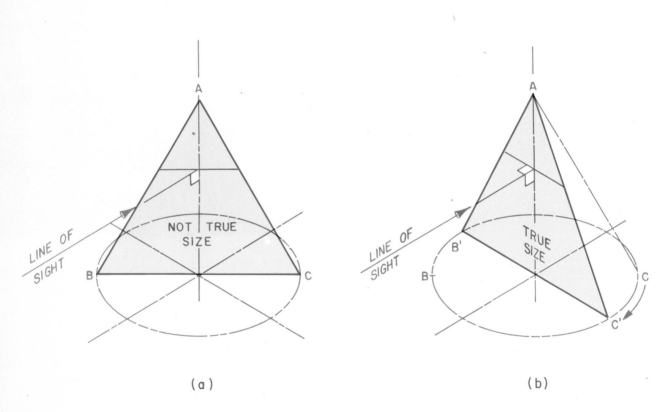

(a)   (b)

Fig. 13-3. Revolving a cone illustrates the principle of revolution in drafting.

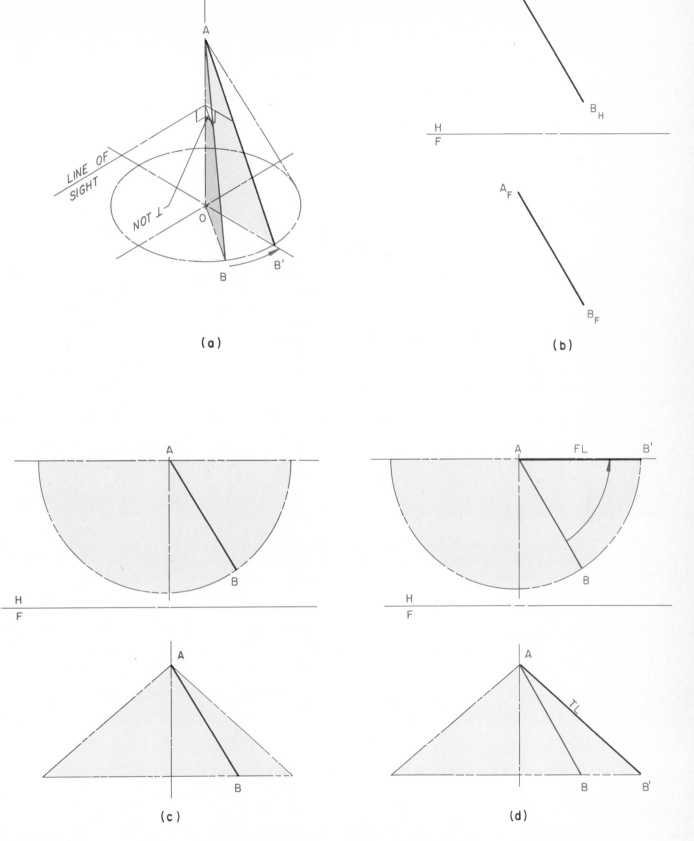

Fig. 13-4. Obtaining the true length of a line by revolution.

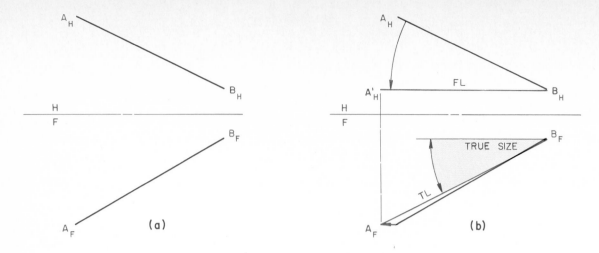

Fig. 13-5. Finding the true size of an angle between a line and a principal plane by revolution.

TRUE SIZE OF AN ANGLE BETWEEN A LINE AND A PRINCIPAL PLANE BY REVOLUTION

The true size of an angle formed between a line and a principal plane may be found by revolving the line to a position parallel to the plane. In Fig. 13-5 (a), the true size of the angle between the front view and the horizontal plane is required. The line is revolved in the top view to a position parallel to the frontal plane and projected to produce a true length line in the front view, (b). The angle between a true length line and a principal plane is a true size angle, the required angle.

REVOLUTION OF A PLANE TO FIND ITS TRUE SIZE

The process of finding the true size of a plane involves the construction of an edge view of the plane in a primary auxiliary. This is done by revolving the edge, then projecting it back to the principal view.

Given the top and front views of plane ABC, proceed as follows:

1. Find a point view of plane by drawing a horizontal line in front view to obtain a true length line in top view, from which a point view primary auxiliary is projected, Fig. 13-6

(a). Plane ABC appears as an edge view in this auxiliary.
2. Revolve edge view of plane ABC to a position parallel to H-1 plane, using point B as radius center, (b).
3. Project points B,C'A' to top view where they intersect with their corresponding projectors which are parallel to H-1 plane, (c).
4. Connect these points of intersection to produce true size plane A'BC'.

A true size of the plane could have been found in the front view in a similar manner.

REVOLUTION OF A PLANE TO FIND THE EDGE VIEW

The edge view of a plane may be found by finding a true length line on the plane in the adjacent view, then projecting a point view of the line to the view where the edge view is desired, Fig. 13-7.

Given the top and front views of the plane ABC, proceed as follows:

1. Draw frontal line AD in top view parallel to H-F plane of projection, Fig. 13-7 (a). Project this line to front view for a true length line AD on plane ABC.
2. Revolve true length line AD to vertical position AD', using

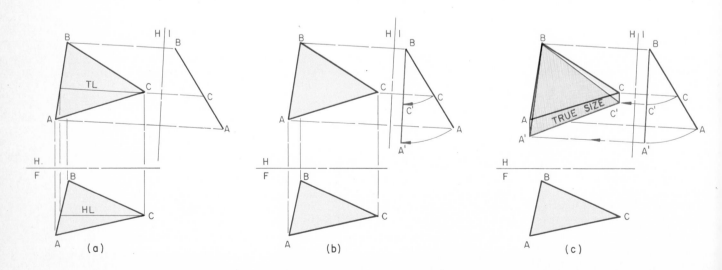

Fig. 13-6. Finding the true size of a plane by revolution.

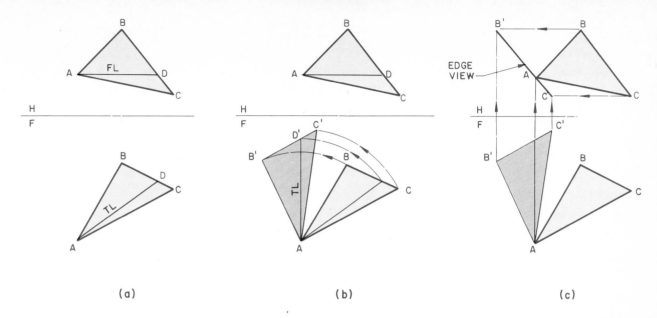

Fig. 13-7. Revolving a plane to find its edge view.

A as a center of radius. Then, by means of a compass and dividers, transfer plane ABC to its corresponding position around line AD', (b).

3. Project points A, B' and C' to top view to intersect with corresponding projectors, (c).

4. Join these points to form an edge view of plane ABC, (c). Points will lie in a straight line if projections and measurements have been accurate.

TRUE SIZE OF AN ANGLE BETWEEN TWO INTERSECTING PLANES BY REVOLUTION

The true size of an angle between two planes, whose line of intersection appears in its true length in a principal view, may be found by drawing a right section through the two planes. This section is then revolved to the horizontal and projected to the adjacent view, Fig. 13-8.

Given the top and front views of two intersecting planes, proceed as follows to find the true size of the angle between

the planes:

1. Draw a right section through view in which line of intersection between two planes appears in its true length, (a).

   Note: Intersecting line AB is parallel to plane of projection in top view and will therefore appear in its true length in front view.

2. Using D as a center, revolve right section line CD to a horizontal position, C'D, on front view, (b). Revolve point where line of intersection of two planes crosses right section line at E.

3. Project points C', E' and D to their corresponding points of intersection in top view.

   Note: Horizontal line C'D in front view projects true length in top view.

4. Connect these points of intersection in top view to form true size angle C'E'D, required angle between two intersecting planes, (c).

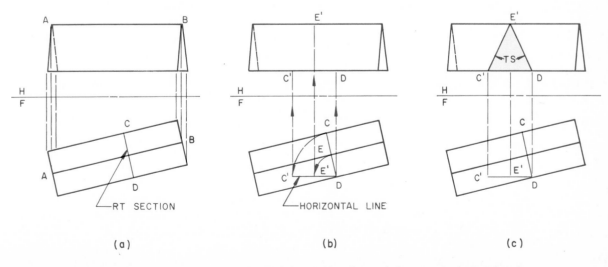

Fig. 13-8. Use of the revolution method to find the true size of an angle between two intersecting planes.

253

TRUE SIZE OF AN ANGLE BETWEEN TWO INTERSECTING OBLIQUE PLANES

The true size of angles between intersecting planes oblique to the principal planes of projection may be found by revolution. First construct a primary auxiliary where the line of intersection between the two planes appears in its true length. Then follow the preceding 4-step procedure.

REVOLUTION OF A POINT ABOUT AN OBLIQUE AXIS TO FIND ITS PATH

Machine designs occasionally include hand cranks which meet the machine surface at an oblique axis. The path of rotation of such a machine part may be traced by revolution with the assistance of auxiliary views.

path of point P.

4. Project point P' back through successive views to top view where it lies on line AB, verifying it as highest point, (c). Position of point P' in views should be established by careful measurements from appropriate reference planes, as shown in (c). Any other position could have been established such as lowest or forward position on path of rotation. This would be done by drawing a line in required direction in appropriate principal view and projecting it into all views.

5. To draw elliptical path of revolution of point in principal views, use an ellipse template of appropriate major diameter and angle. Major diameter is diameter of circular path. Ellipse guide angle is angle formed by line of sight from

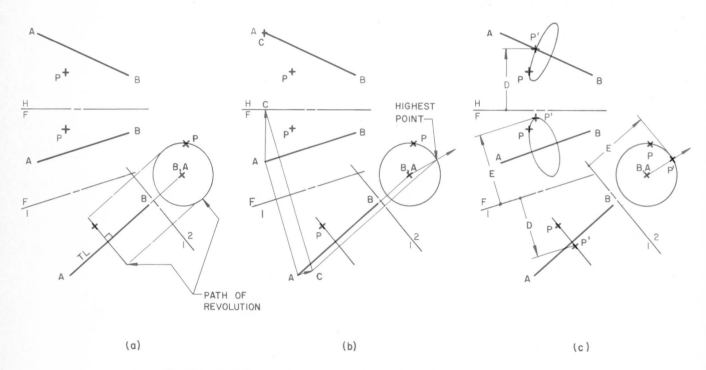

(a)   (b)   (c)

Fig. 13-9. Revolving a point about an oblique axis to find its path and highest point.

Given the top and front views of an oblique axis and a point of rotation, proceed as follows:

1. Develop true length of line AB in a primary auxiliary and project a point view of line AB in a secondary auxiliary, Fig. 13-9 (a).

2. Project point P to primary and secondary auxiliaries and draw path of rotation in secondary auxiliary. As a radius, use distance from point P to axis of line AB. Path of revolution appears as an edge in primary auxiliary, which is perpendicular to line AB and passes through point P.

3. To locate highest point on path of point P, draw vertical line AC to HF plane, then project it as a point in top view, (b). Project this line back through primary and secondary auxiliary views. Where directional arrow crosses circular path in secondary auxiliary, point P', is highest point on

front view with edge view of circular path in primary auxiliary. A horizontal auxiliary view would need to be constructed from top view to find angle of ellipse guide to use in top view.

PRIMARY REVOLUTIONS OF OBJECTS ABOUT AXES PERPENDICULAR TO PRINCIPAL PLANES

Revolutions are made to obtain one or more of three advantages: a clear view of an object; the true length of a line; the true size of a surface. Primary revolutions are drawn perpendicular to one of the principal planes of projection. An object is shown in its normal position in Fig. 13-10 (a). It is revolved in the horizontal plane in (b); in frontal plane in (c); in profile plane in (d). Regular orthographic principles are used in projecting revolved views.

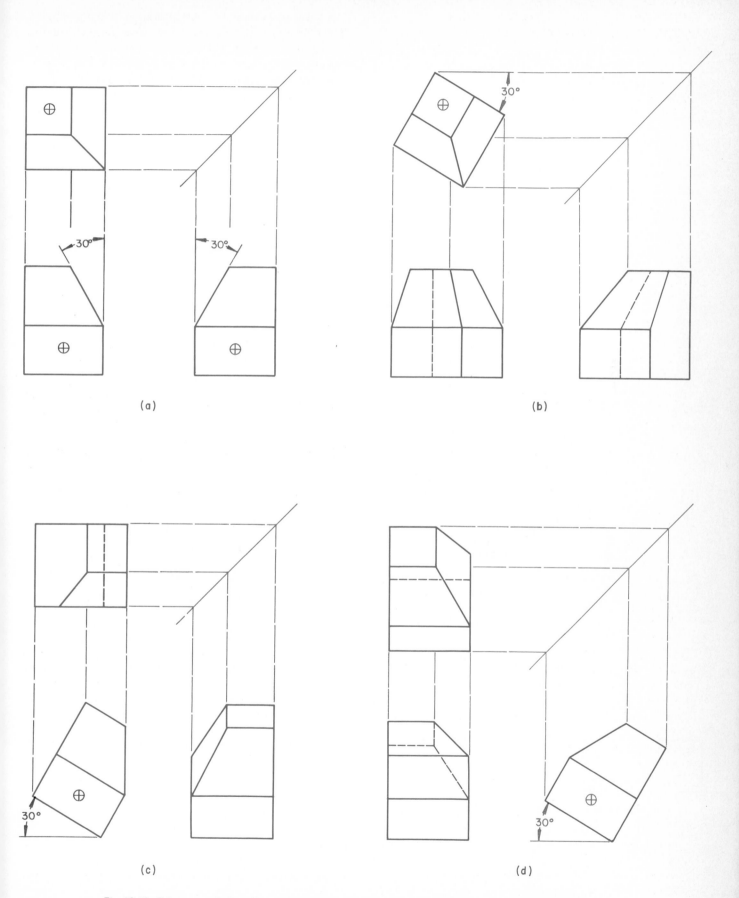

(a)

(b)

(c)

(d)

Fig. 13-10. Primary revolutions of an object around the axes of principal planes of orthographic projection.

SUCCESSIVE REVOLUTIONS OF AN OBJECT WITH AN OBLIQUE SURFACE

The true size of an oblique surface may be obtained by the use of successive revolutions, Fig. 13-11. This method is similar to finding the true size of an oblique surface through successive auxiliary views.

A pictorial view of the object with an oblique surface is shown in Fig. 13-11 (a). The object is shown in its normal orthographic position in (b), with an indication of the revolution to be made in the first of two successive revolutions. The first revolution of the object is performed in (c), while the second revolution, (d), produces the true size view of the oblique surface.

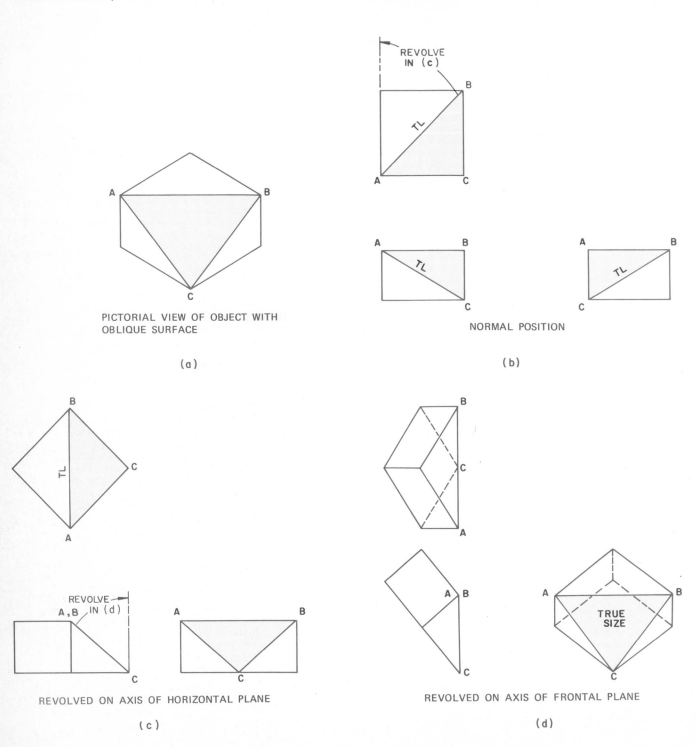

Fig. 13-11. Finding the true size of an oblique surface through successive revolutions.

PROBLEMS IN DRAWING REVOLUTIONS

The following problems are planned to give students proficiency in use of revolution method of solving problems in drafting and design. Size A drafting sheets should be used. Label all reference planes, points and lines. Problems are shown in 1/4 inch section paper to facilitate location on drawing sheet. Use revolution method in solving problems.

1. Refer to Fig. 13-12, then lay out and solve each section as follows: In (a), find true length of line in horizontal view. In (b), find true length of line in front view and indicate angle it makes with horizontal plane. In (c), find true length of line in horizontal view and indicate angle it makes with frontal plane. In (d), find true length of line in horizontal plane and indicate angle it makes in frontal plane.

2. Refer to Fig. 13-13, then lay out and solve each section as follows: In (a), find true size of angle line in profile view makes with frontal plane. In (b), find true size of angle the line in frontal view makes with profile plane. In (c), find true size of plane by projecting an edge view in a horizontal auxiliary and by revolution. In (d), find true size of plane by a primary auxiliary view and by revolution.

3. Refer to Fig. 13-14, then lay out and solve each section as follows: In (a), find the true size of the plane by projecting an edge view in horizontal auxiliary and by revolution. In (b), find the true size of the plane by a primary auxiliary and by revolution. In (c), find true size of angle between intersecting planes. In (d), find true size of angle between intersecting planes.

4. Refer to Fig. 13-15 and find path of revolution of points about lines in sections (a and b). Lay out measurements and projections accurately. Allow approximately one inch space between frontal view of line and primary auxiliary projection plane, and same amount in secondary auxiliary. Locate highest point and project this to all views. Measure and dimension diameter of path of revolution of point in view where path appears as a circle.

5. Refer to Fig. 13-16 and revolve objects about their axes (as indicated) perpendicular to principal planes as follows: In (a), revolve objects about horizontal axes. In (b), revolve objects about frontal axes. In (c), revolve objects about profile axes. Draw three views, starting with view perpendicular to axis of revolution. Place each drawing on a separate A size sheet. Estimate dimensions to retain proportions of object shown in pictorial view. Indicate angle of rotation, but do not dimension further.

6. Refer to Fig. 13-16 (d) and prepare a successive revolution of object to produce a true size view of oblique surface. Use a B size sheet and divide drawing space into four sections. Then prepare in manner shown in Fig. 13-11, starting with a pictorial view. Select an appropriate scale and estimate dimensions to retain proportions shown. Indicate true size surface in view in which it appears, but do not dimension drawing.

7. Find true length of center line of all legs of SUPPORT BRACKET shown in Fig. 13-17.

8. Find true length of all three antenna guy wires, Fig. 13-18.

9. Find true distance between transmission towers A, B and C shown in Fig. 13-19. Each tower is 40 feet above ground level on which it stands.

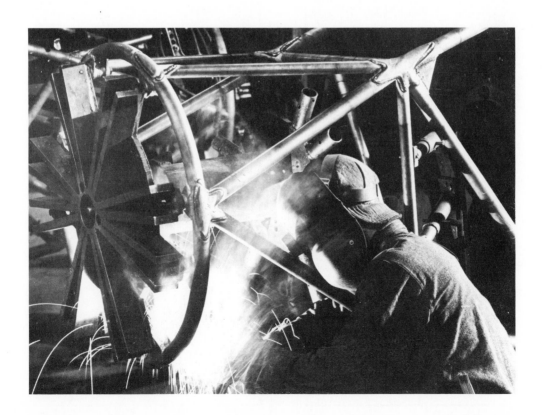

To find the length of the individual members of this steel frame, the projection method of revolution is needed.   (Lincoln Electric Co.)

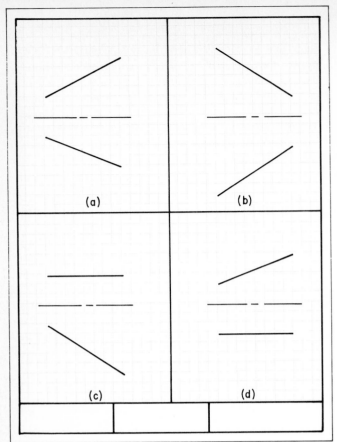

Fig. 13-12. True length of line problems.

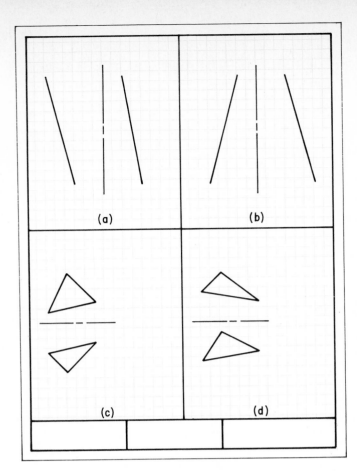

Fig. 13-13. True size of angles and plane problems.

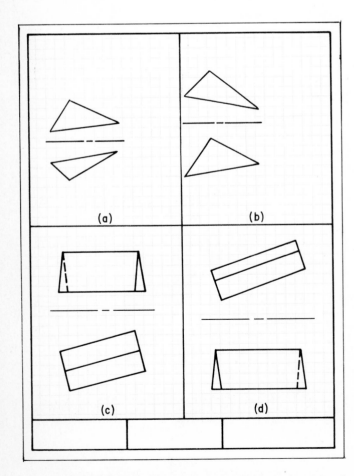

Fig. 13-14. Problems for finding edge views of planes and true size of angle between intersecting planes.

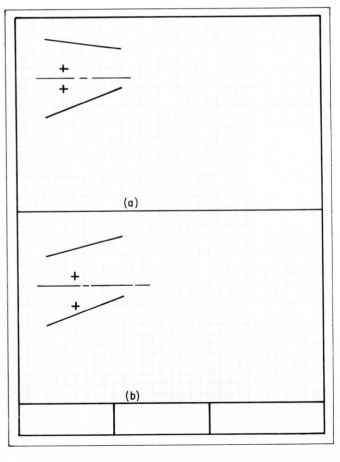

Fig. 13-15. Problems for revolving a point about a line.

Revolutions

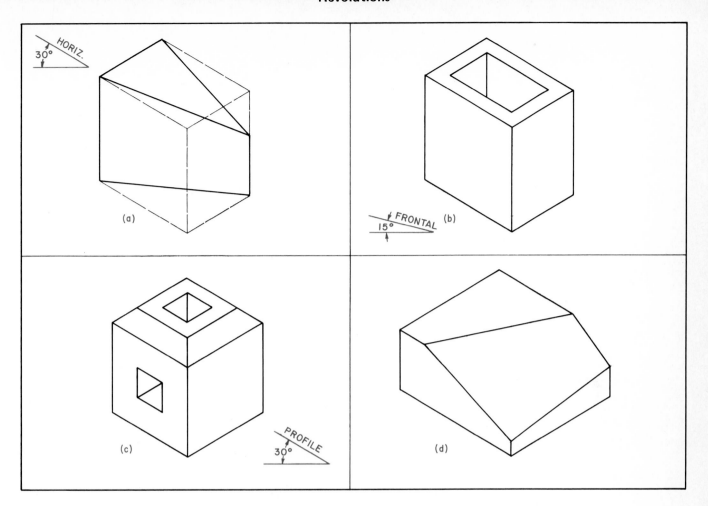

Fig. 13-16. Revolution of objects problems.

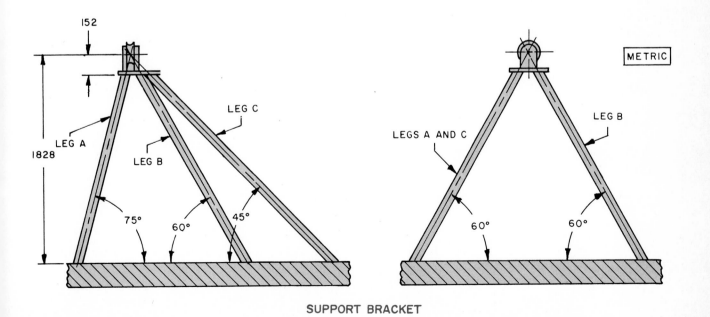

SUPPORT BRACKET

Fig. 13-17. An applied problem in finding true length of a line.

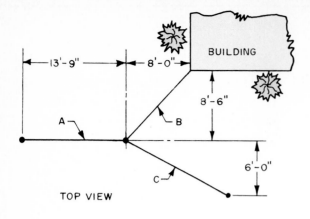

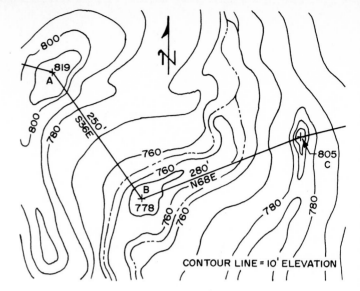

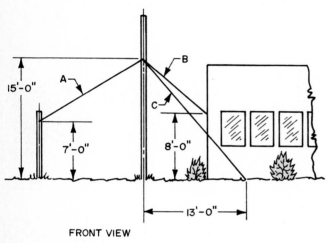

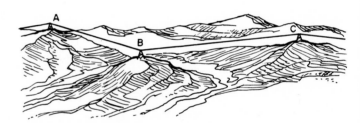

Fig. 13-18. Problem in finding true length of guy wires for an antenna.

Fig. 13-19. A topographic map showing elevation and horizontal distances between three transmission line towers.

Chapter 14
INTERSECTIONS

When two or more objects (such as two planes or a cylinder and a square prism) join or pass through each other, the lines formed at the junction of their surfaces are known as "intersections."

Numerous examples of intersections may be found in industry. Frequently, the design and specification of buildings require architects and engineers to define the intersection of surfaces, Fig. 14-1. The Aerospace and automotive industries also work with intersections of various shapes in the manufacture of instrument panels, body sections and window openings, Fig. 14-2.

This chapter covers basic principles and geometric forms of intersecting objects. Once these principles and the techniques of their application are understood, the solution of most intersection problems is possible.

Fig. 14-2. Defining the intersection of surfaces, such as the fins of this aircraft, requires study and planning for proper assembly. (Fuji Film)

Fig. 14-1. Architects and engineers must resolve intersections created in the design of buildings and clearly specify them on the drawings.

Drafting for Industry

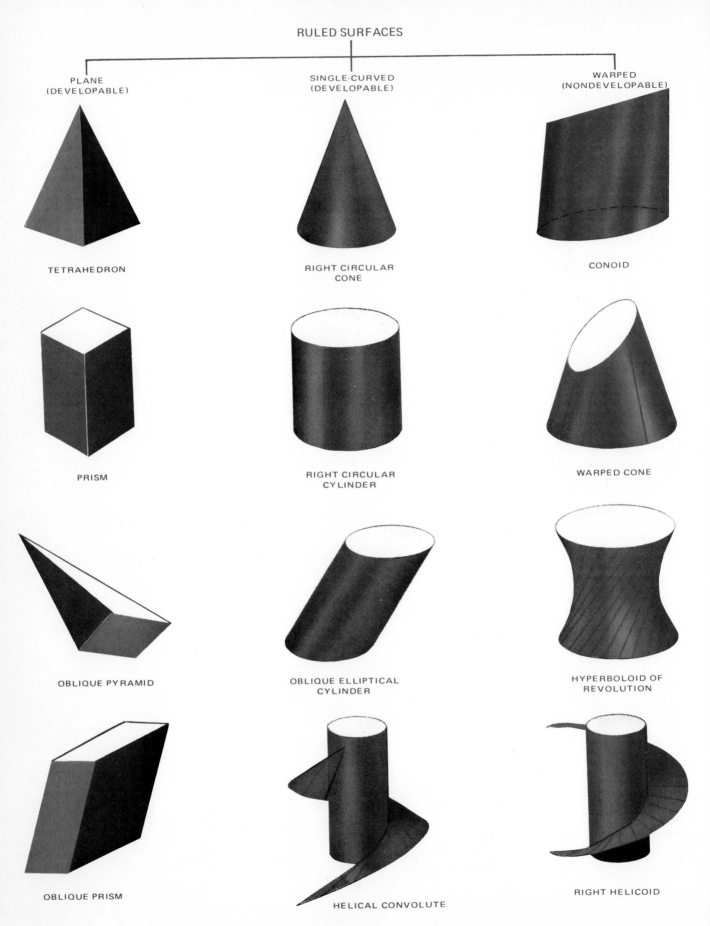

RULED SURFACES

PLANE
(DEVELOPABLE)

SINGLE-CURVED
(DEVELOPABLE)

WARPED
(NONDEVELOPABLE)

TETRAHEDRON

RIGHT CIRCULAR
CONE

CONOID

PRISM

RIGHT CIRCULAR
CYLINDER

WARPED CONE

OBLIQUE PYRAMID

OBLIQUE ELLIPTICAL
CYLINDER

HYPERBOLOID OF
REVOLUTION

OBLIQUE PRISM

HELICAL CONVOLUTE

RIGHT HELICOID

Fig. 14-3. Ruled geometrical surfaces.

Fig. 14-4. A warped surface formed by an explosive process. (Martin Marietta Aerospace)

TYPES OF INTERSECTIONS

Intersections and their solutions are classified on the basis of the types of geometrical surfaces involved. Two broad classifications of geometrical surfaces are RULED SURFACES and DOUBLE-CURVED SURFACES.

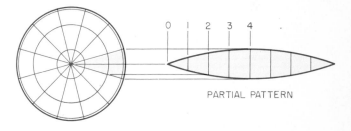

PARTIAL PATTERN

RULED SURFACES

Ruled surfaces are those which are generated by moving a straight line. They may be subdivided into planes, single-curved surfaces and warped surfaces, Fig. 14-3.

Planes and single-curved surfaces are capable of being developed, meaning their ruled surfaces can be unfolded or unrolled to lie in a single plane. Warped surfaces are ruled surfaces which cannot be developed into a single plane. Usually, they are formed to true shape by peening, stamping, spinning or by a vacuum or explosive process, Fig. 14-4. Warped surfaces may be divided into sections and developed, but the result would be an approximate of the true warped surface, Fig. 14-5.

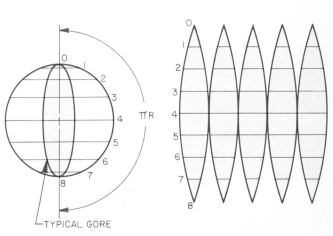

TYPICAL GORE

Fig. 14-5. The gore method of developing an approximate sphere.

DOUBLE-CURVED SURFACES

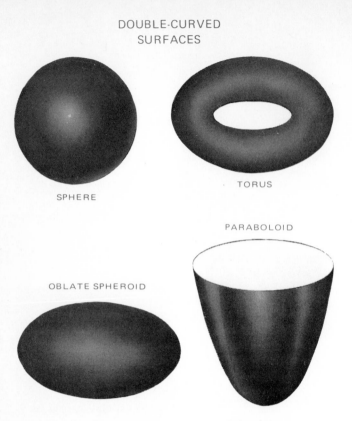

SPHERE

TORUS

OBLATE SPHEROID

PARABOLOID

Fig. 14-6. Double-curved geometrical surfaces.

DOUBLE-CURVED SURFACES

Double-curved surfaces are generated by a curved line revolving about a straight line in the plane of a curve, Fig. 14-6. Some common double-curved surfaces, like warped surfaces, are not capable of being developed into single plane surfaces.

SPATIAL RELATIONSHIPS

In Chapter 7, an introduction was given to normal points,

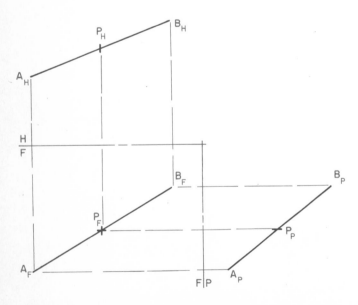

Fig. 14-7. Location of a point on a line in space by orthographic projection method.

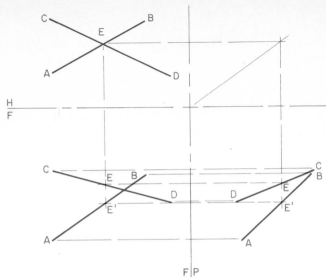

Fig. 14-8. Nonintersecting lines in space.

lines and surfaces in space. Chapter 12 presented the techniques of locating inclined and oblique lines and surfaces in space. Still remaining are a few basic spatial relationships that must be understood before you can solve problems of intersections and development.

POINT LOCATION ON A LINE

Lines are composed of an infinite number of points. To solve problems in space, specific points on lines and surfaces must be located. Line AB is shown in horizontal, frontal and profile views in Fig. 14-7. End points A and B are located in the three views by projectors and any point, P for example, can be located in a similar manner.

INTERSECTING AND NONINTERSECTING
LINES IN SPACE

Lines that cross in space are not necessarily intersecting lines. Intersecting lines have a common point that lies at the exact point of intersection.

If crossing lines AB and CD, Fig. 14-8, are intersecting lines,

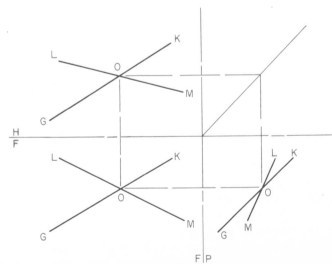

Fig. 14-9. Intersecting lines in space.

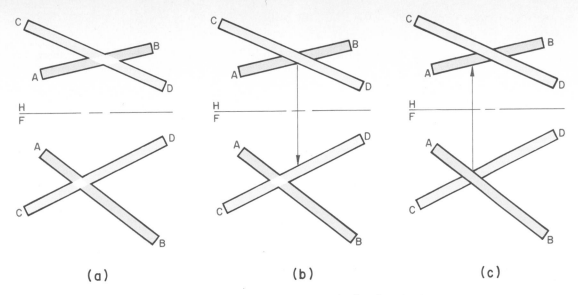

Fig. 14-10. Visibility of crossing lines in space.

they will have a point common to both lines. Locate point E in the horizontal view and project it to intersect lines AB and CD in the frontal view. It is apparent in the front view that point E is not a single point common to both lines. Therefore, the lines do not intersect. This is further shown in the right side or profile view.

In Fig. 14-9, lines GK and LM do intersect. They do have a common point that lies on both lines as revealed by orthographic projection. Two views are sufficient to determine whether crossing lines are intersecting lines. Note that point O is common to both lines in any two views, which verifies the intersection of the lines.

VISIBILITY OF CROSSING LINES IN SPACE

Visibility of crossing lines is established by projecting the crossing point of the lines from an adjacent view, Fig. 14-10 (a). Lines AB and CD are crossing lines.

To determine which line is visible at the point of crossing in the top view, project the crossing point from the top view, Fig. 14-10 (b). The projection line crosses line CD first in the front view, indicating that line CD is higher and, therefore, visible in

the top view at the point of crossing.

To check the visibility of the crossing point in the front view, project the point of crossing from the front view, Fig. 14-10 (c). The projector intersects line AB first, indicating line AB is nearer the frontal viewing plane and visible at the point of crossing of the two lines.

VISIBILITY OF A LINE AND PLANE IN SPACE BY THE ORTHOGRAPHIC PROJECTION METHOD

Determining the visibility of a line and a plane that cross in space is similar to that of two crossing lines. Given line AB and the plane CDE that cross in the horizontal and frontal views, Fig. 14-11 (a), line AB crosses two lines of the plane, lines CE and ED.

Visibility in the horizontal view is determined by projecting the crossing points to the front view, Fig. 14-11 (b). The projectors intersect line AB before they intersect the plane. This indicates that the line is higher than the plane at these points and is visible in the horizontal view.

It is determined that line AB is invisible in the frontal view since the projectors intersect the plane before the line. This

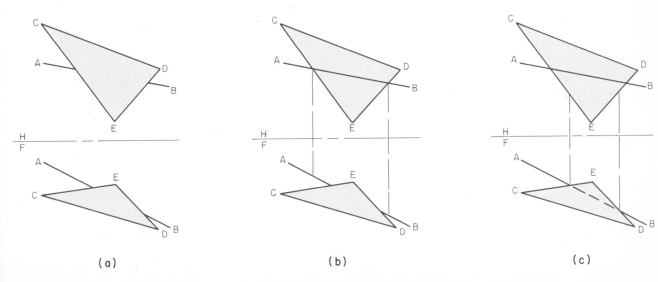

Fig. 14-11. Visibility of a line and plane in space.

Drafting for Industry

indicates that the plane is nearer the frontal view and that line AB crosses behind plane CDE.

Triangular planes and/or prisms have been used here and in other sections to illustrate the intersection of lines and surfaces. This makes the understanding of the principles and procedures less complex. The process described for the intersection of planes is the same regardless of the number of edges a plane contains.

Fig. 14-12. Location of the piercing point of a line and a plane was necessary in the design of this equipment.
(Chicago Bridge and Iron Co.)

266

LOCATION OF PIERCING POINT OF A LINE WITH A PLANE BY THE ORTHOGRAPHIC PROJECTION METHOD

The point of intersection between a plane and a line inclined to that plane is called the piercing point. The location of the piercing point is essential to the solution of many technical problems, such as the intersection of pipe and cables with oblique planes, Fig. 14-12.

Given the top and front views of a line and a plane, Fig. 14-13 (a), proceed as follows to locate the piercing point of a line and a plane:

1. Pass a vertical cutting plane through top view of line AB (see pictorial view) which intersects plane DCE at G and K, as shown in (b).
2. Project G and K to front view.
3. Intersection of the imaginary cutting plane and plane CDE

is represented by trace line GK in front view, as shown in (c).

4. Intersecting line AB lies in imaginary cutting plane and will intersect plane CDE along line GK at point O (front view), the piercing point.
5. Project point O to establish piercing point in top view.
6. Visibility of line segment OB in front view is determined to be invisible by projecting crossing point L to top view where line segment OB is found to be behind line DE.
7. Line segment OB in top view is found to be visible by projecting crossing point K to front view where line segment OB is found to be higher than line DE.
8. Line segment AO is found to be visible in front view and invisible in top view by same process of projecting crossing points M and G.

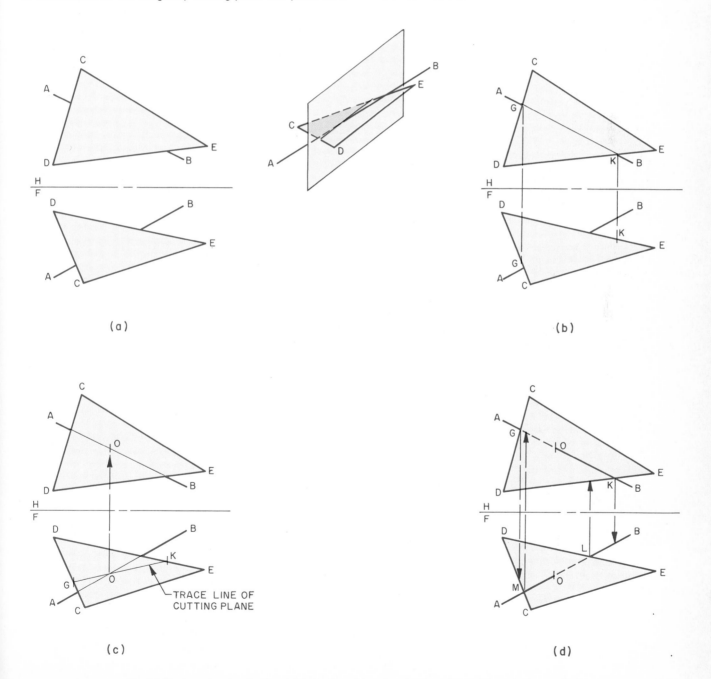

Fig. 14-13. Locating the piercing point of a line with a plane by the orthographic projection method.

LOCATION OF PIERCING POINT OF A LINE WITH A PLANE BY THE AUXILIARY VIEW METHOD

The piercing point of a line and a plane may also be located by means of an auxiliary view as shown in Fig. 14-14. An edge view of the plane ABC is projected in a horizontal auxiliary view. Then the point where the line intersects the edge view of the plane in the auxiliary is projected to the horizontal view. Where this projection intersects line DE is the piercing point in that view. Piercing point is then projected to other view.

Accuracy of the projection may be checked in the front view, Fig. 14-14, by measuring the distance the piercing point lies below the H-F and H-1 projection planes. Visibility of the line and plane is found by determining whether, at the point of crossing, the line or the plane lies nearer the projection plane in the adjacent view. The closest one is visible.

Given the top and front views of plane ABC and line DE which intersect, Fig. 14-14 (a), proceed as follows to locate the piercing point and the visibility of the line and plane:
1. Draw horizontal line AF in front view and project it to top view to produce a true length line, (b).
2. Project this true length line as line of sight and construct auxiliary projection plane H-1 perpendicular to it.
3. Project edge view of plane ABC by finding point view of line AF, (c). Project line DE and where it intersects plane ABC is piercing point O.
4. Project point O to top and front views to intersect with line DE for piercing point in these views, (d). Point O may be checked for accuracy in front view by measuring for H.
5. Visibility is determined by method previously described for orthographic projection.

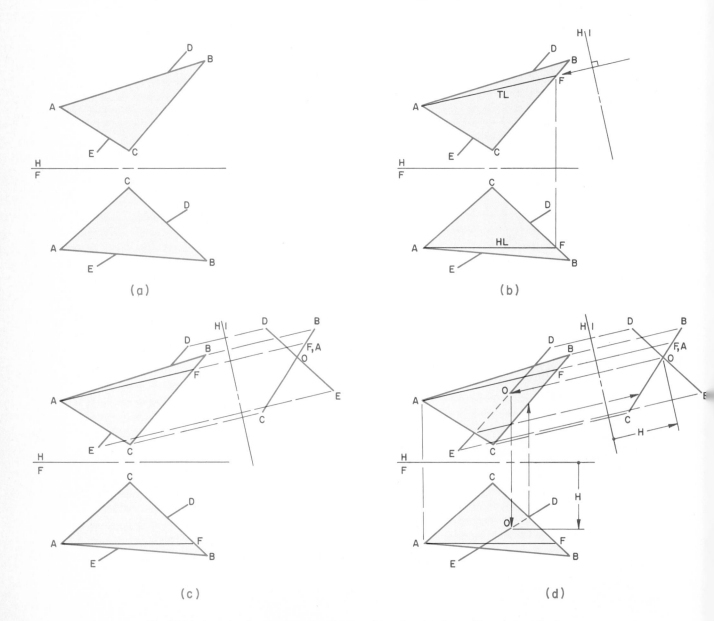

(a) (b) (c) (d)

Fig. 14-14. Locating the piercing point of a line with a plane by the auxiliary view method.

Intersections

A LINE THROUGH A POINT AND PERPENDICULAR TO AN OBLIQUE PLANE

It is sometimes necessary to locate the shortest distance from a point to an oblique plane or construct a perpendicular to the plane. By drawing an auxiliary view, you can show an edgewise view and a perpendicular erected through the point to the plane.

Given the top and front views of a plane and a point in space, Fig. 14-15 (a), proceed as follows:

1. Draw a horizontal line in the front view and project this line to the top view where it appears in true length, as shown in (b).
2. Construct a primary auxiliary view of plane ABC which will appear as an edge.
3. Project point P to auxiliary.
4. Draw line PO perpendicular to edge view of plane ABC in auxiliary, (c).
5. Project piercing point O to top view where it joins line PO. Line PO is parallel to H-1 plane of projection because it is a true length line in auxiliary view. Line PO will also be perpendicular to line of projection which projects true length in top view.
6. Project O to front view where it will intersect line EF, which is projected from a frontal line through point O in top view. Point O will be distance H from H-1 and H-F planes of projection.

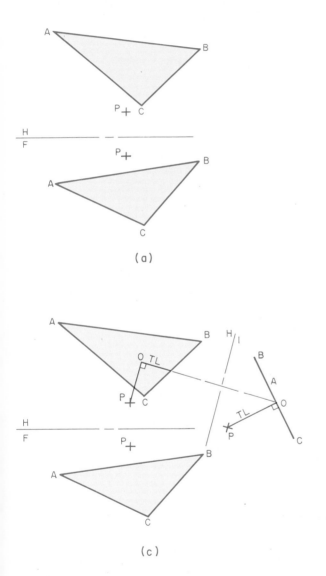

(a)

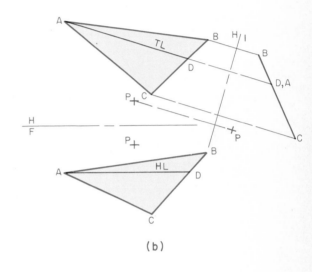

(b)

(c)

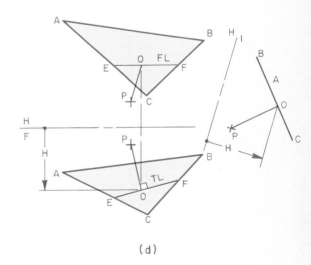

(d)

Fig. 14-15. Constructing a line through a point and perpendicular to an oblique plane.

INTERSECTION OF TWO PLANES BY THE ORTHOGRAPHIC PROJECTION METHOD

The intersection of two planes is a straight line and may be found by locating, on one plane, the piercing points of the lines representing the edges of the second plane. Then connect the two points for the line of intersection, Fig. 14-16.

Given the top and front views of two intersecting planes, Fig. 14-16 (a), proceed as follows to locate their line of intersection:

1. Pass an imaginary cutting plane vertically through line AB in top view to establish points 1 and 2, (b).
2. Project points 1 and 2 to front view where they cross lines DE and FE.
3. Line AB pierces plane DEF at G where it crosses line 1-2. Project point G to top view.
4. In a similar fashion, pass a cutting plane through line AC

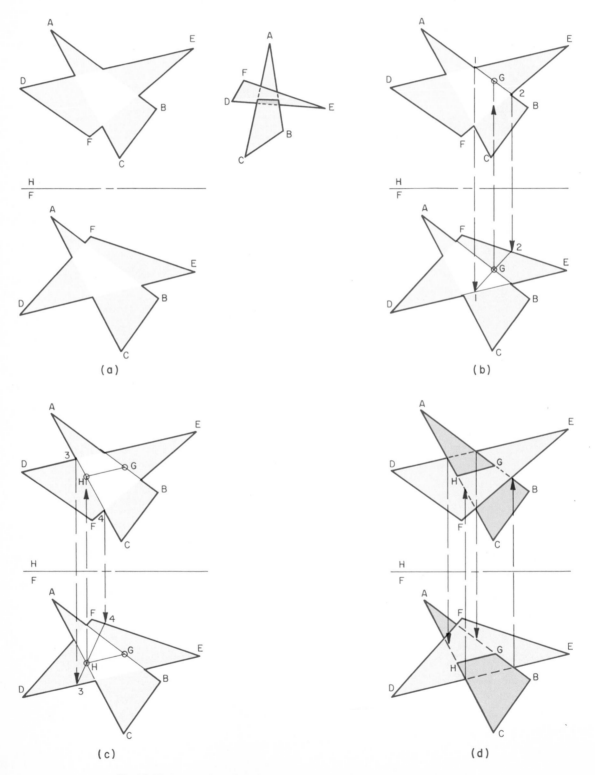

Fig. 14-16. Intersection of two planes by the orthographic projection method.

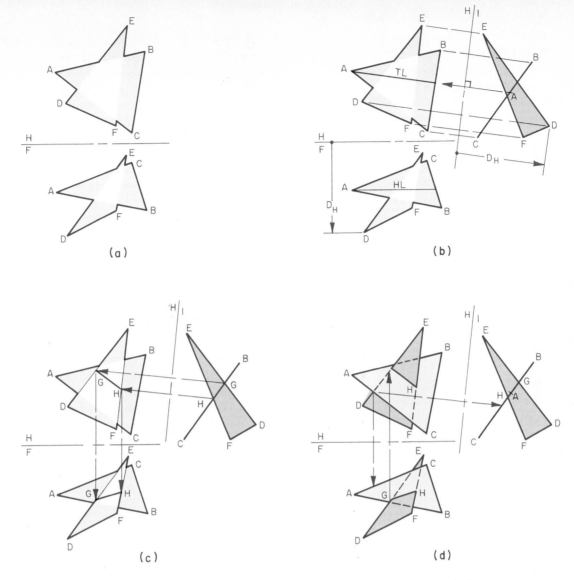

Fig. 14-17. Locating the intersection of two planes by the auxiliary view method.

establishing points which are projected to front view crossing lines DE and FE, (c).

5. Line AC pierces plane DEF at H where it crosses line 3-4. Project point H to top view.

6. Analyze crossing of lines AH and DE in top view for visibility. Line AH is found to be higher, therefore visible in top view, (d).

7. Analyze crossing of AG and DE in top view for visibility. Line AG is found to be higher and visible.

8. GAH is visible in top view.

9. Visibility in front view is determined in a like manner and segment HGBC of plane ABC is found to be visible.

INTERSECTION OF TWO PLANES BY THE AUXILIARY VIEW METHOD

The line of intersection between two planes may be found more readily by the use of an auxiliary view in some instances. One plane is viewed as an edge by constructing a reference plane (H-1) perpendicular to the point view of a true length line on that plane, Fig. 14-17.

Given the top and front views of two intersecting planes,

Fig. 14-17 (a), proceed as follows to locate the line of intersection:

1. Draw a horizontal line in front view and project it to top view to produce a true length line, (b).

2. Project point view of this line to horizontal auxiliary and construct an edge view of plane ABC.

3. Project plane DEF to auxiliary view.

4. Label points of intersection between two planes in auxiliary view, G and H, (c).

5. Project G to top view where it intersects line ED.

6. Project H to its line of intersection, EF. Line GH is line of intersection in top view.

7. Project points G and H to their intersecting lines in front view to establish line of intersection there.

8. Analyze the front view by viewing the top view from crossing in the front view of lines AB and DE. Line DE is found to be nearer and visible in the front view, as shown in Fig. 14-17 (d).

9. Analyze top view for visibility by viewing front view or auxiliary view as shown by crossing lines AC and DE. AC is closest in both views and is visible in top view.

INTERSECTION OF AN INCLINED PLANE AND A PRISM BY THE ORTHOGRAPHIC PROJECTION METHOD

When a plane cuts a prism and appears as an edge in one of the principal views, locating lines of intersection is a matter of finding piercing points of the lines of intersection.

Given at least two principal views, including one with edgewise view of cutting plane, proceed as follows:

1. Label intersections of plane and lateral corners of prism in top and side views, Fig. 14-18 (a).
2. Project these points to their corresponding lines in front view, (b).
3. Join points A, B and C in front view to form line of intersection.
4. Determine visibility of prism and line of intersection by viewing top and side views from front view. Vertical corners of prism which appear in front and below cutting plane are visible in front view. Lateral corners which fall behind plane are invisible in front view. Line of intersection AB in front view is invisible as is a portion of upper edge of plane. This is because their locations are farther away from frontal-profile projection plane than prism.

INTERSECTION OF AN OBLIQUE PLANE AND A PRISM BY THE CUTTING PLANE METHOD

When a cutting plane is oblique to the principal planes of projection (not perpendicular to any one of them), the cutting plane method is used to find the intersection of the plane and prism.

Given two principal views, Fig. 14-19 (a), proceed as follows:

1. Label intersections of plane and lateral corners of prism in top view, (b).
2. Pass cutting planes 1, 2 and 3 through corners of prism in top view and extend through edges of oblique plane. These planes have been drawn horizontally but may be in any direction except perpendicular to the projection plane.
3. Project intersections of cutting planes 1, 2 and 3 with oblique plane edges to front view as shown with line 3 in (b). This line in front view is line of intersection of cutting plane 3 with oblique plane.
4. Project points in top view, where three cutting planes intersect corners of prism, to front view. (That is, point B of line 3 top view to point B, line 3 in front view.)
5. These points in front view represent piercing points of lateral corners of prism. Since prism and oblique plane are intersecting plane figures, their lines of intersection will be straight lines. Join piercing points with light construction lines until visibility is established.
6. Visibility is established in each view by viewing adjacent view for nearness of lines and surfaces, (c).

INTERSECTION OF AN OBLIQUE PLANE AND AN OBLIQUE PRISM BY THE AUXILIARY VIEW METHOD

When neither the cutting plane nor the prism appear as an edge in any of the principal views, construct an auxiliary view to provide an edgewise view of the cutting plane. Once this view is constructed, the solution is similar to that for the inclined plane.

Given two principal views, Fig. 14-20 (a), proceed as follows:

1. Construct primary auxiliary view by taking point view of a line on cutting plane to produce an edgewise auxiliary view of cutting plane 1-2-3-4, (b).

    Note: Either a horizontal or frontal auxiliary could be used.
2. Project prism to this auxiliary view.

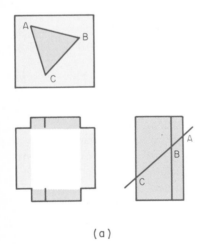

(a)

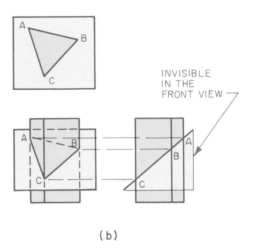

(b)

Fig. 14-18. Locating the intersection of an inclined plane and a prism by the orthographic projection method.

3. Label intersections of plane and lateral corners of prism in auxiliary view, (c).

4. Project these intersections to their corresponding lines in top view. These points are piercing points of lateral corners of prism and cutting plane. Join these points to form line of intersection in top view.

5. Project piercing points from top view to their corresponding lines in front view and draw line of intersection.

6. Visibility is established by checking adjacent views for nearness of surfaces.

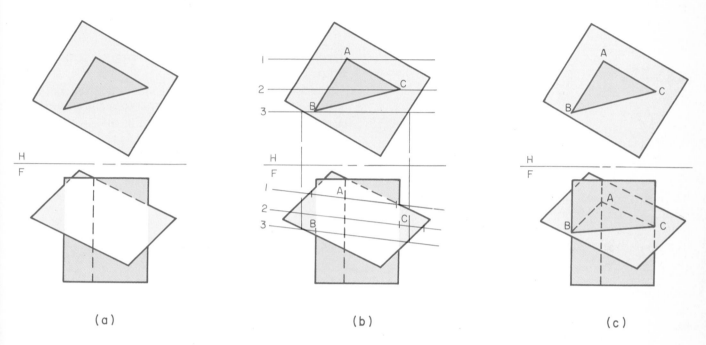

Fig. 14-19. Locating the intersection of an oblique plane and a prism by the cutting plane method.

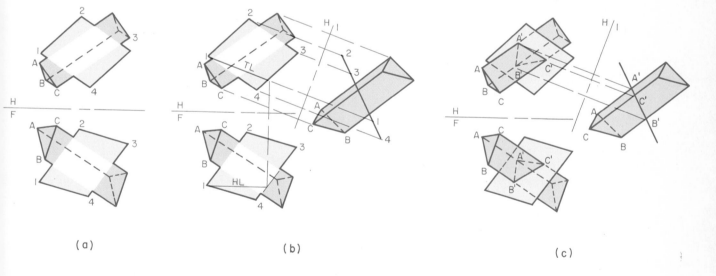

Fig. 14-20. Locating the intersection of an oblique plane and oblique prism by the auxiliary view method.

INTERSECTION OF TWO PRISMS BY THE CUTTING PLANE METHOD

The surfaces of prisms consist of a number of single planes. Therefore, the intersection of two prisms may be thought of as a prism intersecting with one or more single planes. The solution should be approached as outlined in the preceding sections, working with one plane at a time. Two applications are given. First, an illustration of the cutting plane method; second, the auxiliary projection method.

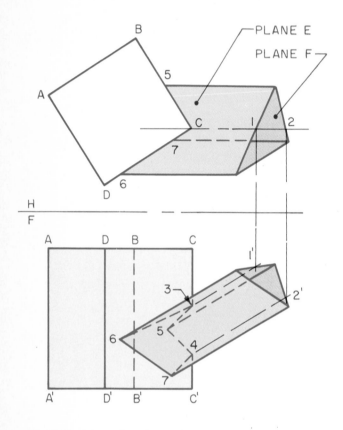

Fig. 14-21. Intersection of two prisms by the cutting plane method.

In Fig. 14-21, two intersecting prisms are shown in top and front views. Two of the lateral edges of the triangular prism intersect the square prism on the front plane and one intersects a rear plane as seen in the top view. The piercing points of all three lateral edges of the triangular prism are shown in the orthographic projection in the top and front views.

The points at which lateral edge CC' (Fig. 14-21) intersects planes E and F cannot be obtained directly by projection in the principal views. However, a vertical cutting plane, parallel to the lateral edges of the triangular prism, can be passed through lateral edge CC' (where the undetermined points lay) and projected to the front view.

Given the top and front views, proceed as follows:

1. Draw a line in top view parallel to edges of triangular prism and through corner C, Fig. 14-21. This line represents vertical cutting plane.

2. Project points 1 and 2, where cutting plane intersects edges of planes E and F, to front view to intersect edges of same planes at 1' and 2'.

3. Project lines parallel to triangular prism through points 1' and 2' to intersect with edge CC' at points 3 and 4, piercing points of edge CC' with planes E and F.

4. Join piercing points 5, 6 and 7 of lateral edges of triangular prism and piercing points 3 and 4 to complete the intersection between the two prisms.

INTERSECTION OF TWO PRISMS BY THE AUXILIARY VIEW METHOD

A second method of finding the intersection between two prisms is the auxiliary view method. The same two prisms used in the cutting plane method are used in the auxiliary view method, Fig. 14-22.

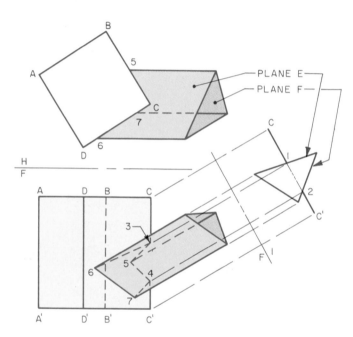

Fig. 14-22. Intersection of two prisms by the auxiliary view method.

The piercing points of the lateral edges of the triangular prism are found by regular orthographic projection. The intersections of lateral edge CC' with planes E and F are found by drawing a frontal-auxiliary view in which a point view is shown of the end of the triangular prism and lateral edge CC' is shown as a line.

Given the top and front views, Fig. 14-22, proceed as follows:

1. Construct a frontal-auxiliary view that shows a point view of triangular prism by using a reference plane that is perpendicular to lateral edges of the prism.

2. Project lateral edge CC' as a line. This is only part of square prism that is necessary to find unknown points of intersection of two prisms (points 1 and 2), since other piercing

points can be located in primary orthographic views.

3. Project points 1 and 2 in auxiliary view to intersect with lateral edge CC' at points 3 and 4, piercing points of edge CC' with planes E and F.
4. Join piercing points 5, 6 and 7 of lateral edges of triangular prism and piercing points 3 and 4 to complete intersection between two prisms.

INTERSECTION OF A PLANE AND A CYLINDER BY THE ORTHOGRAPHIC PROJECTION METHOD

When the intersecting plane appears as an edge view in one of the views, Fig. 14-23, the line of intersection may be found by projection between the principal views.

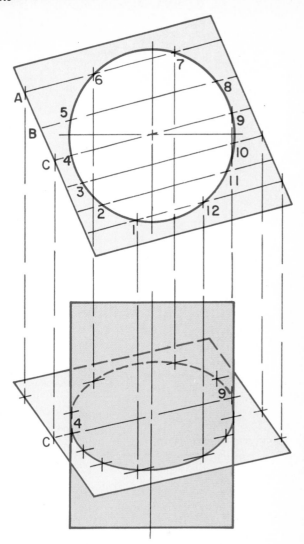

Fig. 14-24. Intersection of a plane and a cylinder by the cutting plane method.

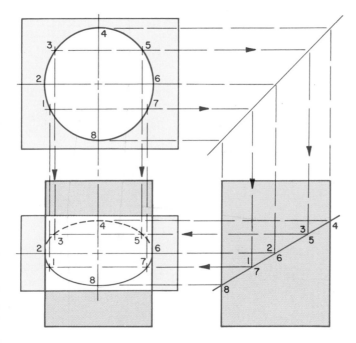

Fig. 14-23. Intersection of a plane and a cylinder by the orthographic projection method.

Given the three principal views, proceed as follows:
1. Project points of intersection of several randomly spaced parallel lines in circular (top) view of cylinder to front and side views, Fig. 14-23.
2. Project points where these lines intersect with edge view of plane, extending them to front view to intersect with corresponding lines of projection from top view.
3. Connect these points of intersection to form line of intersection between plane and cylinder.
4. Visibility is determined by checking adjacent views.

INTERSECTION OF AN OBLIQUE PLANE AND A CYLINDER BY THE CUTTING PLANE METHOD

When the plane is oblique to the principal views, the cutting plane method may be employed to find the line of intersection, Fig. 14-24.

Given the top and front views, proceed as follows:
1. Pass randomly spaced vertical cutting planes through top view of cylinder and plane which run parallel to edges of plane, Fig. 14-24. Select any number of planes at random locations. Greater number produces most accurate line of intersection.
2. Project intersections of cutting plane lines with edges of plane to their corresponding location in plane in front view; example, line C. Connect two intersections of edges of plane with line C in front view. Complete location of cutting plane lines in front view.
3. Project points of intersection of cylinder and cutting plane lines in top view to their corresponding line in front view; example, point 6 and 10. Complete projection of intersecting points which are piercing points of cylinder and the oblique plane in front view.
4. Sketch a light line between these points in front view and finish with an irregular curve.
5. Visibility is determined by checking the two views.

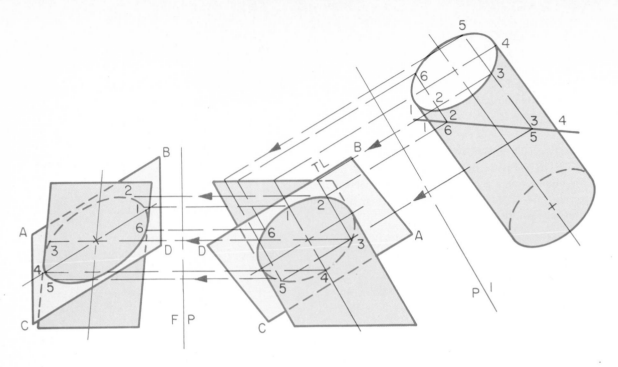

Fig. 14-25. Intersection of an oblique plane and an oblique cylinder by the auxiliary view method.

INTERSECTION OF AN OBLIQUE PLANE WITH AN OBLIQUE CYLINDER BY THE AUXILIARY VIEW METHOD

When a plane and a cylinder which are oblique to the principal views intersect, the line of intersection must be found through an auxiliary view, Fig. 14-25.

Given the frontal and profile views, proceed as follows to find the line of intersection between the oblique plane and oblique cylinder:

1. Construct an edge view of plane in a primary auxiliary, Fig. 14-25. Cylinder will appear foreshortened and elliptical ends are constructed as previously described for orthographic projection.
2. Pass randomly spaced cutting planes through elliptical end of auxiliary view of cylinder to extend to edge view of plane.
3. Project points of intersection of cutting planes from elliptical end and edge view of oblique plane in auxiliary to profile view. This forms line of intersection in this view.
4. Project points of intersection on ellipse in profile view to frontal view and transfer measurements of points from auxiliary view to frontal view. This forms line of intersection in this view.
5. Visibility is determined by checking adjacent views.

INTERSECTION OF A CYLINDER AND A PRISM

The intersection of a cylinder and a prism can be found by approaching the solution as a series of single planes intersecting with a cylinder, Fig. 14-26. Each plane is treated one at a time. Since one or more of the planes of the prism are oblique in the principal views, an auxiliary view to project the true shape of the prism is necessary, Fig. 14-27.

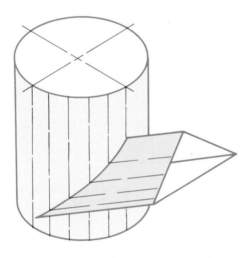

Fig. 14-26. A number of single planes make up the surfaces of cylinders.

Given the top and front views of the cylinder and prism, the procedure is as follows:

1. Construct a primary auxiliary view showing true size and shape of prism, Fig. 14-27.
2. Pass randomly spaced cutting plane lines through top view, intersecting with cylinder and prism.
3. Transfer cutting plane lines to auxiliary view so they are perpendicular to line of sight. Spacing of cutting plane lines must equal that in top view.
4. Project corresponding points of intersection between cutting planes and cylinder and prism in top and auxiliary views to front view. This locates lines of intersection.
5. Check adjacent views for visibility.

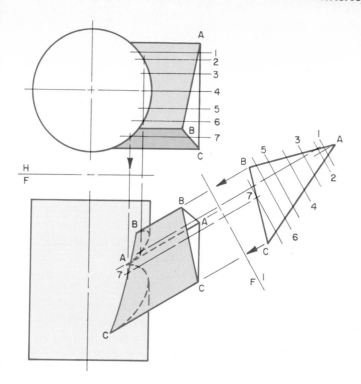

Fig. 14-27. Intersection of a cylinder and a prism.

INTERSECTION OF TWO CYLINDERS

The line of intersection of two cylinders can be found by the cutting plane method, Fig. 14-28. An auxiliary view is constructed to show the true size and shape of the inclined cylinder. Cutting planes are passed through the intersection of the two cylinders to identify piercing points with which to plot the curved line of intersection.

Given the top and front views of the two intersecting cylinders, Fig. 14-28, proceed as follows:

1. Construct an auxiliary view of inclined cylinder to show its true size and shape.
2. Pass randomly spaced cutting plane lines through intersection of two cylinders in top view.
3. Transfer cutting plane lines to auxiliary view so they are perpendicular to line of sight. Spacing of cutting plane lines must equal that in top view.
4. Project corresponding points of intersection between cutting planes and two cylinders to front view from top and auxiliary views. This locates the line of intersection.
5. Check adjacent views for visibility.

INTERSECTION OF AN INCLINED PLANE AND A CONE

The intersection of an inclined plane and a cone may be found in the principal views where one view of the plane is an edge view, Fig. 14-29.

Given the top and front views, proceed as follows:

1. In top view where base of cone appears as a circle, Fig. 14-29, draw a number of randomly spaced diameters intersecting this circle.
2. Project these points of intersection to base of cone in front view and connect them with apex of cone.
3. Lines from base to apex locate piercing points in inclined plane. Project these points to top view to their corresponding diametric lines.
4. Connect these points of intersection in top view to form line of intersection. Line of intersection in front view coincides with inclined plane.

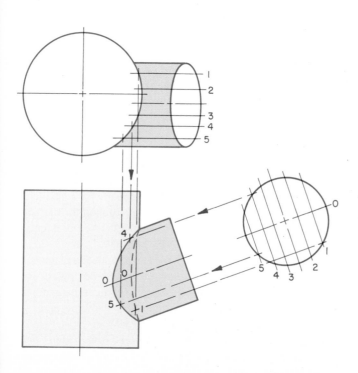

Fig. 14-28. Line of intersection between two cylinders.

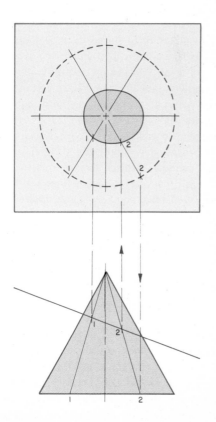

Fig. 14-29. Intersection of a plane and cone.

277

INTERSECTION OF CYLINDER AND CONE

Line of intersection between a cylinder and a cone may be found in principal views when intersecting cylinder is parallel to principal views of orthographic projection, Fig. 14-30.

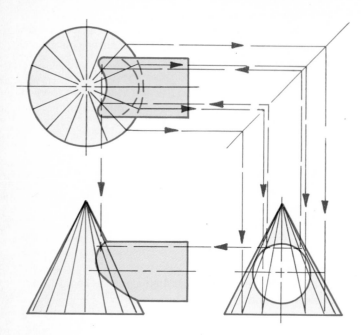

Fig. 14-30. Line of intersection between a cylinder and cone.

Given the three principal views, proceed as follows:

1. In top view, where base of cone appears as a circle, Fig. 14-30, draw a number of randomly spaced diametric lines intersecting circle.

   Note: If a pattern of surface is to be developed later for piece, 12 to 16 diametric lines should be equally spaced.

2. Project points of intersection to front and profile views intersecting with base line and then to apex of cone. If lines in profile view do not coincide with outer surface of cylinder, draw lines that do. Project back to top view where diametric lines should be drawn.

3. Lines in profile view from base to apex locate piercing points of cylinder. Project these points to front view to intersect with their corresponding lines and then on to top view to form piercing points in line of intersection.

4. Connect these points to form line of intersection in top and front views.

SUMMARY OF PROJECTION METHODS APPLICATION IN LOCATING LINES OF INTERSECTION

Three methods of locating lines of intersection between geometric forms have been presented. In selecting the appropriate method to use, consider the following:

1. ORTHOGRAPHIC PROJECTION METHOD should be used when the object is parallel to one or more principal planes of projection and the cutting plane representing the line of intersection is shown as an edge in one of the principal planes of projection.

2. CUTTING PLANE METHOD should be used when plane representing line of intersection is oblique to all principal planes of projection.

3. AUXILIARY VIEW METHOD should be used when both intersecting plane and object are oblique to principal planes of projection.

PROBLEMS IN INTERSECTIONS

Accuracy is extremely important in graphic solution of intersection problems. Use a sharp pencil. Guidelines and construction lines should be drawn very lightly. This will help to increase your accuracy in locating points and intersections as well as minimizing need for erasures.

INTERSECTING AND NONINTERSECTING LINES IN SPACE

1. In problems A through D, Fig. 14-31, ascertain which lines intersect and those that merely cross in space. The lines are shown located on 1/4 inch section paper. Use sheet layout AP1 with only a vertical division and place two problems on sheet. Orthographic projection should be used in solution of these problems. Label your solution. After having solved problems, is there a way of studying views to determine whether lines intersect? Explain your answer.

DRAW LINES THAT INTERSECT IN SPACE

2. Draw the top and front views of two randomly located crossing lines and determine by orthographic projection whether they intersect. If they do not, try to lay out two that do. Check with your instructor if you need assistance.

VISIBILITY OF LINE PROBLEMS

3. In problems A through D, Fig. 14-32, ascertain which is visible in each view; the line or the plane. Complete the line as a visible or hidden line as determined.

PIERCING POINTS OF LINES WITH PLANES

4. In problems A through D, Fig. 14-33, locate piercing point of each line and indicate visibility of line in each view. Use orthographic projection method on problems A and B, and auxiliary view method on problems C and D.

A LINE PERPENDICULAR TO AN OBLIQUE PLANE

5. Construct lines perpendicular to the oblique planes through the points shown in problems A and B, Fig. 14-34. Locate the line in the top and front views.

INTERSECTION OF PLANES

6. In problems A through D, Fig. 14-35, locate the lines of intersection of planes and indicate visibility for each plane in the two views. Use orthographic projection for the first problem and auxiliary views for the last three problems.

INTERSECTION OF A PLANE AND A PRISM

7. Find the line of intersection between the oblique plane and the prism, shown in problem C of Fig. 14-34, by cutting plane method. Lay out the problem a second time and find the line of intersection by the auxiliary view method.

INTERSECTION OF TWO PRISMS

8. Find the line of intersection between the square and triangular prisms shown in problem D in Fig. 14-34, by the auxiliary view method. Indicate by visible and hidden lines, the entire line of intersection.

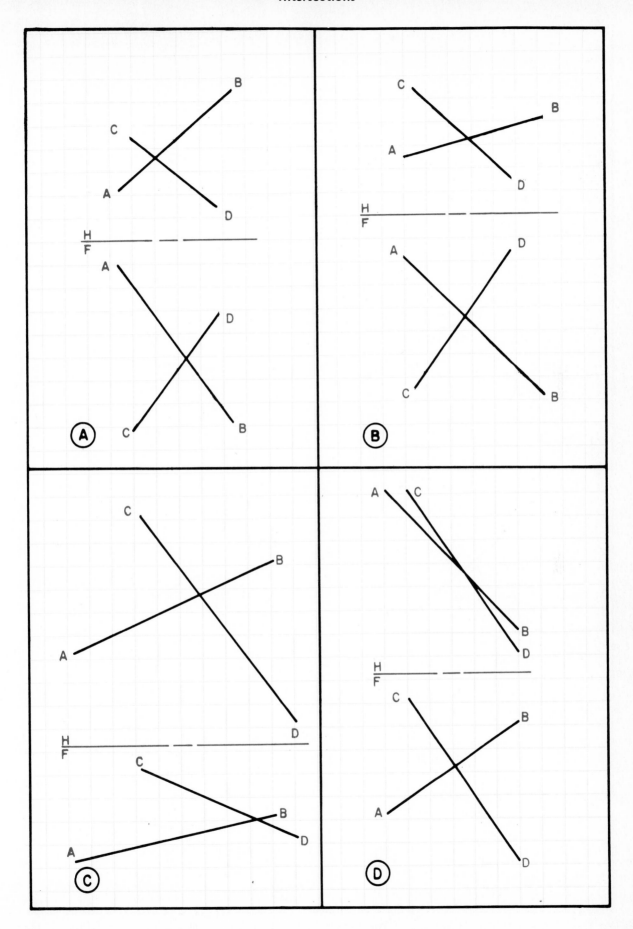

Fig. 14-31. Intersecting and nonintersecting lines problems.

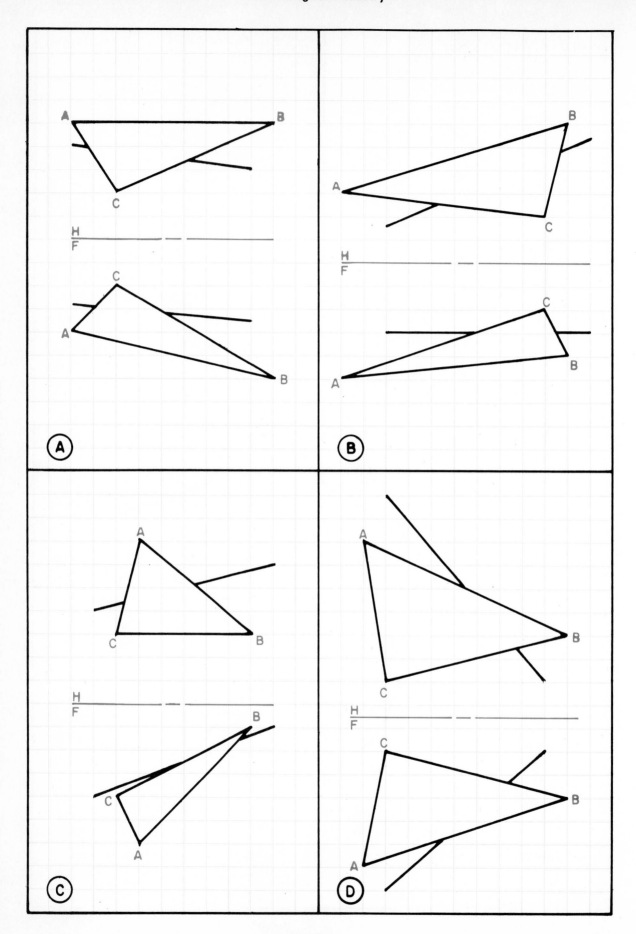

Fig. 14-32. Problems in identifying visibility of lines.

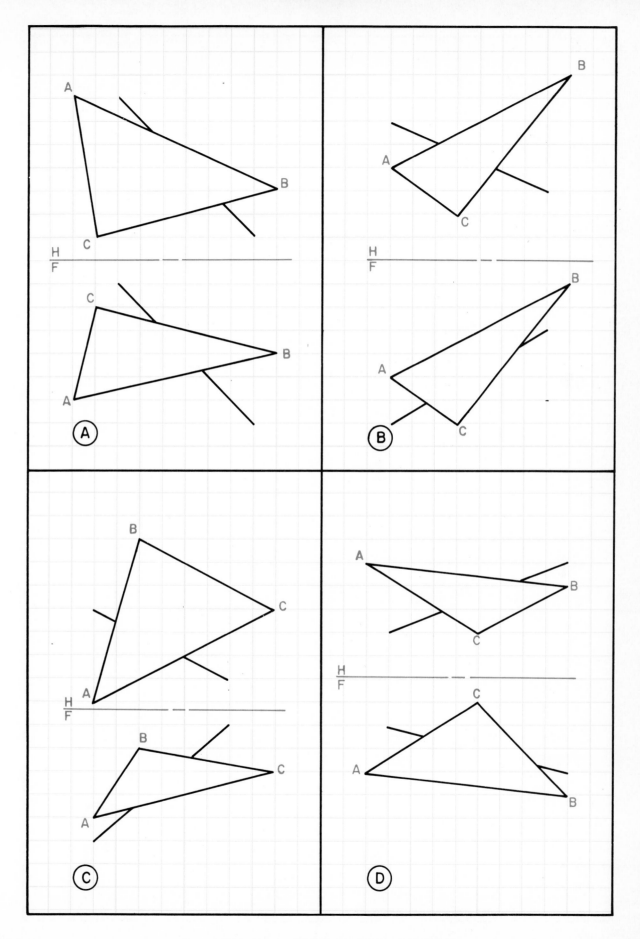

Fig. 14-33. Problems in locating the piercing point of lines.

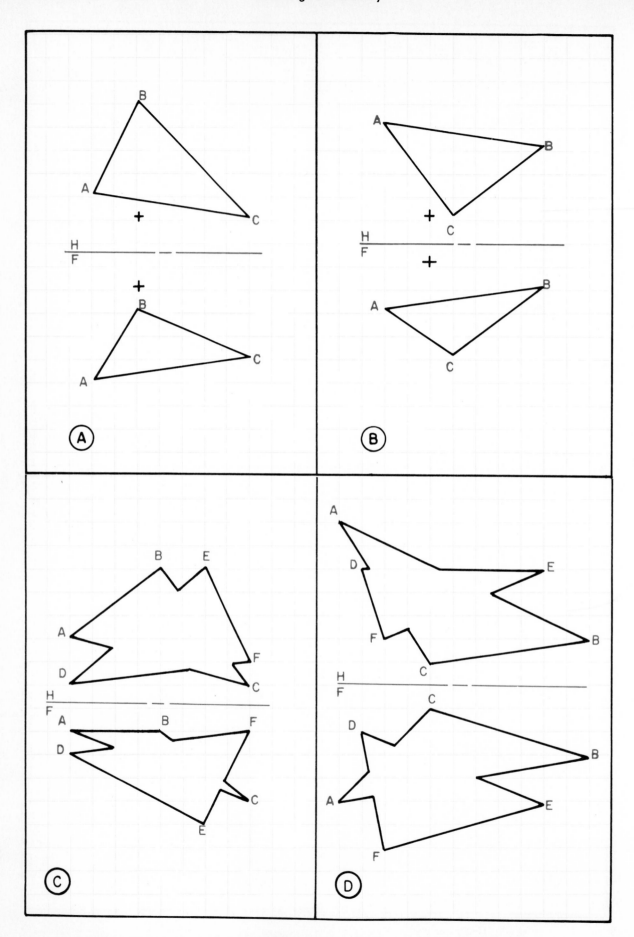

Fig. 14-34. Problems in constructing a line perpendicular to an oblique plane and intersections of prisms.

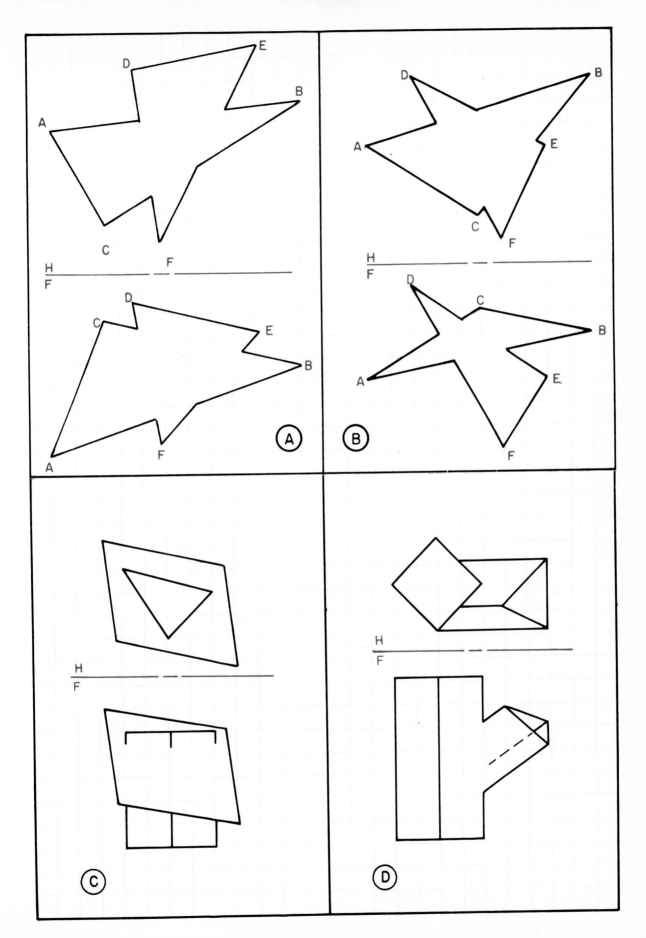

Fig. 14-35. Intersection of planes problems.

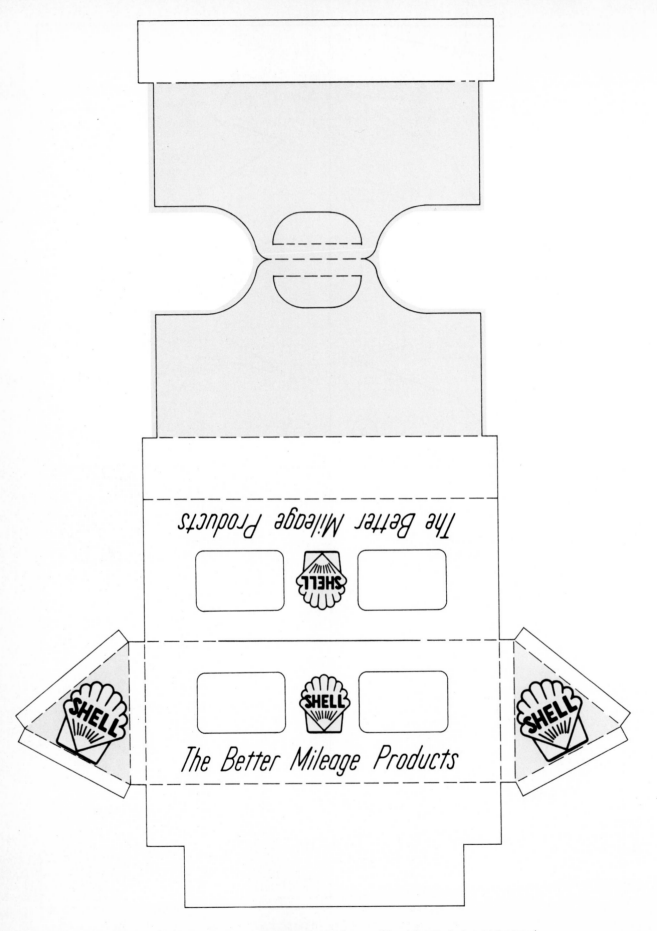

Fig. 15-1. A flat pattern layout for a container package.   (Chet Johnson, Industrial Designer)

Chapter 15
DEVELOPMENTS

A "development" in drafting refers to the layout of a pattern on flat sheet stock. It may be a pattern for a carton, pan, heating and air conditioning duct, hopper or any other manufactured product that requires folding or rolling of sheet materials. A flat pattern development for a package design is shown in Fig. 15-1. The heating and air conditioning industries and the petroleum industries depend heavily upon developments in the design and construction of systems, as shown in Fig. 15-2.

Developments are closely related to the materials on intersections presented in the previous chapter. In many instances, intersections have to be identified before a develop-

ment can be completed.

This chapter presents basic principles, types of developments and their industrial applications.

TYPES OF DEVELOPMENTS

Surfaces may be classified into ruled surfaces and double-curved surfaces, (see page 262). Ruled surfaces are subdivided into planes, single-curved surfaces and warped surfaces. The first two are capable of being developed, while the latter can only be approximated by flat pattern development. Double-curved surfaces cannot be developed into single plane surfaces.

Fig. 15-2. Flat patterns were used in laying out patterns for these pipes, and transition pieces, and the water tower sections. (Central Foundry Division, GMC)

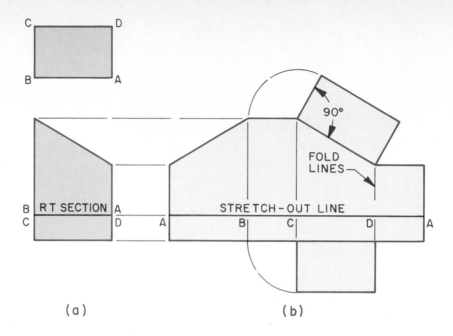

Fig. 15-3. Development of a rectangular prism by the orthographic view method.

DEVELOPMENT OF A RECTANGULAR PRISM

The development for a rectangular prism with an inclined bevel is shown in Fig. 15-3. It is laid out along a stretch-out line which represents, and is parallel to, right section of prism.

Given top and front views of prism, proceed as follows:

1. Draw edge view of right section in front view below bevel line, Fig. 15-3 (a). Lay off a line to one side and parallel to right section to serve as a stretch-out line, (b).
2. Identify corners of right section in top and front views, A, B, C and D, moving in a clockwise order, since stretch-out will be from an inside view. (Most developments are made from an inside view, because they are more easily handled in folding and bending machines. An outside view could be laid out by working in a counterclockwise direction.)
3. True lengths of prism along right section line are shown in top view. Transfer these measurements to stretch-out line starting with A-B, B-C, C-D and D-A. Draw vertical fold lines through these points.
4. Heights of lateral edges of prism are projected from front view where they appear in their true length.
5. Join points to form development or pattern for prism.
6. Project lines at 90 degrees to form bevel surface and bottom if required.
7. Lay off their widths with a compass, using adjacent side as a radius. Join points to complete surfaces.
8. Add material for a seam if required.

DEVELOPMENT OF AN OBLIQUE PRISM

The development of a prism that is oblique to all principal planes requires auxiliary projection to find true size of right section and true length of lateral lines, Fig. 15-4.

1. Construct a primary auxiliary view to find true length of lateral edges, 15-4.
2. Construct a secondary auxiliary to find true size and shape of right section of prism.
3. Draw right section line perpendicular to lateral edges in primary auxiliary and extend it for stretch-out line.

4. Transfer true size measurements from secondary auxiliary to stretch-out line starting with lateral corner A.
5. Draw perpendiculars through these points and project length of each side directly from primary auxiliary.
6. Join points to form development for the oblique prism.
7. If end pieces are required, construct these as indicated in Fig. 15-3.
8. Allow material for a seam if required.

DEVELOPMENT OF A CYLINDER WITH INCLINED BEVEL

The development of a cylinder with an inclined bevel is laid out along a stretch-out line which represents, and is parallel to, the right section of the cylinder, Fig. 15-5.

Given the top and front views with the inclined bevel appearing as an edge, proceed as follows:

1. Draw edge view of right section in front view below bevel cut and extend it to side for stretch-out line.

   Note: Stretch-out line could also coincide with base line if desired, so long as it represents a right section.
2. Divide true size circular view into a number of equal parts (12 in example) and project these to front view for true length lines along height of inclined bevel.
3. Transfer these chord measurements to stretch-out line.

   Note: Figured mathematically, length of stretch-out line is: $\pi \times D$ = circumference. Distance is divided geometrically for greater accuracy, Fig. 15-6 (b).
4. Draw a perpendicular through these points on stretch-out line, Fig. 15-5.
5. Project base line across full length of stretch-out and project height of each true length line in front view to its corresponding lines in stretch-out.
6. Sketch a smooth freehand curve between points and finish with an irregular curve.
7. If a base cover is desired, it is same size as right section. Cover for inclined bevel is constructed as ellipse. See Fig. 12-16 (a).
8. Allow material for a seam if required.

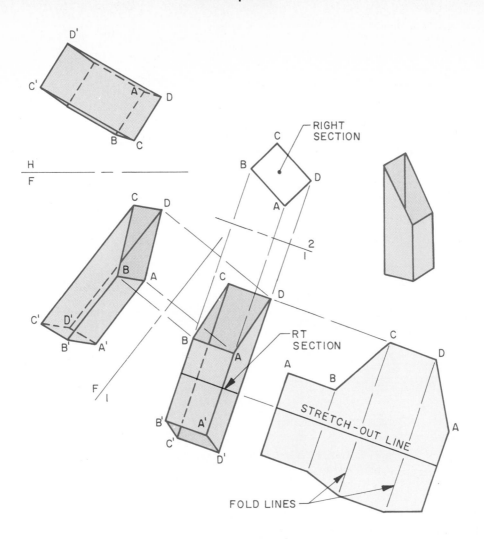

Fig. 15-4. The development of an oblique prism by the auxiliary view method.

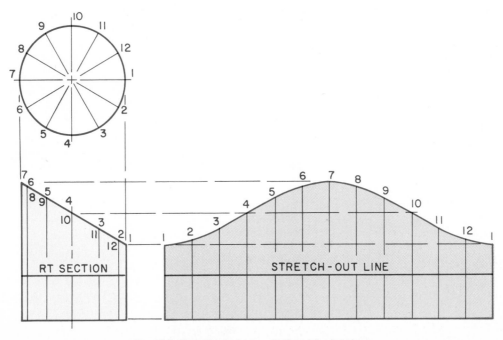

Fig. 15-5. Stretch-out for a cylinder with a bevel cut.

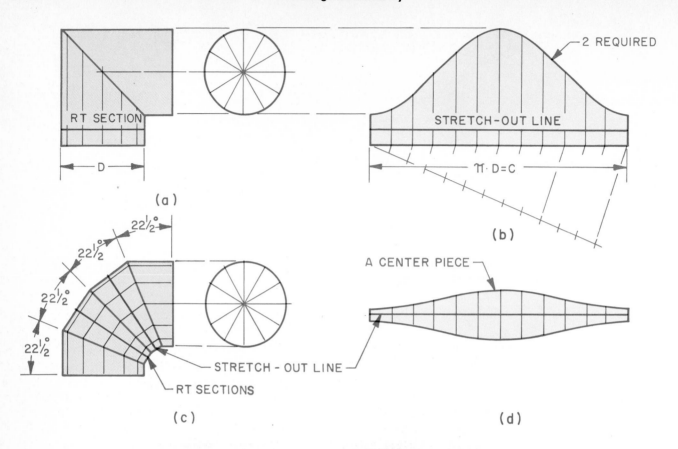

Fig. 15-6. Two-piece and four-piece right elbow pipe developments.

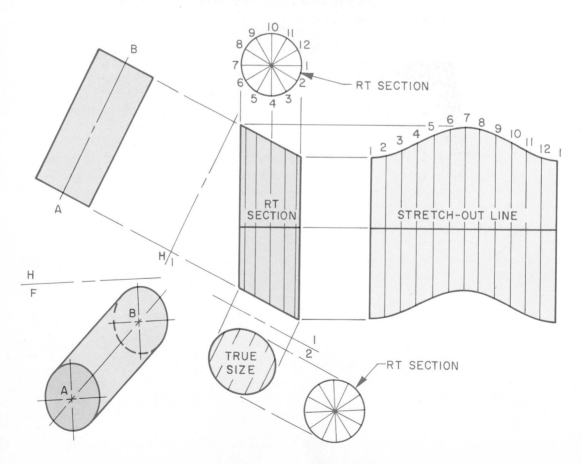

Fig. 15-7. Development of an oblique cylinder.

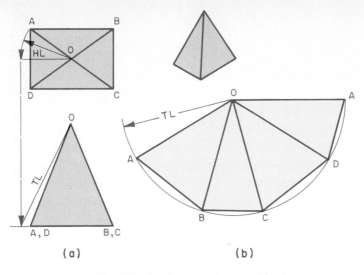

Fig. 15-8. Development of a pyramid.

DEVELOPMENT OF TWO-PIECE AND FOUR-PIECE ELBOW PIPE

The procedure for the development of a two-piece or four-piece right elbow pipe is the same as the bevel cut. Mating pieces are laid out as shown in Fig. 15-6. The stretch-out, (b), for the two-piece elbow was done by figuring the circumference mathematically and dividing distance geometrically.

DEVELOPMENT OF AN OBLIQUE CYLINDER

The procedure for laying out the oblique cylinder is quite similar to that for the inclined bevel-cut cylinder. The chief difference being that the oblique cylinder's true lengths must be found in an auxiliary view, as shown in Fig. 15-7. The end covers, if required, are developed by means of a secondary auxiliary.

DEVELOPMENT OF A PYRAMID

The development of a right pyramid is a type of radial line development and involves finding the true length of the lines and laying these out around a radius point, Fig. 15-8.

Given the top and front views of a right pyramid, proceed as follows:

1. Find true length of corner lines of pyramid by revolving line OA to a horizontal position in top view and projecting

it to front view where it appears in true length, Fig. 15-8 (a). This is true length for all corner lines, since pyramid is a right pyramid.

2. With true length line OA as radius, strike arc OA for stretch-out line of pyramid, (b).

3. Lines AB, BC, CD and DA appear in their true length in top view since base is in a horizontal plane in front view. Lay off line AB as a chord on arc OA and join end points with O to form triangular side OAB.

4. Continue with base line BC, etc., to form remaining triangular sides.

5. Lay out base, if required, adjacent to one of triangular sides.

6. Allow material for a seam if required.

DEVELOPMENT OF A TRUNCATED PYRAMID

The truncated right pyramid shown in Fig. 15-9 is developed in the same manner as any other right pyramid. In addition, the following steps are necessary:

1. Project true lengths of lines from truncated plane horizontally to true length line and transfer these to stretch-out.

2. Join lines along truncated cut.

If a top cover is required for the truncated cut, a primary auxiliary view projected off the front view will produce the desired cover in its true size and shape.

Fig. 15-9. Development of a truncated pyramid.

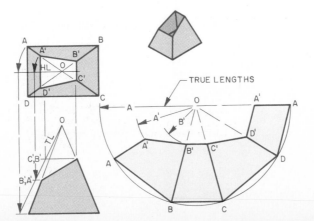

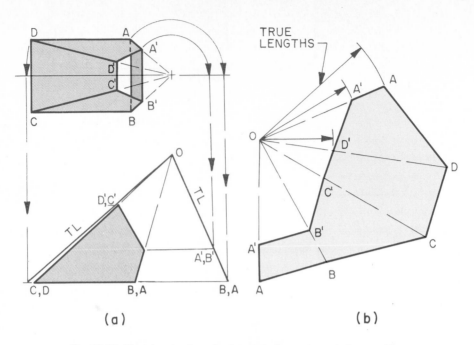

Fig. 15-10. Stretch-out of a development for a truncated pyramid which is inclined to its base.

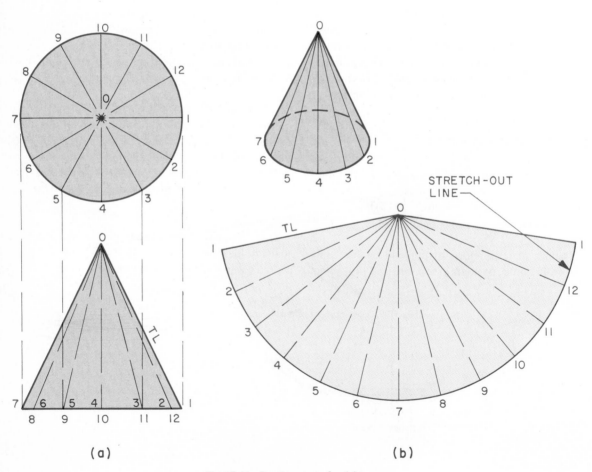

Fig. 15-11. Development of a right cone.

Developments

DEVELOPMENT OF A PYRAMID INCLINED TO ITS BASE

The development of a pyramid which is inclined to its base is shown in Fig. 15-10. The procedure is very similar to the development of a right pyramid except that the sides vary in their true lengths due to the offset of the apex.

The true lengths are found by rotating the corners of the lateral sides in the top view into a horizontal line and projecting these points to the front view, (a). Each surface is laid out in the development as a triangle with three sides given, starting with a side involving the shortest seam, OA'A, (b).

DEVELOPMENT OF A CONE

The stretch-out of cones, like pyramids, involves radial line development. It may be thought of as the development of a series of triangles around a common radius point, Fig. 15-11. The true length line of a side element of a right cone is shown in the frontal view as line O-1, (a). The base circle is divided into a number of equal parts and transferred to the radial arc in the stretch-out to obtain the circular length of the development. The base line of the development is drawn as an arc rather than as chords, since all elements of the lateral surface of the cone are the same length.

Given the top and front views of a right cone, proceed as

follows to lay out development for a cone.
1. Divide base circle into number of equal parts, Fig. 15-11 (a).
2. With true length line O-1 as radius, draw an arc as a stretch-out line, (b).
3. Transfer chord lengths of base circle from top view to stretch-out line to determine circular length of development.
4. Add material for a seam if required.
   If a base is desired, the true size is shown in the top view.

DEVELOPMENT OF A TRUNCATED RIGHT CONE

The development for a truncated right cone is similar to that of the right cone with the additional layout for the truncated part, Fig. 15-12.

Given the top and front views of a truncated right cone, proceed as follows:
1. Lay out development for cone as described earlier.
2. Determine true length of line elements of cone intersecting with inclined surface, Fig. 15-12 (a).
3. Transfer lengths to appropriate line in stretch-out, (b).
4. Sketch a light line through these points and finish line with an irregular curve.

If a cap is required for the inclined surface, the development is achieved through a primary auxiliary as shown.

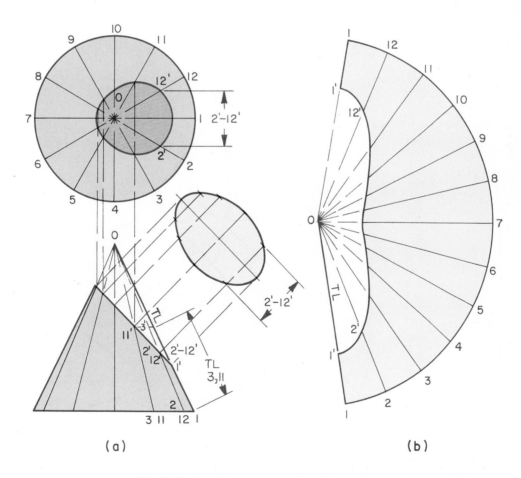

(a)   (b)

Fig. 15-12. Development of a truncated right cone.

DEVELOPMENT OF A CONE INCLINED TO ITS BASE

A cone that is inclined to its base may be developed as shown in Fig. 15-13. The development is sufficiently different from other cones to warrant a careful study of the procedure.

Observe that the cone in Fig. 15-13 is not a right circular cone. The base is a partial true circle, whereas the intersection of an inclined plane with a right cone would appear as an ellipse. The cone shown in Fig. 15-13 actually is an approximate cone.

The development of a cone inclined to its base involves the radial line method. This includes the determination of true lengths of lines and the division of the surface development into triangles as in the development of a right cone. It differs in that the base line of the cone is not laid out along a circular arc. Here, the lateral element lines differ in true lengths rather than all being the same length (as in a right cone). Each triangle on the surface must be constructed by laying off the true lengths of its three sides: two element lines and a chord length.

Given the top and front views, proceed as follows to develop a cone inclined to its base:

1. Divide circular base (top view) into a number of equal parts and project these points to base line in front view, Fig. 15-13 (a).
2. Project these points to apex of cone in top and front views.
3. Construct a true length diagram to find true lengths of lateral element lines, O-1, O-2, etc., (b).
4. Start stretch-out of development by laying off true length of O-1, (c).
5. Lay off from point 1 in stretch-out, chord length arc 1-2 obtained from base circle in top view.
6. Lay off an arc equal to true length line O-2 to intersect with chord arc 1-2 at point 2 on stretch-out, (c).
7. Continue laying out intersecting arcs of true length lateral element lines and chord lengths for remaining points on base circle. Disregard, at this time, vertical cut shown in front view.

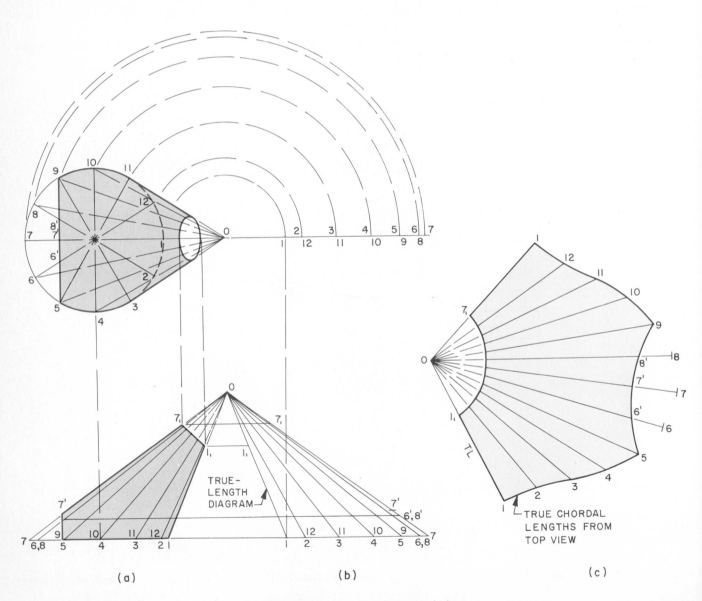

Fig. 15-13. Developing a cone inclined to its base.

8. Project points 6', 7' and 8', which represent vertical cut, from front view to true length diagram.
9. Transfer true lengths 6', 7' and 8' to appropriate lateral in stretch-out.
10. Project points of intersection of cut at upper part of cone and lateral lines to corresponding line in true length diagram, (b).
11. Transfer these true lengths to corresponding lines in stretch-out, (c).
12. Join points of intersection to form required shape of development.
13. Add material for a seam if required.

DEVELOPMENT OF TRANSITION PIECES

Transition pieces are used in gaseous or liquid systems to join pipes or ducts of different cross-section shapes. Some examples of transition pieces are shown in Fig. 15-14.

Fig. 15-14. Transition pieces in ducting system come in a variety of shapes.   (Gleason Works)

A typical transition piece is a section of ducting used to join a square duct to a round one, Fig. 15-15. Transition pieces are developed by dividing surface into triangles, finding true lengths of lateral elements and transferring these to a stretch-out.

B in top view as centers to rotate lengths of lines into a plane parallel to frontal plane (A-B). Then project these lines perpendicularly to height line in front view and join these points with corners A and B where lines appear in

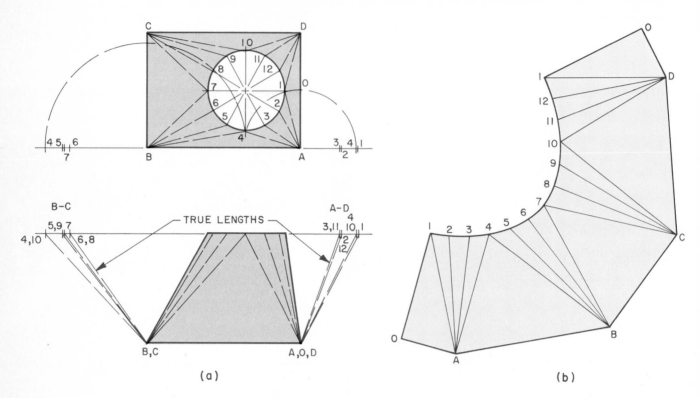

(a)

(b)

Fig. 15-15. Development of a square to round transition piece.

Given the top and front views for the transition piece shown in Fig. 15-15, proceed as follows:

1. Divide circular opening in top view into a number of equal parts, (a).
2. Project these division points to edge view of circular opening in front view.
3. Connect points of four quadrants of circle to adjacent corners of base in top and front views. These lines represent bend lines in transition piece.
4. Determine true lengths of bend lines by using corners A and true length.
5. Make seam in a flat section of development by starting stretch-out with lay out of true length line 1-O, (b).
6. Strike true length intersecting arcs 1-A and O-A (O-A is shown true length in top view) to form triangle 1AO.
7. Strike true length intersecting arcs 1-2 (true length in top view) and A-2 to form adjacent triangle 1A2.
8. Continue with lay out of successive adjacent triangles to complete development.
9. Allow material for a seam if required.

DEVELOPMENT OF A WARPED SURFACE

A warped surface is a ruled surface that cannot be developed into a single plane. However, an approximation can be developed into a single plane by use of triangulation, Fig. 15-16.

equal parts. Project these horizontally to intersect with base circle in top view and from there to other half of top view and to front view where base appears as an edge.

3. Divide top elliptical opening into same number of equal parts and connect these points with points on base circle, forming triangles as shown.

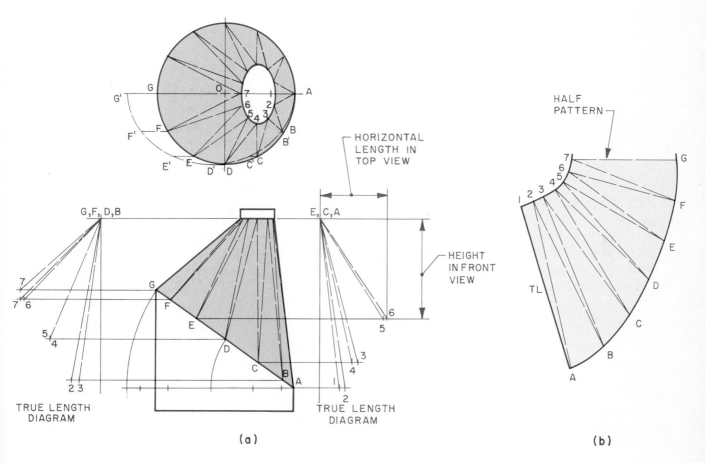

Fig. 15-16. Development of a warped surface through triangulation.

The piece shown is a transition from a right circular cylinder on an incline to an elliptical opening at the top. The surface is divided into a number of triangles whose sides appear as straight lines in the views but will actually be slightly curved when the development is fabricated.

Given the top and front views, proceed as follows to approximate the development of this warped surface.

1. Find true size of elliptical base by rotating major diameter AG, front view in Fig. 15-16, to the horizontal. Project this true length to top view and construct a half ellipse. (Minor diameter OD appears true length in top view.)

2. Using the dividers, divide this half-ellipse into a number of

4. Find true length of lateral lines by constructing two true length diagrams to keep lines separate and identifiable. True length of base line segments (on half-ellipse) and top opening segments are shown in top view.

5. Construct stretch-out by starting with line A1 (shown in its true length in front view) and laying off triangle 1A2, using three-sides-given method, (b).

6. Continue to lay off adjacent triangles in their true length until all are complete.

7. Draw line forming irregular curved lines, connecting points 1 through 7 and A through G.

8. Allow material for a seam, if required.

PROBLEMS IN PATTERN DEVELOPMENT

The following problems provide you with an opportunity to practice the skills and knowledge involved in laying out and developing patterns for various geometrical forms. The problems are represented on 1/4 inch section paper. Use a B size sheet in their development. Transfer your pattern layout to a stiff paper, cut it out, fold or roll it into shape, then test the accuracy of your layout.

1. RECTANGULAR PRISM.
    Lay out the inside patterns for the development of problems 1 through 4, Fig. 15-17.
2. OBLIQUE PRISM.
    Develop the inside patterns for problems 5 and 6 in Fig. 15-17.
3. CYLINDERS WITH INCLINED BEVEL.
    Develop the inside patterns for problems 7 and 8 in Fig. 15-17.
4. MULTI-PIECE ELBOW PIPE.
    Lay out the inside patterns for the development of the pipe elbows in problems 9 through 11, Fig. 15-18.
5. OBLIQUE CYLINDERS.
    Develop the inside patterns for problems 12 and 13 in Fig. 15-18.
6. PYRAMIDS.
    Develop the inside patterns for problems 14 through 16, Fig. 15-18.
7. CONES.
    Lay out the developments for the cones shown in problems 17 through 20, Fig. 15-19.
8. TRANSITION PIECES.
    Lay out the developments for the transition pieces shown in problems 21 through 24, Fig. 15-19.
9. APPLIED PROBLEMS.

Lay out the developments for the Orchard Heater Housing and the Dust Collector Housing shown in Fig. 15-20.

DESIGN PROBLEMS

Make use of the skills and knowledge you have learned in the last four chapters on descriptive geometry, and utilize the techniques of design problem solving studied in Chapter 9, to solve the following problems:

1. Design a horizontal cylinder to hold 1000 gallons of water. The ends of the tank are to be flat. Filling of the tank is accomplished by means of a 4 inch pipe which enters the center top of the tank at an angle that extends directly back. The angle meets the plane in line with the back side of the tank at an elevation of 5 feet above the tank. Make a drawing of the necessary views and prepare a scaled model of the tank and supply pipe.
2. Design a storage device for a liquid, powder or granular material, making use of two or more geometrical shapes studied in these chapters on descriptive geometry. Prepare a scaled model of the design.
3. Design a transition piece to solve some particular problem or need. Draw the necessary orthographic views, develop a full size inside pattern and form the piece out of the required material.
4. Select an item which is commonly sold in stores, such as cosmetics or a gift item, and study the design of its package. Applying the design method studied in Chapter 9, try to improve the design of the package, and then develop a prototype.
5. Select some problem that needs to be solved at school, at home or in the community involving the application of descriptive geometry. Make a drawing and scaled model of the solution which best meets the problem needs.

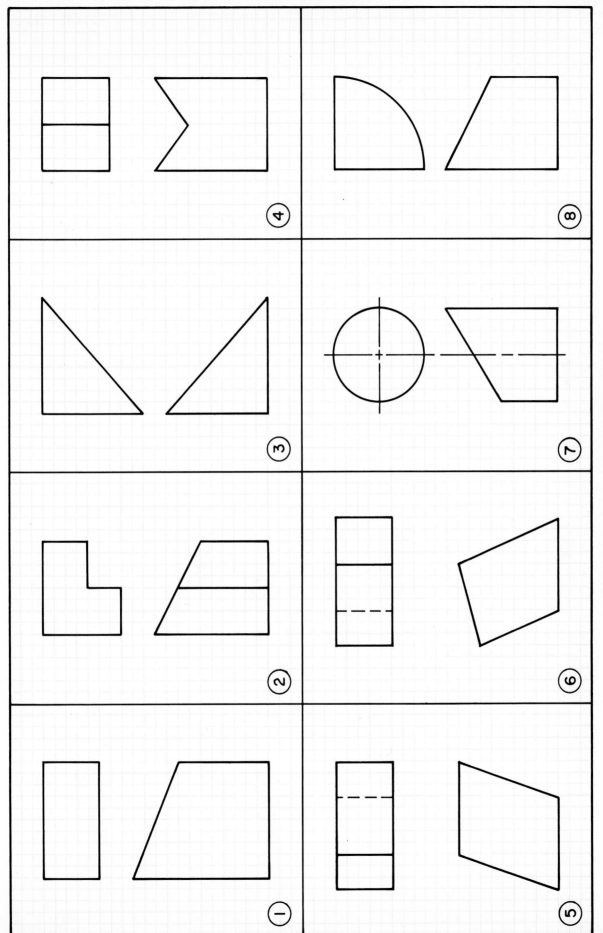

Fig. 15-17. Prism and cylindrical development problems.

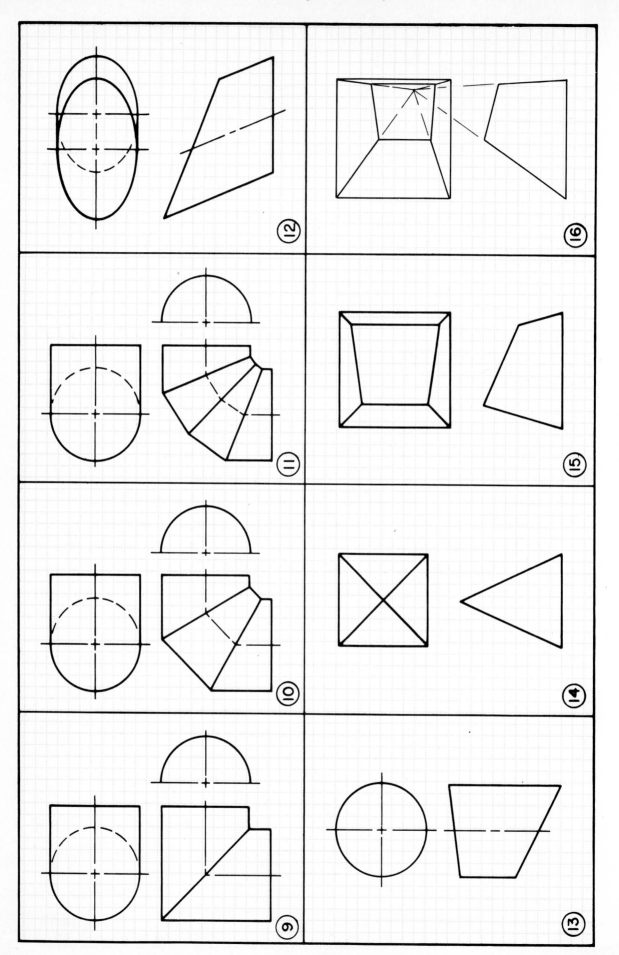

Fig. 15-18. Development problems involving cylinders and pyramids.

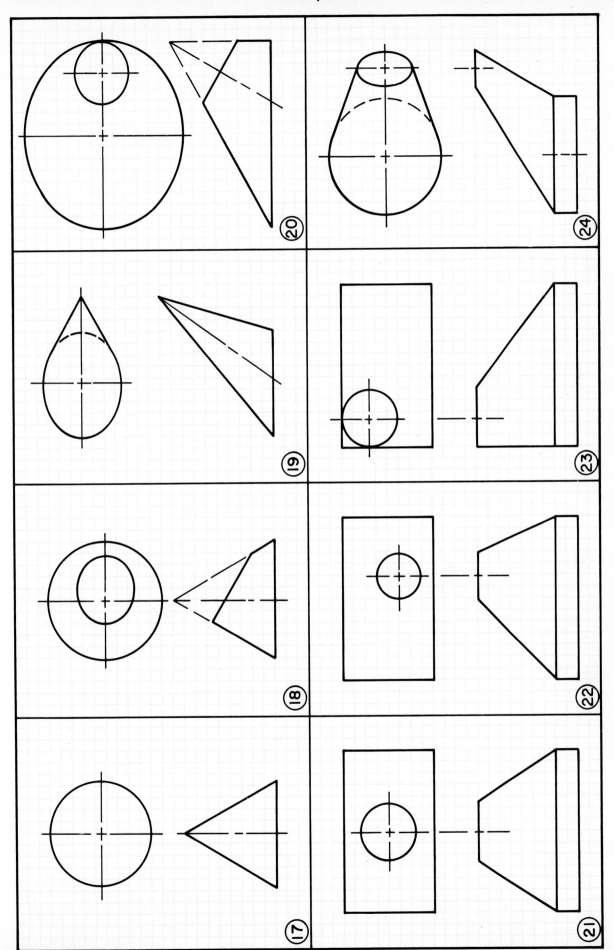

Fig. 15-19. Development problems for cones, transition pieces and warped surfaces.

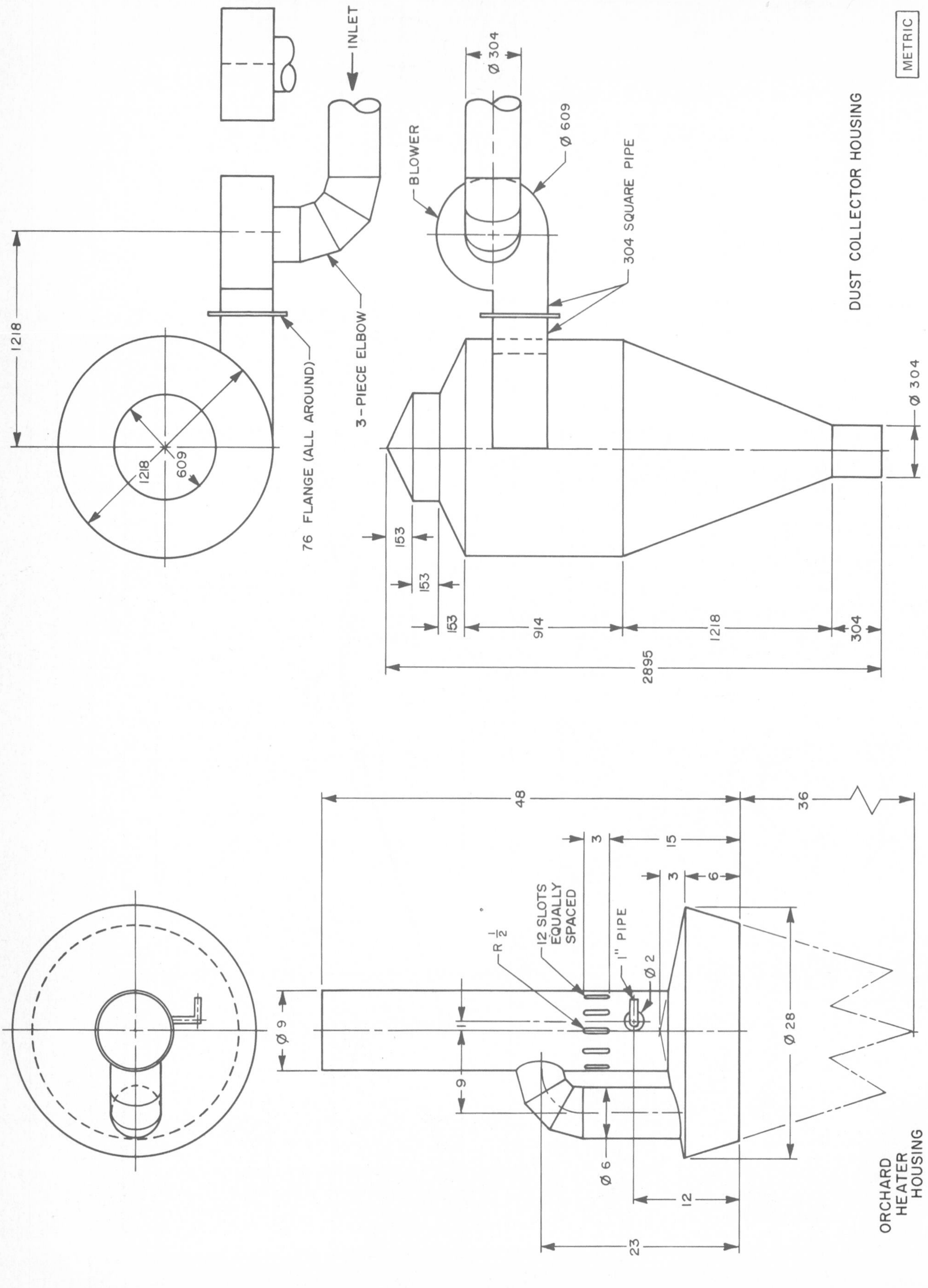

INLET

BLOWER

Ø 304

Ø 609

304 SQUARE PIPE

1218

1218

609

76 FLANGE (ALL AROUND)

3-PIECE ELBOW

DUST COLLECTOR HOUSING

Ø 304

153

153

153

914

1218

304

2895

METRIC

48

36

3

15

R ½

12 SLOTS
EQUALLY
SPACED

1" PIPE

Ø 2

3

6

Ø 9

Ø 28

9

Ø 6

12

23

ORCHARD
HEATER
HOUSING

Fig. 15-20. Applied development problems.

PART III
TECHNICAL DRAFTING

Technical drafting is concerned with the drawing techniques, standard practices and proper terminology used to provide full details and specifications for the manufacture of modern products.

This part of the text covers FASTENERS, MANUFACTURING PROCESSES, PRECISION DIMENSIONING, CAMS, GEARS and SPLINES, WORKING DRAWINGS and COMPUTER GRAPHICS.

The manufacturer of aircraft requires many technical drawings to detail and specify the processes needed.   (McDonnell-Douglas-St. Louis)

Chapter 16
THREADS AND FASTENING DEVICES

The hardware and techniques required by modern industry in the joining of component parts makes fastening one of the most dynamic and fast-growing technologies, Fig. 16-1. To the nontechnical person, fasteners may appear quite simple, and many of them are. However, in high volume assembly work such as in the aerospace, appliance, automotive and electrical industries, speed of assembly, holding capabilities and reliability of the fasteners call for many special types, Fig. 16-2.

The purpose of this chapter is to provide the student of drafting with information about the more common types of fasteners and how they are drawn.

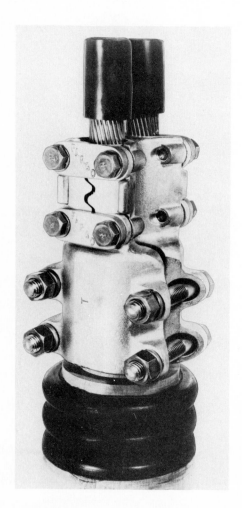

Fig. 16-1. Modern industry depends on fasteners.
(ITT Harper, Inc.)

SCREW THREADS

Threads and threaded fasteners are used on most machine assemblies produced in industry. Standard methods of specifying and representing screw threads are shown in the following sections.

Considerable progress has been made jointly by the United States, Canada and England in standardizing screw threads. The result of this cooperative effort is the Unified Thread Series which is now the American standard for fastening types of screw threads.

Unified threads and the former standard, the American National threads, have essentially the same thread form and are mechanically interchangeable. The chief differences in the two systems lie in the application of allowances, tolerances, pitch diameter and specification.

THREAD TERMINOLOGY

The following thread terminology includes the more important terms, Fig. 16-3.

MAJOR DIAMETER: The largest diameter on an external or internal screw thread.

MINOR DIAMETER: The smallest diameter on an external or internal screw thread.

PITCH DIAMETER: The diameter of an imaginary cylinder passing through the thread profiles at the point where the widths of the thread and groove are equal.

PITCH: The distance from a point on one screw thread to a corresponding point on the next thread measured parallel to the axis. The pitch for a particular thread may be calculated mathematically by dividing one inch by the number of threads per inch. Example:

$$\frac{1 \text{ inch}}{10 \text{thds/inch}} = \text{pitch} = .10 \text{ inch.}$$

CREST: The top surface of thread joining two sides or flanks.

ROOT: The bottom surface of thread joining two sides or flanks.

ANGLE OF THREAD: The included angle between the sides or flanks of the thread measured in an axial plane.

EXTERNAL THREAD: The thread on the outside of a cylinder such as a machine bolt.

INTERNAL THREAD: The thread on the inside of a cylinder such as a nut.

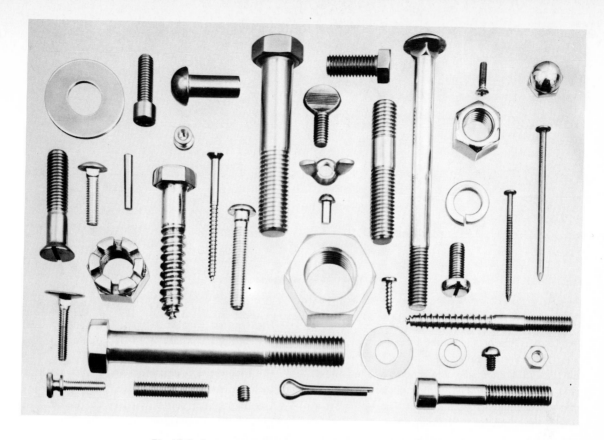

Fig. 16-2. Some examples of the many fasteners used in industry.
(ITT Harper, Inc.)

THREAD FORM: The profile of the thread as viewed on the axial plane. (See Fig. 16-4 for standard thread forms.)

THREAD SERIES: The groups of diameter-pitch combinations distinguished from each other by the number of threads per inch applied to a specific diameter.

THREAD CLASS: The fit between two mating thread parts with respect to the amount of clearance or interference which is present when they are assembled. Class 1 represents a loose fit and class 3 a tight fit.

RIGHT-HAND THREAD: A thread, when viewed in the end view, winds clockwise to assemble. A thread is considered to be right-handed (RH) unless otherwise stated.

LEFT-HAND THREAD: A thread, when viewed in the end view, winds counterclockwise to assemble.

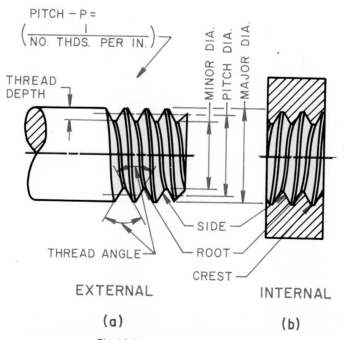

Fig. 16-3. Thread terminology.

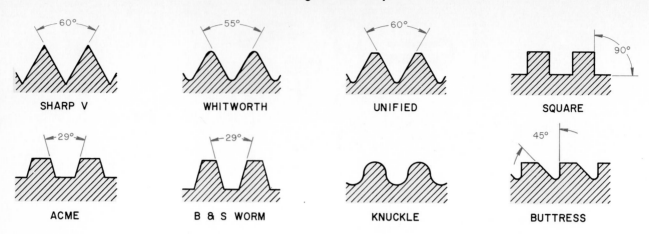

Fig. 16-4. Standard thread forms.

THREAD FORM

The thread form is the profile of the thread as viewed on the axial plane. There are a number of standard thread forms, Fig. 16-4, but the Unified has been agreed upon by the United States, Canada and Great Britain as the standard for such fasteners as bolts, machine screws and nuts. The Unified is a combination of the American National and the British Whitworth and has almost completely replaced the American National form due to fewer difficulties encountered in producing the flat crest and root of the thread.

While the Unified form is used for fasteners, the Square, Acme, Buttress and Worm threads are used to transmit motion and power due to their thread profiles which are more vertical with their axes. Examples of motion and power transmission are steering gears (worm screw) and lead screws (square) on machine lathes. The sharp V is used where friction is desired, such as setscrews. The knuckle form is for fast assembling of parts such as light bulbs and bottle caps.

THREAD SERIES

Thread series designates the form of thread for a particular application. There are four series of Unified screw threads: coarse, fine, extra-fine and constant-pitch series.

The Unified Coarse series is designated UNC and is used for bolts, screws, nuts and threads in cast iron, soft metals or plastic where fast assembly or disassembly is required.

The Unified Fine is labeled UNF and is used for bolts, screws and nuts where a higher tightening of the part is required.

The Fine series is also used where the length of the thread engagement is short and where a small lead angle is desired.

The Extra-Fine series (UNEF) is used for even shorter lengths of thread engagements and for thin-wall tubes, nuts, ferrules and couplings. It is also used for applications requiring high stress resistance.

The Constant-Pitch series is designated UN with the number of the threads per inch preceding the designation, such as 8UN. This series of threads is for special purposes such as high pressure applications and large diameters where other thread series do not meet the requirements.

The 8-thread series (8UN) is also used as a substitute for the Coarse-Thread series for diameters larger than 1 inch.

The 12-thread series (12UN) is used as a continuation of the Fine-Thread series for diameters larger than 1 1/2 inches.

The 16-thread series (16UN) is used as a continuation of the Extra-Fine series for diameters larger than 1 11/16 inches. Dimensions for the Unified series of thread are given in the Reference Section.

THREAD CLASSES

The classes of fit for external and internal threads of mating parts are distinguished from each other by the amount of

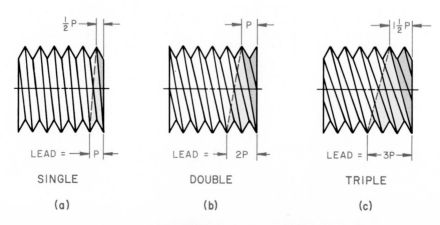

Fig. 16-5. Lead and pitch of single and multiple-threaded screws.

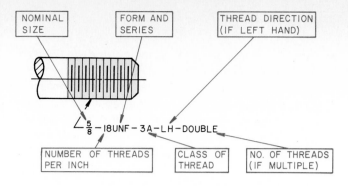

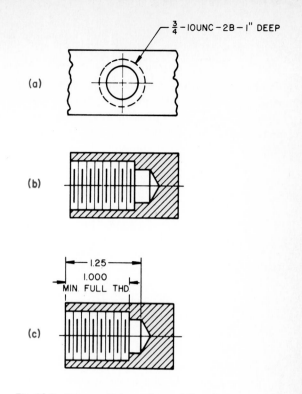

Fig. 16-6. Arrangement of notes specifying threads.

Fig. 16-7. Note specifying an internal thread.

tolerance and allowance permitted for each class. Classes 1A, 2A and 3A indicate external threads and classes 1B, 2B and 3B indicate internal threads.

Classes of fit are identified by numeral "1" for applications requiring minimum binding to permit frequent and quick assembly or disassembly of parts. A class "2" fit is for bolts, screws, nuts and similar fasteners for normal applications in mass production. A class "3" fit is for applications requiring closer tolerances than the other classes for a fit to withstand greater stress and vibration.

SINGLE AND MULTIPLE THREADS

A screw or other threaded machine part may contain single or multiple threads, Fig. 16-5. A screw with a single thread will move forward into its mating part a distance equal to its pitch in one complete revolution (360 degrees). In the case of the single-threaded screw, the pitch (P) is equal to the lead, (a). Notice that the crest line is offset a distance of 1/2P since a single view shows only 1/2 revolution of a thread.

A double-threaded screw has two threads side by side and moves forward into its mating part a distance equal to its lead or 2P, (b). The crest line of a double-threaded screw is offset a distance of P in a single view.

A triple-threaded screw has three individual threads, and it moves forward a distance equal to its lead or 3P, (c).

Single threads are used where considerable pressure or power is to be exerted in the movement of mating parts, such as a bolt and its nut or a machinist vise screw and its jaws. Multiple threads are used where rapid movement between mating parts is desired, such as mating parts of a ball-point pen or water faucet valves.

THREAD SPECIFICATION NOTES

Screw threads are specified by a note, Fig. 16-6, which provides the following information in a standard sequential order: nominal size (major diameter or screw number), number of threads per inch, thread form (UN) and series (F) grouped together (UNF) and the class of fit (3A). The thread is understood to be a right hand, single thread unless noted as left hand (LH) or multiple (DOUBLE) thread.

Thread specifications must be included on all threaded parts by a note and a leader to the external thread, as shown in Fig. 16-6, or to an internal thread in the circular view, Fig. 16-7 (a). The length or depth of the threaded part is given as the last item in the specification, or it may be dimensioned directly on the part, (c). Standard size bolts and nuts may be called out by a letter on the drawing and specified in the materials list.

A thread note providing information on the tap drill size, depth, countersinking and number of holes to be threaded is shown in Fig. 16-8.

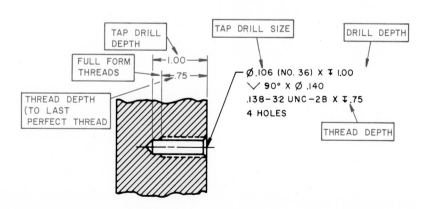

Fig. 16-8. Thread notes may provide information in addition to the thread specifications.

METRIC TRANSLATION OF UNIFIED SCREW THREADS

In drawings which are dual-dimensioned, the thread designation shall specify in sequence the nominal size (expressed in decimal inches), the number of threads per inch, thread series symbol and class symbol. When the pitch diameter (PD) is given, it is presented as follows:

Where inch is primary dimension on dual-dimensioned drawing:

0.375-24 UNF-2A
PD 0.349-0.343
(8.808-8.713 mm)

Where millimeter is primary dimension on dual-dimensioned drawing:

0.375-24 UNF-2A
PD 8.808-8.713
(0.349-0.343 in.)

METRIC THREADS

Metric threads are designated in a manner similar to that used for Unified and American National Standard, but with some slight variations. A diameter and pitch are used to

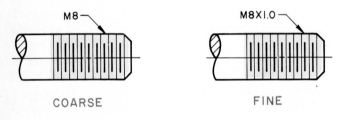

COARSE   FINE

Fig. 16-9. The metric coarse thread designation gives only the diameter, the pitch is understood. The fine thread designation gives the pitch, following the diameter.

designate the metric series, as in the inch system, with the following modifications:

1. Metric coarse threads are designated by simply giving prefix M and diameter. For example, in Fig. 16-9, M8 is a coarse thread designation representing a nominal thread diameter of 8 mm with a pitch of 1.25 mm understood. That is, thread designation is for a coarse thread unless otherwise noted.

2. Metric fine threads are designated by listing pitch as a suffix. A fine thread for a part would be M8 x 1.0, or 8 mm diameter with a pitch of 1.0 mm. Most common metric thread is coarse, which generally falls between the coarse and fine series of inch system measurements for a comparable diameter.

In the inch series, designation of the pitch of a thread is given as the number of threads per inch: 3/4 − 10UNC. The pitch of the thread is actually 1/10 inch. In the metric series, the pitch is really the pitch: M x 1.0. That is, the pitch is actually 1 mm. Tables showing the ISO metric thread series are in the Reference Section. The basic form of the ISO thread is also shown along with information for thread calculations.

The tolerance and class of fit in metric threads are designated by adding numbers and letters in a certain sequence to the callout. The thread designation in Fig. 16-10 calls for a fine thread of 6 mm diameter, 0.75 mm pitch (no pitch is given in the designation for a coarse thread) with a pitch diameter tolerance grade 6 and an allowance "h", crest diameter tolerance grade 6 and allowance "g".

THREAD REPRESENTATION

The conventional methods of representing threads on drawings are: DETAILED, SCHEMATIC and SIMPLIFIED, Fig. 16-11. The detailed convention is a closer representation of the actual thread, and it is sometimes used to show the

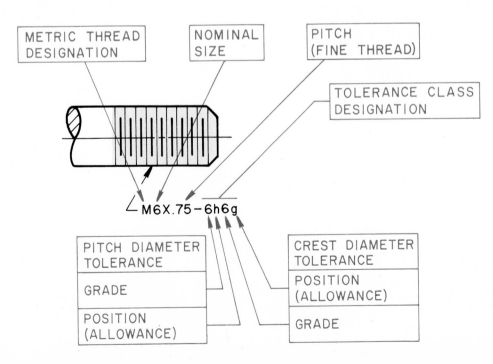

Fig. 16-10. A complete designation for an ISO metric thread.

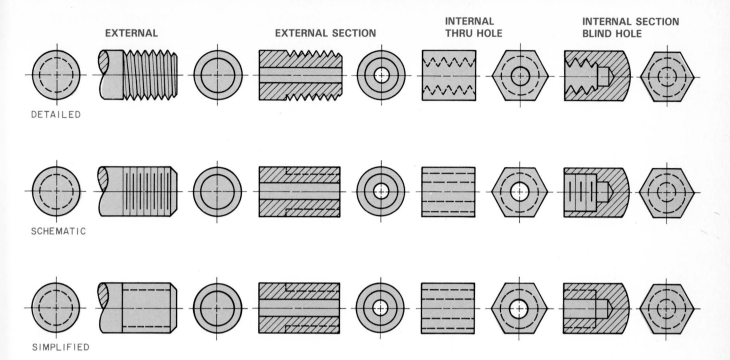

Fig. 16-11. Conventional representations of screw threads.

geometry of a thread form as an enlarged detail. However, the schematic and simplified conventions are most commonly used because they save drafting time and, in many instances, produce a clearer drawing.

DETAILED REPRESENTATION OF V-TYPE THREADS

The construction of the Sharp V, American National or Unified National forms of detail thread representation is shown in Fig. 16-12. The pitch of the thread could be used to lay out thread spacing, (a). However, the conventional practice, especially on small machine parts, is to approximate the pitch spacing for the thread crests so that they appear natural and in keeping with the size of the part, (a).

Start the spacing with a half space, since the first thread crest represents only 1/2 revolution or 180 degrees. Continue the spacing along the bottom edge of the threaded part and draw the crest lines by adjusting a triangle along a straightedge to the correct slope, (b). Actually, the crest and root lines are helix curves (see Fig. 6-21), but they are drawn as straight lines. Notice that the slope for a right-hand thread is shown in Fig. 16-12. The thread advances into its mating part when turned clockwise on its axis. The slope for a left-hand thread would be opposite.

Draw 60 degree sides of the thread form as shown in Fig. 16-12 (c). Join the bottom of these threads to form the root lines. Notice that the root lines are not parallel to the crest lines in this detailed representation due to the difference in diameters. Complete the detail representation by drawing a chamfer on the end of the thread at the minor diameter, (d). Add the callout (note) to provide the thread specification.

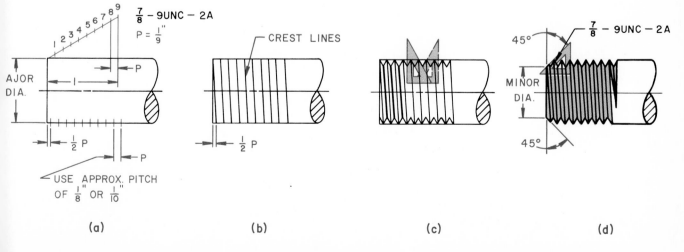

Fig. 16-12. Constructing a detailed representation of a Unified National Thread.

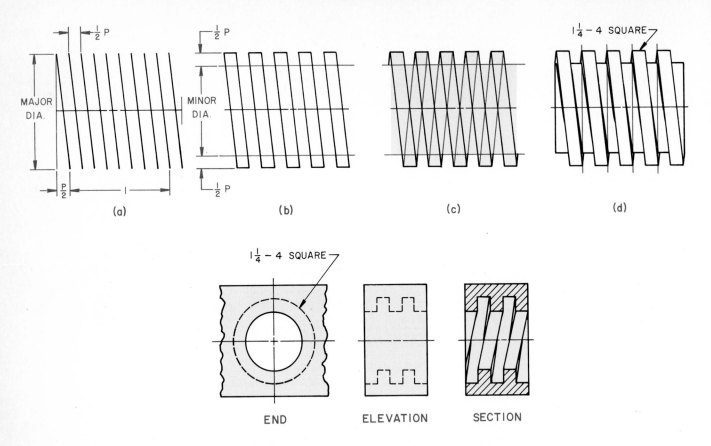

Fig. 16-13. Constructing a detailed representation of an external square thread. A detailed representation of the internal thread is also shown.

DETAILED REPRESENTATION OF SQUARE THREADS

The detailed representation of square threads is shown in Fig. 16-13. This is an approximation of the thread since the thread would appear as a helix curve rather than a straight line. Proceed as follows to draw a detailed thread representation:

1. Construct major diameter and draw crest lines 1/2P apart for single threads, as shown in Fig. 16-13 (a). Pitch of square threads is equal to one over number of threads per inch ($\frac{1}{4}$ in Fig. 16-13).
2. Draw lines for top of threads, (b).
3. Draw light lines representing minor diameter a distance of 1/2P from major diameter.

4. Draw diagonal construction lines connecting tops of thread to represent that portion of thread on back side which is visible. Darken visible thread line outside minor diameter, as shown in (c).
5. Draw a light construction line from inside crest lines to locate points on minor diameter where root lines meet minor diameter, (d).
6. Connect these points with points where adjacent crest line crosses center line to form root lines of thread, (d).
7. Add note to provide thread specification.

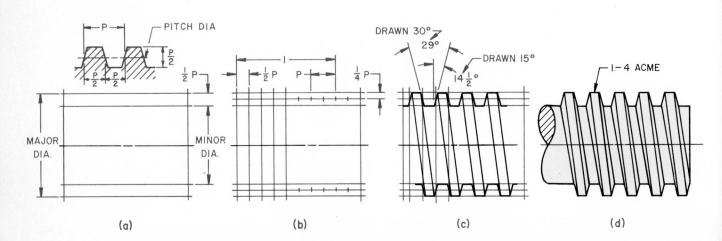

Fig. 16-14. Constructing a detailed representation of an external Acme thread.

DETAILED REPRESENTATION OF ACME THREADS

The Acme thread is similar to the square except the sides of the Acme are drawn to provide an included angle of 30 degrees (actually 29) for the groove and 30 degrees for the thread, Fig. 16-14. The steps in drawing a detailed Acme thread are as follows:

1. Draw center line, lay off length and major diameter, Fig. 16-14 (a).
2. Determine pitch by dividing 1 by number of threads per inch ($\frac{1}{4}$ in our example). Draw minor thread diameter 1/2P from major diameter.
3. Draw pitch diameter halfway between major and minor diameters, (b). Lay off thread pitch along one pitch diameter line by measuring a series of divisions 1/2P (for thread and groove) and projecting these divisions to opposite pitch diameter.
4. Construct sides of threads by drawing lines 15 degrees with vertical and through points marked on pitch diameter, (c). Draw crest and root diameters of thread.
5. Connect thread crests with lines sloping downward 1/2P to right for right-hand threads.
6. Draw root diameter lines to complete thread, (d).
7. Add note to provide thread specification.

The construction of detailed internal Acme threads is as shown in Fig. 16-15.

Two classes of Acme threads are provided: General Purpose and Centralizing. General Purpose classes (2G, 3G and 4G) provide clearances on all diameters for free movement. The variation in classes has to do with the amount of backlash or end play in the threads. Class 2G is the preferred choice and, if less backlash is desired, classes 3G and 4G are provided.

Some examples of Acme thread notes specifying General Purpose classes are:

1/2 - 10 ACME - 2G

2 - 4 ACME - 4G

Centralizing classes of Acme threads (2C to 6C) have

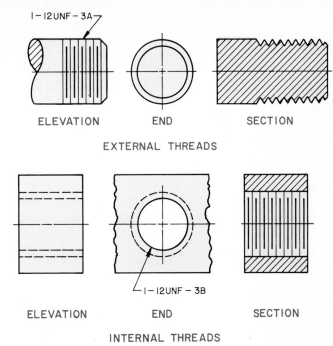

EXTERNAL THREADS

INTERNAL THREADS

Fig. 16-16. Schematic representation of threads.

Acme thread notes specifying Centralizing classes are:

3/8 - 12 ACME - 5C

4 - 2 ACME - 2C

SCHEMATIC REPRESENTATION OF THREADS

Schematic thread symbols are recommended and approved for the representation of all screw threads such as Unified, Square and Acme. These symbols (or simplified representation of threads) are used by industry, along with the thread specifications, on most drawings to represent threaded parts,

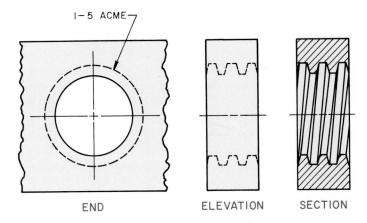

Fig. 16-15. Detailed representation of internal Acme threads.

limited clearances at the major diameters of internal and external threads. This permits a bearing to maintain approximate alignment of the threads and prevents wedging. A class 6C Acme thread is a closer fit than a class 2C. Some examples of

Fig. 16-16. Notice that the external thread in the section view is not a schematic symbol but a detailed representation. Also notice that the internal thread in the elevation is the same symbol used for the simplified internal thread in the elevation.

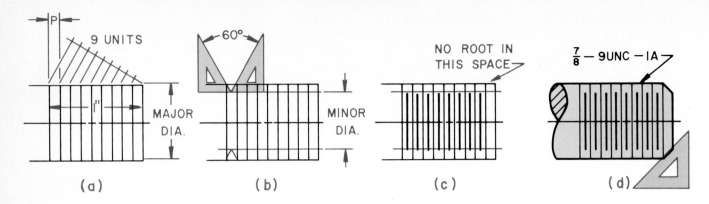

Fig. 16-17. Constructing a schematic thread symbol for an external thread.

The construction of the schematic thread symbol is as follows:

1. Lay out center line and major diameter of thread, Fig. 16-17 (a).
2. Lay off pitch of thread by graphical method or by measurement. Pitch need not be true and can be estimated if laid off uniformly.
3. Draw thin lines across diameter to represent crest lines of thread.
4. To find minor diameter, lay off a 60 degree "V" between crest lines and draw a light construction line along threaded length of part, (b). Repeat on opposite side of piece.
5. Use a heavy line for root line, which is drawn between lines marking minor diameter and spaced uniformily between crest lines, (c). Do not draw a root line in first space next to end.
6. Draw a 45 degree chamfer from last full thread and add note to provide thread specification, (d).

SIMPLIFIED REPRESENTATION OF THREADS

The use of simplified thread symbols is the fastest method of representing screw threads on a drawing, Fig. 16-18. The major diameter is found by direct measurement and minor diameter is found by the 60 degree "V" method or is estimated.

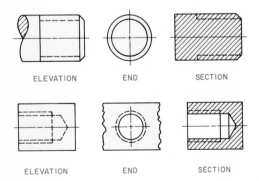

Fig. 16-18. Simplified representation of threads.

REPRESENTATION OF SMALL THREADS

Threaded parts of small diameter are difficult to draw to true or reduced scale dimensions since the small screw pitch crowds the crest and root lines in the schematic method, Fig. 16-19. In the simplified method, clarity of the symbol would be impaired by a crowding of the major and minor diameters. The conventional practice is to exaggerate the space between crests and roots, and major and minor diameters, since accuracy is not as important as clarity of the symbol. The note specifying the thread controls the actual thread characteristics.

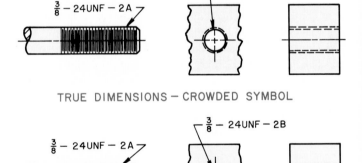

Fig. 16-19. Exaggerate spacings to clarify thread symbols on small size thread drawings.

PIPE THREADS

Three forms of American Standard pipe threads are used in industry: Regular, Dryseal and Aeronautical. The regular pipe thread is the standard for the plumbing trade and it is available in tapered and straight threads.

Tapered pipe threads are cut on a taper of 1 in 16 measured on the diameter. They may be drawn straight or at an angle since the thread note indicates whether the thread is straight or tapered. When the tapered pipe threads are drawn at an angle, they should be exaggerated by measuring 1 unit in 16. These are measured on the radius rather than the diameter, which is an angle of approximately 3 degrees, Fig. 16-20.

Threads and Fastening Devices

Dryseal pipe thread is standard for automotive, refrigeration and hydraulic tube and pipe fittings. The general forms and dimensions of these threads are the same as regular pipe threads except for the truncation of the crests and roots. The common American Standard pipe threads:

NPT — American Standard Taper Pipe Thread
NPTR — American Standard Taper Pipe for Railing Joints
NPTF — Dryseal American Standard Pipe Thread

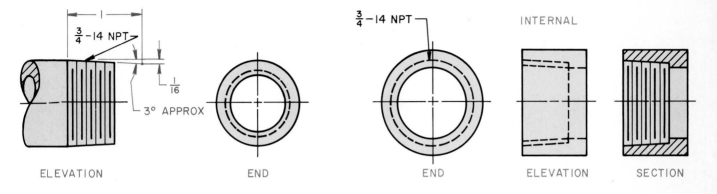

SCHEMATIC

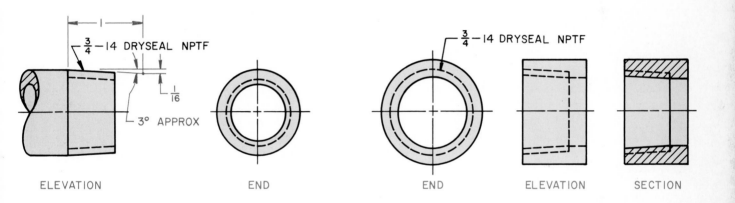

SIMPLIFIED

Fig. 16-20. Tapered pipe threads in schematic and simplified representation.

Dryseal pipe thread form has no clearance since the flats of the crests on the external and internal thread meet, producing a metal-to-metal contact and eliminating the need for a sealer.

Aeronautical pipe thread is the standard in the aerospace industry where the internally threaded part is made of soft light materials (such as aluminum or magnesium alloys) and the screw is made from high-strength steel. An insert, usually of phosphor bronze, is inserted as the bearing part of the internal thread, preventing wear on the light alloy thread, Fig. 16-21.

Regular and aeronautical pipe thread forms require a sealer to prevent leakage in the joint.

The specifications for American Standard pipe threads are listed in sequence: nominal size, number of threads per inch, symbols for form and series. A typical specification, for example, is 1/2 — 14 NPT.

The following symbols are used to designate the more

NPSF — Dryseal American Standard Fuel Internal Straight Pipe Thread
NPSI — Dryseal American Standard Intermediate Internal Straight Pipe Thread

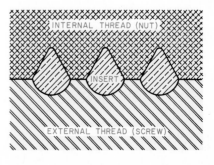

Fig. 16-21. Aeronautical pipe thread showing insert in internal thread.

Fig. 16-22. An anchor fastener that is easy to install.
(Barrett Manufacturing Co.)

TYPES OF BOLTS AND SCREWS

There are many varieties and sizes of bolts, nuts and screws for all kinds of industrial applications, Fig. 16-22. There are five general types of threaded fasteners with which the drafter-designer should be familiar:

A BOLT has a head on one end and is threaded on the other end to receive a nut. It is inserted through clearance holes to hold two or more parts together, Fig. 16-23 (a).

A CAP SCREW is similar to a bolt with a head on one end, but usually a greater length of thread on the other. It is screwed into a part with mating internal threads for greater strength and rigidity, (b).

A STUD is a rod threaded on both ends to be screwed into a part with mating internal threads. A nut is used on the other end to secure two or more parts together, (c).

A MACHINE SCREW is similar to a cap screw except it is smaller and has a slotted head, (d).

A SETSCREW is used to prevent motion between two parts, such as rotation of a collar on a shaft, (e). The range of sizes and exact dimensions for setscrews, and all of these threaded fasteners, are given in the Appendix, starting on page 564.

DRAWING SQUARE BOLT HEADS AND NUTS

The drawing of square bolt heads is identical to that of drawing square nuts except the nut is usually thicker than the bolt head. The method illustrated in Fig. 16-24 is based on the bolt diameter and is an approximation of the actual projection.

The drawing of bolt heads and nuts across corners is the most representative and should be used when a choice is available. This method is shown first. Proceed as follows:

1. Draw bolt diameter (nominal size), Fig. 16-24.
2. Draw bolt head thickness and diameter.
3. Draw square head around diameter at 45 degrees and project to front view, (b).
4. Locate centers for chamfer arcs in front view by projecting lines 60 degrees with horizontal down from center and outside corners of top surface.
5. Complete drawing of square bolt head across corners by drawing a 30 degree chamfer line at outside corners in front view, (c).

The regular square head nut, (d), is 7/8 D in thickness, and the thickness of the heavy duty nut equals the diameter of the bolt it matches. Hidden lines in the front view to represent

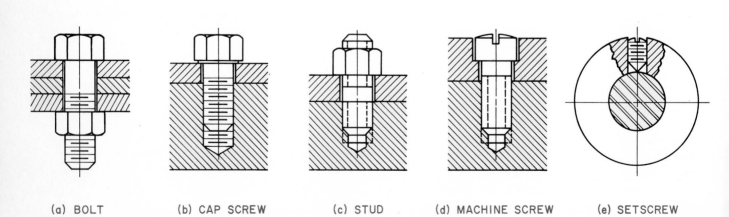

(a) BOLT   (b) CAP SCREW   (c) STUD   (d) MACHINE SCREW   (e) SETSCREW

Fig. 16-23. Five general types of bolts and screws.

threads are normally not shown in application, especially in bolt and nut assemblies. (See Fig. 16-23.)

DRAWING HEXAGONAL BOLT HEADS AND NUTS

The hexagonal bolt head and nut is also most representative when drawn across the corners. This method is shown in Fig. 16-25. Proceed as follows:

1. Draw bolt diameter (nominal size), Fig. 16-25 (a).
2. Draw bolt head thickness and diameter.
3. Lay out hexagonal head around diameter at 60 degrees with horizontal and project to front view, (b).

4. Locate centers for center chamfer arc in front view by projecting lines 60 degrees with horizontal and down from outside corners of top surface.
5. Locate centers for side chamfer arcs by projecting 60 degree lines down from two inside corners to meet other 60 degree lines.
6. Complete drawing of hexagonal bolt head across corners by drawing a 30 degree chamfer line at outside corners in front view, (c).

The construction of a hexagonal nut is shown in Fig. 16-25 (d). The regular nut is drawn 7/8 D in thickness and the heavy

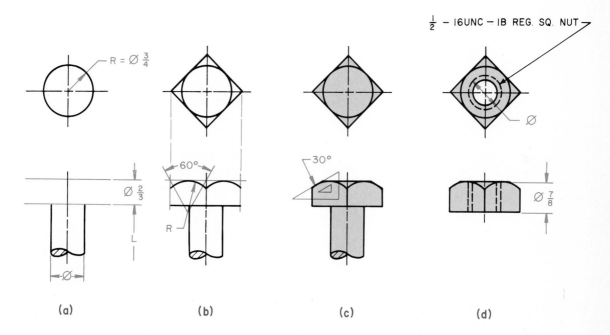

Fig. 16-24. A method of drawing square bolt heads and nuts across corners that approximates true projection.

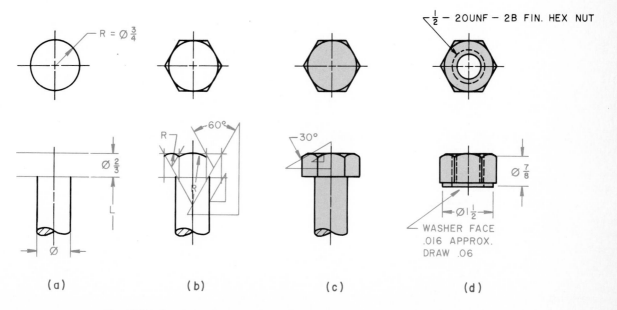

Fig. 16-25. Drawing hexagonal bolt heads and nuts across corners by the approximation method.

duty hex nut is a full D in thickness. Hidden lines may be omitted unless needed for clarity.

Procedures for drawing hexagonal head bolts and nuts across the flats are illustrated in Fig. 16-26.

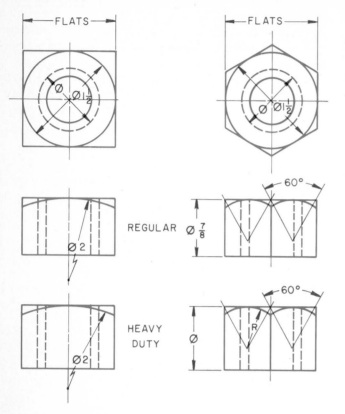

Fig. 16-26. Square and hexagonal nuts are not as representative when drawn across the flats. This position should be avoided whenever possible.

DRAWING CAP SCREWS

The drawing of cap screws (like drawing bolts) is a proportioned, approximate drawing based on the diameter of the screw, Fig. 16-27. Five types of standard cap screws are shown, together with the dimensions for their construction.

Specific dimensions for assembly and thread lengths should be checked in a standards table. Cap screws are specified as follows: 7/16 − 14 UNC − 2A x 2 BRASS HEX CAP SCR. If the cap screw is made of steel, the material term is omitted.

DRAWING MACHINE SCREWS

Four common types of machine screws and their construction are shown in Fig. 16-28. Machine screws are similar to cap screws but, usually smaller in diameter.

Threads on machine screws are either Unified Coarse (UNC) or Unified Fine (UNF), Class 2A. Screws 2 inches in length or less are threaded to within two threads of the bearing surface. The thread length on screws longer than 2 inches is a minimum of 1 3/4 inches. Machine screws are specified as follows:

3/8 − 24 UNF − 2A x 1 OVAL HD MACH SCR

DRAWING SETSCREWS

Setscrews are made in the standard square head and several headless types, Fig. 16-29. Several styles of points are also available with each. When a setscrew is used against a round shaft, a cup point is likely to hold best. A flat or dog point is used where a flat spot on a shaft has been machined. Threads are coarse, fine and 8-thread series, class 2A, except for the square-head setscrews. These are normally stocked in the coarse series and size 1/4 inch or larger. Setscrews are specified as follows:

1/4 − 28 UNF − 2A x 5/8 HEX SOCK CUP PT SET SCR

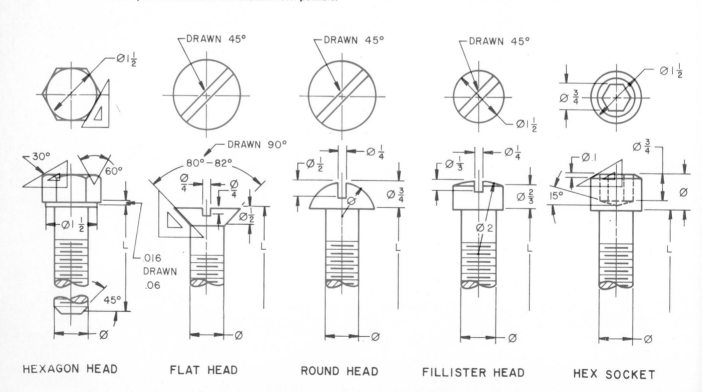

HEXAGON HEAD   FLAT HEAD   ROUND HEAD   FILLISTER HEAD   HEX SOCKET

Fig. 16-27. Representation of cap screws on drawings.

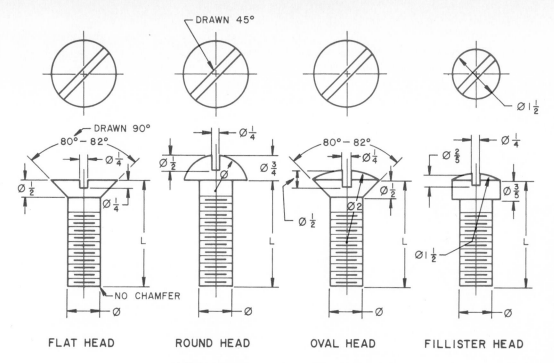

Fig. 16-28. Representation of machine screws on drawings.

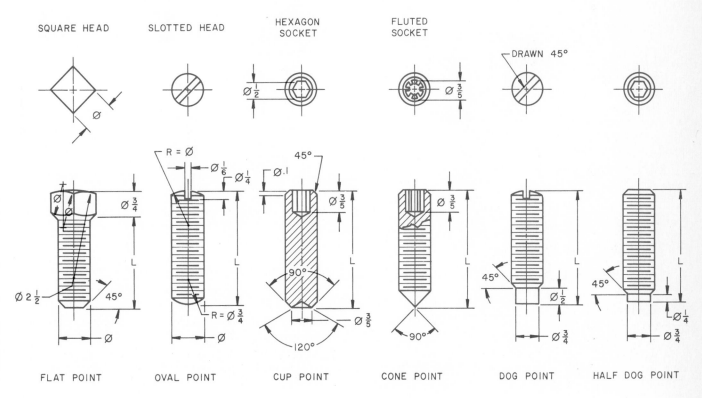

Fig. 16-29. Representation of setscrews on drawings.

SELF-TAPPING SCREWS

Time in assembly work is of great importance in many industrial applications. To meet this condition where threaded fasteners are concerned, self-tapping screws have been developed, Fig. 16-30.

Self-tapping screws are of two major kinds:

1. Thread-cutting screws which act like a tap, cut away material as they enter the hole.

2. Thread-forming screws form threads by displacing the

Fig. 16-30. Self-tapping screws speed the work of industry. (ITT Harper, Inc.)

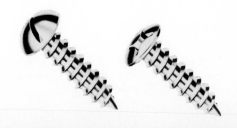

material rather than cutting it.

The thread-cutting screws have flutes or slots in the point to form a cutting edge, and their thread form is similar to standard Unified Threads, Fig. 16-31. This type screw is suitable for applications in metal as well as plastics and can be removed and reassembled without noticeable loss of holding power.

Thread-forming screws (sometimes called sheet-metal screws) are especially suited for thin-gage sheet metal up to .375 inch in thickness, as well as in any soft material such as wood and plastic, Fig. 16-32. The thread form on this screw is a narrow, sharp crest and no chips or waste material is formed in their application. Two patented thread-forming screws have a special shape to displace the metal and form a tight fitting thread. These screws are known as SWAGEFORM ® and TAPTITE ® , Fig. 16-33.

METALLIC DRIVE SCREWS

Some industrial applications call for permanent fasteners not expected to be disassembled while in service. Metallic drive screws designed for this use, Fig. 16-34, have multiple threads with a large lead angle. They are driven by a force in line with their axis, rather than torque.

Once seated, metallic drive screws cannot be removed and reinserted easily. Economy is the main reason for using this kind of screw where circumstances permit. Drive screws are also available for use in assembly of wooden parts.

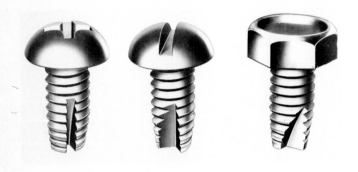

Fig. 16-31. Thread-cutting screws actually cut threads as they are inserted. (ELCO Industries, Inc.)

Fig. 16-32. Thread-forming screws displace the metal to form threads.

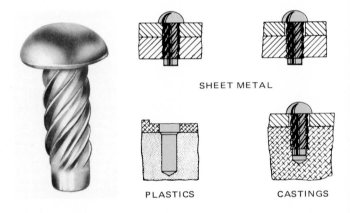

SHEET METAL

PLASTICS   CASTINGS

Fig. 16-34. Drive screws are an economical and permanent type of fastener. (ELCO Industries, Inc.)

WOOD SCREWS

Wood screws are standardized with three head types: flat, round and oval, Fig. 16-35. These are available in slotted or Phillips head drives. The latter is used in most commercially manufactured products where time in assembly is important.

Wood screws range in size from 0 to 32; in diameter from .060 to .372 in. They are specified in a note or in the Materials List as: NO. 9 x 1 1/2 OVAL HD WOOD SCR.

TEMPLATES FOR DRAWING THREADED FASTENERS

A variety of templates are available for the drawing of threads, bolts, screws, nuts and head types, Fig. 16-36. Industry makes use of these templates to speed drafting time to produce more uniform representation of threaded fasteners.

WASHERS AND RETAINING RINGS

Washers are added to screw assemblies for such uses as load distribution, surface protection, insulation, spanning an over-

Fig. 16-33. SWAGEFORM ® and TAPTITE ® threads have special forms to provide a more secure thread engagement. (ELCO Industries, Inc.)

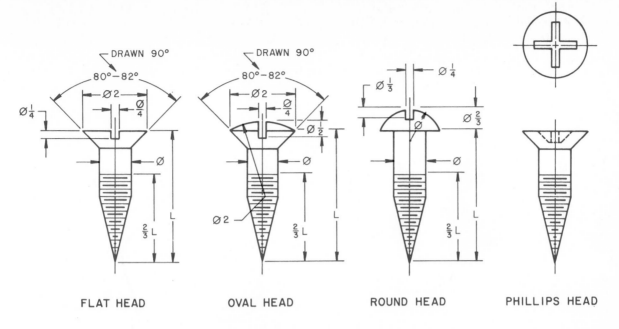

FLAT HEAD　　　OVAL HEAD　　　ROUND HEAD　　　PHILLIPS HEAD

Fig. 16-35. Schematic representation of wood screws.

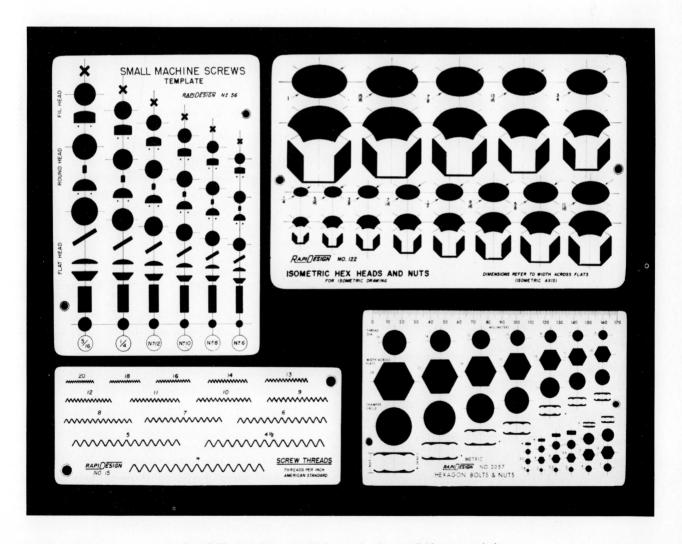

Fig. 16-36. Templates should be used when available to speed the drawing of threaded fasteners.　(RapiDesign)

size clearance hole, sealing, electrical connection, spring-tension takeup and for locking. Washers of the locking type are the split-spring and toothed washers.

A finishing washer distributes the load and eliminates the need for a countersunk hole. It is used extensively for attaching fabric coverings. Flat washers, Fig. 16-37, are used primarily for load distribution.

Fig. 16-37. Examples of finishing, flat and lock washers used in mechanical assemblies. (ELCO Industries, Inc.)

Retaining rings are inexpensive devices, available in a wide variety of designs. They are used to provide a shoulder for holding, locking or positioning components on shafts, pins, studs or in bores, Fig. 16-38. They almost always slip or snap into grooves and are sometimes called "snap" rings.

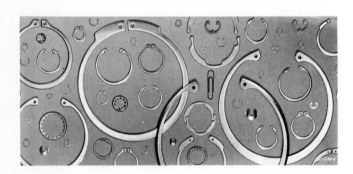

Fig. 16-38. Retaining rings are inexpensive fasteners and can be quickly assembled or removed. (Waldes Kohinoor, Inc.)

Fig. 16-39. The variety of nuts used in industry is seemingly limitless. (Standard Pressed Steel Co.)

NUTS — COMMON AND SPECIAL

There is as much, if not more, variety in nuts as there is in bolts and screws, Fig. 16-39. Nuts are discussed in this section under two broad classes: common and special. Only a few of the special nuts are noted here as representative of those available. A supplier's catalog would contain many hundreds of special use items.

COMMON NUTS

Nuts used on bolts for assemblies are known as common nuts and are generally divided between finished (close tolerances), Fig. 16-40 (a), and heavy (looser fit for large-clearance holes and high loads), (b).

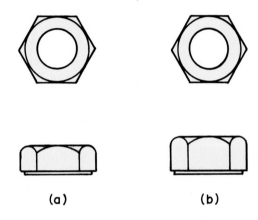

(a)   **(b)**

Fig. 16-40. Types of common nuts.

CAP, WING AND KNURLED NUTS

The cap nut (sometimes called an acorn nut), Fig. 16-41 (a), is used for appearance. The wing nut, (b), and knurled nut, (c), allow for hand tightening. These are typical of the wide variety of nuts for special assembly, fastening or holding purposes.

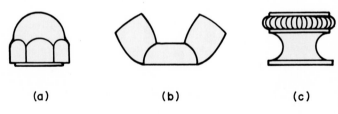

(a)   (b)   (c)

Fig. 16-41. The cap nut and wing nut.

SINGLE-THREAD ENGAGING NUTS

Nuts formed by stamping a thread engaging impression in a flat piece of metal are called single-thread engaging nuts. Shown in Fig. 16-42 is a nut with helical prongs which engage and lock on the screw thread root diameter. A protruding truncated cone nut is stamped into the metal which provides a ramp for the screw to climb as it turns. Single-thread nuts can be formed from nearly any ferrous or nonferrous alloy.

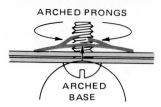
PRE-LOCKED POSITION

ARCHED PRONGS

ARCHED BASE

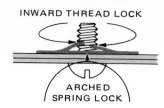
DOUBLE-LOCKED POSITION

INWARD THREAD LOCK

ARCHED SPRING LOCK

Fig. 16-42. An example of a single-thread engaging nut. (Eaton Corp.)

(a)

(b)

(c)

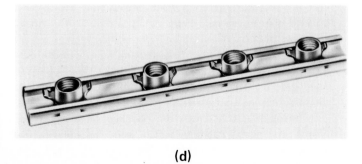

(d)

Fig. 16-43. Some examples of self-retaining nuts. (Esna Corp.)

Usually, however, they are made of high-carbon steel, hardened and drawn to a spring temper. These nuts are often used to reduce assembly costs where lighter-duty applications are involved.

CAPTIVE OR SELF-RETAINING NUTS

Captive or self-retaining nuts are multiple-threaded nuts that are held in place by a clamp or binding device of light gage metal. They are used for applications in which thin materials are used; threaded fasteners are needed at inaccessible or blind locations; or repeated assembly and disassembly is required.

Self-retaining nuts may be grouped according to four means of attachment:
1. Plate or anchor nuts, Fig. 16-43 (a), with mounting lugs that can be screwed, riveted or welded to the assembly.
2. Caged nuts, (b), held in place by a spring-steel cage that snaps into a hole or clamps over an edge.
3. Clinch nuts, (c), designed with a pilot collar clinched or staked into a parent part through a precut hole.
4. Self-Piercing nuts, (d), held in a tool in a punch press that punches a hole in the work piece with the hardened nut. The press then follows through to clinch the nut in place.

LOCKNUTS

There are three groups of locknuts: free-spinning; prevailing-torque; spring-action nuts.

The FREE-SPINNING type grips tightly only when the nut is seated on a surface or when two mating parts are tightened together, Fig. 16-44. There are several types of free-spinning locknuts. Those with two mating parts, (a), clamp the threads of bolt when seated and resist back-off. Locknuts with a recessed bottom and slotted upper portion, (b), cause a spring action when seated, and bind upper threads of the nut. Those

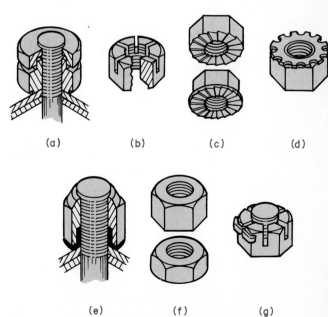

(a) (b) (c) (d)

(e) (f) (g)

Fig. 16-44. Types of free-spinning locknuts.

nuts with a deformed bearing surface, (c), tend to dig in and remain tight when seated.

Some locknuts have a lock washer secured to the main nut, Fig. 16-44 (d). Others have metallic and nonmetallic inserts such as a softer metal or a thermoplastic material, (e). The insert tends to flow around the threads when seated, forming a tight lock and seal. Jam nuts are thin nuts used under common nuts (f), and when seated under pressure, the threads of the jam nut and bolt are elastically deformed, causing considerable resistance against loosening. Slotted nuts, (g), have slots to receive a cotter pin or wire which passes through a drilled hole in the bolt, locking the nut in place.

PREVAILING-TORQUE locknuts start freely, then must be wrenched to their final position due to a deformation of threads or insert in the center or upper portion of the nut. These nuts maintain a constant load against loosening whether seated or not.

SPRING-ACTION locknuts are the single-thread type, usually stamped from spring-steel. They lock in place when driven up against a surface, Fig. 16-42. These nuts are sometimes classed as free-spinning locknuts, but they can be (and frequently are in mass production) jammed onto a thread without spinning.

RIVETS

The manufacture of many assembled products requires a permanent type of fastener. In these cases, rivets often are the answer (on aircraft structures, small appliances and jewelry). Rivet sizes are indicated by the diameter of the shank and, if unusual, by the length of the shank. Rivets are available in a variety of head styles and are grouped into two general types,

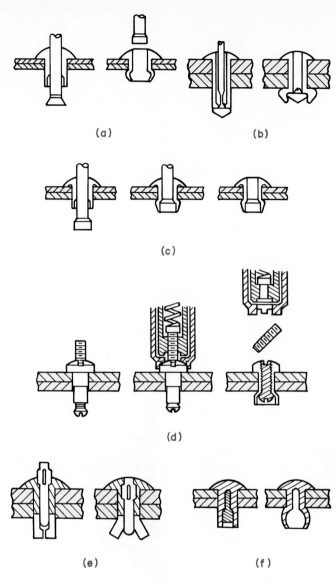

Fig. 16-46. Types of blind rivets.

standard and blind.

STANDARD RIVETS come in several styles, depending on strength, methods of application and other design requirements, Fig. 16-45.

Semitubular rivets are the most widely used type, Fig. 16-45 (a). This type rivet, when properly specified and set, becomes essentially a solid rivet. Semitubular rivets can be used to pierce very thin light metals, although they are not classified as self-piercing rivets.

Full tubular rivets have deeper shank holes and can punch their own holes in fabric, some plastics and other soft materials, (b). The shear strength of full tubular rivets is less than that of semitubular rivets.

Bifurcated or split rivets are punched or sawed to form prongs that enable them to punch their own holes in fiber, wood, plastic or metal, (c). These rivets are also called self-piercing rivets.

Compression rivets consist of two parts: a deep-drilled tubular part; a solid part designed for an interference fit when

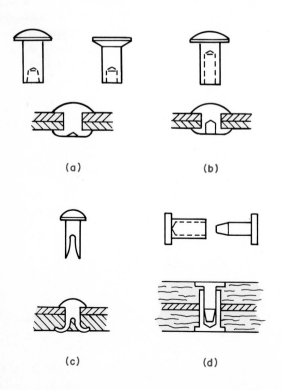

Fig. 16-45. Types of standard rivets.

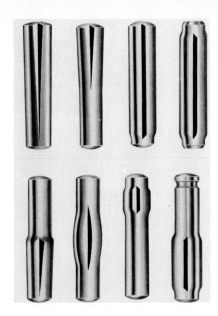

Fig. 16-47. Some examples of pin fasteners used in assembly work. (Groov-Pin Corp.)

set, (d). This type of rivet is used when both sides of a work piece must have a finished appearance, such as the handle of a kitchen knife.

BLIND RIVETS can be installed in a joint that is accessible from only one side. However, they are increasingly being used in applications where standard rivets were formerly used to simplify assembly, reduce cost and to improve appearance. Blind rivets are classified by the methods used in setting. They, too, are available in a variety of head styles.

Pull-mandrel blind rivets (sometimes called "pop rivets") are set by inserting the rivet in the joint and pulling a mandrel to upset the blind end of the rivet, Fig. 16-46. Some rivets have mandrels that pull through leaving a hole in the rivet, (a). Others have break-type mandrels that break during the pull-through process and plug the hole, (b), and a third nonbreak-type mandrel must be trimmed off after the rivet is set, (c).

The threaded-type blind rivet consists of an internally threaded rivet that is torqued or pulled to expand and set the rivet, (d).

The drive-pin blind rivet is like the mandrel type in reverse. The pin is driven into the body to set the blind side of the rivet, (e).

Chemically expanded blind rivets have a hollow end filled with an explosive that detonates when heat or an electric current is applied to set the rivet, (f).

PIN FASTENERS

Where the load is primarily shear, pins can be an inespensive and effective means of fastening. Some representative types include hardened and ground dowel and taper pins, grooved surface, spring or tubular, clevis and cotter pins, Fig. 16-47. The method of representing pins on a drawing is shown in Fig. 16-48.

SPRINGS

The steps in laying out a representation of a coil spring on a drawing are similar to those in representing screw threads. A detail drawing of a coil spring used in conjunction with a check valve is shown in Fig. 16-49.

To lay out a coil spring, mark off the pitch distance along the diameter of the coil, Fig. 16-50. Give the coils a slope of one-half of the pitch for closely wound springs. Note the difference in the representation in a tension and a compression spring and how the different types of ends are drawn in each case. To avoid a repetitious series of coils, phantom lines may be used to represent repeated detail between spring ends.

A schematic representation of various types of coil springs is shown in Fig. 16-50.

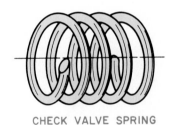

CHECK VALVE SPRING

Fig. 16-49. Detailed representation of coil springs.

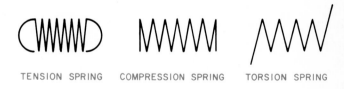

TENSION SPRING   COMPRESSION SPRING   TORSION SPRING

Fig. 16-50. Schematic representation of coil springs.

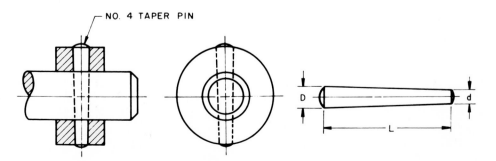

Fig. 16-48. Methods of representing pin fasteners on drawings.

KEYS

Keys are used to prevent rotation between a shaft and machine parts (such as gears, pulleys and rocker arms). Fig. 16-51 illustrates the four most common types of keys (square, gib head, Pratt and Whitney and Woodruff) and the method of dimensioning.

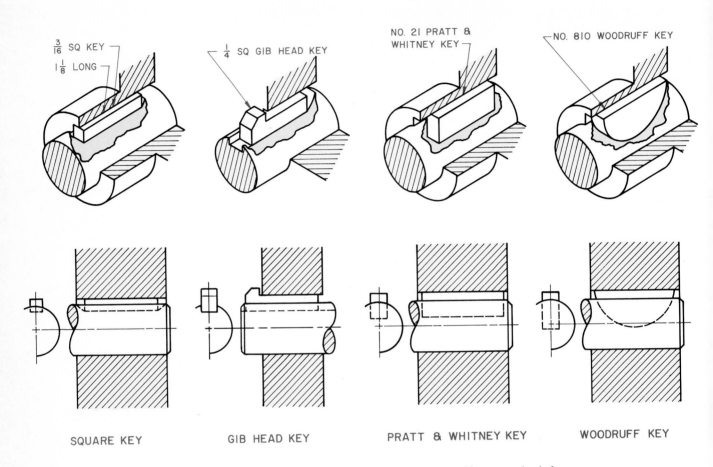

SQUARE KEY   GIB HEAD KEY   PRATT & WHITNEY KEY   WOODRUFF KEY

Fig. 16-51. Standard keys used to prevent rotation between a machine part and a shaft.

Design engineers reviewing a design for a petroleum process plant.
(Standard Oil of California)

PROBLEMS IN THREADS AND FASTENERS

The following problems will provide you with the opportunity to apply the knowledge of threads and fasteners gained in the study of this chapter and to become familiar with the drafting procedures necessary in representing fasteners on drawings.

1. Construct a detail representation of a Unified National Coarse thread showing an external threaded shaft and a sectional view of an internal thread of the same specification. Use a layout similar to the one in Fig. 16-52. The thread specification is: DIA = 1 1/2; 6 THDS/INCH, CLASS 3. Place the thread specification note on the drawing.

2. Make a detailed drawing of an external and internal square thread with the following specifications, using the suggested layout in Fig. 16-52: 1 1/2 − 3 SQUARE. Dimension the thread with a note.

3. Using the same specifications as for problem 2, draw an Acme thread, class 2G.

4. With the layout suggested in Fig. 16-52, draw a schematic representation of threads for a 7/8 class 2 Unified National thread. Check the table in the Reference Section for thread specifications and dimension the thread with a note.

5. This problem is the same as number 4 except the schematic thread is to represent a Square thread with an outside diameter of 3/4 inch.

6. With the layout suggested in Fig. 16-52, draw a simplified thread representation of an Acme thread of the following specification: 1 − 5 ACME − 2G − LH − DOUBLE. Dimension the thread with a note.

7. Draw two views, one a sectional view, of a schematic representation of a threaded hole 2 inches deep in a 1 inch square bar of steel 3 inches long. The hole is to be pilot drilled to a depth of 2 1/4 inches and threaded with a 9/16 Unified National Extra Fine class 3 thread. Dimension the feature with a note.

8. Make a simplified thread drawing of the threaded pieces in problem 1.

9. Using the layout suggested in Fig. 16-52, draw a schematic representation of Unified National Thread Series 8 with an outside diameter of 2 inches.

10. Draw a schematic representation of an American Standard Taper Pipe thread with a nominal pipe size of 1 1/4 inches. Check the Reference Section for the thread specifications and show two views of an external thread and two views, one to be a section, of an internal thread. Dimension the threads with a note.

11. Make a drawing similar to the one in problem 10 for an American Standard Taper Pipe thread with a nominal pipe size 3/8 inch, Dryseal.

12. Draw a regular hexagonal and square head bolt and nut of the following specifications: Nominal size 1/2 inch, length 3 inches, length of thread 2 inches. Show threads in simplified representation and place a dimensional note on the drawing.

13. Make a drawing of a 3/8 inch semifinished hexagonal head bolt 2 inches long, clamping two pieces of 5/8 inch steel plate together and held by a jam nut and a semifinished regular nut. Specify the bolt and nuts by note.

14. Make a two-view drawing of the Spindle Bearing Adjusting Nut, Fig. 16-53 (a), with the circular view a full section and schematic symbol representation for threads.

15. Draw two views of the Special Adjusting Screw, as shown in Fig. 16-53(b). Show the counterbore and full thread as a broken-out section. Use the schematic symbol to represent the thread and dimension the part. (Check the Reference Section for dimensions not furnished on the drawing.)

16. Draw the necessary views to adequately describe the Control Shaft and dimension the drawing, Fig. 16-53 (c). Use the simplified symbol representation for the threads.

17. Draw the views required to adequately describe the Spindle Ram Screw shown in Fig. 16-53 (d). Show a detail representation of the thread by drawing two full threads on each end of threaded portion and indicate the remainder by use of phantom lines at the major diameter. Dimension the part.

18. Draw the necessary views to adequately describe the Pedestal Bracket Shaft, Fig. 16-54 (a), and dimension the drawing. Use schematic thread symbol for threads.

19. Draw a half section of the Shank, Fig. 16-54 (b), and show the threads in schematic form. Dimension the part.

20. Make a front view as shown of the Gear Shaft, Fig. 16-54 (c), and a partial top view sufficient to show both keyseats for the 3/16 inch square key at 1 and a No. 404 Woodruff key at 2. (Check the Reference Section for key specifications.) Use schematic symbols for thread representation. Dimension the drawing.

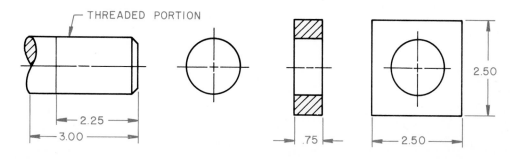

Fig. 16-52. Suggested layout for problems 1-9.

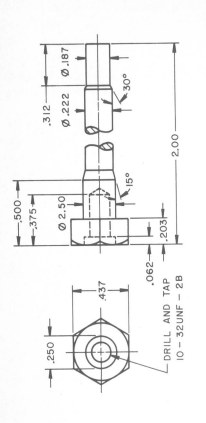

Ø .187
.312
Ø .222
30°

.500
.375
15°
Ø 2.50
.062
.203

DRILL AND TAP
10 – 32UNF – 2B

2.00

.437
.250

SPECIAL ADJUSTING SCREW

(b)

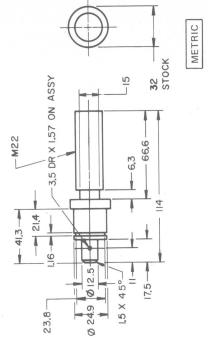

M22
3.5 DR X 1.57 ON ASSY
15
6.3
66.6
114
32
STOCK

41.3
21.4
1.16
Ø 12.5
11
17.5
23.8
Ø 24.9
1.5 X 45°

METRIC

SPINDLE RAM SCREW

(d)

.687
.343

R .062 TYP

1¼ – 12UNC – 2B
TO TAKE THREAD ON
PART NO. 3016

2.375

13/64 TAP DRILL
¼ – 20 UNC TAP

1.265

Ø ¼ DRILL
↧ .312

SPINDLE BEARING ADJ. NUT

(a)

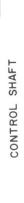

2.7 THD RELIEF
0.7 X 45°
CHAMFER
(Ø 15.8)
M12

R 0.7
3 PLACES
20°TYP

14,3

30° TYP

6.3
TYP
32
52
92

15.8

BREAK SHARP EDGE

3.5

Ø 9.52
Ø 4.70
6.3

CONTROL SHAFT

(c)

METRIC

Fig. 16-53. Problems involving thread representation.

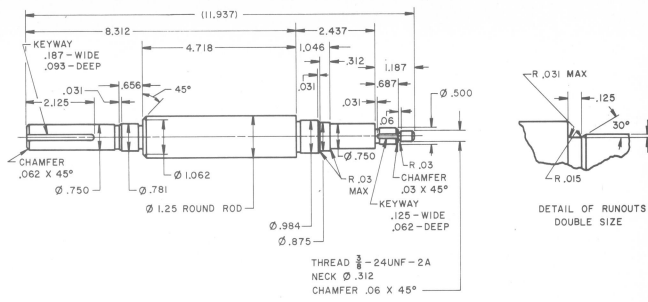

KEYWAY
.187 – WIDE
.093 – DEEP

CHAMFER
.062 X 45°

Ø .750

Ø .781

Ø 1.25 ROUND ROD

Ø 1.062

Ø .984

Ø .875

Ø .750

R .03
MAX

CHAMFER
.03 X 45°

KEYWAY
.125 – WIDE
.062 – DEEP

Ø .500

THREAD $\frac{3}{8}$ – 24UNF – 2A
NECK Ø .312
CHAMFER .06 X 45°

PEDESTAL BRACKET SHAFT

(a)

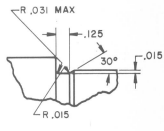

R .031 MAX

.125

30°

.015

R .015

DETAIL OF RUNOUTS
DOUBLE SIZE

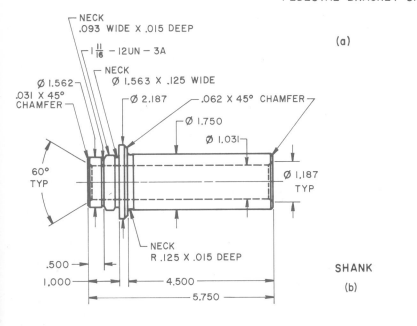

NECK
.093 WIDE X .015 DEEP

$1\frac{11}{16}$ – 12UN – 3A

NECK
Ø 1.563 X .125 WIDE

Ø 1.562

.031 X 45°
CHAMFER

Ø 2.187

.062 X 45° CHAMFER

Ø 1.750

Ø 1.031

60°
TYP

Ø 1.187
TYP

NECK
R .125 X .015 DEEP

.500

1.000

4.500

5.750

SHANK

(b)

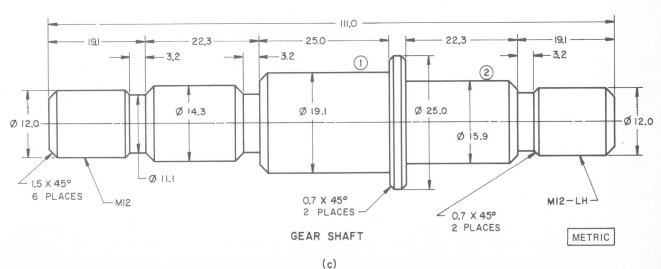

111.0

19.1

22.3

25.0

22.3

19.1

3.2

3.2

①

②

3.2

Ø 12.0

Ø 14.3

Ø 19.1

Ø 25.0

Ø 12.0

Ø 15.9

1.5 X 45°
6 PLACES

Ø 11.1

M12

0.7 X 45°
2 PLACES

0.7 X 45°
2 PLACES

M12–LH

METRIC

GEAR SHAFT

(c)

Fig. 16-54. Thread representation problems.

Chapter 17
MANUFACTURING PROCESSES

An important aspect of the drafter's and designer's work in the preparation of a drawing is specifying the features on a machine part. This is done by notes and/or symbols which are known as CALLOUTS. This chapter presents those manufacturing processes used most frequently, with the exception of threads, fasteners and weldments, which are covered in other sections of the text.

MACHINE PROCESSES

A feature on a drawing such as a hole may simply be dimensioned by giving its diameter, Fig. 17-1 (a). However, those features which are to be machined in a certain way (in order to produce a desired surface texture or hold a certain tolerance) must be specified as to the machine process. Following are the more common machine processes and their callouts.

DRILL

Drilled holes are produced by a drill bit, usually chucked in a drill press or portable power drill (depending on nature of

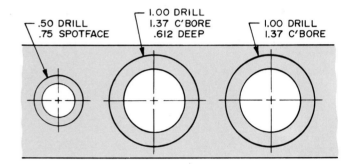

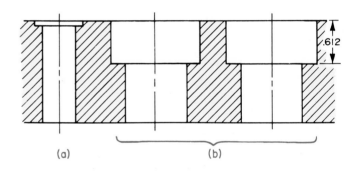

Fig. 17-2. Spotfaced and counterbored holes.

piece and accuracy required), Fig. 17-1 (b). Note that the specification of the drilled hole may be entirely by a callout or by a callout and a dimension for depth on the feature.

SPOTFACE

Spotfacing is when a surface around a hole needs to be cleaned up or leveled to provide a bearing for a bolt head or nut, Fig. 17-2 (a). A spotface may be specified by note only and need not be shown on the drawing.

COUNTERBORE

Counterboring is deeper than spotfacing to allow fillister and socket head screws to be seated below the surface, as shown in Fig. 17-2.

COUNTERSINK

Countersinking is cutting a chamfer in a hole to allow a flat head screw to seat flush with the surface, Fig. 17-3 (a). Outside diameter of countersunk feature on the surface of the part and the angle of countersink are dimensioned.

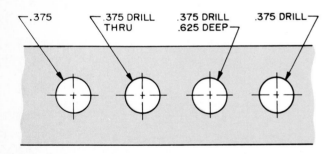

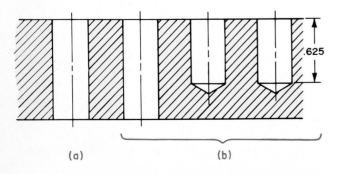

Fig. 17-1. Methods of representing and dimensioning drilled holes.

COUNTERDRILL

Smaller holes may be partially drilled with a larger drill to allow room for fastener or feature of mating part, Fig. 17-3 (b).

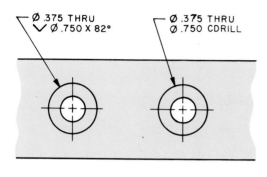

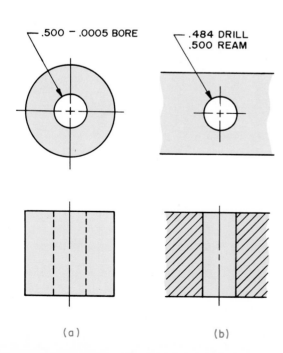

Fig. 17-3. Representation of countersunk and counterdrilled holes.

BORE

When an extremely accurate hole with a smooth surface texture is required, boring is usually specified as the machine process, Fig. 17-4 (a). This may be done on the lathe or boring mill.

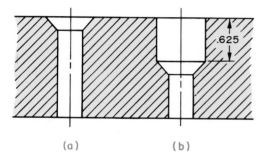

Fig. 17-4. Callouts for bored and reamed holes.

REAM

A hole may be reamed after drilling for greater accuracy and a smooth surface texture, Fig. 17-4 (b). The hole is drilled slightly undersize and then reamed to the desired size.

BROACH

Broaching is the process of pulling or pushing a tool over or through the workpiece to form irregular or unusual shapes. A very simple type of broach is one used to cut a keyway on the inside of a pulley or gear hub. A more complex broach is used to cut an internal spline, Fig. 17-5. The broach is a long tapered tool with cutting teeth that get progressively larger so that at the completion of a single stroke the work is finished.

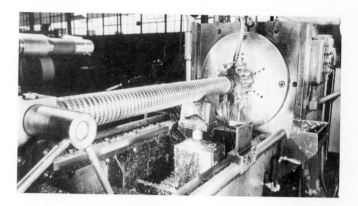

Fig. 17-5. The internal splines of a pinion being broached on a 40 ton horizontal broaching machine.
(Milwaukee Gear)

STAMPING OPERATIONS

Circular or irregular holes and other features are often prepared with dies in a punch press, Fig. 17-6. Stamping operations include perforation, blanking, shearing, bending

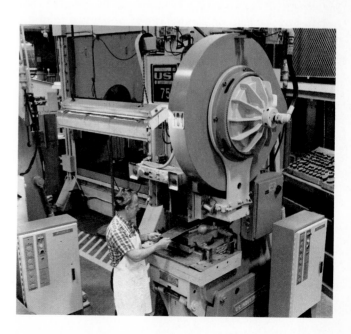

Fig. 17-6. A 75-ton mechanical press used for punching and forming sheet metal. (Rockwell International Corp.)

and forming. These are normally used on sheet metals. The features may be dimensioned directly, placed in a callout, Fig. 17-7 (a), or specified in a tabular dimension table on the drawing, (b).

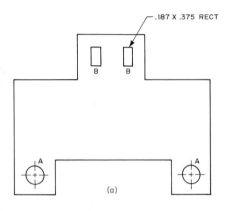

(a)

DESCRIPTION OF HOLES		
SIZE	DESCRIPTION	QTY
A	φ .375 THRU	2
B	.187 x .375 RECT	2

(b)

Fig. 17-7. Blanking features which are regular or irregular in shape is common practice on light gage metals.

KNURL

The process of knurling is forming straight-line or diagonal-line (diamond) serrations on a part to provide a better hand grip or interference fit between parts. See Fig. 17-8 (a). The diametral pitch (DP) type, grade (coarse, medium and fine) and length of knurl should be specified, (b). The knurled surface may be fully or partially drawn or omitted from the drawing since the callout provides clarity, (c).

NECKS AND UNDERCUTS

It is sometimes necessary on machine parts to provide a groove on a shaft to terminate a thread, Fig. 17-9 (a), or to cut a recess at a point where the shaft changes size and mating parts such as a pulley must fit flush against a shoulder, (b). These necks and undercuts are specified as noted in (a). When too small to detail on the part itself, they should be drawn as an enlarged detail, (c).

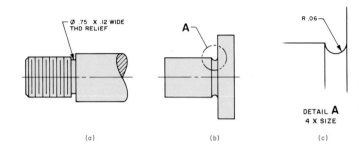

Fig. 17-9. Specifying necks and undercuts on drawings.

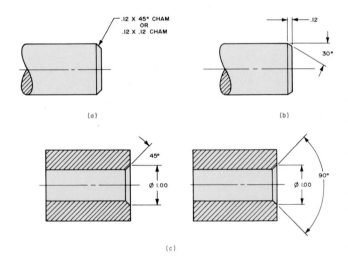

Fig. 17-10. Representing and dimensioning chamfers on a drawing.

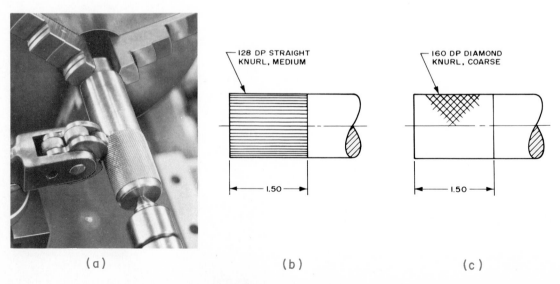

Fig. 17-8. Representing knurling patterns on drawings.

CHAMFERS

Small beveled edges are usually cut on the ends of holes and shafts, as well as on threaded fasteners, to facilitate their assembly. These "chamfers" are dimensioned as shown in Fig. 17-10. When the chamfer angle is 45 degrees, the dimension should be noted as at (a). The use of the word "chamfer" is optional. Angles other than 45 degrees must be included in the dimension note, (b). Internal chamfers are dimensioned as shown at (c).

TAPERS—CONICAL AND FLAT

The conical surface on a shaft or in a hole is called a conical taper, Fig. 17-11. Standard machine tapers are used on various machine tool spindles and mating tapers on drill bits and tool shanks used in the machine. A flat taper increases or decreases in size at a uniform rate to assume a wedge shape.

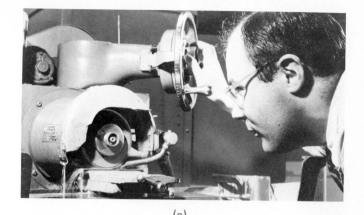

(a)

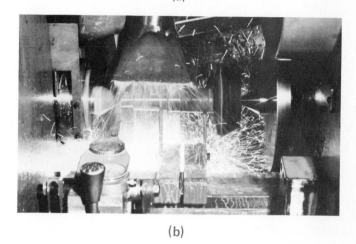

(b)

Fig. 17-12. Finish grinding, (a), and abrasive machining, (b), on grinders. (Norton Co.)

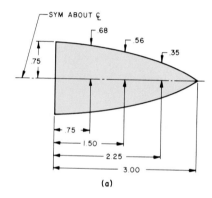

(a)

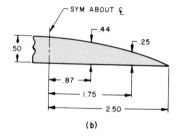

(b)

Fig. 17-11. Dimensioning of tapers. (ANSI)

a cylinder). They are rotated rather slowly and moved backwards and laterally. Lapping is quite similar to honing except a lapping plate or block is used with a very fine abrasive, in paste or liquid form, between the metal lap and work surface.

Fig. 17-13. A cylindrical machine part being ground in a centerless-cylindrical grinder. (Bryant Grinder Corp.)

GRIND, HONE AND LAP

The process of removing metal by means of abrasives is known as grinding. This is usually a finishing operation, Fig. 17-12 (a), but some actual machining and shaping of parts is done by grinding which is also called machining, (b). Abrasive wheels used for most grinding operations come in a variety of sizes, shapes and coarseness of abrasives.

Grinding for purposes of producing a finished surface may be done on a surface grinder for flat work, on a horizontal spindle machine for grinding precision tool and die work, or on a lathe or cylindrical grinder for internal or external grinding of cylindrical parts, Fig. 17-13.

The surface texture of a machine part is usually produced with a grinding operation, particularly finer finishes.

Honing is done with blocks of very fine abrasive materials under light pressure against the work surface (such as inside of

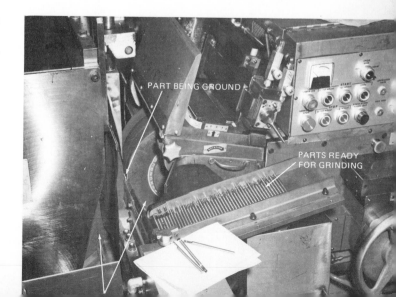

Fig. 17-14. Computer numerical controlled (CNC) machines is a part of modern manufacturing today. Transporter carts automatically deliver and remove workpieces from the CNC machines.   (Kearney & Trecker Corp.)

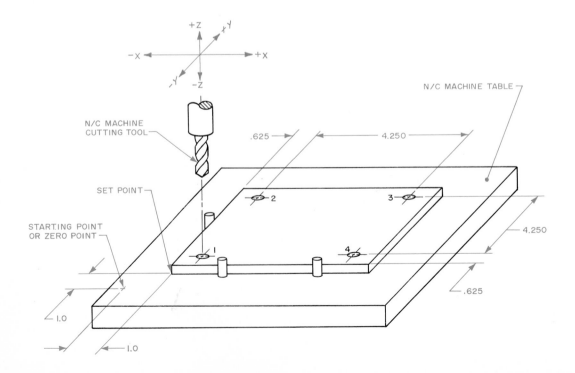

Fig. 17-15. A workpiece located on a N/C machine.

NUMERICAL CONTROL MACHINING

Numerical control (N/C) machining is a means of controlling machine tools, usually with perforated tape, Fig. 17-14. It has been applied extensively to milling, drilling, lathe, punch press work and wire wrapping.

Numerical control machining is very flexible and can be used for machining long or short run production items. There is a great reduction in conventional tooling and fixturing made possible by the programmed instructions given in the tape control.

Two systems of machine control have been added to basic N/C machining. These systems, operated from computer "software," are computer numerical control (CNC) and direct numerical control (DNC). When a machine is operated by its own computer, it is called CNC. When several CNC machines are controlled by a central computer which is directly wired to the machines, the system is called DNC.

DRAWINGS FOR NUMERICAL CONTROL MACHINING

There are two reference point systems of positioning the cutting tool of a numerical control machine for work on a part, incremental and absolute. Drawings used in programming N/C machines are much the same as those used for more traditional machining. However, the dimensioning system used should be compatible with the reference point system of the N/C machine.

INCREMENTAL POSITIONING

The incremental (continuous path) system of positioning the cutting tool in relation to the workpiece is based upon programming the machine a specific distance and direction from its previous position rather than from a fixed zero reference point, Fig. 17-15. That is, each move the tool must make is given as a distance and direction from the previous location or point.

The first dimension is given as a distance and direction from the starting point to the first location where the tool will perform its work. In the example in Fig. 17-15, the distance and direction from the starting point to the first hole is X = 1.625 and Y = 1.625. The second hole is X = 0.0 and Y = 4.250 from the previous location; the third, X = 4.250 and Y = 0.0; the fourth, X = 0.0 and Y = 4.250. The programming to return the N/C tool to its starting point would be X = 5.875 and Y = 1.625.

INCREMENTAL DIMENSIONS should be applied as successive (chain) dimensions, Fig. 17-16. The programmer can read these directly without having to calculate individual settings for preparing the documents needed to punch the tape which feeds information into the N/C machine control unit. These dimensions are the same as BASIC dimensions in that they are untoleranced. The tolerances which can be held between features in N/C machining are built into the machine. Toleranced dimensions on the drawing would not change the machined part.

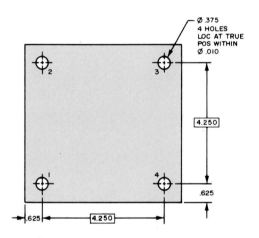

Fig. 17-16. Successive dimensions, sometimes called chain dimensions, are used for dimensioning drawings for N/C machines having incremental positioning systems.

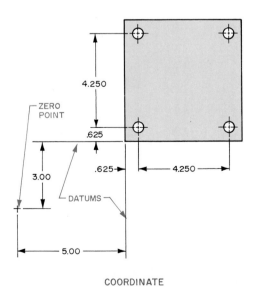

COORDINATE

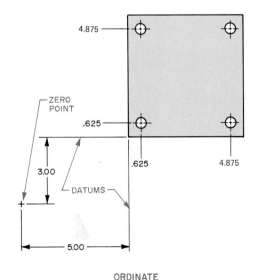

ORDINATE

Fig. 17-17. Coordinate or ordinate dimensions measured from datums are used for dimensioning drawings for N/C machines having absolute positioning systems.

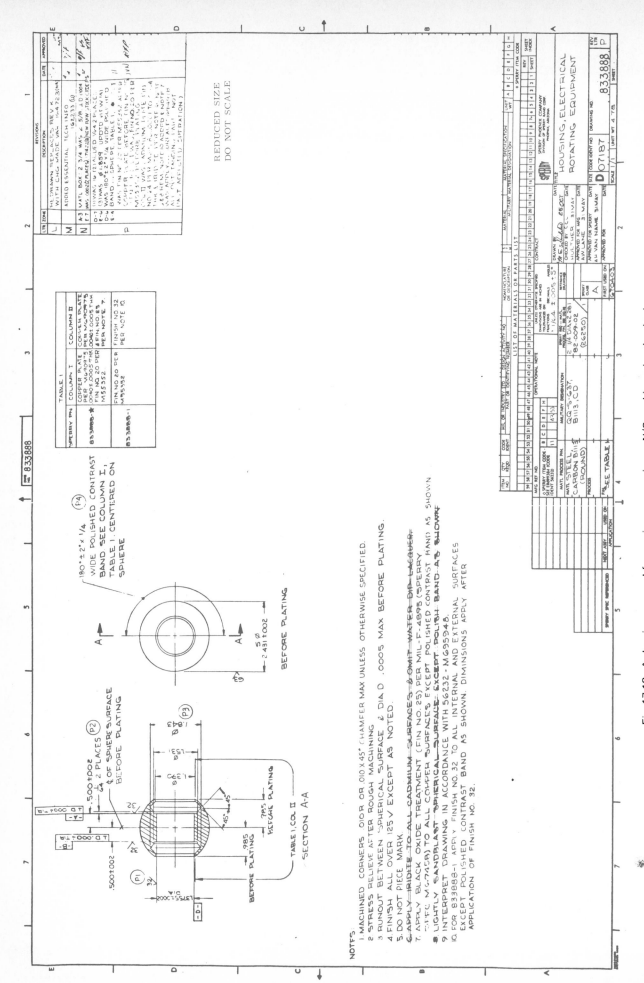

Fig. 17-18. A drawing prepared for use in programming an N/C machine using the absolute positioning system. Note the datum dimensioning and zero point. (Sperry Flight Systems)

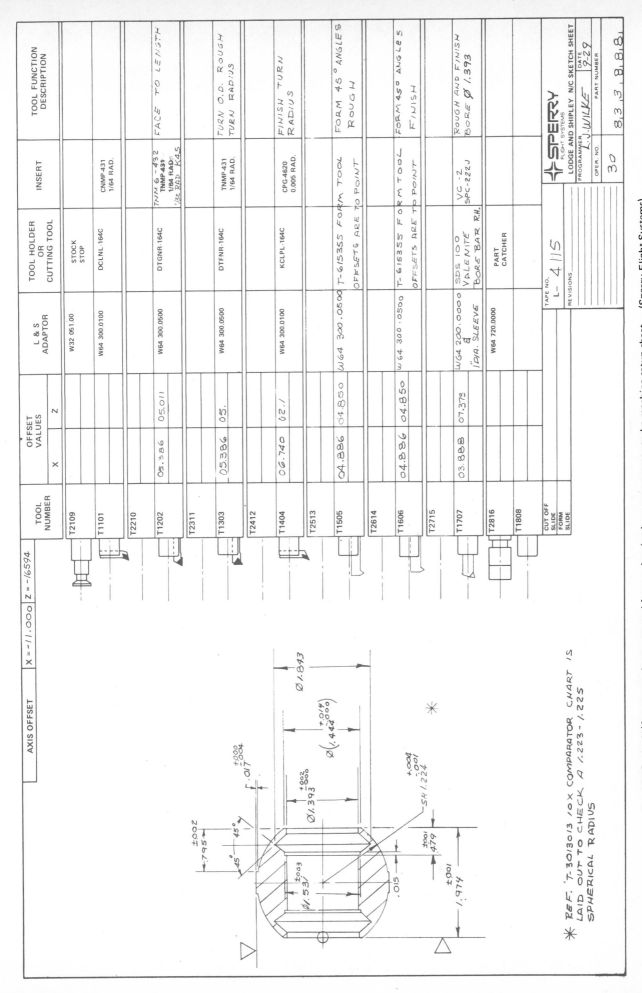

Fig. 17-19. An N/C program sheet prepared from a drawing by a programmer and a machine setup sheet. (Sperry Flight Systems)

333

ABSOLUTE POSITIONING

Many N/C machines use the absolute reference point system of positioning the cutting tool. In this system, all locations should be given as distances and directions from datums. That is, each move the tool makes is given as a distance and direction from the zero point, Fig. 17-17. The first X dimension is 5.000 + .625 or +5.625. The Y dimension is 3.000 + .625 or +3.625. The dimension for the second hole is X = 5.625; Y = 7.875. The remaining holes are located in a similar manner.

ABSOLUTE DIMENSIONS on drawings for N/C machining should be of the rectangular datum dimensioning type. Two types, coordinate and ordinate, are shown in Fig. 17-17. The coordinate type is typical of rectangular dimensioning, but each dimension is measured from a datum plane. Ordinate dimensions are also measured from datums and are shown on extension lines without the use of dimension lines or arrowheads. Fig. 17-18 shows a drawing dimensioned in a manner suitable for N/C machining with the absolute reference point system of positioning.

PROGRAM SHEET AND MACHINE SETUP SHEET

The program sheet for N/C machining, Fig. 17-19, is prepared from the drawing by a technician called a "programmer." The programmer must be able to read and interpret prints made from drawings and be thoroughly familiar with machine processes and capabilities. A punch tape will be prepared by a tape punch machine operator working from the program sheet. Tapes may also be prepared on a tape punch machine connected to a digitizer, a measuring device which converts a drawing to a series of digits or coordinate points.

Fig. 17-20. A computer controlled machine center.
(Kearney & Trecker Corp.)

After the program sheet has been carefully written, a machine setup sheet is prepared, as shown in Fig. 17-19, indicating how the part is to be located and secured on the machine worktable and selecting the proper tools which are to be loaded into the CNC or N/C machine. Data from the program sheet is then input to the N/C tape or computer. When programmed, the machine center, shown in Fig. 17-20, can operate on several types of complex parts simultaneously, selecting them randomly from a pre-loaded queue of palletized parts.

INTERPRETING AN N/C PROGRAM

It will help you in drawing for N/C machining if you understand the operational movement of an N/C machine.

An N/C machine is wired either for a FIXED ZERO set point or a FLOATING ZERO set point. In the fixed zero system, the machine refers to this point as zero and parts to be machined are located with reference to this point. In the floating zero system, the N/C programmer may establish zero at any convenient point by coding it into the punched tape.

Dimensional instructions for an N/C program are given on a two or three dimensional coordinate plane as in Fig. 17-15. When the operator is facing a vertical spindle machine, table movement to the left or right is the "X" direction or axis. Table movement away from the operator or toward the operator is the "Y" direction or axis. Vertical movement of the working tool is called the "Z" axis. Machining of some parts requires the use of only the X and Y axes; others require X, Y and Z.

It is easier to understand the direction of movement if you assume the table remains stationary and the tool moves over the work. When the work is located on the table with the datum zero point at the zero point of the machine, movement of the tool along the X axis to the right is in + X direction and to the left is in the − X direction. Movement of the tool into the work away from the operator is the + Y direction and toward the operator is the − Y direction. Tool movement down into the work is − Z and up from the work is + Z.

Tool movements are assumed to be in the PLUS direction unless marked MINUS. It is therefore desirable to establish the datum zero point on a drawing at the furthermost point on the lower left-hand corner or at a point just off the part to be machined (see Fig. 17-17).

When the zero point is located in this manner, all datum dimensions are plus dimensions and DO NOT need to be indicated as such. This also eliminates the possibility of errors in working with plus and minus dimensions.

SURFACE TEXTURE

The measurement of the smoothness of a surface texture, or finish, is done by a profilometer. This instrument measures the roughness of a surface in microinches or micrometres. The symbol used on drawings for micro (millionths) is the Greek letter μ (microinch μ in; micrometre μ m). The surface texture of a machine part should be specified on the drawing as part of the design specifications and not left to the machine operator.

The surface texture value is used in conjunction with the American Standard Surface Texture symbol, Fig. 17-22.

Machine processes such as milling, shaping and turning can produce surface textures in the order of 125 to 8 μ inches (3.2 to 0.2 μ meters). Grinding operations can produce surface textures in the range of 64 to 4 μ inches. This depends on the coarseness of the wheel and rate of feed.

Honing and lapping remove only very small amounts of metal and surface textures as fine as 2 μ inches are possible.

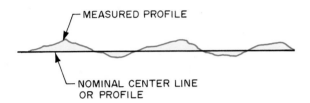

Fig. 17-21. The nominal profile or assumed true surface and the measured profile as measured by a profilometer (greatly exaggerated).

SURFACE TEXTURE SYMBOL

The American Standard Surface Texture symbol is used to designate the classifications of roughness, waviness and lay.

Roughness of a surface refers to the finer irregularities in a surface. Included are those which result from action of the machine production process such as traverse feed marks. Surface roughness is measured for height and width.

Surface roughness height can deviate from 1 to 1000 microinches, as measured along a nominal center line, Fig. 17-21. The preferred series of roughness height values, however, are shown in Fig. 17-22. Surface roughness height is designated above vee in surface texture symbol, Fig. 17-23. Horizontal extension bar is added to symbol where values other than roughness are specified.

Roughness width is the distance between successive peaks or ridges of the predominate pattern of roughness. This characteristic is measured in inches or millimeters, Fig. 17-23. The roughness-width cutoff is the greatest spacing of ir-

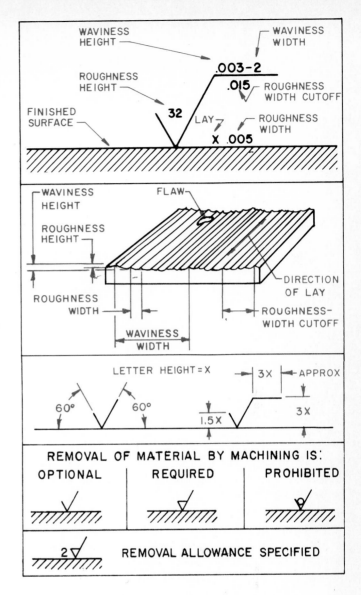

Fig. 17-23. Surface texture symbol.

regularities in the measurement of roughness height.

Waviness is the widely spaced component of the surface and

ROUGHNESS HEIGHT RATING		SURFACE DESCRIPTION	PROCESS
MICROMETERS	MICROINCHES		
25.2	1000	Very rough	Saw and torch cutting, forging or sand casting.
12.5	500	Rough machining	Heavy cuts and coarse feeds in turning, milling and boring.
6.3	250	Coarse	Very coarse surface grind, rapid feeds in turning, planning, milling, boring and filing.
3.2	125	Medium	Machining operations with sharp tools, high speeds, fine feeds and light cuts.
1.6	63	Good machine finish	Sharp tools, high speeds, extra fine feeds and cuts.

ROUGHNESS HEIGHT RATING		SURFACE DESCRIPTION	PROCESS
MICROMETERS	MICROINCHES		
0.8	32	High grade machine finish	Extremely fine feeds and cuts on lathe, mill and shapers required. Easily produced by centerless, cylindrical and surface grinding.
0.4	16	High quality machine finish	Very smooth reaming or fine cylindrical or surface grinding, or coarse hone or lapping of surface.
0.2	8	Very fine machine finish	Fine honing and lapping of surface.
0.05 0.1	2-4	Extremely smooth machine finish	Extra fine honing and lapping of surface.

Fig. 17-22. Description of roughness height values.

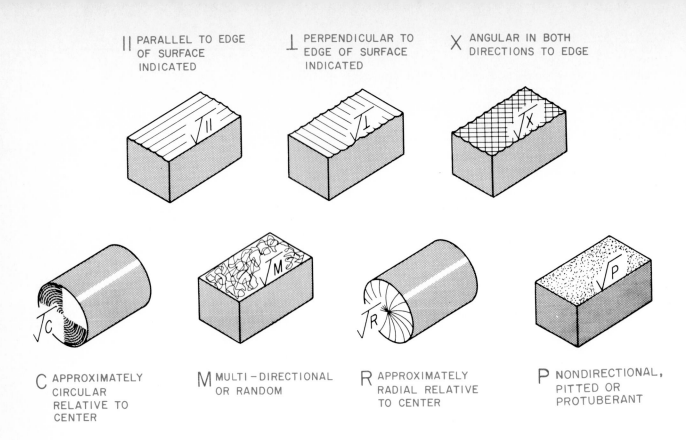

‖ PARALLEL TO EDGE OF SURFACE INDICATED	⊥ PERPENDICULAR TO EDGE OF SURFACE INDICATED	X ANGULAR IN BOTH DIRECTIONS TO EDGE

C APPROXIMATELY CIRCULAR RELATIVE TO CENTER

M MULTI−DIRECTIONAL OR RANDOM

R APPROXIMATELY RADIAL RELATIVE TO CENTER

P NONDIRECTIONAL, PITTED OR PROTUBERANT

Fig. 17-24. Designation and interpretation of lay symbols for surface texture.

is wider spaced than roughness-width cutoff, Fig. 17-23. Roughness may be thought of as occurring on a "wavy" surface. Waviness width is the spacing from one wave peak to the next, and waviness height is the distance from peak to valley, measured in inches or millimetres. See Fig. 17-21.

Lay is the direction of the predominant surface pattern (such as parallel, perpendicular or angular to line representing surface to which symbol is applied), Fig. 17-24.

SHEET METAL FABRICATION

The fabrication of sheet metal varies considerably from the construction industry to the electronics and instrumentation industries. The difference is primarily in the degree of tolerances held on the fabrication of objects produced.

SHEET METAL WORK IN THE CONSTRUCTION INDUSTRY

The development of pattern layouts for sheet metal products in the construction industry was discussed in Chapter 14. The flat-pattern layouts for these products vary from quite simple to very complex. These objects usually have considerable margin in fit, so tolerances are not held closely. Nor is bend allowance normally figured in laying out patterns.

Some common types of sheet metal hems and joints used in air conditioning sheet metal work are shown in Fig. 17-25.

Fig. 17-25. Sheet metal hems and joints.

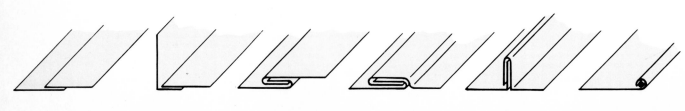

SINGLE HEM	DOUBLE HEM	SINGLE FLANGE	DOUBLE FLANGE	SINGLE SEAM	DOUBLE SEAM

LAP SEAM	OUTSIDE LAP SEAM	PLAIN FLAT SEAM	GROOVED SEAM	STANDING SEAM	WIRED EDGE

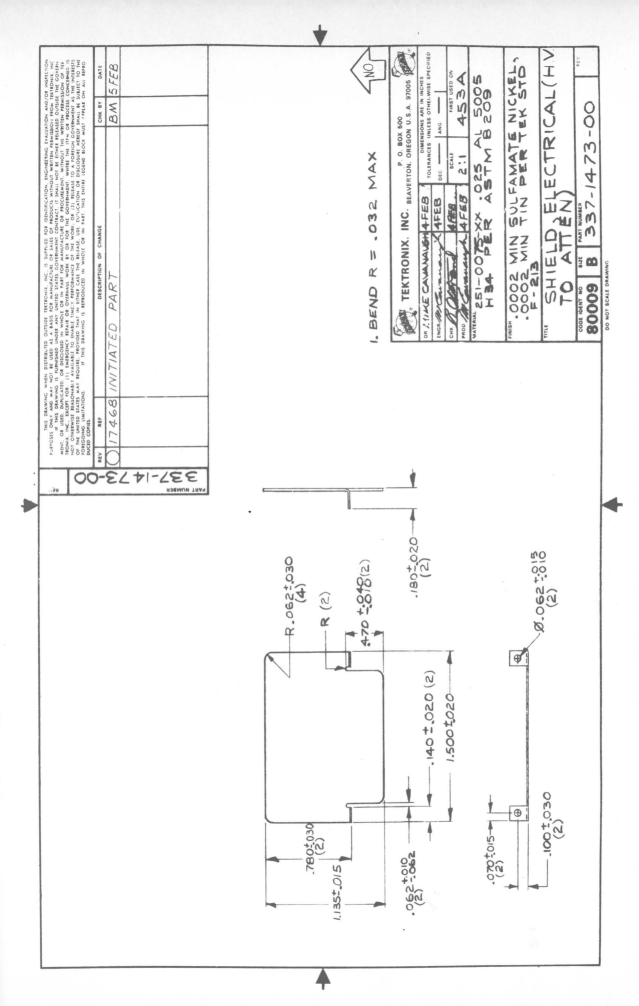

Fig. 17-26. A precision sheet metal drawing for an instrument housing.

SHEET METAL WORK IN ELECTRONIC AND INSTRUMENTATION INDUSTRIES

Sheet metal work in the electronic and instrumentation industries is frequently referred to as precision sheet metal work because of the close tolerances to which the parts are held. Precision sheet metal may be defined as "working thin-gaged metal to machine shop tolerances."

Precision sheet metal parts are machined in the flat, folded to shape and must hold the tolerances between related features, Fig. 17-26. The design drafter must calculate the bend allowance or obtain these from charts in order to properly dimension the layout for a flat pattern.

DEFINITIONS

BEND ALLOWANCE: Length of material required for a bend from bend line to bend line, Fig. 17-27.

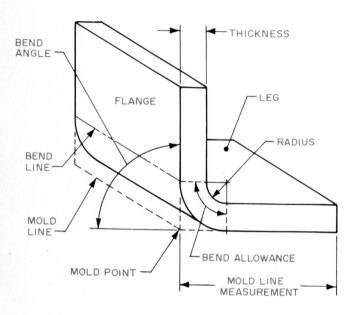

Fig. 17-27. Precision sheet metal terms.

BEND ANGLE: Full angle through which sheet metal is bent; not to be confused with angle between flange and adjacent leg.

BEND LINE: Tangent line where bend changes to a flat surface. Each bend has two bend lines.

BLANK: A flat sheet metal piece, approximately correct size, on which a pattern has been laid out and is ready for machining and forming.

CENTER LINE OF BEND: A radial line, passing through bend radius, which bisects included angle between bend lines.

DEVELOPED LENGTH: Length of flat pattern layout. This length is always shorter than sum of mold line dimensions on part.

FLAT PATTERN: Pattern used to lay out sheet metal part on blank.

MOLD LINE: Line of intersection formed by projection of two flat surfaces.

SET-BACK: Amount of deduction in length resulting from a bend to be developed in a flat pattern.

FORMULA

One of two formulas is used for calculating the lineal length of the bend on precision sheet metal parts, depending on the size of the bend radius and the thickness of the metal. After finding the lineal length, the developed length (length in the flat) can be calculated.

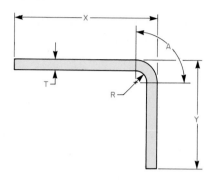

Fig. 17-28. Diagram for calculating developed length, using formula.

To find A (lineal length of a 90 degree bend), Fig. 17-28, when R (inside radius) is LESS THAN TWICE THE STOCK THICKNESS, use the following formula:

$$A = 1/2\,\pi\,(R + .4T)$$

To find A, where R = 1/16 or .0625 and T = .064 (inside radius is less than twice stock thickness):

$$A = 1/2\,\pi\,(R + .4T)$$
$$= 1/2 \times 3.1416\,(.0625 + .4 \times .064)$$
$$= 1.5708\,(.0625 + .0256)$$
$$= 1.5708\,(.0881)$$
$$A = .1384 = \text{Lined length of bend}$$

When the bend radius is MORE THAN TWICE THE STOCK THICKNESS, the following formula should be used to find A (lineal length of bend):

$$A = 1/2\,\pi\,(R + .5T)$$

To find the developed length of the part after the lineal bend length (A) has been found, refer to the diagram in Fig. 17-28 and use the formula which follows:

Developed length = $X + Y + A - (2R + 2T)$

Where: X = Outside distance of one side
      Y = Outside distance of other side
      A = Lineal length of bend
      R = Bend radius
      T = Material thickness

Developed length = $1.00 + .75 + .1384 - (.125 + .128)$
$$= 1.8884 - .253$$
$$= 1.6354 \text{ or } 1.64$$

SET-BACK CHARTS

Set-back charts are available in most industries where precision sheet metal work is done. A portion of a chart for 90 degree bends is shown in Fig. 17-29. A more complete chart is shown in the Reference Section. These charts will save time and errors in calculations in the shop.

The set-back figure is found by following across the row representing the bend radius until it meets the vertical column representing the thickness of the sheet metal to be bent. For

90 DEG BEND RADIUS	PRECISION SHEET METAL SET–BACK CHART						
	MATERIAL THICKNESS						
	.016	.020	.025	.032	.040	.051	.064
1/32	.034	.039	.046	.055	.065	.081	.097
3/64	.041	.046	.053	.062	.072	.086	.104
1/16	.048	.053	.059	.068	.079	.093	.110
5/64	.054	.060	.066	.075	.086	.100	.117
3/32	.061	.066	.073	.082	.092	.107	.124
7/64	.068	.073	.080	.089	.099	.113	.130
1/8	.075	.080	.086	.095	.106	.120	.137
9/64	.081	.087	.093	.102	.113	.127	.144
5/32	.088	.093	.100	.109	.119	.134	.150
11/64	.095	.100	.107	.116	.126	.140	.157
3/16	.102	.107	.113	.122	.133	.147	.164

Fig. 17-29. Precision sheet metal set-back chart for 90 degree bends.

example, a bend radius of 1/8 inch on metal .040 inch thick would require a set-back figure of .106 inch, Fig. 17-29. The diagram in Fig. 17-30, and the formula which follows, show the application of the set-back figure in calculating the DEVELOPED LENGTH (length in the flat to produce desired folded size) of a precision sheet metal part.

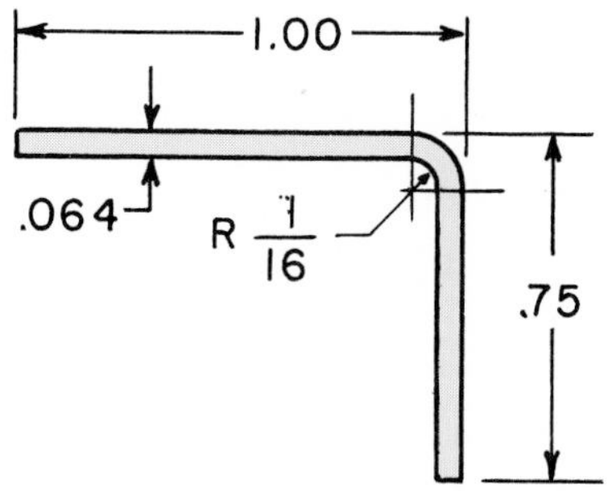

Fig. 17-30. Diagram for calculating developed length from set-back chart.

Using the set-back chart:

Developed length = X + Y − Z

Where: X = Outside distance of one side

Y = Outside distance of other side

Z = Set-back allowance for the 90 degree bend (from chart, Fig. 17-29)

Example, Fig. 17-30:

Developed length = X + Y − Z

= 1.00 + .75 − .110

= 1.75 − .110

Developed length = 1.64

Note that the 90 degree bend precision sheet metal part was the same part in examples used for the formula calculation and with the set-back chart. The same answer was obtained for the developed length in each example. However, the process is much shorter using the set-back chart.

Allowances for bends of angles other than 90 degrees may be easily calculated by multiplying the developed length for 90 degree bends by a factor representing the number of degrees in the bend desired over 90 degrees. For instance, suppose in the above example the bend had been 60 degrees. The bend length for the 60 degree bend would be calculated by taking the reading for the 90 degree bend from the set-back chart and multiplying it by $\frac{60}{90}$.

$$60^{\circ} \text{ Bend length} = \text{Reading for } 90^{\circ} \cdot \frac{60}{90}$$

$$60^{\circ} \text{ Bend length} = .110 \cdot \frac{60}{90} = .073$$

VARIATIONS IN BEND LENGTHS OF DIFFERENT METALS

When sheet metal is bent, the median line along the interior of the metal remains true length throughout the bending process, Fig. 17-31. The metal on the outside of this line is stretched and the metal on the inside of the bend is compressed. For most metals, the median line is approximately 44 percent of the distance from the interior face.

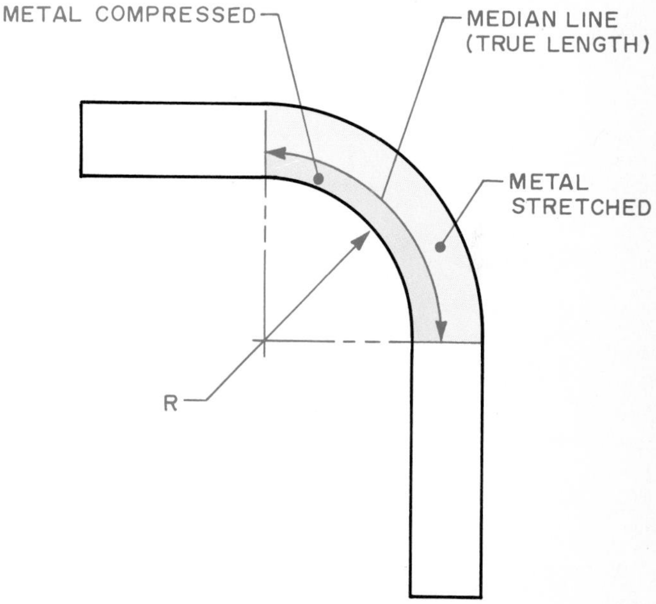

Fig. 17-31. When bent, metal tends to stretch on the outside portion of the bend and compress on the inside. The median line remains true length.

The location of the median line forms the basis for bend allowance calculations. The harder the metal, the greater the bend length. In actual practice, some experimentation may need to be done with the different metals and different shipments of the same metal alloy to arrive at the correct allowance. Most industries will lay out the flat pattern on two identical blanks. One blank will be formed and measurements checked. Any adjustments needed can then be layed out on a third blank, using the first blank as a reference.

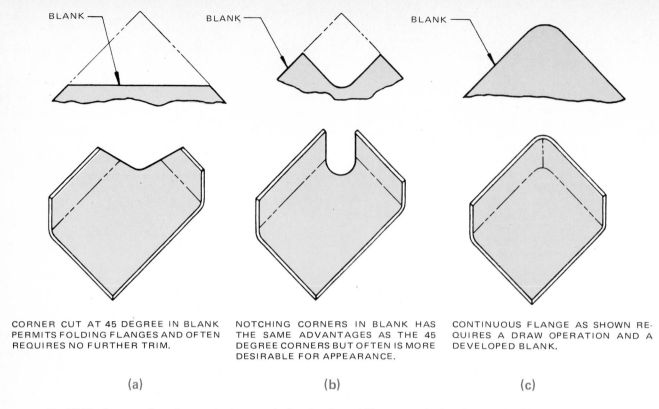

CORNER CUT AT 45 DEGREE IN BLANK PERMITS FOLDING FLANGES AND OFTEN REQUIRES NO FURTHER TRIM.

(a)

NOTCHING CORNERS IN BLANK HAS THE SAME ADVANTAGES AS THE 45 DEGREE CORNERS BUT OFTEN IS MORE DESIRABLE FOR APPEARANCE.

(b)

CONTINUOUS FLANGE AS SHOWN RE-QUIRES A DRAW OPERATION AND A DEVELOPED BLANK.

(c)

Fig. 17-32. Cutouts relieve the stress in sheet metal where bends would intersect and otherwise cause buckling and wrinkling. (General Motors Drafting Standards)

BEND RELIEF CUTOUTS

Wherever sheet metal bends intersect, a relief (notch) must be cut out to prevent the part from buckling or wrinkling, as shown in Fig. 17-32. The size and configuration of the sheet metal relief cutout may vary as long as it extends at least .03 inch beyond the intersection of the bend lines. The cutout usually has a radius, although it is not required for proper bend results.

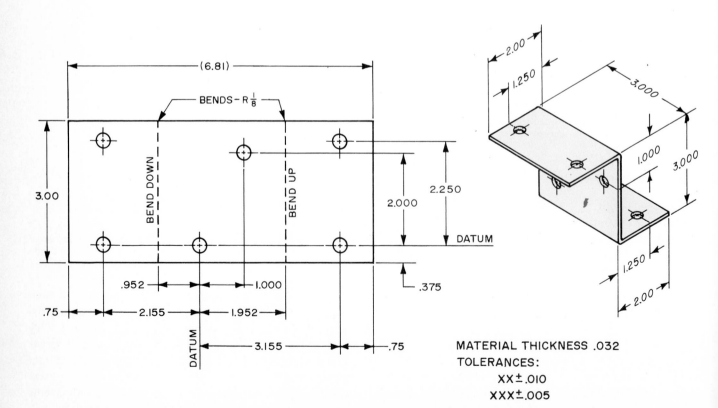

MATERIAL THICKNESS .032
TOLERANCES:
XX±.010
XXX±.005

Fig. 17-33. A precision sheet metal drawing of a flat pattern layout.

PRECISION SHEET METAL DRAWINGS

Because most precision sheet metal work will be machined on numerical control equipment, drawings should be datum dimensioned. Usually, a flat pattern and a pictorial of the folded pattern are drawn, showing toleranced dimensions which must be held. The direction of bend on a flange or leg is indicated by BEND UP or BEND DOWN shown on the affected part, Fig. 17-33.

UNDIMENSIONED DRAWINGS

An undimensioned drawing is a precise, full-scale drawing of a flat pattern on environmental stable material. It is used as a direct pattern for layout or machining.

METAL FORMING AND HEAT TREATMENT PROCESSES

Manufacturing processes other than the shaping of metals by cutting action should also be understood by the design drafter. Most drawings refer to materials that have been formed or must be heat treated by one of these processes. A comprehensive treatment of these processes is beyond the scope of this text, but the presentation here will help the design drafter become familiar with each and a more intensive study can be found in manufacturing processes texts.

CASTING

Casting, as used in the metal industries, is the process of pouring molten metals into molds where it hardens into the desired form as it cools, Fig. 17-34. Casting processes differ primarily in the type of materials used for molds and the type of molds.

Fig. 17-34. A mechanized molding line with 3-piece combination shell and hot box cores, four per mold, set in green sand drag molds. Cope molds are precisely set by the mechanized handling equipment in the background. (Central Foundry Div., General Motors Corp.)

The principal casting processes and their uses are:

SAND CASTING: All castable metals of any size from a few ounces to large pieces weighing many tons may be cast by the sand casting process, Fig. 17-34. Intricate details in castings and low cost are the advantages of this process.

Fig. 17-35. Ceramic shell mold for casting turbine blades. (Garrett-AiResearch Casting Div.)

Disadvantages are rough surface and low accuracy of castings.

SHELL-MOLD CASTING: The shell-mold process is used mainly for smaller castings up to 25 to 50 pounds, although some larger castings are being poured. The molds are made in the form of thin shells, Fig. 17-35. Sharp reproduction of details on original pattern, smooth finish and fairly close accuracy are possible. Nearly all castable alloys can be shell molded. Molds are more expensive than sand casting, but this is offset by time and labor savings where the casting has to be machined.

PLASTER-MOLD CASTING: Nonferrous metals (aluminum, brass and bronze) can be cast with a very smooth finish and good accuracy, Fig. 17-36. Small castings from a fraction of an ounce to 10 pounds are normal with pieces up to 200

Fig. 17-36. These turbocharger wheels have been cast, using an aluminum alloy, with a rubber pattern in plaster. (Garrett-AiResearch Casting Div.)

Fig. 17-37. Molten aluminum is poured into an iron permanent mold. These molds have a long life. (Garrett-AiResearch Casting Div.)

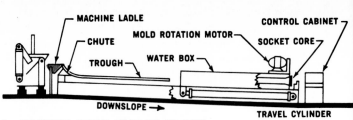

1. SOCKET CORE PUT IN PLACE

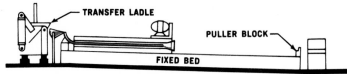

2. START OF CONTINUOUS CAST IN SPINNING MOLD

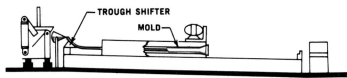

3. CAST HALF COMPLETE

4. CAST COMPLETED

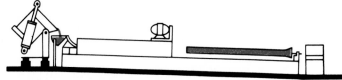

5. PIPE REMOVED—LADLE REFILLED

Fig. 17-38. Pipe is cast in centrifugal molds. (United States Pipe and Foundry Co.)

Fig. 17-39. Small, intricate parts are cast by the investment casting process. (Garrett-AiResearch Casting Div.)

pounds possible. The Antiock process, which is a mixture of plaster and sand, provides excellent results with large and complex pieces such as tire molds, wave guides for the electronics industry and torque converters. Ceramicast Ⓡ is a special ceramic material that can be used for a mold in casting certain ferrous alloys such as carbon and stainless steels.

PERMANENT MOLD CASTING: Permanent molds are cast iron or steel molds used for casting lower melting point alloys such as aluminum, magnesium, zinc, tin, lead and some of the copper base alloys, Fig. 17-37. Accuracy is exceptionally good in these castings, which are best adapted to medium or large quantities of small or medium-sized parts. Slush mold casting (a variation of permanent mold casting) uses zinc, lead or tin alloys. The alloy is poured into the permanent mold and left just long enough to form a shell on the inside of the mold and the remaining molten metal poured out.

CENTRIFUGAL CASTING: Tubular or cylindrical parts, such as cast iron or aluminum pipe, are cast in whirling metal molds. This forces the metal against the wall of the mold, leaving a hole in the center, Fig. 17-38.

INVESTMENT CASTING: The investment casting process, sometimes called lost wax or precision casting is used primarily for small, intricate parts requiring high accuracy and excellent finish, Fig. 17-39. The process is relatively expensive, but the time and labor saved in machining offset the cost.

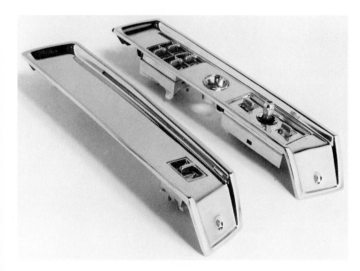

Fig. 17-40. Die casting produces items of high accuracy and excellent finish. (Fisher Body Div., GMC)

DIE CASTING: Sometimes called pressure casting, the die casting process is particularly well suited to mass production, Fig. 17-40. Molten aluminum, zinc or magnesium is forced under pressure into a metal die or mold and allowed to harden. The mold is opened, the part ejected and the cycle repeated. Very good accuracy and an excellent finish are obtainable. This process involves expensive dies and is suited only to quantity processing.

Fig. 17-41. Automatic drop forging eliminates man-handling of steel bars. Stock is heated in electrical resistance heaters (foreground) and then progressively manipulated through multiple stages of forging die, mounted on machine's horizontally oriented impellers. (Chambersburg Engineering Co.)

FORGING

The working of heated (not molten) metal into shape by means of pressure, usually a hammering or squeezing action, is known as forging, Fig. 17-41. This process develops the greatest strength and toughness possible into steel, bronze, brass, copper, aluminum and magnesium parts. Huge mechanically operated presses are used to forge machine parts that, for some applications, are not otherwise attainable.

EXTRUSION OF METALS

Extrusion is the process of producing long lengths of rod, tubing and other shapes from aluminum, brass, copper and magnesium. The extrusions are formed by placing a billet of hot metal in the cylinder of an extruding press and forcing it through a die with an opening of the desired shape, Fig. 17-42.

Fig. 17-42. Metal is formed into various shapes by the extrusion process. (Aluminum Assoc.)

343

Tremendous pressure is created by a hydraulic ram that squeezes the plastic-like metal out through the die and onto a conveyor to cool. The extrusions are cut to length and straightened by a stretching operation. Extrusions are used extensively in the aerospace, automotive and structural industries.

COLD HEADING

Metals may be formed while cold if sufficient pressure and the correct type of dies are applied. Cold heading (also called chipless machining, cold extrusion, cold forging or impact forging) is the forcing of metal under tremendous pressure to flow upward into a narrow space around a punch and die or to form an enlarged head or flange on a rod or bar.

The cold heading process is used in making bolts, screws and other fasteners where large rolls of "wire" stock are fed to automatic cold heading machines, Fig. 17-43.

Fig. 17-43. Threaded fasteners are made in these cold heading machines. (Elco Industries, Inc.)

HEAT TREATING AND CASEHARDENING

The heat treatment of steels is done by heating the steels to high temperatures and cooling at various rates to produce qualities of hardness, ductility and strength. Annealing is a form of heat treatment to reduce the hardness of a metal to make it machine or form more easily. Normalizing is a form of heat treatment aimed at relieving stresses caused by previous hot or cold working of a part.

Casehardening is the hardening of an outer layer on a piece and leaving the inner core more ductile. This produces a part that is very hard, wear-resistant and, at the same time, resistant to breaking under impact. Casehardening is sometimes called carburizing, since the part is usually heated to a high temperature for an extended period in contact with materials from which the steel absorbs more carbon or nitrogen. Casehardening is generally applied to lower-carbon steels and used for such parts as automobile wrist pins and races of ball

and roller bearings.

Flame hardening is similar to casehardening. It is a method of producing surface or localized hardening by immediately quenching the heated piece before the heat has had a chance to penetrate far below the surface. This process is used to harden gears, splines and ratchets, Fig. 17-44.

Fig. 17-44. Flame hardening of gear teeth.

HARDNESS TESTING

The hardness of a piece of metal is measured by the indentation impression or height of rebound on machines designed for the purpose, Fig. 17-45. Steels must be carefully selected to give the desired qualities of strength and hardness.

Fig. 17-45. Measuring the hardness of a piece of steel in a hardness tester. (Wilson Instrument Div. ACCO)

SOME RECENT METALWORKING PROCESSES

The space age has created the need and developed new metal alloys which are far tougher and more difficult to work than any used before. The requirements for working these alloys has also developed some sophisticated techniques and processes.

ELECTRICAL DISCHARGE MACHINING (EDM)

The working of metals by eroding the material away with an electric spark is known as electrical discharge machining, Fig. 17-46. The process is a magnification of the pitting or burning that occurs when a charged electrical wire momentarily contacts a piece of metal that is grounded. A pitting of the metal occurs.

Fig. 17-46. An electrical discharge machine with a machine part in place for processing. (AiResearch Mfg. Co.)

The electrode in an EDM machine, usually graphite or a brass material, is the cutting tool. It is surrounded with a coolant, normally an oil, that serves as a dielectric barrier between the electrode and the workpiece at the arc gap. A servo-mechanism accurately controls movement of the electrode and maintains the proper gap. The electrical discharge occurs at a rate of 20,000 to 30,000 times per second, and with each spark a small amount of metal is eroded.

The dielectric flushes the eroded particles away from the cutting action and keeps the electrode and workpiece cool. This machining process is very effective on metals that would be difficult to machine in any other way. Accuracies within .0005 inch are possible with a very fine surface texture as well.

ELECTROCHEMICAL MACHINING (ECM)

The process of electrochemical machining is the reverse of electroplating. In the electroplating process, a thin layer of metal is added by means of a direct current through a liquid solution to another metal. In electrochemical machining, a very high current density is used. The workpiece is the anode or positive part of the circuit; the electrode is the cathode or negative part.

The electrolyte or solution, flows at high pressure between the shaped electrode and workpiece. This results in the metal electrolytically eroding from the workpiece and being washed away rather than being deposited on the electrode, Fig. 17-47.

The electrochemical method of machining metal can be used on materials that are very difficult to machine by other methods. No strains are set up in the workpiece since no contact occurs between it and the electrode. Considerable work has been done in perfecting the shape of electrodes to produce desired shapes on workpieces. ECM is a much faster method of machining than Electrical Discharge Machining (EDM) and is used more in production operations.

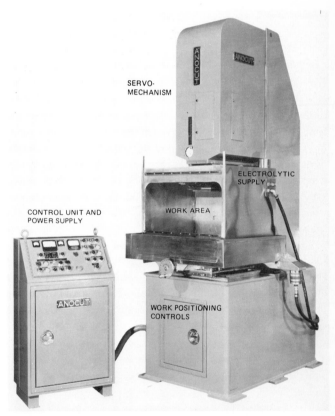

Fig. 17-47. Electrochemical machining equipment.

CHEMICAL MILLING (CHEM MILLING)

A method of removing material by etching with a chemical is called chemical milling. The process, which can be used on metals as well as plastics and glass, is also referred to as "chem milling" or "contour etching." Chemical milling works on that portion of the metal which is not protected by a mask.

On large pieces, such as aircraft structural members which are chem milled to reduce their weight, the masking is done by dipping in a plastic material. The portion to be etched is then cut away by hand. This process is also used extensively with parts made from thin metals, printed circuits and integrated circuits. The masking for these is usually done by the use of a photosensitive resist material.

Chemical milling of metal has several advantages. It is relatively inexpensive, produces a surface texture in the range of 30 to 125 μ inches, tolerances can be held within a few thousandths, the metal is not stressed as it would be in blanking. Also, heat-treated or hardened parts can be worked without affecting the characteristics of the metal. See Fig. 17-48.

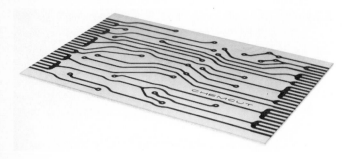

Fig. 17-48. A printed circuit board which has been chem milled. (Chemcut Corp.)

ULTRASONIC MACHINING

Certain materials change dimensions slightly when subjected to a strong magnetic field. Ultrasonic machining is based on this property of a nickel-alloy rod. When placed in a strong magnetic field that fluctuates rapidly, a nickel rod changes its length about .004 inch. An alternating current is used to vary the magnetic field about 25,000 to 30,000 cycles per second. The "cutting tool," a rod of brass or soft steel of the desired shape, is attached to the nickel rod.

In operation, the cutting tool rests on the workpiece and a solution of water and an abrasive flows between the points of contact. The abrasive is driven into the workpiece by ultrasonic vibration of the cutting tool. The shaped-rod cutting tool works its way through the workpiece leaving a clean hole of the desired shape, Fig. 17-49.

Fig. 17-49. Ultrasonic machining is done at frequencies of 25,000 to 30,000 cycles per second, well above the sounds heard by the human ear. (Raytheon Co.)

Materials such as tungsten carbide, quartz, glass and ceramics can be machined with amazing speed. Tolerances as close as .0005 inch and surface textures of 10 to 15 μ inches may be obtained. The process is used for specialized cutting operations and for making carbide dies for extrusion, drawing and stamping.

LASER APPLICATIONS

One of the recent developments that holds considerable promise in science, research, machining, welding and measurement is the laser (pronounced lay-zer), Fig. 17-50. Laser is an abbreviation for Light Amplification by Stimulated Emission of Radiation. The laser is a means of generating a narrow beam of monochromatic light of extremely high intensity in very short pulses. Its principal industrial applications are in metal removal, welding and measurement. Because of its capability of producing an extremely narrow beam of light and temperatures up to 75,000 degrees, Fahrenheit, laser can be used for perforating holes in stainless steel, carbide, ceramics and even diamonds. Actually, the holes are not drilled. The material is melted and vaporized with each burst of energy from the laser beam. The laser machine can work to very close tolerances.

Fig. 17-50. A two-axis, numerically controlled laser cutting machine trimming a titanium workpiece. (Grumman Aerospace Corp.)

HIGH ENERGY RATE FORMING (HERF)

The forming of large diameter sheet metal parts requires tremendous power as well as a press of sufficient size. Often neither are available. However, a rather ingenious and inexpensive method has been developed to process these kinds of jobs. It is known as High Energy Rate Forming, HERF for short, and is sometimes called explosive forming. A die of the desired shape is prepared, the sheet metal piece is cut or fabricated and clamped in place with a retaining ring. The area around and over the die is filled with water, a vacuum pulled between the piece and the die, and an explosive material of the correct amount is set off above the piece of sheet metal to be formed, Fig. 17-51. The piece is formed in a fraction of a second and retains its new shape within acceptable tolerances. A piece formed by HERF is shown in Fig. 17-52.

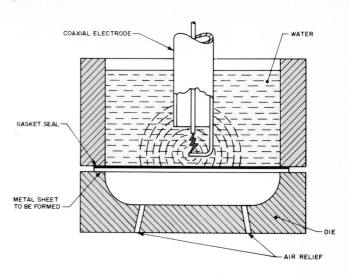

Fig. 17-53. Hydrospark forming used to form a sheet metal part.

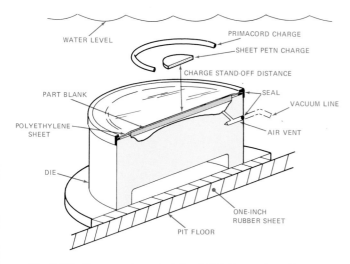

Fig. 17-51. The operating principle of High Energy Rate Forming. (Grumman Aerospace Corp.)

ELECTROHYDRAULIC FORMING

Two methods of forming sheet metal parts and bulging metal tubular parts by electrohydraulic forming are the HYDROSPARK PROCESS and the EXPLODING BRIDGE WIRE. The hydrospark process discharges an electric spark under water to produce a high velocity shock wave which forms the part, Fig. 17-53. The shock wave is created by stored electrical energy generated by a high-voltage power supply and a coaxial electrode. The force of the shock wave can be changed by varying the voltage.

The exploding bridge wire is similar to the electrospark process. It uses equipment of the same type, except a wire is used between the electrodes. The stored electrical energy vaporizes the wire and, as the vapor expands, the surrounding water is compressed against the sheet metal and forces it against the die wall.

The electrospark method is faster since the wire need not be replaced as in the exploding bridge wire method. The wire method does lend itself to better control by shaping the wire to fit the cavity. The electrohydraulic forming method is safer, more precise and lower in cost than conventional hydraulic forming.

PLASTIC PRODUCTION PROCESSES

Plastics have become one of the significant materials of modern industry and are used for everything from electronic components to machine tool parts. The broad field of plastics can be divided into two general groups, thermoplastics and thermosetting plastics.

Thermoplastic materials become soft when heated, harden when cooled. They can be resoftened repeatedly by heating. Included in this group are the styrenes, vinyls, acrylics, polyethylene, nylon and Teflon.

Thermosetting plastics cannot be resoftened by heating since they change chemically during the curing process. Included in the thermosets are the phenolics, epoxies, ureas, melamine and polyester.

Fig. 17-52. The part on the left was formed in a fraction of a second by the explosive force used in High Energy Rate Forming. The part on the right was hot-spun and required considerably more time. (Lockheed Aircraft Corp.)

Many of the production processes for working plastics are the same as for metalworking. However, there are a few techniques for processing plastics that are different, and these are discussed in the following sections.

INJECTION MOLDING

Many different processes and techniques are used to form plastics into useful products. Injection molding is the most widely used. Plastic granules are loaded into the hopper of the injection-molding machine, Fig. 17-54, heated, and then the softened plastic is forced through a nozzle into the mold to form and cool. After cooling, the part is ejected and the feeder gates trimmed to finish the part.

Injection molding lends itself to a diversity of products which can be produced, Fig. 17-55. Thermoplastics are primarily used with the injection molding process.

EXTRUSION OF PLASTICS

The extrusion of plastics is similar to that of metals. Continuous shapes of pipe, rod, special shapes and sheets can be formed through dies of the machine. Thermoplastic resin is heated and forced through the die in its finished shape and picked up on a conveyor belt as it leaves the machine to cool, Fig. 17-56.

Fig. 17-56. A plastic extrusion machine.
(Rohm and Haas)

Fig. 17-54. An injection molding machine for forming plastic parts.
(Union Carbide Corp.)

BLOW MOLDING

Thin-wall hollow plastic parts are formed in a blow mold, using a thermoplastic resin, Fig. 17-57. A tube or cylinder of heated plastic, called a PARISON, is extruded and placed between the split mold. The mold is closed, pinching the ends of the plastic tube and a blast of air forces the thermoplastic against the mold. When the plastic cools, it becomes stable and retains the shape of the mold.

The technology of plastic blow molding has advanced to the place where it is possible to produce unique products at a high rate of speed. Objects such as fuel tanks for automobiles and boats, containers for soap, medicine, instruments and tools have been blow molded.

COMPRESSION MOLDING

Compression molding is one of the most common processes used in forming thermosetting plastics. The correct amount of plastic resin is placed in the open heated mold. The mold is closed and pressure is applied to force the plastic into the shape of the mold cavity. The plastic first transforms into a liquid and then into a permanently hard material.

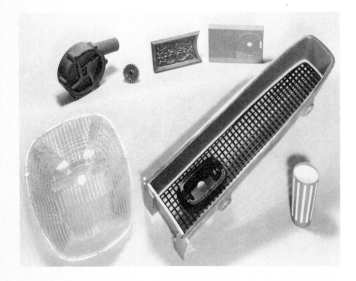

Fig. 17-55. Some examples of plastic products which have been injection molded. (Cincinnati Milacron)

Fig. 17-57. A blow molding machine with four die heads automatically producing gallon milk bottles. (Uniloy)

Compression molding presses come in varying sizes. Those commonly used generate temperatures of 270 to 360 degrees F., and pressures from 300 to 8000 psi, Fig. 17-58.

TRANSFER MOLDING

The plastic resin in transfer molding is not fed directly into the mold as it is in compression molding. Rather, the resin is placed in a separate chamber where it is heated under the pressure of a plunger until molten. Higher pressures are then exerted, forcing the softened resin through runners and gates into the mold cavities, Fig. 17-59.

Fig. 17-58. Forming plastic in a compression mold. (Union Carbide Corp.)

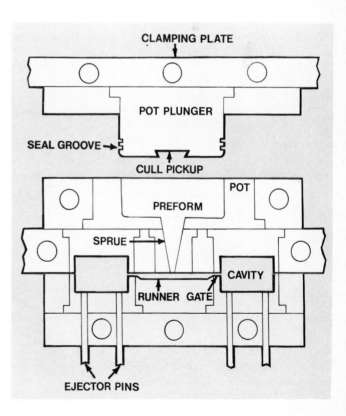

Fig. 17-59. Schematic drawing of a transfer mold showing the separate chamber or transfer pot, runner, gates and mold cavity. (Durez Div.)

CALENDERING

Plastic sheet and film stock are produced by a process known as CALENDERING, Fig. 17-60. The plastic resin is fed through a hopper, heated and passed through a series of rollers. The rollers reduce the resin to the desired thickness. It is cooled and the finished sheet trimmed to width. Various materials can be mixed with the plastic resin to give it the color and qualities desired.

Fig. 17-61. A vacuum forming machine producing plastic window shutters. (B. F. Goodrich Chemical Co.)

Fig. 17-60. A plastic calendering machine forming sheet material. (Union Carbide Corp.)

ROTATIONAL MOLDING

In rotational molding, plastic resin in the form of powder or liquid is placed in a mold of the desired form. The mold rotates in a manner so as to spread the powder or liquid evenly over the interior mold surface. As the resin melts, a solid coating is formed over the mold to produce the required shape. Some products of rotational molding are plastic garbage cans, ice chests, footballs and helmets.

POST-FORMING OF THERMOPLASTIC SHEETS

In the post-forming process, thermoplastic sheets from .010 inch to 5/8 inch or more can be formed into desired shapes by heating them and forcing them into a mold by a pair of matched molds; or forced with air pressure over or into a half mold; or vacuum formed over male or female molds, Fig. 17-61.

The vacuum process is useful with thinner sheets. Air pressure and matched molds are used for heavier forming such as aircraft noses, pilot domes and similar products.

CASTING

Plastic in a liquid form may be cast from thermoplastic or thermosetting materials. The materials may be cured by chemical action or heating. Sheets, rods and special shapes may be formed by casting. It is suitable for small production runs or for casting parts for a prototype. Molds for casting plastic may be made from such inexpensive materials as lead, plaster or plastisols. Pressure is not required, but a vacuum is sometimes used to eliminate bubbles and voids in castings.

PLASTISOLS

Plastisols are a mixture of plastic resins and plasticizers (chemicals) to improve their workability and reduce brittleness. Plastisols are cured by heating to about 350 degrees F., at which time they become tough, flexible, solid materials. They are generally used in slush molding and dip molding processes. Common uses are coatings for such articles as wire dish racks, electroplating racks, and for coating the inside of drums and tanks.

SLUSH MOLDING AND DIP MOLDING

Slush molding of plastics is similar to the slush molding of metals except the mold is heated and filled with plastisol. After a short period, a layer of the plastisol is formed on the inside of the mold and the remainder is poured out. The mold is placed in the oven to cure the plastic. Such products as toy doll parts, syringe bulbs and spark plug covers are molded by this process.

Dip molding is similar to slush molding except a heated male mold is used and the plastisol fuses to the outside surface of the mold, Fig. 17-62. This is further cured in an oven and then stripped from the mold. The detail of the male mold is reproduced on the interior of the plastisol piece. Typical products are toy doll parts, boots and spark plug covers. Tool handles may also be coated by this process. Sometimes the dip mold product is stripped and used as a mold for casting other plastic parts.

Fig. 17-62. Plastisol dip molding system for making insulating "boots"
for high voltage switchgear. Thickness of cured plastic is .125 inch.
(W. S. Rockwell Co.)

PRESSURE LAMINATES

When two or more layers of material are bonded together they are known as a laminate. The individual layers making up the laminate may be sheet plastic, cloth, paper and/or wood. The layers are impregnated with a plastic resin and heat and pressure applied. When the pressures exceed 1000 psi, the process is known as high-pressure laminating. Typical high-pressure laminates are used for the familiar kitchen cabinet tops and furniture made with a plastic laminate overlay.

Low-pressure laminating is used for forming boat hulls, automobile bodies, luggage, credit card plates and component housings of various types.

FOAMED PLASTIC

Plastic resins to which air or gas have been added to form a sponge-like substance are known as "foamed plastics." Two types are the rigid and flexible.

The rigid type is used as core material for sandwich panels in aircraft and construction work. The panels are strong and have good heat and sound insulating qualities. This type of foam plastic can be formed, removed from the mold and used as packing for machine parts or used in slab form. It can also be made to adhere to the insides of refrigerator doors, aircraft assemblies or wherever its insulating qualities are needed, Fig. 17-63.

Flexible foam plastic resembles foam rubber and is being used in many applications to replace rubber.

Plastic resins commonly used for foam plastics are the styrenes, phenolics, polyurethane, epoxies and silicones.

Fig. 17-63. Rigid foam plastic can be formed in a variety of ways to provide insulation for temperature control and for use in packing shipments of machine parts. (Sinclair-Koppers Co.)

QUESTIONS FOR DISCUSSION

1. Which machine process, drilling or reaming, is likely to produce a hole with greater accuracy? Why?

2. How does boring differ from drilling or reaming?

3. Explain the difference between counterboring and spotfacing. Make a sketch to illustrate the difference.

4. What is broaching? Can you give examples of machine parts or objects which have been produced by broaching?

5. How does chamfering differ from countersinking? What is the function of each?

6. Give a definition of numerical control machining in terms which you understand.

7. Does a drawing to be used for preparing a tape for N/C machining differ from a standard drawing? Why, or why not?

8. What other documents must be prepared from the drawing for N/C machining prior to making the N/C tape? Who prepares these?

9. Differentiate between a "fixed zero" and a "floating zero" set point in N/C work.

10. How does honing and lapping differ from grinding?

11. Explain the meaning of surface texture and state the purpose of designating this feature on a drawing. How does surface texture differ from surface finish?

12. There are two general types of sheet metal drafting. Explain how these two differ.

13. Give some examples of metal parts which have been cast. Which process of casting was used (or likely used)?

14. What is forging? Give examples of forged parts.

15. How is the extrusion of metal parts done? List some examples of parts which have been or could be extruded.

16. Differentiate between heat treatment and casehardening.

17. Explain the process of electrical discharge machining. How does electrochemical machining differ from EDM?

18. What is chemical milling and where is it used?

19. What is ultrasonic machining and how does it work? Where is it used?

20. What uses does the laser have in modern industrial work?

21. Explain the meaning and process of high energy rate forming. How does this differ from electrohydraulic forming?

22. Name the two general types of plastics and explain how they differ.

23. How does the injection molding process of forming plastics operate?

24. What is blow molding in the plastics industry? What are its uses?

25. How does transfer molding differ from compression molding of plastics?

26. What uses are made of calendering in the production of plastic products?

27. What are plastisols? What uses do they have?

28. Explain the difference in slush and dip molding of plastics?

29. What is meant by plastic laminates? Give some examples of high and low-pressure laminated plastic products.

30. Name the two types of foam plastics and their chief uses?

Chapter 18
PRECISION DIMENSIONING

The manufacture of a product today, such as an automobile, usually requires the assembly of parts supplied by a number of different industries in widely separated locations. In such cases, it is necessary to control dimensions very closely to assure that parts will fit properly. This capability in modern industry is know as interchangeable manufacture. Interchangeability is also essential in the replacement of parts on equipment already in service.

TOLERANCING

The control of dimensions to achieve interchangeable manufacturing is known as tolerancing. A toleranced dimension means that the dimension has a range of permissible sizes within a ''zone.'' The size of this zone depends on the function of the part. To achieve an exact size (non-toleranced dimension) is not only very expensive but virtually impossible under normal conditions. Therefore, tolerances are set as liberal as possible and still achieve the function of the part.

Industrial designers and engineers establish tolerances based on industry standards and practice. Drafters, too, must understand the application of these tolerances to engineering drawings.

TYPES OF TOLERANCES

Tolerances are used to control the size of the features of a part. In addition to tolerances on size, there are two general types of tolerances: (1) Positional tolerances controlling the location of features of a part. (2) Form tolerances controlling the form or the geometric shapes of features of a part. These basic types and uses of tolerances are presented here as they pertain to dimensioning drawings.

DEFINITION OF TERMS

Standard terms have been adopted by industry to effectively communicate information related to tolerancing.

BASIC DIMENSION
A BASIC dimension is a theoretically exact, untoleranced value used to describe the size, shape or location of a feature. Basic dimensions are used as a base from which permissible

variations are established by tolerances or other associated dimensions.

Basic dimensions are not directly toleranced. Any permissible variation in the feature located by the basic dimension is contained in the tolerance on the dimension associated with the basic dimension, such as the tolerance on the hole diameter, Fig. 18-1 (a).

Basic dimensions are indicated on the drawing by the word BASIC or the abbreviation BSC following or immediately below the dimension figure, or by enclosing the dimension figure in a rectangular frame, or by a general note; UNTOLERANCED DIMENSIONS LOCATING TRUE POSITION ARE BASIC.

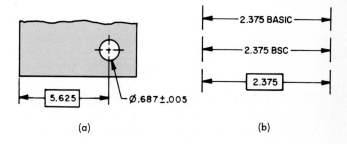

(a)   (b)

Fig. 18-1. Indicating BASIC dimensions.

REFERENCE DIMENSION
Reference dimensions are placed on drawings for the convenience of engineering and manufacturing personnel and are indicated by enclosing the dimension within parentheses, Fig. 18-2. They are not required for the manufacturing of a part or in determining its acceptability. Referenced dimensions are untoleranced dimensions with values equal to the nominal sum or difference of the associated toleranced dimensions. They may be rounded off to any degree desired.

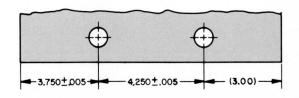

Fig. 18-2. Indicating reference dimensions.

DATUMS

Datums are theoretically exact planes, lines or points from which other features are located. A datum is usually a plane or point on the part, but it may be a plane or surface on the machine being used. Care should be exercised in the selection of datums on drawings to make certain they are recognizable, accessible and useful for measuring. Corresponding features on mating parts should be selected as datums to assure ease of assembly.

Datums are indicated by the word DATUM, Fig. 18-3 (a), or symbolically by a capital letter in a rectangular frame, (b).

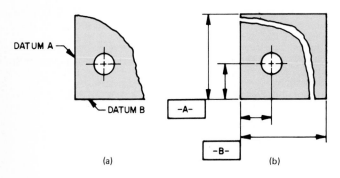

Fig. 18-3 Identifying datums.

A machined part may require more than one datum plane in its dimensioning. Letters such as A, B, and C are assigned to each datum. Datums may be considered to be primary, secondary, and tertiary depending on the design of the part, Fig. 18-18. Preference for the datums would appear in that order, and dimensions on the part would be established in the given sequence.

NOMINAL SIZE

The nominal size is a classification size given commercial products such as pipe or lumber. It may or may not express the true numerical size of the part or object. For example, a seamless wrought-steel pipe of 3/4 inch nominal size has an actual inside diameter of 0.824 and an actual outside diameter of 1.050 inches, Fig. 18-4 (a). In the case of cold-finished low-carbon steel round rod, the nominal 1 inch size is within .002 inch of actual size, (b).

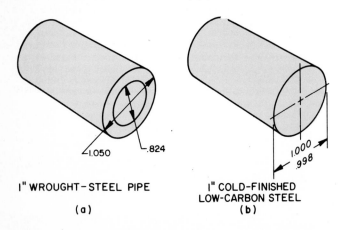

1" WROUGHT-STEEL PIPE
(a)

1" COLD-FINISHED LOW-CARBON STEEL
(b)

Fig. 18-4. Nominal sizes of products do not necessarily reflect actual sizes.

BASIC SIZE

The basic size is the size of a part determined by engineering and design requirements. It is from this size that allowances and tolerances are applied. For example, strength and stiffness of a shaft may require one inch diameter material. This basic one inch size (with tolerance) is usually applied to the hole size and allowance to the shaft, Fig. 18-5.

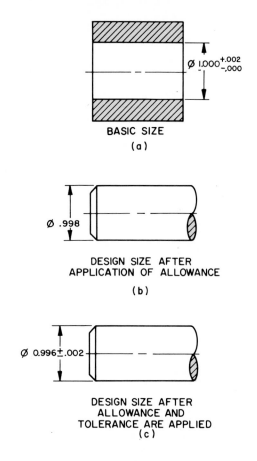

BASIC SIZE
(a)

DESIGN SIZE AFTER APPLICATION OF ALLOWANCE
(b)

DESIGN SIZE AFTER ALLOWANCE AND TOLERANCE ARE APPLIED
(c)

Fig. 18-5. Basic size, allowance and design size.

ACTUAL SIZE

Actual size is the measured size of a part or object.

ALLOWANCE

Allowance is the intentional difference in the dimensions of mating parts to provide for different classes of fits. It is the minimum clearance space or maximum interference, whichever is intended, between mating parts. In the example shown in Fig. 18-5 (b), an allowance of .002 has been made for clearance (1.000 − .002 = .998).

DESIGN SIZE

The design size of a part is the size after an allowance for clearance has been applied and tolerances have been assigned. The design size of the shaft shown in Fig. 18-5 (b) is shown in (c) after tolerances are assigned.

LIMITS OF SIZE

Limits are the extreme permissible dimensions of a part resulting from the application of a tolerance. Two dimensions are always involved, a maximum size and a minimum size. For example, the design size of a feature may be 1.625. If a tolerance of plus or minus two thousandths (± .002) is applied, then the two limit dimensions are maximum limit 1.627 and minimum limit 1.623, Fig. 18-6 (a).

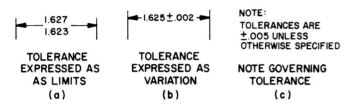

Fig. 18-6. Means of indicating tolerances on drawings.

TOLERANCE

Tolerance is the total amount of variation permitted from the design size of a part. Tolerances may be expressed as limits, Fig. 18-6 (a), or as the design size followed by the tolerance, (b). Tolerances should always be as large as possible, and still produce a usable part, to reduce manufacturing costs.

Tolerances may be specific and applied directly to the dimension value, Fig. 18-6 (a and b), or they may be given in the title block, or in a note, (c), and apply to all dimensions unless otherwise noted.

UNILATERAL TOLERANCE

A unilateral tolerance is one in which variation is in only one direction from the specified dimensions, Fig. 18-7 (a).

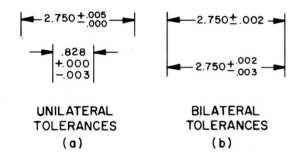

Fig. 18-7. Types of tolerances.

BILATERAL TOLERANCE

Bilateral tolerances are given as variations permitted in both directions from the specified dimension, Fig. 18-7 (b).

FIT

Fit is a general term referring to the range of tightness or looseness resulting from the application of a specific combination of allowances and tolerances in the design of mating parts. There are three general types of fits: clearance, interference and transition.

CLEARANCE FIT

A clearance fit is one in which a positive allowance, or air space, occurs. The limits of size are so prescribed that a clearance always results when mating parts are assembled, Fig. 18-8 (a).

INTERFERENCE FIT

An interference fit is one in which an interference, or negative allowance, always occurs when mating parts are assembled, Fig. 18-8 (b).

TRANSITION FIT

In a transition fit, the limits of size are so prescribed that the result may be either a clearance fit or an interference fit. That is, the smallest shaft sizes permitted within its tolerance will fit within the largest hole size within the hole tolerance and a clearance will result. However, the largest shaft size permitted will interfere with the smallest hole size permitted and the two mating parts will have to be forced together, Fig. 18-8 (c).

BASIC SIZE SYSTEMS

In the design of mating cylindrical parts, it is necessary to assume a basic size for either the hole or shaft, then calculate design sizes of mating parts by applying allowance to this basic size. Manufacturing economy determines which mating part, the hole or shaft, becomes the standard size.

If standard tools can be used to produce the holes, the basic hole size system is used. If a machine or an assembly requires several different fits on a cold-finished shaft, the

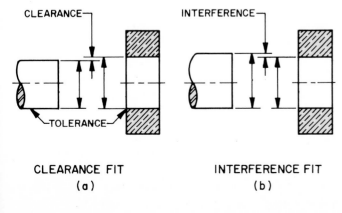

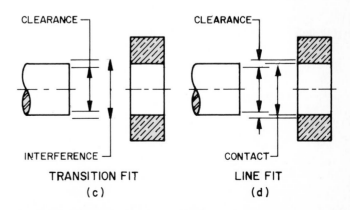

Fig. 18-8. Types of fits used in the design of mating parts.

basic shaft size is most economical in manufacturing and is the one used. When standard parts (such as ball bearings) are inserted in castings, the basic shaft size also applies.

BASIC HOLE SIZE

In the basic hole size system, the basic size of the hole is the design size, and the allowance (for clearance or interference) is applied to the shaft. Basic hole size is the minimum hole size produced by standard tooling, such as reamers and broaches. Tolerances are specified for this basic or design size to produce the type fit desired.

An example of a basic hole size is shown in Fig. 18-9 (a), where the basic hole size is the minimum size, .500''. An allowance of .002'' is subtracted from the basic hole size and applied to the shaft for clearance, providing a maximum shaft size of .498''. A tolerance of .002'' is then applied to the basic hole size and .003'' to the shaft size to provide a maximum hole size of .502'' and a minimum shaft size of .495''.

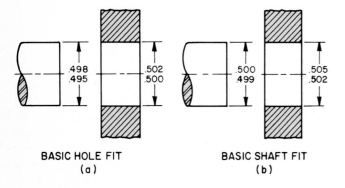

BASIC HOLE FIT
(a)

BASIC SHAFT FIT
(b)

Fig. 18-9. Basic hole and basic shaft systems.

The tightest fit (minimum clearance) is .500'' (smallest hole size) − .498'' (largest shaft size) = .002''.

The fit giving maximum clearance is .502'' (largest hole size) − .495'' (smallest shaft size) = .007''.

These examples have provided a clearance fit in mating parts. To obtain an interference fit in the basic hole size system, add the allowance to the basic hole size and assign this value as the largest shaft size.

BASIC SHAFT SIZE

When the basic shaft size system is used, the design size of the shaft is the basic size and the allowance is applied to the hole. The basic shaft size is the maximum shaft size. Tolerances are then specified for this basic or design size to produce the desired fit.

In Fig. 18-9 (b), the maximum shaft size of .500'' is taken as the basic (design) size. An allowance of .002'' is added to the basic shaft size and applied to the hole size, providing a minimum hole size of .502''. A tolerance of .001'' is specified for this basic shaft size and .003'' for the hole size. This provides a minimum shaft size of .499'' and a maximum hole size of .505''. The minimum clearance provided is .502'' (smallest hole size) − .500'' (largest shaft size) = .002''. The maximum clearance provided is .505'' (largest hole size) − .499'' (smallest shaft size) = .006''.

The example has provided a clearance fit in the mating parts. Interference fits may be obtained in the basic shaft size system by subtracting the allowance from the basic shaft size and assigning this value as the minimum hole size.

MAXIMUM MATERIAL CONDITION (MMC)

The maximum material condition is present when the feature contains the maximum amount of material. MMC exists when internal features, such as holes and slots, are at their minimum size, Fig. 18-10 (a), and when external features, such as shafts and bosses, are at their maximum size (b).

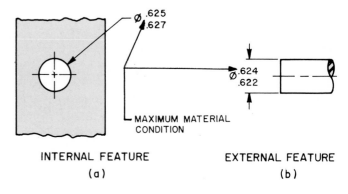

INTERNAL FEATURE
(a)

EXTERNAL FEATURE
(b)

Fig. 18-10. Maximum material condition.

MMC is applied to the individual tolerance, datum reference, or both. The positioned or form tolerance increases as the feature departs from MMC by the amount of such departure.

LEAST MATERIAL CONDITION (LMC)

Least material condition is present when the feature contains the least amount of material within the stated limits of size. LMC exists when holes are at maximum size and shafts are at minimum size.

REGARDLESS OF FEATURE SIZE (RFS)

Regardless of feature size means that geometric tolerances or datum references must be met, irrespective of where the feature lies within its size tolerance. Where RFS is applied to a positional or form tolerance, the tolerance must not be exceeded regardless of the actual size of the feature. RFS applies with respect to individual tolerance, datum reference, or both, where no symbol is specified.

MMC, LMC, and RFS must be specified on the drawing as applicable.

DIMENSIONS NOT TO SCALE

All drawings (with the exception of diagrammatic, schematic, etc.) should be drawn to scale. However, on a drawing revision, the correction of a dimension in scale may necessitate an unjustified amount of drafting. It is permissible, providing the drawing remains perfectly clear, to change the dimension and underline it with a straight, thick wavy line to indicate "not-to-scale," Fig. 18-11.

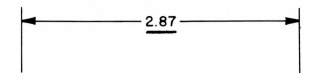

Fig. 18-11. Indication of a not-to-scale dimension.

Original drawings should not be issued with dimensions out-of-scale, and such dimensions should be kept to an absolute minimum. Where there is the slightest chance of misinterpretation of an out-of-scale dimension on a revised drawing, the drawing should be corrected.

APPLICATION OF TOLERANCES

In manufacturing items for assemblies and subassemblies requiring interchangeability of parts, tolerancing of all dimensions is required except those labeled BSC, REF, MAX or MIN. Tolerances are normally expressed in the same number of decimal places as the dimension (see Fig. 18-7). These tolerances are applied to the dimensions either as limit dimensioning or as plus and minus tolerancing.

LIMIT DIMENSIONING

In tolerancing a dimension by limit dimensioning, only the maximum and minimum dimensions are given, Fig. 18-12. The limit numerals shall be arranged by one of the following methods:

1. For dimensions given directly (not by note), maximum (high) limit is always placed above minimum (low) limit, (a). For dimensions given in note form (single line), minimum limit always precedes maximum limit, (b).
2. For positional dimensions of features given directly, maximum limit is placed above and minimum limit below, (a). For size dimensions of features given directly, numeral representing maximum material condition (MMC) is placed above numeral representing minimum material condition, (c). For dimensions given in note form, MMC numeral precedes other, (d).

PLUS AND MINUS TOLERANCING

In plus and minus tolerancing, the tolerances generally are placed to the right of the specific dimension as a plus and minus expression of the permissible variation of the size or location of a feature, Fig. 18-6 (b).

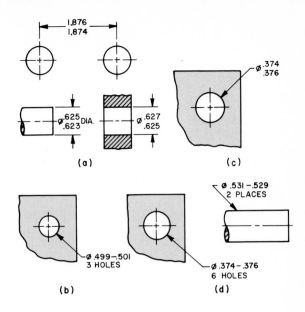

Fig. 18-12. Limit dimensioning.

CHECKING TOLERANCES

Drawings are checked for analysis of tolerances to see that all parts will assemble without interference. Two methods of computing the tolerance between two features are shown in Fig. 18-13.

SELECTIVE ASSEMBLY

For the manufacture of mating parts with allowances and tolerances of a fairly wide range, interchangeability of parts is easily obtained. However, for assemblies requiring mating parts with close fits, very small allowances and tolerances are required in order to maintain interchangeability of parts. Manufacturing costs of producing parts with very small allowances and tolerances is often prohibitive.

Selective assembly is a process of selecting mating parts by inspection and classification into groups according to actual sizes. Small size features (such as cylindrical shafts and holes) are grouped for matching with small size features of mating parts, medium size features with medium size mating features, large with large.

The cost of manufacturing is reduced considerably in selective assembly due to less restricted allowances and tolerances. This method is usually more satisfactory in achieving a transition fit of the desired function in mating parts than in interchangeable assembly.

FEATURE	NORMAL METHOD	GRAPHIC METHOD
HOLE 1.000 $^{+\ .002}_{-\ .000}$	1.002 max 1.000 min	1.000 $^{+\ .002}_{-\ .000}$
SHAFT .996 $^{+\ .002}_{-\ .000}$	.996 min .998 max	.996 $^{+\ .002}_{-\ .000}$
Find resultant difference	.006 max .002 min	.004 $^{+\ .002}_{-\ .002}$
		$^{+\ .006\ max}_{-\ .002\ min}$

Fig. 18-13. Chart shows normal and graphic methods of computing tolerance between hole and shaft. (Sperry Flight Systems Div.)

CHAIN DIMENSIONING

The dimensioning of a series of features, such as holes, from point to point is known as chain dimensioning, Fig. 18-14 (a). When these dimensions are toleranced, overall variations in position of the features may occur that exceed the tolerances specified, (b). The possible variations are equal to the sum of the tolerances on the intermediate dimensions, (b). For example, the variation in position between features A and B range from 3.006'' (5.504'' − 2.498'' = 3.006'') to 2.994'' (5.496'' − 2.502'' = 2.994'') a difference of .012'' instead of the intended ± .001 or a total variation .004'', (c).

Where the distance between two features, such as A and B, must be closely controlled for assembly purposes, the overall accumulation of tolerances can be avoided by dimensioning the features individually, (c).

Chain dimensioning, Fig. 18-14 (a), is used on drawings prepared for numerical-controlled machining of the incremental positioning type (see Chapter 17) where the tolerances are built into the machine and dimensions are given as BASIC dimensions. Tolerancing of these dimensions would not change the part being machined.

DATUM DIMENSIONING

In datum dimensioning (sometimes called base line dimensioning), features are dimensioned individually from a datum, Fig. 18-14 (c). This system of dimensioning avoids accumulation of tolerances from feature to feature. Where the distance between two features must be closely controlled, without the use of an extremely small tolerance, datum dimensioning should be used.

ANGULAR SURFACE TOLERANCING

Angular surfaces may be dimensioned and toleranced by a combination of linear and angular dimensions or by linear dimensions alone. A dimension and its tolerance specify a tolerance zone within which the surface must lie, Fig. 18-15 (a). The tolerance zone widens as it moves away from apex of the angle. Where a tolerance zone with parallel boundaries is desired, a basic angle may be specified as in Fig. 18-15 (b). The surface controlled must lie within the tolerance zone.

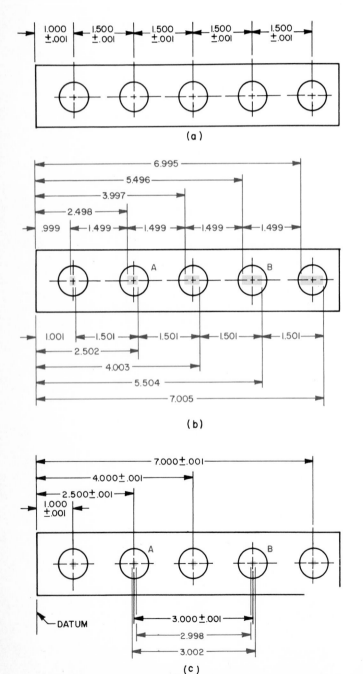

Fig. 18-14. Tolerance accumulation in chain dimensioning.

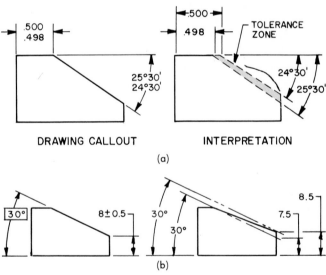

Fig. 18-15. Tolerancing an angular surface.

SELECTION OF FITS

Tables of recommended tolerances have been worked out by many companies as well as the American National Standards Institute (ANSI) Committee. The designer or drafter should consult these when it is necessary to select tolerances for a specific size feature and mating part.

The use required from a piece of equipment determines the limits of size of mating parts and the selection of type of fit.

Precision Dimensioning

STANDARD FITS

A number of types and classes of fits are prescribed in the Reference Section. Any fit of mating parts will usually be required to perform one of three functions: running or sliding fit, locational fit, force fit. These fits are further divided into classes and assigned letter symbols for the purpose of study. They are not shown as such on a drawing, rather, fits of mating parts are specified in the tolerances. These types and classes of fits are described here.

RUNNING AND SLIDING FITS (RC)

Running and sliding fits are a type of fit designed to provide similar running performance, with suitable lubrication, throughout the range of sizes:

RC1: Close sliding fits are designed for accurate location of parts which must assemble without perceptible play.

RC2: Sliding fits are intended for accurate location but with greater maximum clearance than RC1. Parts move and turn easily, but are not intended to run freely.

RC3: Precision running fits are about closest fits which can be expected to run freely at slow speeds and light journal (shaft) pressure.

RC4: Close running fits are designed for running fits on accurate machinery with moderate surface speeds and journal pressures, where accurate location and minimum play is desired.

RC5 and RC6: Medium fits are designed for higher running speeds and/or heavy journal pressures.

RC7: Free running fits are designed for use where accuracy is not essential or where temperature variations are likely to occur.

RC8 and RC9: Loose running fits are designed for use where wide commercial tolerances may be necessary, together with an allowance on the external member.

LOCATIONAL FITS (LC)

Locational fits relate to the location of mating parts. They are subdivided into three classes based on design requirements:

LC: Locational clearance fits are designed for parts which can be freely assembled or disassembled. Classes of fits run from snug fits for parts requiring accuracy of location, through medium clearance fits for parts such as ball bearing race and housing, to looser fits for fastener parts requiring considerable freedom of assembly.

LT: Locational transition fits are a medium fit between clearance and interference fits where accuracy of location is important but some clearance or interference is permissible.

LN: Locational interference fits provide accuracy of location for parts requiring rigidity and alignment with no special requirements for bore pressure. Such fits are not intended for parts designed to transmit frictional loads from one part to another by virtue of tightness of fit, as these conditions are covered by force fits.

FORCE FITS (FN)

Force or shrink fits are a type of interference fit, normally characterized by maintenance of constant bore pressures throughout the range of sizes. The interference varies almost directly with the diameter:

FN1: Light drive fits are those requiring light assembly pressures and produce more or less permanent assemblies. They are used for thin sections or long fits, or in cast iron external members.

FN2: Medium drive fits are designed for ordinary steel parts or for shrink fits on light sections. They usually are tightest fits that can be used with high-grade cast iron external members.

FN3: Heavy drive fits are suitable for heavier steel parts or for shrink fits in medium sections.

FN4 and FN5: Force fits are designed for parts which can be highly stressed or for shrink fits where heavy pressing forces required are impractical.

GEOMETRIC DIMENSIONING AND TOLERANCING

Geometric dimensioning and tolerancing is a system of dimensioning and tolerancing drawings with emphasis on the

SYMBOL FOR:	ANSI Y14.5
STRAIGHTNESS	—
FLATNESS	▱
CIRCULARITY	○
CYLINDRICITY	⌭
PROFILE OF A LINE	⌒
PROFILE OF A SURFACE	⌓
ALL AROUND – PROFILE	⌖
ANGULARITY	∠
PERPENDICULARITY	⊥
PARALLELISM	//
POSITION	⊕
CONCENTRICITY/COAXIALITY	◎
SYMMETRY	NONE
CIRCULAR RUNOUT	*↗
TOTAL RUNOUT	*↗↗
AT MAXIMUM MATERIAL CONDITION	Ⓜ
AT LEAST MATERIAL CONDITION	Ⓛ
REGARDLESS OF FEATURE SIZE	Ⓢ
PROJECTED TOLERANCE ZONE	Ⓟ
DIAMETER	⌀
BASIC DIMENSION	[30]
REFERENCE DIMENSION	(30)
DATUM FEATURE	-A-
DATUM TARGET	Ⓐ⌀6
TARGET POINT	✕
DIMENSION ORIGIN	⟜
FEATURE CONTROL FRAME	⊕ ⌀0.5Ⓜ A B C
CONICAL TAPER	▷
SLOPE	◺
COUNTERBORE/SPOTFACE	⌴
COUNTERSINK	⌵
DEPTH/DEEP	↧
SQUARE (SHAPE)	□
DIMENSION NOT TO SCALE	15
NUMBER OF TIMES/PLACES	8X
ARC LENGTH	⌒105
RADIUS	R
SPHERICAL RADIUS	SR
SPHERICAL DIAMETER	S⌀

*MAY BE FILLED IN

Fig. 18-16. Geometric characteristic symbols for positional and form tolerances. (ANSI — American National Standards Institute)

actual function and relationship of part features where interchangeability is critical. This system does not replace the coordinate dimensioning system but may be used in conjunction with it, particularly where interchangeability and mating features are required. It should not be assumed that positional and form tolerancing imply tighter tolerance. Rather, it permits the use of maximum tolerance while maintaining 100 percent interchangeability.

Because of its clarity and preciseness in communicating specifications, geometric dimensioning and tolerancing has become the system used by most industries. Every drafter, designer and engineer should understand its use.

SYMBOLS FOR POSITIONAL AND FORM TOLERANCES

Geometric characteristic symbols for tolerances reduce the number of notes required on a drawing. They are compact, recognized internationally and reduce misinterpretation.

The standard symbols used to denote geometric characteristics of part features are shown in Fig. 18-16 (a). Modifying symbols are shown in Fig. 18-16 (b). The meaning and application of these symbols are discussed later in this chapter.

Templates are available for use in drawing the symbols for geometric dimensioning and tolerancing, Fig. 18-17. The template shown features symbols, specifications for surface characteristics and a complete alphabet.

Fig. 18-17. Symbols template for geometric dimensioning and tolerancing.   (RapiDesign)

DATUM IDENTIFYING SYMBOL

Datums are identified on the drawing by a reference letter (any letter except I, O and Q), which is preceded and followed by a dash and enclosed in a frame, Fig. 18-18 (a). Where more than one datum is used on a drawing, the desired order or precedence of datums is shown from left to right in the feature control symbol (b).

DATUM TARGETS

MATERIAL CONDITION APPLICABILITY FOR TOLERANCES OF POSITION AND FORM

The applicability of material condition symbols, RFS, MMC, and LMC is limited to features subject to variations in size. These may be any feature or datum feature whose axes or center is controlled by geometric tolerances. The following general rules apply:

1. Tolerances of Position—RFS, MMC, or LMC must be specified on the drawing for an individual tolerance,

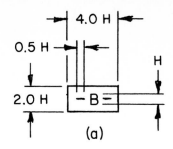

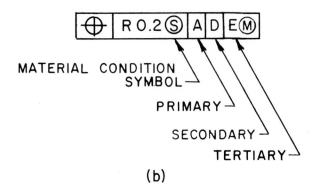

Fig. 18-18. Datum identifying symbol and order of datum precedence. (ANSI)

datum reference, or both, as applicable.
2. All other Geometric Tolerances—RFS applies, for an individual tolerance, datum reference, or both, where no modifying symbol is specified. MMC must be specified on the drawing where it is required.

The symbol Ⓜ is used to designate "Maximum Material Condition." The symbol Ⓛ specifies the "Least Material Condition." The symbol Ⓢ designates that "Regardless of Feature Size" the tolerance must not be exceeded. These symbols may be used only as modifiers in feature control frames, Fig. 18-18 (b). The abbreviations MMC, LMC, and RFS should be used in notes and dimensions rather than the symbols.

Where MMC, RFS, or LMC is specified, the appropriate symbol follows the specified tolerance and applicable datum reference in the feature control frame, Fig. 18-18 (b).

FEATURE CONTROL FRAME

The feature control frame is the means by which a geometric tolerance is specified for an individual feature, Fig. 18-19. The frame is divided into compartments containing, in order from the left, the geometric characteristic symbol followed by the tolerance. Where applicable, the tolerance is preceded by the diameter or radius symbol and followed by a material condition symbol, Ⓜ , Ⓛ , or Ⓢ .

If the tolerance is related to a datum(s), the datum reference letter(s) follows in the next compartment. Where applicable, the datum reference letter is followed by a material condition symbol. Datum reference letters are entered in the desired order of precedence, from left to right, and need not be in alphabetical order, Fig. 18-18 (b).

The feature control frame is associated with the feature(s) by:
1. Attaching a side or end of the frame to an extension line from the feature, provided it is a plane surface, Fig. 18-19.

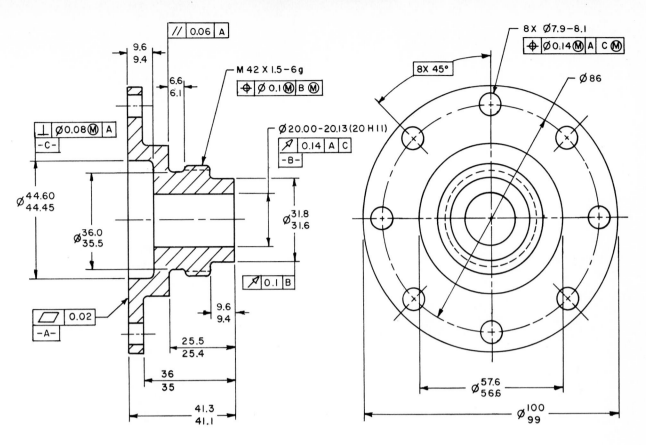

Fig. 18-19. Feature Control Frame Placement.

2. Attaching a side or end of the frame to an extension of the dimension line pertaining to a feature of size.
3. Running a leader from the frame to the feature.
4. Placing the frame below or attached to a leader-directed callout or dimension controlling the feature.

TOLERANCES OF LOCATION

Tolerances assigned to dimensions which locate one or more features in relation to other features or datums are known as positional tolerances. There are three basic types of positional tolerances: true position, concentricity (coaxiality) and symmetry.

TRUE POSITION TOLERANCE

The term "true position" has a meaning similar to basic dimension. It describes the exact (true) location of a point, line or plane (usually the center) of a feature in relation to another feature or datum. A true position is located by basic dimensions, Fig. 18-20. A feature control frame is added to the note used to specify the size and number of the feature, (c).

A true position tolerance is the total permissible amount a feature may vary around its true position. Note in Fig. 18-20 (b) that the coordinate system of tolerancing results in a square tolerance zone, and the actual variation from true position may exceed the specified variation. The actual variation along the diagonal is .014 or 1.4 times the tolerance specified. By utilizing a true position tolerance (circular zone), the larger tolerance can be specified and interchangeability of parts still maintained, (c).

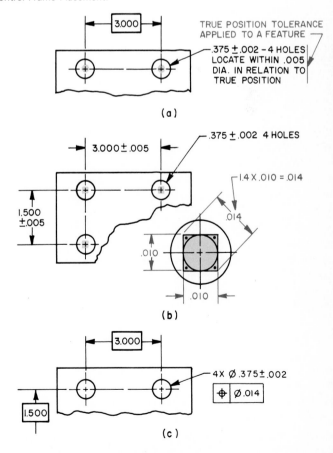

Fig. 18-20. True position tolerancing compared with the coordinate system.

361

The tolerance zone is represented as a circle in one view and is assumed to be a cylindrical zone for the full depth of the hole, Fig. 18-21. (Figs. 18-21 through 18-39 courtesy of ANSI — American National Standards Institute.) The axis of the feature hole) must be within the tolerance zone.

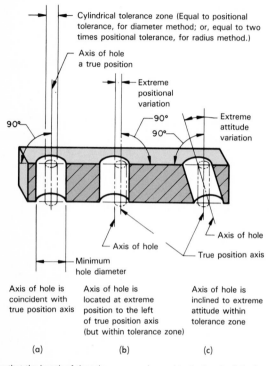

Note that the length of the tolerance zone is equal to the length of the feature, unless otherwise specified on the drawing.

Fig. 18-21. Hole axes in relation to positional tolerance zone.

PROJECTED TOLERANCE ZONE

A projected tolerance zone is specified where the variation in perpendicularity of threaded or press-fit holes could cause fasteners such as screws, studs, or pins to interfere with mating parts. The application of a projected tolerance zone to a positional tolerance is shown in Fig. 18-22. The extent of the projected tolerance zone may be indicated in the feature control frame, (a), or shown as a dimensioned heavy chain line drawn closely to the center line of the hole, (c).

TRUE POSITION FOR NONCIRCULAR FEATURES

Noncircular features such as slots, tabs, and elongated holes may be toleranced for position by using the same basic principles used for circular features. Positional tolerances usually apply only to surfaces related to the center plane of the feature, Fig. 18-23.

Where the feature is at maximum material condition, its center plane must fall within a tolerance zone having a width equal to the true position tolerance for the diameter method (or twice the tolerance when the radius method is used). Note that the tolerance zone also defines the variation limits of squareness of the feature.

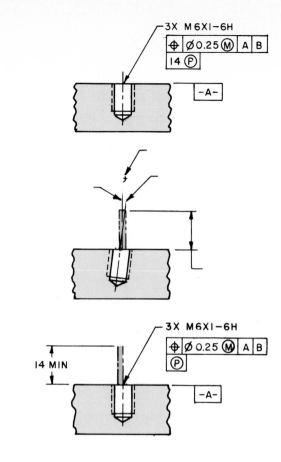

Fig. 18-22. Projected tolerance zone specified.

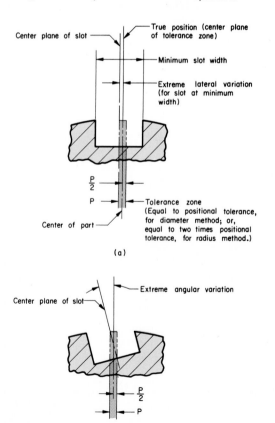

Fig. 18-23 Tolerance zone for center plane of slot at MMC.

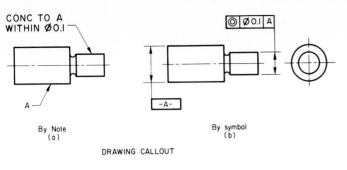

By Note (a)   By symbol (b)

DRAWING CALLOUT

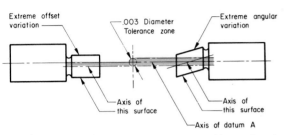

The feature must be within a cylinder zone, regardless of feature size, whose axis coincides with the datum axis.

(c)   (d)

INTERPRETATION

Fig. 18-24. Concentricity callout and interpretation.

CONCENTRICITY TOLERANCE

Concentricity is the condition of two or more surfaces of revolution having a common axis. Concentricity tolerance callout and interpretation are shown in Fig. 18-24. In cases where it is difficult to find the axis of a feature (and where control of axis is not necessary to part's function), it is recommended that the control be specified as a runout tolerance or true position tolerance.

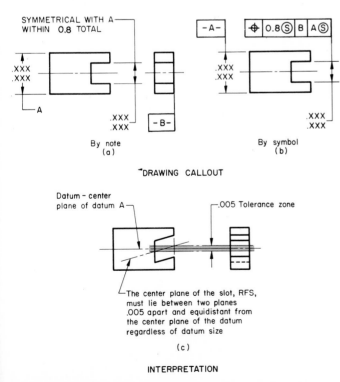

By note (a)   By symbol (b)

DRAWING CALLOUT

The center plane of the slot, RFS, must lie between two planes .005 apart and equidistant from the center plane of the datum regardless of datum size

(c)

INTERPRETATION

Fig. 18-25. Symmetry callout and interpretation.

SYMMETRY

A symmetrical feature or part has the same contour and size on opposite sides of a central plane or datum feature. A symmetry tolerance may be specified by using a positional tolerance at MMC, as shown in Fig. 18-25. The true position symbol is recommended where a feature is to be located symmetrically about a datum plane and the tolerance is expressed on an MMC or RFS basis, depending upon the design requirements.

TOLERANCES OF FORM, PROFILE, ORIENTATION AND RUNOUT

Tolerances which control the form of the various geometrical shapes and free-state variations of features are called form tolerances. Form tolerances are used to control the conditions of straightness, flatness, roundness, cylindricity, profile of a surface or line, angularity, parallelism, perpendicularity and runout. A form tolerance specifies a tolerance zone within which the particular feature must lie.

STRAIGHTNESS TOLERANCE

The characteristic of straightness is where an element of an axis or surface is a straight line. Straightness is specified by a tolerance zone of uniform width along a straight line within which all elements of the line must lie. Fig. 18-26 indicates the drawing callout by geometric symbol, by note and interpretation. All elements of this feature must lie within a tolerance zone of .010 inch for the total length of the part. Straightness may be applied to control line elements in a single direction or in two directions.

FLATNESS TOLERANCE

The characteristic of flatness of a surface is where all elements of a surface lie in one plane. Flatness is specified by a tolerance zone between two parallel planes within which the surface must lie. The specification for a flatness tolerance is shown by symbol and by note, together with its interpretation,

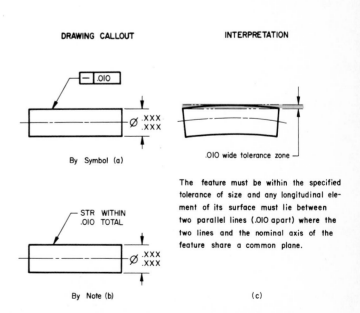

DRAWING CALLOUT   INTERPRETATION

By Symbol (a)

.010 wide tolerance zone

The feature must be within the specified tolerance of size and any longitudinal element of its surface must lie between two parallel lines (.010 apart) where the two lines and the nominal axis of the feature share a common plane.

By Note (b)   (c)

Fig. 18-26. Specifying straightness.

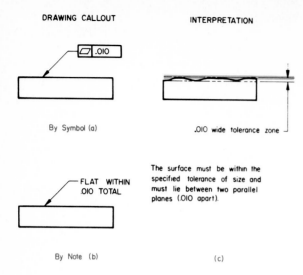

Fig. 18-27. Specifying flatness.

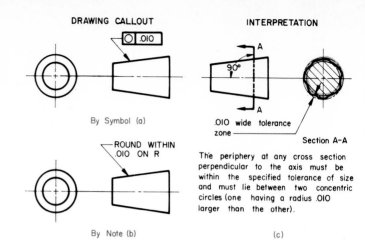

Fig. 18-29. Specifying circularity for a cone.

in Fig. 18-27. All elements of this surface must lie within a tolerance zone of .010 inch for the total length and width of the part. The feature control frame is placed in a view where the surface to be controlled is represented by a line.

CIRCULARITY (ROUNDNESS) TOLERANCE

The characteristic of circularity of a surface of revolution such as a cylinder or cone, is where all elements of the surface intersected by any plane are equidistant from the axis. For a sphere, all elements of the surface intersected by any plane passing through a common center are equidistant from that center. The intersection plane is perpendicular to a common axis for the cylinder and cone and it passes through a common center for the sphere. A circularity tolerance is specified by two concentric circles confining a zone within which the periphery must lie. The permissible circularity tolerance is specified for a cylinder in Fig. 18-28, a cone in Fig. 18-29, and a sphere in Fig. 18-30. Note that the tolerance zone for each of these is established by a radius.

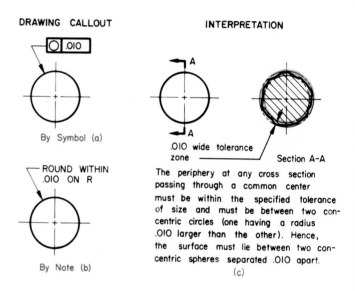

Fig. 18-30. Specifying circularity for a sphere.

CYLINDRICITY TOLERANCE

Cylindricity is characterized by a surface of revolution in which all elements of the surface are equidistant from a common axis. A cylindricity tolerance zone is defined as the annular space between two concentric cylinders within which the specified surface must lie. Note in Fig. 18-31 that the tolerance zone is established by a radius, and that the cylindricity tolerance controls roundness, straightness and parallelism of the surface elements.

PROFILE TOLERANCING

The elements of profiles consist of straight lines and curved lines. The curved lines may be either arcs or irregular curves. The elements in a profile line or surface are located with basic dimensions, and the profile tolerance zone is established by applying a specified amount of permissible variation to these dimensions, Fig. 18-32.

The profile tolerance zone may be specified as bilateral (to both sides of true profile) or unilateral (to either side of true profile). The tolerance zone is shown along the profile in a con-

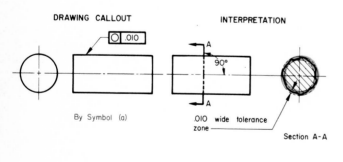

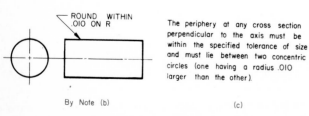

Fig. 18-28. Specifying circularity for a cylinder.

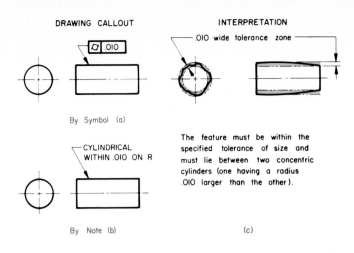

Fig. 18-31. Specifying cylindricity.

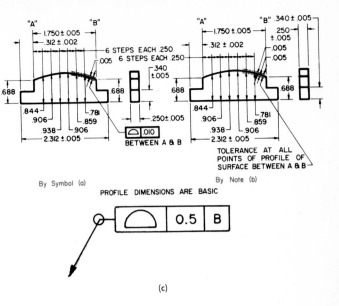

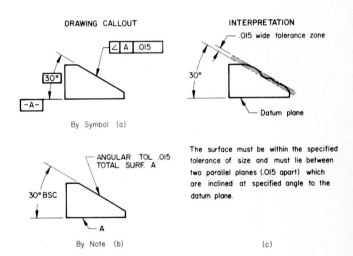

Fig. 18-33. Specifying profile of a surface.

spicuous place by one or two phantom lines at a distance greater than the actual tolerance for drawing clarity.

A dimensional part is shown with a profile tolerance specification for a surface in Fig. 18-33. A profile tolerance for a line is specified in the same manner except the symbol is different. (See Fig. 18-16.)

When a profile tolerance applies to surfaces all around the part, a circle is located at the junction of the feature control frame leader, Fig. 18-33 (c).

ANGULARITY TOLERANCE

The means of controlling the specific angle of a surface or axis (other than 90 degrees) with respect to a datum plane or axis is known as angularity tolerance. This is achieved by specifying a tolerance zone confined by two parallel planes, inclined at the required angle to a datum plane or axis. The parallel planes establish the zone within which the toleranced surface or axis must lie, Fig. 18-34.

Fig. 18-34. Specifying angularity.

PARALLELISM TOLERANCE

Parallelism is characterized by a surface or line which is equidistant at all points from a datum plane or axis. Parallelism is specified for a plane surface by a tolerance zone confined by two planes parallel to a datum plane within which the specified feature (axis or surface) must lie, Fig. 18-35.

Parallelism is specified for a cylindrical surface by a tolerance zone parallel to a datum feature axis within which the axis of a feature must lie, Fig. 18-36.

PERPENDICULARITY TOLERANCE

Perpendicularity is characterized by surfaces, axes or lines which are at right angles to a datum plane or axis. Perpendicularity tolerance for a surface is specified by a zone confined by two parallel planes (or cylindrical tolerance zone) perpendicular to a datum plane or axis within which the controlled surface of the feature must lie, Fig. 18-37.

A method of tolerancing a cylindrical feature for perpendicularity is shown in Fig. 18-38.

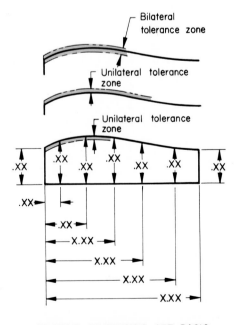

PROFILE DIMENSIONS ARE BASIC

Fig. 18-32. Profile tolerance zones.

THIS ON THE DRAWING

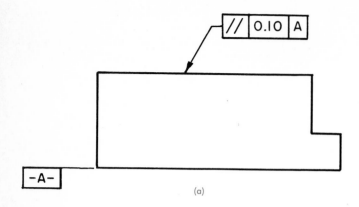

(a)

MEANS THIS

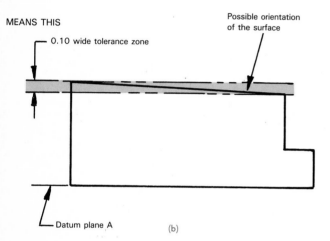

Datum plane A

(b)

The surface must lie between two planes 0.10 apart which are parallel to datum plane. A. Additionally, the surface must be within the specified limits of size.

Fig. 18-35. Specifying parallelism for a plane surface.

THIS ON THE DRAWING

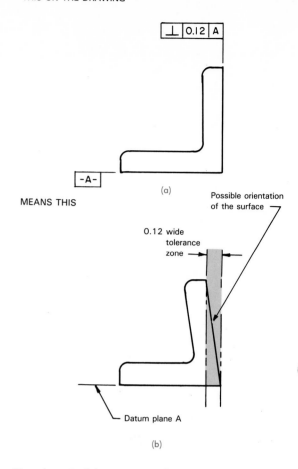

MEANS THIS

(a)

Possible orientation of the surface

0.12 wide tolerance zone

Datum plane A

(b)

The surface must lie between two parallel planes 0.12 apart which are perpendicular to datum plane A. Additionally, the surface must be within the specified limits of size.

Fig. 18-37. Specifying perpendicularity.

DRAWING CALLOUT

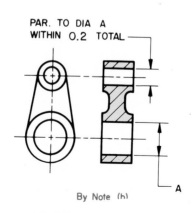

By Symbol (a)

PAR. TO DIA A
WITHIN 0.2 TOTAL

By Note (b)

INTERPRETATION

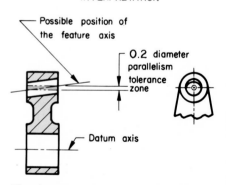

Possible position of the feature axis

0.2 diameter parallelism tolerance zone

Datum axis

The feature axis must be within the specified tolerance of location. Regardless of the actual size of the feature, its axis must lie within a cylindrical zone (0.2 diameter) which is parallel to the datum axis.

(c)

Fig. 18-36. Specifying parallelism for a cylindrical surface.

DRAWING CALLOUT

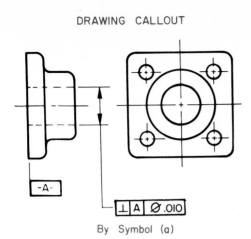

-A-

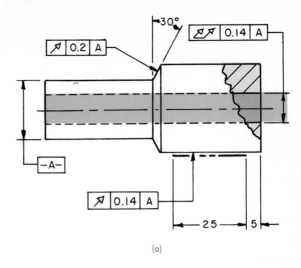

(a)

By Symbol (a)

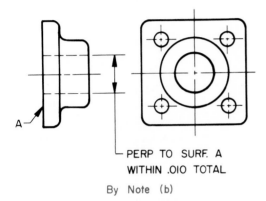

PERP TO SURF. A
WITHIN .OIO TOTAL

By Note (b)

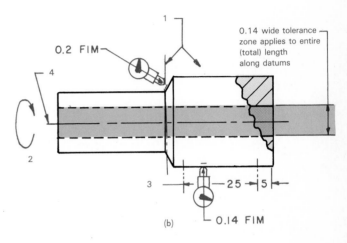

Fig. 18-39. Interpretation of runout tolerance zone.

INTERPRETATION

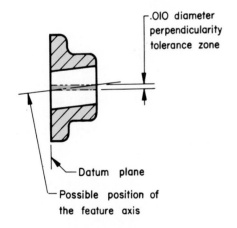

.OIO diameter
perpendicularity
tolerance zone

Datum plane

Possible position of
the feature axis

Regardless of the actual size of the
feature, its axis must lie within a
cylindrical zone (.OIO diameter) which
is perpendicular to the datum plane.

(c)

Fig. 18-38. Specifying perpendicularity for a cylindrical feature.

RUNOUT TOLERANCE

Runout is a composite tolerance used to control those surfaces constructed around a datum axis and those at right angles
to a datum axis. Runout control is of two types, circular and
total.

Circular runout tolerance is applied independently and controls the elements of circularity and coaxiality of a surface. The
measurement is taken as the part is rotated 360°, Fig. 18-39
(b). Where applied to surfaces constructed at right angles, to
the datum axis, circular elements of a plane surface (wobble)
are controlled. Where the runout tolerance applies to a specific
portion of a surface, the extent is shown by a chain line adjacent to the surface profile.

Total runout is applied to all (total) circular and profile surface elements as the part is rotated 360°.

ROUND-OFF RULES

The general rules for rounding off the converted millimeter
and inch values are shown in Fig. 18-40.

Two methods of rounding tolerances may be used, depending on the degree of accuracy which must be maintained in
the part:

METHOD A involves rounding to the nearest rounded value
of the limit so that, on the average, the converted tolerances
remain statistically identical to the original tolerance. For
example:

$$1.282 = 32.5628 \text{ rounded to } 32.563$$
$$1.273 = 32.3342 \text{ rounded to } 32.334$$

The limits converted by this method may, in some instances,

ROUND-OFF RULES

Total Tolerance in Inches		Millimeter Conversion Rounded to
At Least	**Less Than**	
.00001	.0001	5 Decimal Places
.0001	.001	4 Decimal Places
.001	.01	3 Decimal Places
.01	.1	2 Decimal Places
.1	1	1 Decimal Place

Total Tolerance in Millimeters		Inch Conversion Rounded To
.005	.05	5 Decimal Places
.05	.5	4 Decimal Places
.5	5.0	3 Decimal Places
5.0 and Over		2 Decimal Places

Fig. 18-40. General rules for rounding off converted millimeter and inch values. (General Motors Drafting Standards)

be outside the original tolerance (as in case of upper limits in example). Where this variance is acceptable for interchangeability of parts, this method of rounding tolerances is the basis for inspection.

METHOD B rounding is done systematically toward the interior of the tolerance zone, so that the converted tolerances are never larger than the original tolerances. For example:

1.282 = 32.5628 rounded (inward) to 32.562
1.273 = 32.3342 rounded (inward) to 32.335

Where tolerance limits must be respected absolutely (when parts made to converted limits are to be inspected by means of original gages), Method B should be used.

LIMIT DIMENSION

Where tolerances are expressed on dual dimensioned drawings as limit dimensions, the dimension is expressed as a limit dimension for the decimal inch as well as the metric:

$$\begin{bmatrix} 1.135 & 28.829 \\ 1.125 & 28.575 \end{bmatrix}$$

PLUS AND MINUS TOLERANCING

The round-off practice described earlier cannot be applied directly to plus and minus tolerances. The inch dimensions and tolerance should first be changed into a limit dimension, then converted to millimeters as illustrated:

$$1.130 \pm .005 = \begin{bmatrix} 1.135 & 28.829 \\ 1.125 & 28.575 \end{bmatrix}$$

DIMENSIONING FOR NUMERICALLY CONTROLLED EQUIPMENT

Numerical control is a system of controlling the movements of machines such as drills, mills, welding, flame cutting, filament winding, paint spraying and drafting machines. The control is accomplished by numbers coded into a tape or disc. In the case of programs involving considerable calculations in the dimensional location of features, the tape is punched by a computer from the program sheets prepared directly from the engineering-design drawing.

N/C PROGRAM

The N/C program is a translation of requirements shown on the drawing to numerical control language. The program is prepared by a "part programmer" who knows how to read drawings, material characteristics, machine processes, tool requirements and the particular N/C machine operation and capabilities.

DIMENSIONING REQUIREMENTS

Dimensioning requirements for N/C machining are not significantly different from those required for manual machine operation. The drawing must be fully descriptive of the part and its features. Dimensional requirements must be in terms of distance from a point of origin (datum) or from point to point along two or three mutually perpendicular axes. The rectangular coordinate system of dimensioning should be used to facilitate programming time and to reduce the possibility of error.

The N/C machine receives its numerical instruction based upon the rectangular coordinates shown in Fig. 18-41. The point of origin is the "zero" point (reference plane) on the machine and should be identified on the N/C machine drawing. If desired, the point of origin or zero point may be positioned off the left front corner of the part, making all X and Y dimensions positive. Any position on the part can be described in terms of the X, Y and Z axes.

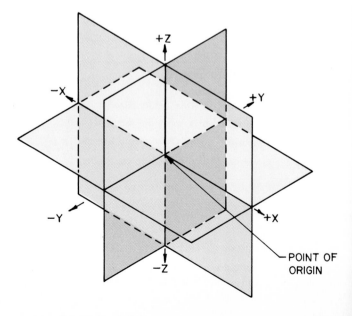

Fig. 18-41. N/C machine rectangular coordinates.

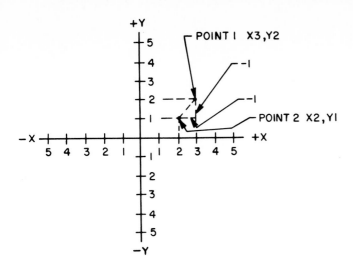

Fig. 18-42. Comparison between absolute and incremental methods of describing N/C machine movement.

Within the rectangular coordinate system, two methods of indicating movement to points or positions are used: absolute (address) and incremental:

1. Absolute Method. Machine positions are given in terms of distance from the origin of the coordinate axes (datum dimensioning).

    Example: Refer to Fig. 18-42. With center of machine spindle or cutter positioned at point 1 and ready for movement to point 2, absolute system command would be, "move to point 2, plus X2 plus Y1." This is in terms of zero point of origin.

2. Incremental Method. Machine positions are given in terms of distance from the preceding point or present machine location (chain dimensioning).

    Example: Refer to Fig. 18-42. With center of machine spindle or cutter positioned at point 1 and ready for movement to point 2, incremental system command would be "move from point 1, minus X1, minus Y1."

TOLERANCING REQUIREMENTS FOR N/C MACHINING

Drawings for N/C machining should include tolerances based upon the design needs rather than N/C machine capabilities. N/C machines only recognize exact commands (dimensions) which, in themselves, imply no degree of tolerance. However, tolerances should be assigned for the following reasons:

1. Accuracy varies between N/C machines and cannot be controlled by operator manipulation.
2. Parts may need to be produced on conventional machines with manual operator.
3. Maximum tolerances permit programming flexibility.

N/C DIMENSIONING RULES

To facilitate the use and accuracy of drawings in numerical controlled machining, the following rules should be observed:

1. Dimensioning must fully describe part and its features.
2. Sufficient dimensions must be supplied to eliminate necessity of part programmer making assumptions or calculations.

3. All dimensions should be in decimal and/or metric units.
4. Rectangular coordinate dimensions must be used to describe distances from datum planes or surfaces.
5. Angles should be specified by rectangular coordinate position dimensions.
6. Reference planes (machine zero) may be common with datum plane of drawing or they may be assigned outside part. They should be clearly marked on drawing.
7. Tolerances should be based on design requirements rather than on N/C machine capability.
8. Geometric position and form tolerances may be used with N/C machining.

N/C DESIGN CONSIDERATIONS

The following design techniques are recommended for use on drawings for parts and assemblies regardless of manufacturing method to be employed (courtesy of Allis-Chalmers, ENGINEERING STANDARDS):

1. Draw principally machined surfaces as primary (front) view. This will keep drawing and major machining operation in same relationship.
2. Design for symmetry so that initial programming steps can be reused.
3. As many part features as practical should be located on a common X or Y axis. Two or more holes, bolt circles, slots, etc., located on a common center line can reduce total programmed movement with less chance of error than random location.
4. Aim for a minimum number of machine setups. Wherever practical, design for machining from one side only.
5. Drilled or tapped holes and counterbore depths should always be dimensioned from a finished surface (datum). A finished surface (pad or boss) should be provided on rough castings for this purpose.
6. Avoid blind-tapped holes and back spotfacing operations. This will eliminate "down-time" to clear chips and to add special cutters.
7. A preferred tool list should be followed whenever possible. For example, Standard 534 lists preferred sizes of tap drills. This recommendation is based on a reduction of tool inventories to essential items compatible with machine capacities and consistent with economical design considerations. It is not practical to call for .250-20, .312-18 and .375-16 tapped holes in same part when one size could satisfy all three applications.
8. Reduce number of different hole sizes. (Drilled, tapped, punched, etc.) An N/C machine can store just so many different cutting tools; operations beyond this number require special handling.
9. Specify same radius for fillets and rounded curves, where possible, to reduce number of different cutting tools required.

    NOTE: Consideration should be given to standardized design characteristics that can be programmed on a reusable basis. Specific families of parts that follow an established set of rules are adaptable to computer programming where optimum design can be selected electronically.

    Typical flow of N/C information is shown in Fig. 18-43.

Fig. 18-43. Typical flow of N/C information.

PROBLEMS IN PRECISION DIMENSIONING

The following problems provide you with an opportunity to apply the information presented in this chapter on precision dimensioning.

BILATERAL TOLERANCING AND LIMIT DIMENSIONING

1. Draw the front view and a left-side section of the DIFFERENTIAL SPIDER, Fig. 18-44. Dimension using bilateral tolerances. Delete the note on tolerances but include the other two notes on the drawing in an appropriate space.

2. Draw two views of the SHAFT BEARING CARTRIDGE, Fig. 18-44, including the right-side view as a section. Use bilateral tolerances and identify the surface roughness as indicated.

3. Draw two views of the TRUNNION IDLER, Fig. 18-44, with the upper portion of the profile view as a broken-out section to show hole detail. Show bilateral toleranced

dimensions as limit dimensions and tolerance of other dimensions as a note. Include notes on the drawing.

4. Draw the SLIDE NUT, Fig. 18-45, as a full section. Show threads in simplified form and dimension the drawing using limit dimensions.

READING GEOMETRIC DIMENSIONS AND TOLERANCES

5. Refer to the drawing in Fig. 18-45 and answer the following questions:

   a. What is the name of the part and the drawing number?
   b. What feature of the part is datum A?
   c. Interpret the feature control symbol at B.
   d. What type of dimension is at (X) ? What relationship must exist on the finished part between surface Y and datum A?
   e. What relationship is specified for feature (Z) and datum A?
   f. Which feature calls for the smoother surface texture, V or W? Explain.

6. Refer to the drawing in Fig. 18-47 and answer the following questions:

   a. Give the name and number of the part.
   b. What general tolerances are specified for the part?
   c. What surface roughness is called for at X? At Z?
   d. What specification is given for hole D?
   e. Identify datums A, B and C.
   f. What relationship must exist between hole B and datums A, B and C?
   g. What relationship is specified between hole D and hole C?
   h. State the dimension at Y as a limit dimension.

GEOMETRIC DIMENSIONING AND TOLERANCING

7. Draw the necessary orthographic views of the BEARING SUPPORT, Fig. 18-45, including a sectional view. Delete the notes on geometric tolerancing and include these as "feature control symbols" on the drawing.

8. Draw the necessary views of the SLIDE NUT, Fig. 18-45, and dimension in metric. Change the general tolerance note to metric.

9. Draw the necessary views of the objects shown in Fig. 18-48, and dimension. Use geometric "feature control symbols" to replace notes where possible.

DIMENSION FOR N/C MACHINING

10. Draw the necessary views of LIGHT SWITCH MOUNTING PLATE, Fig. 18-49, and dimension for incremental N/C machining.

11. Draw the necessary views of the MOUNTING PLATE, Fig. 18-49, and dimension for the absolute method of N/C machining, using ordinate dimensioning.

SELECTED ADDITIONAL READINGS

1. DIMENSIONING AND TOLERANCING FOR ENGINEERING DRAWINGS, ANSI Y14.5M-1982, American National Standards Institute, 1430 Broadway, New York, NY 10018.

2. DRAWING REQUIREMENTS MANUAL, Mil-D-1000/Mil-Std-100, Global Engineering Documentation Services, Inc., P.O. Box 2060, Newport Beach, CA 92660.

3. GEOMETRIC DIMENSIONING AND TOLERANCING, Madsen, David A., Goodheart-Willcox Co., Inc.

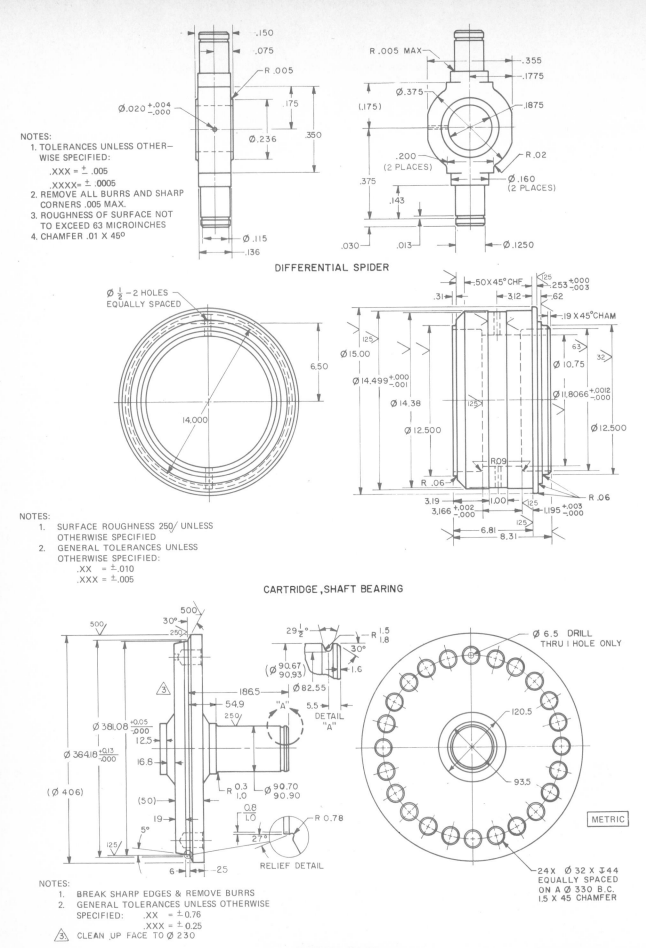

NOTES:
1. TOLERANCES UNLESS OTHER-
   WISE SPECIFIED:
   .XXX = ± .005
   .XXXX= ± .0005
2. REMOVE ALL BURRS AND SHARP
   CORNERS .005 MAX.
3. ROUGHNESS OF SURFACE NOT
   TO EXCEED 63 MICROINCHES
4. CHAMFER .01 X 45°

DIFFERENTIAL SPIDER

NOTES:
1. SURFACE ROUGHNESS 250/ UNLESS
   OTHERWISE SPECIFIED
2. GENERAL TOLERANCES UNLESS
   OTHERWISE SPECIFIED:
   .XX = ±.010
   .XXX = ±.005

CARTRIDGE, SHAFT BEARING

NOTES:
1. BREAK SHARP EDGES & REMOVE BURRS
2. GENERAL TOLERANCES UNLESS OTHERWISE
   SPECIFIED: .XX = ±0.76
                  .XXX = ±0.25
3. CLEAN UP FACE TO Ø 230

TRUNNION IDLER

Fig. 18-44. Bilateral tolerance dimensioning problems.

372

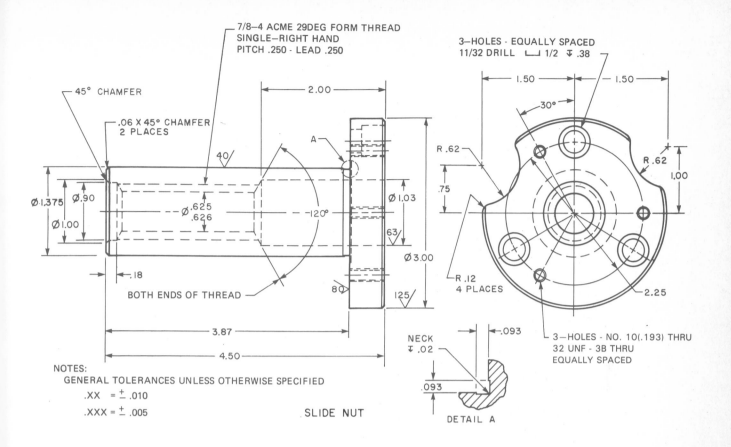

7/8-4 ACME 29DEG FORM THREAD
SINGLE-RIGHT HAND
PITCH .250 - LEAD .250

3-HOLES - EQUALLY SPACED
11/32 DRILL ⌴ 1/2 ⊼ .38

45° CHAMFER

.06 X 45° CHAMFER
2 PLACES

2.00

A

40/

Ø.90

Ø1.375

Ø.625
.626

Ø1.00

120°

Ø1.03

63/

Ø3.00

.18

BOTH ENDS OF THREAD

80/

125/

3.87

4.50

NOTES:
GENERAL TOLERANCES UNLESS OTHERWISE SPECIFIED
.XX = ± .010
.XXX = ± .005

SLIDE NUT

1.50   1.50

30°

R .62

.75

R .62

1.00

R .12
4 PLACES

2.25

3-HOLES - NO. 10(.193) THRU
32 UNF - 3B THRU
EQUALLY SPACED

NECK
⊼ .02

.093

.093

DETAIL A

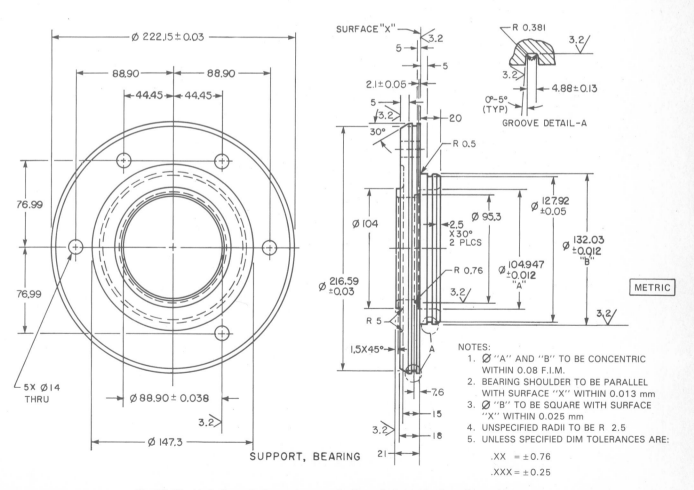

Ø 222.15 ± 0.03

88.90   88.90

44.45   44.45

SURFACE "X"

3.2

5

5

2.1 ± 0.05

5

3.2

30°

20

R 0.381

3.2

3.2

4.88 ± 0.13

0°-5°
(TYP)

GROOVE DETAIL-A

76.99

Ø 104

R 0.5

Ø 127.92
±0.05

Ø 132.03
±0.012
"B"

76.99

Ø 216.59
±0.03

2.5
X30°
2 PLCS

Ø 95.3

R 0.76

3.2

Ø 104.947
±0.012
"A"

METRIC

5X Ø14
THRU

Ø 88.90 ± 0.038

3.2

Ø 147.3

1.5X45°

R 5

A

7.6

15

3.2

18

21

SUPPORT, BEARING

NOTES:
1. Ø "A" AND "B" TO BE CONCENTRIC
   WITHIN 0.08 F.I.M.
2. BEARING SHOULDER TO BE PARALLEL
   WITH SURFACE "X" WITHIN 0.013 mm
3. Ø "B" TO BE SQUARE WITH SURFACE
   "X" WITHIN 0.025 mm
4. UNSPECIFIED RADII TO BE R 2.5
5. UNLESS SPECIFIED DIM TOLERANCES ARE:

   .XX = ± 0.76
   .XXX = ± 0.25

Fig. 18-45. Limit dimensioning and geometric dimensioning and tolerancing problems.

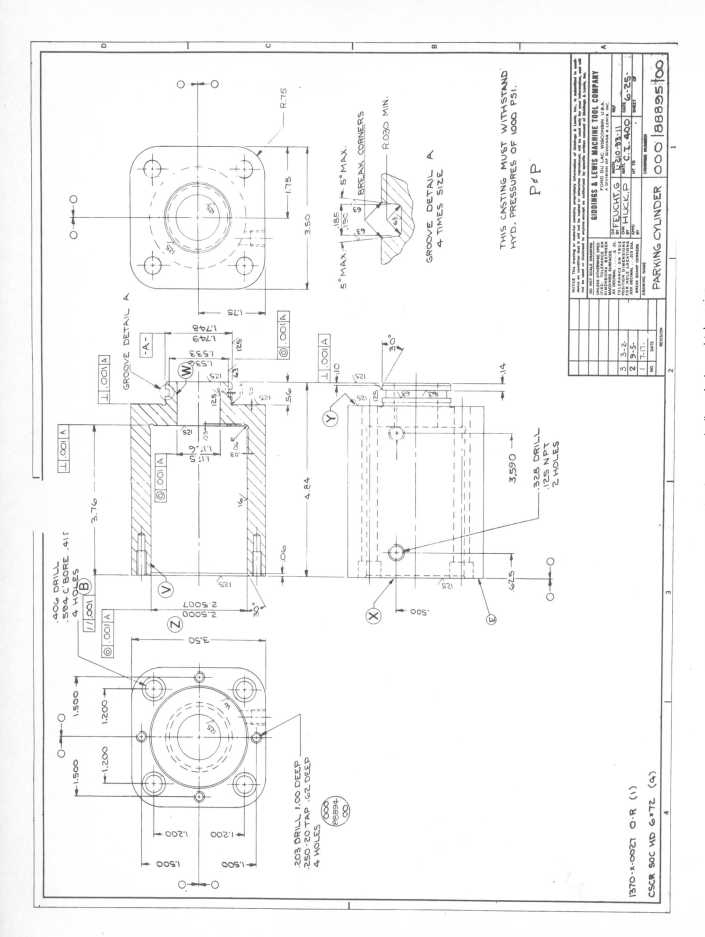

Fig. 18-46. Blueprint reading problem in geometric dimensioning and tolerancing.

Precision Dimensioning

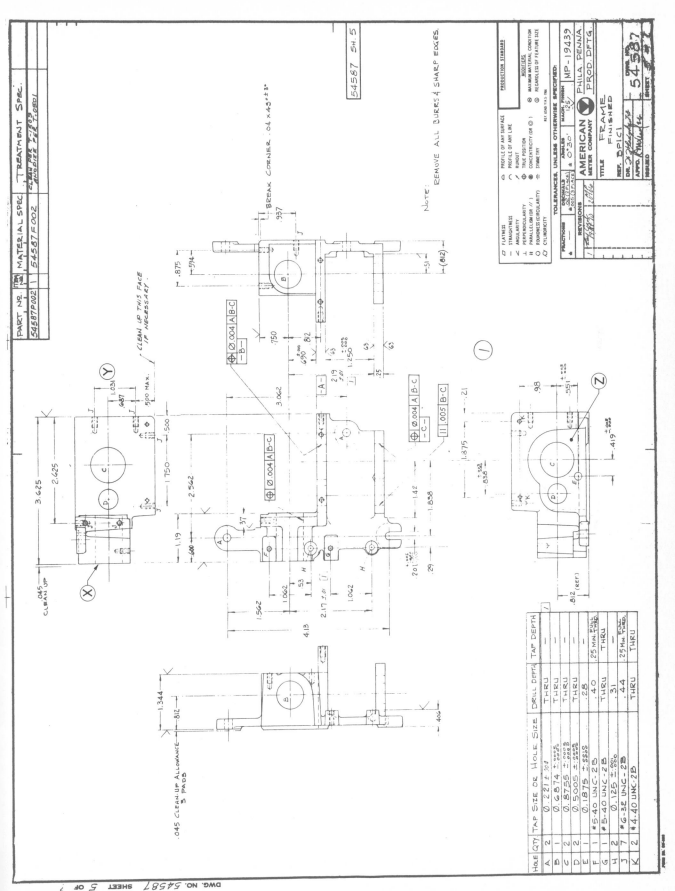

Fig. 18-47. Blueprint reading problem in geometric dimensioning and tolerancing.

375

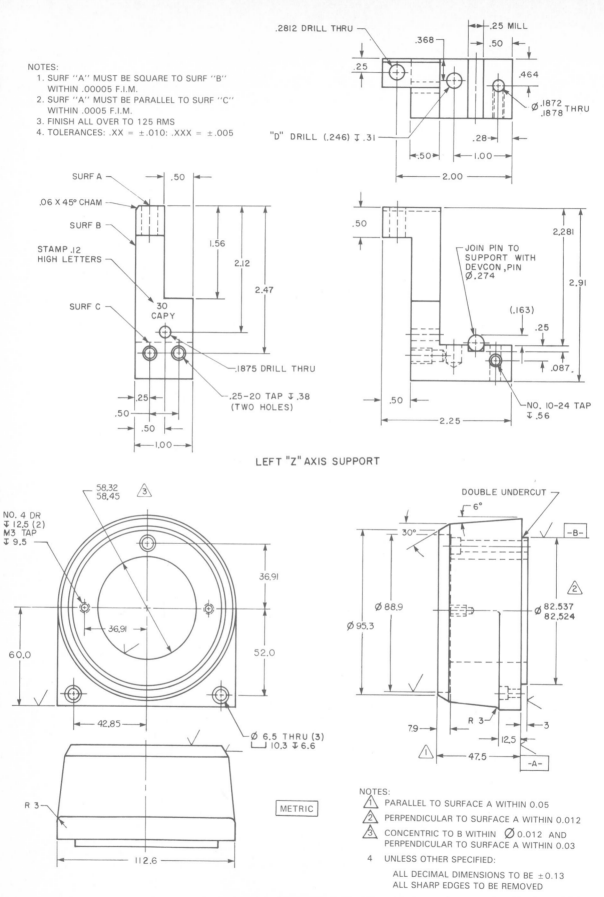

NOTES:
1. SURF "A" MUST BE SQUARE TO SURF "B" WITHIN .00005 F.I.M.
2. SURF "A" MUST BE PARALLEL TO SURF "C" WITHIN .0005 F.I.M.
3. FINISH ALL OVER TO 125 RMS
4. TOLERANCES: .XX = ±.010: .XXX = ±.005

.2812 DRILL THRU

.368

.25 MILL

.50

.25

.464

Ø.1872 / .1878 THRU

"D" DRILL (.246) ↧ .31

.28

.50

1.00

2.00

SURF A

.50

.06 X 45° CHAM

SURF B

STAMP .12 HIGH LETTERS

1.56

2.12

2.47

SURF C

30 CAPY

.1875 DRILL THRU

.25-20 TAP ↧ .38 (TWO HOLES)

.25

.50

.50

1.00

.50

2.281

JOIN PIN TO SUPPORT WITH DEVCON, PIN Ø.274

(.163)

.25

2.91

.087

.50

NO. 10-24 TAP ↧ .56

2.25

LEFT "Z" AXIS SUPPORT

58.32 / 58.45 △3

NO. 4 DR ↧ 12.5 (2) M3 TAP ↧ 9.5

36.91

36.91

52.0

60.0

42.85

Ø 6.5 THRU (3) ⌴ 10.3 ↧ 6.6

R 3

METRIC

112.6

DOUBLE UNDERCUT

6°

30°

-B-

△2

Ø 88.9

Ø 95.3

Ø 82.537 / 82.524

R 3

3

7.9

12.5

△1

47.5

-A-

NOTES:
△1 PARALLEL TO SURFACE A WITHIN 0.05
△2 PERPENDICULAR TO SURFACE A WITHIN 0.012
△3 CONCENTRIC TO B WITHIN Ø 0.012 AND PERPENDICULAR TO SURFACE A WITHIN 0.03
4 UNLESS OTHER SPECIFIED:
ALL DECIMAL DIMENSIONS TO BE ± 0.13
ALL SHARP EDGES TO BE REMOVED

BRACKET, Y-AXIS DRIVE COVER

Fig. 18-48. Precision dimensioning problems.

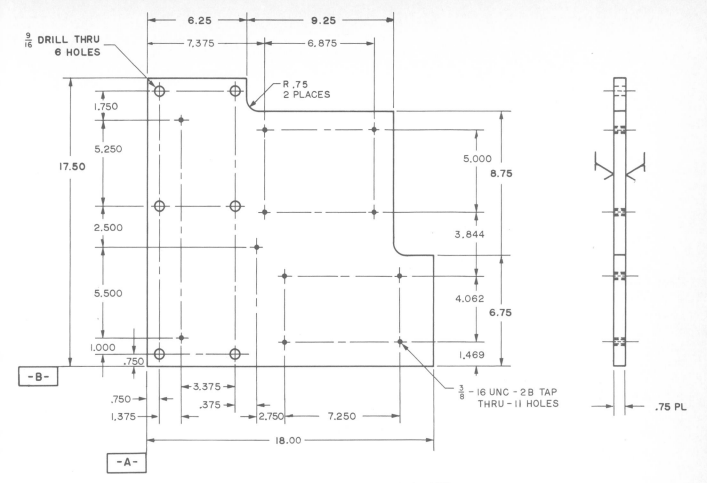

LIMIT SWITCH MOUNTING PLATE

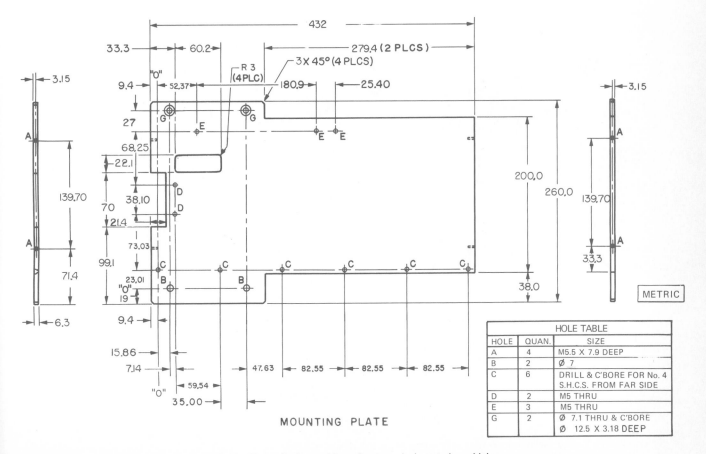

HOLE TABLE		
HOLE	QUAN.	SIZE
A	4	M5.5 X 7.9 DEEP
B	2	Ø 7
C	6	DRILL & C'BORE FOR No. 4 S.H.C.S. FROM FAR SIDE
D	2	M5 THRU
E	3	M5 THRU
G	2	Ø 7.1 THRU & C'BORE Ø 12.5 X 3.18 DEEP

MOUNTING PLATE

Fig. 18-49. Dimensioning problems for numerical control machining.

Chapter 19
CAMS, GEARS AND SPLINES

Modern machines usually require mechanisms to transfer motion and power from one source to another without slippage as might occur with belts. It is also necessary in some instances to convert rotary motion to reciprocal motion at a certain rate of speed for related parts. An example is the firing action of a four-stroke cycle internal combustion engine, as shown in Fig. 19-1.

In four-stroke cycle engine operation, there must be a definite timing of the opening and closing of valves in relation to the cycling of the piston. This is achieved with gears and cams that are capable of maintaining the desired mechanical relationship. This chapter presents information on some basic types of cams, gears and splines, and how these features are represented on drawings.

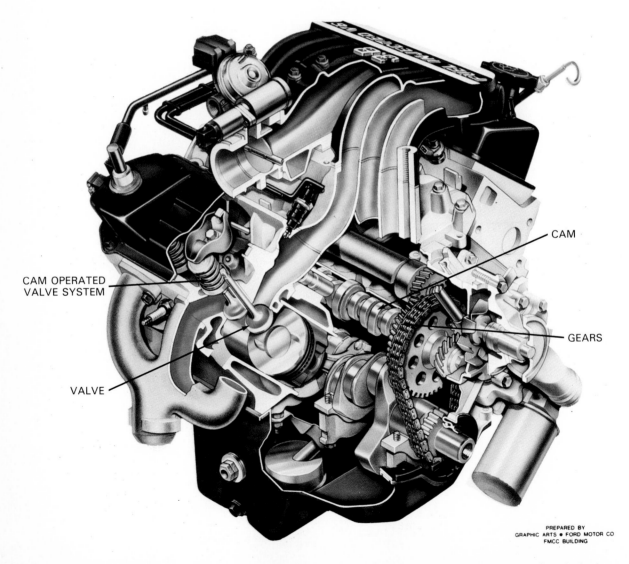

Fig. 19-1. Gears and cams are used to maintain the precise relationship between valves and pistons to produce the intake, compression, power and exhaust strokes of an internal combustion engine. (Ford Motor Co.)

Cams, Gears and Splines

CAMS

A cam is a mechanical device for changing the motion of a uniformly rotating shaft into a reciprocating motion of varying speed, Fig. 19-2. Three types of cams are in common use: plate cam, Fig. 19-2 (a), groove cam (b) and cylindrical cam (c).

Contact with the surface or groove of the cam is made with a follower held against the cam by gravity, spring action or by the groove in the groove cam, Fig. 19-3. Basic types of cam followers are: knife edge, Fig. 19-3 (a), flat face, (b), and roller, (c).

Cams may be designed to provide a number of different types of motion and displacement patterns. MOTION refers to the rate of speed or movement of the cam follower in relation to the uniform or constant speed of rotation of the cam, Fig. 19-4. DISPLACEMENT refers to the distance the cam follower moves in relation to the rotation of the cam. The three basic types of cam follower motions will be discussed later in this chapter.

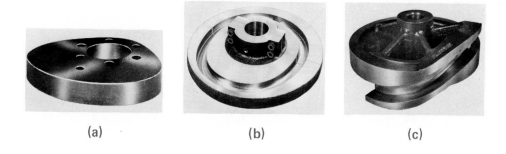

(a)   (b)   (c)

Fig. 19-2. Cam types commonly used in mechanisms. (Ferguson Machine Co.)

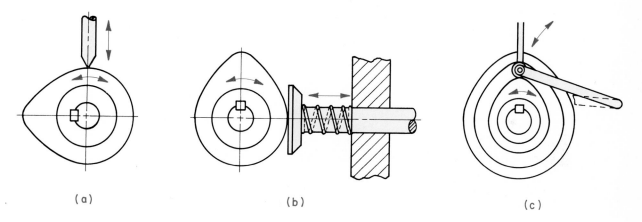

(a)   (b)   (c)

Fig. 19-3. Cam followers pick up the rotating motion of the cam and change it to reciprocating motion.

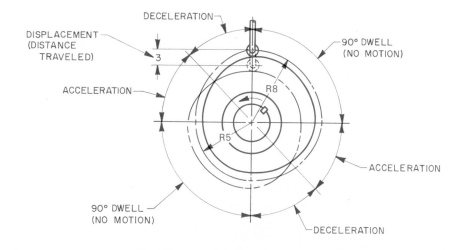

Fig. 19-4. Cams can be designed to provide a variety of motion and displacement patterns.

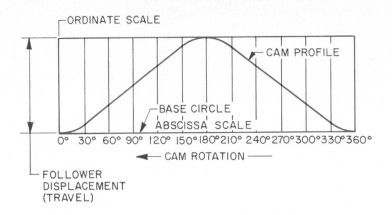

Fig. 19-5. A cam displacement diagram showing the abscissa and ordinate scales, and a graph of the cam profile.

CAM DISPLACEMENT DIAGRAMS

A displacement diagram is a graph or drawing of the displacement (travel) pattern of the cam follower caused by one rotation of the cam, Fig. 19-5. Construction of a displacement diagram is usually the first step in the design of a cam.

In Fig. 19-5, divisions on the abscissa scale or angular sectors around the base circle represent time intervals of the revolving cam. (The abscissa is the horizontal coordinate of a point in a plane Cartesian coordinate system obtained by measuring parallel to the X axis.) It has been established that when the speed of rotation of a cam is constant, the time intervals are uniform. Note in Fig. 19-5 that the distance from the base circle on the ordinate scale represents distance of travel or displacement of the cam follower. When the time intervals are kept constant, the rate of speed of the cam follower varies as the angle or incline of the cam changes.

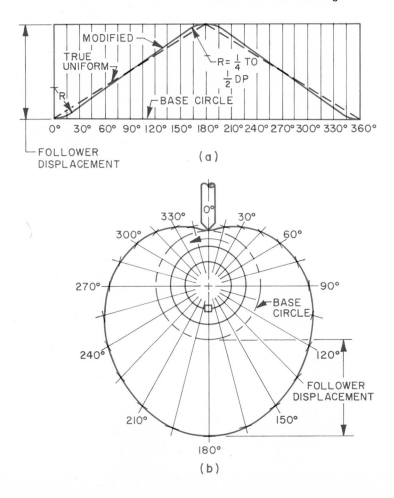

Fig. 19-6. Displacement diagram and cam layout for a uniform motion cam.

The divisions along the abscissa scale may approximate the actual spaces on the cam base circle since the displacement diagram is only representative of the motion of the cam. Displacement on the ordinates must be accurate since these are used in laying off measurements on the cam layout itself.

TYPES OF CAM FOLLOWER MOTIONS

The three principal types of motion for cam followers are uniform motion, simple harmonic motion and uniformly accelerated motion.

UNIFORM MOTION is produced when the cam moves the follower at the same rate of speed from the beginning to the end of the displacement cycle. The shape of a uniform motion cam is a straight line as shown in the displacement diagram, Fig. 19-6 (a).

However, with a straight line cam design, the starting and stopping of the follower would be very abrupt due to instantaneous changes in velocity. So cam shape is usually modified with arcs (R) having a radius of one-fourth to one-half the follower displacement to smooth out the beginning and ending of the follower stroke. The uniform motion cam is satisfactory for machinery operating at a slow rate of speed.

Proceed as follows to lay out a uniform motion cam:

1. Lay out base circle with a radius equal to distance from cam axis to lowest follower position as shown at zero degree position, Fig. 19-6 (b).
2. Draw a convenient number of equally spaced radial lines dividing base circle into intervals representing angular motion of cam. (Number of divisions must equal divisions along base circle of displacement diagram for cam.) In our example, 24 increments of 15 degrees have been used.
3. Starting with zero degree position of displacement diagram, transfer with dividers distances modified cam profile line lies above base circle on each 15 degree line to corresponding line on cam layout beyond base circle. Note cam rotates counterclockwise and plotting progresses in opposite direction, (b).
4. When all points have been located, sketch a smooth curve through points and finish with an irregular curve.

HARMONIC MOTION moves the follower in a smooth continuous motion based on the successive positions of a point moving at constant velocity around the circumference of a circle, Fig. 19-7 (a). The harmonic cam is satisfactory for machinery operating at moderate speeds.

Proceed as follows to lay out a harmonic motion cam:

1. Lay out displacement diagram by constructing a semicircle

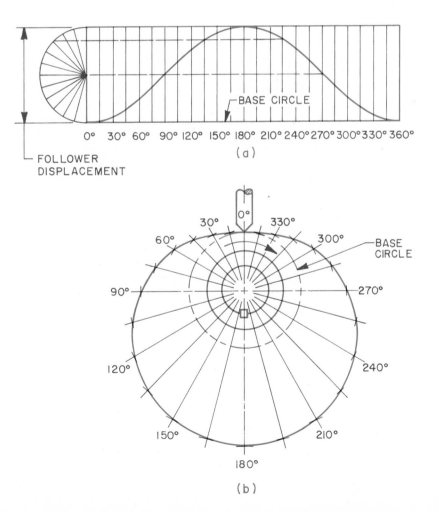

Fig. 19-7. Displacement diagram and cam layout for a harmonic motion cam.

381

whose diameter is equal to desired follower displacement. Divide semicircle into same number of equal parts as there are angular divisions for one-half of cam layout, Fig. 19-7 (a). Project these divisions to their corresponding angular ordinate and draw curve representing displacement diagram for harmonic motion cam.

2. Lay out base circle of cam with a radius equal to distance from cam axis to lowest follower position as shown at zero degree position, Fig. 19-7 (b).

3. Draw a convenient number of equally spaced radial lines dividing base circle into sectors representing angular motion of cam. (Number of divisions must equal divisions along base circle of displacement diagram.)

4. Starting with zero position of displacement diagram, transfer distances harmonic curve lies off of base circle at each ordinate to its corresponding radial line in cam layout, (b). Note that cam rotates clockwise and plotting progresses in opposite direction.

5. When all points have been located, sketch a smooth curve through points and finish with an irregular curve.

UNIFORMLY ACCELERATED MOTION is designed into a cam to provide constant acceleration or deceleration of the follower displacement, Fig. 19-8. The displacement on the ordinates varies at the end of successive uniform intervals on the abscissa scale or angular sectors around the base circle such as 0, $1^2 = 1$, $2^2 = 4$, $3^2 = 9$, and so on.

Divisions along the abscissa scale, Fig. 19-8 (a), represent equal intervals of time. Divisions on the ordinate scale represent distances which are the squares of each successive time interval.

During the first one-half revolution of the cam, the follower rises with constant acceleration from 0° to 90°. From 90° to 180°, the cam still rises but with constant deceleration. Note on the displacement diagram, Fig. 19-8 (a), that there is a reversal of the ordinate scale at 90° (or midway).

During the second half revolution of the cam, the follower falls with constant acceleration from 180° to 270°. It continues to fall from 270° to 360° with constant deceleration, returning the follower to its lowest or zero point, Fig. 19-8 (a). This type of cam motion is suited for high speed cam operation.

Proceed as follows to lay out a uniformly accelerated motion cam:

1. Lay out displacement diagram by drawing an inclined line and laying off squares of successive intervals of time, Fig. 19-8 (a). Note that squares of intervals increase through interval 3 (90°) and decrease in same manner from interval 3 to height of full displacement (180°).

Project these divisions to ordinate lines at 0° and from there to their corresponding angular ordinate. Draw curve representing displacement diagram for uniformly accelerated motion cam.

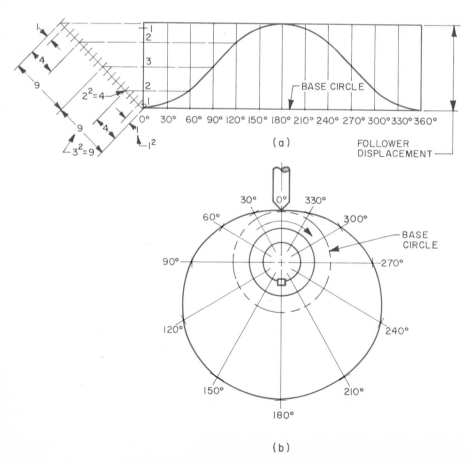

Fig. 19-8. Displacement diagram and cam layout for a uniformly accelerated motion cam.

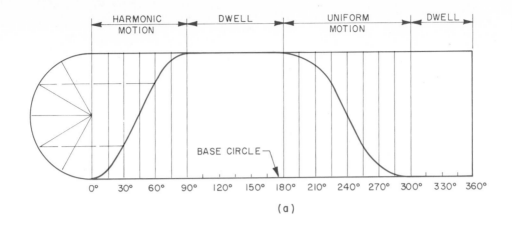

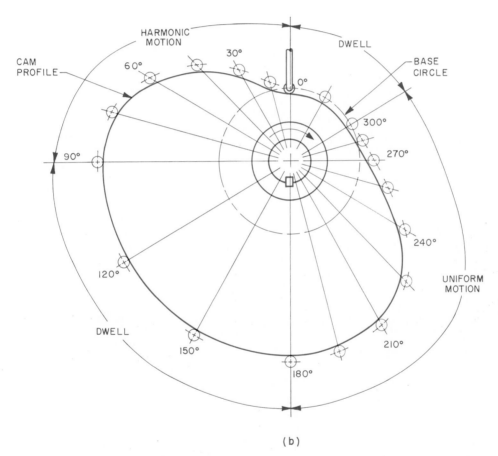

Fig. 19-9. Displacement diagram and cam layout for a cam with a combination of displacement motions.

2. Lay out base circle of cam with a radius equal to distance from cam axis to lowest follower position as shown at 0° position, Fig. 19-8 (b).

3. Draw a convenient number of equally spaced radial lines dividing base circle into sectors representing angular motion of cam. (Number of divisions must equal divisions along base circle of displacement diagram.)

4. Starting with zero position of displacement diagram, transfer distances uniformly accelerated curve lies off of base circle at each ordinate to its corresponding radial line in cam layout, (b). Note that this cam rotates clockwise and plotting progresses in opposite direction from base circle

outwards.

5. When all points have been located, sketch a smooth curve through points and finish with an irregular curve.

COMBINATION MOTION may be designed for a single cam in order to achieve the follower displacement desired, Fig. 19-9. Note the cam follower is the roller type, and the center of the roller is assumed to start on the base circle for layout purposes. Transfer displacement distances from the diagram to their respective radial lines in the layout in the usual manner. Lay off an arc whose radius is equal to that of the roller from these points. The cam profile is drawn tangent to the roller positions on the radial lines.

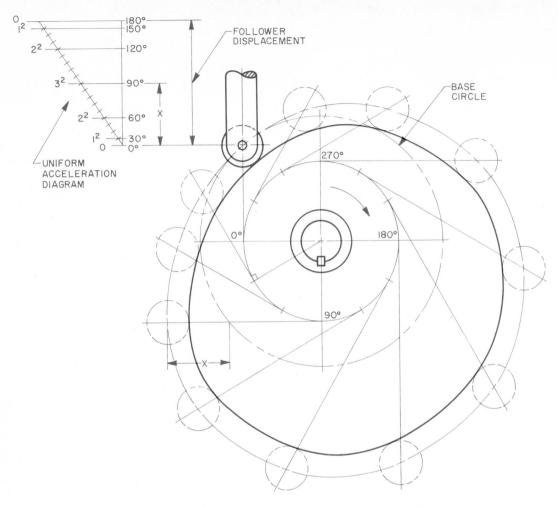

Fig. 19-10. A uniformly accelerated motion cam with an offset follower.

CAM WITH OFFSET ROLLER FOLLOWER

A uniformly accelerated cam with an offset roller follower is shown in Fig. 19-10. Since the motion is uniformly accelerated throughout, it can be plotted directly from the follower without the necessity of drawing a displacement diagram.

Note in Fig. 19-10 that the center of the roller follower is located on the base circle. Draw a circle with its center at the center of the base circle and tangent to the extended center line of the roller follower. Divide this circle into twelve 30 degree sections and tangents drawn at the section points. Next, transfer distances from the uniform acceleration diagram to these tangent lines that extend from the base circle outward as shown at the 90 degree radial line. Then draw circles representing the roller at each of these locations, and draw a smooth curve tangent to the 12 positions of the roller to form the profile of the offset roller uniformly accelerated motion cam.

GEARS

Gears are machine parts used to transmit motion and power by means of successively engaging teeth. Gear teeth are shaped

so contact between the teeth of mating gears is continually maintained while rotation is occurring. Teeth with the involute curve are the type most commonly used for gears. A variety of gear types are shown in Fig. 19-11.

The purpose of this section is to provide an introduction to the terminology, representation and specification of basic gear types on drawings. The spur gear, rack and pinion, bevel gear and worm gear are covered.

SPUR GEARS

Spur gears are used to transmit rotary motion between two or more parallel shafts, Fig. 19-12. The teeth of a spur gear may be cut parallel to the gear axis, in which case the gear is a straight spur gear as shown in Fig. 19-12. These gears are satisfactory for low or moderate speeds but tend to be noisy at high speeds. Modifications of the spur gear for heavier loading and higher speeds are achieved through helical and herringbone toothed gears, Fig. 19-13.

When mating spur gears of different size are in mesh, the larger one is called the gear, the smaller one is the pinion. Only straight spur gears are discussed here. Information on other types may be found in gear standards bulletins and manufacturer's catalogs.

Fig. 19-11. A variety of types of gears, all of which have the involute form of gear tooth. (Western Gear Corp.)

(a)   (b)

Fig. 19-13. The teeth of spur gears may be helical in shape (a) or herringbone (b).

Fig. 19-12. Many spur gears are used to transmit motion and power in this truck transmission. (International Harvester Co.)

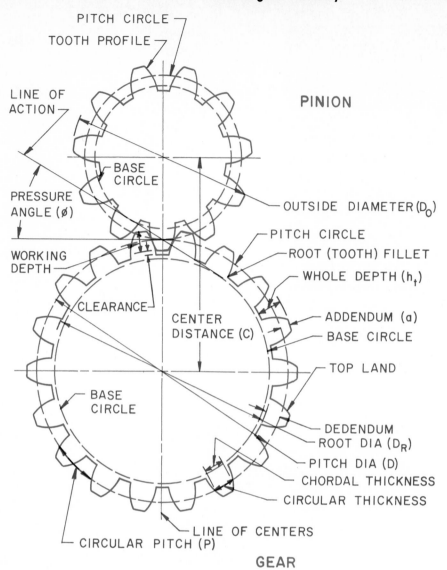

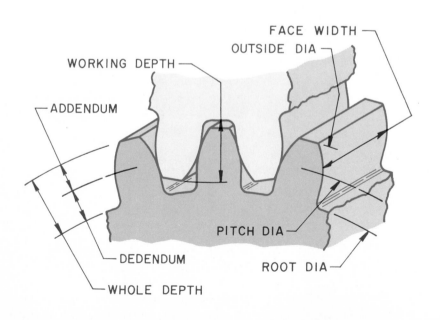

Fig. 19-14. Spur gear terminology.

SPUR GEAR TERMINOLOGY

The design drafter must know and understand gear terminology in order to properly specify and represent gears on drawings. Some essential terms are defined here. Formulas are given, when appropriate, for finding various gear measurements.

NUMBER OF TEETH OR THREADS (N, n): The number of teeth in the gear or pinion or number of threads in the worm.

DIAMETRAL PITCH (P): The number of teeth (N) in a gear per inch of pitch diameter. A gear having 48 teeth and a pitch diameter of 3 inches has a diametral pitch of 16.

$$P = \frac{N}{D}$$

PITCH CIRCLE: An imaginary circle located approximately half the distance from the roots and tops of the gear teeth. It is tangent to the pitch circle of the mating gear, Fig. 19-14.

PITCH DIAMETER (D): The diameter of the pitch circle.

$$D = \frac{N}{P}$$

ADDENDUM (a): The radial distance between the pitch circle and the top of the tooth.

$$a = \frac{1}{P} = 0.5\,(D_o - D)$$

DEDENDUM (b): The radial distance between the pitch circle and the bottom of the tooth.

$$b = \frac{1.157^*}{P} = 0.5\,(D - D_R)$$

*A constant for involute gears.

OUTSIDE CIRCLE OR ADDENDUM CIRCLE: The diameter of the pitch circle plus twice the addendum (same as outside diameter).

OUTSIDE DIAMETER (D_o): The diameter of a circle coinciding with the tops of the teeth of an external gear (same as addendum circle).

$$D_o = D + 2a = \frac{N}{P} + 2\left(\frac{1}{P}\right) = \frac{N+2}{P}$$

ROOT CIRCLE OR DEDENDUM CIRCLE: The circle which coincides with the bottom of the gear teeth.

ROOT DIAMETER (D_r): The diameter of the root circle. It is equal to the pitch diameter minus twice the dedendum.

$$D_r = D - 2b = \frac{N}{P} - \frac{2(1.157)}{P} = \frac{N-2.314}{P}$$

CENTER DISTANCE (C): The center-to-center distance between the axes of two meshing gears.

$$C = PR_1 + PR_2 = \frac{N_1 + N_2}{2P}$$

Where PR_1 and PR_2 are the respective pitch radii and N_1 and N_2 are the respective number of teeth of the two meshing gears.

CLEARANCE (c): The radial distance between the top of a tooth and the bottom of the tooth space of a mating gear.

$$c = b - a = \frac{1.157}{P} - \frac{1}{P} = \frac{0.157}{P}$$

CIRCULAR PITCH (p): The length of the arc along the pitch circle between similar points on adjacent teeth.

$$p = \frac{\pi D}{N} = \frac{\pi}{P}$$

CIRCULAR THICKNESS (t): The length of the arc along the pitch circle between the two sides of the tooth.

$$t = \frac{p}{2} = \frac{\pi D}{2N}$$

FACE WIDTH (F): The width of the tooth measured parallel to the gear axis.

CHORDAL ADDENDEM (a_c): The radial distance from the top of the tooth to the chord of the pitch circle.

$$a_c = a + \frac{D}{2}\left[1 - \cos\left(\frac{90^o}{N}\right)\right]$$

CHORDAL THICKNESS (t_c): The length of the chord along the pitch circle between the two sides of the tooth.

$$t_c = D \sin\left(\frac{90^o}{N}\right)$$

WHOLE DEPTH (h_t): The total depth of a tooth (addendum plus dedendum).

$$h_t = a + b = \frac{1}{P} + \frac{1.157}{P} = \frac{2.157}{P}$$

WORKING DEPTH: The sum of the addendums of two mating gears.

PRESSURE ANGLE (ϕ): The angle of pressure between contacting teeth of meshing gears. Two involute systems, the 14 1/2° and 20°, are standard with the 20° gradually replacing the older 14 1/2°. It may be observed that the pressure angle determines the size of the base circle to which it is tangent, Fig. 19-14.

BASE CIRCLE: The circle from which the involute profile is generated. The diameter of the base circle is determined by the pressure angle of the gear system.

SPUR GEAR REPRESENTATION

The normal practice in representing gears on industrial drawings is to show the gear teeth in simplified conventional form, Fig. 19-15, rather than to draw them in detail form.

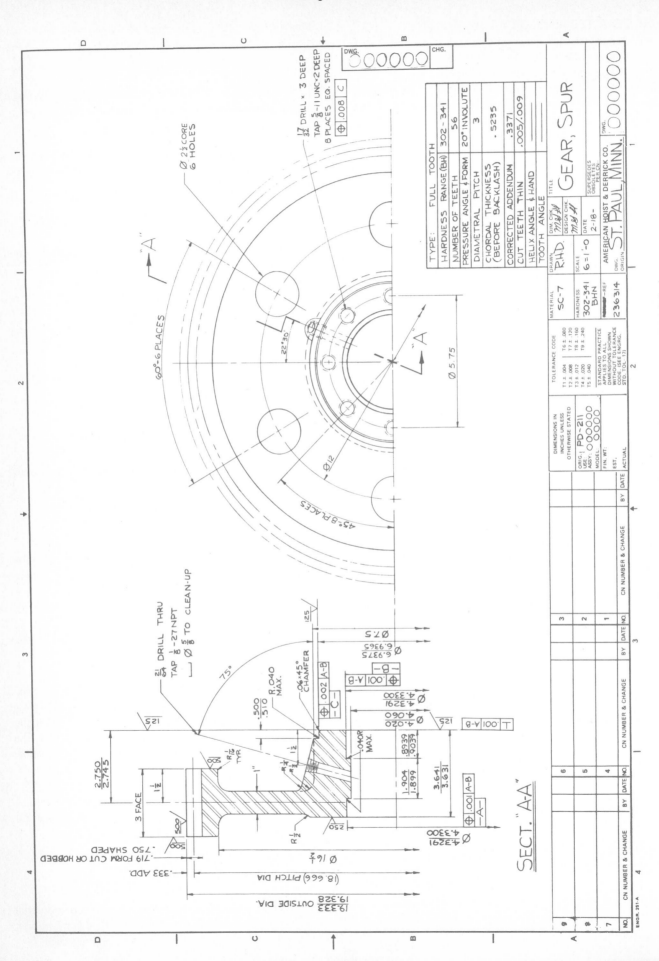

Fig. 19-15. An industrial drawing of a spur gear. (American Hoist & Derrick Co.)

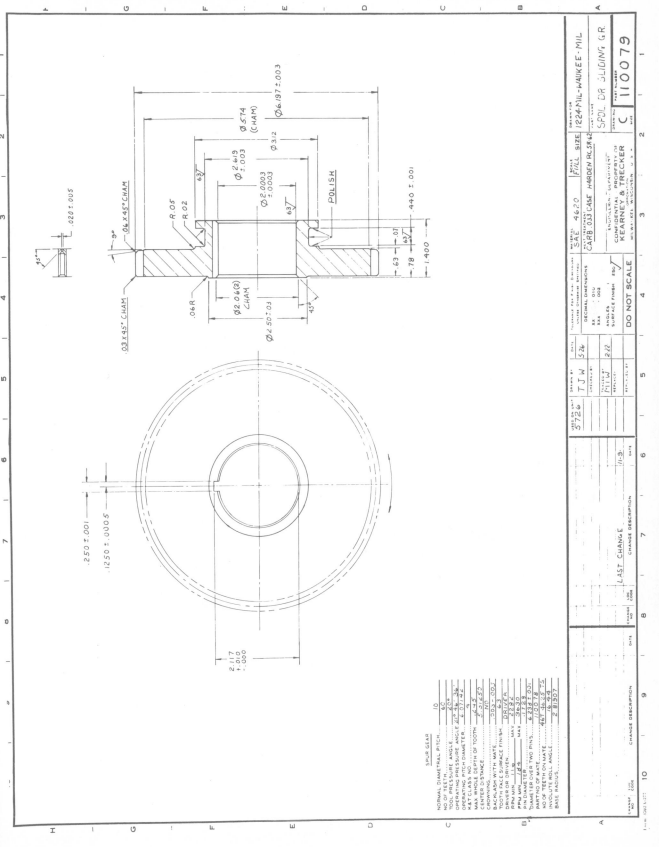

Fig. 19-16. A detail drawing of a spur gear and table of gear data. (Kearney & Trecker Corp.)

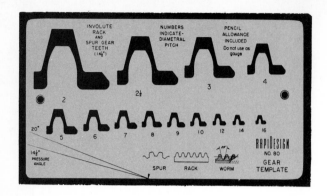

Fig. 19-17. A template for drawing spur rack and spur gear teeth. (RapiDesign)

The circular view may be omitted unless needed to more fully describe the gear. A table of gear data is included on the drawing to supply the specifications needed to manufacture the gear, Fig. 19-16. Note that a phantom line is used to represent the outside and root diameters and a center line is used to represent the pitch circle.

Where it is necessary to show tooth profiles for clarity, a gear template should be used, Fig. 19-17.

RACK AND PINION

A rack is a spur gear with its teeth spaced along a straight pitch line, Fig. 19-18. The rack and pinion have a number of uses in machinery and equipment such as lowering and raising the spindle of a drill press.

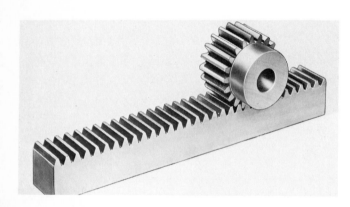

Fig. 19-18. A rack and pinion.   (Boston Gear Div.)

BEVEL GEARS

Bevel gears are used to transmit motion and power between two or more shafts whose axes are at an angle (usually 90°) and would intersect if extended, Fig. 19-19. Bevel gears of same size and at right angles are called miter gears. Straight toothed bevel gears are discussed here, but helical toothed bevel gears are often used for quieter and smoother operation.

BEVEL GEAR TERMINOLOGY

Some of the terms used for bevel gears are the same as for spur gears. These are noted in the list that follows. Also given, where appropriate, are formulas for straight bevel gear measurements:

Fig. 19-19. A pair of bevel gears. The smaller gear is called a pinion.

DIAMETRAL PITCH (P_d): Same as for spur gears.

PITCH DIAMETER (D): The diameter of the pitch circle at the base of the pitch cone, Fig. 19-20.

$$D = \frac{N}{P_d}$$

CIRCULAR PITCH (p): Same as for spur gears.

CIRCULAR THICKNESS (t): Same as for spur gears, but measured at large end of tooth.

OUTSIDE DIAMETER (D_o): Diameter of the crown circle of the gear teeth.

$$D_o = D + 2a \cos\Gamma$$

CROWN HEIGHT (χ): Distance from the cone apex to the crown of the gear tooth measured parallel to the gear axis.

$$\chi = \frac{1}{2}\, D_o/\tan\Gamma_o$$

BACKING (Y): Distance from the back of the gear hub to the base of the pitch cone measured parallel to the gear axis.

CROWN BACKING (Z): Distance from the back of gear hub to the crown of the gear measured parallel to the gear axis.

$$Z = Y + a \sin\Gamma$$

MOUNTING DISTANCE (MD): The distance from a locating surface of a gear (such as end of hub) to the center line of its mating gear. It is used for proper assembling of bevel gears.

$$MD = Y + \frac{1}{2}D/\tan\Gamma$$

ADDENDUM (a): Same as for spur gears, but measured at large end of tooth.

ADDENDUM ANGLE (a): The angle between elements of the face cone and pitch cone. It is the same for the gear and pinion.

$$a = \tan^{-1}\frac{A}{CD}$$

DEDENDUM (b): Same as for spur gears, but measured at large end of tooth.

DEDENDUM ANGLE (δ): The angle between elements of root cone and pitch cone and is the same for gear and pinion.

$$\delta = \tan^{-1}\frac{D}{CD}$$

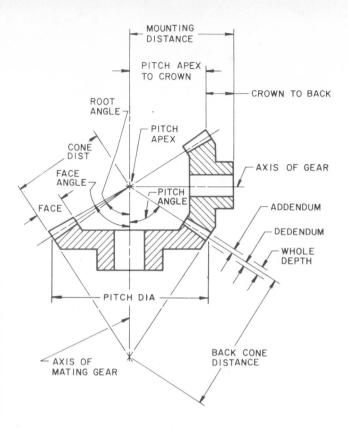

Fig. 19-20. Bevel gear terminology.

FACE ANGLE (Γ_o, γ_o): The angle between an element of the face cone and the axis of the gear or pinion.

$$\Gamma_o = \Gamma + \delta_P$$

$$\gamma_o = \gamma + \delta_G$$

PITCH ANGLE (Γ, γ): The angle between an element of the pitch cone and its axis.

$$\text{Gear: } \Gamma = \tan^{-1}\frac{N}{n} = \tan^{-1}\frac{D}{d}$$

$$\text{Pinion: } \gamma = \tan^{-1}\frac{n}{N} = \tan^{-1}\frac{d}{D}$$

ROOT ANGLE (Γ_R, γ_R): The angle between an element of the root cone and its axis.

$$\text{Gear: } \Gamma_R = \Gamma - \delta_G$$

$$\text{Pinion: } \gamma_R = \gamma - \delta_P$$

SHAFT ANGLE (Σ): The angle between the shaft of the two gears, usually $90°$.

PRESSURE ANGLE (ϕ): Same as for spur gears.

CONE DISTANCE (A_o): The distance along an element of the pitch cone and is the same for the gear and pinion.

$$A_o = \frac{D}{2\sin\Gamma}$$

WHOLE DEPTH (h_t): Same as for spur gears, but measured at large end of tooth.

CHORDAL THICKNESS (t_c): The length of the chord subtending a circular thickness arc.

$$\text{For Bevel Gear: } t_c = D\sin\left(\frac{90° \cos\Gamma}{N}\right)$$

$$\text{For Pinion: } t_c = D\sin\left(\frac{90° \cos\gamma}{n}\right)$$

CHORDAL ADDENDUM (a_c): The distance from the top of the tooth to the chord subtending the circular thickness arc.

$$a_c = a + \frac{D}{2\cos\Gamma}\left[1 - \cos\left(\frac{90° \cos\Gamma}{N}\right)\right]$$

BEVEL GEAR REPRESENTATION

The construction of a bevel gear for purposes of representation on a drawing is shown in Fig. 19-21. The teeth are normally drawn in simplified conventional form. Proceed as follows to draw the gear:

1. Lay out pitch diameters and axes of gear and pinion, Fig. 19-21 (a).

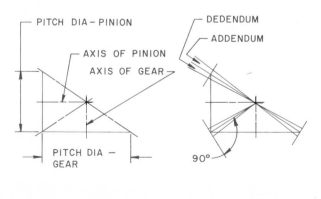

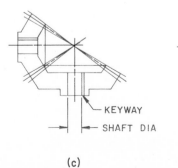

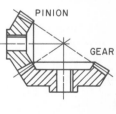

(a)   (b)   (c)   (d)

Fig. 19-21. Construction of a bevel gear drawing.

Fig. 19-22. A typical worm mesh (worm gear and worm).

2. Show whole tooth depth by drawing light construction lines for addendum and dedendum, (b).
3. Lay off face width and other features using dimensions specified or dimensions from gear data tables, (c).
4. Erase construction lines and complete drawing of bevel gear and pinion, (d).

WORM GEAR AND WORM

The worm mesh (worm gear and worm) is a gear type used for transmitting motion and power between nonintersecting shafts usually at 90°, Fig. 19-22. They are characterized by a high velocity ratio of worm to gear and are capable of carrying greater loads than the cross helical gears. The driving member of the worm mesh is the worm.

The worm is actually an Acme-type thread that in section appears much like a gear rack, Fig. 19-23. To increase the contact of the worm mesh, the worm gear is made in a throated (concave) shape to wrap around the worm. (See Fig. 19-22). One revolution of a single-threaded worm advances the worm gear one tooth space, called the "lead." Worms may be either right or left hand thread depending on the rotation desired. Worms may have single, double or triple threads.

The speed ratio of a worm mesh depends on the number of threads on the worm and the number of teeth on the gear. A worm with a single thread meshed with a gear having 48 teeth must revolve 48 times to rotate the gear 1 time; a ratio of 48:1. To gain the same speed reduction with a pair of spur gears would require a gear with 480 teeth and a pinion with 10 teeth. A double-threaded worm would require 24 revolutions to rotate the 48 tooth gear once; a ratio of 24:1.

WORM AND WORM GEAR TERMINOLOGY

The following terms are used in reference to worm gears and worms.

AXIAL PITCH (p_x): The distance between corresponding sides of adjacent threads in a worm. The axial pitch of the worm and the circular pitch of its mating worm gear are the same, Fig. 19-23.

LEAD (ℓ): The axial advance of the worm in one complete revolution. The lead is equal to the pitch for single-thread worms, twice the pitch for double-thread, and three times the pitch for worms with triple threads.

LEAD ANGLE (λ): The angle between a tangent to the helix of the thread at the pitch diameter and a plane perpendicular to the axis of the worm.

$$\lambda = \tan^{-1} \frac{\ell}{\pi D_\omega}$$

PITCH DIAMETER OF WORM (D_ω): The diameter of the pitch circle of a worm thread and may be calculated using the formula below. This is a recommended value, but it may be varied.

$$D_\omega = 2.4 p_x + 1.1$$

ADDENDUM OF THREAD (a_ω): Same as for spur gears.
$$a_\omega = 0.318 p_x$$

DEDENDUM OF THREAD (b_ω): Same as for spur gears.
$$b_\omega = 0.368 p_x$$

WHOLE DEPTH OF THREAD ($h_{t\omega}$): Same as for spur gears.
$$h_{t\omega} = 0.686 p_x$$

OUTSIDE DIAMETER OF WORM ($D_{o\omega}$): The pitch diameter of the worm plus twice the addendum.
$$D_{o\omega} = D_\omega + 0.636 p_x$$

FACE LENGTH OF WORM (F_ω): The overall length of the worm thread section.

$$F_\omega = p_x \left(4.5 + \frac{N_\omega G}{50} \right)$$

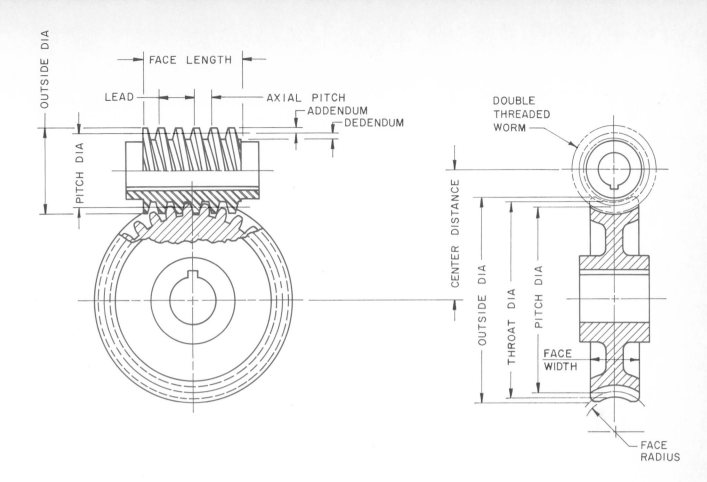

Fig. 19-23. Worm gear and worm terminology.

NUMBER OF TEETH ON WORM GEAR ($N_{\omega}G$): Determined by the desired speed ratio* between the worm and worm gear.

$$N_{\omega}G = SR^* \times \text{No. of Threads}$$

CIRCULAR PITCH OF WORM GEAR: Same as for spur gears, and must be the same as the axial pitch of the worm.

PITCH DIAMETER OF WORM GEAR ($D_{\omega}G$): Same as for spur gears, and the following formula is recommended.

$$D_G = \frac{p_x(N_{\omega}G)}{\pi}$$

ADDENDUM ($a_{\omega}G$): Must equal the addendum of the worm thread.

$$a_{\omega}G = 0.318p\chi$$

WHOLE DEPTH ($h_{t\omega}G$): Must equal the whole depth of the worm thread.

$$h_{t\omega}G = 0.696p\chi$$

THROAT DIAMETER OF WORM GEAR (D_t): The outside diameter of the worm gear measured at the bottom of the tooth arc. It is equal to the pitch diameter of the gear plus twice the addendum.

$$D_t = \frac{p_x(N_{\omega}G)}{\pi} + 0.636p\chi = p\chi\frac{N_{\omega}G + 1.113\pi}{\pi}$$

FACE RADIUS OF WORM GEAR (F_r): The radius of the outside arc of the teeth of the worm gear that curves around the worm.

$$F_r = \frac{D_{\omega}}{2} - 0.318p\chi$$

OUTSIDE DIAMETER OF WORM GEAR ($D_{o\omega}G$): The outside diameter of the worm gear measured at the top of the tooth arc.

$$D_{o\omega}G = D_t + 0.477p\chi$$

WORM GEAR AND WORM REPRESENTATION

The manner of representing worm gears and worms on drawings is shown in Fig. 19-24. The gear teeth and worm thread are usually drawn in simplified, conventional form. Specifications for machining the gear and worm are given in table form on the drawing.

SPLINES

Splines are like multiple keys on a shaft which prevent rotation between the shaft and its related member. The teeth on a spline may have parallel sides, but splines with involute teeth are increasing in use, Fig. 19-25.

A drawing of an external and internal spline is shown in Fig. 19-26. Note the specifications given for each spline on the drawing. Terminology for involute splines is the same as for spur gears.

WORM

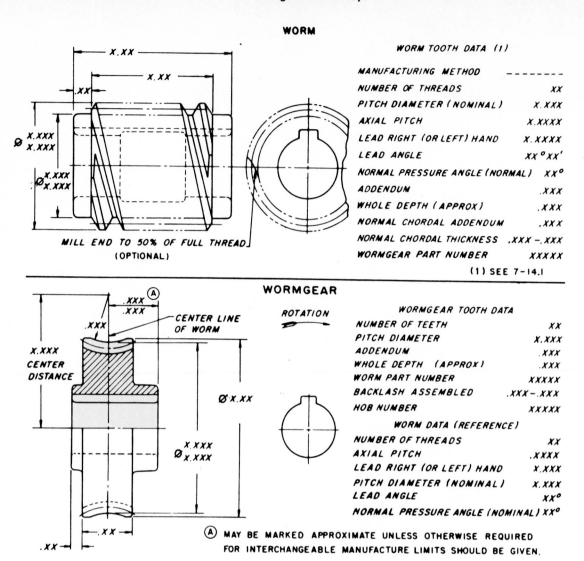

X.XX

X.XX

.XX

Ø X.XXX
X.XXX

Ø X.XXX
X.XXX

MILL END TO 50% OF FULL THREAD
(OPTIONAL)

WORM TOOTH DATA (1)	
MANUFACTURING METHOD	--------
NUMBER OF THREADS	XX
PITCH DIAMETER (NOMINAL)	X.XXX
AXIAL PITCH	X.XXXX
LEAD RIGHT (OR LEFT) HAND	X.XXXX
LEAD ANGLE	XX°XX'
NORMAL PRESSURE ANGLE (NORMAL)	XX°
ADDENDUM	.XXX
WHOLE DEPTH (APPROX)	.XXX
NORMAL CHORDAL ADDENDUM	.XXX
NORMAL CHORDAL THICKNESS	.XXX −.XXX
WORMGEAR PART NUMBER	XXXXX

(1) SEE 7-14.1

WORMGEAR

ROTATION

CENTER LINE OF WORM

.XXX
.XXX

.XXX

X.XXX
CENTER
DISTANCE

Ø X.XX

X.XXX
Ø X.XXX

.XX

WORMGEAR TOOTH DATA	
NUMBER OF TEETH	XX
PITCH DIAMETER	X.XXX
ADDENDUM	.XXX
WHOLE DEPTH (APPROX)	.XXX
WORM PART NUMBER	XXXXX
BACKLASH ASSEMBLED	.XXX−.XXX
HOB NUMBER	XXXXX
WORM DATA (REFERENCE)	
NUMBER OF THREADS	XX
AXIAL PITCH	.XXXX
LEAD RIGHT (OR LEFT) HAND	X.XXX
PITCH DIAMETER (NOMINAL)	X.XXX
LEAD ANGLE	XX°
NORMAL PRESSURE ANGLE (NOMINAL)	XX°

(A) MAY BE MARKED APPROXIMATE UNLESS OTHERWISE REQUIRED
FOR INTERCHANGEABLE MANUFACTURE LIMITS SHOULD BE GIVEN.

Fig. 19-24. Representing and specifying worm gears and worms on drawings. (ANSI)

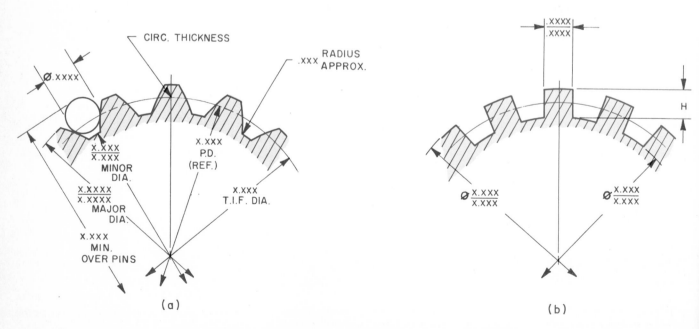

CIRC. THICKNESS

RADIUS
.XXX APPROX.

Ø.XXXX

X.XXX
X.XXX
MINOR
DIA.

X.XXXX
X.XXXX
MAJOR
DIA.

X.XXX
MIN.
OVER PINS

X.XXX
P.D.
(REF.)

X.XXX
T.I.F. DIA.

(a)

.XXXX
.XXXX

H

Ø X.XXX
X.XXX

Ø X.XXX
X.XXX

(b)

Fig. 19-25. Involute (a) and parallel (b) splines are used to prevent rotary
motion between a shaft and coupling or gear mounted on the shaft.

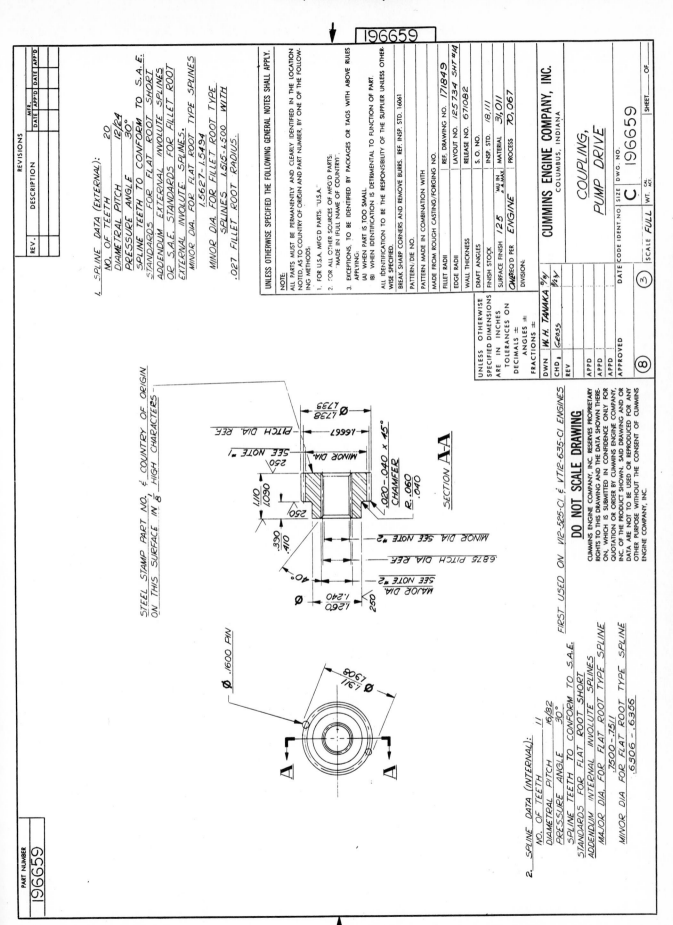

Fig. 19-26. An industrial drawing using simplified convention to represent splines.   (Cummins Engine Co., Inc.)

PROBLEMS IN CAMS, GEARS AND SPLINES

In laying out the cams in the problems which follow, use size B sheets and arrange the required features to make good use of the space available.

The following dimensions, where used, are standard for all cam problems: base circle, 3.50''; shaft diameter, 1''; hub, 1.50'' diameter; keyway, 1/8'' x 1/16''; knife edge follower, 0.625'' round stock; roller follower, 0.875'' diameter. Follower is aligned vertically over the center of the base circle and the cam rises in 180° and falls in 180° unless otherwise noted.

Make a displacement diagram and cam layout for the cams specified:

1. Modified uniform motion cam (use arc of one-quarter of rise to modify uniform motion in displacement diagram) with a rise of 1.375''. Cam rotates clockwise, and knife edge follower is used.

2. Modified uniform motion cam (use arc of one-third of rise to modify uniform motion in displacement diagram) with a rise of 1.250''. Cam rotates counterclockwise, and a knife edge follower is used.

3. Harmonic motion cam with a rise of 1.50''. Cam rotates counterclockwise and a knife edge follower is used.

4. Harmonic motion cam with a rise of 1.125'' in 120°, dwell for 90°, fall 1.125'' with harmonic motion in 120° and dwell for 30°. Cam rotates clockwise, and a knife edge follower is used.

5. Uniformly accelerated motion cam with a rise of 1.250''. Cam rotates clockwise, and a knife edge follower is used.

6. Uniformly accelerated motion cam with a rise of 1.375'' in 90°, dwell for 90°, fall 1.375'' with uniformly decelerated motion in 90° and dwell for 90°. Cam rotates counterclockwise, and a roller follower is used.

7. Uniformly accelerated motion cam with a rise of 1.125'' in 120°, dwell for 60°, fall 1.125'' in 120° with uniformly decelerated motion and dwell for 60°. Cam rotates counterclockwise, and the roller follower is offset .50'' to left of vertical center line.

8. Uniformly accelerated motion cam with a rise of 1.50'' in 180°, dwell for 60° and fall 1.50'' with harmonic motion in 120°. Cam rotates clockwise and has a roller follower.

CAM DESIGN PROBLEMS

9. Design a cam which will, in one revolution, open and close a valve on an automatic hot-wax spray at a car wash. To open valve, cam follower must move 1.125''. Valve is to open in 20° of cam rotation, remain open for 320°, close in 10° and remain closed for 10°. Cam operates at moderate speed. You are to select appropriate cam motion, size of base circle and type of cam follower. Make a full-size working drawing of displacement diagram and cam.

10. Design a cam which will raise a control lever, permitting a workpiece to be fed to a machine. Lever must be raised a distance of 1'', remain open, and close in equal segments of cam revolution. Cam operates at a relatively high speed with moderate pressure on cam follower, which must be offset to right of center .75''. Select appropriate cam motion, size of base circle, type of cam follower. Make full-

size working drawing of displacement diagram and cam.

GEARS

11. Make a working drawing of a spur gear having 40 teeth, diametral pitch of 8, a pressure angle of 20°, shaft diameter .75'', hub diameter 1.50'', hub width 1.00'', face width .50'' and keyway 1/8'' x 1/16''. Compute values for pitch diameter, circular thickness and whole depth. Include these in a table on drawing. One view should be a sectional view.

12. Make an assembly drawing of a spur gear having 48 teeth and a pitch diameter of 3.00'', shaft diameter .625'', face width .75'', and keyway 1/8'' x 1/16''. Also draw a pinion having 24 teeth and a pitch diameter of 1.250''. Other dimensions of pinion are same as for spur gear. Pressure angle of gear and pinion is 20°. Include a table of specifications for gear and pinion on drawing.

13. Make a detail drawing of a bevel gear having 36 teeth, diametral pitch of 12, pressure angle of 20°, face width .53'', shaft diameter 1.00'', hole length 1.25'', mounting distance 1.875'', hub diameter 2.125'' and 1/8'' x 1/16'' keyway. Make a sectional view for clarity and compute values for pitch diameter, circular pitch, whole depth, addendum and dedendum. Include these in a table on drawing.

14. Make an assembly drawing of a 64 tooth bevel gear and a 16 tooth pinion assembled at a 90° shaft angle. Diametral pitch is 16, pressure angle 20°, face width .48''. For gear: shaft size is .625'', hub diameter 2.250'', keyway 1/8'' x 1/16'' and mounting distance (MD) is 1.375''. For pinion: shaft size is .375'', hub diameter .8125'', keyway 1/8'' x 3/64'' and mounting distance is 1.50''. Draw assembly in section and compute values for pitch diameter, circular pitch and whole depth. Include these in a table on drawing.

15. Make an assembly drawing of a worm mesh which has following specifications. For gear: pitch diameter is 5.80'', pressure angle 20°, number of teeth 29, face width 1.375'', shaft diameter 1.250'', hub diameter 2.750'', keyway 1/4'' x 1/8'', outside diameter 6.40''. For worm: pitch diameter is 2.30'', face length 3.0'', shaft diameter 1.125'', hub diameter 1.837'', keyway 1/4'' x 1/8''.

   Note that axial pitch of worm may be found by computing circular pitch of gear. Other values may be found by using formulas given in section on worm gears and worms. Show noncircular view as a sectional view. Include following specifications either in table form or as direct dimensions on drawing.

   For gear: number of teeth, pressure angle, pitch diameter, outside diameter and face width.

   For worm: lead, pitch diameter, outside diameter, length of face and whole depth of thread.

16. Design a gear assembly involving two gears or a worm gear and worm to achieve a definite ratio. Obtain basic specifications for gears from a machinist's handbook or from a gear catalog. Make an assembly drawing of gears and add the necessary dimensions and specifications.

Chapter 20
WORKING DRAWINGS

Working Drawings are drawings that provide all the necessary information to manufacture, construct, assemble or install a machine or structure. Usually, a working drawing is the product of a team of engineers or architects, designers, technicians and drafters who add their special talents to the solution of production problems, Fig. 20-1. Literally thousands of hours go into the preparation of industrial drawings used in modern industrial production.

Fig. 20-1. A typical industrial drafting room where the efforts of many specialists are combined to produce the working drawings essential to manufacturing and construction. (Standard Oil of California)

TYPES OF WORKING DRAWINGS

Working drawings may be divided into a number of subtypes, depending on their use. The first type of working drawing is a freehand sketch; the remainder are instrument drawings to serve various purposes.

FREEHAND SKETCHES OF DESIGN PROTOTYPES

The engineer-designer frequently will prepare a freehand sketch on sectional paper that will serve as the working drawing for tooling, jigs and fixtures or test equipment setups, Fig. 20-2. These drawings are also used on occasion for prototype, or for experimental or research parts and assemblies.

Modifications to the drawing are made during the construction by the designer or technician. Because of the nature and use of these sketches, only the basic information for fabrication of a part or an assembly is included. It should be emphasized that these working drawings are for limited use and are not released for general production.

DETAIL DRAWINGS

A detail drawing describes one part which is to be made from a single piece of material. Information is provided through views and by notes, dimensions, tolerances, material, and finish specifications. Also included are any other requirements sufficient to fabricate, finish and inspect the part.

The views of the part usually are in orthographic projection as normal views or sections. Pictorial views may be included for clarification when necessary. An example of a detail working drawing of a machine part is shown in Fig. 20-3. Some industries permit detailing of several parts on one detail drawing when parts are closely related and space permits.

Before starting a detail drawing of a part, the drafter should study and consider the methods by which the part is to be processed until it is finally assembled on a machine. Sufficient information should be included on the drawing to purchase or make the part and to design the tools used for its manufacture. The drafter should decide how many views will be necessary, and locate the views to allow plenty of space for dimensions and necessary notes.

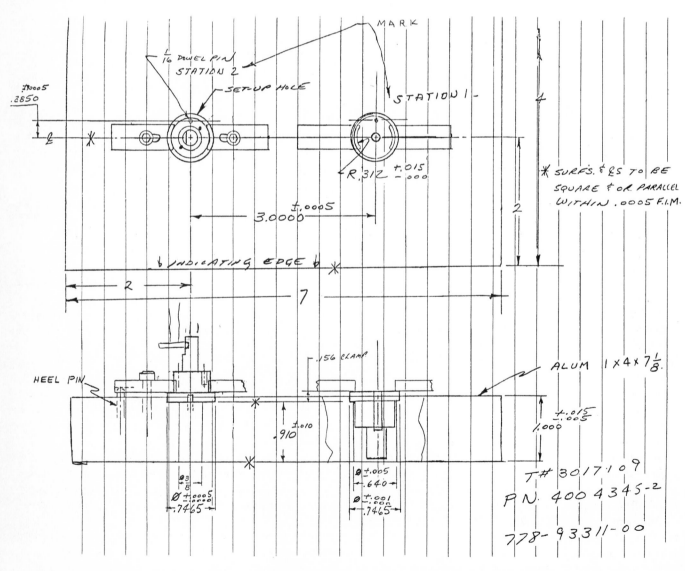

Fig. 20-2. A freehand sketch used as a working drawing in the production of a tooling fixture.
(Sperry Flight Systems Div.)

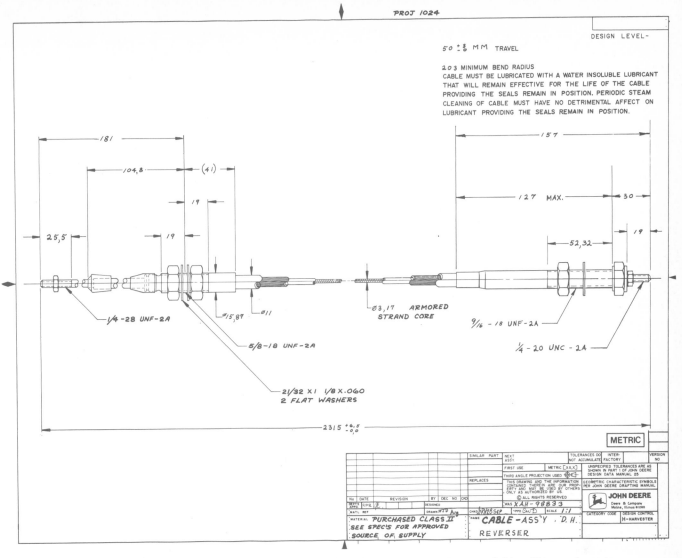

PROJ 1024

DESIGN LEVEL-

50 ± 3/0 MM TRAVEL

203 MINIMUM BEND RADIUS

CABLE MUST BE LUBRICATED WITH A WATER INSOLUBLE LUBRICANT
THAT WILL REMAIN EFFECTIVE FOR THE LIFE OF THE CABLE
PROVIDING THE SEALS REMAIN IN POSITION. PERIODIC STEAM
CLEANING OF CABLE MUST HAVE NO DETRIMENTAL AFFECT ON
LUBRICANT PROVIDING THE SEALS REMAIN IN POSITION.

181

104,8

(41)

19

157

127 MAX.

30

25,5

19

52,32

19

1/4-28 UNF-2A

Ø15,87   Ø11

Ø3,17 ARMORED
STRAND CORE

9/16 - 18 UNF-2A

5/8-18 UNF-2A

1/4 - 20 UNC - 2A

21/32 X1 1/8 X .060
2 FLAT WASHERS

2315 +6,5 -0,0

METRIC

			SIMILAR PART	NEXT ASSY		TOLERANCES DO NOT ACCUMULATE	INTER- FACTORY		VERSION NO
				FIRST USE	METRIC □X,X□	UNSPECIFIED TOLERANCES ARE AS SHOWN IN PART 1 OF JOHN DEERE DESIGN DATA MANUAL 25			
				THIRD ANGLE PROJECTION USED ⊕⊏⊟					
				REPLACES		GEOMETRIC CHARACTERISTIC SYMBOLS PER JOHN DEERE DRAFTING MANUAL			

THIS DRAWING AND THE INFORMATION CONTAINED THEREIN ARE OUR PROPERTY AND MAY BE USED BY OTHERS ONLY AS AUTHORIZED BY US.
© ALL RIGHTS RESERVED

WAS XAH-98833

JOHN DEERE
Deere & Company
Moline, Illinois 61265

No	DATE	REVISION	BY	DEC NO	CKD
DEPT'S APPR.	APR		DESIGNED		
MAT'L REF			DRAWN WJD Aug	CHK'D RKJ 28 SEP	'PPD CWD	SCALE 1:1

MATERIAL PURCHASED CLASS II
SEE SPEC'S FOR APPROVED
SOURCE OF SUPPLY

'NAME CABLE-ASS'Y .D.H.
REVERSER

CATEGORY CODE   DESIGN CONTROL
H-HARVESTER

Fig. 20-3. A detail working drawing provides complete information necessary to fabricate, finish and inspect the part.
(John Deere)

TABULATED DRAWING

A tabulated drawing is a type of detailed working drawing which provides information needed to fabricate two or more items which are basically identical but vary in dimensions, material, finish or other characteristics, Fig. 20-4 (a). The fixed characteristics such as dimensions, materials or finish should be detailed only once, either on the body of the drawing, in the material block or in the tabulation block. Characteristics, such as variable dimensions, are expressed on the drawing with letter symbols, and the value for each symbol is given in the tabulation block. Stock sizes for various parts are given in a materials list, Fig. 20-4 (b).

A tabulated drawing avoids the necessity of preparing separate drawings of parts which are basically alike.

ASSEMBLY DRAWINGS

An assembly drawing depicts the assembled relationship or position of two or more detail parts, or of parts and subassemblies required to comprise a unit, Fig. 20-5. The views of the object are usually orthographic. Isometric or other pictorial views are permissible if needed for clarity. Only those

views, sections and details necessary to adequately describe the assembly should be used. A list of parts is detailed in tabulated form in the materials block on the drawing, or on a separate sheet, and referenced to the parts in the assembly by numbers.

Assembly drawings should be drawn by referring to the detail drawings of the respective parts. This provides an excellent check for fits, clearances and interferences. Only those operations performed on the assembly in the condition shown should be specified. No detail dimensions should be shown on assembly drawings except to cover operations performed during assembly or to locate detail parts in an adjustable assembly.

A special type of assembly drawing is the EXPLODED ASSEMBLY drawing, Fig. 20-6. This type of drawing is most useful in assembling various components. These are usually drawn in pictorial form with an axis line showing the sequence of assembly.

Another type of special assembly drawing is the OUTLINE ASSEMBLY drawing, Fig. 20-7. These drawings are used for the installation of units and provide overall dimensions to show size and location of points necessary in locating and

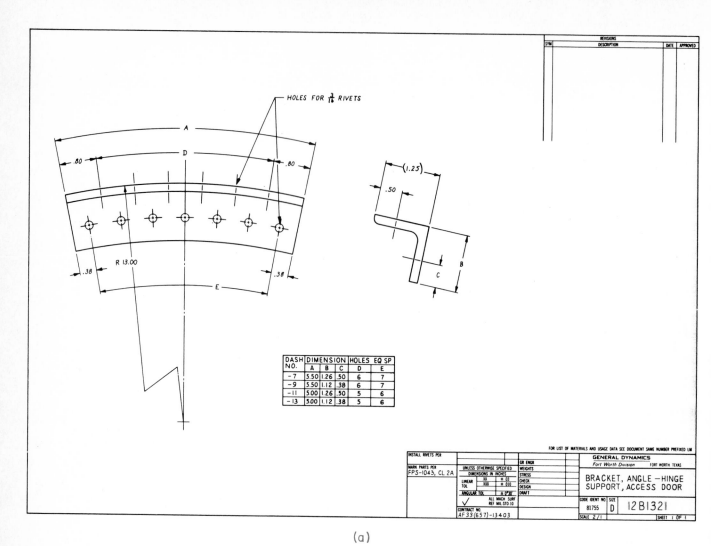

DASH NO.	DIMENSION A	B	C	HOLES EQ SP D	E
-7	5.50	1.26	.50	6	7
-9	5.50	1.12	.38	6	7
-11	5.00	1.26	.50	5	6
-13	5.00	1.12	.38	5	6

(a)

(b)

Fig. 20-4. A tabulated working drawing is used for the fabrication of parts which are nearly identical.
(Convair Aerospace Div., General Dynamics)

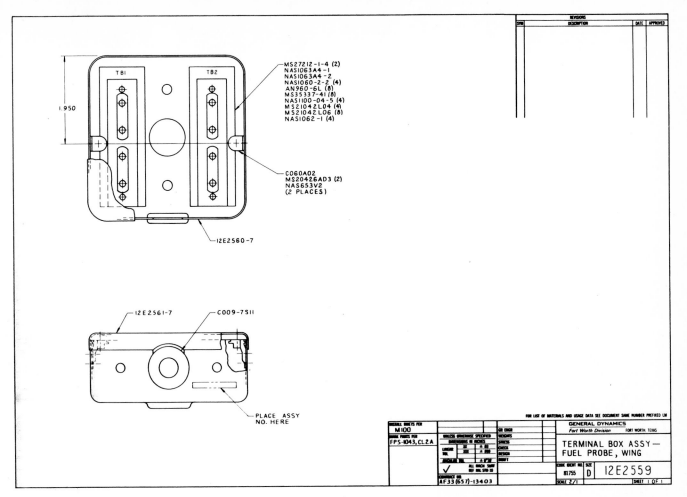

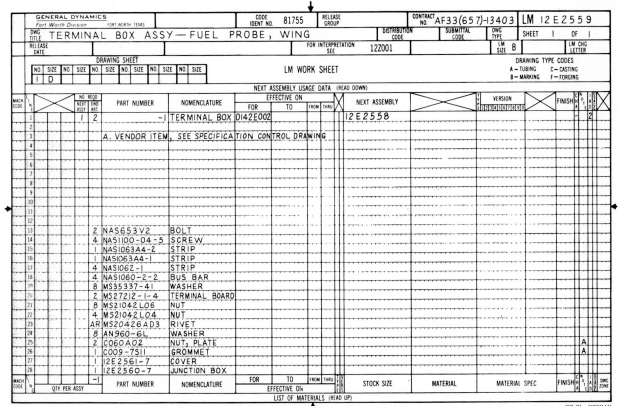

Fig. 20-5. An assembly working drawing and materials list.
(Convair Aerospace Div., General Dynamics)

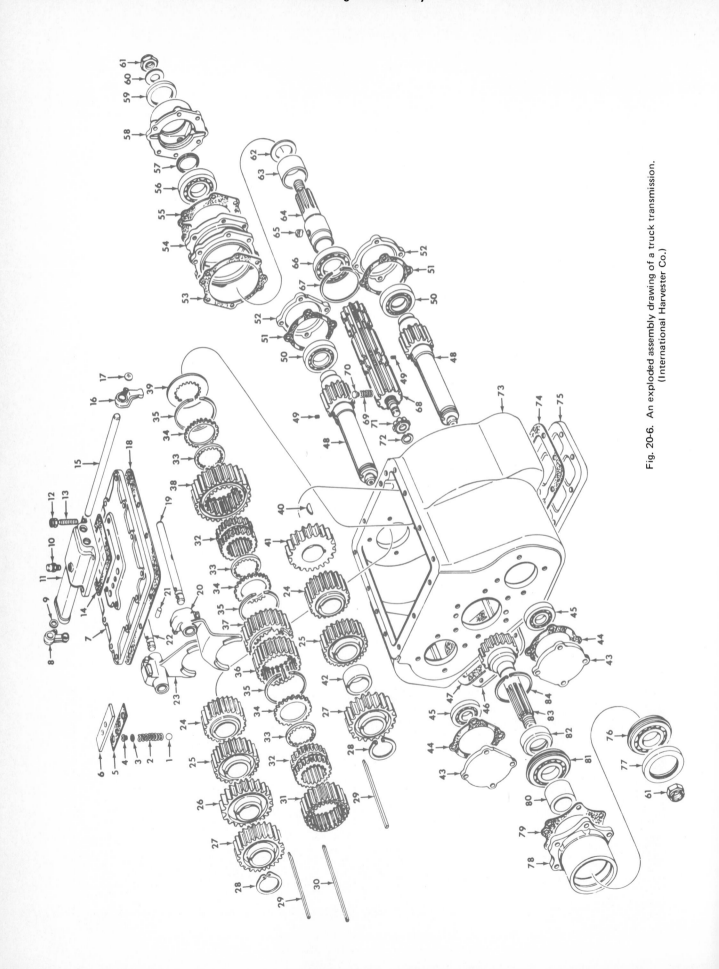

Fig. 20-6. An exploded assembly drawing of a truck transmission. (International Harvester Co.)

DIMENSIONS

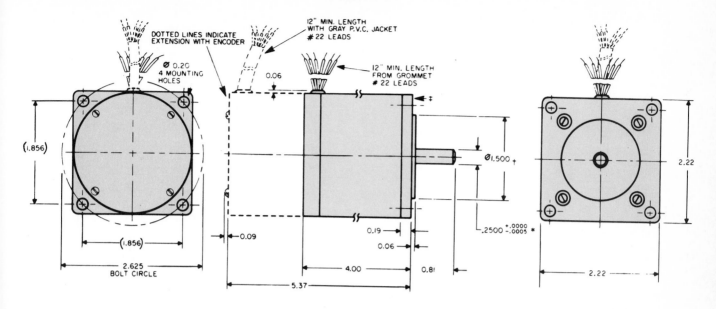

M063 MOTORS

* SHAFT RUNOUT .002 MAX.

† DIAMETER TOLERANCE ±.002
DIAMETER CONCENTRIC TO MOTOR AXIS WITHIN .003 F.I.M.

‡ SURFACE SQUARE TO MOTOR AXIS WITHIN .003 F.I.M.

Fig. 20-7. An outline assembly drawing which provides the necessary dimensions for the installation of a motor on a numerical control machine.   (Superior Electric Co.)

fastening each unit in place. Amount of clearance required to operate and service the unit is given in outline assembly drawing. Dimensions essential to making electrical, air and other connections are provided in outline assembly drawing.

Notes may be included to indicate the weight of the unit, the electrical and cooling requirements, and also include any special notes of caution. These drawings are sometimes termed INSTALLATION drawings.

Complex structures require many working drawings.

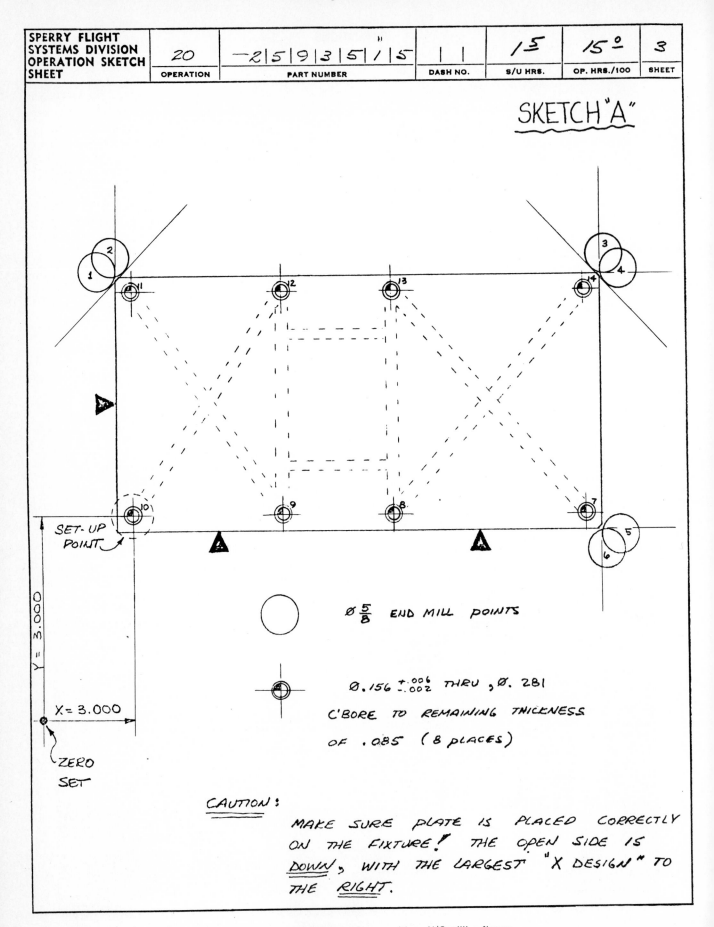

SPERRY FLIGHT SYSTEMS DIVISION OPERATION SKETCH SHEET	20	—2|5|9|3|5|1|5	1 1	1⁵	15°	3
	OPERATION	PART NUMBER	DASH NO.	S/U HRS.	OP. HRS./100	SHEET

SKETCH "A"

SET-UP POINT

Y = 3.000

X = 3.000

ZERO SET

Ø 5/8 END MILL POINTS

Ø.156 +.006 −.002 THRU, Ø.281

C'BORE TO REMAINING THICKNESS

OF .085 (8 PLACES)

CAUTION:

MAKE SURE PLATE IS PLACED CORRECTLY ON THE FIXTURE! THE OPEN SIDE IS DOWN, WITH THE LARGEST "X DESIGN" TO THE RIGHT.

Fig. 20-8. An operation drawing for use with an N/C milling fixture.
(Sperry Flight Systems Div.)

Working Drawings

PROCESS OR OPERATION DRAWINGS

In addition to the principal types of working drawings discussed previously, there is another type that concerns production methods. These drawings usually provide information for only one step or operation in the making of a part, as shown in Fig. 20-8.

These process or operation drawings are used by machine operators when they make particular machine setups and perform single operations such as drilling a hole or milling a slot. These process or operation drawings usually are accompanied by the machine setup specifications and specific steps for performing the operation in the machine shop. The preparation of process drawings and operation sheets is usually done by a person knowledgeable in drafting and machining operations.

LAYOUT DRAWINGS

A layout drawing is often the original conception of a design of a machine or of placement of units. It is not a production drawing, but rather serves to record developing design concepts. It is used to obtain approval of a particular design or to check clearances and interrelation of component parts, Fig. 20-9. Layout drawings are used by experimental shops in the construction of models or prototypes, and by design drafters as a reference in preparing detail drawings of various parts.

Although layout drawings may look like assembly type drawings, the purposes of the two are entirely different. Layout drawings are used in the early concept and design stage of a product, while assembly drawings are used near the end for final fabrication of the product.

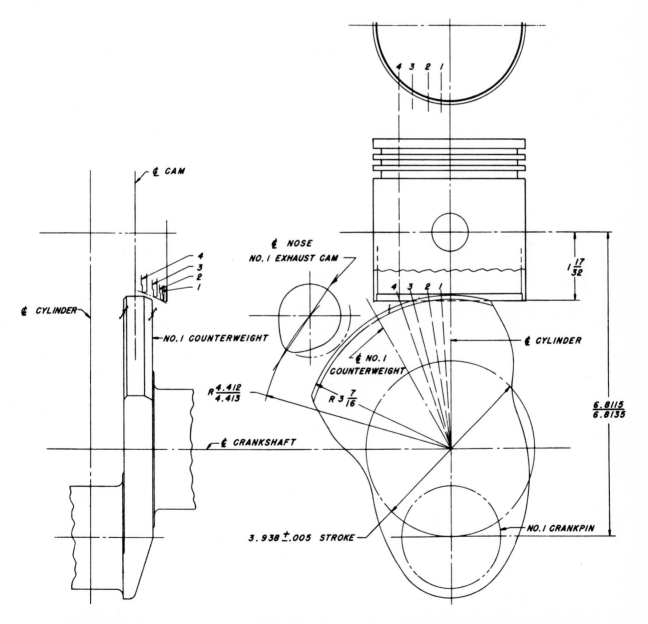

Fig. 20-9. An industry layout drawing to determine the largest radius crankshaft counterweight which can be used and still maintain clearance between the counterweight and the cam and piston.   (General Motors Engineering Standards)

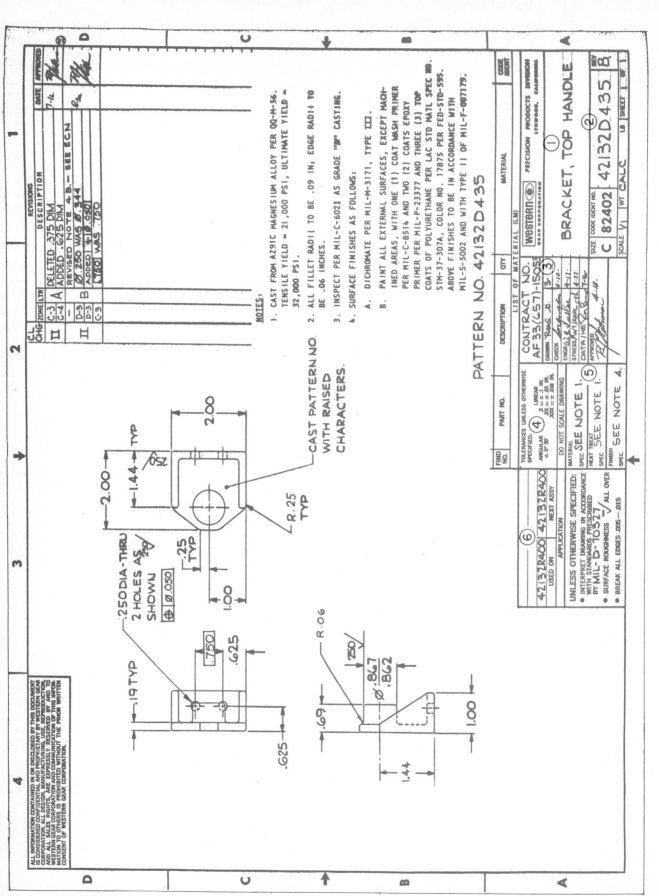

Fig. 20-10. Elements of a drawing title block. (Western Gear Corp.)

REQ'D	REQ'D	ITEM	PART NO.	DESCRIPTION	SPECIFICATION	NOTES
REF		11	500040-1	WIRE LIST (DSTS)		
122		10	270193-1	CLIP, TERMINAL		
AR		9	220226-9	WIRE, ELECTRICAL		
✕	1	8	750098-2	ELECTRONIC CMPNT ASSY, CONTROL		
	1	7	803239-1	CKT CARD ASSY, BCD TO DEC CONV		
	1	6	803235-1	CKT CARD ASSY, BCD TO DEC CONV		
	1	5	803231-1	CKT CARD ASSY, BCD TO DEC CONV		
	1	4	803230-1	CKT CARD ASSY, FIXED DATA CONV		
4		3	MS51957-21	SCREW, MACHINE-PAN HEAD	FF-S-92	
1		2	750044-1	CONNECTOR, PLATE ASSY		
	1	1	850064-1	BRACKET ASSY, CONTROL		
-2	-1			LIST OF PARTS		

Fig. 20-11. A materials block lists all of the parts required to complete the assembly shown on the drawing.

FORMATS FOR WORKING DRAWINGS

All formal industrial drawings have a title block, materials block and change block in which essential information is recorded in an organized manner. Certain basic information is common to these blocks in nearly all industries. The style and location of the block on the drawing may vary with individual industries. The following sections discuss the information usually recorded in these blocks.

TITLE BLOCK

The title block, Fig. 20-10, usually is placed in the lower right-hand corner of the drawing. It includes ① the title (name) of the part or assembly, ② the drawing number (same as the part number), ③ name of the drafter and checker, and signatures of those responsible for engineering, materials and production approvals. Other items usually given in the title block are ④ general tolerances, ⑤ specifications for material, heat treatment and finish, ⑥ an application block that indicates the equipment in which the part or assembly is used.

MATERIALS BLOCK

The materials block is a tabular form that usually appears immediately above the title block on assembly and installation drawings, Fig. 20-11. This block is sometimes called a Parts List, List of Materials, Bill of Materials or Schedule of Parts. This block lists the different parts which go into the assembly shown on the drawing, the quantity of each part required, name or description of the part and specification of material. If the part is to be purchased, the supplier should be identified by name or identification code in the materials block.

CHANGE BLOCK

Once prints of a drawing have been released to production, it is sometimes necessary to make changes because of design improvement, production problems or errors found in the drawing. All changes to the original drawing must have approval of the proper authority.

When changes have been approved and made on the drawing, a record is made in the change block, Fig. 20-12. A brief description of the change is given, an identifying letter referencing it to the specific location on the drawing, the date, initials or signature of drafter making the change, and those of the person approving the change.

The change block is usually located in the upper right-hand corner of the drawing and is sometimes titled Alterations, Notice of Change or Revisions.

INDUSTRIAL STANDARDS FOR DRAFTING

All industrial drawings should be drawn according to the National Standards published for that industry. These standards are available from the American National Standards Institute and are prepared by various industrial groups. The American Society of Mechanical Engineers has prepared the Y.14 series of standards for drafting. Most companies have their own drafting standards, contained in a Drafting Room

LAYOUT OR REF. NO. AM-89303 #44, AM-94524 #44					
CHK.	DATE	CHANGE	REV.	RELEASE NO.	
CP 46.8	14OCT	X-6643-4062 DIM 'A' WAS 8, DIM 'D' WAS 3.4, PROD N° ADDED	A	27990 T	
CP 8.0	10AUG	W 4" & WERE CI: B DIMS WERE 8.9 W 5" PIPE & D DIMS WERE 3.0 W 4" & 3.2 W 5" PIPE	B	65030-A " "	

Fig. 20-12. The change block is a record of changes made to the original drawing. (International Harvester Co.)

Manual, which are in conformance with the National Standards with modifications or additions to fit the particular needs of that industry.

There is a strong trend toward the adoption of certain international symbols (see page 597 of the Appendix), particularly by those companies engaged in international marketing.

INDUSTRIAL DESIGN AND DRAFTING PROCESS

The production of an industrial drawing begins with an expressed need for something to be produced, Fig. 20-13. This need is given to the design department where the concept is developed, researched, and where original designs of the product are sketched or drawn. The designers give the sketches to their drafters for further development of the design ideas in layout drawings, or for preparation of detail and assembly drawings. Prints are made of these original drawings and are carefully checked by experienced design drafters for details in meeting the original need.

When the drawing meets with the approval of design engineers, cost analysts, production engineers and management, it is released to the reproduction department where prints are made. These are sent to production where tool designers

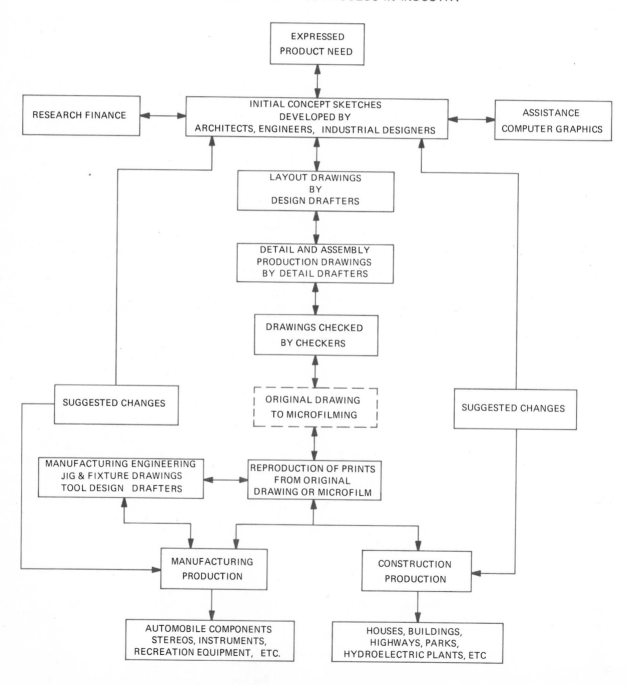

Fig. 20-13. The industrial drafting process from product conception to production.

prepare drawings for the jigs and fixtures to be used with the machines in producing the product. Once prints of the production drawings are in use, changes may be made to the prints only by the design department under authority of the project engineer for the product.

APPLICATIONS OF WORKING DRAWINGS

The working drawing is the vital link of communication in industry, making it possible to produce individual parts of a machine in widely separated plants. Some applications of working drawings in major industries are discussed in the following sections.

AEROSPACE DRAFTING

Because of the product manufactured, aerospace drawings are perhaps the most elaborate of those of any industry. Precision dimensioning, tolerancing and rigid specifications characterize these drawings, since many of the parts and components are made by subcontracting industries and brought together for assembly by the prime contractor.

Literally thousands of detail and assembly working drawings are used in the production of one model of an aerospace vehicle, Fig. 20-14. Every type of drawing is employed in the aerospace industry.

Fig. 20-14. Launching of an aerospace vehicle is made possible by our advanced technology and ability to communicate this knowledge through working drawings. (NASA)

AUTOMOTIVE DRAFTING

The automotive industry, together with its subcontractors, constitutes one of the largest industrial groups in America. Perhaps more has been done in this industry to perfect the drafting process than any other. As in the aerospace industry, every type of drawing is employed in the production of automobiles and trucks. Most of these industries have produced extensive drafting room procedures in supplement to the National Standards in order to perfect the communication process.

The automotive industry is also one of the leaders in the use of computer control which necessitates special drafting procedure, Fig. 20-15.

Fig. 20-15. A computer-integrated manufacturing system controls movement of these car chassis. To program these machines requires a skilled programmer and the use of many detail working drawings. (Cincinnati Milacron)

ARCHITECTURAL DRAFTING

Architects, engineers, technicians and construction workers rely heavily on the working drawing in planning and construction of a residence or commercial building. The nature and types of architectural drawings are discussed in Chapter 22. This is a field in which creative design and individuality are expressed in nearly every well planned structure. In order to achieve the desires of the owner and the design solution planned by the architect, careful working drawings and specifications are necessary.

ELECTRICAL-ELECTRONICS DRAFTING

Working drawings in the electrical industry differ from those in other industries as discussed and illustrated in Chapter 23. Instead of detail and assembly drawings of machine parts, various diagrams and line drawings with numbers of graphic symbols are used. The complex circuitry of modern electronic devices such as computers, process controllers, aerospace vehicles as well as that of many household appliances has increased the need for drawings in the electrical industries.

FORGING AND CASTING DRAWINGS

Forging is the forming of heated metal by a hammering or squeezing action, as discussed in Chapter 17. A forged part should be shown on one drawing with the outline shown in phantom lines, Fig. 20-16. Forging outlines for machining should not be dimensioned, other than a note indicating the allowance, unless the amount of finish cannot be controlled by

the machining symbol. Where the forging is complex, and where the outline of the rough forging must be maintained for tooling purposes, separate dimensioned drawings for the forged and machined piece should be made. Material specification and heat treatment must be called out on the drawing and the part number indicated and located on the piece.

In the casting drawing, the rough and machined versions of the casting should be combined in the same views on one drawing. The material to be removed by machining is shown in phantom, Fig. 20-17. For complex castings, where the combined rough and machined castings in a single view would cause confusion, separate dimensioned drawings showing the rough casting and machined piece should be drawn on one

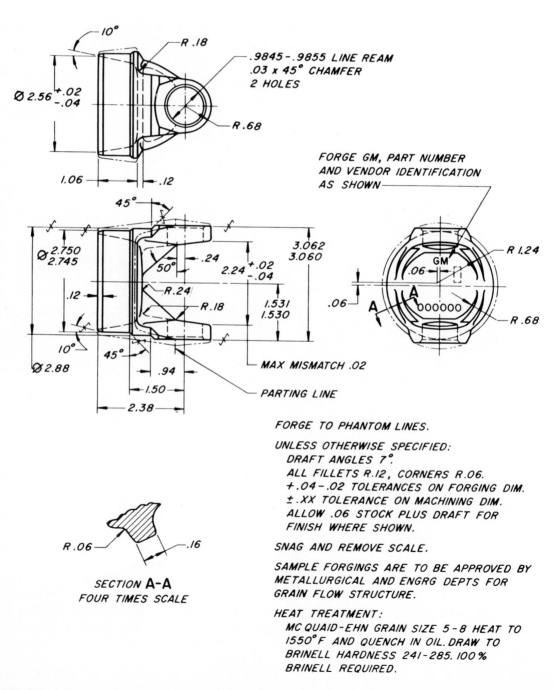

FORGE TO PHANTOM LINES.

UNLESS OTHERWISE SPECIFIED:
DRAFT ANGLES 7°.
ALL FILLETS R.12, CORNERS R.06.
+.04 −.02 TOLERANCES ON FORGING DIM.
±.XX TOLERANCE ON MACHINING DIM.
ALLOW .06 STOCK PLUS DRAFT FOR
FINISH WHERE SHOWN.

SNAG AND REMOVE SCALE.

SAMPLE FORGINGS ARE TO BE APPROVED BY
METALLURGICAL AND ENGRG DEPTS FOR
GRAIN FLOW STRUCTURE.

HEAT TREATMENT:
MC QUAID − EHN GRAIN SIZE 5 − 8 HEAT TO
1550°F AND QUENCH IN OIL. DRAW TO
BRINELL HARDNESS 241 − 285. 100%
BRINELL REQUIRED.

Fig. 20-16. A composite forging drawing with machining dimensions given and the rough forging shown in phantom lines. (General Motors Corp.)

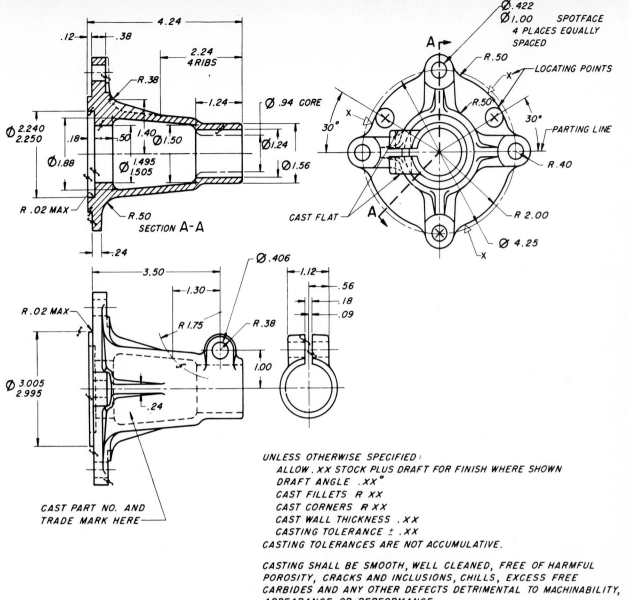

Fig. 20-17. A composite casting drawing with finished machining dimensions given and locating points identified.
(General Motors Corp.)

sheet when possible. A detail drawing of a casting should give complete information on the following:

Material specification.
Hardness specification if required.
Machining allowances.
Kind of finish.
Draft angles.
Limits on draft surfaces that must be controlled.
Locating points for checking the casting.
Parting line.
Part number and trademark.

PIPING DRAWINGS

Piping drawings are a type of assembly drawing using either double-line or single-line drawings to represent the piping layout. They use either pictorial or graphic symbols to represent the various fittings. The usual practice is to show piping layouts as single-line drawings in isometric or oblique because of the difficulty of reading orthographic projection views. A typical piping layout is shown in Fig. 20-18.

STRUCTURAL STEEL DRAWINGS

Structural steel drawings are of two types: design drawings, and shop or working drawings. Design drawings, which show the overall design and dimension of the structure and specify sizes and types of material used are prepared by structural engineers. Working drawings for the actual fabrication of steel members are prepared by the fabricator under the direction and approval of the design engineer. Working drawings which are sent to the job site for erection purposes must detail

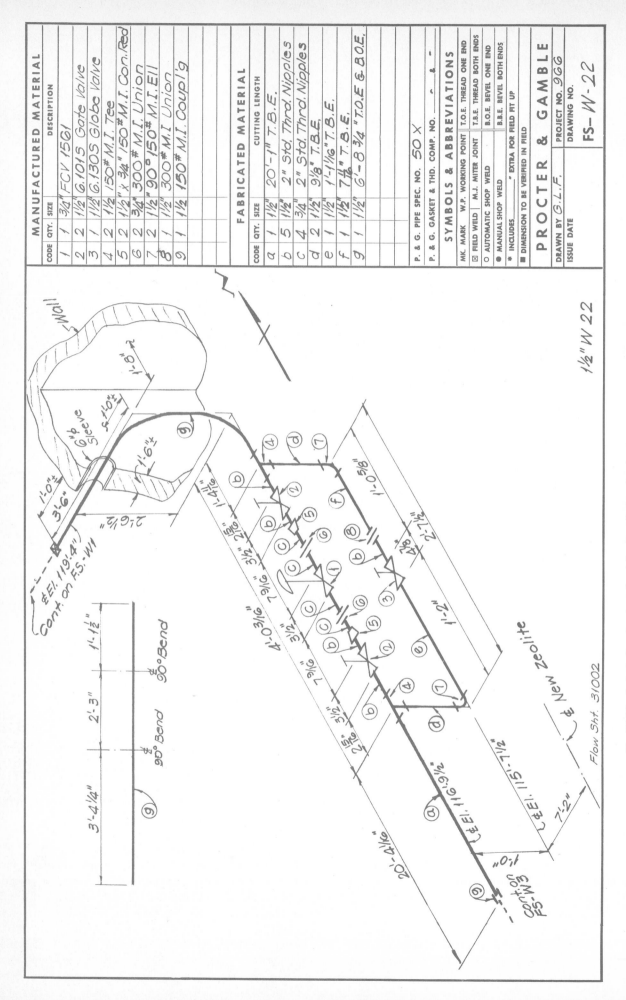

Fig. 20-18. A single-line drawing of a piping installation showing location of valves and other fittings. (Proctor & Gamble Co.)

412

Fig. 20-19. This large steel girder has been fabricated as a subassembly in the shop and shipped to the job site. (Kaiser Steel)

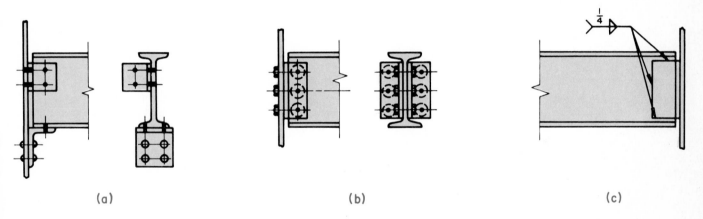

(a)   (b)   (c)

Fig. 20-20. Methods of illustrating steel beam connections with (a) rivets, (b) bolts and (c) welds.

connections which are to be made in the field.

Units shipped to the job site from the fabricator are called subassemblies, Fig. 20-19. These are fastened together on the job either by rivets or bolts or are welded in place, Fig. 20-20.

REINFORCED CONCRETE DRAWINGS

Reinforced concrete construction is achieved by placing steel reinforcing rods or beams strategically in the forms and pouring the concrete around these, Fig. 20-21. This construction is very strong structurally and is fire resistant.

Drawings for reinforced concrete construction must show the dimension and shape of the concrete members, and the

Fig. 20-21. Steel reinforcing rods are used
to form reinforced concrete members.
(Allison Steel)

size and location of the reinforcing steel, Fig. 20-22. The symbol for the reinforcing rod is the Greek letter phi (ϕ). The rods are represented by long dashes in the elevation view and as darkened circles in the sectioned view. The American Concrete Institute publishes an approved Manual of Standard Practice for Detailing Reinforced Concrete Structures which is helpful in preparing drawings for reinforced concrete structures such as buildings, bridges and walls.

accuracy of the drawing nor the quality of lettering and line delineation.

Every industry is continually seeking ways of improving their drafting communications. This is caused by the amount of drafting time involved in planning and development of most industrial production projects, and time spent in reading and interpreting industrial drawings by industrial personnel. Functional drafting techniques have been successful in reducing

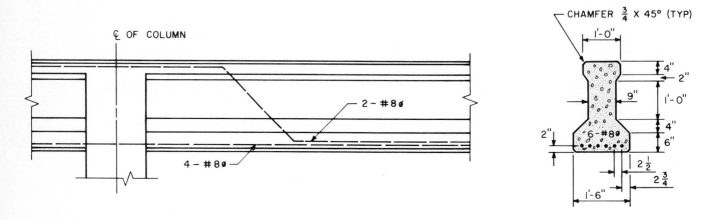

Fig. 20-22. A detail of a reinforced concrete member showing size and location of steel reinforcing rods.

WELDING DRAWINGS

The welding drawing is a type of assembly working drawing showing the various parts or members of an assembly in position to be welded rather than as separate parts. Specification of the type of welds to be used on various joints has become standard procedure on welding drawings. A series of welding symbols has been prepared by the American Welding Society. These symbols and the procedure for preparing welding drawings are discussed in Chapter 24.

PATENT DRAWINGS

When a patent is sought for a machine or other device which lends itself to illustration by a drawing, the applicant is required to furnish the drawing. Very specific instructions on the preparation of patent drawings are available from the Superintendent of Documents, Washington D.C. These include the type of paper and ink, size, line density, etc., and the manner in which the drawings are to be sent to the Patent Office.

Patent drawings follow the American Drafting Standards, but it is not necessary that the views appear in a strict orthographic projection relation to one another or even be placed on the same sheet. Isometric or other pictorial views are acceptable if they clearly represent the object for which a patent is desired.

FUNCTIONAL DRAFTING TECHNIQUES

Functional drafting may be defined as making a drawing which includes the essential lines, views, symbols, notes and dimensions to completely clarify the construction of an object or part. Application of this technique should not reduce the

drafting time in industry and have been welcomed in most industries.

The following functional drafting techniques have been discussed in other sections of this text as standard drafting practices. They are listed here as a summary of standard practices.

1. Use minimum number of views.
2. Use partial view where adequate.
3. Eliminate a view where thickness note will suffice.
4. Use symmetry to reduce drawing time.
5. Eliminate superfluous detail such as detail thread representation; use schematic or simplified form.
6. Omit drawing of standard parts such as bolts, nuts and rivets. Locate them conventionally or by a note, and list them in the materials block.
7. Avoid unnecessary repetition of detail.
8. Omit cross hatching except where clarity demands its use; use outline sectioning.
9. Use standard graphic symbols such as piping and welding symbols.
10. Use templates to draw ellipses, circles, symbols, etc.
11. Save time by using mechanical lead holders, pencil pointers and thin-lead holders.
12. Photograph usable sections of drawings to be reworked and start with this.
13. Use photo-drafting of models, electronic circuits and other assemblies to which notes can be added to speed the drafting process.
14. Use adhesive-backed appliques for information used repeatedly on drawings such as change blocks and gear data blocks.
15. Use tabulated semicomplete drawings of common shaped

items which require only the addition of dimensions and/or specifications.

16. Use a general tolerance note in title block for indicating tolerances on drawing when possible.

17. Avoid freehand lettering of parts list and other material which can be typed on adhesive-backed appliques.

PROBLEMS IN PREPARING WORKING DRAWINGS

The problems in this chapter are to be drawn and should included all dimensions, specifications and notes required to release the drawing to production.

DETAIL WORKING DRAWINGS

1. Make detail working drawings of those parts in Fig. 20-23 as assigned by your instructor. Change all dimensions to decimal limit dimensions with the following tolerances: .XXX = ± .003; .XX = ± .010. Delete all unnecessary dimensions; indicate all flat surfaces as 125 microinches (3.2 micrometers) and all bored and counterbored holes as 63 microinches (1.6 micrometers) in texture.

2. Prepare a three-view detail working drawing of the BODY PITOT OVERRIDE, Fig. 20-24, with one view as a section to clarify interior detail. Dimension the drawing using geometric dimensioning and tolerancing. Delete the notes where they are replaced by geometric dimensioning symbols and add the remaining necessary notes.

3. Draw two views, one a section, of the HYDRAULIC DECHUCK PISTON, Fig. 20-24. Change the number of equally spaced holes from 24 to 18 and dimensions for N/C machining. Use chart in Reference Section on page 596 for calculating equally spaced locations. Dimension drawing, using geometric feature control symbols.

PRECISION SHEET METAL DRAWINGS

4. Refer to Figs. 20-25 and 20-26 and calculate the flat pattern size and make flat pattern dimensioned drawings of those parts shown.

ASSEMBLY WORKING DRAWINGS

5. Make a detail working drawing of the ROLLER FOR BRICK ELEVATOR, Fig. 20-27. Add bilateral tolerances to those parts requiring fits as indicated in the notes.

6. Make an exploded assembly drawing of the ROLLER FOR BRICK ELEVATOR, Fig. 20-27.

7. Make detail drawings of the component parts and an assembly drawing of the WHEEL, IMPELLER, Fig. 20-28. Include all notes and material list.

8. Prepare detail working drawings of the various parts of the STEERING QUADRANT ASSEMBLY, Fig. 20-29. Add material and revision items to the individual detail drawings either as callouts or notes. The drawings may be placed on one sheet or on separate sheets.

9. Make an assembly drawing of the STEERING QUADRANT ASSEMBLY, Fig. 20-29. Add material and revision blocks and callouts to your drawing.

10. Make detail drawings of component parts (except standard bolts, nuts and washers) and an assembly drawing of WIRE STRAIGHTENER, Figs. 20-30 and 20-31. Include all notes and material list.

DESIGN PROBLEMS

The following are suggested problems which can be solved by the design method discussed in Chapter 9. With the approval of your instructor, you may select any of these or one of your own for which you desire a solution. Prepare detail and assembly working drawings of the solutions.

1. Stereo component cabinets.

2. Quick-action, easy operated automobile tire jack.

3. Water safety device useful in hunting or fishing.

4. Desk caddy to help organize paper clips, rubber bands, pens, pencils, notes and other items used at your desk.

5. Storage device for your favorite sporting equipment.

6. Jig or fixture for holding a workpiece in a machine tool.

7. Special tool or clamp which combines the function of two or more separate tools.

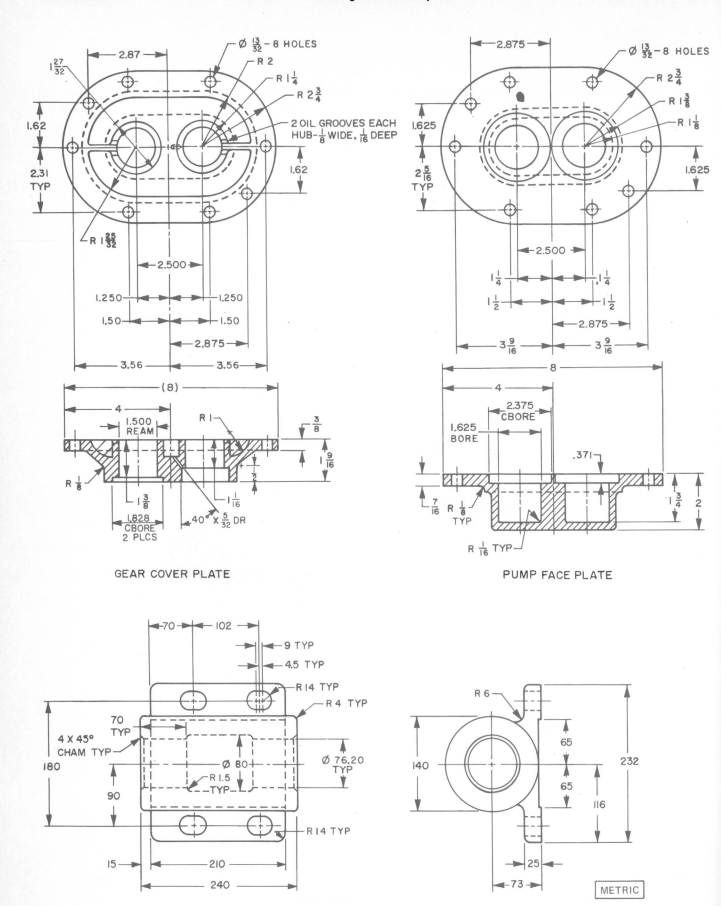

GEAR COVER PLATE

PUMP FACE PLATE

ROLLER SHAFT HOUSING

METRIC

Fig. 20-23. Parts for which detail working drawings are to be prepared.

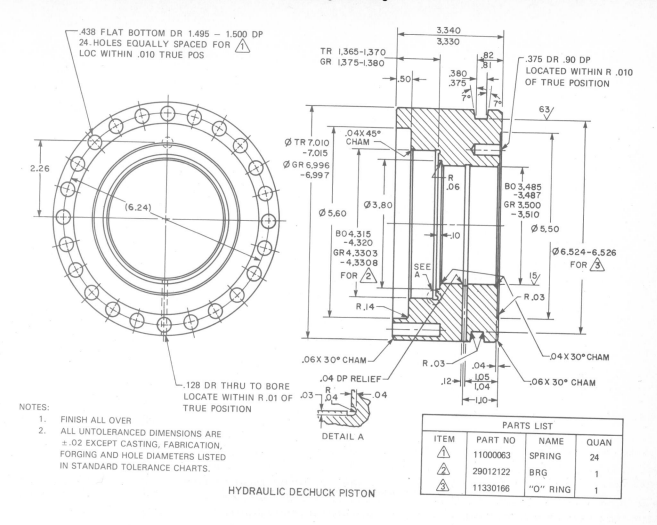

.438 FLAT BOTTOM DR 1.495 – 1.500 DP
24 HOLES EQUALLY SPACED FOR △1
LOC WITHIN .010 TRUE POS

2.26

(6.24)

.128 DR THRU TO BORE
LOCATE WITHIN R .01 OF
TRUE POSITION

TR 1.365-1.370
GR 1.375-1.380

.50

Ø TR 7.010
-7.015
Ø GR 6.996
-6.997

Ø 5.60

BO 4.315
-4.320
GR 4.3303
-4.3308
FOR △2

.04 X 45°
CHAM

Ø 3.80

R .14

.06 X 30° CHAM

.04 DP RELIEF
R .04   .04
.03

DETAIL A

3.340
3.330

.82
.81

.380
.375

7°   7°

R
.06

.10

R .03

63

.375 DR .90 DP
LOCATED WITHIN R .010
OF TRUE POSITION

BO 3.485
-3.487
GR 3.500
-3.510

Ø 5.50

Ø 6.524-6.526
FOR △3

15

R .03

.04
1.05
1.04

.12   1.10

.04 X 30° CHAM

.06 X 30° CHAM

SEE
A

NOTES:
1. FINISH ALL OVER
2. ALL UNTOLERANCED DIMENSIONS ARE
   ±.02 EXCEPT CASTING, FABRICATION,
   FORGING AND HOLE DIAMETERS LISTED
   IN STANDARD TOLERANCE CHARTS.

HYDRAULIC DECHUCK PISTON

PARTS LIST			
ITEM	PART NO	NAME	QUAN
△1	11000063	SPRING	24
△2	29012122	BRG	1
△3	11330166	"O" RING	1

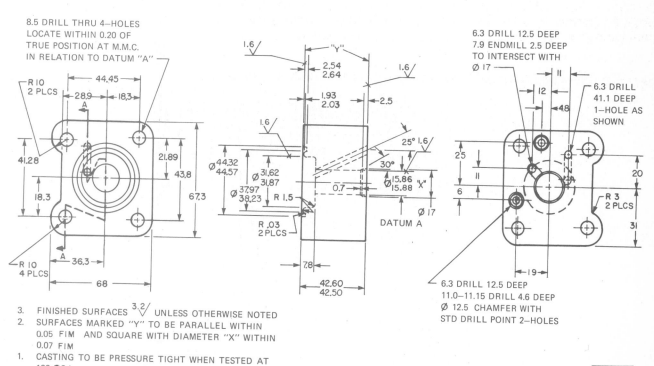

8.5 DRILL THRU 4 HOLES
LOCATE WITHIN 0.20 OF
TRUE POSITION AT M.M.C.
IN RELATION TO DATUM "A"

R 10
2 PLCS

44.45

28.9   18.3

A

41.28

21.89

43.8

18.3

67.3

R 10
4 PLCS

A   36.3

68

1.6

"Y"

2.54
2.64

1.93
2.03

1.6

1.6

Ø 44.32
44.57

Ø 31.62
31.87

Ø 37.97
38.23

R 1.5

R .03
2 PLCS

7.8

42.60
42.50

1.6

2.5

25° 1.6

30°

0.7

Ø 15.86 "X"
15.88

Ø 17

DATUM A

6.3 DRILL 12.5 DEEP
7.9 ENDMILL 2.5 DEEP
TO INTERSECT WITH
Ø 17

11

12

4.8

6.3 DRILL
41.1 DEEP
1 HOLE AS
SHOWN

25

11

6

20

R 3
2 PLCS

31

19

6.3 DRILL 12.5 DEEP
11.0–11.15 DRILL 4.6 DEEP
Ø 12.5 CHAMFER WITH
STD DRILL POINT 2 HOLES

3. FINISHED SURFACES ³·² √ UNLESS OTHERWISE NOTED
2. SURFACES MARKED "Y" TO BE PARALLEL WITHIN
   0.05 FIM AND SQUARE WITH DIAMETER "X" WITHIN
   0.07 FIM
1. CASTING TO BE PRESSURE TIGHT WHEN TESTED AT
   100 P.S.I.
NOTES:

BODY PITOT OVERRIDE

UNLESS OTHERWISE SPECIFIED:
ALL DIMENSIONAL TOLERANCES ±0.25
ALL ANGULAR TOLERANCES ±1°

METRIC

Fig. 20-24. Parts for which detail working drawings are to be prepared.

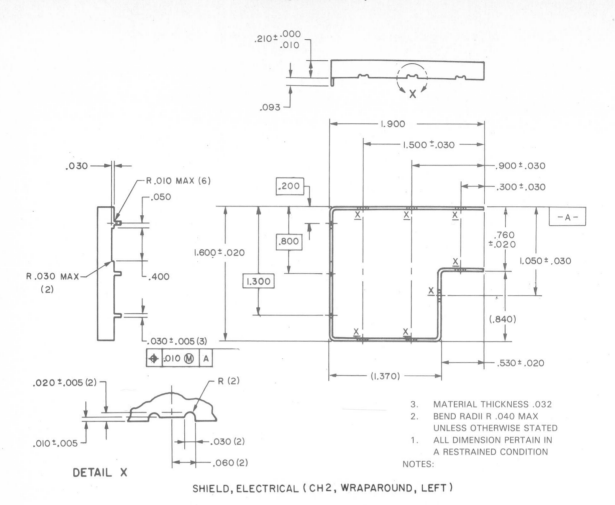

SHIELD, ELECTRICAL (CH 2, WRAPAROUND, LEFT)

3.   MATERIAL THICKNESS .032
2.   BEND RADII R .040 MAX
     UNLESS OTHERWISE STATED
1.   ALL DIMENSION PERTAIN IN
     A RESTRAINED CONDITION
NOTES:

DETAIL X

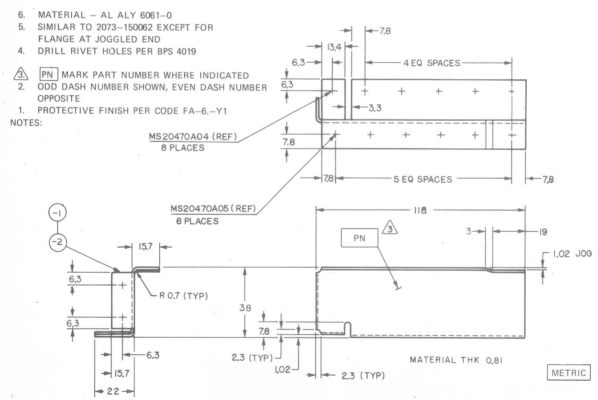

6.   MATERIAL — AL ALY 6061—0
5.   SIMILAR TO 2073—150062 EXCEPT FOR
     FLANGE AT JOGGLED END
4.   DRILL RIVET HOLES PER BPS 4019

⚠3. PN MARK PART NUMBER WHERE INDICATED
2.   ODD DASH NUMBER SHOWN, EVEN DASH NUMBER
     OPPOSITE
1.   PROTECTIVE FINISH PER CODE FA—6.—Y1
NOTES:

CLIP—SHEAR UPPER WATERTIGHT BULKHEAD HYDROSKIMMER

Fig. 20-25. Precision sheet metal parts for which flat pattern dimensioned drawings are to be prepared.

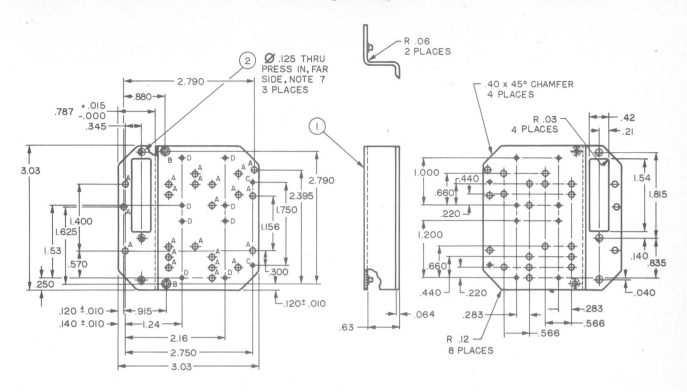

3	X	MIL−C−5541, GR C, CL 1	M690278−1	CHEMICAL FILM TRTMT
2			1752101−11	NUT, SELF LOCKING
1	1		4005071−2	CHASSIS
ITEM NO.	QTY REQD		STANDARD	NAME OR DESCRIPTION
LIST OF MATERIALS				

HOLE TABLE		
HOLE CODE	HOLE SIZE	QTY REQD
A	$\varnothing$ .125 THRU	22
B	$\varnothing$ .125 THRU 100° CSK TO $\varnothing$ .204	2
C	$\varnothing$ .096 $^{+.004}_{-.001}$ THRU	2
D	$\varnothing$ .116 $^{+.004}_{-.001}$ THRU	8

NOTES:
1. INTERPRET DRAWING PER ANSI Y 14.5M−1982
2. REMOVE BURRS AND SHARP EDGES, UNLESS OTHERWISE SPECIFIED
3. FINISH ALL OVER 125
4. DIMENSIONS, TOLERANCES AND SURFACE FINISH VALUES APPLY BEFORE THE APPLICATION OF THE FINISH
5. DO NOT APPLY PIECE MARK
6. FINISH AS FOLLOWS:
   6.1   CHEMICAL FILM TREATMENT, ITEM 1 ONLY, PER ITEM 3
7. DO NOT CHAMFER OR BREAK ENTRANCE EDGE OF MOUNTING HOLE RECEIVING INSERT. PRESS IN UNDER STEADY LOAD UNTIL UNDERSIDE OF SHOULDER IS FLUSH WITHIN .005 OF ADJACENT SURFACE
8. TOLERANCES, UNLESS OTHERWISE SPECIFIED:
   2 PLACE DECIMALS = ±.02
   3 PLACE DECIMALS = ±.005
   ANGLES = ±2°
9. MATERIAL THICKNESS .064

CHASSIS   ASSEMBLY

Fig. 20-26. Precision sheet metal part for which a flat pattern dimensioned drawing is to be prepared.

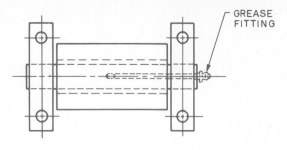

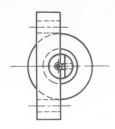

GREASE
FITTING

ASSEMBLY VIEWS

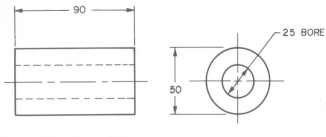

90

50

25 BORE

ROLLER – C.R.S.

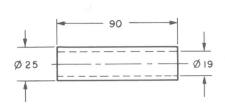

90

Ø 25

Ø 19

BUSHING – BRONZE

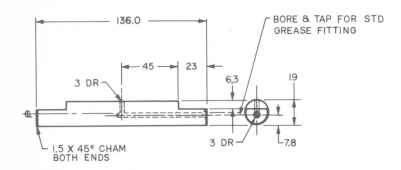

136.0

45   23

3 DR

6.3

19

BORE & TAP FOR STD
GREASE FITTING

1.5 X 45° CHAM
BOTH ENDS

3 DR

7.8

SHAFT – C.R.S.

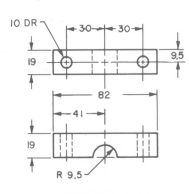

10 DR

30   30

19

9.5

82

41

19

R 9.5

KEEPER – C.R.S.

NOTES:
2. FINISH 125/ ALL OVER
1. BUSHING TO BE A LIGHT DRIVE FIT
    IN ROLLER & RUNNING FIT ON SHAFT

METRIC

ROLLER FOR BRICK ELEVATOR

Fig. 20-27. A subassembly for which an assembly drawing and an exploded assembly drawing are to be prepared.

Working Drawings

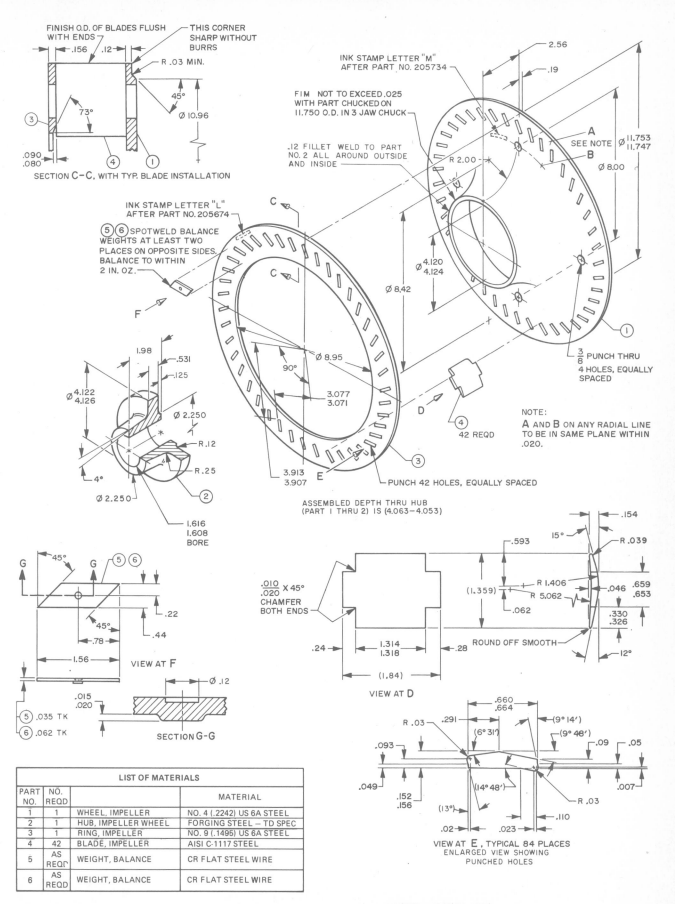

SECTION C-C, WITH TYP. BLADE INSTALLATION

FINISH O.D. OF BLADES FLUSH WITH ENDS

THIS CORNER SHARP WITHOUT BURRS

R .03 MIN.

45°

73°

Ø 10.96

.156 .12

.090 .080

INK STAMP LETTER "M" AFTER PART NO. 205734

FIM NOT TO EXCEED .025 WITH PART CHUCKED ON 11.750 O.D. IN 3 JAW CHUCK

.12 FILLET WELD TO PART NO. 2 ALL AROUND OUTSIDE AND INSIDE

2.56

.19

R 2.00

A SEE NOTE B

Ø 11.753 11.747

Ø 8.00

Ø 4.120 4.124

INK STAMP LETTER "L" AFTER PART NO. 205674

⑤ ⑥ SPOTWELD BALANCE WEIGHTS AT LEAST TWO PLACES ON OPPOSITE SIDES. BALANCE TO WITHIN 2 IN. OZ.

Ø 8.42

Ø 8.95

90°

3.077 3.071

3.913 3.907

1.98

.531

.125

Ø 4.122 4.126

Ø 2.250

R .12

4°

R .25

Ø 2.250

1.616 1.608 BORE

② ③ ④

3/8 PUNCH THRU 4 HOLES, EQUALLY SPACED

42 REQD

NOTE:
A AND B ON ANY RADIAL LINE TO BE IN SAME PLANE WITHIN .020.

PUNCH 42 HOLES, EQUALLY SPACED

ASSEMBLED DEPTH THRU HUB (PART I THRU 2) IS (4.063-4.053)

VIEW AT F

45°

⑤ ⑥

.22

.44

.78

1.56

.015 .020

⑤ .035 TK
⑥ .062 TK

SECTION G-G

Ø .12

.010 .020 X 45° CHAMFER BOTH ENDS

.593

.062

.24

1.314 1.318

.28

(1.84)

VIEW AT D

.154

15°

R .039

R 1.406

R 5.062

.046

.659 .653

.330 .326

12°

(1.359)

ROUND OFF SMOOTH

.660 .664

R .03

.291

(6° 31')

(9° 14')

(9° 48')

.09 .05

.093

.049

.152 .156

(13°)

(14° 48')

.02

.023

.110

R .03

.007

VIEW AT E, TYPICAL 84 PLACES ENLARGED VIEW SHOWING PUNCHED HOLES

LIST OF MATERIALS

PART NO.	NO. REQD		MATERIAL
1	1	WHEEL, IMPELLER	NO. 4 (.2242) US 6A STEEL
2	1	HUB, IMPELLER WHEEL	FORGING STEEL — TD SPEC
3	1	RING, IMPELLER	NO. 9 (.1495) US 6A STEEL
4	42	BLADE, IMPELLER	AISI C-1117 STEEL
5	AS REQD	WEIGHT, BALANCE	CR FLAT STEEL WIRE
6	AS REQD	WEIGHT, BALANCE	CR FLAT STEEL WIRE

WHEEL, IMPELLER

Fig. 20-28. Exploded assembly view of IMPELLER WHEEL.

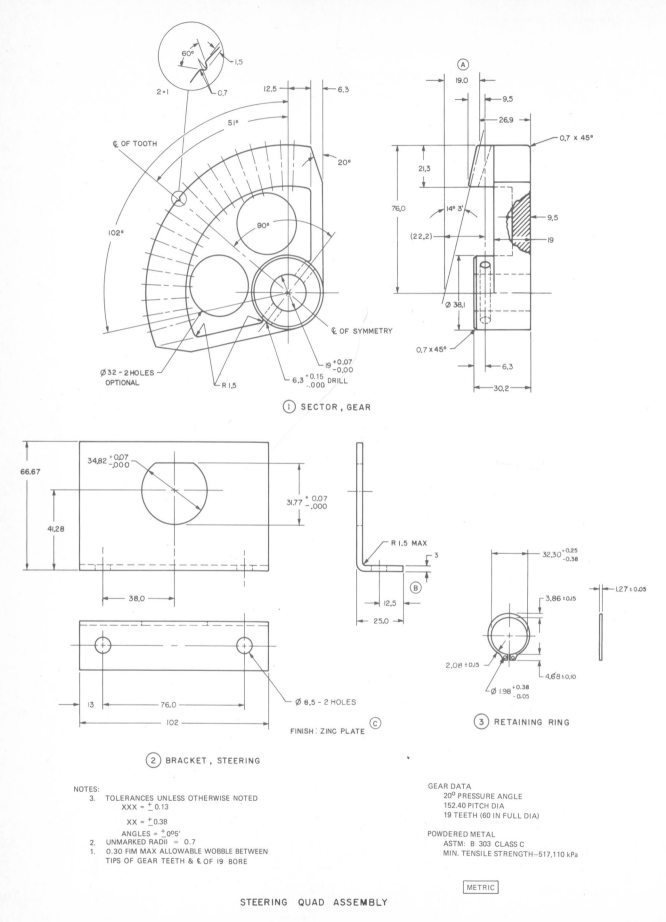

STEERING QUAD ASSEMBLY

Fig. 20-29. An assembly for which detail working drawings and an assembly drawing are to be prepared.

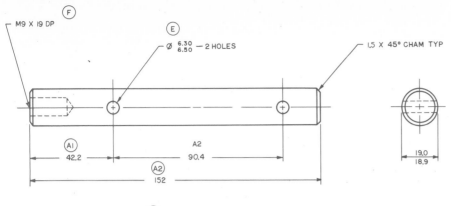

REVISIONS			
LTR	DESCRIPTION	DATE	APPD
A	DIM CORRECTED		
B	WAS 11		
C	WAS PAINT		
D	MATL SPEC ADDED		
E	WAS 6.30 DIA		
F	2B ADDED		

(4) SHAFT, SECTOR

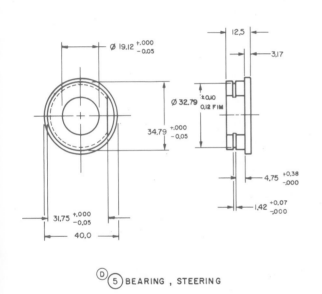

(D) (5) BEARING , STEERING

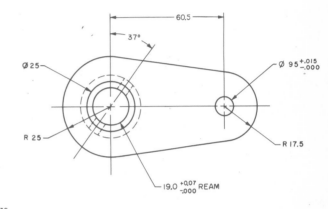

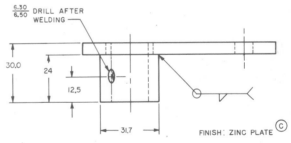

(6) STEERING ARM ASSY,
SECTOR SHAFT

METRIC

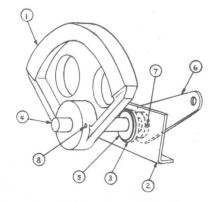

8	1	456722	ROLL PIN (PURCHASE)
7	1	454565	ROLL PIN (PURCHASE)
6	1	39563	ARM. 19 HRS; HUB C1018 STEEL
5	1	39580	ASTM B−202−60T, TYPE 2 CLASS B SINTERED IRON, OIL IMPREGNATED
4	1	39564	19 O.D. C1018 CRS
3	1	39492	TRU−ARC NO. 5100-137
2	1	40200	11GA H.R.P & O
1	1	41088	POWERED METAL ASTM: B 303 CLASS C
ITEM	REQD	PART NO.	MATERIAL
LIST OF PARTS			

Fig. 20-29. (Continued)

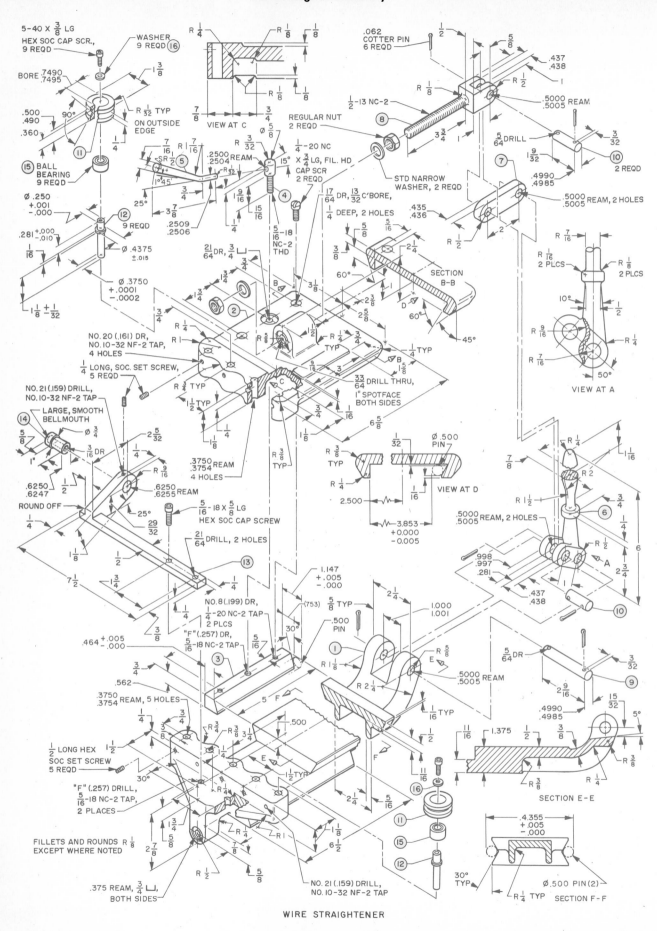

WIRE STRAIGHTENER

Fig. 20-30. Exploded assembly view of WIRE STRAIGHTENER.

PART NO.	NO. REQD.	PART NAME	MATERIAL
		MATERIAL LIST	
1	1	LOWER ADJUSTER BRACKET	CAST IRON
2	1	LEFT ADJUSTER SLIDE BRACKET	CAST IRON
3	1	GIB FOR SLIDE	AISI C-1018 CRS
4	1	CLAMP SCREW FOR SLIDE GIB	AISI C-1018 CRS
5	1	HANDLE FOR SLIDE GIB CLAMP SCREW	AISI C-1018 CRS
6	1	WIRE SET CLAMP HANDLE	AISI C-1018 CRS
7	1	LINK FOR HANDLE	AISI C-1018 CRS
8	1	ADJUSTING FORK SCREW	AISI C-1018 CRS
9	1	PIVOT PIN	AISI C-1018 CRS
10	2	PIVOT PIN	AISI C-1018 CRS
11	9	WIRE STRAIGHTENER ROLLER	CARBON HDN
12	9	WIRE STRAIGHTENER ROLLER SHAFT	CARBON HDN
13	1	STOCK GUIDE BUSHING HOLDER	CARBON HDN
14	1	STOCK GUIDE BUSHING	SAE W1 TOOL STL HDN 60-62R_c
15	9	BALL BEARING	NEW DEPARTURE
16	9	WASHER	MILD STEEL

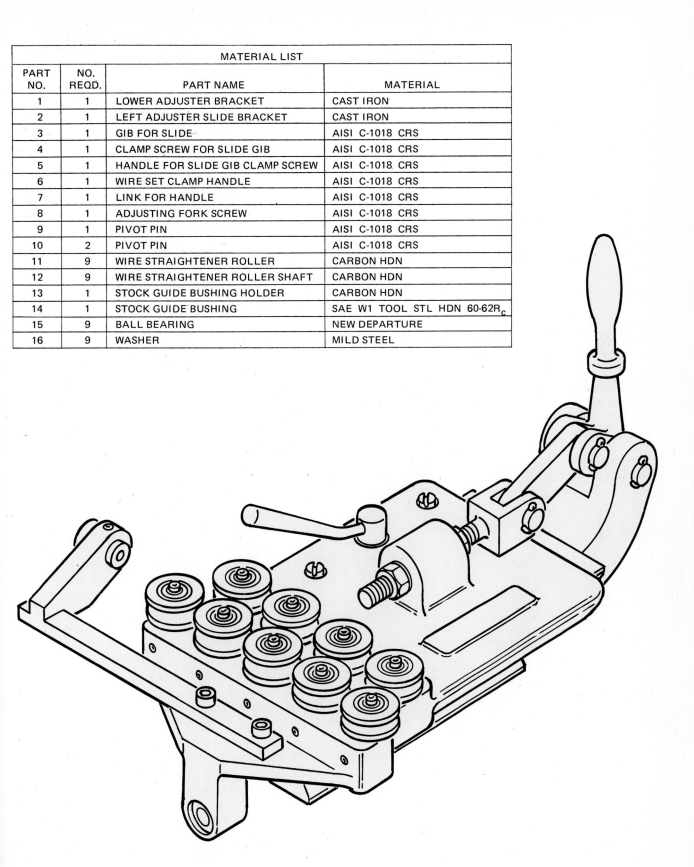

Fig. 20-31. Pictorial view and material list for the WIRE STRAIGHTENER.

Chapter 21

REPRODUCTION OF DRAWINGS

Although making blueprints or whiteprints of original drawings is a major portion of reproduction activities of most industrial drafting departments, it is not the only type of reproduction work performed. Other important aspects include microfilming, Fig. 21-1, photodrafting and scissors drafting, all of which have done much to improve the drafting process.

REPRODUCTION OF DRAWINGS AND THE DRAFTING PROCESS

Elements of reproduction processes are presented in this chapter to provide the drafting student with an understanding and appreciation of their importance to the drafting function.

Fig. 21-1. Microfilm copies of drawings are also a type of reproduction in drafting. (3M Co.)

DEFINITION OF TERMS

An understanding of the following terms will be helpful in the reproduction of drawings:

APERTURE CARD: An electronic accounting machine card with a rectangular hole designed to hold a single frame of microfilm.

AUTOPOSITIVE: A print made on paper or film by means of a positive-to-positive silver type emulsion.

BLOWBACK: An enlarged print made from a micro-image.

GENERATION: The blowback made from a microfilm of an original drawing is a first generation print. A blowback made from a microfilm of this first generation print would be a second generation print, etc. The term "generation" is used to express the quality required of an original drawing being microfilmed. It is usually one that is capable of being reproduced as a clearly readable fourth generation print. Fourth generation quality is necessary because drawing changes are made on a reproduction copy of the original drawing. This revised copy is reproduced and changes are made on this revision for four or more reproduction sequences. Fourth generation quality must be present in the original drawing so that all lines, notes and dimensions are clearly readable on the fourth generation print, Fig. 21-2.

HARD COPY: An enlarged print, on paper, cloth or film, made from an original drawing or microfilm image.

INTERMEDIATE: A translucent reproduction made on vellum, cloth or film from an original drawing to serve in place of the original for making other prints.

NEGATIVE PRINT: A print, usually on opaque material, which is opposite to the original drawing; that is, light lines on a dark background.

POSITIVE: A print which is similar to the original drawing; that is, dark lines on a light background.

REPRODUCIBLE: Capable of being used as a master for making prints by the action of radiant energy on chemically treated media.

SENSITIZED: A reproduction material coated with a light-sensitive emulsion.

TRANSLUCENT: A material that permits the passage of light; partially transparent.

MICROFILMING

The technique of microfilming has been known since the early 1800s, but it was not until World War II that it was

Reproduction of Drawings

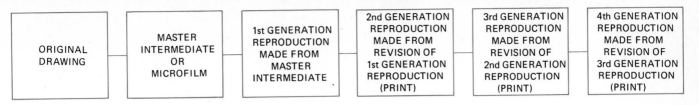

| ORIGINAL DRAWING | MASTER INTERMEDIATE OR MICROFILM | 1st GENERATION REPRODUCTION MADE FROM MASTER INTERMEDIATE | 2nd GENERATION REPRODUCTION MADE FROM REVISION OF 1st GENERATION REPRODUCTION (PRINT) | 3rd GENERATION REPRODUCTION MADE FROM REVISION OF 2nd GENERATION REPRODUCTION (PRINT) | 4th GENERATION REPRODUCTION MADE FROM REVISION OF 3rd GENERATION REPRODUCTION (PRINT) |

Fig. 21-2. Fourth generation reproducibility of a drawing.

applied to drafting for storage of security copies of original drawings at a separate location in the event of a disaster. More recently, microfilming has been developed as an active working technique in creating new drawings, updating old drawings, reducing storage space requirements and in finding drawings once they have been filed, Fig. 21-3.

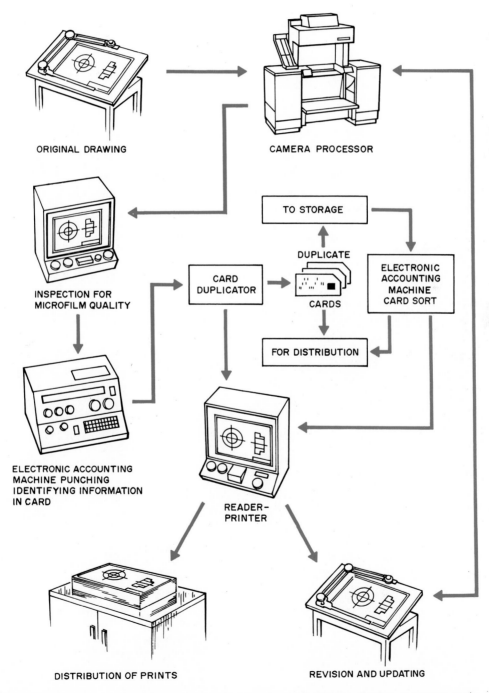

Fig. 21-3. The microfilm system from original drawing to distribution of prints or aperture cards.

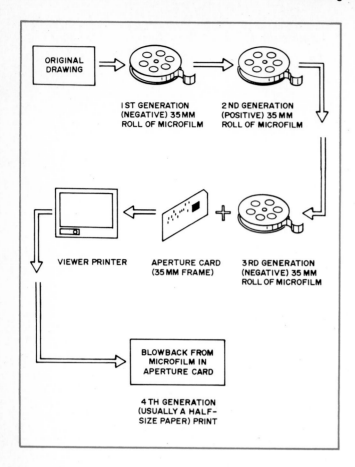

ORIGINAL DRAWING

1ST GENERATION (NEGATIVE) 35 MM ROLL OF MICROFILM

2 ND GENERATION (POSITIVE) 35 MM ROLL OF MICROFILM

VIEWER PRINTER

APERTURE CARD (35 MM FRAME)

3 RD GENERATION (NEGATIVE) 35 MM ROLL OF MICROFILM

BLOWBACK FROM MICROFILM IN APERTURE CARD

4 TH GENERATION (USUALLY A HALF-SIZE PAPER) PRINT

Fig. 21-4. A flow chart illustrating a fourth generation print from microfilm.

The microfilm system has done more to revolutionize drafting operation than any other single factor. The principal advantages are:

1. Less storage space required for original drawings and copies of prints made up ahead of time of use.
2. Near instantaneous retrieval time of aperture card for viewing or for hard copy prints.
3. Increased use of data in drawings because of ready availability.
4. Reduces handling and helps to preserve original drawings.

In meeting these requirements, microfilming has placed more rigid quality control demands on the original drawing. The microfilm process reduces a 34 x 44 inch (size E) drawing 30 times. To appreciate this reduction, consider that the recommended height of letters and dimensions on a size E drawing is .20 inch. This will be reduced on microfilm to a height of .007 inch, or about half the height of a period on the original drawing.

If the microfilm copy is of a quality to produce a fourth generation drawing, it must have sufficient clarity to maintain reproducibility through three generations of microfilm and the subsequent generation of a hard copy print which is clearly readable, Fig. 21-4.

Some industries expect the original drawing input to be of sufficient quality to produce seven generations of updating before touch-up is required.

APERTURE CARD

The microfilm aperture card, an electronic accounting machine (EAM) tabulating card, is a punched card with a single frame microfilm insert that contains the image of the

Fig. 21-5. The aperture card is an effective means of storing, retrieving and disseminating drawing information.

Fig. 21-6. A drawing image projected on the screen of a microfilm reader-printer from an aperture card. (3M Co.)

original drawing, Fig. 21-5. The card may be retrieved quickly in an electronic card sort, inserted in a reader-printer and viewed on a screen, Fig. 21-6, or used to produce a hard copy.

Additional copies of the aperture card may be made for distribution to other offices within the plant or to sub-contractors. Most industries using microfilm maintain a security file in a separate location.

DRAFTING TECHNIQUES FOR MICROFILM QUALITY

The quality of any drawing reproduction depends on the quality of line work used for delineation, and on clear, legible, well-formed letters and numerals used for notes and dimensions. For prints of drawings reproduced by the microfilm process, these qualities are especially important.

Microfilming results in a great reduction of the original drawing and blowback of the print. In addition to the normal procedures in drafting relating to cleanliness, clarity and observing standard drafting practices, the following steps are important when working on drawings which are to be microfilmed:

LINE WORK

1. Use graphite lead pencils on vellum or cloth.
2. Use plastic pencils on film.
3. All lines should be a minimum of .01 inch in width. Thin or light lines tend to "burn out." Also avoid thick lines.
4. Open space between two adjacent parallel lines should be consistent and no closer than .06 inch, Fig. 21-7. Draw slightly out of scale if necessary or show removed detail view at a larger scale.
5. Show hidden lines only when necessary for clarity.
6. Never use shading.

LETTERING AND DIMENSIONING

1. The recommended minimum height for letters and numerals is as follows for various size drawings: The drawing number on all size drawings should be .40 inch. Titles, dimensions and notes on A, B and C sized drawings should be .10 inch in height; for D and E sizes, 0.12 and .20 inch, respectively.
2. A modified form of lettering which is more open and legible should be used on drawings to be microfilmed, Fig. 21-8.
3. The open space between adjacent and parallel letter features in a horizontal line should be a minimum of .06 inch; the space must average .06 inch between nonparallel features; the space between lines of notes must be .10 inch, Fig. 21-9.
4. Each element of a fraction must have the same height and spacing as whole numerals, Fig. 21-9. Space between the fraction bar and the fraction element is .06 inch, minimum.
5. All lettering and numerals should be single stroke with .01 inch minimum line width.
6. Decimal points should be located in line with bottom of associated digits; they should be uniform and dense.

BACKGROUND CONTRAST

1. The background of the drawing should be kept uniformly clean so as to contrast sharply with line work. Dark or smudged areas result in a camera exposure reading which

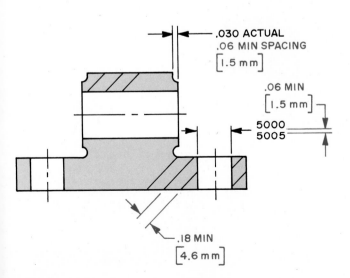

Fig. 21-7. Minimum open space for adjacent parallel lines for drawings to be microfilmed.

Unit	Comments	Not This	Possible Error
B	Upper part small, but not too small.	B	8
H	Bar above center line.	H	——
M	Center portion extends below center line. Slight slant on uprights.	M or M	——
S	Lower part large, ends open. Slight angle on center bar.	S	8
T	Horizontal bar shall be full width of letter "E."	T	7
U	Full width.	V	V
V	Sharp point.	V	U
Z	All lines straight.	Z	2
I	Full height and heavy enough to be identified. No serifs.	1	7
2	Upper section curved with open hook. Bottom line straight.	2	8 or Z
3	Upper portion smaller than lower. Never flat on top.	3 or 3	8 or 5
4	Body large, ends extended.	4	7 or 9
5	Body large, curve dropped to keep large opening. Top fairly wide.	5 or 5 or 5	6 or 3 or S
6	Large body, stem curved but open.	6	8
8	Lower part larger than upper, full and round to avoid blur.	8 or 8	B
9	Large body, stem curved and open.	9	8

Fig. 21-8. Style of lettering recommended for drawings to be microfilmed. (Massey-Ferguson Co.)

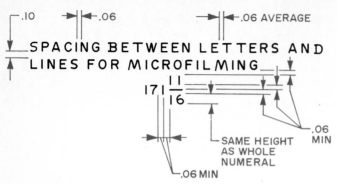

Fig. 21-9. Letter, numeral and fractional spacing for microfilm.

adversely affects the camera lens opening and the quality of microfilm.

2. High concentrations of line work such as large cross-hatched areas should be avoided (see Fig. 21-7). Use a minimum of cross-hatching sufficient only to clarify sectional view.

3. Where erasures have left a "clean" spot which contrasts sharply with the remainder of an old drawing, dull the contrast by smudging slightly with the fingers.

ERASURES

1. On polyester film, use a vinyl eraser crosswise to the line for plastic pencil work; for ink, use the vinyl eraser and moisten slightly.

2. On vellum or cloth, use a vinyl or soft rubber eraser for graphite pencil leads; for ink, use an electric erasing machine with a soft rubber eraser.

3. On drawings that are photographic reproductions, use a dampened soft rubber or vinyl eraser. Do not use a steel knife eraser on this media.

4. When using an electric erasing machine, use only soft rubber erasers. Take care to exert only light pressure to prevent "ghosting" the surface and creating an image which cannot be eliminated.

5. Place a hard surface material, such as a plastic triangle, under the drawing area to be erased. This improves erasability.

6. Use an erasing shield to protect surrounding area.

7. Erase lines thoroughly to avoid "ghost" lines in resultant prints.

8. Remove all eraser crumbs from the drawing surface and from your drawing table with a brush to avoid smudging pencil lines.

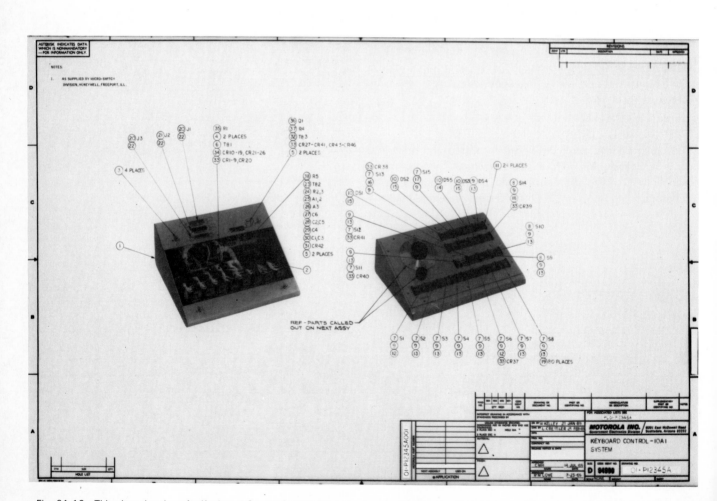

Fig. 21-10. This photodrawing of a Keyboard Control System illustrates how a photograph of an object can be reproduced on drafting film or paper and notes and callouts added to complete the drawing.   (Motorola Inc.)

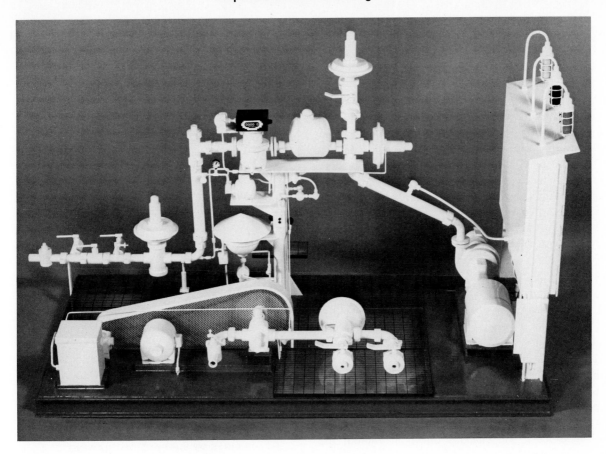

Fig. 21-11. A scale model used in making a photodrawing of an automatic custody transfer unit. (Cities Service Oil Co.)

PHOTODRAWINGS

A photodrawing is a photograph of a model or the actual object. It is made on reproducible drafting film or paper to which the necessary lines, dimensions and notes have been added, Fig. 21-10. Photodrawings have been used for some time in aerial mapping of land areas for highway and other construction projects. They have also been widely used in the electronic industry in wire assembly work.

Where a drawing would require a considerable number of hours in layout work, and where a pictorial presentation would aid in interpreting the drawing, a photodrawing could be more descriptive than the usual drawing and could mean considerable time saving as well.

Photodrawings begin with a photograph of a machine part, model, building, etc., prepared as a continuous tone or half-tone print, Fig. 21-11. The latter type produces the best quality diazo prints.

Follow these steps in making photodrawings:

1. Make a photograph with good detail and little or no shadow effect. Use a good view camera or a 4 x 5 Graphic and flat diffused light.

2. Make a half-tone positive of photograph, using an 85 or a 100 line screen or lithographic film.

3. Position half-tone on drawing format (vellum or film) and mortise the two together by cutting through both at same

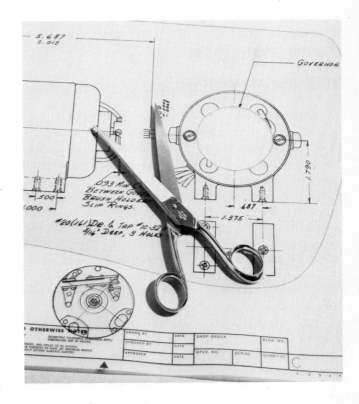

Fig. 21-12. Scissors drafting makes use of photodrafting techniques taking parts of original drawings and splicing in other parts or notes for a "second-original" drawing. (Eastman Kodak Co.)

time. This insures a tight fit. Tape half-tone in place with transparent tape.

4. Make an autopositive print of composite on photographic drafting film. This reproduction serves as master drawing.
5. Add necessary callouts, notes and dimensions to complete photodrawing.

The term "photodrafting" is also used around industrial drafting departments. This term includes photodrawing, but also refers to the process of making a drawing where no photographs are used. In this case, sections of one or more drawings are combined in a new or revised drawing by use of photographic techniques. This latter technique is discussed in the next section on "scissors drafting."

SCISSORS DRAFTING

When part (or all) of one drawing is combined into another reproduction media to serve as part of a "second-original" drawing, this process is known as scissors drafting. Modern reproduction methods make it possible to merge parts of several drawings onto one drawing media and add to this the necessary line delineation, notes and dimensions.

Scissors drafting is similar to photodrawing except the source of the "add-on" material is from other drawings or prints, parts catalogs and charts, Fig. 21-12. Unwanted sections on a drawing can be cut or blanked out during the reproduction process and changes drawn in without the necessity of erasing.

When the parts have been combined onto one reproducible drawing, an autopositive print is made and serves as the "second original" drawing. Callouts, notes and dimensions are then added to complete the drawing. Considerable drafting time can be saved when the parts added on represent complicated and detailed objects.

REPRODUCTION PRINTS

Today, there are numerous processes by which prints can be made from original drawings or from a microfilm copy of an original drawing. The prints most commonly used require a translucent drawing (or microfilm copy) and a paper, cloth, film or other medium which has been coated with a chemical sensitive to light (the electrostatic process excepted). After exposure to a light source in conjunction with the translucent drawing, the materials can be developed chemically into legible reproductions or prints.

Fig. 21-13 illustrates how the lines and lettering on the original drawing, or other master copy, prevent the light from acting on the sensitized coating of the reproduction base material. This unexposed material will react during the chemical treatment process to develop the image of the original drawing. The type of reproduction print made depends upon the type of sensitized material used and its subsequent processing.

Several types of reproductions are made to serve different purposes. One type of reproduction, called an intermediate, is developed on a suitable medium such as vellum, film or photographic paper to serve as a drawing medium in preparing

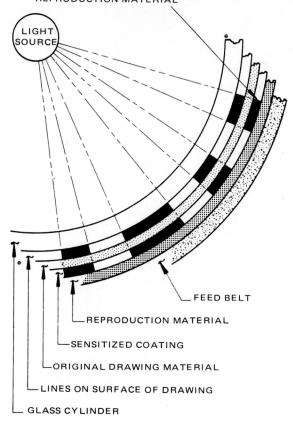

DRAWING LINES ARE IMAGED BY INHIBITING LIGHT ACTION ON BLUEPRINT, DIAZO OR SILVER EMULSION COATINGS OF REPRODUCTION MATERIAL

LIGHT SOURCE

FEED BELT
REPRODUCTION MATERIAL
SENSITIZED COATING
ORIGINAL DRAWING MATERIAL
LINES ON SURFACE OF DRAWING
GLASS CYLINDER

Fig. 21-13. An illustration of the operating principles of a typical reproduction machine. After exposure to the light source, the original returns to the operator and the sensitized material continues through a chemical treatment to develop the print. (General Motors Corp.)

"second-original" drawings. These were discussed in the previous paragraphs on photodrawing and scissors drafting. Another type intermediate is developed on a translucent material and serves as the "tracing" for use in making additional prints, saving the wear and tear on the original drawing. The largest number of reproduction prints used are the opaque type used by the workmen who will actually produce the object described on the drawing.

Some of the more common types of opaque prints used today are discussed in the following sections.

BLUEPRINT

The term "blueprint" is used loosely to refer to all types of hard copy prints. For many years it was the only type of reproduction made. This print has white lines on a blue background and is developed from a material coated with light-sensitive iron salts (see page 159). The blueprint is called a "negative" print because reproduction is opposite in tone to the original drawing. Dark lines on a light background produce light lines on a dark background. It is possible to produce a "positive" print with the blueprint process by first producing a negative translucent intermediate and then making a blue line print fron this negative.

A blueprint is made by placing the original drawing over sensitized paper and exposing both to an intense light source. The lines and lettering on the original drawing protect chemical sensitizers on the sensitized paper. When the original drawing is removed and the exposed paper is carried through a water wash, the chemical sensitizers are removed from those protected portions. A rinse in an oxidizing agent (usually potassium dichromate), a final water wash and a drying operation complete the print.

DIAZO PROCESS

A more recent and widely used process of making prints is the diazo process which produces positive type prints with dark lines on a light background. These prints may have blue, black, brown, red or other colored lines and are often referred to as whiteprints, directline prints, black line, brown line, etc. This process utilizes the light sensitivity of certain diazo compounds and is available in dry, moist and pressurized form.

The original translucent drawing and sensitive diazo paper are inserted into a diazo type print machine, Fig. 21-14, and exposed to the light source which destroys the unprotected diazo compound. The original drawing is returned to the operator and the exposed sensitized paper is carried through the developing section of the machine. Here it is exposed to a chemical which develops the print. The moist form transfers an ammonia solution to the print to cause the development. The print is delivered in a somewhat moist or damp state.

Fig. 21-14. A pressure diazo print machine that produces positive prints with dark lines and light background. (Teledyne Rotolite)

Dry form of diazo print making utilizes an ammonia vapor to develop the exposed copy and the resultant print is relatively dry. The pressure form of the diazo process utilizes a thin film of a special activator delivered under pressure to the exposed copy to complete the development.

ELECTROSTATIC PROCESS

The electrostatic process is a means of producing paper prints and intermediate transparencies from original drawings or microfilm, Fig. 21-15. This process, commonly referred to

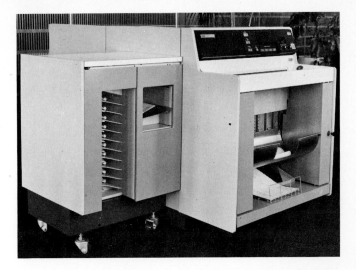

Fig. 21-15. The Xerox machine is an electrostatic print making machine.

as Xerography, will develop a print on unsensitized paper. It will produce either enlarged or reduced copies of the original drawing. The process works by forming an image electrostatically on a selenium coated drum or plate and then transferring the image onto most any type of material. The exposed surface is dusted with a dark colored powder which is affixed quite permanently by heat or solvent action.

Reproduction by the electrostatic process has found increasing use as a means of producing prints and other documents because of its convenience and simplicity. Intermediate transparencies may also be produced by this process.

PROBLEMS AND ACTIVITIES

The following problems and activities will provide an opportunity for you to gain further knowledge and understanding of the reproduction processes.

1. Select problems as directed by your instructor and prepare working drawings to microfilm quality standards.
2. Prepare a photodrawing utilizing a suitable photograph. Dimension the drawing and add necessary notes.
3. Prepare second-original drawing, using scissors drafting technique. Select earlier drawing to be modified or combined with another. Estimate time saved by this technique over preparing an entirely new drawing.
4. Make a print of one of your drawings prepared on a translucent medium. Use either the blueprint process or the diazo process. Study the reproduction qualities of the drawing as revealed in the print.
5. Make an electrostatic copy of one of your drawings prepared earlier. Compare the quality of this reproduction with the one in problem No. 4 above.
6. Visit a local blueprint service company and find out the types of reproduction processes they perform. Which type of reproduction is in greatest demand? What are the costs for the various types of reproduction prints?
7. Visit a drafting equipment supply company and inquire about the type of reproduction equipment they sell. Find out about new processes or trends developing in this field.

PART IV
DRAFTING APPLICATIONS

Drafting applications relate to the many areas in which advanced skill in drawing and knowledge of the field are essential to the preparation of acceptable technical drawings for industry.

This part of the text covers ARCHITECTURAL DRAWINGS, ELECTRICAL AND ELECTRONICS DRAWINGS, WELDING DRAWINGS, MAPS AND SURVEYS, TECHNICAL ILLUSTRATIONS, and GRAPHS AND CHARTS.

The construction and allied fields depend heavily on industrial drawings.
(International Paper Co.)

Chapter 22
ARCHITECTURAL DRAWINGS

Architecture and architectural drafting is the area of design and drafting which specializes in the preparation of drawings for the construction of houses and commercial buildings. It also deals with the design of churches and entire facility complexes, such as schools and colleges, shopping centers, municipal parks and golf courses, Fig. 22-1. That phase of architectural drafting relating to the design and preparation of drawings for residences and light commercial buildings is covered in this chapter.

STYLES OF RESIDENTIAL ARCHITECTURE

Various styles of residential architecture have developed in America as the result of influences from Europe and because of climatic conditions in various regions. It was natural that early American houses were modeled after ones familiar to settlers in their home land.

Consequently, there was considerable influence from English architecture in the New England states and Virginia

Fig. 22-1 Architects may work with a single residence or several buildings and surrounding grounds. Shown is the Central Library on the campus of the University of California at San Diego.
(Architects—William Pererira and Associates. Fischbach and Moore, Electrical Engineers)

Fig. 22-2. Traditional styles of residential architecture in America. (a) Elizabethan, (b) Georgian, (c) Southern Colonial, (d) Dutch Colonial, (e) Cape Cod and (f) Spanish.

where the English Puritans settled. Dutch settlers in New York and Pennsylvania and the Spanish in Florida, the southwest, and southern California influenced the style of architecture in those regions.

TRADITIONAL ARCHITECTURE

The traditional styles of architecture, imported as well as native forms, represent a wide range of design and construction techniques. Some of the traditional styles that have had considerable influence on architecture in America since colonial times are presented in this section.

The Elizabethan style house of English origin is characterized by prominent gables and half-timber construction, Fig. 22-2 (a). Brick, stone or stucco is usual siding material while timber frame is exposed. These houses of one-and-one-half or two stories are characterized by several large chimneys.

The Georgian style, with its very formal proportions and balance in exterior walls, represents another English influence on the architectural styles of America, Fig. 22-2 (b). Houses of this style vary from one-and-one-half to two-and-one-half stories, with gable, hip or gambrel roofs, dormer windows and chimneys.

Georgian houses built in New England and Southern Colonies were usually frame with wood or brick exterior. The exteriors of those built in the middle colonies were largely of stone because of the bountiful supply of this building material.

The Southern Colonial is a one-and-one-half story house built around a central hall. It has a gable roof with dormer windows and large exterior chimneys at each end of the house, Fig. 22-2 (c). The house was first made of wood frame construction with white clapboard siding. Later, brick was used for the exterior walls.

The Dutch Colonial is a style of architecture that was introduced in America by the Dutch settlers in the Hudson Valley of New York. The houses were from one-and-one-half to two-and-one-half stories with gambrel roofs, Fig. 22-2 (d). The second story frequently projected out over the first, and a combination of exterior stone for the first story and wood siding for the second was used.

The Cape Cod is a single or one-and-one-half story of wood frame and clapboard siding, Fig. 22-2 (e). Earlier Cape Cod houses did not have dormer windows. Later versions, and ones being built at present, usually include two dormer windows in front and a shed type dormer in the rear roof. The roof is a gable type with little overhang and windows just below the eaves. This roof tends to make the house appear small and low.

The chimney of old Cape Cod houses was centrally located and served several flues. Present day versions of the Cape Cod usually locate the chimney on one end of the house.

The Spanish style house was fashioned after the Spanish mission buildings, particularly those in the Southwest. These mission buildings had walls of adobe brick (a mixture of native clay and straw dried in the sun). Flat or shed type roofs were formed by placing heavy hewn logs as beams. Log planking was placed over these and covered with a thick layer of adobe clay to form a solid slab roof, Fig. 22-2 (f).

Later, building materials included lime and sand to protect the walls from rain erosion. In some areas, burned bricks were available and used to form the door and window arches so characteristic of Spanish architecture. Red tiles, formed into half cylinders, replaced the adobe slab roofs. Houses frequently were constructed in a "U" shape, with all rooms opening

Fig. 22-3. Some examples of present day architectural styles in America. (a) Modern, (b) Ranch, (c) Contemporary and (d) Modified Spanish.

(a)

(b)

(c)

(d)

onto a central patio.

Today's Spanish style house retains many of the features characteristic of early Spanish architecture in America: stucco walls; flat or shed type tile roofs; covered patios, featuring out of doors living.

PRESENT DAY ARCHITECTURE

Architectural styles in America today tend to emphasize a freer, more relaxed arrangement between space within the house and out of doors. The styles most representative of residential architecture in America today are presented here.

The Modern style house developed in this country (as well as in Europe) has emphasized the interior design and arrangement of rooms. The purpose has been to make the interior of the house "more livable" for the occupants. The exterior appearance, while pleasing in its simplicity, has become a secondary consideration, Fig. 22-3 (a).

In the modern style of architecture, related areas such as living and dining rooms, or family room, dining area and kitchen tend to flow together to provide feeling of spaciousness. Generous use of glass "captures" the out of doors.

The Ranch style of residential architecture features a long and sometimes angular or "U" shaped one-story house with a low roof line that seems to tie the structure to the ground, Fig. 22-3 (b). This style of house started in the Southwest, but it has spread throughout the country in modified form. Patios and out of doors living, so common to the Southwest and California, are a part of the design of this style of house.

The Contemporary style, like modern, features functional living interiors. There is considerable openness in kitchen, dining, family and living areas, yet it retains privacy for individual members. The house design blends indoor and outdoor living with entire walls of glass and covered patios, Fig. 22-3 (c).

In a Contemporary house, more attention is given to exterior design (wider variety than Modern) and to fitting the house to its surroundings. This design is equally applicable to single or two-story houses as well as to the split-level.

Modified Spanish is a version of Spanish architecture that presently is very popular in the Southwest, Fig. 22-3 (d). This style of house retains the flat, or shed type, roof and stucco walls of the traditional Spanish style. However, it has a more simple exterior design and a more functional arrangement of the interior of the house.

Like the Contemporary house design, the rooms in a Modified Spanish house have an openness to them. Also, indoors and outdoors blend together through walls with large glass sections leading onto covered patios with appropriate plantings. This architectural style may be found in individual houses or row houses of condominiums.

TYPES OF ROOFS

There are a variety of roof styles found on residential structures today. Most can be classified as one of the following basic types, or a modification of one.

The gable type roof is perhaps the most common of all roofs, Fig. 22-4. The hip roof is a modification of the gable, where the ends are given a pitch and roofed the same as the sides. The hip roof is more difficult to construct, but it reduces the maintenance of painting gable ends.

The flat roof is the most easily constructed of all roof styles. However, care must be taken to adequately seal against leaks. The shed roof is a flat roof with a one-way pitch to provide for drainage. These roofs are found in the dryer climates on the Modern, Spanish and Modified Spanish style houses. When the shed type roof is used, the ceiling of the house frequently follows the same roof pitch giving a spacious feeling to rooms.

The gambrel and mansard roof types are used on present-day structures, but more often on traditional houses.

DESIGNING OF A HOUSE

Designing a house involves a study of the needs of the family planning to occupy the house. Certain information must be obtained if the architect is to adequately plan to meet the needs of his client.

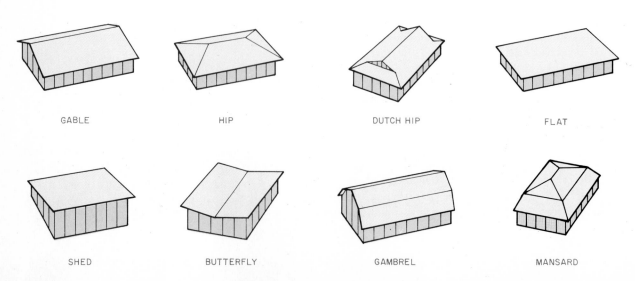

GABLE   HIP   DUTCH HIP   FLAT

SHED   BUTTERFLY   GAMBREL   MANSARD

Fig. 22-4. Basic types of roofs.

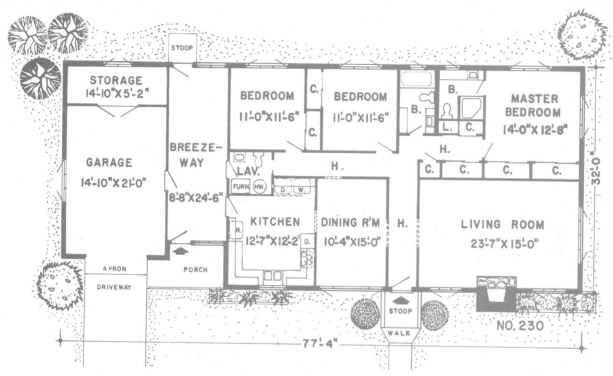

Fig. 22-5. Presentation drawings usually consist of a basic floor plan and pictorial rendering of elevations.
(Garlinghouse, Inc.)

THE FAMILY GROUP AND SPECIAL NEEDS

The number of members in the family, their ages, desires for separate bedrooms, special needs of children and adults and family customs must all be considered. Do their occupations and hobbies require special rooms (a study, workshop, playroom, sewing room, office) or special adaptations of rooms (nursery, invalid care, gourmet kitchen, dark room)? What special built-in features (bookshelves, a storage wall for off-season clothing, recreational and hobby equipment) are needed? Is there to be a basement? How many cars? Garage or car port? Separate room for laundry? What type heating and cooling is preferred? What colors are preferred?

STYLE AND CONSTRUCTION

What general type of architectural style is wanted? Is the house to be a single-story, two-story, split-level or other style? What type construction is best for this family — wood frame, brick, stone or other? What style roof and roofing material is preferred?

THE LOT AND ITS FEATURES

What is the size and shape of the lot and what features does it have? Are certain trees or rocks to be retained? What views are to be planned for in the house design? Where may patios, recreation areas, drives and entries to the house be located? What bearing does the sun's path have on orientation of the house design?

FINANCIAL CONSIDERATIONS

What financial resources does the family have? Is their present and anticipated future income in keeping with their building aspirations? What items or features of the architectural design should be considered as essentials? What items or features can be considered as alternatives to be eliminated if funds are short?

INITIAL PLANNING AND PRESENTATION DRAWINGS

Once the architect has the information needed to establish objectives, study can begin regarding the relationship of eating, living, recreation and sleeping areas. Preliminary freehand sketches are usually prepared on graph paper, taking into consideration all of the factors on needs, desires and financial resources. After carefully studying several preliminary sketches of the interior room arrangement, exterior design and plot plan location, the architect and drafters will prepare presentation drawings for the client's review, Fig. 22-5.

439

Presentation drawings consist of a floor plan showing room arrangement and size and one or two elevation views on a perspective rendering. These are not detail dimensioned drawings. Rather, they serve as further communication between the client and architect in arriving at the final plan for the house. Scaled models are sometimes prepared for making presentations to clients, especially on large commercial or government buildings.

ARCHITECTURAL LETTERING AND DIMENSIONING

Lettering and dimensioning techniques learned in earlier chapters are applicable, with minor exceptions, to architectural drawings.

LETTERING STYLE AND SIZE

The lettering style used on architectural drawings is the single-stroke Gothic capital letters used on other type drawings, but it tends to be a little more free style and individualized by the drafter, Fig. 22-6.

Fig. 22-6. The style of architectural lettering is not as formal as on other type drawings.

Clarity and neatness are just as essential in architectural lettering and dimensioning as they are in preparing drawings for manufacturing. The letters are usually 1/8 inch on the drawing proper and 1/4 inch on subheadings such as room identification, window and door schedules. Lettering in title blocks is 5/16 inch for main title and 3/16 inch for remainder, Fig. 22-7. Use guidelines for all letters and numerals.

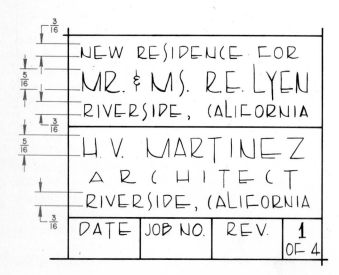

Fig. 22-7. A title block for architectural drawings.

DIMENSIONING TECHNIQUES

Floor plans and elevations usually are drawn to a scale of 1/4" = 1' − 0". Details of wall sections and built-ins, such as cabinets, utilize a scale of 1/2" = 1' − 0". Plot plans usually are drawn to a greatly reduced scale of 1" = 10'.

Architectural drawings usually are dimensioned with an unbroken dimension line terminating with arrowheads, Fig. 22-8 (a), or with slashes, (b). The dimension figure is located above the dimension line when read from the bottom or right hand side of the sheet (aligned system).

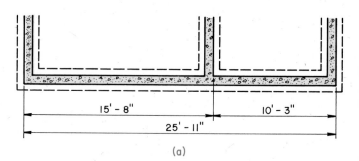

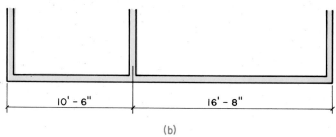

Fig. 22-8. Architectural dimensioning.

It is customary to dimension to the outside of the foundation wall on the foundation plan and to the outside of the stud wall on wood frame houses. Dimensions through the house are to partition wall centers. This makes it easier for the workers on the job to lay out the features of the house.

ARCHITECTURAL WORKING DRAWINGS

Working drawings in architectural drafting consist of roof and plot plans, foundation/basement plans, floor plans, elevations and framing details. Together, they are called a set of plans.

FLOOR PLAN

The floor plan is a fully dimensioned and graphic description of the layout of one floor, Fig. 22-9. It includes the location of all features such as walls, doors, windows and built-ins, Fig. 22-10. It may include the layout and location of the electrical system, plumbing fixtures and references to detail sections contained in other sheets.

Note how symbols are used in Fig. 22-9 to represent doors,

Architectural Drawings

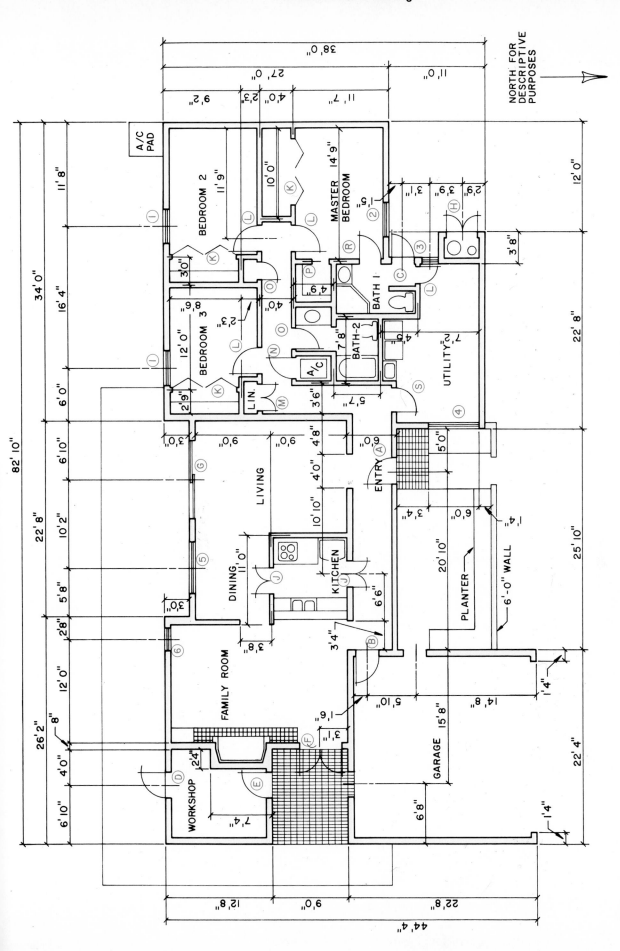

FLOOR PLAN LIVING AREA 2320 SQ. FT.

Fig. 22-9. An architectural floor plan for a residence. The door and window schedules for this plan are on the next page.

441

DOOR SCHEDULE

SYMBOL	QUANTITY	TYPE	DOOR SIZE	REMARKS
A	1	Panel	3' – 0'' x 6' – 8'' x 1 3/4''	Fir
B	1	Flush	2' – 6'' x 6' – 8'' x 1 3/4''	Birch, solid core
C	1	Panel	2' – 6'' x 6' – 8'' x 1 3/4''	Fir w/2 hammered glass lites
D	1	Panel	2' – 6'' x 6' – 8'' x 1 3/4''	Fir w/ventilating lite
E	1	Panel	2' – 6'' x 6' – 8'' x 1 3/4''	Fir
F	2	French	2' – 6'' x 6' – 8'' x 1 3/4''	Fir
G	1	Glass Sliding	9' – 0'' x 6' – 10''	Aluminum
H	2	Flush	2' – 0'' x 6' – 8'' x 1 3/4''	Masonite, w/louvres
J	4	Louvre	1' – 4'' x 6' – 8'' x 1 3/8''	Fir
K	3	Bifold Louvre	5' – 0'' x 6' – 8'' x 1 3/8''	Fir, track at top only
L	4	Panel	2' – 6'' x 6' – 8'' x 1 3/8''	Fir
M	2	Panel	1' – 10'' x 6' – 8'' x 1 3/8''	Fir
N	1	Flush	2' – 6'' x 6' – 8'' x 1 3/8''	Birch, hollow core w/louvre
O	2	Panel	2' – 4'' x 6' – 8'' x 1 3/8''	Fir
P	1	Pocket Louvre	2' – 0'' x 6' – 8'' x 1 3/8''	Fir
R	1	Panel	2' – 0'' x 6' – 8'' x 1 3/8''	Fir
S	1	Panel Dutch	2' – 6'' x 6' – 8'' x 1 3/8''	Fir

WINDOW SCHEDULE

1	2	Sliding	4' – 0'' x 4' – 0''	Aluminum
2	1	Sliding	4' – 0'' x 3' – 0''	Aluminum
3	1	Sliding	2' – 0'' x 3' – 0''	Aluminum
4	4	Ventilating	6' – 0'' x 6' – 10''	Aluminum
5	1	Fixed	6' – 0'' x 6' – 10''	Aluminum
6	1	Ventilating	2' – 8'' x 6' – 10''	Aluminum

Fig. 22-9. (Continued) Door and window schedules for the floor plan shown on page 441.

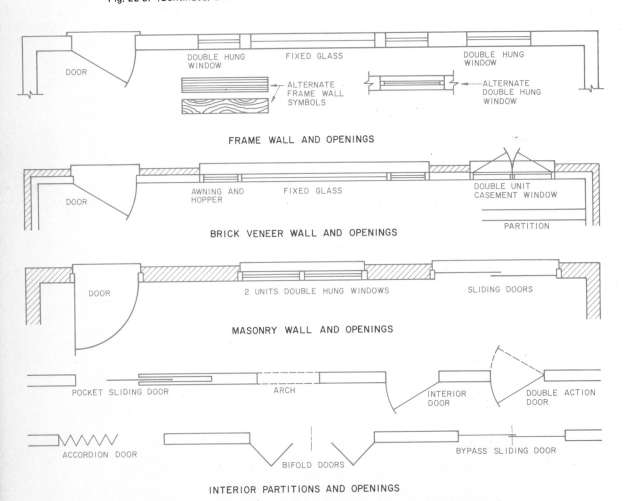

FRAME WALL AND OPENINGS

BRICK VENEER WALL AND OPENINGS

MASONRY WALL AND OPENINGS

INTERIOR PARTITIONS AND OPENINGS

Fig. 22-10. Representation of wall openings in the plan view.

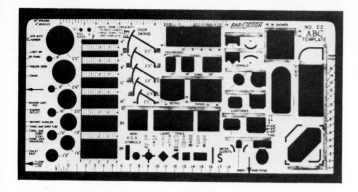

Fig. 22-11. This template is useful in drawing architectural symbols on floor plans. (RapiDesign)

windows, electrical and plumbing details. Architectural symbol templates are available for use with floor plan layouts, as illustrated in Fig. 22-11.

ROOF AND PLOT PLAN

The roof and plot plan shows the location of the house on the plot (lot), Fig. 22-12. It is usually drawn to a scale of 1″ = 10′ in order to get it on the sheet. Compass direction NORTH is indicated by an arrow. Overall dimensions of the plot and those locating the house, other buildings, walks, drives and prominent features such as shrubs and trees are given. It is also general practice to give the elevation of finish grade at each corner of the house, other buildings and at the street curb or crown.

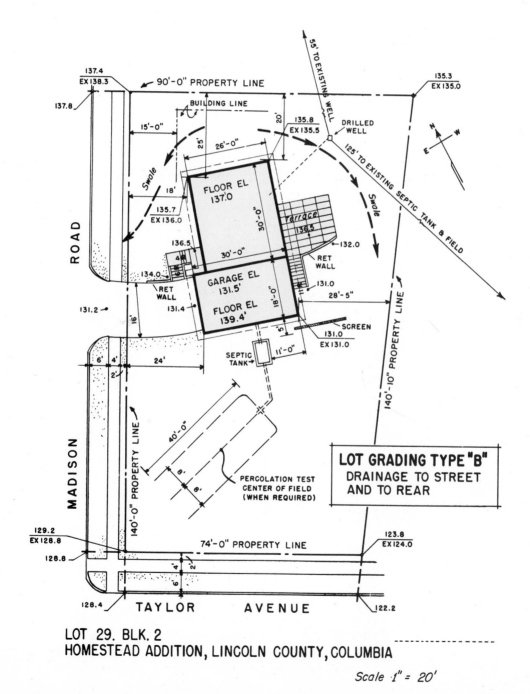

Fig. 22-12. A plot plan showing roof outline and location of principal features on the lot. (FHA)

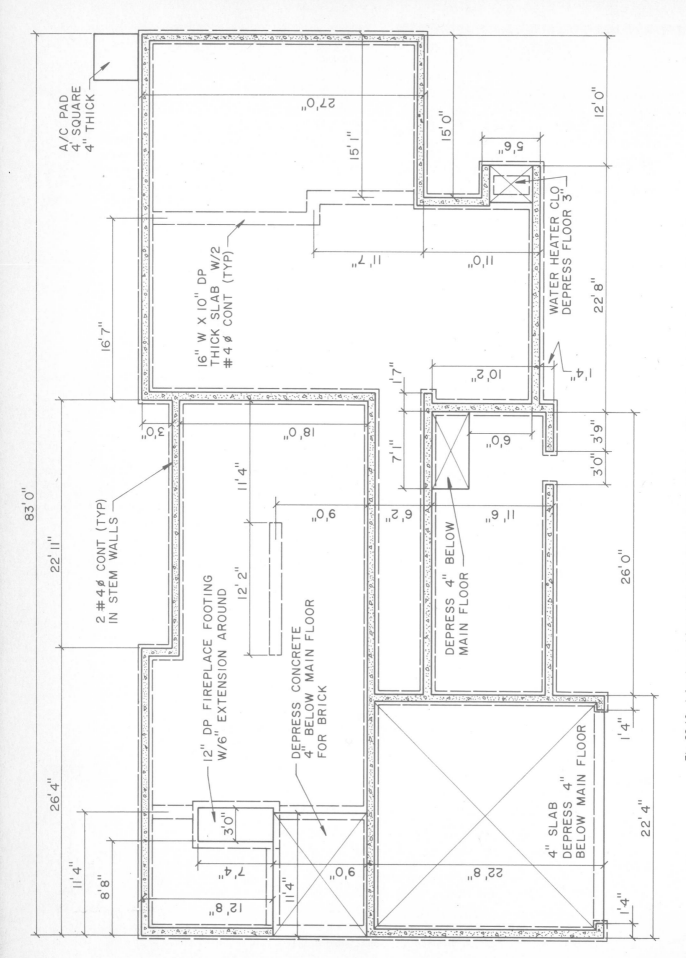

Fig. 22-13. A footing and foundation plan showing the foundation wall in section and the footings as broken dash lines.

A/C PAD
4' SQUARE
4" THICK

27'0"

15'1"

15'0"

12'0"

5'6"

WATER HEATER CLO.
DEPRESS FLOOR 3"

11'7"

11'0"

22'8"

16" W X 10" DP
THICK SLAB W/2
#4 Ø CONT (TYP)

16'7"

10'2"

1'4"

1'7"

3'0"

7'1"

9'0"

6'0"

3'0" 3'9"

83'0"

2 #4 Ø CONT (TYP)
IN STEM WALLS

22'11"

18'0"

11'4"

6'2"

11'6"

26'0"

12'2"

9'0"

12" DP FIREPLACE FOOTING
W/6" EXTENSION AROUND

DEPRESS CONCRETE
4" BELOW MAIN FLOOR
FOR BRICK

DEPRESS 4" BELOW
MAIN FLOOR

26'4"

3'0"

1'4"

22'4"

11'4"

8'8"

7'4"

9'0"

4" SLAB
DEPRESS 4"
BELOW MAIN FLOOR

12'8"

1'4"

22'8"

1'4"

444

Architectural Drawings

The roof plan is the view from above, and the roof extension beyond the walls of the house is shown in phantom lines. Fig. 22-12 shows the roof outline.

FOOTING AND FOUNDATION PLAN

The footing and foundation plan can be drawn by placing vellum or tracing paper directly over the floor plan and tracing the outline of the floor plan. The foundation wall is shown in a visible line; the footing outline is represented by a broken dash line, Fig. 22-13. Beams below the floor level (used to support bearing walls) or footings for piers and columns are also shown on the plan as broken dash lines.

When a concrete slab foundation is used (no footings), the inner line of the slab foundation wall and other beams below the floor are shown in hidden lines, Fig. 22-14. The foundation and slab floor are poured at the same time.

BASEMENT PLAN

The basement plan may also be prepared by tracing the floor plan, including whatever details are required. Openings for doors and windows are shown and called out for listing in a schedule as they are on plans for other floors, Fig. 22-15.

ELECTRICAL PLANS

The wiring diagram for the electrical service in a house may be drawn on the regular floor plan, Fig. 22-16. Or, it may be drawn on a separate undimensioned tracing of the floor plan when the amount of information tends to be crowded and

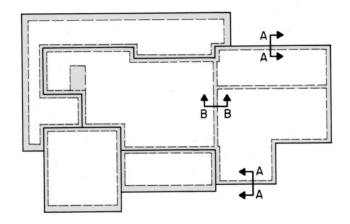

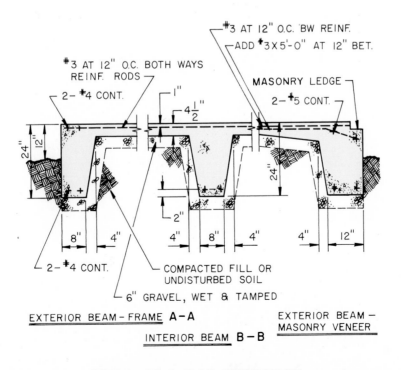

DETAIL OF A SLAB FOUNDATION

Fig. 22-14. A slab foundation plan showing thickness of beams.

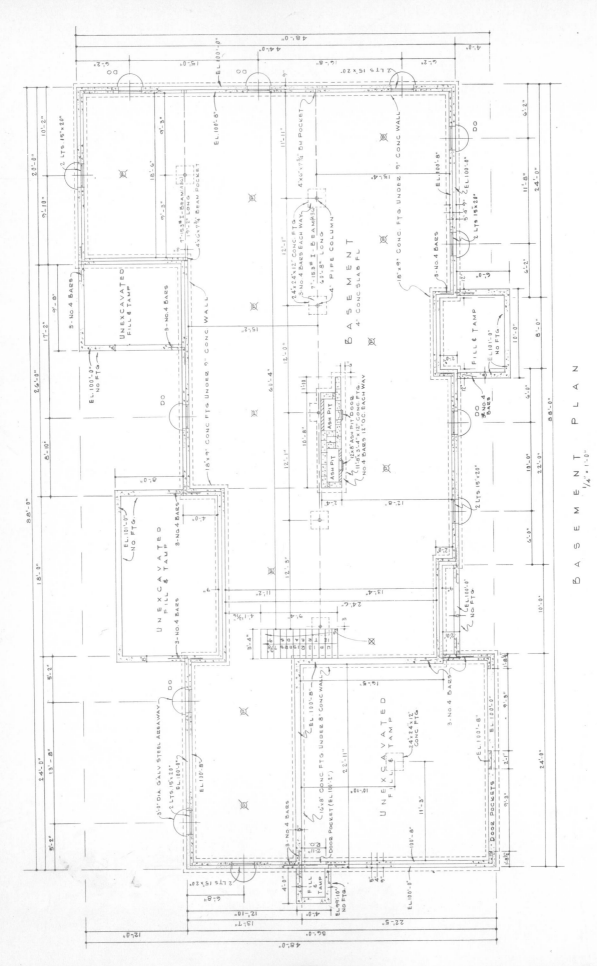

BASEMENT PLAN $\frac{1}{4}" = 1'-0"$

Fig. 22-15. A basement floor plan. (Garlinghouse, Inc.)

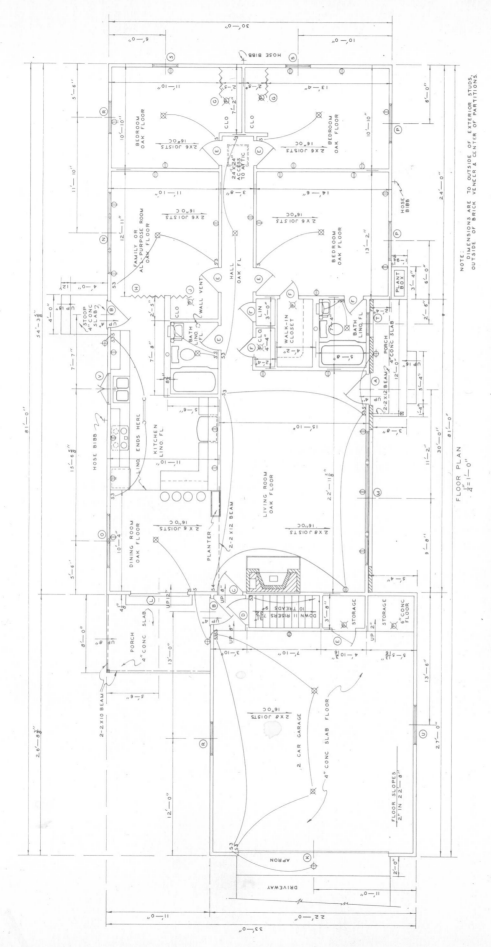

Fig. 22-16. An electrical floor plan showing the wiring diagram.

ELECTRICAL SYMBOLS

CEILING OUTLETS FOR FIXTURES

WALL FIXTURE OUTLET

CEILING OUTLET WITH PULL SWITCH

WALL OUTLET WITH PULL SWITCH

DUPLEX CONVENIENCE OUTLET

WEATHERPROOF CONVENIENCE OUTLET

CONVENIENCE OUTLET 1,3, I = SINGLE 3 = TRIPLE

RANGE OUTLET

CONVENIENCE OUTLET WITH SWITCH

220 VOLT OUTLET

JUNCTION BOX

SPECIAL PURPOSE (SEE SPECS.)

CLOCK OUTLET

FLOOR OUTLET

CEILING LIGHT FIXTURE

PULL CHAIN LIGHT FIXTURE

EXTERIOR LIGHT FIXTURE

RECTANGULAR RECESSED LIGHT (SIZE VARIES)

POWER PANEL

SINGLE-POLE SWITCH

DOUBLE-POLE SWITCH

THREE-WAY SWITCH

FOUR-WAY SWITCH

WEATHERPROOF SWITCH

SWITCH WITH PILOT LIGHT

PUSH BUTTON

BELL

CHIME

ELECTRIC DOOR

OUTSIDE TELEPHONE CONNECTION

TELEVISION OUTLET

EXTERIOR CEILING FIXTURE

SWITCH WIRING

FLUORESCENT LIGHTING

Fig. 22-17. Electrical symbols used in wiring diagrams.

cluttered. Electrical symbols are used. See Fig. 22-17. Lights or outlets to be controlled by switches are indicated by connecting the symbols with a dash line, Fig. 22-17.

FRAMING PLANS

The framing of a house requires considerable planning and detail, Fig. 22-18. Framing plans should be prepared for floor, ceiling, walls and roof, particularly where special framing details are involved. Where double framing is to be included (floor framing of double joists beneath a wall) or around openings in a floor, ceiling or wall, these double members should be drawn. Either use double lines to show thickness of framing lumber or include a note when the framing lumber is represented by a single line, Fig. 22-19. The framing members should be located 16 inches on center (16" o.c.) unless otherwise required. The framing plan for the roof shows the

Fig. 22-18. A scale model of a house showing framing details.   (Jon Salvesen)

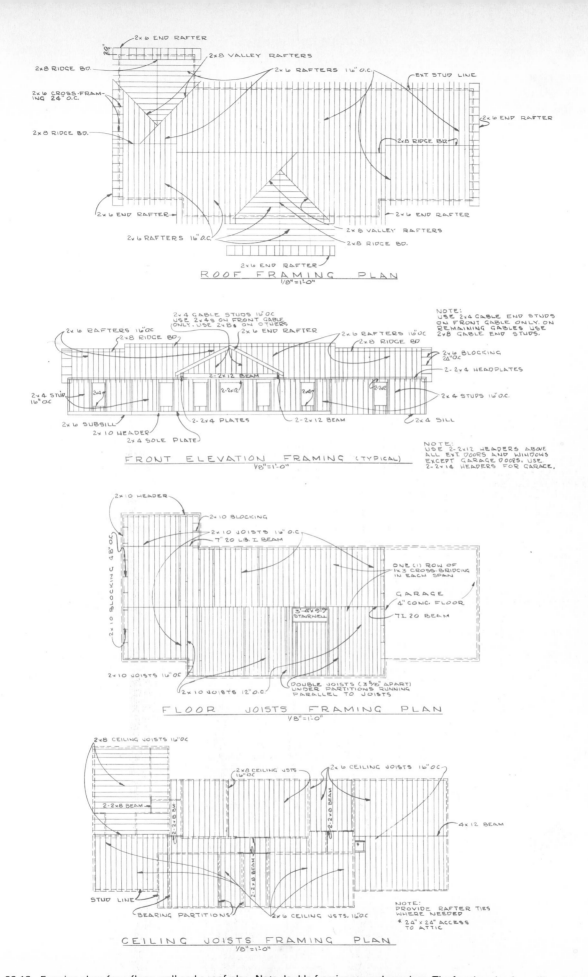

ROOF FRAMING PLAN
1/8"=1'-0"

FRONT ELEVATION FRAMING (TYPICAL)
1/8"=1'-0"

FLOOR JOISTS FRAMING PLAN
1/8"=1'-0"

CEILING JOISTS FRAMING PLAN
1/8"=1'-0"

Fig. 22-19. Framing plans for a floor, wall and a roof plan. Note double framing around openings. The framing plan for ceilings is prepared in a similar manner. (Garlinghouse, Inc.)

449

ridge board and rafter arrangement, plus double framing at all openings. Any special framing such as decking or beam construction should be shown, detailed or noted. Material symbols should also be used on any plan which will serve to provide clarity to the drawing, Fig. 22-20.

Fig. 22-21 shows a pictorial representation of two types of wall framing, platform (western) and balloon framing in which the studs run from sill to top plate.

MATERIAL	PLAN	ELEVATION	SECTION
EARTH	NONE	NONE	
CONCRETE			SAME AS PLAN VIEW
CONCRETE BLOCK			
GRAVEL FILL	SAME AS SECTION	NONE	
WOOD	FLOOR AREAS LEFT BLANK	SIDING PANEL	FINISH FRAMING
BRICK	FACE COMMON	FACE OR COMMON	SAME AS PLAN VIEW
STONE	CUT RUBBLE	CUT RUBBLE	CUT RUBBLE
STRUCTURAL STEEL		INDICATE BY NOTE	SPECIFY
SHEET METAL FLASHING	INDICATE BY NOTE		SHOW CONTOUR
INSULATION	SAME AS SECTION	INSULATION	LOOSE FILL OR BATT BOARD
PLASTER	SAME AS SECTION	PLASTER	STUD LATH AND PLASTER
GLASS			LARGE SCALE SMALL SCALE
TILE			

Fig. 22-20. Architectural material symbols used on drawings.

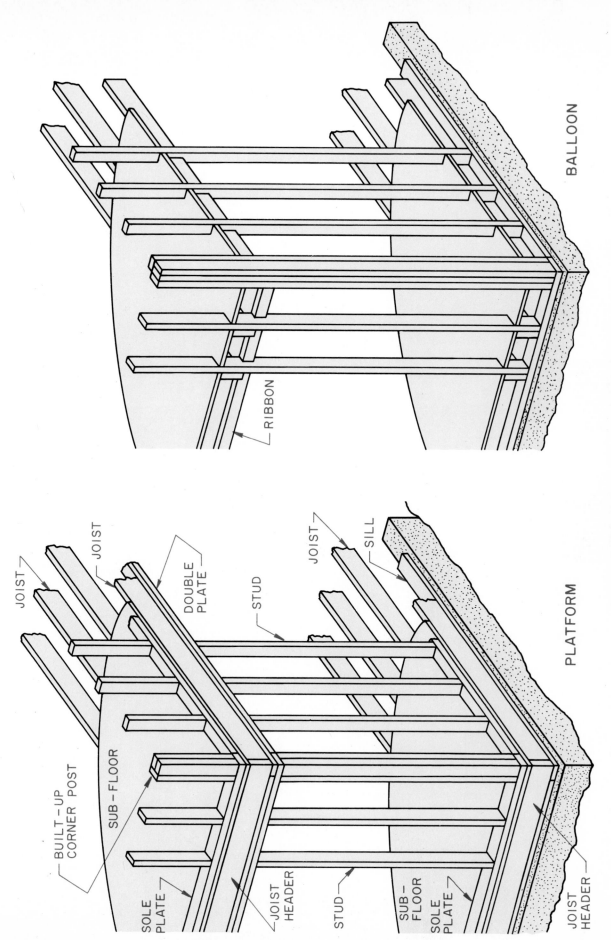

BALLOON

RIBBON

JOIST

JOIST

BUILT–UP
CORNER POST

SUB–FLOOR

SOLE
PLATE

JOIST HEADER

STUD

DOUBLE
PLATE

STUD

JOIST

SILL

SUB–
FLOOR

SOLE
PLATE

JOIST
HEADER

PLATFORM

Fig. 22-21. Platform and balloon framing of exterior walls.

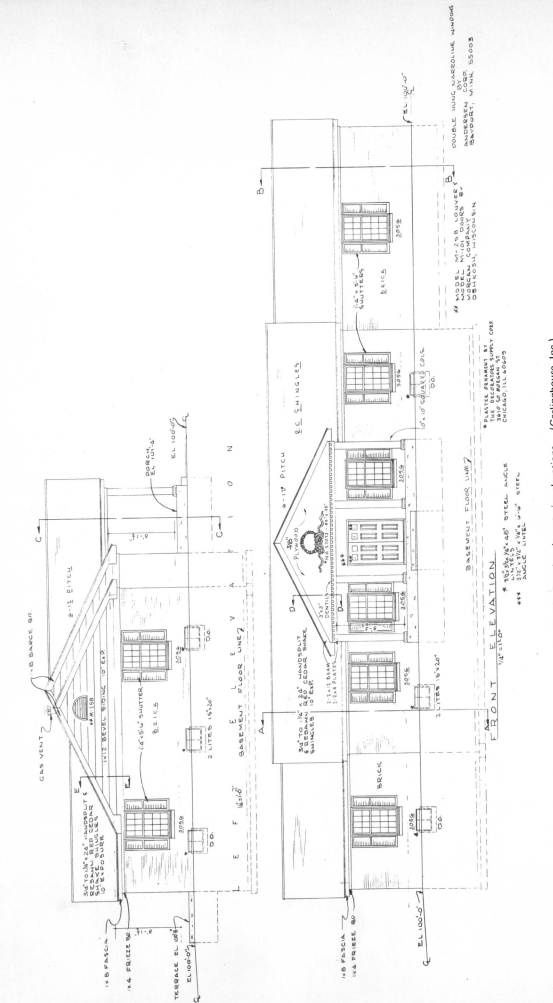

Fig. 22-22. Drawing the exterior elevations. (Garlinghouse, Inc.)

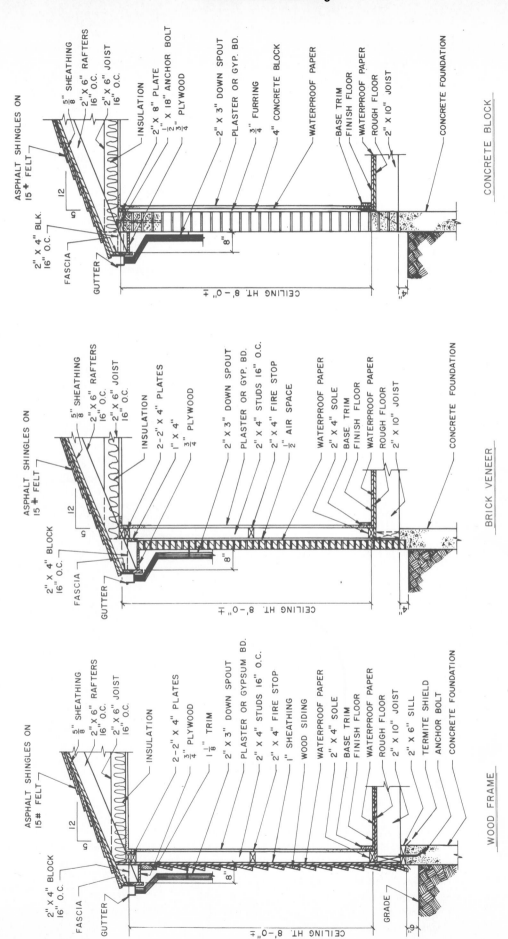

Fig. 22-23. Detail sections of three types of walls.

CONCRETE BLOCK

ASPHALT SHINGLES ON 15# FELT
5/8" SHEATHING
2" X 6" RAFTERS 16" O.C.
2" X 6" JOIST 16" O.C.
INSULATION
2" X 8" PLATE
1/2" X 18" ANCHOR BOLT
3/4" PLYWOOD
2" X 3" DOWN SPOUT
PLASTER OR GYP. BD.
3/4" FURRING
4" CONCRETE BLOCK
WATERPROOF PAPER
BASE TRIM
FINISH FLOOR
WATERPROOF PAPER
ROUGH FLOOR
2" X 10" JOIST
CONCRETE FOUNDATION
2" X 4" BLK. 16" O.C.
FASCIA
GUTTER
CEILING HT. 8'-0" ±
8"
4"

BRICK VENEER

ASPHALT SHINGLES ON 15# FELT
5/8" SHEATHING
2" X 6" RAFTERS 16" O.C.
2" X 6" JOIST 16" O.C.
INSULATION
2-2" X 4" PLATES
1" X 4"
3/4" PLYWOOD
2" X 3" DOWN SPOUT
PLASTER OR GYP. BD.
2" X 4" STUDS 16" O.C.
2" X 4" FIRE STOP
1/2" AIR SPACE
WATERPROOF PAPER
2" X 4" SOLE
BASE TRIM
FINISH FLOOR
WATERPROOF PAPER
ROUGH FLOOR
2" X 10" JOIST
CONCRETE FOUNDATION
2" X 4" BLOCK 16" O.C.
FASCIA
GUTTER
CEILING HT. 8'-0" ±
8"
4"

WOOD FRAME

ASPHALT SHINGLES ON 15# FELT
5/8" SHEATHING
2" X 6" RAFTERS 16" O.C.
2" X 6" JOIST 16" O.C.
INSULATION
2-2" X 4" PLATES
3/4" PLYWOOD
1 1/8" TRIM
2" X 3" DOWN SPOUT
PLASTER OR GYPSUM BD.
2" X 4" STUDS 16" O.C.
2" X 4" FIRE STOP
1" SHEATHING
WOOD SIDING
WATERPROOF PAPER
2" X 4" SOLE
BASE TRIM
FINISH FLOOR
WATERPROOF PAPER
ROUGH FLOOR
2" X 10" JOIST
2" X 6" SILL
TERMITE SHIELD
ANCHOR BOLT
CONCRETE FOUNDATION
2" X 4" BLOCK 16" O.C.
FASCIA
GUTTER
GRADE
CEILING HT. 8'-0" ±
8"
6"

Drafting for Industry

EXTERIOR ELEVATIONS

Elevations are the views looking directly at the sides of the house. All sides must be shown unless they are identical and this information is noted on the drawing. All elevations may be included on the same sheet or each on a separate sheet, depending on the space available. When two or more views are included, they should be aligned either vertically or horizontally to permit projections from one view to another.

Front and rear elevations are often aligned vertically or horizontally. When a side elevation is shown with a front or rear elevation, it should appear as is (left side elevation to left of front view, if aligned horizontally). The rules of orthographic projection are not as strict in architectural drafting but should be observed when feasible.

Elevations show the floor line, ceiling line and first floor joist line as center lines, Fig. 22-22. The floor line represents the top surface of the subfloor, and the line at the lower edge of the first floor joists represents the basement ceiling or lower edge of joists over unexcavated areas. The grade line may be drawn lightly with instruments, then darkened freehand to indicate irregularities. Footings and foundation walls are shown by hidden lines below grade level. Only visible edges are shown above grade level.

WALL SECTIONS

Wall sections are drawings of imaginary cuts through a typical wall of a building to show more clearly the construction details, Fig. 22-23. They are usually drawn at a scale of 1/2" = 1' 0". Wall sections are an important part of a set of plans because they clearly communicate to the contractor and his craftsmen how the building is to be constructed. Any wall that is to be constructed in a different manner from that shown in the above mentioned section should also be detailed. An example might be a wall which joins the house and an adjacent building, such as a garage.

Three types of poured concrete foundations are in common use for houses. One is the slab foundation (see Fig. 22-14), in which the foundation and the floor are poured at the same time. Another is the foundation wall, which is the same representation as the basement foundation wall except for height. The third type is the foundation pier. Detail sections for these foundations are shown in Fig. 22-24.

FIREPLACE PLAN VIEW AND DETAIL SECTION

Fireplaces may be constructed of solid masonry or a commercial, double-walled, sheet-metal unit may be installed and "framed" in masonry. Freestanding fireplaces which

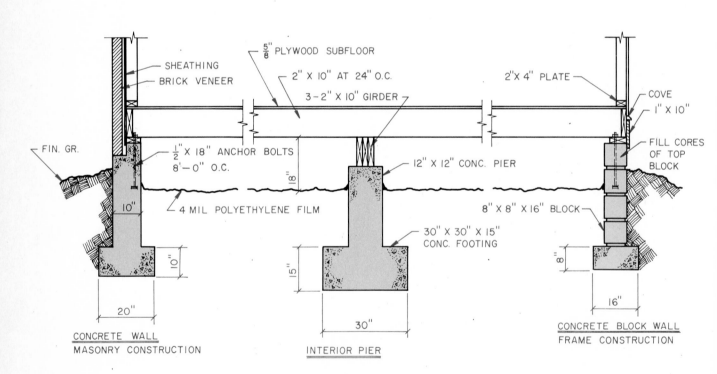

Fig. 22-24. Detail sections of poured concrete foundations.

Start the elevation drawings by beginning with the floor line on the elevation with the largest gable end. Locate the ceiling line on the exterior wall to indicate plate height on which the bird cut in the rafters will rest. The pitch (slope) of the rafters is the rise divided by the run, in our example 6:12 or one-half pitch. That means for every 12 feet of run the rafter rises 6 feet. This will locate the roof height for the elevation drawing.

require no masonry construction have become popular.

The methods of representing a solid masonry fireplace in the plan view and in a detail section are shown in Fig. 22-25. Dimensions are given for the fireplace shown, but these should be checked for the particular size and style of fireplace desired. Where a commercial unit is to be used, the number of the unit and method of detailing should be obtained from the manufacturer's literature.

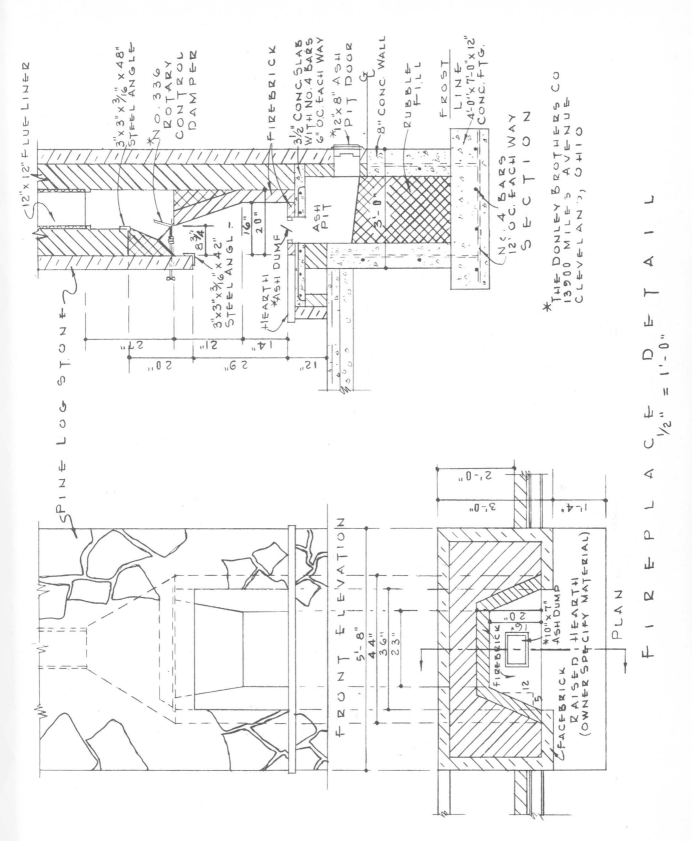

Fig. 22-25. Methods of representing a solid masonry fireplace in the plan view and detail section.

STAIRWAY PLAN VIEW AND DETAIL SECTION

The method of representing stairways on the plan view and on the elevation are shown in Fig. 22-26. Stairways that are functionally well designed have minimum size standards that should be observed. These are:

WIDTH OF STAIRWELL	MINIMUM	PREFERRED
Main stairs	2' – 8''	3' – 2'' to 3' – 6''
Service stairs	2' – 6''	3' – 0''

RISE PER STEP* (RISER HEIGHT)	MAXIMUM	PREFERRED
Interior stairs	8 1/4''	7'' to 7 5/8''
Exterior stairs	7 1/2''	7''

RUN PER STEP* (TREAD WIDTH)	MINIMUM	PREFERRED
Interior stairs	9'' + 1/2'' nosing	10 1/2'' to 11'' + 1/2'' nosing
Exterior stairs	10'' + 1'' nosing	11'' + 1'' nosing
Step without nosing	11''	12''

HEADROOM**	MINIMUM	PREFERRED
Main stairs	6' – 8''	7' – 6''
Service stairs	6' – 4''	7' – 0''

STAIR RAIL HEIGHT		STANDARD
At rake angle of stairs		30''
At stair landings		34''

  * The ideal proportion of tread width to rise is 17 to 18 inches when the two are added together, or 70 to 75 inches when the two are multiplied.
** Measured vertically at front corner of step or nosing.

To calculate the number of treads and risers for a stairway, divide the preferred riser size into the total rise of the stairway. In the example in Fig. 22-26, the total rise is 107 inches. Divide this by an ideal riser size of 7 1/2 inches, which equals 14.26 risers. All steps must be equal, so divide by the number of the closest full step, which is 14. This gives 14 risers of 7.64 inches.

The number of treads is always one less than the number of

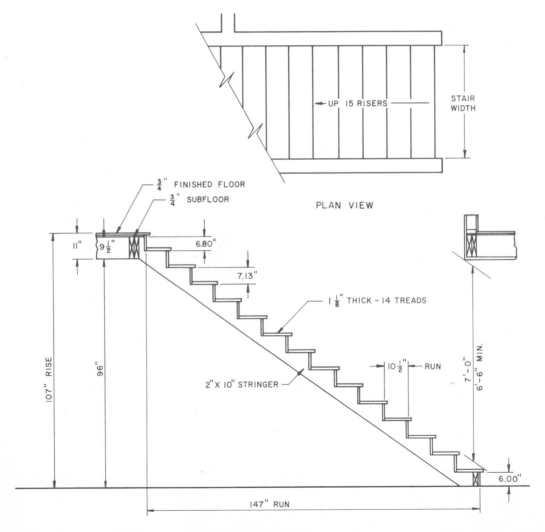

PLAN VIEW

STAIRWAY DETAIL

Fig. 22-26. Stairway representation in plan view and detail section.

DOOR SCHEDULE

CODE	QUAN.	SIZE	THK.	ROUGH OPENING	MASONRY OPNG.	JAMB SIZE	TYPE	DESIGN	REMARKS
A	1	3'-0"x 6'-8"	1$\frac{3}{4}$"	3'-3"x 6'-10"	3'-4" x ——	1$\frac{5}{16}$"x 4 1/2"	HINGED	1-LT. FLUSH SOLID CORE	FRONT ENTRANCE DOOR
B	2	2'-8"x 6'-8"	1$\frac{3}{4}$"	2'-11"x 6'-10"	3'-0" x ——	1$\frac{5}{16}$" x 4 1/2"	HINGED	3-LTS., 1 PANEL	EXTERIOR DOORS TO BREEZEWAY
C	1	9'-0"x 7'-0"	1$\frac{3}{8}$"	9'-3"x 7'-1 1/2"	9'-4" x ——	3/4" x 5 5/8"	OVERHEAD	2-LTS., 8 PANELS	GARAGE DOOR
D	1	(2) 3'-0"x 6'-8"	1$\frac{3}{8}$"	6'-2 1/2"x 6'-9 3/4"		3/4" x 4 5/8"	HINGED	FLUSH HOLLOW CORE	STORAGE DOORS IN GARAGE
E	2	2'-8"x 6'-8"	1$\frac{3}{8}$"	2'-10 1/2"x 6'-9 3/4"		3/4" x 4 5/8"	HINGED	FLUSH HOLLOW CORE	INTERIOR DOORS
F	3	2'-6"x 6'-8"	1$\frac{3}{8}$"	2'-8 1/2" x 6'-9 3/4"		3/4" x 4 5/8"	HINGED	FLUSH HOLLOW CORE	INTERIOR DOORS
G	1	2'-4"x 6'-8"	1$\frac{3}{8}$"	2'-6 1/2"x 6'-9 3/4"		3/4" x 4 5/8"	HINGED	FLUSH HOLLOW CORE	INTERIOR DOOR
H	3	2'-0"x 6'-8"	1$\frac{3}{8}$"	2'-2 1/2"x 6'-9 3/4"		3/4" x 4 5/8"	HINGED	FLUSH HOLLOW CORE	INTERIOR DOORS
J	2	(2)3'-0"x6'-8"	1$\frac{3}{8}$"	6'-1 1/2" x 6'-10 1/2"		3/4" x 4 5/8"	SLIDING	FLUSH HOLLOW CORE	BY-PASSING CLOSET DOORS
K	4	(2)2'-6"x6'-8"	1$\frac{3}{8}$"	5'-1 1/2" x 6'-10 1/2"		3/4" x 4 5/8"	SLIDING	FLUSH HOLLOW CORE	BY-PASSING CLOSET DOORS
L	1	2'-0"x 6'-8"	1$\frac{3}{8}$"	4'-2" x 6'-11 1/4"			SLIDING	FLUSH HOLLOW CORE	RECESSED DOOR
M	2	4'-0"x 6'-8"		4'-2 1/2"x 6'-9 3/4"		3/4" x 4 5/8"			CASED OPENING
N	1	3'-0"x 6'-8"		3'-2 1/2"x 6'-9 3/4"		3/4" x 4 5/8"			CASED OPENING
O	1	2'-6"x 6'-8"		2'-8 1/2"x 6'-9 3/4"		3/4" x 4 5/8"			CASED OPENING
P	1	6'-1 1/2"x 6'-8"		6'-0 3/8"x 6'-9"		WALLBOARD	FOLDING	WOOD SLATS	*PELLA WOOD FOLDING DOORS

WINDOW SCHEDULE

CODE	QUAN.	NO. LTS.	GLASS SIZE	SASH SIZE	ROUGH OPENING	MASONRY OPENING	REMARKS
R	1	1	60" X 48"	9'-3" X 4'-4"	9'-4 1/2"x 4'-7"	9'-8 3/8"x ——	*1 - NO.64 PELLA FIXED CASEMENT UNIT FLANKED EACH
		2	20" X 48"				SIDE BY ONE (1) NO. 24 PELLA CASEMENT WINDOW UNIT.
S	1	3	20" X 60"	5'-11" X 5'-4"	6'-0 1/2"x 5'-7"	6'-4 3/8"x ——	*NO. 325 PELLA CASEMENT WINDOW UNIT
T	2	3	16" X 60"	4'-11" X 5'-4"	5'-0 1/2"x 5'-7"	5'-4 3/8"x ——	*NO. 325N PELLA CASEMENT WINDOW UNIT
U	2	3	20" X 48"	5'-11" X 4'-4"	6'-0 1/2"x 4'-7"	6'-4 3/8"x ——	*NO. 324 PELLA CASEMENT WINDOW UNIT
V	1	4	20" X 36"	7'-11" X 3'-4"	8'-0 1/2"x 3'-7"	8'-4 3/8"x ——	*NO. 423 PELLA CASEMENT WINDOW UNIT
W	2	3	20" X 36"	5'-11" X 3'-4"	6'-0 1/2"x 3'-7"	6'-4 3/8"x ——	*NO. 323 PELLA CASEMENT WINDOW UNIT
X	1	2	20" X 36"	3'-11" X 3'-4"	4'-0 1/2"x 3'-7"	4'-4 3/8"x ——	*NO. 223 PELLA CASEMENT WINDOW UNIT
Y	1	2	20" X 24"	3'-11" X 2'-4"	4'-0 1/2"x 2'-7"	4'-4 3/8"x ——	*NO. 222 PELLA CASEMENT WINDOW UNIT
Z	2	1	20" X 36"	1'-11" X 3'-4"	2'-0 1/2"x 3'-7"	2'-4 3/8"x ——	*NO. 23 PELLA CASEMENT WINDOW UNIT
A'	2	4	20 1/2"x 8"	(2) 2'-0"x 4'-2"	4'-7 1/2" X 4'-5 3/8"	4'-8"x ——	**NO. 2042 ANDERSEN DOUBLE HUNG WINDOW UNIT 1-4"MULLION

*ROLSCREEN CO.
PELLA, IOWA

**ANDERSEN CORP.
BAYPORT, MINN.

Fig. 22-27. Door and window schedules. (Garlinghouse, Inc.)

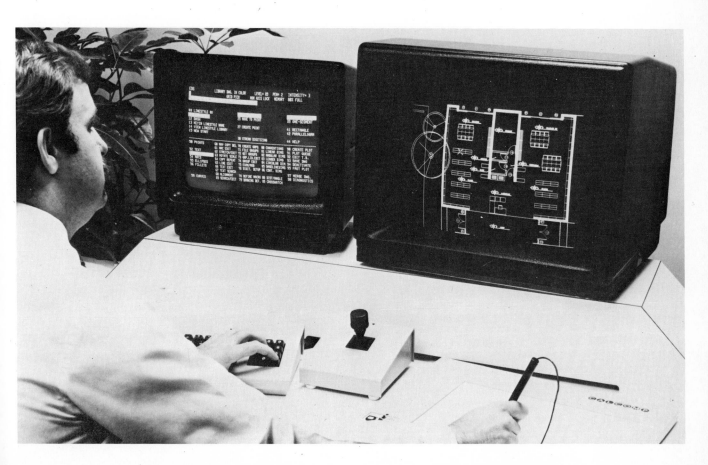

Architects are finding computer graphics a valuable tool in their work. (California Computer Products, Inc.)

risers (the top landing serves as a tread). An ideal size tread width is 10 1/2 inches. To check these proportions, add the riser to the tread (7.64 + 10.5 = 18.14 inches). This sum exceeds the ideal proportion of 17 to 18 inches. When 7.64 is multiplied by 10.5, the product is 80.22 inches, which also exceeds the ideal proportion of 70 to 75 inches.

Since the riser height is at the upper limits of ideal (7 5/8 inches), 15 risers divided into the total rise will give you a riser height of 7.13 inches. Using the same tread width of 10 1/2 inches, check this against the ideal proportion by adding the two figures (17.63), then by multiplying (74.87). Both are satisfactory. To obtain the total run, multiply the tread width

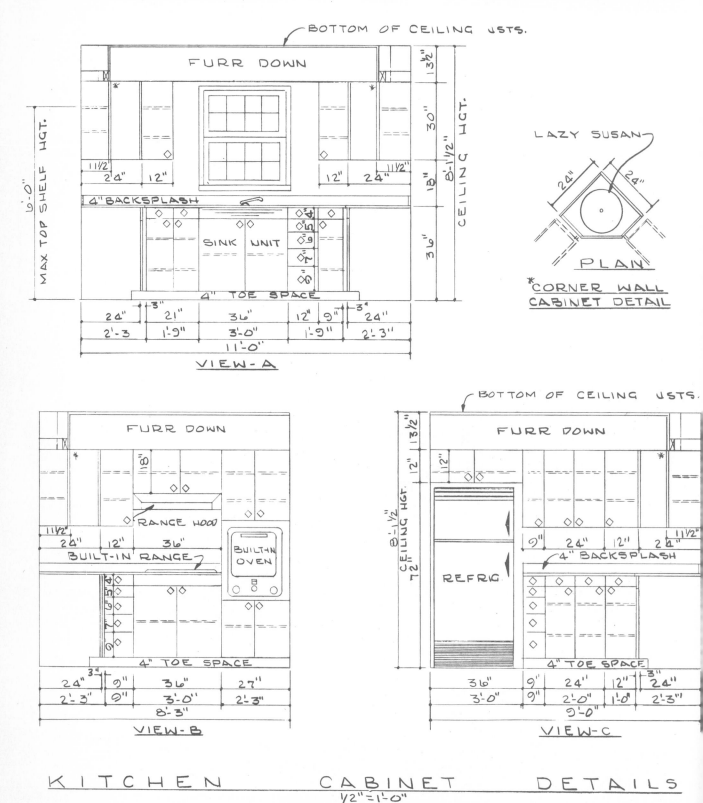

Fig. 22-28. An elevation showing kitchen cabinet arrangement.

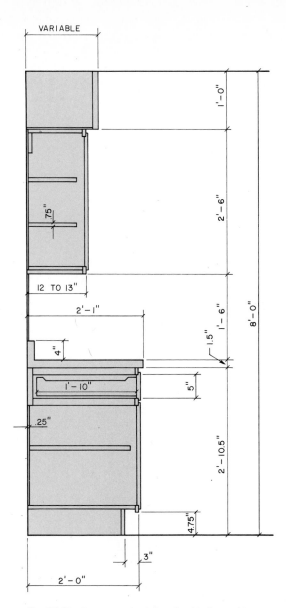

Fig. 22-29. Some standard sizes for kitchen cabinets.

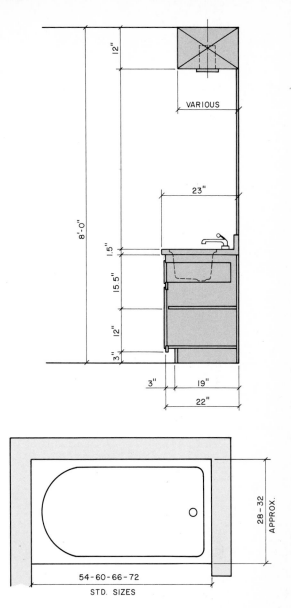

Fig. 22-30. Some standard sizes for bathroom cabinets.

of 10.5 inches by 14 (one less than the number of risers) to get a total run of 147 inches.

Using these dimensions, the headroom shown in the detail section, Fig. 22-26, is satisfactory.

DOOR AND WINDOW SCHEDULES

The schedules for doors and windows to be used in a house are listed in tabular form, usually on the sheet with the detail drawings or on a separate sheet. The key number or letter referencing each item to the drawing is shown. The size, material, description and manufacturer's number, if specified, is given, Fig. 22-27.

CABINET AND SHELF DETAILS

Kitchen and bathroom cabinets are represented on the floor plan, but elevations of these should be drawn and dimensioned or specified by the manufacturer's number. An elevation of kitchen cabinets is shown in Fig. 22-28. Some standard sizes for kitchen cabinets are given in Fig. 22-29 and for bathroom

cabinets in Fig. 22-30.

Sometimes the plan calls for shelves to be built into the walls of the den or study, Fig. 22-31. Such shelves should be located and noted on the plan drawing and detailed in an elevation drawing. Other fixtures such as showers, lavatories and built-in units should also be shown on the plan drawing using the correct symbols, Fig. 22-32.

ARRANGING THE VIEWS

The plan views and the elevations may all be included on the same sheet or each on a separate sheet, depending on the space available. When two or more related views are included on the same sheet, they should be aligned vertically or horizontally to permit projections from one view to another.

The floor plan and basement plan can be aligned vertically, as can the front and rear elevations. Side elevations can be aligned horizontally with the front and rear elevations or vertically with each other.

The normal practice is to group the plan views on one or

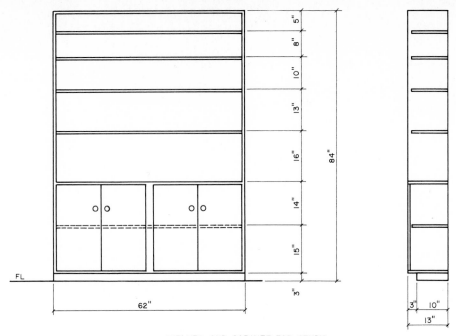

SHELVES AND CABINET FOR STUDY

Fig. 22-31. An elevation drawing of shelves to be built into the wall.

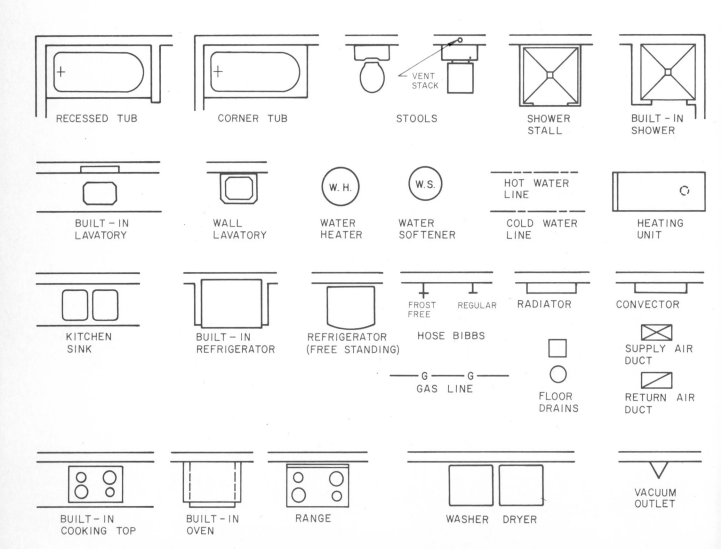

Fig. 22-32. Symbols for plumbing, appliances and mechanical equipment.

Fig. 22-33. An architectural rendering of a house improves the appearance of the drawing and helps in the presentation. The elevation drawings for this home are shown in Fig. 22-22. (Garlinghouse, Inc.)

more sheets, the elevations and sections on one or more sheets, and the details of cabinets, shelving, stairways and other such items together on one or more sheets.

CHECK LIST FOR A SET OF ARCHITECTURAL DRAWINGS

A complete set of architectural working drawings would include, usually on four or more sheets, the following drawings and items:

PLAN DRAWINGS
Floor plan
Roof and plot plan
Footing and foundation plan
Basement plan (if basement is included)

Electrical plan
Framing plans

ELEVATIONS AND SECTION DRAWINGS
Exterior elevations
Wall sections
Fireplace sections
Stairway elevation

DETAIL DRAWINGS AND SCHEDULES
Door and window schedules
Cabinet and built-in shelf details

The number of sets of prints needed depends on the number of individual subcontractors, government and finance offices, and suppliers requiring prints in addition to the owner.

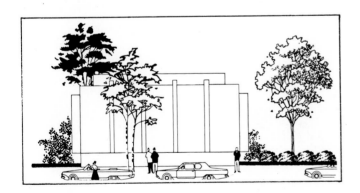

Fig. 22-34. Examples of symbols which can be transferred to assist in preparing architectural renderings. (Para-Tone Inc.)

ARCHITECTURAL RENDERINGS AND SCALED MODELS

The presentation and promotion of a solution to an architectural design problem can be considerably enhanced by the use of renderings and models.

RENDERING ARCHITECTURAL DRAWINGS

Rendering is the process of giving a drawing a more realistic appearance. It is usually done by shading and sketching in shrubs and trees on a perspective view of a house, Fig. 22-33. This can be done by charcoal pencil, soft lead pencil or airbrush. Appliques and transfer symbols of trees and shrubs are also available for use in making renderings, Fig. 22-34.

ARCHITECTURAL MODELS

Scale models of architectural designs help the client to visualize how the completed project will appear, Fig. 22-35. Model making in some architectural and engineering firms is a full time job for skilled persons in this field.

Fig. 22-36. The lot is rectangular with a frontage of 80 feet and a depth of 115 feet. Indicate the north direction.

3. Make a footing and foundation plan for the house shown in Fig. 22-36. There is no basement, and the foundation is a 36 inch stem wall on a footing. On the same sheet, prepare a detail section of the foundation showing any beams and piers necessary in the foundation to support the floor and interior walls.

4. Trace the floor plan which you made of the house in Fig. 22-36 and prepare an electrical wiring plan for the house. Check the local electrical code, if one is available, for the requirements on spacing wall outlets. Show lines to switches on all outlets controlled by switches.

5. Prepare floor, ceiling and roof framing plans for the house shown in Fig. 22-36. Show double framing members where required and dimension as a working drawing.

6. Make a front and side elevation for the house shown in Fig. 22-36. Also make a wall section to show details of construction. Add necessary dimensions and notes to the elevations and wall section.

Fig. 22-35. Scaled model of an automobile service station.
(Cities Service Oil Co.)

PROBLEMS IN ARCHITECTURAL DRAFTING

The following problems are planned to give you experience with the basics of architectural drafting. You will have an opportunity to make various types of drawings used in architectural work and to try your problem solving skills in the field of architectural design.

1. Prepare a scaled working drawing of the floor plan of the house shown in Fig. 22-36. Include all necessary dimensions and notes, but do not include the electrical wiring plan.

2. Prepare a scaled roof and plot plan of the house shown in

7. Prepare interior plan and elevations of the kitchen cabinets and window and door schedules, Fig. 22-36. Check a builder's supply catalog for specifications.

8. Make a freehand sketch of the floor plan of your own home. Grid paper may be used if desired. When the sketch is approved by your instructor, make a presentation drawing of the floor plan showing room sizes and overall dimensions.

9. Design and prepare a working drawing of the floor plan for the house shown in Fig. 22-37. The house has no basement.

10. Based on the floor plan design in 9, prepare elevation

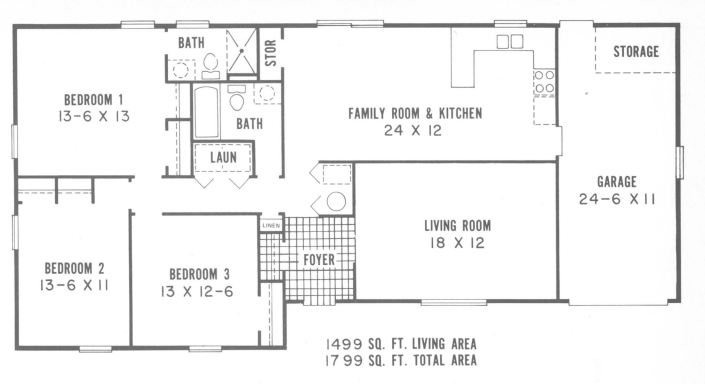

1499 SQ. FT. LIVING AREA
1799 SQ. FT. TOTAL AREA

Fig. 22-36. A presentation drawing of a three bedroom house.
(National Homes Corp.)

drawings for all elevations. You may select the exterior wall material and the type of foundation suitable for your area. Include a detail of a typical wall section and the foundation wall.

11. Design the interior elevations for the kitchen and bathrooms for the house in Fig. 22-37 and prepare the working drawing.

12. Select one of the designs of the houses shown in Fig. 22-38 and prepare a full set of working drawings. You may select the lot size and direction of orientation of the house.

13. Select one of the floor plan designs in Fig. 22-39 and prepare a full set of working drawings. You may select the exterior wall material, roof style and material, lot size and house orientation.

14. Design a house to meet the needs of a particular family. Prepare a form to gather information on the items discussed in Designing Of A House. Interview the family and make note of the items discussed. Study the information collected and design and sketch a floor plan which meets their needs. Prepare a presentation drawing of the floor plan and a two-point perspective rendering of the house to show to the family.

15. After you and the family have discussed the presentation drawing and rendering, prepare a full set of working drawings for the house.

16. With several of your class members, organize a team to study and plan the solution to some community or school

Fig. 22-37. A perspective rendering of a four bedroom house.
(National Homes Corp.)

463

need such as: converting a vacant lot into a playground or park; a weekend cabin for a boys' or girls' club; shelter for spectators at a ball park; swimming pool shower and locker room or a chip-and-putt golf course.

Check your proposed study and plan of attack with your instructor and get his or her suggestions on persons to interview for information and ideas. You may want to divide the responsibilities for interviewing and planning various aspects of the project and assign these to various members of your design team.

Study the information gathered and, using the problem solving Design Method discussed in Chapter 9, prepare the presentation drawings and make a formal presentation of your solution to the proper authorities. With their approval, prepare the working drawings and budget estimate necessary to complete the project.

17. Make a scaled model of one of the projects developed in the earlier problems in this chapter. Check with your instructor on sources of supply to construct and complete the decor of the model.

Fig. 22-38. Views of houses for which a set of working drawings are to be prepared. (Garlinghouse, Inc.)

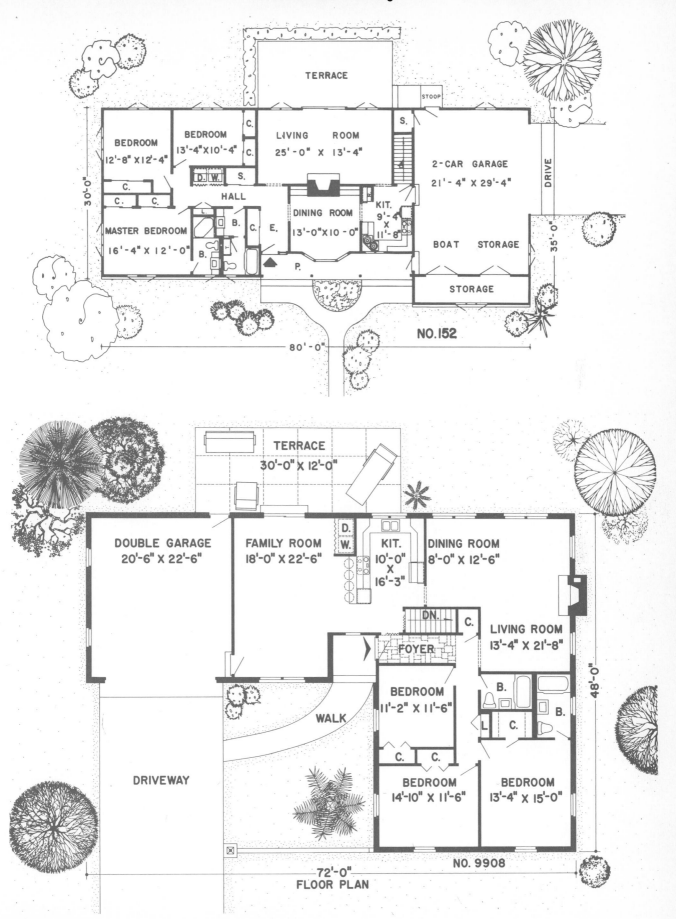

TERRACE

BEDROOM
12'-8" X 12'-4"

BEDROOM
13'-4" X 10'-4"

C.

C.

C.

LIVING ROOM
25'-0" X 13'-4"

S.

STOOP

2-CAR GARAGE
21'-4" X 29'-4"

DRIVE

30'-0"

D.W.

S.

HALL

C.

L.

B.

B.

C. E.

DINING ROOM
13'-0" X 10'-0"

KIT.
9'-4"
X
11'-8"

BOAT STORAGE

35'-0"

MASTER BEDROOM
16'-4" X 12'-0"

C.

C.

C.

P.

STORAGE

80'-0"

NO. 152

TERRACE
30'-0" X 12'-0"

DOUBLE GARAGE
20'-6" X 22'-6"

FAMILY ROOM
18'-0" X 22'-6"

D.
W.

KIT.
10'-0"
X
16'-3"

DINING ROOM
8'-0" X 12'-6"

DN.

C.

LIVING ROOM
13'-4" X 21'-8"

FOYER

BEDROOM
11'-2" X 11'-6"

B.

B.

L.

C.

48'-0"

DRIVEWAY

WALK

C.

C.

BEDROOM
14'-10" X 11'-6"

BEDROOM
13'-4" X 15'-0"

72'-0"
FLOOR PLAN

NO. 9908

Fig. 22-39. Floor plan designs. (Continued)   (Garlinghouse, Inc.)

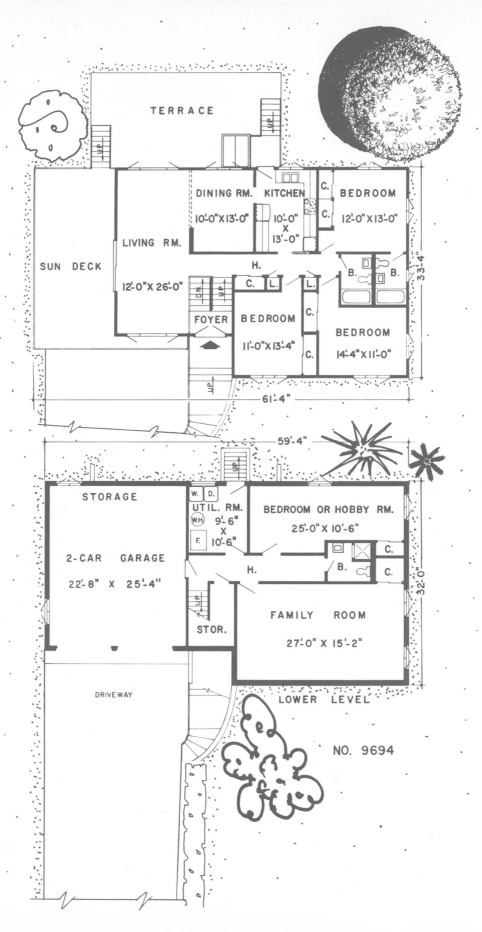

Fig. 22-39. (Continued) Floor plan designs. (Garlinghouse, Inc.)

466

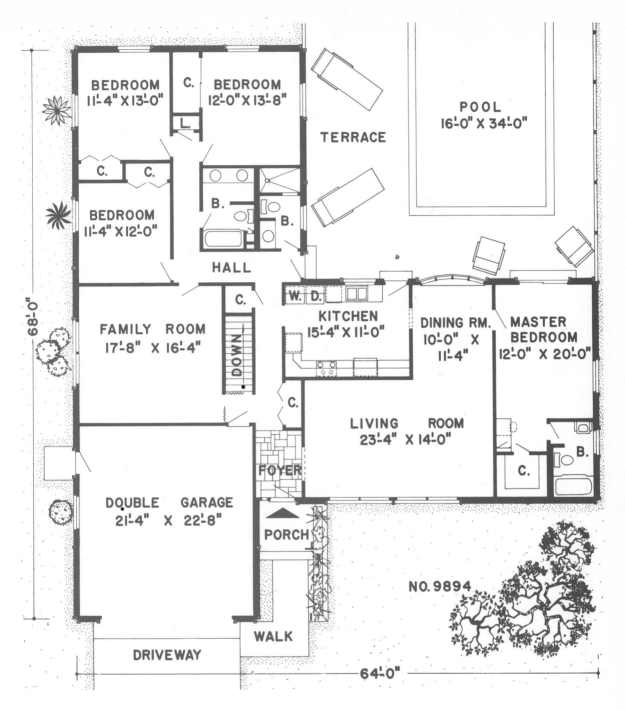

BEDROOM
11'-4" X 13'-0"

C.

BEDROOM
12'-0" X 13'-8"

L.

POOL
16'-0" X 34'-0"

TERRACE

C.   C.

BEDROOM
11'-4" X 12'-0"

B.

B.

HALL

FAMILY ROOM
17'-8" X 16'-4"

C.

DOWN

W. D.

KITCHEN
15'-4" X 11'-0"

DINING RM.
10'-0" X
11'-4"

MASTER
BEDROOM
12'-0" X 20'-0"

C.

LIVING   ROOM
23'-4" X 14'-0"

B.

C.

FOYER

DOUBLE   GARAGE
21'-4"   X   22'-8"

PORCH

NO. 9894

WALK

DRIVEWAY

68'-0"

64'-0"

Fig. 22-39. (Continued) Floor plan designs.   (Garlinghouse, Inc.)

Chapter 23
ELECTRICAL AND ELECTRONICS DRAWINGS

Fig. 23-1. This single chip, the size of a dime, is the first full 32-bit microprocessor. It contains 150,000 transistors and was made possible by computer graphics and microelectronics. (Bell Labs)

Growth of the electronics industries in recent years has brought an increased demand for drafters who are capable of preparing various kinds of electrical and electronic circuit drawings. Electrical and electronics drafting involves the same basic drawing principles used in other types of drawings. The difference is in the special symbols that have been developed to represent electrical circuits and wiring diagrams.

Major requirements for electrical and electronics drawings have been standardized by industry through the American National Standards Institute and by the military through the Military Standard Publications. These two sets of standards are almost identical.

Most electrical drawings begin as rough engineering sketches, or as written or verbal instructions from electrical engineers. This information must be interpreted and checked for completeness by the drafter before further work on the drafting assignment is accomplished.

Because of the complex devices and their relationship to each other in an electrical circuit, it is recommended that you acquire a basic knowledge of electricity/electronics and under-

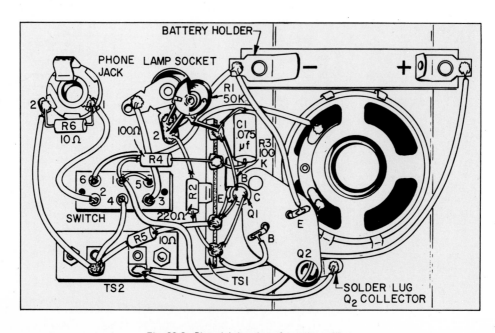

Fig. 23-2. Pictorial drawing of a code oscillator.

stand how an electrical or electronic circuit operates if you wish to specialize in this type of drafting.

ELECTRICAL AND ELECTRONIC COMPONENTS

The components used in electronics vary in size from large transformers at an electrical generating plant to microscopic integrated semiconductor circuits, Fig. 23-1. Most components may be unfamiliar, but the drafter should try to associate the actual component with its graphic symbol. (See Fig. 23-4.)

PICTORIAL DRAWINGS

Pictorial symbols, Fig. 23-2, are sometimes used to illustrate component parts in electrical and electronics drawings. Pictorials are particularly useful for assembly line workers, do-it-yourself hobbyists and other personnel not trained in

reading graphic symbols in electrical and electronics drawings. Fig. 23-2 illustrates a pictorial drawing of several electrical components and their relative position on a circuit board.

PHOTODRAWINGS

Another drafting technique used in electronics work is the photodrawing. This type of drawing is produced by photographing electronic components or assemblies and adding line work to complete the drawing, Fig. 23-3. Photodrawing is discussed in detail in Chapter 28.

GRAPHIC SYMBOLS

According to the definition in ANSI Standard Graphic Symbols for Electrical and Electronics Diagrams, graphic symbols are:

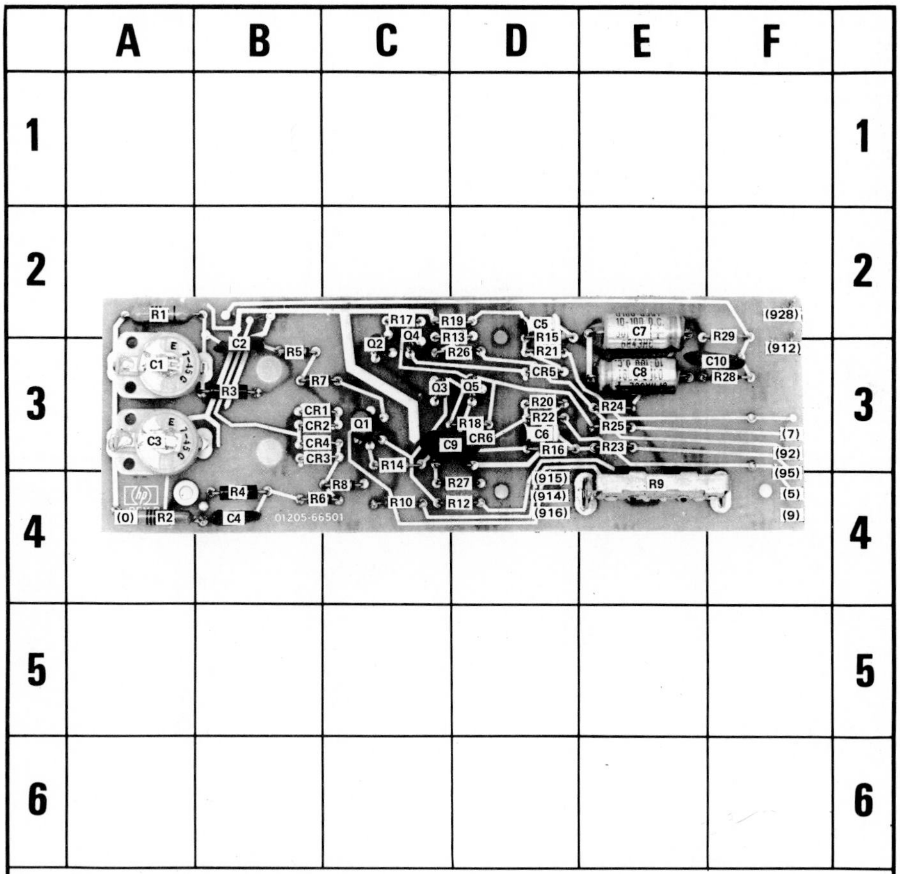

REF DESIG	GRID LOC	REF DESIG	GRID LOC	REF DESIG	GRID LOC	REF DESIG	GRID LOC	REF DESIG	GRID LOC	REF DESIG	GRID LOC	REF DESIG	GRID LOC
C1	A-3	C8	E-3	CR5	D-3	R1	A-2	R8	C-4	R16	D-3	R23	E-3
C2	B-3	C9	C-3	CR6	D-3	R2	A-4	R9	E-3	R17	C-2	R24	E-3
C3	A-3	C10	F-3	Q1	C-3	R3	B-3	R10	C-4	R18	D-3	R25	E-3
C4	B-4	CR1	B-3	Q2	C-3	R4	B-4	R12	D-4	R19	D-2	R26	D-3
C5	D-2	CR2	B-3	Q3	C-3	R5	B-3	R13	B-3	R20	D-3	R27	D-4
C6	D-3	CR3	B-3	Q4	C-2	R6	B-4	R14	C-3	R21	D-3	R28	F-3
C7	E-2	CR4	B-3	Q5	D-3	R7	B-3	R15	D-3	R22	D-3	R29	F-2

Fig. 23-3. Vertical preamplifier, A1A1, component identification. (Hewlett-Packard)

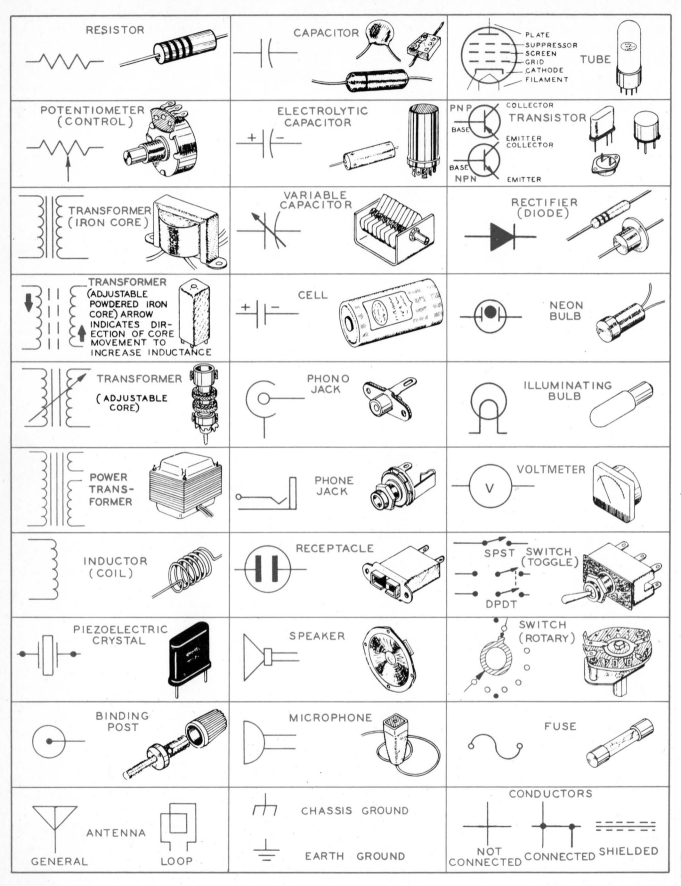

(Reprinted with permission of Heath Company)

Fig. 23-4. Typical components and their symbols.

... a shorthand used to show graphically the functioning or interconnections of a circuit. A graphic symbol represents the function of a part in the circuit. Graphic symbols are used on single-line (one-line) diagrams, on schematic or elementary diagrams, or as applicable on connection or wiring diagrams. Graphic symbols are correlated with parts lists, descriptions or instructions by means of designations.

The key word in the definition is "shorthand," since graphic symbols are capable of conveying a tremendous amount of information about the function of an electrical circuit. However, the language of electronics is constantly evolving, so it becomes the responsibility of the drafter to select the proper symbols and use them correctly on a drawing.

The current editions of ANSI Standard Graphic Symbols for Electrical and Electronics Diagrams should be used as guides when selecting a particular symbol. A few of the many symbols are shown with their corresponding parts in Fig. 23-4.

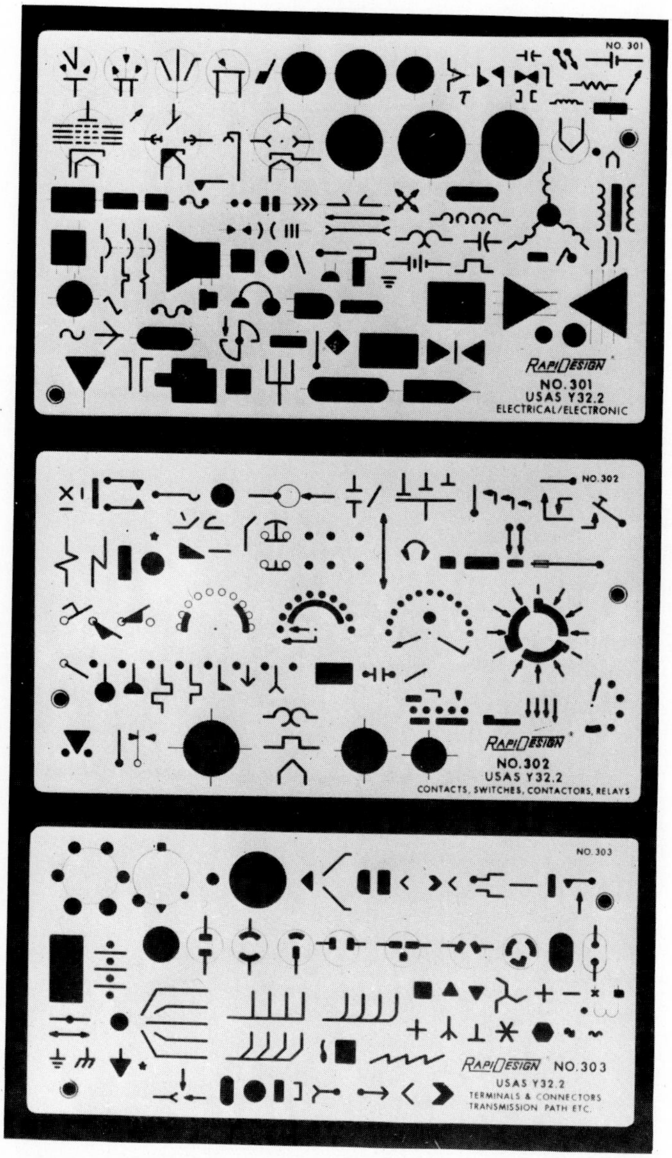

A more complete selection may be found in the Reference Section, page 598.

The symbols in the ANSI guidebook are shown in their relative sizes but they may be reduced or enlarged as long as the relative size is maintained on a particular drawing. The drafter should select a scale based on the size of the final print. Neither the line width nor the orientation affects the meaning of a symbol, but a wider line may be used on part of a drawing for emphasis.

Most electronic drafting today is accomplished with the use of drafting aids such as templates, preprinted symbol appliques and special typesetting equipment, Fig. 23-5. The scale of a drawing is frequently determined by the drafting aids that are available to the drafter.

COMPONENT DESIGNATIONS

A graphic symbol may be repeated many times in a single drawing, therefore it becomes necessary to identify each component by a reference designator, Fig. 23-6. Military

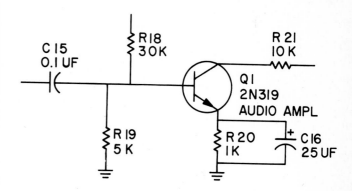

Fig. 23-6. Reference designators.

Standard Electrical and Electronic Reference Designations defines a complete reference designation as:

> ... a unique combination of letters and numbers which identifies a part, subassembly or unit of a set on equipment diagrams, drawings, parts list, technical manuals, etc. The letters in a reference designation identify the class of item such as a resistor, coil or electron tube, or identify a subassembly .... the number differentiates between parts or subassemblies of the same class. A reference designation is not an abbreviation for the name of an item.

Identification letters used on a drawing are contained in ANSI Standard Electrical and Electronics Reference Designations. A partial list of designators is contained in the Reference Section. After the correct designator for a class of item has been selected, the number portion is assigned. The lowest number should always be in the upper left-hand corner of the drawing and proceed consecutively from left to right and from top to bottom layer if two or more layers are used, Fig. 23-7.

UNIT AND SUBASSEMBLY DESIGNATIONS

If two or more units are contained in a set, the unit designations will start with the number "1" and increase

Fig. 23-5. Electrical and electronics drafting templates. (RapiDesign)

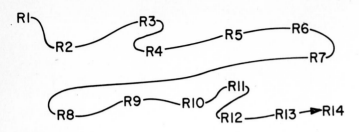

Fig. 23-7. Reference designation number assignments; two or more layers.

HIGHEST REFERENCE DESIGNATIONS		
R 67	C 15	L 2
REFERENCE DESIGNATIONS NOT USED		
R34 , R41	C 3	

Fig. 23-9. Table indicating omitted and highest numerical reference designations.

consecutively for all units of the set. Subassemblies and parts within a unit or subassembly will be assigned reference designations in the following order: unit number; the letter "A" identifying a subassembly; a number identifying a specific subassembly; a letter(s) identifying the class of items to which the part belongs; a number identifying the specific part, as shown in Fig. 23-8.

If some parts are eliminated as the result of a revision, it is not necessary to renumber the remaining parts, but the deleted items should be listed in a note or table, Fig. 23-9. This is done to account for all numbers and to prevent subsequent errors in equipment lists and other documents related to the unit or subassembly.

In addition to the reference designation, it is frequently necessary to indicate the specific type of item within a class, the circuit function, the item location, numerical values, rating and contact or terminal designations. This information should be located as near as possible to the component symbol and centered immediately under the reference designation as shown in Fig. 23-6.

If abbreviations are used, they should be selected from those listed in ANSI Standard Abbreviations for Use on Drawings, or other nationally recognized sources. A word should always be spelled out if there is any doubt about its correct abbreviation.

SINGLE-LINE DIAGRAMS

Single-line diagrams are simplified representations of complex circuits or entire systems. The definition for a single-line diagram in ANSI Standard Electrical and Electronics Diagrams

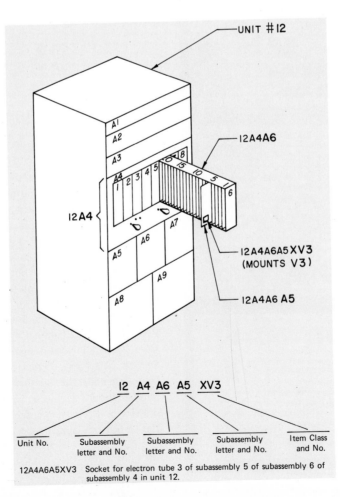

12A4A6A5XV3 Socket for electron tube 3 of subassembly 5 of subassembly 6 of subassembly 4 in unit 12.

Fig. 23-8. Application of reference designations to a unit.

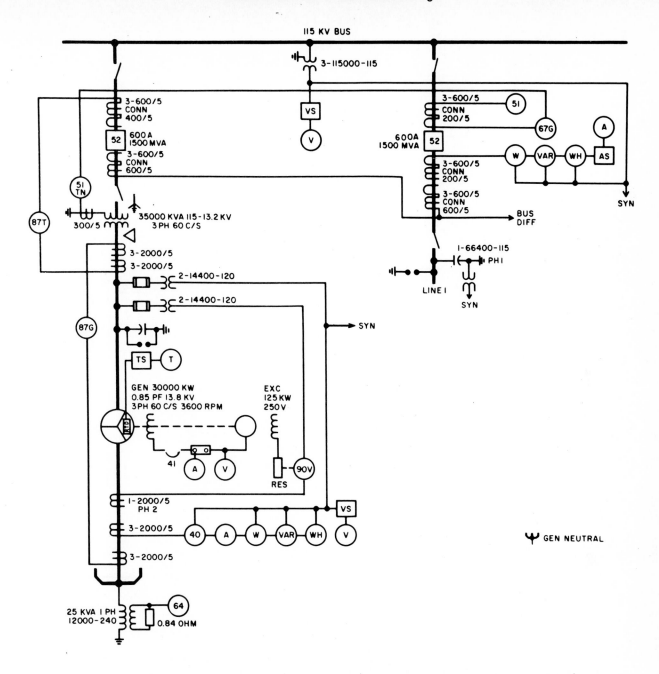

Fig. 23-10. Typical, power switchgear, single-line diagram with complete device designations.   (ANSI)

states that a single-line or one-line diagram is:

> ... a diagram which shows, by means of single lines and graphic symbols, the course of an electric circuit or system of circuits and the component devices or parts used therein.

Single-line diagrams are used primarily in the electrical power and industrial control areas with some limited applications in electronics and communications. However, it is usually one of the first drawings made in the design of a large electrical power system, because it contains the basic information that will serve as a guide in the preparation of more detailed plans.

Fig. 23-10 is a typical, single-line diagram used in the electrical power field. The thick connecting lines on the drawing indicate primary circuits, and the medium lines indicate connections to current or potential sources. In either case, a single line may be used to represent a multiconductor circuit.

In single-line diagrams, it is standard practice to use either horizontal or vertical connecting lines with the highest voltages at the top or left of the drawing and successively lower voltages toward the bottom or right of the drawing. An effort should be made to maintain a logical sequence while avoiding an excessive number of line crossings.

Small circles and rectangles are used to depict components which are identified by abbreviations and letter combinations. The use of graphic symbols and component designations previously described apply to single-line diagrams, but the designations used in a specific field should be selected from the standards listed in this chapter under ELEMENTARY DIAGRAMS.

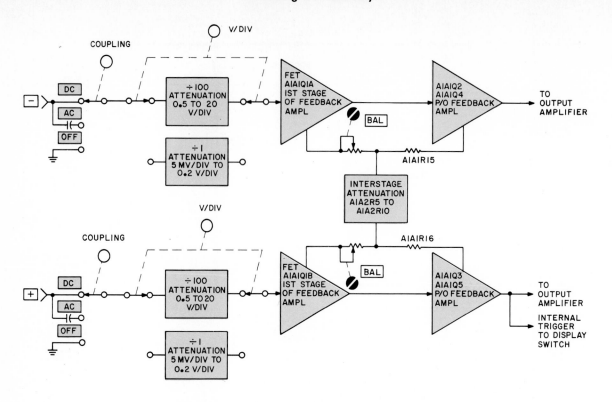

Fig. 23-11. A typical block diagram. (Hewlett-Packard)

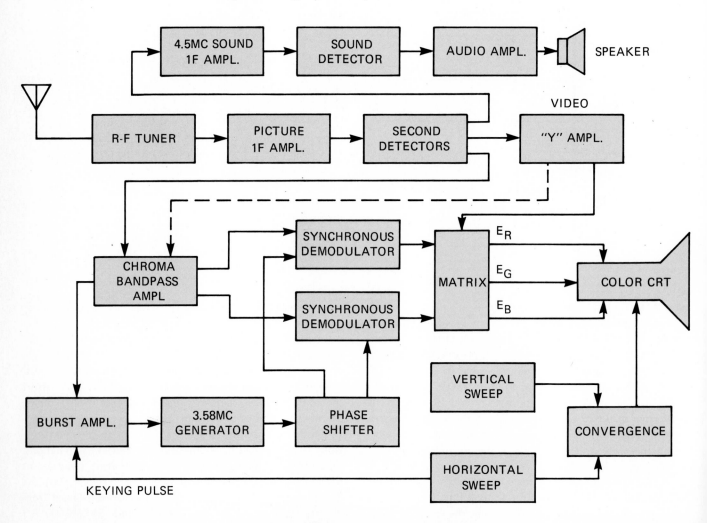

Fig. 23-12. Block diagram of a color television receiver.

BLOCK DIAGRAMS

Block diagrams are closely related to single-line diagrams. Each contains basic information which represents the overview of a system in its simplest form. Squares and rectangles are primarily used on block diagrams, but an occasional triangle or circle may be used for emphasis. Graphic symbols are rarely used except as input and output devices.

The blocks should be arranged in a definite pattern of rows and columns with the main signal path progressing from left to right whenever possible, Fig. 23-11. Auxiliary units, such as power supply or oscillator circuits, should be placed below the main diagram. Each block should contain a brief description or function of the stage it represents. Additional information may be placed elsewhere on the drawing. The block that requires the greatest amount of lettering usually determines the size of all the blocks. However, two block sizes on one drawing is not objectionable.

A heavy line should be used to show the signal path. In a complex circuit or system, more than one line may lead into or away from a block, Fig. 23-12. Arrows should be used to show the direction of the signal flow. The overall appearance should be a consistent and organized pattern that is well balanced and easy to read.

ELEMENTARY DIAGRAMS

Functional drawings prepared for industrial control and electrical power switchgear equipment are referred to as elementary diagrams. The definition given in ANSI Standard

Computer graphics plays a large part in the design and layout of electronic diagrams. (California Computer Products, Inc.)

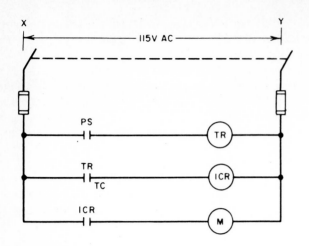

Fig. 23-13. Example of circuit arrangement in functional sequence.

Electrical and Electronics Diagrams includes the schematic diagram, but this diagram will be discussed later in this chapter.

A schematic or elementary diagram is:

. . . a diagram which shows by means of graphic symbols, the electrical connections and functions of a specific circuit arrangement. The schematic diagram facilitates tracing the circuit and its functions without regard to the actual physical size, shape or location of the component device or parts.

The graphic symbols used in industrial control and electrical power drawings are seldom used in electronics and communications drawings. Even the component designations are different and care should be taken in selecting the applicable standard when preparing drawings in a specific area. Standards not previously mentioned which apply to elemen-

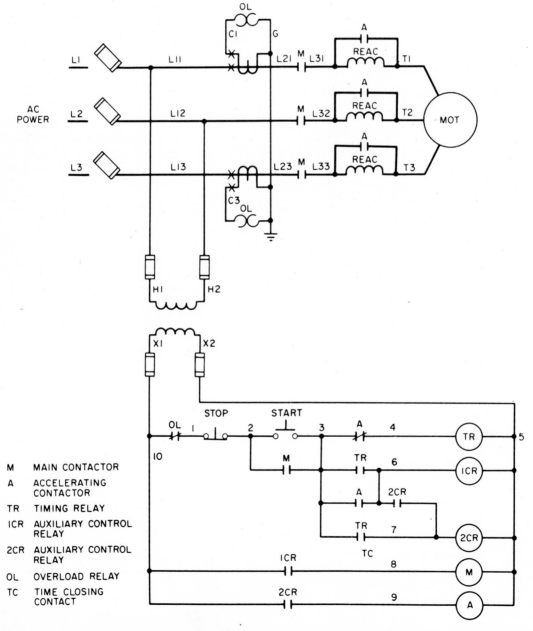

M	MAIN CONTACTOR
A	ACCELERATING CONTACTOR
TR	TIMING RELAY
ICR	AUXILIARY CONTROL RELAY
2CR	AUXILIARY CONTROL RELAY
OL	OVERLOAD RELAY
TC	TIME CLOSING CONTACT

Fig. 23-14. Typical industrial control schematic diagram. (ANSI)

476

tary diagrams are ANSI Standards:

C37.2 — Manual and Automatic Station Control, Supervisory, and Associated Telemetering Equipments

C37.11 — Requirements for Power Circuit Breaker Control

C37.20 — Switchgear Assemblies Including Metal Enclosed Bus

NEMA Standards (National Electrical Manufacturers Association, Inc.)

1C1 — Industrial Control

SG4 — High Voltage Power Circuit Breakers

SG5 — Power Switchgear Assemblies

The circuits should be arranged in a functional sequence from left to right or from top to bottom. This should not be rigidly followed if excessive line crossings reduce the clarity of the drawing. For example, in Fig. 23-13, the sequence of operation would begin by closing the pressure switch PS. This would activate relay TR and close the time-closing contact labeled TR on the next lower level. With contact TR closed, relay 1CR will energize and close contact 1CR which will, in turn, apply power to device M.

Fig. 23-14 is a more detailed drawing of the same circuit function. These diagrams are characterized by the prominent use of contactor, relay, switch and overload symbols. Another feature is the extensive use of notes to specify the operational sequence and other pertinent data.

SCHEMATIC DIAGRAMS

The most frequently used drawing in the electronics field is the schematic diagram. It serves as the master drawing for production drawings, parts lists and component specifications. It is used by the engineering group for circuit design and analysis, and by the technical personnel for installation and maintenance of the finished product. Elements of a schematic diagram may be combined with other types of diagrams to provide a more comprehensive and useful drawing.

A schematic diagram frequently originates as an engineer's or technologist's rough sketch which may be incomplete, contain obsolete symbols or represent nonstandard drafting practices, Fig 23-15 (a). The drafter takes the rough sketch and makes the necessary additions and/or corrections according to standard drafting practices. A freehand sketch is then prepared showing a tentative layout of the schematic diagram. This is usually done on vellum with nonreproducible lines and with the aid of templates and a straightedge.

Throughout, the layout is carefully developed so that the signal path is from upper left to lower right and connecting lines are either horizontal or vertical with as few bends or crossovers as possible. Avoid long lines or groups of lines by interrupting them at convenient points and indicate destinations and line identifiers.

Component designations, notes and related information are then added. When the sketch is complete, it is checked for accuracy, completeness and clarity. The formal schematic diagram is traced or copied from the sketch, Fig. 23-15 (b). A maintenance type schematic diagram illustrating many of the practices described in previous sections is shown in Fig. 23-16, and an equipment schematic diagram in Fig. 23-17.

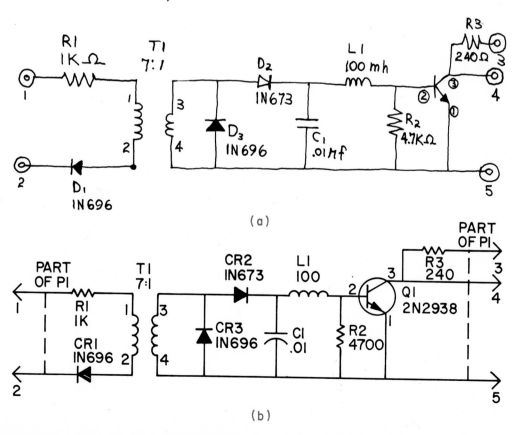

Fig. 23-15. Engineer's sketch of a circuit (a), finished drawing (b).

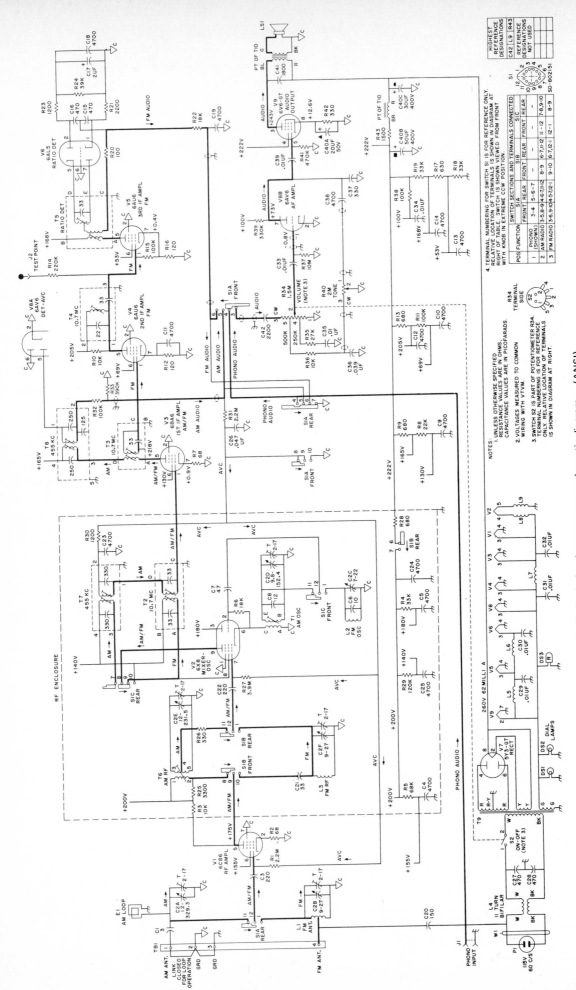

Fig. 23-16. Typical maintenance type schematic diagram. (ANSI)

478

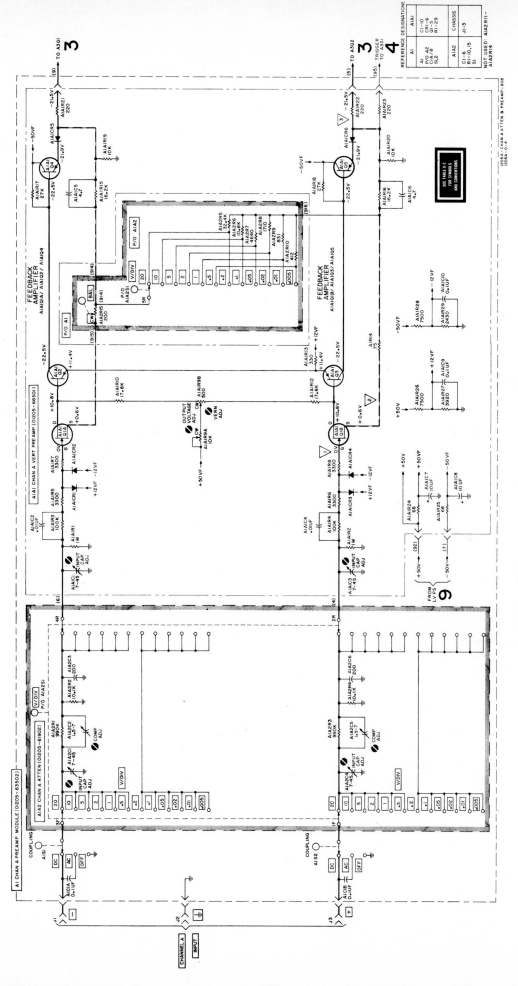

Fig. 23-17. Preamplifier Module Schematic. (Hewlett-Packard)

479

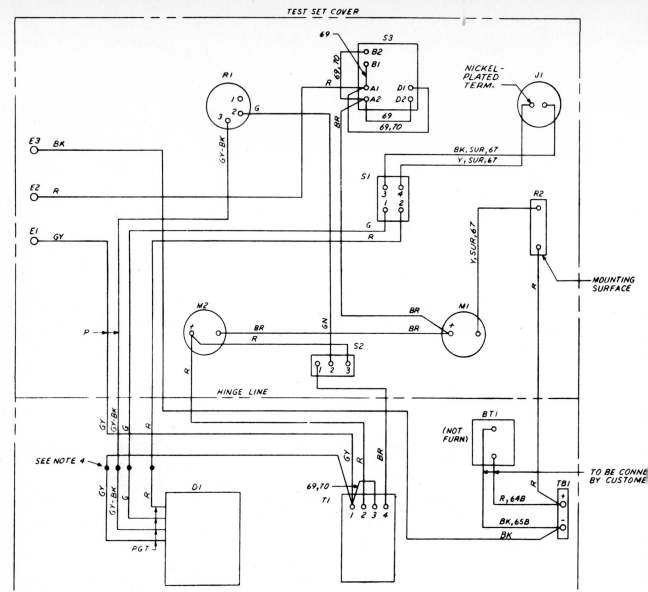

Fig. 23-18. Typical continuous line (point-to-point) connection diagram. (ANSI)

NOTES:

1. UNLESS OTHERWISE SPECIFIED, ALL WIRES ARE INCLUDED IN THE CABLE ASSEMBLY XXXXX.

2. ITEM NUMBERS REFERRED TO ARE SHOWN IN PARTS LIST OF ASSEMBLY DRAWING XXXXX.

3. ALL SOLDERING SHALL BE IN ACCORDANCE WITH QQ-S-524 METHOD C.

4. SPLICE AND SOLDER AND WRAP WITH ONE LAYER OF TAPE ITEM 58 AND TWO LAYERS OF TAPE ITEM 60.

5. SUR-WIRING-WIRE TO BE DRESSED BACK AND RUN ALONG THE MOUNTING SURFACES IN THE MOST CONVENIENT MANNER.

6. PGT - LEADS FURNISHED WITH PART.

CONNECTION AND INTERCONNECTION WIRING DIAGRAMS

Connection and interconnection wiring diagrams are supplementary drawings to schematic diagrams. These drawings contain information used in the manufacture, installation and maintenance of electrical and electronic equipment. They graphically represent the conducting paths (wires or cables) between component devices.

Connection diagrams and interconnection diagrams are very similar. The definitions contained in ANSI Standard Electrical and Electronics Diagrams may help to differentiate between the two types.

Connection or wiring diagram is:

. . . a diagram which shows the connections of an installa-

tion or its component devices or parts. It may cover internal or external connections, or both, and contains such detail as is needed to make trace connections that are involved. The connection diagram usually shows general physical arrangement of the component devices or parts. Interconnection diagram is:

. . . a form of connection or wiring diagram which shows only external connections between unit assemblies or equipment. The internal connections of the unit assemblies or equipment are usually omitted.

Wiring diagrams are divided into three major classifications: continuous line, Fig. 23-18; interrupted line, Fig. 23-19; tabular types, Fig. 23-20. Each type indicates the method used to show the connections between component parts or devices.

Fig. 23-18 is a type of continuous line diagram that is often

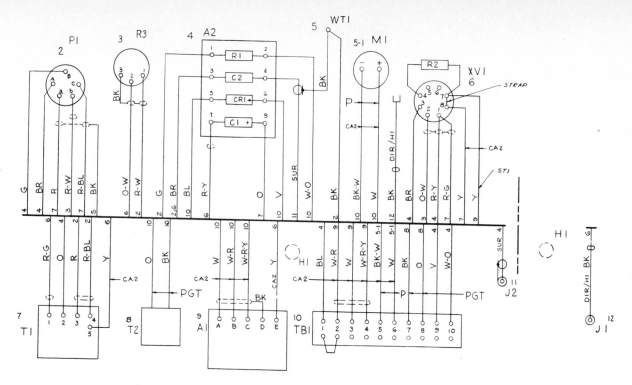

Fig. 23-19. Typical interrupted line (base-line) connection diagram. (ANSI)

Fig. 23-20. Typical tabular type connection diagram. (ANSI)

REV		WIRE					FROM				TO			
SYM	TRAN	COLOR	AWG	SYMBOL	METHOD OR PATH	NOTE	AREA LOC	TERMINAL	LEVEL	NOTES	AREA LOC	TERMINAL	LEVEL	NOTES
		W-R	ST1		CA2		TB1	2			A1	B		
		MS1			CA1		TB1	2			TB1	1		
		W	ST1		CA2		TB1	3			A1	A		
		W-R-Y	ST1		CA2		TB1	4			A1	C		
		BK-W	P1		CA2		TB1	5			M1	NEG		
		W	P1		CA2		TB1	6			M1	PØS		
		BK			PGT		TB1	7			T2			
		Ø			PGT		TB1	8			T2			
		V			CA1		TB1	9			A2	6		
		W-Ø			CA1		TB1	10			A2	2		
		R-G	SS4		CA1		T1	1			XV1	1		
		Ø			CA1		T1	2			A2	8		
		R	SP1		CA1		T1	3			P1	A*		
		R-BL	SP1		CA1		T1	4			P1	C		
		Y			CA2		T1	5			XV1	8		
		BK			PGT		T2				TB1	7		
		Ø			PGT		2				TB1	8		
		BK	H1		DIR		V1	CAP			J1			
		BK			SUR		WT1				A2	4$		
		BK			CA1		WT1				P1	C$		
		R-Y	SS3		CA1		XV1	1			A2	7		
		R-G	SS4		CA1		XV1	1			T1	1		
		CØM			CA1		XV1	1$			XV1	1$		
		CØM			CA1		XV1	1$			XV1	1$		
		CØM			CA1		XV1	1$			XV1	2$		
		Ø-W	SS2		CA1		XV1	2			R3	2		
		CØM			CA1		XV1	2$			XV1	1$		
		BR			CA1		XV1	3			A2	3		
		R2			PGT		XV1	4			XV1	7		
							XV1	5						
							XV1	6						
		R2			PGT		XV1	7			XV1	4		
		STRAP					XV1	7			XV1	8		
		Y			CA2		XV1	7(ST1						
		W	ST1		CA2		A1	A			A1	E		
		W-R	ST1		CA2		A1	B			TB1	2		

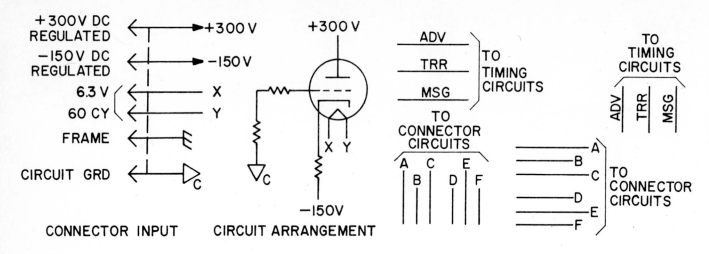

Fig. 23-21. Symbols used to indicate connecting paths.

used to show the point-to-point connections of a device. The diagram in Fig. 23-16 is more complicated, but it is also a point-to-point connection diagram.

Not shown is another type of continuous line diagram where groups of lines are combined into one-line paths or highways. Feeder lines leading from the terminals to the highway must have routing designations so that the user can locate the other ends of the connections.

An interrupted line connection diagram may contain symbols shown in Fig. 23-21 to indicate the connecting paths between component parts or devices. Another type, the base-line connection diagram, Fig. 23-19, also may be used to reduce the number of lines on a drawing. A section of the base line may be interrupted near a component and another section of the same base line drawn near the destination.

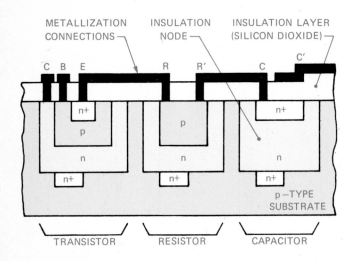

Fig. 23-22. Cross sectional view of a typical integrated circuit.

The tabular type of connection diagram, Fig. 23-20, is sufficient for many wiring operations. It is a simple "from-to" list, but it may be expanded to show additional information such as wire lengths, sizes or types.

The component parts in a connection diagram are represented by pictorial, block or graphic symbols with only outline and terminal circles shown. A circular symbol should not be used unless the component device or part closely approximates a circle. On most drawings there are numerous designations in coded or abbreviated form which identify component parts, terminals, wire sizes, wire colors, destinations and other related information.

SOLID-STATE ELECTRONICS

The greatest development in the field of electronics in recent years has been in solid-state electronics. The accompanying problems of miniaturization have placed unusual demands upon the drafter. These have led to the increased use of sophisticated drafting equipment such as artwork generators, automatic drafting machines and computer-aided design equipment.

Solid-state electronics began with the invention of the transistor and the subsequent development of the printed circuit board. Both involve an etching process where very precise masks, or windows, are used to control the areas eaten or etched away by a chemical solution.

The next step in the development of solid-state electronics was to construct thin film circuits in which passive components and connections were deposited in extremely thin films on a supporting substrate or base. At the present time there are numerous active devices interconnected on a supporting substrate, Fig. 23-22.

The current phase of development in solid-state technology is very large scale integration (VLSI) of complex circuits and functions. These have made possible a "computer-on-a-chip" device within a single package, Fig. 23-23.

The drawings for solid state devices are usually photographically reduced from 2 to 40 times size and several drawings must be successively aligned to an accuracy of one ten-thousandth of an inch (0.0001). They are typically plotted on a special film base, Fig. 23-24. This replaces the manual method of cutting a peelable opaque overlay which is then removed to form the circuit pattern.

Fig. 23-23. A one-inch square hybrid microcomputer. (Motorola, Inc.)

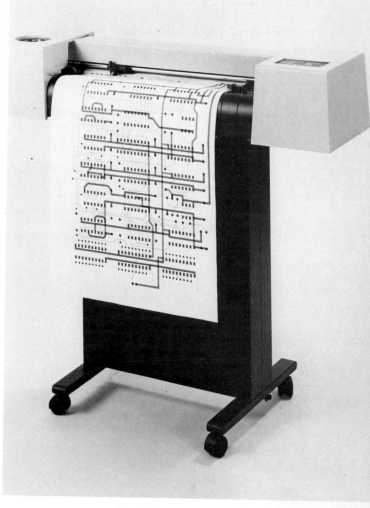

Fig. 23-24. Plotting a solid state schematic on a film base. (Hewlett-Packard Co.)

PROBLEMS IN ELECTRICAL AND ELECTRONICS DRAFTING

PICTORIAL DRAWINGS

1. Select one of the components from Fig. 23-4 and make an enlarged, freehand sketch of it on a size A sheet. Estimate proportions.
2. Make a pictorial drawing of a small transistor radio or similar electronic device.
3. If electrical or electronic components are available, make a scaled drawing of the components.

GRAPHIC SYMBOLS

4. Draw and label the following component symbols: (Use a template or draw each component to same relative size.)
   a. Battery, 9 volts, BT1.
   b. Switch, single-pole, single-throw, S1.
   c. Ammeter, M1.
   d. Resistor, 4700 OHMS, R1.
   e. Lamp, incandescent, dial lamp, DS1.
5. Draw the symbols shown to the right and completely identify the component. On the line to the right of the symbol, give the name of the component.

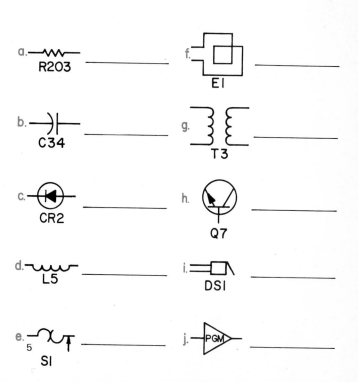

SINGLE LINE DIAGRAMS

6. Replace the blocks with the correct single line symbols in the following diagram of an electrical power system.

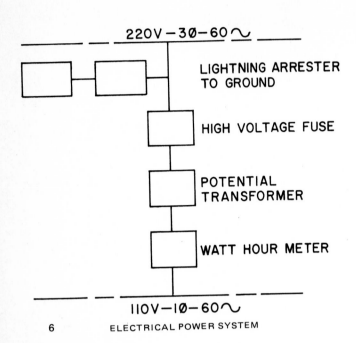

6   ELECTRICAL POWER SYSTEM

BLOCK DIAGRAMS

7. Draw a block diagram of a noise level meter with the following stages:
   a. Input Microphone.
   b. Audio Amplifier, Q1.
   c. Audio Amplifier, Q2.
   d. Audio Amplifier, Q3.
   e. Decibel Meter.

8. Draw a block diagram of a television receiver. Use 5 rows and 7 columns to represent the following signal paths.
   a. From antenna to RF amplifier, to first IF amplifier, to second IF amplifier, to third IF amplifier, to video detector, to video amplifier, to picture tube.
   b. From output side of video amplifier to sound amplifier, to FM detector, to audio amplifier, to audio output, to loudspeaker.
   c. From output side of video amplifier to sync separator, to vertical oscillator, to vertical output, to deflection yoke.
   d. From output side of sync separator to horizontal oscillator, to deflection yoke.
   e. From output side of horizontal output to high voltage power supply.
   f. A single block showing the low voltage power supply.

9. Draw a block diagram of the circuit shown in Fig. 23-16. The heavy line indicates the signal path.

ELEMENTARY DIAGRAMS

10. Complete the following elementary diagram so that the operation will cease after a complete sequence of operations.

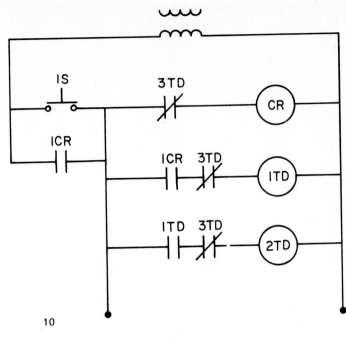

10

SCHEMATIC DIAGRAM

11. Redraw the diagram below by replacing the numbered blocks with the correct graphic symbols. Show a battery at Block 1. A voltmeter goes in Block 2 and an ammeter in Block 5. Resistors go in Blocks 3, 4 and 6.

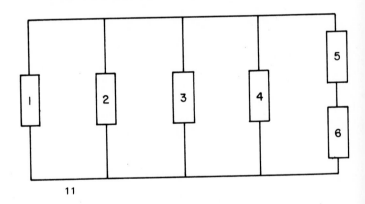

11

12. Draw the schematic diagram for the device shown in Fig. 23-2. The switch used is a double-pole double-throw (DPDT). Refer to Fig. 23-4 for schematic clarification.

13. Make a 4X enlargement of the diagram of the RF amplifier shown on the top of the next page by replacing the numbered blocks with the correct schematic symbols. Retain correct proportion between symbols.

     Oscillator, 1
     Capacitor, Fixed, 3, 11, 12
     Resistor, Fixed, 4, 6, 8, 9
     Transistor, PNP, 7
     Battery, Multicell, 15
     Connector, Male, 14
     Switch, Single-Pole, Single-Throw, 16
     Common Connection Designated C, 2, 5, 10, 13, 17

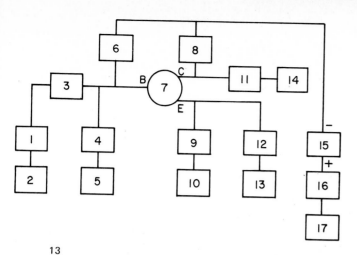

14. Redraw the sketch below and add the information listed below. Avoid crowding and wasted spaces.

R_1, R_2, 220K, 1/2W	C_1, 0-365 pF
R_3, R_6, 1K, 1/2W	C_2, .01 μF
R_4 100K POT	C_3, 10 μF, 25V
R_5 100K, 1/2W	Q_1, 2N663
CR_1, 1N63	

13

CONNECTION WIRING DIAGRAM

15. Redraw the continuous line connection diagram, shown below, as an interrupted line connection diagram. For each lead, show the subassembly number and terminal to which it is going and the color code. For example, Lead 3 on Unit A1 would be labeled A4/4-GN-BK meaning it is a green lead with a black tracer stripe that goes to Unit A4, Terminal 4.

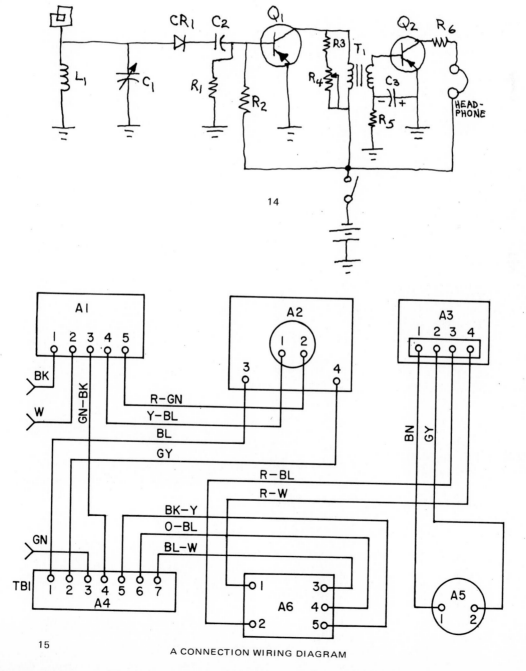

14

15   A CONNECTION WIRING DIAGRAM

Chapter 24
WELDING DRAWINGS

Fig. 24-1. Modern industry depends upon welding processes for many jobs. A drawing must clearly specify the engineering designer's intent for each weld if the part is to be properly fabricated.
(Lincoln Electric)

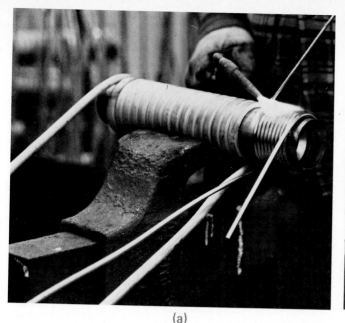

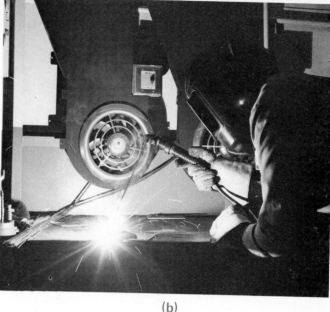

(a)   (b)

Fig. 24-2. Gas welding (a) and arc welding (b) of metal by addition of filler metals from rods.
(McDonnell Douglas)

Welding has become one of industry's principal means of fastening parts together or building up surfaces of parts, Fig. 24-1. Modern technology has developed welding processes and materials which can meet nearly any metal fabricating need. This capability has placed a major responsibility on the design-drafting department to adequately specify welds required for a particular structure or machine part.

WELDING PROCESSES

Numerous welding processes have been developed to meet the need for joining like and unlike types of metals. Those processes which are in common use and those developed for welding the "exotic" metals of the space industry are discussed in this chapter, along with specification of welds on drawings.

BRAZING

Brazing is the joining of metals by adhesion with a low melting point alloy (a copper base with tin, zinc and/or lead) that does not melt the parent metal.

FUSION WELDING

The basic types of fusion welding processes include oxyacetylene, arc, TIG and MIG welding. The heat generated by the flame or arc causes the parent metal and a feeder metal rod to melt and fuse into one piece, joining the metals by cohesion, Fig. 24-2.

TIG stands for tungsten inert gas welding and is a gas-shielded arc welding process. The tungsten electrode maintains an intense heat and a metal filler rod may or may not be added, depending on the requirements of the joint. An inert gas which does not chemically combine with the weld, usually argon and helium, surrounds the weld and produces a clean weld, Fig. 24-3.

The primary use of the TIG welding process is in joining lightweight (less than 1/4 inch in thickness) nonferrous metals

including aluminum, magnesium, silicon-bronze, copper and nickel alloys, stainless steel and precious metals. The gas-shielded arc gives an unobstructed view of the slag-free weld.

MIG is the abbreviation for metal inert gas welding. It is a gas-shielded arc welding process similar to TIG welding. In MIG welding, the electrode is the filler wire which is fed into the weld automatically. MIG is used for welding metals 1/4 inch or more in thickness.

Fig. 24-3. TIG welding showing a filler rod being used.
(Lockheed Aircraft Corp.)

Fig. 24-4. A spot welder is a type of resistance welder.
(Grumman Aerospace Corp.)

RESISTANCE WELDING

Resistance welding is an effective and economical means of fastening metal parts, Fig. 24-4. An electric current is the source of heat, and pressure is applied to bring the parts together at the point of weld.

Resistance welding is based on the principle that resistance to current flow causes metal to become hot. Resistance is greatest at the joint between the pieces. Therefore, when the current is properly adjusted, the metal pieces melt and fuse at the joint.

Types of resistance welding are:
1. Spot welding, where the metal is fluxed only in the contact spots.
2. Butt or seam welding of an entire joint or seam.
3. Flash welding is where the ends of two metal parts are brought together under pressure and resistance welded.

INDUCTION WELDING

Induction welding is similar to resistance welding. However, in induction welding, the heat generated for the weld is produced by the resistance of the metal parts to the flow of an induced electric current. The welding action may occur with or without pressure.

ELECTRON BEAM WELDING (EBW)

The source of heat in electron beam welding is a high-intensity beam of electrons focused in a small area at the surface to be welded, as shown in Fig. 24-5 (a). Although this electron beam welding is a fusion welding process, it is unlike the common fusion welding processes which are often used.

Electron beam welding is done in a vacuum which practically eliminates contamination of the weld from the atmosphere, as shown in Fig. 24-5 (b). There is a minimum of distortion of the workpiece because the heat is concentrated in a small area. No rod, gas or flux is needed in electron beam welding. EBW is used in welding metals such as titanium, beryllium and zirconium, which are difficult to weld by other welding processes.

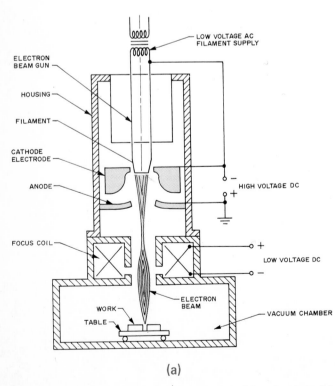

(a)

(b)

Fig. 24-5. A diagram of an electron beam welder (a) and a photo (b) showing a view of the electron beam welding a piece of metal inside the vacuum chamber. (United Aircraft Corp.)

TYPES OF WELDED JOINTS

The welding process lends itself to a variety of joints in fastening metal parts. The basic types of joints used in welding, and welds applicable to each type, are shown in Fig. 24-6.

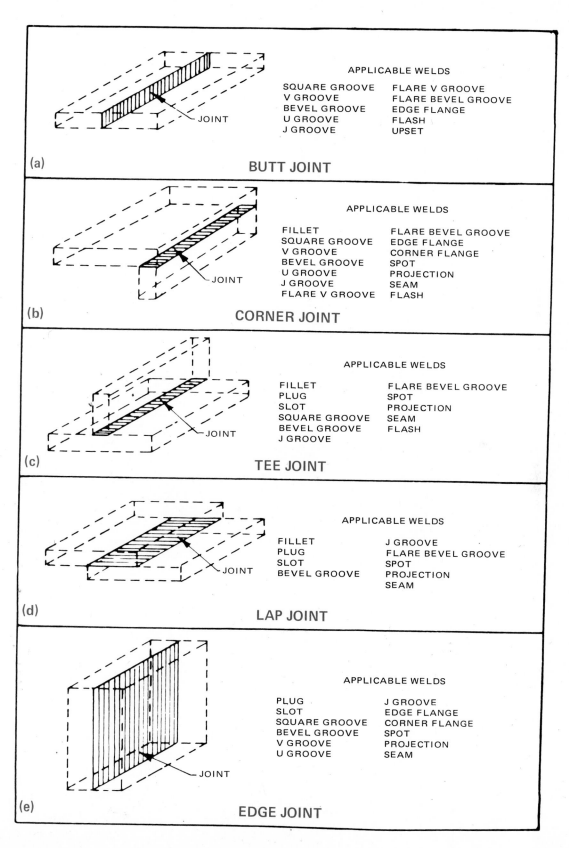

(a)

APPLICABLE WELDS

SQUARE GROOVE   FLARE V GROOVE
V GROOVE   FLARE BEVEL GROOVE
BEVEL GROOVE   EDGE FLANGE
U GROOVE   FLASH
J GROOVE   UPSET

JOINT

BUTT JOINT

(b)

APPLICABLE WELDS

FILLET   FLARE BEVEL GROOVE
SQUARE GROOVE   EDGE FLANGE
V GROOVE   CORNER FLANGE
BEVEL GROOVE   SPOT
U GROOVE   PROJECTION
J GROOVE   SEAM
FLARE V GROOVE   FLASH

JOINT

CORNER JOINT

(c)

APPLICABLE WELDS

FILLET   FLARE BEVEL GROOVE
PLUG   SPOT
SLOT   PROJECTION
SQUARE GROOVE   SEAM
BEVEL GROOVE   FLASH
J GROOVE

JOINT

TEE JOINT

(d)

APPLICABLE WELDS

FILLET   J GROOVE
PLUG   FLARE BEVEL GROOVE
SLOT   SPOT
BEVEL GROOVE   PROJECTION
   SEAM

JOINT

LAP JOINT

(e)

APPLICABLE WELDS

PLUG   J GROOVE
SLOT   EDGE FLANGE
SQUARE GROOVE   CORNER FLANGE
BEVEL GROOVE   SPOT
V GROOVE   PROJECTION
U GROOVE   SEAM

JOINT

EDGE JOINT

Fig. 24-6. Basic types of joints used in welding.
(American Welding Society)

Drafting for Industry

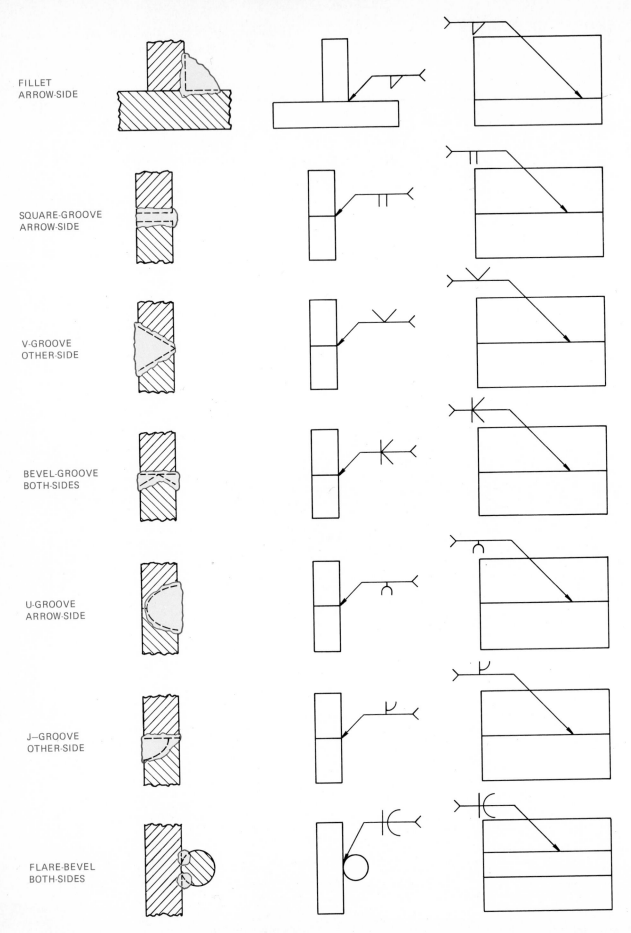

FILLET
ARROW-SIDE

SQUARE-GROOVE
ARROW-SIDE

V-GROOVE
OTHER-SIDE

BEVEL-GROOVE
BOTH-SIDES

U-GROOVE
ARROW-SIDE

J—GROOVE
OTHER-SIDE

FLARE-BEVEL
BOTH-SIDES

Fig. 24-7. Types of welds.

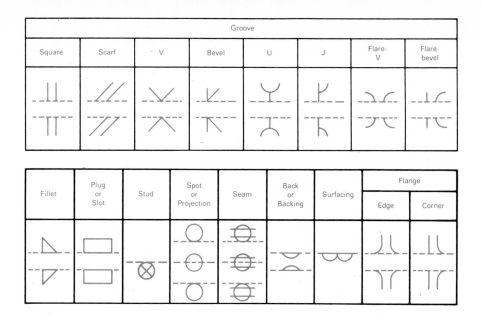

Fig. 24-8. Basic weld symbols. (American Welding Society)

TYPES OF WELDS

The term "weld" refers to the basic design of the weld itself, Fig. 24-7. Design selection, basically, is determined by the thickness of the metals to be joined and the penetration of the weld into the joint for the strength required. Type of metal also has a bearing on the weld design selected.

BASIC WELD SYMBOLS

The American Welding Society (AWS) has developed a set of standard symbols for use in specifying types of fusion and resistance welds on drawings, Fig. 24-8. These weld symbols are available to designers, drafters and welders and should be understood by persons in industries using welding processes.

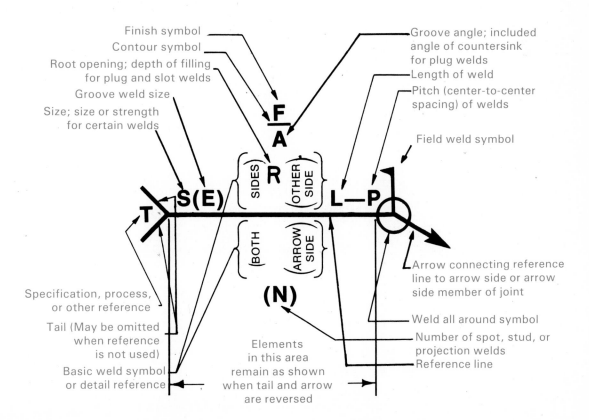

Fig. 24-9. Standard location of elements of a welding symbol.

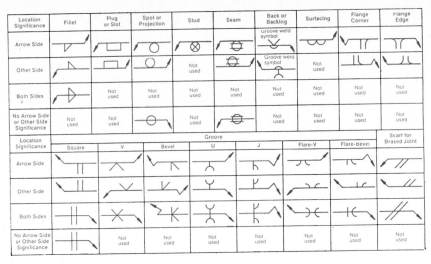

Fig. 24-10. Some basic weld symbols and their location significance.

Weld symbols should be used only as a part of the welding symbol discussed in the next section.

STANDARD SYMBOL

The standard welding symbol is a composite symbol that carries all pertinent information for a particular weld. It indicates type of weld, size, location and welding process if specified, Fig. 24-9. Note that the elements along the reference line of the symbol remain the same when the tail and arrow are reversed, Fig. 24-10.

A template for use in preparing standard welding symbols is shown in Fig. 24-11.

Weld symbols attached to the reference line are shown in an "upright" position when on the far side (top side) of the line; in an "up-side-down" position when on the near side (lower side) of the line. The weld symbols are never reversed. For example, the perpendicular leg of the fillet and groove weld symbols always are shown on the left.

When no specification, welding process or other reference is given, the tail section of the symbol may be omitted.

The location of welds with respect to a joint is controlled by the placement of the weld symbol on the reference line of the welding symbol. Welds which are to be located on the arrow side of the joint are shown by placing the weld symbol on the side of the reference line toward the reader, Fig. 24-12 (a). Welds which are to be on the side opposite the arrow are considered to be on the other side of the joint, so the weld symbol is shown on the side of the reference line away from the reader, (b).

When the joint is to be welded on both sides, the weld symbol is shown on both sides of the reference line, Fig. 24-12 (c). Note in the second example of (c), that a different weld may be called out for each side of the joint and that a combination of welds may also be specified.

SUPPLEMENTARY SYMBOLS

The dimensions of welds, the contour of the weld surface, welds which are to melt through, etc., may be specified by adding the appropriate information or symbol to the welding symbol. Refer to the welding symbols chart in the Reference Section, and to the bulletin, Standard Symbols for Welding, Brazing, and Nondestructive Examination (ANSI/AWS A2.4-86) published by the American Welding Society.

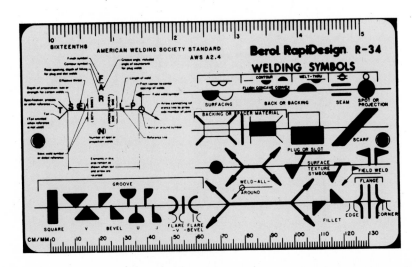

Fig. 24-11. This template will speed the application of welding symbols.   (RapiDesign)

Welding Drawings

SYMBOL INTERPRETATION

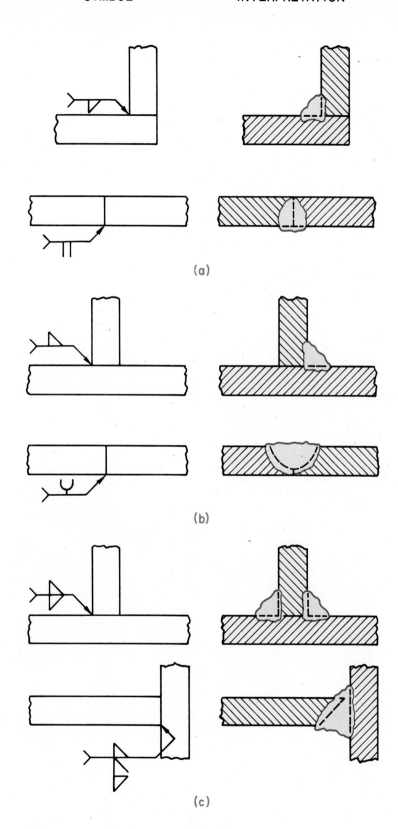

(a)

(b)

(c)

Fig. 24-12. Significance of arrow in welding symbol in locating welds.

QUESTIONS FOR DISCUSSION

1. How does brazing of a metal joint differ from a fusion welded joint?
2. Check a welding text or reference book for the meaning and interpretation of the terms adhesion and cohesion. Report the findings to your class.
3. What are TIG and MIG welding, and how do they differ from regular electric arc welding?
4. What is the principle on which resistance welding is based, and how does it work? Give some examples of products which have been resistance welded.
5. How does induction welding operate?
6. What is electron beam welding, and what metals is it used on? Where is it done?
7. Differentiate between a weld symbol and the welding symbol.
8. Of what significance is the arrow in the welding symbol in determining the location of the weld?

PROBLEMS AND ACTIVITIES

1. Make working drawings, including the specification of welds, for objects shown in Figs. 24-13, 24-14 and 24-15 as assigned by your instructor.
2. Design a piece of furniture requiring welded parts for use inside or outside the home. Use the design method for arriving at the final design and make a working drawing of the piece.
3. Using the design method, design a tool, jig or fixture requiring welded parts for some problem which needs solving around home, school or your place of work. Make the working drawing and construct a scaled model or prototype of the item.

SELECTED ADDITIONAL READING

1. WELDING SYMBOLS, ANSI, Y32.3-1969, American Welding Society, 2501 N.W. Seventh St., Miami, FL 33125.

Industry photo. View of "Clam Shell" chamber showing gun, gantry, and tooling of electron beam welder. Before welding process begins, the chamber is closed and a vacuum is created in the chamber. (Grumman Aerospace Corp.)

Welding Drawings

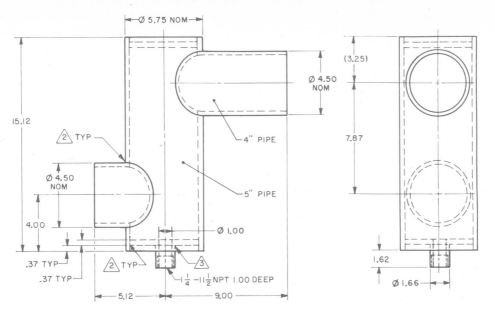

15.12

Ø 5.75 NOM

Ø 4.50 NOM

(3.25)

7.87

2 TYP

4" PIPE

Ø 4.50 NOM

5" PIPE

4.00

.37 TYP

2 TYP

Ø 1.00

.37 TYP

3

1.62

1 1/4 –11 1/2 NPT 1.00 DEEP

Ø 1.66

5.12

9.00

1. TOLERANCES UNLESS OTHERWISE NOTED:

    .XXX = ± .015
    .XX   = ± .030

2. .31 FILLET WELD ARROW SIDE ALL–AROUND

3. .18 FILLET WELD ARROW SIDE ALL–AROUND

LIQUID SUMP ASSEMBLY FOR CTV WITH RCC MOTOR

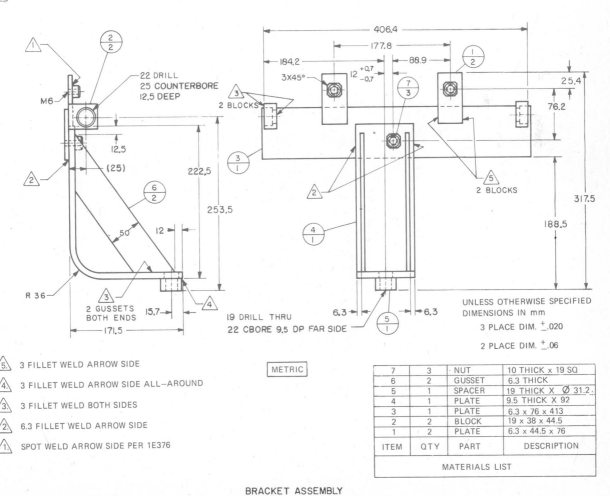

1

2/2

22 DRILL
25 COUNTERBORE
12.5 DEEP

406.4

177.8

184.2

88.9

1/2

M6

3X45°

12 +0.7 / -0.7

7/3

25.4

2 BLOCKS

3

76.2

12.5

(25)

3/1

222.5

2

253.5

6/2

2

5

50

12

2 BLOCKS

4/1

317.5

188.5

R 36

3

2 GUSSETS BOTH ENDS

15.7

4

19 DRILL THRU
22 CBORE 9.5 DP FAR SIDE

6.3

5/1

6.3

UNLESS OTHERWISE SPECIFIED DIMENSIONS IN mm

    3 PLACE DIM. ±.020

    2 PLACE DIM. ±.06

171.5

5. 3 FILLET WELD ARROW SIDE

4. 3 FILLET WELD ARROW SIDE ALL–AROUND

3. 3 FILLET WELD BOTH SIDES

2. 6.3 FILLET WELD ARROW SIDE

1. SPOT WELD ARROW SIDE PER 1E376

METRIC

ITEM	QTY	PART	DESCRIPTION
7	3	NUT	10 THICK x 19 SQ
6	2	GUSSET	6.3 THICK
5	1	SPACER	19 THICK X Ø 31.2
4	1	PLATE	9.5 THICK X 92
3	1	PLATE	6.3 x 76 x 413
2	2	BLOCK	19 x 38 x 44.5
1	2	PLATE	6.3 x 44.5 x 76

MATERIALS LIST

BRACKET ASSEMBLY

Fig. 24-13. Parts for which working drawings are to be prepared, including the specification of welded joints.

495

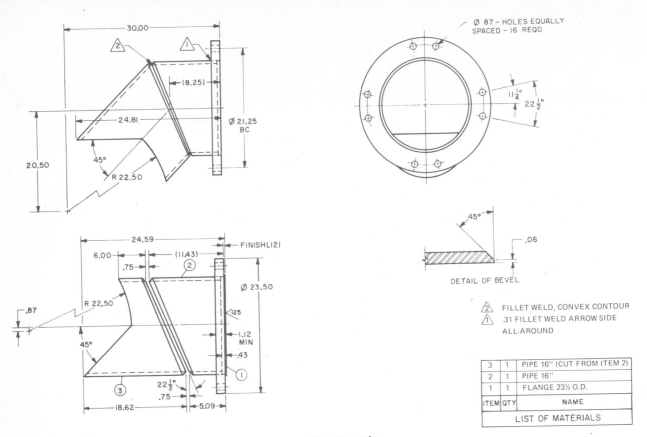

DISCHARGE CONN. ASS'Y.

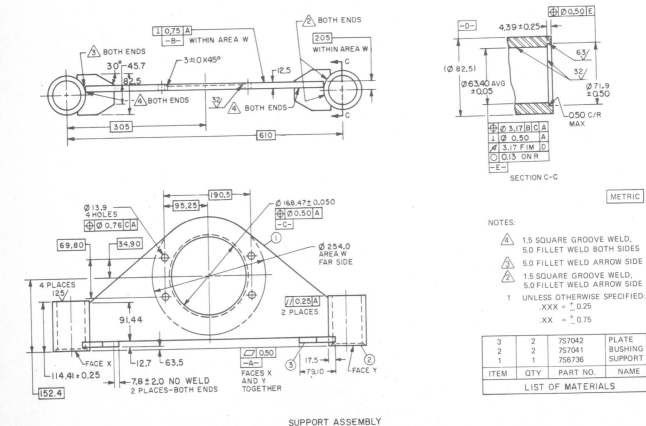

SUPPORT ASSEMBLY

Fig. 24-14. Parts for which working drawings are to be prepared, including the specification of welded joints.

Welding Drawings

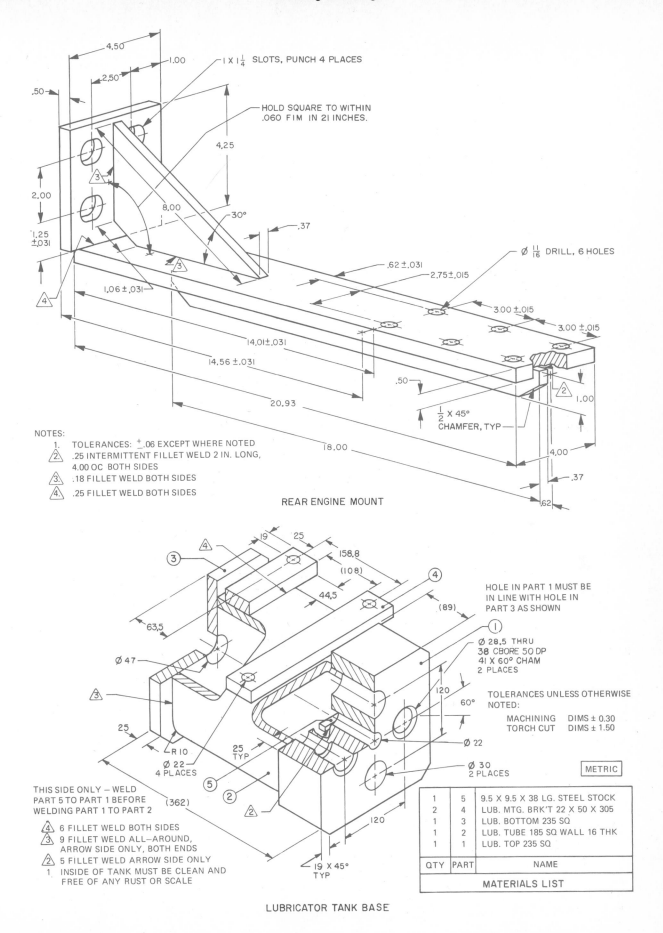

REAR ENGINE MOUNT

1 X 1¼ SLOTS, PUNCH 4 PLACES

HOLD SQUARE TO WITHIN .060 FIM IN 21 INCHES.

Ø 11/16 DRILL, 6 HOLES

½ X 45° CHAMFER, TYP

NOTES:
1. TOLERANCES: ±.06 EXCEPT WHERE NOTED
2. .25 INTERMITTENT FILLET WELD 2 IN. LONG, 4.00 OC BOTH SIDES
3. .18 FILLET WELD BOTH SIDES
4. .25 FILLET WELD BOTH SIDES

HOLE IN PART 1 MUST BE IN LINE WITH HOLE IN PART 3 AS SHOWN

Ø 28.5 THRU
38 CBORE 50 DP
41 X 60° CHAM
2 PLACES

TOLERANCES UNLESS OTHERWISE NOTED:

MACHINING   DIMS ± 0.30
TORCH CUT   DIMS ± 1.50

Ø 22

Ø 30
2 PLACES

METRIC

THIS SIDE ONLY — WELD PART 5 TO PART 1 BEFORE WELDING PART 1 TO PART 2

4. 6 FILLET WELD BOTH SIDES
3. 9 FILLET WELD ALL—AROUND, ARROW SIDE ONLY, BOTH ENDS
2. 5 FILLET WELD ARROW SIDE ONLY
1. INSIDE OF TANK MUST BE CLEAN AND FREE OF ANY RUST OR SCALE

R 10
Ø 22
4 PLACES

19 X 45° TYP

LUBRICATOR TANK BASE

1	5	9.5 X 9.5 X 38 LG. STEEL STOCK
2	4	LUB. MTG. BRK'T 22 X 50 X 305
1	3	LUB. BOTTOM 235 SQ
1	2	LUB. TUBE 185 SQ WALL 16 THK
1	1	LUB. TOP 235 SQ
QTY	PART	NAME
		MATERIALS LIST

Fig. 24-15. Parts for which working drawings are to be prepared, including the specification of welded joints.

Chapter 25
MAP AND SURVEY DRAWINGS

Cartography is the science of map making, and special kinds of drafting are required in the preparation of copy for maps. Drafting techniques used for some of the more common types of maps are presented in this chapter.

DEFINITION OF TERMS

The following terms are basic to an understanding of surveying and map drafting.

AZIMUTH: The angle a line makes with a north and south line measured clockwise from the north, Fig. 25-1.

BACK AZIMUTH: In surveying, the angle measured clockwise from the north to a line running in the opposite direction from the azimuth measurement. The back azimuth is always equal to the azimuth plus or minus 180°, Fig. 25-1.

BEARING: An angle having a value of 0 to 90° measured from either the north or south. A line with a bearing of 20° to the west of south would be stated as South 20° West or S20°W, Fig. 25-1.

BACKSIGHT: A line in surveying; sighting with the transit back on to the last station occupied, Fig. 25-2.

FORESIGHT: A line in surveying; when occupying a new station, and with the transit sighting on the previous station, revolving the transit 180° gives the foresight, Fig. 25-2.

DEFLECTION ANGLE: The angle of a line in surveying laid out to the right of the foresight for the right deflection angle or to the left for the left deflection angle, Fig. 25-2.

HORIZONTAL CURVE: A change of direction in the horizontal or plan view which is achieved by means of a curve (see Fig. 25-7).

INTERPOLATION: To locate, by proportion, intermediate points between grid data given in contour plotting problems.

MOSAIC: An assembly of a series of aerial photographs or radar imagery, taken with intentional overlaps of adjacent land areas and fitted together to produce a larger picture.

NORTH: The direction normally indicated on a map; magnetic north as indicated by a magnetic compass is satisfactory for most maps, but is subject to local deflection errors affecting the magnetic compass. True north, as determined by sighting on Polaris (North Star) is considered to be most accurate.

PLANIMETRIC: Suitable for measurement of plane areas.

PLAT: A plot of ground, usually small.

SURVEY: The use of linear and angular measurements and calculations to determine boundaries, position, elevation, profile, etc., of a part of the earth's or other planets' surfaces.

STATIONS: The turning points in a map traverse (see traverse), Fig. 25-2. In highway construction surveys, points at 100 foot intervals on the center line in the plan view are also called stations and are located by stakes with station numbers on them. Points of change between stakes are given the number of the last station with a plus number equal to the distance beyond that stake (see Fig. 25-7).

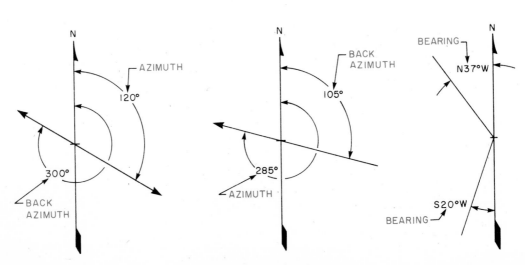

Fig. 25-1. The azimuth and bearing of a line.

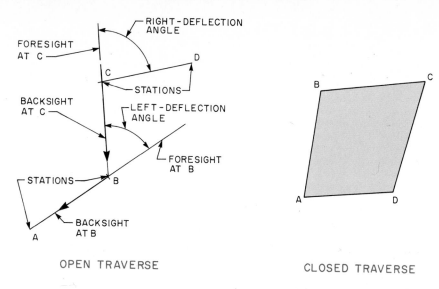

Fig. 25-2. Illustration of terms used in laying out map traverses.

TRAVERSE: Measuring or laying out a line, such as a property line, by means of angular and linear measurements, Fig. 25-2. A closed traverse is one that returns to its point of origin in a previously identified point. An open traverse neither returns nor ends at a previously identified point.

VERTICAL CURVE: A change of direction in the grade shown in a profile view which is achieved by means of a curve, usually a parabolic curve (see Fig. 25-7).

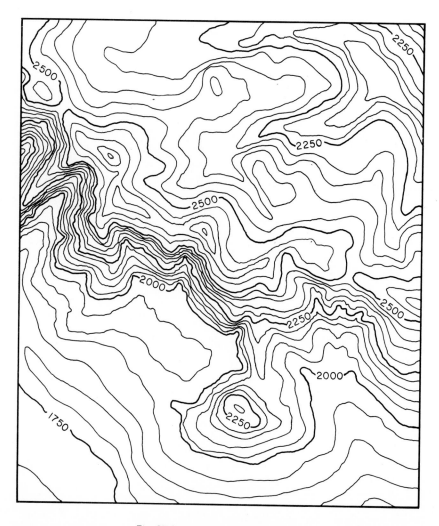

Fig. 25-3. A geologic surface map.

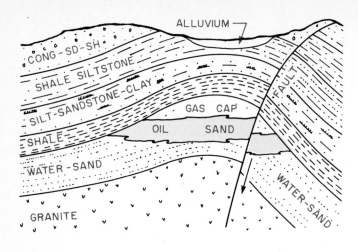

Fig. 25-4. A geologic cross section map showing a section of the earth's interior structure.

Industry photo. Checking highway right-of-way. (Virginia Department of Highways and Transportation)

TYPES OF MAPS

All maps are representations, on flat surfaces, of a part of the earth's surface (or any planet's surface). Although all maps have much in common, they may be classified according to types based upon their intended use.

GEOGRAPHIC MAPS

The geographic map is familiar to most students as the type contained in social studies texts. It illustrates, usually by shading or color variation, such elements as climate, soil, vegetation, land use, rivers, population, cities and topographic features. The geographic map normally represents a large area and must be drawn to a very small scale.

GEOLOGIC MAPS

Geology is the study of the earth's surface, its outer crust and interior structure, and the changes which have taken and are taking place. Geologic maps report this information pictorially. Maps showing topographic surface, Fig. 25-3, and geologic cross sections of the subsurface, Fig. 25-4, are used.

TOPOGRAPHIC MAPS

Contrasted with a geographic map, a topographic map gives a detailed description of a relatively small area. Depending on the intended use of topographic maps, they may include natural features, boundaries, cities, roads, pipelines, electric lines, houses and vegetation, Fig. 25-5. Contour lines are normally used to show elevation. Standard map symbols may be used to show natural or man-made features.

CADASTRAL MAPS

Cadastral maps are drawn to a scale large enough to accurately show locations of streets, property lines, buildings, etc. They are also used in the control and transfer of property, Fig. 25-6. Cadastral maps are used to show city addition plats and to identify property owners along a road right-of-way.

ENGINEERING CONSTRUCTION MAPS

Maps under this category range from the simple plot map for a residence to such major engineering projects as commercial buildings, electrical transmission lines, bridges and hydroelectric dams. An engineering map for a highway construction project is shown in Fig. 25-7. The horizontal curve is plotted in the aerial or plan view. The vertical curve is shown on the same sheet in the profile view. The two are referenced by common check points.

Another engineering map that requires careful study and detailing is one that locates a dam in relation to the elevation and configuration of the surrounding land. The location and cross section of Morrow Point Dam (Colorado) is shown in Fig. 25-8.

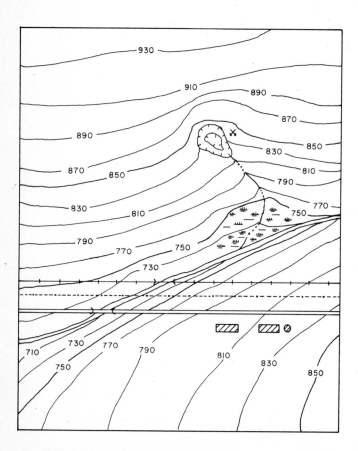

Fig. 25-5. A topographic map showing contour elevations and both natural and man-made features.

Fig. 25-6. A cadastral map aids in locating property lines.

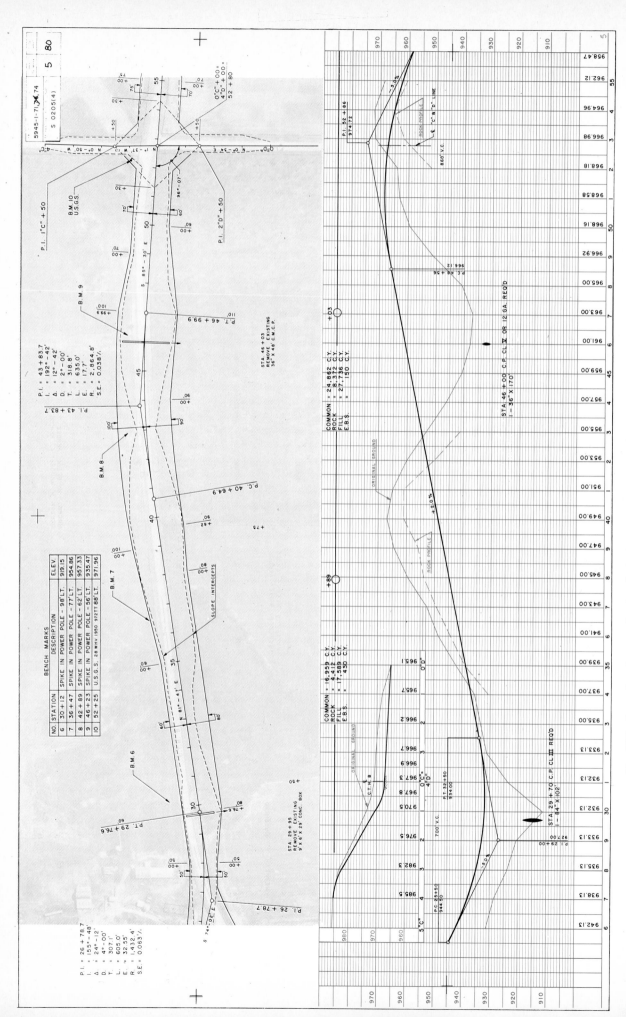

Fig. 25-7. An engineering map showing the horizontal and vertical curve of a section of highway construction. (Wisconsin Department of Transportation)

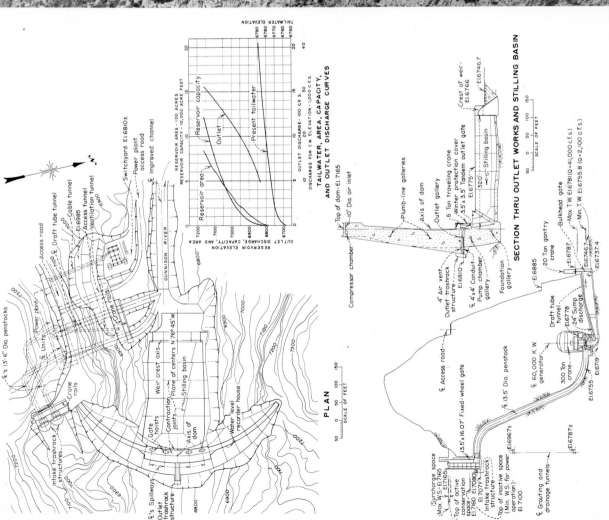

Fig. 25-8. Morrow Point Dam (Colorado) construction site was located by means of an engineering map.
(U. S. Department of the Interior)

MAP FORMAT

There are perhaps as many different layouts for maps as there are types of maps. However, each type of map has a title, scale, lettering and notes, symbols or other standard data for which certain guide lines are recognized.

TITLE

The title of a map is a statement of what the map is, its location, when it was prepared and the individual, company or government agency for whom it was prepared, as shown in Fig. 25-9. The title is usually placed in the lower right-hand corner of the sheet when possible; otherwise in an area that affords clarity.

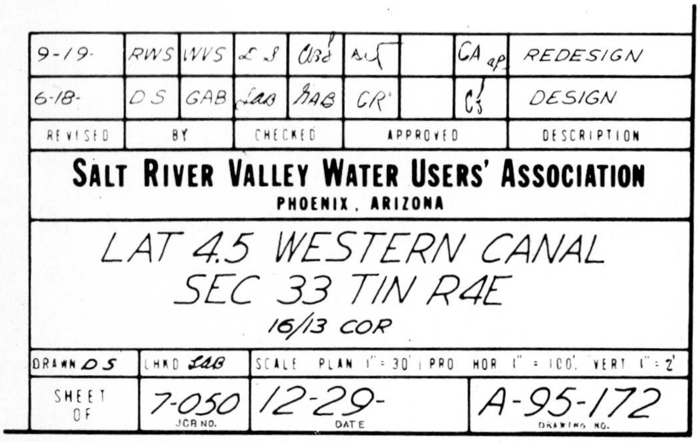

SCALE

The scale of a map should be indicated just below or near the title. Most maps are laid out on a base ten ratio with a civil engineer's scale. The scales range from very large (0.1 inch = 1 foot), such as on a plot for a residence or commercial building, to greatly reduced scales (1 inch = 400 miles) on geographic maps.

Scales also may be indicated as a ratio such as 1:250,000. In this case, 1 inch equals 250,000 inches, or nearly 4 miles. The decimal scale marked "50" could be used for this scale by letting each major unit equal 50,000 similar units.

For a scale of 1 inch = 400 feet, the "40" scale would be used with the smallest subdivision equal to 10 feet. Some maps use the graphic scale, Fig. 25-10, which is quickly and easily interpreted.

Fig. 25-10. A type of graphic scale used on some maps.

LETTERING AND NOTES

Lettering on engineering maps is done with single-stroke capital letters, either vertical or inclined. The two are never used on the same map. Titles of maps are sometimes done in Roman style letters with a little more flair than the single-stroke Gothic.

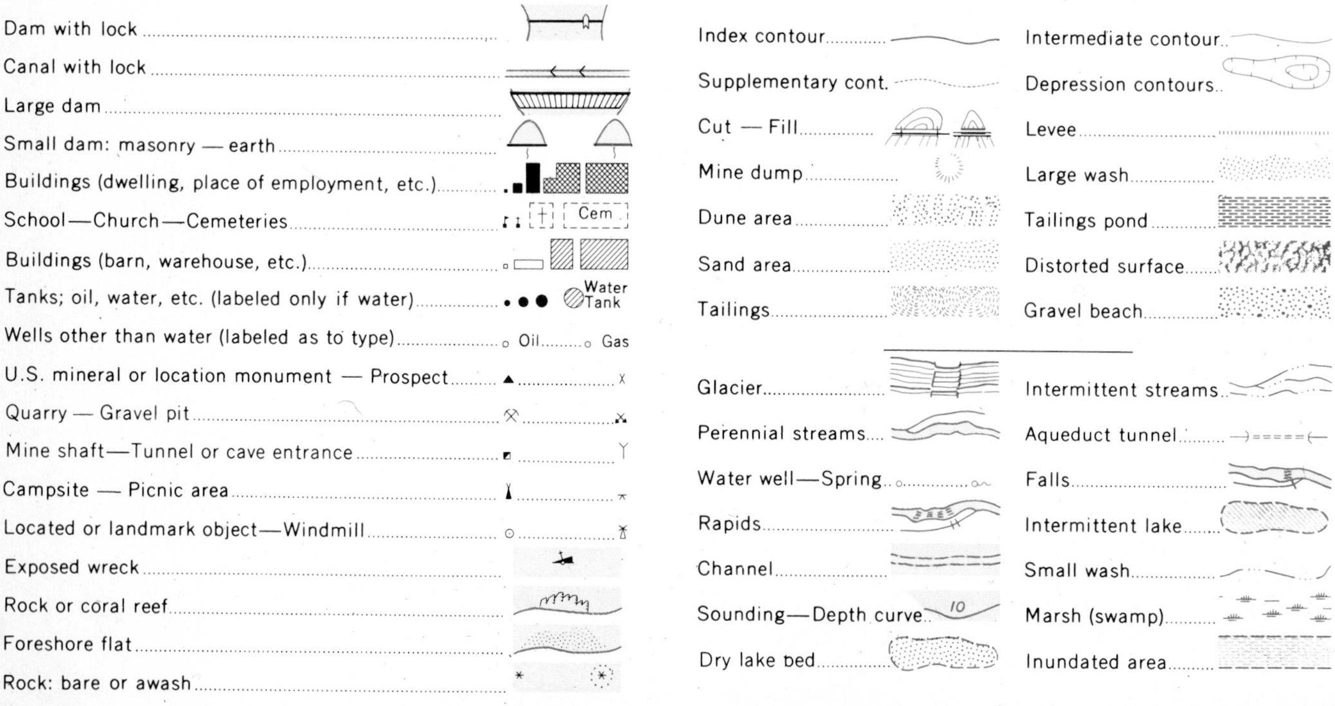

Fig. 25-9. A map title should be located in the lower right-hand corner of the sheet and give data pertinent to the map.

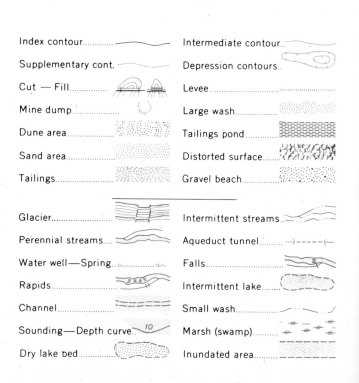

Fig. 25-11. Standard symbols for use on maps.
(U. S. Department of the Interior)

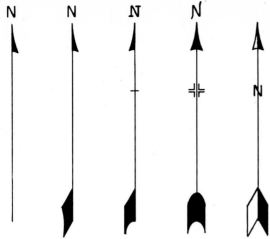

Fig. 25-12. Some sample arrows indicating direction NORTH.

Regardless of the scale of the map, symbols are drawn essentially the same size. Symbols important to a particular type map should be emphasized by making them darker than the less important symbols. Additional map symbols are shown on page 601 in the Reference Section.

NORTH INDICATION

Maps must be properly oriented to be useful. This is done with the direction arrow NORTH, which indicates true north unless otherwise stated. The main feature of the arrow should be its body line with the arrowhead clearly indicating north. Some examples are shown in Fig. 25-12. Typical NORTH symbols are available in the pressure-sensitive transfers.

Some drafters use lower case letters in notes on maps. Notes and tabular materials are frequently typed on adhesive back overlays and added to the drawing. All lettering and notes are placed to read from the bottom or right-hand side of the sheet.

SYMBOLS

Because of the necessity of using very small scales on most maps, not all features can be shown. Many that can be shown must be represented by symbols, Fig. 25-11. Details of symbols should be carefully studied so they may be drawn correctly on maps. Avoid the mistake of crowding symbols on maps by using too many of one symbol. (For example, too many symbols for grass.)

METHODS OF GATHERING MAP DATA

Surveying is the means of collecting data for use in making maps. This is accomplished in a variety of ways, the most important of which are discussed in this section.

FIELD SURVEY CREWS

For years, survey crews equipped with the surveyor's chain (a distance measuring device), transit and level have gathered data in the field for use in making maps. This is a time-consuming and laborious task, particularly where the terrain is rough and equipment has to be carried manually.

Many of these surveying devices have been replaced today, particularly on large scale projects, with faster and more accurate instruments. However, transit-equipped survey crews are still used to gather map-making information on small tracts of land, in highway construction, and on geological exploration projects. Fig. 25-13.

Fig. 25-13. A highway survey crew at work staking out horizontal and vertical curves which were designed from earlier field data. (Virginia Department of Highways)

Industry photo. Highway department employee taking a reading through a transit theodolite. More women each year are applying engineering principles in the field. (Colorado Department of Highways)

PHOTOGRAMMETRY

The tremendous expansion in state and interstate highway programs has caused a need for new and improved techniques of gathering survey information. One of these techniques, photogrammetry, which had been in limited used for a number of years, is now used extensively.

Photogrammetry is the use of photography, either aerial or land-based, to produce useful data for the preparation of contour topographic, planimetric and orthophoto cross section profile maps.

Once the area to be mapped has been identified, control points to be used in controlling photographic stereo models (3-dimension viewing) are placed. Next, ground control surveys are made as checks. Then aerial photographs are taken for translation into photomaps, orthophoto cross section maps and topographic maps. This is accomplished by means of a stereoplotter for reading elevations from a flat surface, as shown in Fig. 25-14.

Automated digitizers for recording horizontal coordinates and control points are used, along with computers. The data are returned to drafters and a map is drawn from the survey data. Or, the data are processed and fed to a plotter in the form of a computer program on cards. The plotter then produces the maps.

Photogrammetry represents a considerable savings of time over field survey methods in the collection of survey data.

RADAR IMAGERY

A recent development in gathering information for map drafting is radar imagery, Fig. 25-15. This is accomplished by a high-resolution, side-looking, airborne radar that records a "photo-like" image of the terrain. It is called side-looking because the electronic signals are sent out at right angles to the aircraft's path.

Radar imagery mapping can be done day or night in any kind of weather, even when heavy clouds block out the use of aerial photography. Unlike aerial photography, the scale of the radar imagery is constant and without distortion, regardless of the range or altitude of the aircraft. This makes it ideal for mosaics and other mapping purposes.

A radar system works by radiating an electronic beam, then reads and records the reflections received from the surface being surveyed. As the aircraft travels forward, successive strips of terrain are exposed to the beam and bounced back to the aircraft. The radar is sensitive to variations in intensity of the reflections and these variations produce images of the terrain covered.

Both natural and man-made features are recorded by radar imagery. "Shadow" effects in the imagery are not from the sun, but from a lack of electromagnetic return from an area shielded from radar by mountains. Water imagery appears dark since the radar beam is not returned, but deflected away by the surface of the water.

The radar imagery method of surveying has applications in the fields of geology, geophysics, hydrology, topographic mapping, highway construction and agriculture. Output is planimetric, but three-dimensional qualities are provided when prepared in duplicate, overlapping sets. A radar system was used in mapping the moon surface and subsurface (1.8 miles) on the Apollo 17 flight.

MAP DRAFTING TECHNIQUES

Once the map data has been gathered, the plotting of the data and actual drawing of the map may begin. However, there are several techniques which are somewhat special to map drafting. These should be understood and are presented here.

CONTOURS

Contours are the irregular shaped lines found on topographic and other plan view maps to indicate changes in elevation of the terrain. Contours may be thought of as the line produced when an imaginary horizontal plane meets the earth's surface, Fig. 25-16. Every point on a single contour line is at the same elevation, and every contour line closes when extended far enough. (This point may be off the particular map being drawn.)

Intervals (vertical distances) of spacing between contour

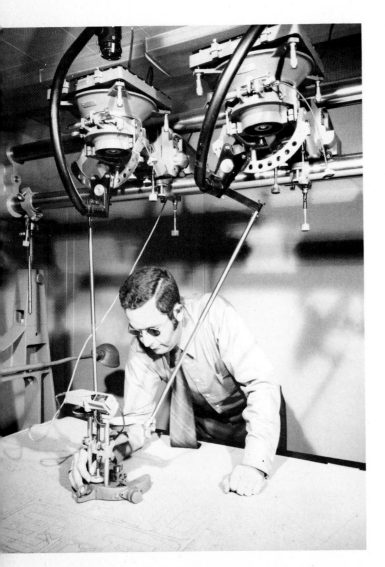

Fig. 25-14. A stereoplotter produces vertical readings from a flat surface. (Virginia Dept. of Highways)

Fig. 25-15. A radar imagery of an Indonesian Island coastal terrain. (Goodyear Aerospace Corp. and International Aero Service Corp.)

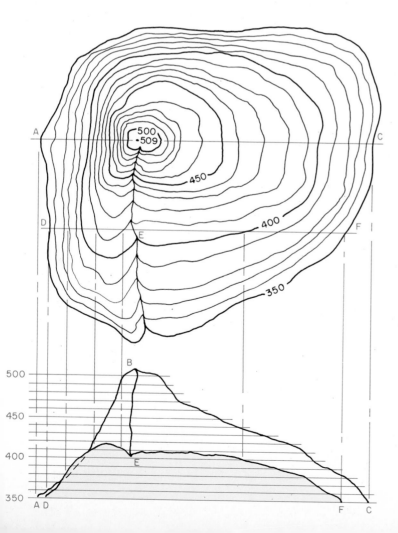

Fig. 25-16. A contour line is an imaginary
line on the earth's surface which represents
a perfectly horizontal plane.

lines may be any distance, depending somewhat on the scale of the map and the characteristic of the terrain. The interval is usually 5, 10 or 20 feet on maps where the terrain is reasonably flat. However, they may be 100 to 200 feet, or more, in mountainous terrain.

The elevation is given for each contour interval or for alternate intervals. Elevation figures should appear parallel to the contour lines, in rows for consecutive intervals, and read from the bottom of the map where possible, Fig. 25-16. The elevation of a peak or depression is represented by a point, and the elevation figure is given. The usual practice is to show every fifth contour in a heavier line to assist a reader in following contours.

Contour lines are plotted from data gathered in the field. The field data are usually taken in some pattern such as a coordinate grid, Fig. 25-17 (a). Or, the data are taken radially from traverse points, and close enough together so that interpolations between the measured points may be made.

Where the terrain takes a decided change in elevation, such as a steep bank or cliff, measurements may be taken at the point of change and recorded for use in plotting. This adds to the accuracy in plotting par-

Drafting for Industry

ticular features of an area.

A grid, drawn to the same scale as the map, and with the elevations recorded, may be placed beneath tracing paper or vellum on which the map is being drawn. Then the contour lines can be plotted. Since contour lines are drawn at regular intervals of previously determined elevations, it is necessary to locate these intervals between the grid intersections by interpolation.

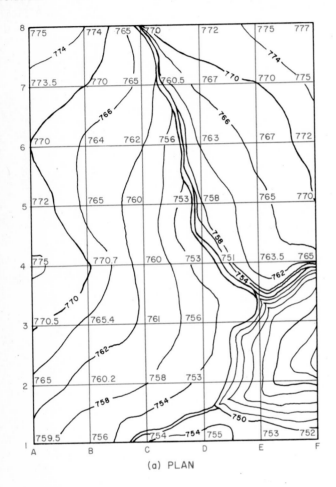

(a) PLAN

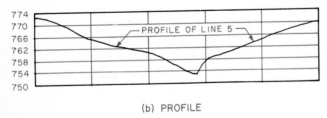

(b) PROFILE

Fig. 25-17. A rectangular-coordinate grid system for use in plotting contour lines on maps.

In making interpolations, it is assumed that the slope of the terrain is uniform between measured points. The horizontal distance between measured points on the grid may be divided proportionally to arrive at a point on the contour line between grid intersections, Fig. 25-17 (a).

When points for the various contour lines have been plotted along the grid lines, the contours may be sketched in lightly and then drawn in freehand to the correct line weight.

PROFILES

In regular orthographic projection, a profile view is one of the side views. In map drafting, however, a profile view is any view of a vertical plane passing through a section of the earth's surface, Fig. 25-17 (b).

A profile may be drawn to the same scale as the contour map from which it is taken. Or, the scale may be exaggerated to emphasize changes in elevation (see Fig. 25-7). Profiles are used to detail vertical curves and elevations of cuts and fills for highways, canals and similar construction projects.

TRAVERSES

Map traverses are straight, intersecting lines which may be laid out in several ways, depending upon the time available and accuracy required. For most maps, a protractor (no smaller than 6 inches and preferably a vernier protractor) and a scale are satisfactory in laying out traverses.

The drafting machine with vernier protractor is also satisfactory for drawing map traverses. For map traverses requiring greater accuracy, use trigonometric calculations.

The layout procedure starts at a station from a known backsight line, Fig. 25-18.

1. Point B represents station 2, and line BA is backsight line.
2. Extend line AB far enough through C to permit protractor to be aligned with protractor center at B.
3. Assume station 3 has a deflection of 82° 30' to right (R) of foresight line BC and a length of 127 feet.
4. Locate protractor to right side of line AC with its center B at station 2, and lay off 82° 30' at D.
5. Extend BD and lay off scale measurement of 127 feet. This is station 3.
6. Subsequent lines and stations to complete closed traverse of land plot are located in a like manner.

SCRIBING IN MAP DRAFTING

Until recent years the usual practice in map drafting was to ink the original in order to have camera ready copy. Many map drafting groups still ink their drawings, but there is a trend toward producing camera ready copy by a technique called "scribing."

Scribing is the process of removing colored coating from a transparent polyester film based material with scribing instruments, Fig. 25-19. These instruments resemble conventional drafting instruments except the working points are scribing points instead of pencils or pens.

Scribe points are made to fit drafting instruments and mechanical lead holders. They are made of carboloy and ground to a shape and size to produce sharp lines of various widths without damaging the drawing surface.

The chief advantages of scribing are that it is more accurate, gives sharper line work, is faster than inking and the copy serves as the film negative which can be used for final composite positives.

Layouts may be done lightly in pencil and then traced, using a scribe point. Corrections may be made by rubbing an orange crayon at right angles across the scribed line or by adding a fluid touch-up. A new line may then be scribed through the area. Lettering may also be done with a scriber.

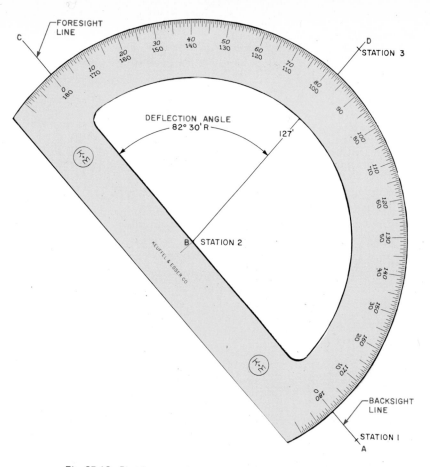

Fig. 25-18. Plotting a map traverse with the use of a protractor.

Fig. 25-19. A cartographic drafter uses a scribing instrument in preparing an original drawing of a map on scribe-coat film.   (Keuffel & Esser Co.)

PROBLEMS IN MAP DRAFTING

The following problems are designed to provide you with the opportunity to apply knowledge gained in your study of map drafting and to help you become familiar with the procedures used.

1. Select an appropriate scale and contour interval and plot the contours for the map shown in Fig. 25-20.

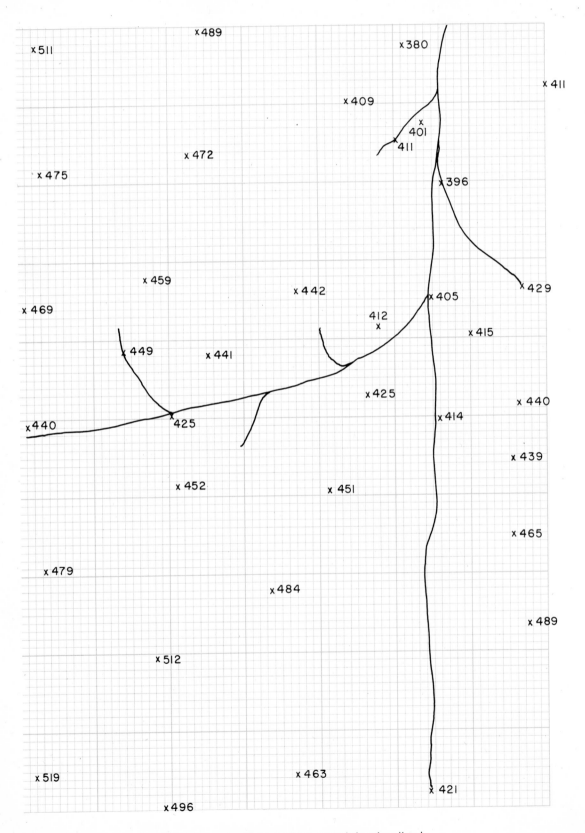

Fig. 25-20. A topographic map with measured elevations listed.

2. Select an appropriate scale and contour interval and plot the contours, and natural and constructed features, for the map shown in Fig. 25-21.

3. Draw a profile map showing the shape of the terrain at lines 1 and 4 through the map section shown in Fig. 25-21. Use the same scale as used for the contour map.

4. Draw a profile map showing the shape of the terrain at line 3 through the map section shown in Fig. 25-21, and another perpendicular to line 3 at D.

5. Lay out the map traverse shown in Fig. 25-22. Indicate a NORTH point on the map and orient the first station and backsight line with it.

6. Draw an irregular shaped (not rectangular), four-sided plot to any convenient size. With a protractor or vernier protractor, select a scale suitable for producing a reasonable size lot, and measure the angle and length of each traverse. Label each station and measurement appropriately.

7. With the assistance and approval of your instructor, select a teammate and survey a plot on the school campus. Record your data and draw a map traverse, using the data collected. Locate the natural and constructed features as directed by your instructor.

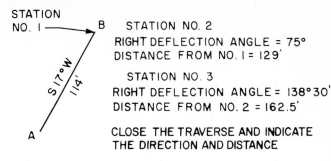

STATION NO. 2
RIGHT DEFLECTION ANGLE = 75°
DISTANCE FROM NO. 1 = 129'

STATION NO. 3
RIGHT DEFLECTION ANGLE = 138°30'
DISTANCE FROM NO. 2 = 162.5'

CLOSE THE TRAVERSE AND INDICATE THE DIRECTION AND DISTANCE

Fig. 25-22. A map traverse with three stations.

8. With the approval of your instructor, organize a team of 3 to 5 students. Then arrange for the necessary equipment and survey a plot of ground, including boundary lines and elevation.

Select a grid for measuring the elevation at regular intervals and at each corner of any structure on the plot. Record your data and prepare a topographic map, showing property lines or boundary of property, contour lines and their elevation (assume elevation of lowest point as 0). Locate principal natural and constructed features.

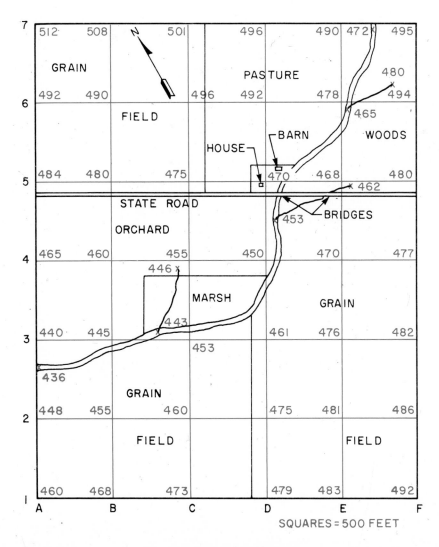

SQUARES = 500 FEET

Fig. 25-21. Map drafting problem.

Chapter 26
TECHNICAL ILLUSTRATION

Industrial processes and products have become so technical that most items sold are accompanied with one or more technical illustrations explaining their operation and use, Fig. 26-1. Technical illustration is the preparation of drawings, usually pictorial, with three-dimensional effect. Usually, they are shaded or finished in multicolor to give more realism and understanding to the object or process.

Technical illustrations are used by technical personnel to supplement working drawings and to clarify complex assembly and operational procedures. They are indispensable to nontechnical personnel and to those who use technical equipment but have difficulty reading working drawings.

Technical illustrations are widely used in industry to speed the production processes. They are also used in service manuals and in do-it-yourself kits.

matched and are not
care when inserting the plunger
the barrel.

Fuel Injection Pump Installation

The installation of fuel injection pumps requires that the lifter be at a low point and the fuel rack be centered or at "zero" position. To center or "zero" the rack, install the 7S7113 Rack Setting Gauge and set it at .000 in., retract the speed limiter, and move the rack in "fuel on" direction until it contacts the gauge. To install the pump, sight down the pump and align notches in bonnet and barrel with slot in the pump gear segment. Slot is 180° from pump gear segment center tooth.

RACK SETTING GAUGE INSTALLED

1. 7S7113 Rack Setting Gauge. A. 9S240 Rack Positioning Tool Group can also be used.

Position the notches in bonnet and barrel to align with dowels in the housing. Install the pump. Keep a downward force (by hand) on the pump

Fig. 26-1. Technical illustrations are a valuable part of equipment service manuals. (Caterpillar Tractor Co.)

TYPES OF TECHNICAL ILLUSTRATIONS

Many of the techniques and types of technical illustrations have been covered in previous chapters. Their special application and use is discussed in this chapter.

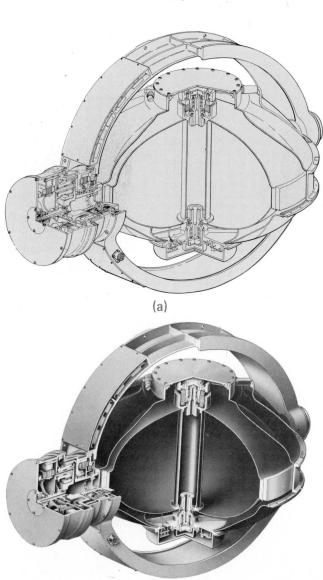

(a)

(b)

Fig. 26-2. The pictorial drawing (a) is a basic type of technical illustration that can be changed into a rendering (b) of the object and used for a presentation drawing. (Sperry Flight Systems Div.)

Technical Illustration

Technical illustration work may be classified into two general fields: engineering-production illustrations; publication illustrations.

Engineering-production illustrations are used for engineering design, contract proposals and production work. More emphasis is given to technical accuracy than to their styling.

Publication illustrations are used in service manuals, parts

PICTORIAL

The pictorial drawing is perhaps the most elementary type of technical illustration. It may be a line drawing in any one of the standard pictorial projections: isometric, dimetric, trimetric, oblique or perspective, Fig. 26-2 (a). Or, it may be a sophisticated rendering of the same drawing that closely resembles a photograph of the object, (b).

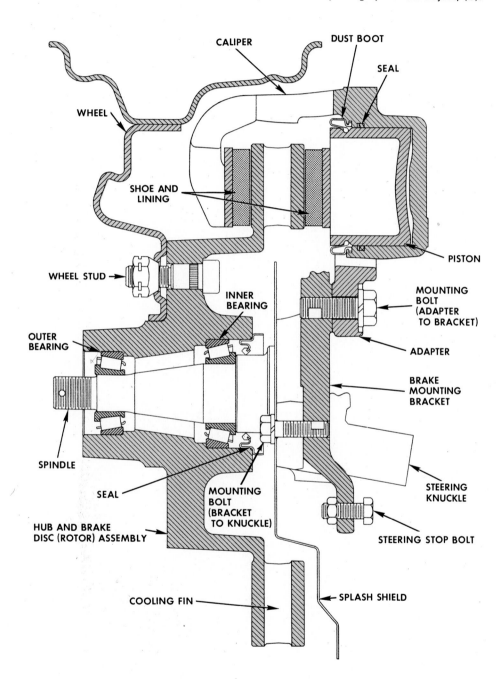

Fig. 26-3. The cutaway assembly illustrates the assembled relationship of the various parts. (International Harvester Co.)

catalogs, operational handbooks and in sales and advertising brochures and catalogs. Publication illustrations frequently include cartoon sketches, shading effects, life drawings and other commercial art techniques. In both fields of illustration, the following types of illustrations are used.

CUTAWAY ASSEMBLY

The cutaway assembly drawing helps to clarify multiview drawings of complex assemblies, Fig. 26-3. Used in production operations, cutaway assembly drawings frequently are found in service manuals.

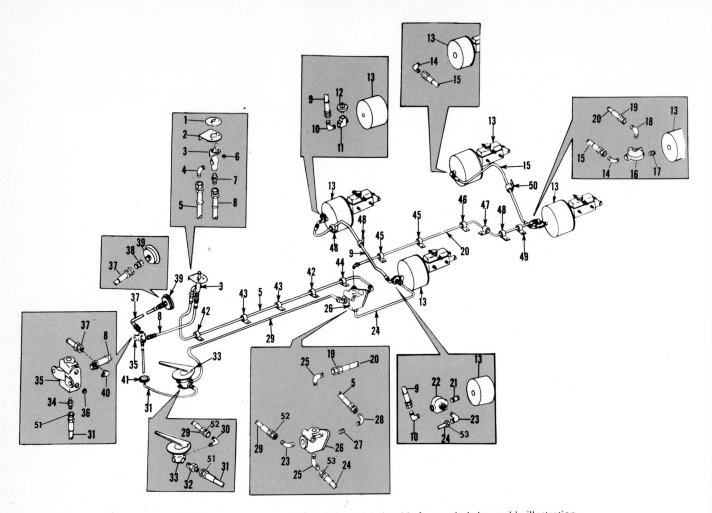

Fig. 26-4. Assembly operations are easy to follow with the aid of an exploded assembly illustration.
(International Harvester Co.)

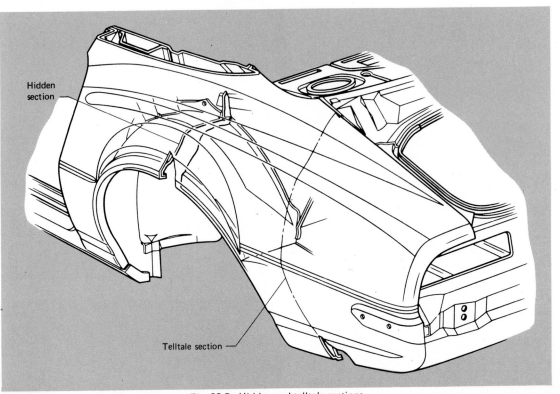

Fig. 26-5. Hidden and telltale sections.
(General Motors Corp.)

EXPLODED ASSEMBLY

A type of technical illustration used frequently in manufacturing assembly, service manuals and customer purchase instructions is the exploded assembly, Fig. 26-4. Few Illustrations are as effective as the exploded assembly in clarifying a procedure for assembly. Note that a center line joining parts leaves no doubt about the order of assembly.

HIDDEN AND TELLTALE SECTIONS

Hidden and telltale sections permit the illustrator to display sheet metal relationships when it is desired to show the outside piece in its entirety, Fig. 26-5.

PEELED SECTION

Several layers of material in a built-up part may be shown by use of a peeled section, Fig. 26-6. The use of this type of section calls for artistic judgment on the part of the illustrator and checker.

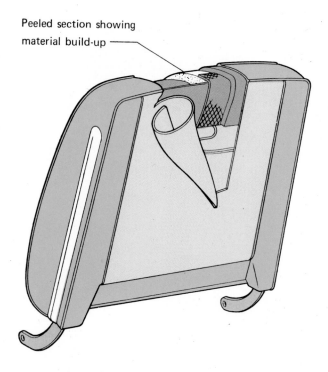

Peeled section showing material build-up

Fig. 26-6. A peeled section showing an automobile seat construction. (General Motors Corp.)

FILM SLIDES AND TRANSPARENCIES

In some industries, much of the technical illustrator's work is the preparation of artwork used in making film slides and overhead projection transparencies. The artwork for these visuals is, in many cases, the same as that used for other purposes. Normally, however, it includes more color work.

BASIC ILLUSTRATION TECHNIQUES

The following techniques are widely used in industry to prepare all types of technical illustrations. It is assumed that the student has skill in the basic techniques of orthographic projection, dimensioning, sectioning and pictorial drafting.

OUTLINE SHADING

A multiview or a pictorial drawing may be given more realism by a technique called outline shading, Fig. 26-7. The light direction is assumed, and the lines that are on the far side of the light source are considered to be in the shade. For holes, the side nearest the light is assumed to be in the shade.

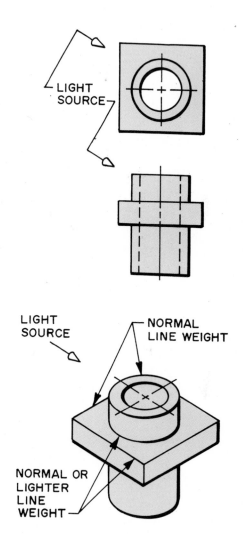

LIGHT SOURCE

LIGHT SOURCE — NORMAL LINE WEIGHT

NORMAL OR LIGHTER LINE WEIGHT

Fig. 26-7. Outline shading technique applied to a multiview drawing and to a pictorial.

The remaining lines that outline the part are done in normal visible line weight. The lines inside may be drawn either normal weight or lighter weight for greater emphasis.

In multiview drawings, the line is widened on one side away from the light approximately three times its normal width. Notice that cylindrical features are shifted on an axis parallel to the light source.

In pictorial drawings, the line is widened on both sides to approximately one-and-one-half times its normal width for a total of three line widths. Notice that cylindrical features, which appear as ellipses in pictorial views, are shifted along the axis which most nearly aligns with the light source.

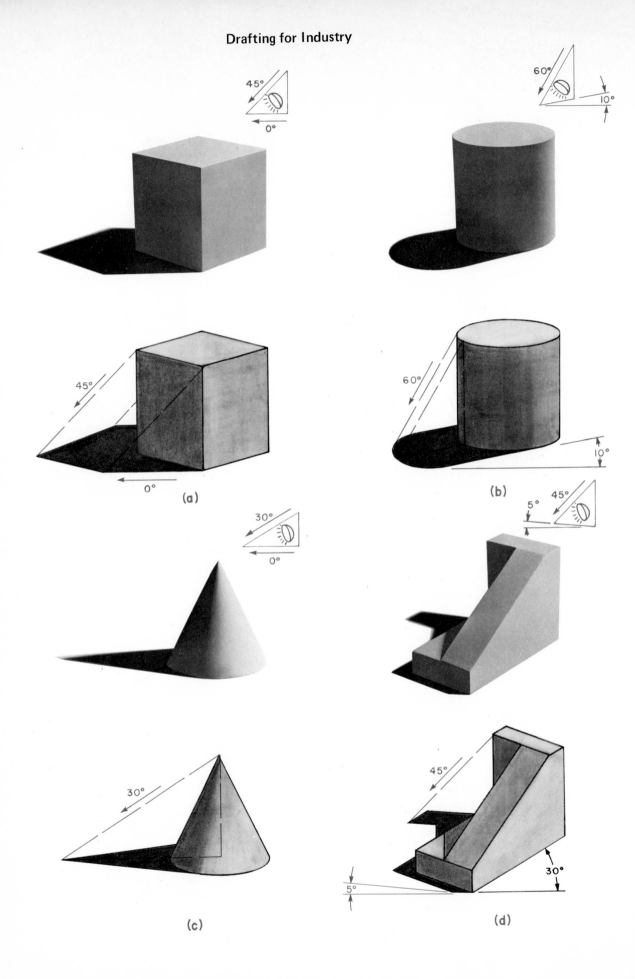

Fig. 26-8. Locating shadows of geometric figures with various directions of light source.

LOCATING SHADOWS OF GEOMETRIC FORMS

Locating shadows of geometric forms helps in understanding the shading of these forms in technical illustrations. Four geometric figures are shown in Fig. 26-8, along with the procedures used in locating their shadows. Notice the direction of the light source and how it affects the shading on the objects. Carefully study the projection of the shadows and how the light source affects the direction of projection.

LINE SHADING

Shading to produce a three-dimensional effect on an object may be accomplished by the use of lines of varying spacing and length, Fig. 26-9. A light source should be assumed. The shading lines are then drawn more closely together near the back of surfaces exposed to light or those shaded from the light. Note how the lines may be shortened or omitted on surfaces where the light seems to fall.

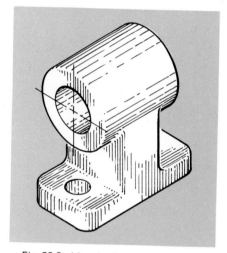

Fig. 26-9. Line shading of a machine part.

On external cylindrical surfaces, note that the lines appear closer together as they move to the shaded side of the piece. Where the light falls on the far side on internal cylindrical walls, fewer lines are drawn. It is best to arrange the light so that it does not fall in the center of a cylindrical surface.

Shaded flat surfaces are darkest in the foreground where they contrast sharply with lighted surfaces. Or, they tend to lighten in the background. Lighted flat surfaces are lightest, or void of line shading, in the foreground and tend to darken somewhat in the background.

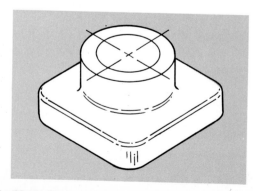

Fig. 26-10. Shading fillets and rounds with straight line shading.

FILLETS AND ROUNDS

Fillets and rounds usually catch the light source and tend to appear light. When line shading is used, fillets and rounds may be treated either with curved lines as shown in Fig. 26-9 or with straight lines, Fig. 26-10. The latter are preferred because they may be applied more readily and present a neater appearing drawing.

THREADS

Threads are represented by a series of ellipses unless they appear in the frontal plane of an oblique drawing. There they are represented as circles. They may be shaded solid except for the highlights, as shown in Fig. 26-11.

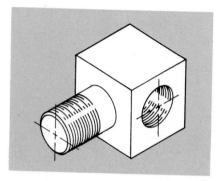

Fig. 26-11. Shading of external and internal threads.

SMUDGE SHADING

When a tone shading that flows smoothly from dark to light and back to dark is desired, smudge shading is commonly used, Fig. 26-12. This is done by going over the area to be shaded with a soft lead pencil, 3B to HB. Then, to produce the desired tone, the graphite is rubbed into the texture of the paper with a stub of paper, a piece of soft cloth, or the finger. Where a darker portion is needed, more graphite may be added with the pencil and rubbed.

A protective spray coating should be applied to the rendering when it is completed.

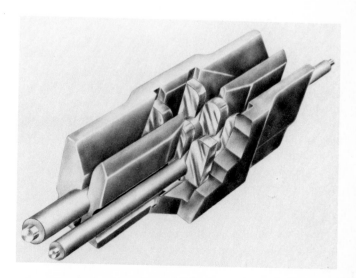

Fig. 26-12. A smudge shading of a telephone wire device. (Bell Laboratories)

Fig. 26-13. An architectural rendering in pencil.

PENCIL SHADING

A soft lead pencil, 3B to HB, used on paper that has a slight texture is an effective means of shading a drawing. Variations in rendering are achieved by using the side of the lead as well as the point, and by using leads of different degrees of hardness. This technique produces soft tones especially suited to architectural renderings, Fig. 26-13.

A spray coating should be applied to the rendering to protect it from smears during handling and storage.

INKING

Technical illustrations planned for publications are usually inked to achieve desired reproduction qualities. Not only is ink well suited for line work, but it is adaptable to stippling, by use of a fine pen point or sponge.

Sponge shading is done with small piece of sponge used with stamp pad or film of printer's ink, Fig. 26-14. Light touches should be used first, darker areas may then be reworked to desired tone. Protect surrounding areas with paper overlay.

SHADING WITH THE AIRBRUSH

One of the most effective tools used in illustration work is the airbrush, Fig. 26-15. The airbrush operates off compressed air or a carbon dioxide cylinder. It sprays a mist of ink or water color onto the drawing.

Parts of the drawing not to be sprayed can be protected by paper templates or a frisket, which is a special paper with an adhesive back. The frisket is laid over the entire drawing and "windows" are cut and removed over areas to be worked.

Some practice is needed to develop skill in the use of the airbrush. Once the technique is mastered, illustrations more effective than an actual photograph are possible, Fig. 26-16.

Fig. 26-14. An example of sponge shading with ink. (Bell Laboratories)

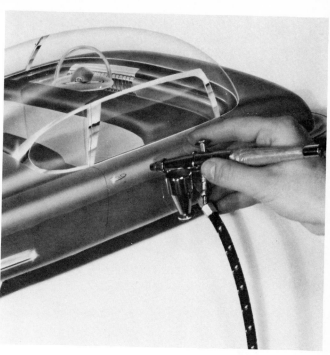

Fig. 26-15. An airbrush is a valuable tool in technical illustration work. (Paasche Airbrush Co.)

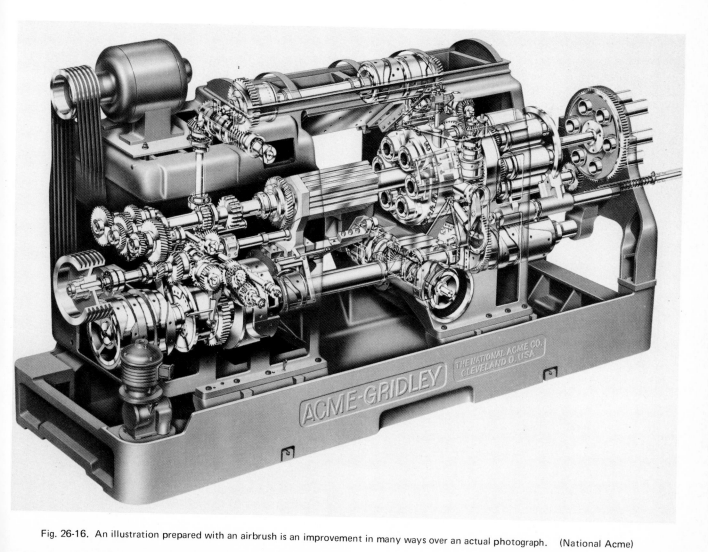

Fig. 26-16. An illustration prepared with an airbrush is an improvement in many ways over an actual photograph. (National Acme)

(a)   (b)

Fig. 26-17. A photograph before (a) and after (b) retouching.
(Mack Trucks, Inc.)

PHOTO RETOUCHING

Photographs are a fast and inexpensive means of producing a technical illustration. However, the photograph often fails to bring out detail that is desired. Sometimes, only a part of the object photographed is to be emphasized, while the rest is subdued.

The process of reworking photographs to emphasize or sharpen certain details and hold others back is known as photo retouching. This may be done by use of the airbrush or hand brush, as on any other illustration. Lines may be added by scratching the negative or may be removed by using a special lacquer. Fig. 26-17 shows a photograph before and after retouching to bring out the detail.

OVERLAY FILM

Overlay films for shading or color work in technical illustration are available commercially in a wide assortment. (Fig. 26-8 shows use of shading film.) These films come in glossy finish for illustrations that are to be reproduced by the diazo (dark line) print process or by photography. When the original artwork is to serve as the finished illustration, overlay film with a matte finish should be used.

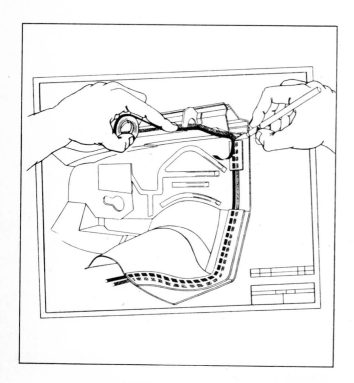
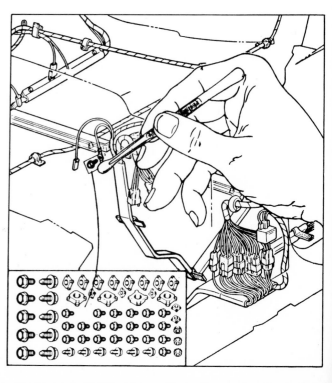

Fig. 26-18. Transfer sheets and tapes of standard parts are a valuable aid to technical illustrators.   (General Motors Corp.)

Overlay film has an adhesive back and can be applied to almost any working surface. The film is laid in place and lightly pressed. The part to be overlayed is cut in place and left, while the remainder is removed. Some films require burnishing to set in place while others do not.

STANDARD PARTS TRANSFERS AND TAPES

Where a number of standard parts or a number of like components are needed, adhesive back or pressure sensitive transfers and tapes should be used. These are available commercially and represent a considerable savings in time for technical illustrators, Fig. 26-18.

COMPUTER-ASSISTED ILLUSTRATION

The computer has become a valuable tool for the technical illustrator. The speed and ease of the computer, along with its great diversity, offer the creative illustrator many design possibilities not attainable by traditional methods.

As in computer graphics applications for standard drafting procedures, the computer-assisted illustration provides for a stored library. This means immediate access to symbols and illustrations of parts for creating new illustrations. Changes and revisions can be performed faster, leaving the illustrator more time for creativity.

PROCEDURE FOR PREPARING TECHNICAL ILLUSTRATIONS

The following steps should be observed in preparing technical illustrations:
1. Know purpose illustration is to serve.
2. Familiarize yourself with part to be illustrated and study blueprint, proposed drawings or model.
3. Make a freehand sketch of your proposed solution and check it against steps 1 and 2.
4. Accurately lay out illustration and check it against blueprint.
5. Trace illustration on vellum, cloth, film or high grade of illustration paper.
6. Add shading, overlays and notes to finish illustration.

PROBLEMS IN TECHNICAL ILLUSTRATION

The following problems afford you with the opportunity to try several of the techniques used in technical illustration. With the guidance and approval of your instructor, select as many types of illustrations and techniques as time and equipment permit. Then prepare the technical illustrations.
1. Select an object from Fig. 7-40, page 139, and prepare a multiview illustration using outline shading in pencil.
2. Select an object from Fig. 11-62, page 225, and prepare an isometric illustration using outline shading in pencil.
3. Select an object from Fig. 7-41, page 140, and prepare a multiview illustration using outline shading in ink.
4. Select an object from Fig. 11-63, page 226, and prepare an isometric illustration using outline shading in ink.
5. Select an object from Fig. 11-64, page 227, and prepare an isometric illustration using line shading in pencil.
6. Select an object from Fig. 11-64, page 227, and prepare an oblique illustration using line shading in ink.
7. Prepare a smudge shading illustration of an object in Fig. 11-64, page 227. The illustration is to be a dimetric drawing using one of the axes suggested in Fig. 11-32, page 205. Select the axes which will present the best view of the object's features.
8. Prepare a pencil shading rendering of the front elevation of one of the houses you drew in Chapter 22. You may pencil in the shrubs or use transfer appliques.

Computer graphics will increasingly be used by the technical illustrator to produce pictorial drawings. (Integraph)

Chapter 27
GRAPHS AND CHARTS

Drafting departments are called on from time to time to prepare graphs and charts as a means of presenting data in graphic form for contract proposals, analysis of data and marketing. As one familiar with drafting procedures, you possess many of the skills for this work.

Graphs are diagrams showing relationship between two or more factors. Charts may be defined as a means of presenting information in a tabulated form, or sometimes in graphic form as a line diagram.

TYPES OF GRAPHS AND CHARTS

Many forms of graphs and charts are used to analyze and clarify data. The major types and the techniques of their construction are presented in this chapter.

LINE GRAPH

One of the simplest graphs to construct is the line graph, Fig. 27-1. Line graphs are used to show relationships of

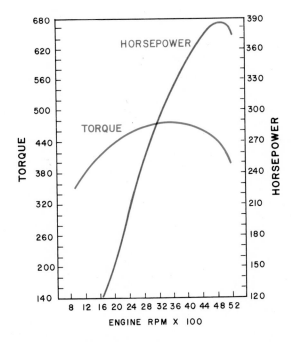

The data represented by these curves were obtained through Standard Test No. 20 of the General Motors Automotive Engine Test Code. This test establishes a uniform method of determining the gross power output of the bare engine. It is run with distributor and carburetor adjusted for maximum power at each speed. Data are corrected to 60°F. using an SAE correction factor.

Fig. 27-2. Smooth curves in a line graph reflect a continuous change rather than a sharp change for each interval. Data are gathered for each interval and plotted. The curve approximates these points.

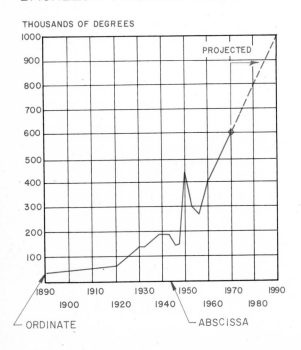

Fig. 27-1. A line graph showing bachelor degrees granted.

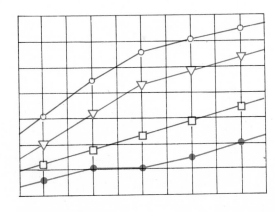

Fig. 27-3. Symbols may be used to differentiate curves if several are to be plotted on the same graph.

MINIMUM FORGEABLE RIB THICKNESS

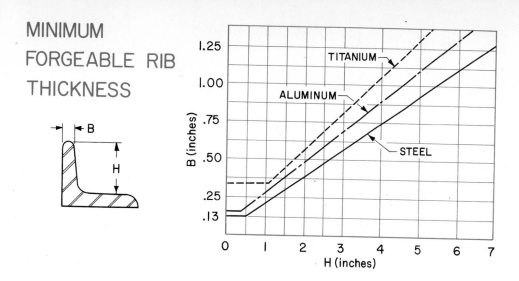

Fig. 27-4. A line graph used to compare dimensional features of metal stock.

quantities to a time span. The horizontal axis, called the abscissa or X axis, usually contains the time element. The vertical axis, called the ordinate or Y axis, expresses the other factor in terms of numbers or percentages.

After the data are plotted in the line graph, the line may be a broken line curve (drawn point-to-point), Fig. 27-1, or a smooth curve, Fig. 27-2. When it is desired to show an actual condition or status for each of the time periods, use the broken line curve. However, when the change is continuous, the smooth curve approximating the actual data for each time interval is more meaningful.

Symbols at each of the data points or variations in the form of the line itself are sometimes used to represent differences in factors or methods of treatment, Fig. 27-3.

The line graph may also be used for comparison of two design factors of material as shown in Fig. 27-4, or to show a set of characteristic curves for an electronic transistor. Fig. 27-5 illustrates this technique.

CONSTRUCTION OF LINE GRAPHS

Given the data to be plotted, proceed as follows for constructing a line graph:

1. Select a scale (uniform spacing), based on data given, which will fit allotted space and effectively present the furnished data, Fig. 27-6.
2. Use a prepared coordinate grid paper, or prepare a grid, and lay off scales on X and Y axes.
3. Plot data at X and Y coordinates. Points may be a symbol if this is to be used to distinguish between sets of data.
4. Join points with a light weight, broken line since progression of data is not continuous.
5. Darken line graph, place a border around graph if desired and add identifying notes and title.

COMMON EMITTER OUTPUT CHARACTERISTICS FOR A 2N1234

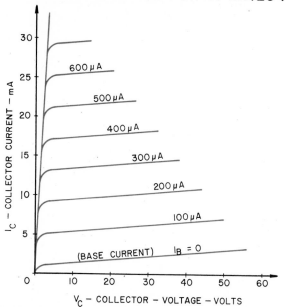

Fig. 27-5. A family of output characteristic curves for a hypothetical silicon bipolar transistor in a common emitter connection.

COMPANY HOURLY LABOR COST INDEX
TEN YEAR PERIOD

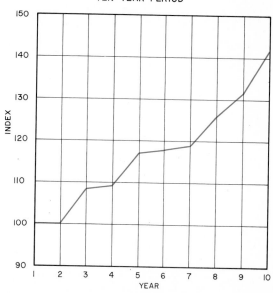

Fig. 27-6. Constructing a line graph.

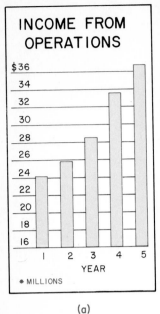

INCOME FROM OPERATIONS

$36
34
32
30
28
26
24
22
20
18
16

1 2 3 4 5
YEAR

* MILLIONS

(a)

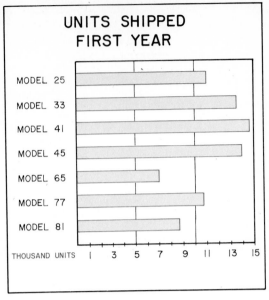

UNITS SHIPPED FIRST YEAR

MODEL 25
MODEL 33
MODEL 41
MODEL 45
MODEL 65
MODEL 77
MODEL 81

THOUSAND UNITS 1 3 5 7 9 11 13 15

(b)

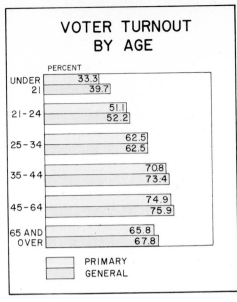

VOTER TURNOUT BY AGE

PERCENT

UNDER 21   33.3 / 39.7
21-24   51.1 / 52.2
25-34   62.5 / 62.5
35-44   70.8 / 73.4
45-64   74.9 / 75.9
65 AND OVER 65.8 / 67.8

□ PRIMARY
□ GENERAL

(c)

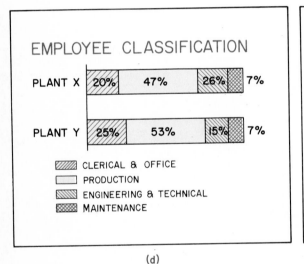

EMPLOYEE CLASSIFICATION

PLANT X   20%   47%   26%   7%

PLANT Y   25%   53%   15%   7%

▨ CLERICAL & OFFICE
□ PRODUCTION
▨ ENGINEERING & TECHNICAL
▨ MAINTENANCE

(d)

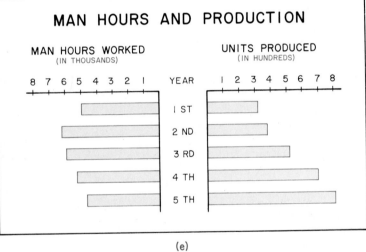

MAN HOURS AND PRODUCTION

MAN HOURS WORKED (IN THOUSANDS)   UNITS PRODUCED (IN HUNDREDS)

8 7 6 5 4 3 2 1   YEAR   1 2 3 4 5 6 7 8

1 ST
2 ND
3 RD
4 TH
5 TH

(e)

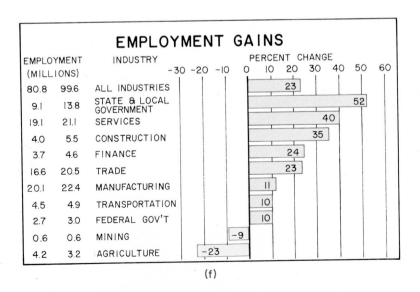

EMPLOYMENT GAINS

EMPLOYMENT (MILLIONS)		INDUSTRY	PERCENT CHANGE
80.8	99.6	ALL INDUSTRIES	23
9.1	13.8	STATE & LOCAL GOVERNMENT	52
19.1	21.1	SERVICES	40
4.0	5.5	CONSTRUCTION	35
3.7	4.6	FINANCE	24
16.6	20.5	TRADE	23
20.1	22.4	MANUFACTURING	11
4.5	4.9	TRANSPORTATION	10
2.7	3.0	FEDERAL GOV'T	10
0.6	0.6	MINING	-9
4.2	3.2	AGRICULTURE	-23

-30 -20 -10 0 10 20 30 40 50 60

(f)

Fig. 27-7. Variations in preparing index bar graphs.

BAR GRAPHS

Bar graphs are a popular form of presenting statistical data, because they are easily understood by lay persons. They are used to show relationships between two or more variables, Fig. 27-7, but have fewer plotted values for each variable than the line graph.

The data presented in bar graphs are usually for a total period of time rather than for various periods such as those in a line graph. For example, the bar graph shows production per hour, day or year, but usually not successive periods of production.

There are two basic types of bar graphs: index bar and range bar. Index bar graphs have a common base where the bars originate, Fig. 27-7. Range bar graphs are individual bars, representing segments of the whole, which are plotted within the range of the total project time schedule (see Fig. 27-8).

sizable element should be plotted first (next to index line). Follow this with the next in importance and so on. The same order of elements should be retained when two or more subdivided bars are used, regardless of the variation in importance or size in successive bars.

The percentage bar graph is a type of subdivided bar which is particularly easy to read when the percentages are included, Fig. 27-7 (d).

The paired bar graph is useful to compare two sets of factors on different scales, Fig. 27-7 (e). A common use is to show total value of raw products purchased as contrasted with the percentage of this amount purchased from one supplier.

The deviation bar graph provides a comparison between a number of factors and their deviation from a "break-even" point. This type of graph lends itself well on such comparisons as profit and loss, or increase and decrease. On a horizontal bar

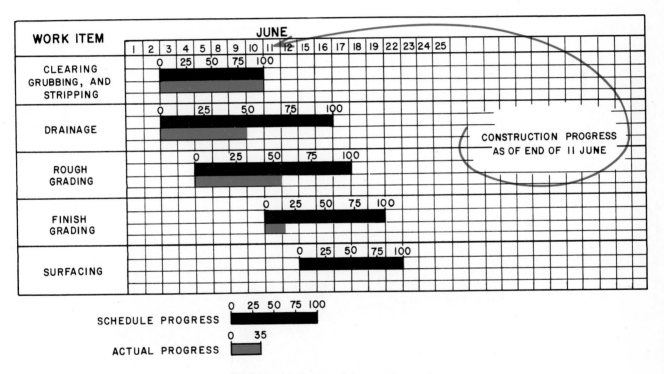

Fig. 27-8. Application of the range bar graph.

A number of variations are possible with the index bar graphs. Variations include: vertical bar, Fig. 27-7 (a), horizontal bar (b), grouped bar (c), subdivided bar (d), paired bar (e) and deviation bar (f).

Vertical and horizontal bar graphs are most commonly used. However, the grouped bar permits the inclusion of other variables in an effective manner. When the grouped bar method is used, the sequence of the elements should be maintained throughout and each element should be distinctively shaded or colored.

The subdivided bar graph is effective when there are fewer than five subdivisions. The graph loses its value when too many divisions make it difficult to appraise the relative value of each.

When the subdivided bar is used, the most important or

graph, Fig. 27-7 (f), the bars are drawn from a zero index line with positive values running to the right and negative values to the left. On a vertical deviation bar graph, positive values should appear on top and negative values below the index line.

The range bar variation of the bar graph normally plots the items against a time line. A typical example would be to plot a production schedule where each phase would be plotted as a time range within the time schedule shown for the entire project, Fig. 27-8. The range bar graph also may be used to show progress.

CONSTRUCTION OF BAR GRAPHS

Given the data to be plotted, proceed as follows for constructing a bar graph:

1. Review data and select most appropriate type of bar graph

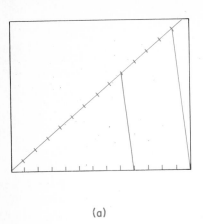

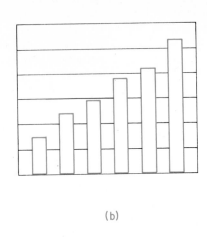

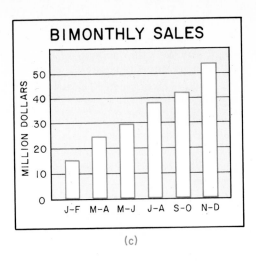

(a)   (b)   (c)

Fig. 27-9. Construction of the bar graph.

for displaying material.

2. Select a pleasing rectangular proportion for graph based on number of bars to be used. Bars should make good use of space available without appearing crowded, Fig. 27-9.

3. To space bars equally, multiply number of bars by two and add one. (6 x 2 + 1 = 13 in our example.) Divide space available by this number, (a). This can be done geometrically.

4. Locate bars between these division marks, starting with second space and skipping every other one, (b).

5. Lay off scale on other axis.

6. Draw bars and shade or otherwise distinguish each set.

7. Draw a border around graph if desired and add identifying notes and title, (c).

SURFACE OR AREA GRAPHS

A surface or area graph is an adaptation of the line or bar graph. The area between the curve and the abscissa axis is shaded for emphasis, Fig. 27-10 (a). Variations in the surface graph are shaded-zone graph, (b) and pictorial-surface graph, (c).

The computer is very useful in preparing charts and graphs. (Tektronix)

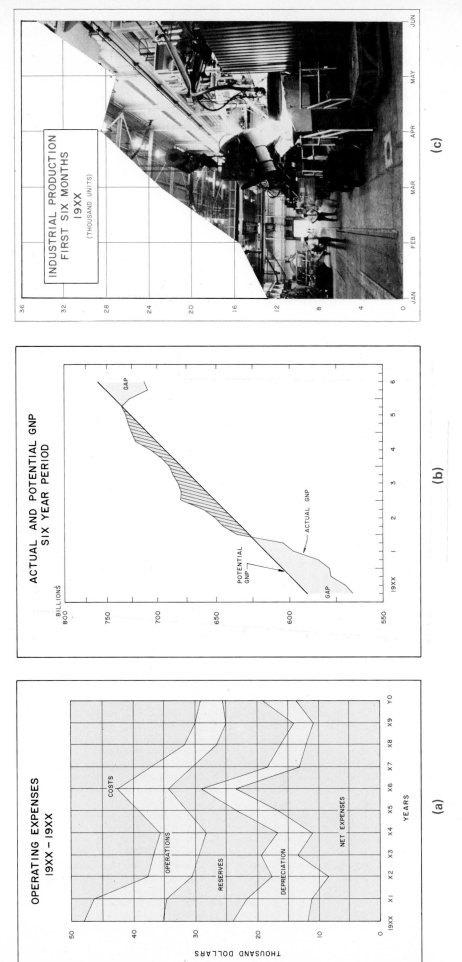

Fig. 27-10. Types of surface or area graphs.

PIE GRAPHS

Pie graphs, sometimes called circle or sector graphs, are frequently used to contrast individual segments, or parts, with the whole. A typical example is the graph shown in Fig. 27-11 (a), which shows the distribution of the labor force in one area.

CONSTRUCTION OF PIE GRAPHS

Given the data to be plotted, proceed as follows for constructing pie graphs:

For a regular pie graph:

1. Select size and draw circle for graph, Fig. 27-12 (a).

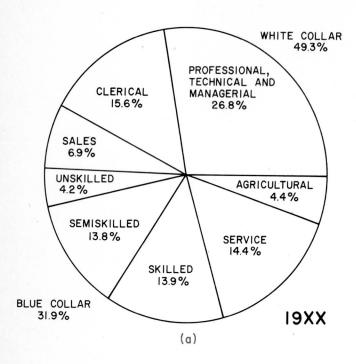

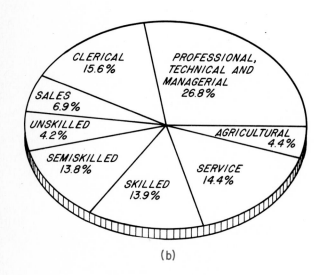

EMPLOYMENT
BY OCCUPATIONAL GROUP

Fig. 27-11. A pie graph (a) and a pictorial pie graph (b).

One variation of the pie graph is to draw it as a pictorial, (b). This type is sometimes used to represent cost expenditures. The pictorial graph shown resembles a silver dollar divided into the various categories of expense.

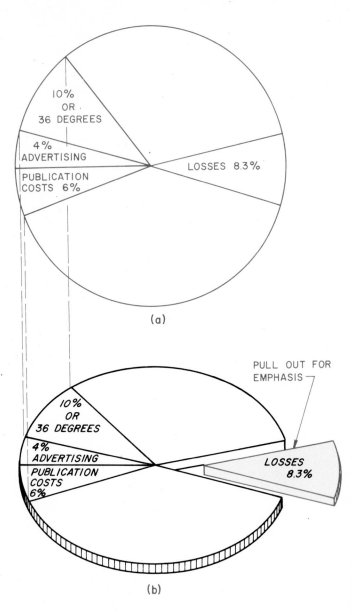

Fig. 27-12. Construction of a pie graph.

2. Lay off various sectors, using a circular percentage grid sheet or a percentage protractor (Fig. 2-63, page 43). Or, calculate degrees for various sectors, using following formula (lay off sectors with a protractor).

Formula:　N times 3.6 = degrees for sector.

Where:　　N = percentage one sector is of whole.

Example:　One sector is 10% of whole.

　　　　　　10 x 3.6 = 36 degrees.

3. If possible, start laying off smallest sectors in a horizontal position to facilitate labeling.

4. Complete sectors and add identifying notes and title.
5. Shade sectors or add colors if desired.
6. Draw a border around graph if desired.
   For a pictorial pie graph:

1. Select a template of size desired and draw an ellipse.
2. To find sector divisions of ellipse, draw a circle with a diameter equal to the major diameter of the ellipse you constructed. On a piece of scrap paper (or lightly on your working sheet), divide the same as in step 2 for a regular pie graph and project to ellipse, Fig. 27-12 (b).
3. Remaining steps are same as 3 through 6 for constructing

regular pie graph.

4. A sector may be "pulled out" for greater emphasis if desired, Fig. 27-12 (b).

FLOW CHARTS

Flow charts are a graphic means of describing the sequence of technical processes that would be difficult to describe in narrative form, Fig. 27-13. The flow of various processes and materials in the line of manufacturing production can be clearly detailed in a flow chart. Pictures, symbols and diagrams should be used when they aid in understanding flow charts.

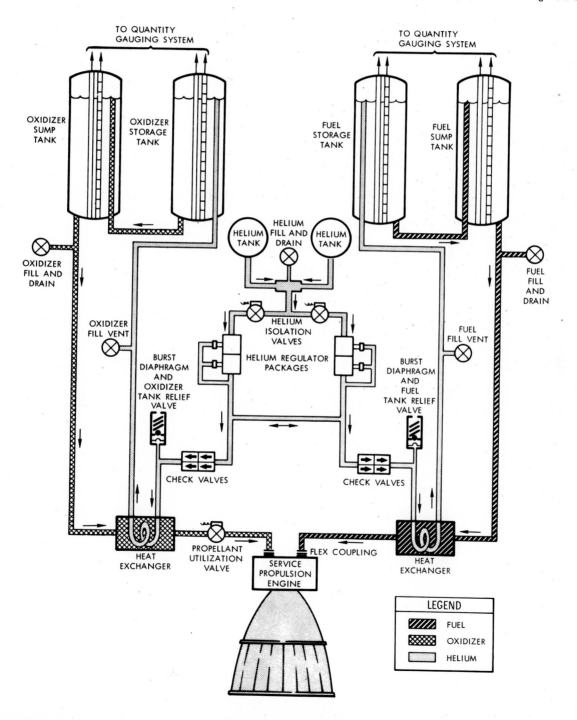

Fig. 27-13. Flow chart showing how fuel and oxidizer get to the combustion chamber of a rocket engine used on Saturn V.
(Rockwell International)

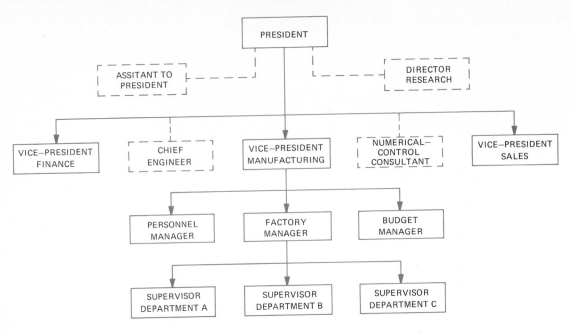

Fig. 27-14. Organizational charts show lines of authority and responsibility as well as those personnel who serve in a staff or support position.

SPEEDS FOR MACHINING STEEL
WITH CARBIDE TOOLS

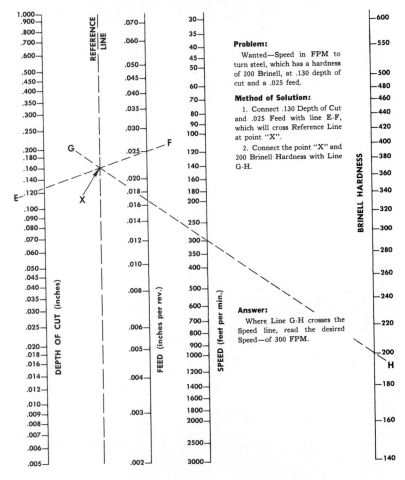

Problem:

Wanted—Speed in FPM to turn steel, which has a hardness of 200 Brinell, at .130 depth of cut and a .025 feed.

Method of Solution:

1. Connect .130 Depth of Cut and .025 Feed with line E-F, which will cross Reference Line at point "X".

2. Connect the point "X" and 200 Brinell Hardness with Line G-H.

Answer:

Where Line G-H crosses the Speed line, read the desired Speed—of 300 FPM.

Fig. 27-15. A nomograph designed to permit the rapid determination
of cutting speeds when using carbide cutting tools.
(Carboloy Div., G. E. Co.)

ORGANIZATIONAL CHARTS

Organizational charts do for personnel groups what flow charts do for processes and materials. They show relationships between individuals within a company or organization and the operations or services each performs, Fig. 27-14.

The organizational chart also shows the relationship between line personnel and staff personnel within an organization. Line personnel, such as supervisors or department heads, have authority to direct an operation or a group. Lines in the chart clearly show this authority.

Staff personnel, such as consultants for numerical control machines, may suggest and recommend procedures and types of equipment to the manufacturing manager. However, they cannot direct that the suggestions or recommendations be carried out. The lines on the organizational chart indicate these responsibilities.

NOMOGRAPHS

A type of graph useful in solving a succession of nearly identical problems is the nomograph, Fig. 27-15. This graph usually contains three parallel scales graduated for different variables. When a straight line connects values of any two scales, the related value may be read directly from the third at the point intersected by the line.

GRID PAPER FOR GRAPHS AND CHARTS

Graphs and charts may be drawn directly on grid or plain paper. A better appearance can be obtained by drawing on plain paper and including only the grid lines for major divisions on graphs or lines of a chart. In laying out a graph or chart on plain paper, it is helpful to insert a grid paper below the translucent paper as a guide.

Grid papers are available in a wide variety of patterns and scales for all types of graphs and charts.

LETTERS AND SYMBOLS

Titles, notes and symbols may be added to graphs and charts freehand. Often, however, it is done by instrument or transfer sheet. Most graphs and charts are a part of a promotional presentation to management or to a customer. They must be well done to be effective. Special type styles of letters are available in various sizes on transfer sheets and tapes, Fig. 27-16.

COLOR AND SHADING FILMS

Overlay films for coloring or shading areas of graphs, charts and other illustrations are available in a wide variety of colors and "fill-in" designs, Fig. 27-17. To apply color and shading

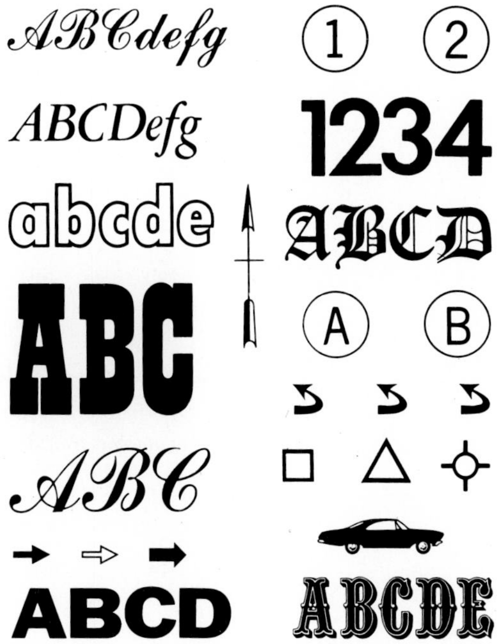

Fig. 27-16. Some examples of type styles and symbols which are available on transfer sheets and tapes for preparing graphs and charts. (Chartpak)

Fig. 27-17. Color and overlay shading film help to put action in graphs and charts. (Chartpak)

MATERIALS AND AIDS FOR GRAPHS AND CHARTS

A number of aids are available to the drafter for use in making graphs and charts. Most have been presented in other chapters, but aids referenced here have special use in constructing graphs and charts.

film, lay a sheet over the area to be treated and cut around the area, leaving some extra film. Remove the backing sheet and press the film lightly in place. Trim around the edges of the area with an X-acto type knife and remove surplus. Place backing sheet over the film and burnish firmly to assure adhesion. (Note: Some film overlays do not require burnishing. Check manufacturer's recommendation.)

A variety of pressure sensitive tapes are available in widths from 1/64 to 2 inches and in many styles and colors, Fig. 27-18. These tapes can be used for line graphs, bar graphs and labeling purposes.

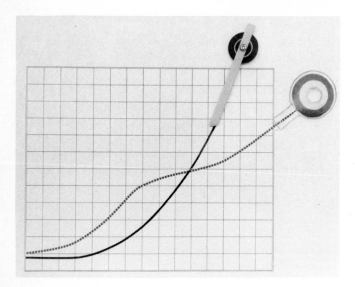

Fig. 27-18. Pressure sensitive tapes speed the process of preparing graphs and charts.

INSTRUMENT AIDS

In addition to the instruments used for drawing horizontal, vertical and other straight lines, there are several instruments that speed the process of making graphs and charts.

Templates are valuable for drawing circles, ellipses, diagram boxes for flow and organizational charts; also for drawing various symbols used in graphs and charts. Proportional dividers are useful in making special divisions of lines, and for dividing circles and areas into proportional parts. Irregular curves are useful in establishing smooth curves on graphs and charts. (See Chapter 2.)

PROBLEMS IN GRAPHS AND CHARTS

The following problems are planned to provide understanding and skill in the use of basic drafting techniques related to constructing graphs and charts.

LINE GRAPH PROBLEMS

1. Prepare a line graph contrasting the number of persons employed in the United States in service producing industries with those in goods producing industries. Use the following data. Let the present represent the current year and show other dates accordingly.

2. Plot the following data as a line graph. Assign year dates and show monthly changes.

CHANGES IN THE UNEMPLOYMENT RATE
5-YEAR PERIOD

		YEARS				
		1	2	3	4	5
	J	5.6	4.6	4.0	3.8	4.0
	F	5.5	4.8	4.0	3.8	3.9
	M	5.4	4.7	3.8	3.8	3.8
	A	5.4	5.0	3.8	3.8	3.7
MONTHLY	M	5.2	4.8	3.8	3.8	3.6
PERCENT OF	J	5.1	5.0	3.8	3.9	3.5
UNEMPLOYMENT	J	5.2	4.6	3.8	3.9	3.6
	A	4.9	4.4	3.8	3.8	3.7
	S	5.1	4.3	3.7	3.8	3.6
	O	5.1	4.2	3.7	3.9	3.5
	N	5.0	4.2	3.6	4.0	3.6
	D	4.8	4.1	3.8	4.1	3.7

3. Collect data on a subject of interest to you and construct a line graph. Magazines in your school, public or drafting library are good sources of data.
4. Prepare a line graph that compares the increase in average horsepower of automobiles since 1930 to the present with the highway speed limits during the same period.

BAR GRAPH PROBLEMS

5. Select a type of bar graph and show the melting temperatures of various metals. Investigate if there is a way in which the metals could be arranged, other than alphabetically, which would make the graph more meaningful.

METAL OR COMPOSITION AND
MELTING POINT – DEG. F.

Aluminum	1220	Copper	1981	Nickel	2651
Brass	1823	Gold	1945	Silver	1761
Bronze	1841	Lead	621	Steel	2500
Cadmium	610	Magnesium	1204	Tin	449
Chromium	2939	Manganese	2300	Zinc	788

6. The specific gravity of the metals listed in the above problem is shown below. Select a different type of bar graph to contrast this factor of the various metals.

Aluminum	2.70	Copper	8.89	Nickel	8.80
Brass	8.60	Gold	19.30	Silver	10.45
Bronze	8.78	Lead	11.34	Steel	7.80
Cadmium	8.65	Magnesium	1.74	Tin	7.29
Chromium	6.93	Manganese	7.30	Zinc	7.10

7. Construct a paired bar graph to compare the two sets of factors listed for the metals in the two previous problems.
8. Collect data on a subject of interest to you and prepare a bar graph of a type not yet used.

		PRIOR		PRESENT USE CURRENT YEAR	FUTURE (ESTIMATED)	
		10 YEARS	5 YEARS		5 YEARS	10 YEARS
MILLIONS OF WORKERS	SERVICE	24.8	27.2	34.0	47.6	59.7
	GOODS	27.3	26.0	24.8	28.5	30.0

SURFACE GRAPH PROBLEMS

9. Prepare a surface or area graph to show sources of steel imports over an eight year period as shown by the data below.

MILLIONS OF NET TONS*

	JAPAN	BELGIUM-LUXEMBOURG	ALL OTHER SOURCES
1	0.1	0.3	0.4
2	0.1	0.4	0.3
3	0.2	0.5	0.5
YEARS 4	0.3	0.8	1.4
5	0.4	0.4	0.5
6	0.7	0.6	0.9
7	0.9	0.7	0.8
8	1.1	0.7	1.4

*Data modified for illustration purposes

10. Construct a pictorial surface graph illustrating traffic safety and accident statistics. Gather data from driver training, safety classes or an insurance company. Select an appropriate photo from a magazine or take one yourself.

PIE GRAPH PROBLEMS

11. Gather information from an annual report of a local or state government, organization or business and prepare a pie graph to graphically contrast the budgeted dollars.

12. Draw a pictorial pie graph to represent your personal budget of time, earnings or expenditures.

CHARTING PROBLEMS

13. Construct a flow chart illustrating the sequence followed in an industrial process, such as preparing and spray painting a metal surface, anodizing aluminum or making an electronic-circuit board.

14. Gather information on your city government and prepare an organizational chart showing line and staff organization.

15. Collect as many different types of graphs and charts from magazines, brochures and other sources available to you. Mount these on a suitable display board or in a notebook and classify each.

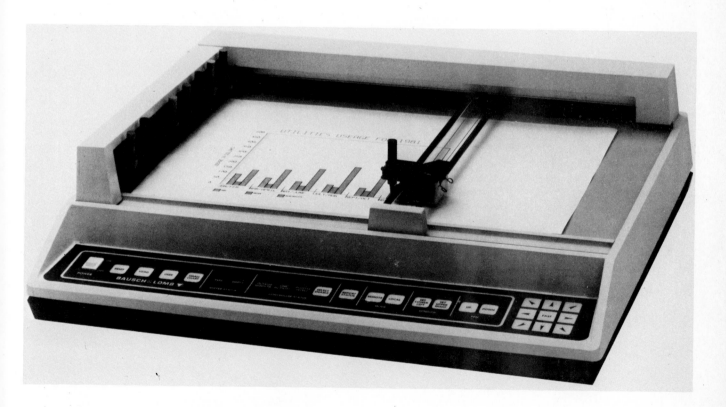

The computer graphics flat-bed plotter is capable of preparing graphs.   (Bausch & Lomb)

PART V

COMPUTER-AIDED DRAFTING AND DESIGN

The role of the drafter has changed dramatically with the utilization of CAD in the workplace. Drafters and designers are now able to design, create, and evaluate a product before it is produced. Industry is also able to produce the parts with limited human intervention.

This part of the text covers LINKING DESIGN AND MANUFACTURING, INTRODUCTION TO COMPUTER-AIDED-DRAFTING AND DESIGN, THE CAD WORKSTATION, CONSTRUCTING AND EDITING DRAWINGS, AND DETAILING AND ADVANCE CAD FUNCTIONS.

Integrating CAD and CAM has enabled industry to become more efficient and productive. (Caterpillar, Inc.)

Chapter 28
LINKING DESIGN AND MANUFACTURING

National and international competition in manufacturing is causing industrial leaders to re-think their production plans. Recent developments in computers and manufacturing have prompted these leaders to look at new strategies for remaining competitive and improving quality in their products. Developments in the computer industry are having a profound impact on the manufacturing industries from design to machine processing as well as the management and marketing components.

The purpose of this chapter is to assist the student to develop a knowledge and understanding of these new strategies in manufacturing. Also covered in this chapter is the effect they are likely to have on the design-drafting component of industry now and in the future. The impact of computer technology on design and manufacturing is presented first as it relates to design-documentation—computer-aided design (CAD). Then, the study of computer-aided manufacturing (CAM) and how it is linked to CAD is covered. Other innovative manufacturing plans, such as flexible manufacturing system (FMS), are included and finally a look at computer-integrated manufacturing (CIM) which is still being developed today.

COMPUTER-AIDED DESIGN AND DRAFTING (CAD)

CAD refers to the computer-aided design and/or drafting process. The term CADD is sometimes used but the more recent practice is to refer to the design process with the assistance of the computer as CAD. It is assumed that the documentation of the design process will result in some computer-generated drawings (drafting) and other documents essential to the entire manufacturing cycle, Fig. 28-1.

DEVELOPMENT OF CAD

In the early stages of CAD, the principal use was to produce and maintain drawings. There was considerable value even in the generation of drawings in that the process could be accelerated with the ready application of symbols, dimensioning elements, and projection of views. But it was realized that CAD had much more to offer in the design of a product. Related data such as alternate design of a product, materials analysis including stress analysis, costs, etc., could be stored in a data bank ready for instant recall. Further, as the concept of the automated factory developed, the information generated in the design process was extended for use in manufacturing processing.

NEED FOR DATA BASE IN TOTAL MANUFACTURING SYSTEM

In the highly competitive manufacturing market, the designer does not have the luxury of re-designing or re-working of a product once the manufacturing process has started. The design must be the best selection among several which have

Fig. 28-1. Computer-generated drawings are essential to the design process. (Versatec)

been proven by thoroughly analyzing alternate designs, material options, machine processes, and labor costs. Anything less than top performance in product design contributes to problems in the manufacturability of the product, cost overruns, product failure in service, and lack of customer confidence in the company. Poorly designed products eventually contribute to a company's failure.

Perfection in the manufacturability of a product requires design capability to thoroughly examine design alternatives. Knowledge gathered from previous experience can be stored in the computer data base for ready use in design work. This information will be available for the designer to make wise decisions in the design process.

HOW THE CAD PROCESS WORKS AS A SUB-SYSTEM OF CIM

CAD is one sub-system of computer-integrated manufacturing (CIM), Fig. 28-2. With the assistance of CAD equipment, designers are able to analyze, test, and discuss each design decision, Fig. 28-3. There is input to the design process

Fig. 28-3. This advanced workstation enables designers to optimize and verify their designs and generate drawings, parts lists, and layout plotting instruction. (Gerber Scientific, Inc.)

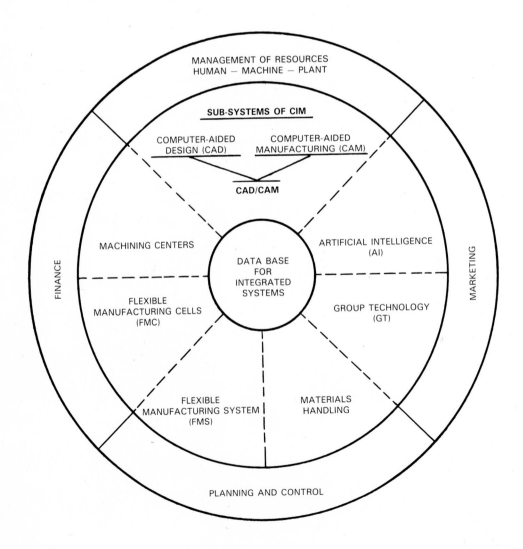

Fig. 28-2. Components of computer-integrated manufacturing. These components of computer-integrated manufacturing (CIM) are described in this chapter.

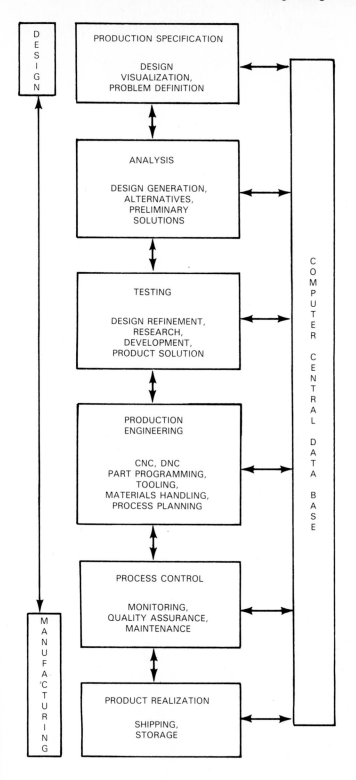

Fig. 28-4. Linking design and manufacturing.

ADVANTAGES OF CAD

In addition to making the design-drafting department more interesting and creative for persons working there, CAD has many other advantages. Following are a few of these:
1. Removes tedious calculations by designers and drafters.
2. Generates notations, bills of materials, and symbols to be placed on the drawing, saving valuable time.
3. Eliminates time-consuming tasks of manual drafting, such as drawing lines, geometric shapes, and measuring distances.
4. Provides time and essential data to review alternate design solutions.
5. Requires the input from other industry personnel such as manufacturing processes and sales, thus eliminating problems later.
6. Provides more reliability in design work by having relevant information available to design personnel.
7. Generates CAD data base in the design and documentation of the product which can be used in other sub-systems of CIM.
8. Reduces the number of drawings required by the ability to retrieve design models when needed.

CAD DESIGNER-DRAFTER QUALIFICATIONS

The design-drafter has available more assistance in the way of design information, analysis, and testing than at any previous time. To make the full use of this capability, the designer-drafter must acquire the knowledge and skill necessary to fully utilize the CAD systems that will be in the design departments of tomorrow, Fig. 28-5. In Chapters 29 through 32, the fundamentals of computer-aided design and drafting are presented to assist the student in getting started in computer graphics.

Fig. 28-5. This CAD system is used today in three-dimensional wireframe modeling, three-dimensional surface modeling, three-dimensional solids modeling, finite element analysis, and numerical control output. (Gerber Scientific, Inc.)

by other specialists in materials, tooling, manufacturing processing, as well as in sales marketing. Once the design decision has been made, information is entered into the central data base for use and adaptation to other sub-systems in the manufacturing cycle, Fig. 28-4. When the accepted design for a product leaves the CAD department, it is assumed to meet all requirements of manufacturability and customer needs.

Fig. 28-7. A designer compares a finished part to the original CAD data using product design graphics. (Ford Motor Company)

Fig. 28-6. Computer numerical control (CNC) machining of a component. (Ford Motor Company)

COMPUTER-AIDED MANUFACTURING (CAM)

The computer-aided manufacturing (CAM) facility is a natural extension of the technology of CAD. CAM can be defined as a plan for utilizing numerically controlled machines such as mills, lathes, drills, punches, and other programmable production equipment controlled by a computer, Fig. 28-6. These machines are known as computer numerical control (CNC) machines and can be programmed to perform a wide variety of machine processes with great speed while holding close tolerances. Robots are another type of programmable equipment that is essential to a CAM environment.

CAD/CAM is linked together by the data base developed in the design of the product, Fig. 28-7. This same data base is used by production engineering to program computer-aided manufacturing equipment.

Computers are also used in a CAM facility to control production scheduling and quality control, as well as the business functions of manufacturing such as purchasing, financial planning, and marketing.

NUMERICAL CONTROLLED (NC) MACHINES

When the system of numerical control (NC) was first introduced to program and control production machine tools, punch cards, and later, punched paper tape were used. These methods of numerical control were subject to damage in a hostile machine tool environment. Paper tape was replaced by the more durable mylar tape. With the advances of microelectronics technology, computers were introduced as the control system for NC machines, hence the term CNC.

In CNC, a computer is used to write, store, edit, and control an NC program. The next stage in manufacturing was direct numerical control (DNC) where the computer is the control unit for one or more NC machines. A more accelerated stage today is called distributive numerical control (DNC) in which a main host computer controls several intermediate computers that are coupled to certain machine tools, robots, and inspection stations.

ROBOTS

Robots have been performing tasks of varying degrees for some time in industry, Fig. 28-8. The single-purpose robots, such as ones which transfer a part from one machine to

Fig. 28-8. Robotics vision-assisted insertion of gauges into the cluster housing. (Ford Motor Company)

another and cannot be re-programmed to perform other tasks, are not considered in the definition of robot as stated by the Robot Institute of America.

A robot is a programmable multifunctional manipulator designed to move material, parts, tools, or specific devices through variable motions for the performance of a variety of tasks.

This defines a "flexible" tool capable of functioning in a number of industrial settings just as any other programmable machine.

ADVANTAGES OF CAM

Many of the advantages claimed for CAM come from the linkage with CAD as CAD/CAM. As manufacturing becomes more computer based and as it moves toward full integration in all facets of design, production engineering, process control, and marketing, the advantages will become even more pronounced. Advantages claimed for CAM include:

1. Communications are improved by the direct transfer of documentation from design to manufacturing.
2. Production is increased and more efficient.
3. Errors are reduced with the same data base used by design and manufacturing.
4. Materials handling and machine processing are more efficient.
5. Quality control is improved.
6. Lead times are reduced, improving market response.
7. Work environment is safer and more humane.

CAM WORKS AS A SUB-SYSTEM OF CIM

The scope of CAM may be limited to a few machines in a machining cell or it may be expanded to include an entire department or facility, to become a computer-integrated manufacturing (CIM) facility, Fig. 28-9. The following sections describe other components usually found in a more comprehensive CIM industry.

MACHINING CENTERS

A machining center is a machine tool which is capable of performing a variety of metal removal operations on a part such as drilling, milling, or boring, usually under computer numerical control (CNC). Often times these machines are equipped with automatic tool changing and storage capabilities, and a variety of part delivery or shuttle mechanisms. CNC turning centers, CNC grinding centers, etc., are also available for stand-alone machining or systems integration of these processes.

Operations scheduled for machining centers are numerically controlled by computers and sensors are built into the system to protect the equipment from overload and maintain product quality. These sensors enable the controller to monitor the plant, process, and product, Fig. 28-10. The flow of lubricant, coolant, tool life, and tool breakage are also monitored.

Machining centers require a minimum of operator supervision and work in process is limited only by pallet storage and the number of tools stored in the tool magazine. Machining

Fig. 28-9. The systems operator in the control room can monitor diagnostics via the color graphics monitor. The supervisory unit also accumulates information at the same time. (Allen-Bradley)

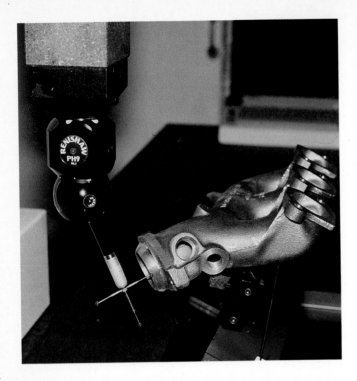

Fig. 28-10. Automatic gauging of a finished part. (Ford Motor Company)

Fig. 28-11. Control room for overall FMS/CIM manufacturing operations. (Ford Motor Company, Kentucky Truck Plant)

centers are usually installed as integral parts of flexible manufacturing systems (FMS) and are considered the smallest building block of FMS.

FLEXIBLE MANUFACTURING CELLS (FMC) OR CENTERS

A flexible manufacturing cell (FMC) or center consists of a grouping of machine tools organized into a cell. Cells are usually configured to provide most of the machining processes to produce a part or family of parts without leaving the cell. Another form of the FMC is a grouping of like machines dedicated to a particular type of machining process, such as a small group of horizontal machining centers. The equipment in an FMC can be linked by automated material handling equipment, like a robot or robot cart system. The more "intelligent" FMCs will also be directed by a "cell controller" or cell computer. In many ways, an FMC is just a mini-version of an FMS (flexible manufacturing system).

FLEXIBLE MANUFACTURING SYSTEMS (FMS)

Flexible manufacturing systems (FMS) is a production system of highly automated and computer controlled machines, assembly cells, robots and inspection equipment, together with computer-integrated materials handling and storage systems. The automated manufacturing operation is monitored and controlled from a central location, Fig. 28-11. The term "flexible" refers to the capability of the system to process a variety of similar products as well as to reroute or reschedule production in the event of equipment failure. Some FMS can also select, at random, parts needing special

materials, all controlled by the computer program. These systems are designed to deliver quality output in a cost effective manner and to respond quickly to changing demands.

ADVANTAGES OF FMS

FMS is the form of manufacturing that closely approaches computer-integrated manufacturing (CIM). FMS offers the following advantages:
1. Increases volume of production.
2. Decreases costs of production.
3. Manufactures completed single parts and/or batches in random order.
4. Produces parts "on order" rather than being warehoused, thereby reducing inventory.
5. Improves quality control by 100% inspection.
6. Decreases hazardous and repetitive work, making the work environment more humane.

TRENDS IN FMS

When this system is more fully understood and as world manufacturing competition increases, FMS will be seen as a more cost effective means of production, permitting increased product design changes and variation in production.

GROUP TECHNOLOGY (GT)

Group technology (GT) is a manufacturing philosophy of organizing parts to be manufactured into families of parts to be produced in cells of machine tools. These parts are similar in design and/or manufacturing requirements. The design characteristics are similar in materials, dimensions and tolerances, shape, and finish. Manufacturing characteristics include such factors as tool and machine processes, fixtures necessary, and sequence of operations. GT is also an essential element in the implementation of computer-integrated manufacturing (CIM).

ADVANTAGES OF GT

Although there is expense involved in coding and organizing products by group technology, there are also some distinct advantages:
1. Improves product design.
2. Provides standardization and families of parts, thereby reducing costs.
3. Reduces tooling and set-up costs.
4. Reduces work-in-progress (WIP).

JUST-IN-TIME (JIT) MANUFACTURING

One of the most rewarding concepts in manufacturing in recent years has been the introduction of "Just-In-Time" (JIT) philosophy. In this plan of operation, the goal is to reduce work-in-progress (WIP) to an absolute minimum. This involves the reduction of: (1) lead times in getting WIP, (2) actual WIP inventories, and (3) set-up times for WIP, Fig. 28-12.

The conventional approach in industry has been to automate existing processes of machining and assembly which has resulted in "isolated cells" of production. There have been costly periods of wasting time in between these cells of production rather than a smooth efficient flow of material between production processes. The most efficient system moves the product to be manufactured from the firm's suppliers to its customers in a continuous manner with few or no rejects.

The JIT system regards production processes as the only means of adding value to a manufactured product. All other tasks such as transportation, inspection, and storage are defined as "wastes" to be eliminated wherever possible. An efficient production system requires a highly consistent, short-cycled process with minimal inventory in process.

JIT manufacturer's suppliers are expected to function as extended storage facilities of the company. Material and parts are purchased in small lot sizes (just-in-time for use on the line) as opposed to the purchase of large amounts and storing a portion until needed for manufacture. The latter requires capital to be invested in nonproductive (at the time) supplies and storage facilities.

ARTIFICIAL INTELLIGENCE (AI) AND EXPERT SYSTEMS

Artificial intelligence (AI) is the attempt to program or to place in the computer's "memory," knowledge (data) that would cause the computer to recognize problems to be solved, and make decisions normally associated with human intelligence. Speech recognition, language interpretation, and computer vision (scene interpretation) are examples of this branch of computer science, Fig. 28-13.

Fig. 28-12. Automatic storage and retrieval system for finished rails. Small inventory buffer is needed to match assembly line speed and mix of vehicles. (Ford Motor Company, Kentucky Truck Plant)

Fig. 28-13. An automated optical inspection (AOI) system used in the PCB manufacturing industry for inspection of printed circuit boards and artwork. (Gerber Scientific, Inc.)

Expert systems technology is a branch of AI. It is the design of computer systems that exhibit characteristics associated with human intelligence. An expert system's software uses knowledge and inference procedures to solve problems. A fundamental concept is the ability for the program to "learn" rather than being programmed with firm, preprogrammed decisions.

This is an over simplification of the topics of AI and expert systems but until further progress is made in these areas, the human interaction with automated manufacturing is likely to remain.

COMPUTER-INTEGRATED MANUFACTURING (CIM)

Computer-integrated manufacturing (CIM) is the full automation and joining of all facets of industry: design, documentation, materials selection and handling, machine processing, quality assurance, storing and/or shipping, management, and marketing. There are no such CIM installations at this time in operation. There are, however, partial CIM systems such as CAD/CAM, FMS and FMCs existing today that have been discussed in this chapter and sometimes are referred to as CIM. These partial CIM facilities vary in size and complexity, Fig. 28-14.

Fig. 28-14. A CIM facility for manufacturing electrical parts.
(Allen-Bradley)

ADVANTAGES OF CIM

The advantages of CIM in today's manufacturing enterprise are inherent somewhat in all of the sub-systems of CIM, and these may be summarized as follows:
1. Improves productivity and efficiency in manufacturing.
2. Increases quality control and improves product reliability.
3. Reduces costs of production and makes industry more competitive.
4. Makes the work environment more safe and humane.

FUTURE DEVELOPMENTS IN CIM

Over time, computer systems have developed in capacity and speed since their inception and have decreased in costs. Likewise, computer-integrated manufacturing is not likely to become obsolete, but rather will be improved in its adaptive characteristics. At the present time there are too many subjective decisions which CIM users have to make. There is a need for CIM to be more "intelligent," with the capability to self-determine which decision to make. This will be accomplished when computer designers develop computers with memories to respond with the kind of knowledge a worker gains from experience.

Progress is likely to occur in the area of artificial intelligence and application of this new computer technology will further enhance CIM. Regardless, CIM will require individuals who understand CIM and the design-manufacturing problems to be solved.

SUMMARY

The trend in manufacturing today is to link the design process with the machine processes by utilizing the same data base. Progress will continue to be made in computer technology that will lead to further integration of design and manufacturing, becoming one continuous process from product conception to product shipment.

QUESTIONS FOR DISCUSSION

1. In the early stages of computer-aided design and drafting (CAD), the computer was used primarily for generation of drawings. What brought about the expanded use of CAD and what is included?
2. How does CAD contribute to an industry's ability to remain competitive? To improve quality in products?
3. Explain the term "Central data base."
4. Enumerate the advantages of a CAD system.
5. Define computer-aided manufacturing (CAM). How is it related to CAD?
6. How does a machining center differ from an FMS cell or center?
7. Explain the term "flexible manufacturing cell (FMC)." How does this differ from the "flexible manufacturing system (FMS)?"
8. What is "group technology?"
9. Explain the meaning of the term "Just-In-Time" manufacturing.
10. What is meant by "artificial intelligence" as it relates to computers?
11. What is computer-integrated manufacturing (CIM) and what are its capabilities? How does CIM relate to the fully automated factory?

Chapter 29

INTRODUCTION TO COMPUTER-ASSISTED DRAFTING AND DESIGN

The growing use of computer graphics has affected all areas of our technological society. Medical technicians use computer-generated images to track biological functions. Cartographers use computer graphics to map out land contours and terrain. Graphic artists use computers to lay out designs for printed materials. No less affected by computers is industry. Computer-controlled systems are revolutionizing the methods used to design and manufacture a product. At the front end of the manufacturing process, designers use computer graphics to create and document product designs, Fig. 29-1. During the production process, computers may control machine tool movement and product transport. This chapter will focus on a specific subset of computer graphics, CAD.

The term CAD commonly refers to computer-assisted drafting. Often you may see CADD, standing for computer-assisted drafting and design. CADD means the use of a computer not only to draw, but also to analyze and test product prototypes. Each of these acronyms refer to one concept, the use of computers to assist the design-drafter in developing an idea into a product. The key word is assist. The computer cannot create a drawing without human skill. The process of drafting still requires the knowledge and talents of the drafter. However, with CAD, the speed of a computer is matched with the skills of the drafter. The result is an increase in productivity, both in the quality and amount of drawings completed as compared to traditional drafting.

TOOLS OF CAD: THE CAD WORKSTATION

Like a pencil, scale, and T-square, a CAD system is a tool of the drafter. The combination of equipment to operate a CAD system is referred to as the workstation. It includes a computer, CAD software, display screen, input device, and hardcopy device, Fig. 29-2.

The *computer* is the heart of any CAD workstation. Equally important is CAD software, the list of instructions which operate the computer. Advancements in both computer technology and software programs now allow personal computers to perform functions which only five years ago were limited to larger mainframe systems.

Input and output devices allow the drafter to interact with the computer system. *Input devices* are used to enter commands. They are also used to locate positions on the drawing when constructing and editing lines, circles, and other entities. *Output devices* allow you to see the created drawing. The display screen, much like a TV shows an image of the drawing. Hardcopy devices prepare a paper or film copy of the drawing.

Various equipment comprising the CAD workstation is discussed in Chapter 30.

HISTORY OF COMPUTER GRAPHICS

As discussed earlier, computer-assisted drafting is a part of the broader area of computer graphics. The development of computer graphics has progressed greatly since its beginning at the Massachusetts Institute of Technology.

In the the 1950s, MIT developed the APT program, which stood for Automatically Programmed Tools. The APT program, still used today, is a method of generating computer code to define the geometry of simple products. The code was then used by numerical control (NC) equipment to control machine movement while producing a part. Although this program

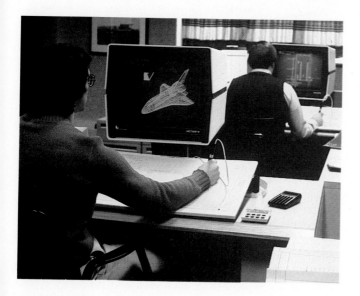

Fig. 29-1. Design engineering problem on a display screen. The growing application of computer-assisted drafting/computer-aided manufacturing (CAD/CAM) technology used by engineers and drafters. (3M Company)

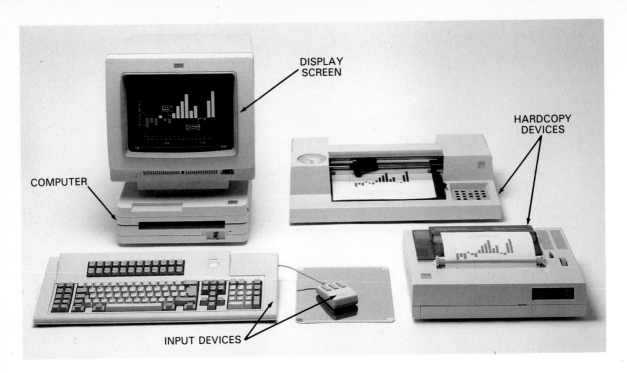

DISPLAY SCREEN

HARDCOPY DEVICES

COMPUTER

INPUT DEVICES

Fig. 29-2. The CAD workstation. (IBM)

wasn't a true interactive graphics program, it raised the idea of using computer code to define a product, rather than using pencil lead on a piece of paper.

Later that decade, the Air Force developed the SAGE (Semi-Automatic Ground Environment) project. It was a simulation program that displayed radar images about enemy attacks. Military personnel would view the screen, then point to a sector of the screen to indicate interceptor aircraft. This was the first limited use of interactive graphics, the ability to communicate with the computer.

The 1960s saw the first true application of interactive graphics as we know it today. In 1962, a doctoral student, Ivan Sutherland developed software which allowed a light pen to draw on a display screen. The light pen could not only point (as it did in the SAGE project), it could also draw images. The sketchpad project was refined the next year by T.E. Johnson to draw multiview and perspective views.

Throughout the 1960s, automotive, aerospace, defense, and computer industries began further research and development of computer design and engineering systems. Most of these systems were specific to the company; they were not developed to be sold. Today many of these companies now market their systems such as *Unigraphics* by McDonnell-Douglas and *CADAM* by Lockheed. Later in the 60s, companies specializing in CAD systems appeared. They developed and sold "turnkey" systems, a workstation which includes all of the hardware (computer and related equipment) and software needed to run the system.

These graphics systems introduced some of the first highly sophisticated examples of CADD. They were capable of advanced work, such as solid modeling. Solid models are representations of an object in three dimensions, Fig. 29-3.

Fig. 29-3. Solid models represent an object in three dimensions. (IBM)

The computer "thinks" of the object as a solid; it recognizes that the object has not only a surface, but material on the inside as well. Using the three-dimensional concept, a product design can be translated into code for use with numerical control machinery.

The 1970s saw advances in hardware technology and the introduction of more CAD systems. The primary developments in hardware were the progressive use of miniature circuits and displays which were less expensive and produced clearer images. The 1970s also saw the growth of CAD software vendors. These companies specialize in CAD programs while computer companies specialize in developing computers.

The 1980s are most noted for the development of microcomputer CAD systems, Fig. 29-4. Although companies continue to market mainframe and minicomputer graphics systems, the PC has revolutionized CAD. The low cost of PC CAD has allowed most industries and schools to run full-fledged computer drafting and design programs. Because of further advances in computer hardware, it is possible for microcomputers to perform most of the CAD functions of larger systems.

There is little doubt that the future of CAD is bright. As new technology upgrades hardware and software, industry will combine CAD with all aspects of manufacturing, construction, and electronics design. The use of computer graphics will grow as industrial and graphic engineers realize the great benefits of CAD.

Fig. 29-4. Microcomputer CAD systems perform many of the tasks of mainframe systems. (Calcomp, Inc., A Lockheed Company)

BENEFITS OF CAD

The productivity gained by using a CAD system lies in many areas. Five of these include: drawing speed, drawing quality, quick modifications, better communication, and analysis tools.

DRAWING SPEED

Research has indicated that skilled CAD users are up to three times faster than drafters using traditional tools. This can mostly be attributed to the removal of tedious work. A traditional drafter spends a large amount of time "laying lead" to create each image on paper. Using a CAD system, most of your input is points. Instead of drawing a line, you select endpoints, then the computer completes the line between the endpoints. Another example is a circle. The CAD user simply specifies the center and radius, Fig. 29-5.

Another factor in drawing speed is repetition. A rule of thumb with CAD is never to draw the same item twice. Instead, the system is commanded to copy. The object to copy could be as simple as a rectangle or as complex as

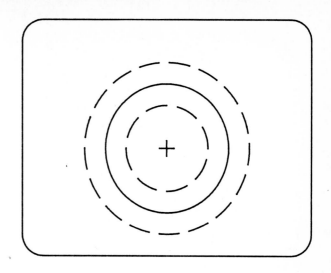

Fig. 29-5. Drawing speed is increased with a CAD system. Here the center point and radius are specified and the computer creates the circle.

subassembly. In addition, commonly used symbols and parts can be held in a symbol library. Basically, a symbol is a miniature drawing, which can be inserted into the current version in any position. Symbol libraries may be sold with the system or marketed separately. For example, an architectural symbol library would contain all the symbols for doors, windows, plumbing, electrical, and heating equipment.

QUALITY

Quality gained using CAD includes line quality and accuracy. Using the computer, all additions and changes are entered into memory and viewed on the display screen. Only when the drawing is complete is it sent to a hardcopy device for reproduction. This eliminates the constant drawing and erasing commonly done on a traditional drawing.

Another key to quality is the precision of the drawing. Lines, circles, and other entities can be positioned with accuracy greater than one-thousandth of an inch because all calculations of distance are done in computer memory.

MODIFICATIONS

Modifying a drawing is one of the most time consuming tasks of the drafter. There may be one way to draw a line, but a hundred ways to change it. A full range of CAD editing functions make modifications quick and easy. Since modifications are done in computer memory, the original is not destroyed. A revised drawing is simply replotted using a hardcopy device.

COMMUNICATIONS

Application of computers in the production process has increased communications among departments of a company. First of all, information about the drawing (including the drawing itself) can be stored in a main computer. Each department can access this information with their terminal to verify the progress of the product, Fig. 29-6.

Communications are also improved because guidelines can be standardized. These might include linetypes, style of lettering, and layout of the working drawings. Also, drafters use the same data base as symbols and subassemblies previously drawn. Mistakes in documentation are greatly reduced and confusion is kept at a minimum.

ANALYSIS TOOLS

Sophisticated CAD systems are capable of a wide variety of analysis and testing functions. Applications programs can use the drawing data created by the CAD system to study many aspects including durability, strength, and potential failures of the product. The end result is a better quality product without having to build expensive prototype models for destructive testing.

QUALIFICATIONS OF THE CAD DRAFTER

The introduction of computer-assisted design has changed several of the required skills of the drafter. Tedious tasks such as linework and lettering are eliminated. More emphasis is being placed on analysis and problem solving. Even though a CAD system is simply a tool of the drafter, as systems become more advanced the task of actually drawing objects will decline. More significance will be placed on the drafter's ability to specify design features.

The qualifications of the CAD operator include drafting discipline skills, software proficiency, and hardware knowledge. Discipline skills are mentioned only to stress that a CAD system is only as effective as the person using it. An operator with poor knowledge of drafting standards will create poor drawings.

Knowing the capabilities of the software is an important step to becoming proficient with the system. You should be able to interact freely with the computer. Know the command you want, and know how to find it. Your drafting speed will increase as you become familiar with the system.

Most employers will not require that you know a high level computer programming language. However, having some programming experience is an advantage. In time you will encounter *parametric programming,* which is listing a series of CAD commands used to create a certain object. These programs automate the drafting process. You need only give certain values. For example, to draw a screw, you need only give the diameter, length, and threads per inch. These values are the parameters. Using these values, the CAD system draws the necessary entities to describe the screw.

As a CAD operator, you will be expected to understand the function of different hardware. Although you may use a specific workstation, take time to study the different types of computers, input devices, display screens, and output devices, as well as the proper care of this equipment. At some time, you may be required to recommend equipment purchases. Know what aspects make one piece of equipment better than another.

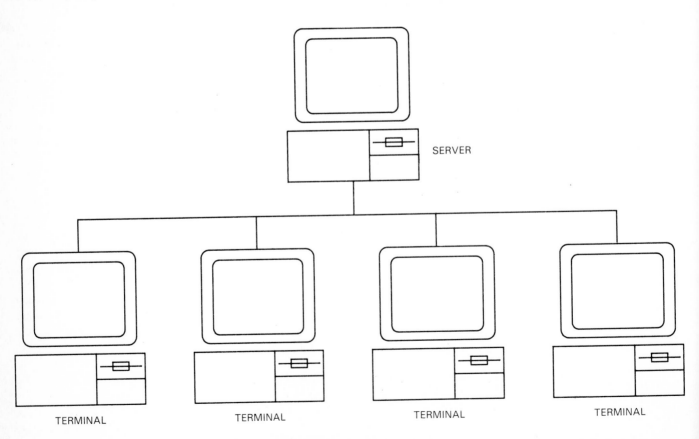

Fig. 29-6. Server is accessible from all terminals.

CAD DRAFTING POSITIONS

Drafting positions have remained relatively stable with the introduction of CAD. The position of layout and detail drafters had generally changed job title to CAD operator. Design drafters are now referred to as graphic designers. The biggest growth has been in positions related only to operation of the CAD system. *Systems managers* load software, start the system, and notify operators that the system is ready for use. During operation of the system, managers report hardware problems and software bugs. They might also schedule operation of the system. *CAD programmers* develop and maintain the functions of the CAD software. They load new software updates as they become available and correct bugs. They may also research new applications of CAD software. The programmer may design interface software which allows increased communication between the CAD system and production machinery.

THE DRAFTING ENVIRONMENT

The environment of drafting departments has changed rather drastically in companies using CAD. Modern facilities include comfortable settings. Tables and stools are replaced by ergonomically designed workstations with ample room for equipment. The atmosphere (temperature and humidity) is carefully controlled to insure the reliable operation of the computer.

Because of the high initial cost of a CAD system, most companies have drafters work in shifts. The number of drafters generally does not decline. However, most companies invest in fewer drafting stations and to reduce payback time, run the system from 14 to 24 hours a day. For companies just switching to computers, old drawings are digitized into CAD drawing files almost constantly. This is in addition to creating, modifying and revising existing drawings.

JOB OUTLOOK

The drafting department has been recognized as the bottleneck of work flow. The time required to create and revise drawings often exceeds that for design and production. Computer-assisted drafting has lessened much of the bottleneck. Currently, the department of labor has indicated a need for approximately 50,000 additional CAD operators per year. This presents a great opportunity for all persons, especially those who are handicapped. The physical ease of running a CAD system means those who are physically handicapped are less hindered by CAD environment than by manual drafting.

APPLICATIONS OF COMPUTER-ASSISTED DRAFTING AND DESIGN

Computer-assisted drafting and design has become a vital part of various drafting disciplines. In addition to a basic drafting package, meaning the functions are generic, additional software is added for special applications. Customized CAD packages are generally programmed for three areas:

MECHANICAL

Mechanical CADD includes machine and product design related to manufacturing. Mechanical designers might have special software to analyze a product design and generate code for numerical control machinery.

ARCHITECTURAL ENGINEERING AND CONSTRUCTION (AEC)

AEC applications include architecture, facilities planning, landscape architecture, and structural design. AEC CAD systems speed the layout of space and materials in residential and commercial structures.

ELECTRONICS

Electronic CADD involves design and layout of components for circuit boards and electronic products. Electrical engineers use programs which automatically place electrical components and their connections.

These three areas make up more than 90 percent of the CAD applications. Other areas include civil drafting, business graphics, publishing, and any company which has reason to draw. This section will discuss those CAD functions which enhance a certain drafting discipline.

MECHANICAL APPLICATIONS

Mechanical design and drafting remains the most prominent use of CAD systems. The accuracy of CAD increases the precision and dimensioning of mechanical designs since measurements are taken from the drawing data, not measured by the drafter. This precision is necessary if the drawing data will be sent directly to machinery for production. The drafter can also magnify the connection of two parts to inspect how they match. Motion may also be analyzed to make sure movable parts do not interfere with each other.

MODELING

A foremost use of CADD in mechanical applications is modeling. *Modeling* is creating a three-dimensional view of an object. The drafter identifies faces, or surfaces, of an object from which the computer creates the model.

Three-dimensional models help drafters visualize the product. Models can be shaded for appearance and rotated to help the designer see the object from different angles. Certain types of models are used to test products for strength, durability, etc. There are three types of models: wireframe, surface, and solid.

Wireframe models are created by connecting points of an object. The points refer to intersections of lines. Connecting these positions makes the object appear three dimensional, Fig. 29-7.

Surface models are constructed by connecting edges. Basically, a surface model begins with a wireframe model. Then each plane is converted to a surface as if a sheet of plastic covered the wire frame. Many CAD systems allow you to assign color to surfaces to enhance the appearance, Fig. 29-8. Also, because the model has solid surfaces, the drafter can specify shading as if there were a light source at

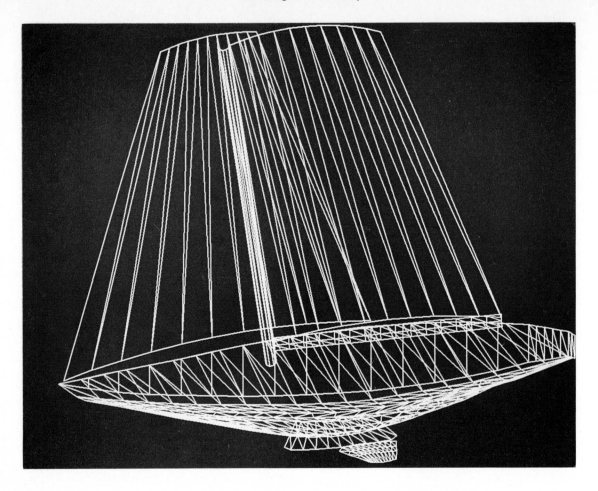

Fig. 29-7. Wireframe models are transparent views. (CADKEY, Inc.)

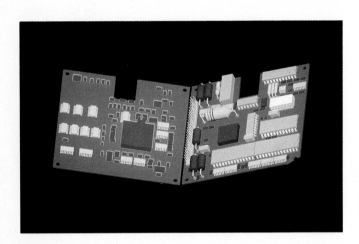

Fig. 29-8. Electronics design utilizing surface modeling techniques. (Intergraph)

of the object as solid material, Fig. 29-9. Solid models are created in computer memory using very small elements, called primitives. *Primitives*, including cubes, spheres, cones,

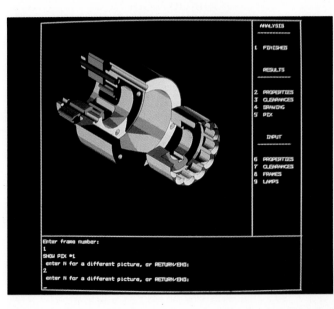

some angle. A special application of surface models is sheet metal design. After making the surface model of the product, it can be "unfolded" to view the pattern required to make the sheet metal product. This ability saves much time and effort for pattern makers.

Solids modeling is the the most refined type of modeling. Instead of recognizing only surfaces, the computer "thinks"

Fig. 29-9. Solids modeling technique. (CADAM, Inc.)

pyramids, and other shapes, form the building blocks of the model.

Because solid models are made of small elements, like a real object is made of atoms, a solid model can be cut, pressed, pulled, and twisted much like a real object. This type of testing is referred to as *finite element analysis*. If you put force on one area of the model, you can analyze the effect of that force in all areas of the model.

COMPUTER-AIDED MANUFACTURING

The use of computers is seen throughout the manufacturing process. Mechanical CADD is part of a much broader concept, computer-aided manufacturing (CAM). At first, computer-aided design systems are used to define the geometry of the object. This geometry is then converted to numerical code (NC) by an NC processor. The code is read by a controller, connected to a milling machine or lathe, to direct the speed and direction of the machine tool, Fig. 29-10. The code may be sent directly to the controller or stored in a data storage device for later use. CAM speeds up the manufacturing process since the same information used to create the design is used by machinery to machine the part.

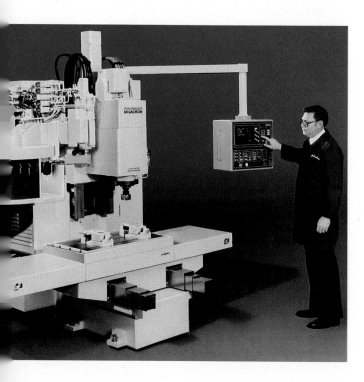

Fig. 29-10. The machine tool is numerically controlled by the computer. (Cincinnati Milacron)

Before being sent to a machine, the numerical code can be used for machine tool simulation. Using this type of program, the computer displays the tool path on the screen, Fig. 29-11. Errors in tool movement can be detected before a part is machined. If the tool moves too fast or too deep, the designer can edit the tool path. Without CAD/CAM, most errors are caught when the tool breaks because of a wrong move.

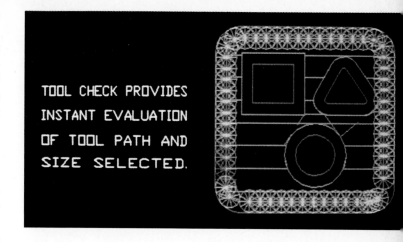

Fig. 29-11. Machine tool simulation. (NC Microproducts, Inc.)

COMPUTER-INTEGRATED MANUFACTURING

CAM systems generally control one or several machines, called machining cells. A *computer-integrated manufacturing* (CIM) system controls the entire product assembly line, Fig. 29-12. CIM interfaces machining cells with robots and transport systems which load, process, remove, and transport parts as they pass through the production line. A CIM system reduces the amount of human labor required in the total production process.

ARCHITECTURAL ENGINEERING AND CONSTRUCTION APPLICATIONS

Architectural engineering firms are increasing their use of CADD in the design and layout of residential and commercial buildings. *AEC applications* include facilities design, plans, elevations, details, structural design, and landscape architecture.

FACILITIES DESIGN

A growing use of computers in architecture is *facilities design*. These programs help determine space and room arrangements the buyer needs. Using facilities design software, an engineer identifies the type of activities in certain departments. Also specified is the relationship of these departments, whether it be high or low. The computer then calculates a proposed arrangement of rooms and space for optimum communication and traffic.

PLANS

Plans include layouts for foundation, floor, electrical, plumbing, and HVAC systems. When drawing floor plans, the CAD functions allow a designer to move, copy, and mirror rooms. If one room shape is commonly used, the designer may choose to make an array. An array copies one shape into numerous copies in a rectangular or circular fashion, Fig. 29-13.

Fig. 29-12. CIM systems involve the use of machining cells, robots, and transport systems. (Cincinnati Milacron)

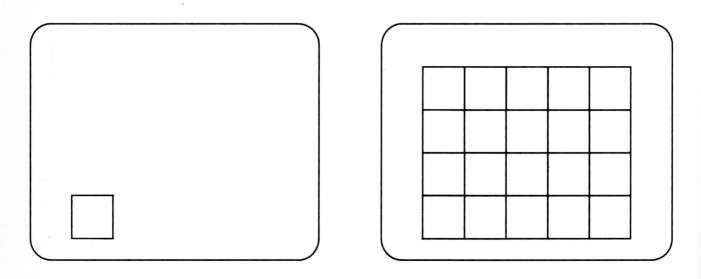

Fig. 29-13. An array command allows you to duplicate entities in a rectangular or circular arrangement.

Special inquiry commands are used to calculate properites of space allocations. Area can be found to report the square footage of each room, or of the total building. The volume of rooms can be calculated to determine requirements for heating and cooling.

An important feature of CAD when producing plans is symbol libraries. The most commonly used symbols for construc-

tion, electrical, and HVAC systems are held in a library. When inserting a door, window, outlet, vent, etc., the designer needs only to call in the appropriate symbol. Many CAD systems also keep track of the number of items inserted into the drawing. When the plan is complete, a bill of materials can be listed automatically, based on the number of components inserted into the plan.

ELEVATIONS

The presentation of a proposed building is critical to contract sales. Certain CAD capabilities, such as color and three-dimensional imaging have greatly changed presentation graphics. Instead of a two color, two-dimensional drawing of the building, the architect can create a realistic perspective view complete with color, Fig. 29-14. In addition, the architect can rotate the image, showing clients views of the proposed structure from different angles.

SECTION VIEWS

Section views in both AEC and mechanical applications are quickened by the use of hatching. Instead of drawing single section lines, the CAD operator simply defines a border around the area to be hatched. Then by selecting the type of hatch, section lines within the border are automatically drawn at the correct angle and spacing.

STRUCTURAL DESIGN

A CAD system capability of analysis facilitates the work of structural designers. Members, such as girders, columns, and beams can be placed to provide maximum strength for the structure. With three-dimensional commands, the structure can be viewed from various angles. Clearance between members can be analyzed. This information is then used by other AEC engineers when adding other systems, such as heating and cooling ducts.

PIPING PLANS

Another growing use of CAD in AEC applications is the design and layout piping systems consisting of motors, pumps, and valves to transport water or other fluids. The initial layout is made in two dimensions. Symbols for pipes, pumps, valves, regulators, etc. are arranged to provide proper flows. Once the entire system is laid out, the designer makes a three-dimensional model of the design, Fig. 29-15. The three-dimensional model helps the designer check whether any pipes or components interfere with each other. Advanced piping systems can even indicate errors in flow direction and valve or pump selection.

LANDSCAPE ARCHITECTURE

Applications programs in *landscape architecture* allow the drafter to create and easily change the layout of trees, shrubs, and other ground cover surrounding a building, Fig. 29-16. Symbols are used to represent each of these items. The features of each tree or other foliage are included when symbols are entered into the drawing. Cost estimates can automatically be produced from the quantity of each item in the landscape design. The ease of modification allows landscape architects to edit layouts according to client's needs and wants.

Fig. 29-14. Computer-generated model of an office facility.

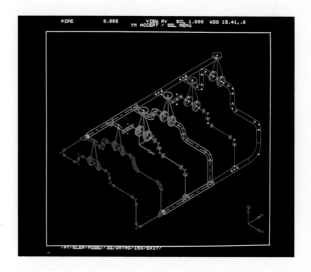

Fig. 29-15. Three-dimensional piping layout can disclose errors in the design. (CADAM)

ELECTRONIC APPLICATIONS

The evolution of many modern electronic components can be attributed to use of CAD electronic application programs. This software allows designers to more easily lay out and test the logic of circuits.

The initial layout is done with use of a symbol library. The library contains a list of prepared symbols which refer to standard electronic components. Once all components are identified, pin connections can be identified for automatic tracing, the task of connecting the wires from one component to another, Fig. 29-17. The computer finds the shortest route which doesn't interfere with other traces. This job typically

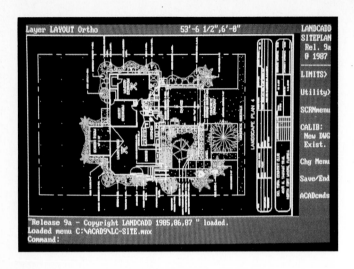

Fig. 29-16. Landscape architecture benefits from the use of applications programs. (Landcadd, Inc.)

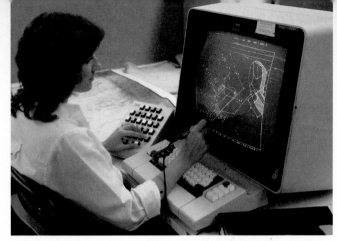

Fig. 29-18. The cathode ray tube (CRT) permits the designer to alter a design and see the results immediately. (McDonnell-Douglas, St. Louis)

Fig. 29-17. Computer-assisted design of integrated circuits. (Bell Labs)

took weeks manually; using the computer the same practice can be done in minutes. Routes for different circuits are given color to clarify the design.

Once tracing is complete logic testing is performed. Historically, *logic testing*, was done by breadboarding the components manually. Automatic logic testing simulates the operation of the circuit board. Tests are done for open circuits, shorts, or other improper connections. Errors can be easily edited on the display screen.

OTHER APPLICATIONS

Any industry which has reason to draw is a candidate for CAD. Each year, a wider variety of CAD applications are found to replace traditional drafting techniques. Applications dis-cussed in this chapter cover only a few of the many areas where computer graphics might be used.

CIVIL DRAFTING

Civil drafting involves making surveys, charts, and maps that describe land terrain, road systems, utility systems, etc. using CAD, Fig. 29-18. Changes due to construction can be easily modified on the existing map. Data concerning mineral deposits can be analyzed on the screen to determine mining and drilling operations.

BUSINESS GRAPHICS

In addition to product drawings, companies use CAD to create charts and graphs for sales and market data to production flow and process sequences. Various CAD features, such as color and hatching, allow graphic artists to create appealing presentations.

PUBLISHING

The benefits of computer-assisted drafting make it cost effective to replace T-squares and technical pens for layout of artwork. More recently, CAD drawings are being transferred directly into electronic page layouts. When corrections are necessary, the illustration can easily be modified using CAD editing features. Artwork is then produced using a hardcopy device, such as a pen plotter. Also, entire pages, including copy and imported CAD artwork can be reproduced using laser typesetting machines.

QUESTIONS FOR DISCUSSION

1. Present several benefits of computer-aided drafting and design over manual drafting.
2. List the components of a CAD workstation.
3. Briefly summarize the history of computer graphics.
4. Compare the qualifications of a CAD drafter versus those of a manual drafter.

5. Describe the changes that occur in the environment of a drafting department that has converted to computers.
6. Describe the impact computers have made on mechanical design and manufacturing.
7. Describe the impact computers have made on the architectural engineering and construction field.
8. What do you believe the future holds in store for computer graphics applications?

ACTIVITIES

1. Review current literature on the applications of computer graphics for drafting. Check your school or community library for magazines containing current articles. Noted magazines are *Computer Graphics World*, the *S. Klein Computer Graphics Review*, *Design Graphics World*, and *Plan and Print*. Note the trends in the field and prepare a report for presentation to your class.
2. Plan a visit to the computer graphics department of a local industry. Make a list of the equipment they are using. Ask whether the department has experienced any increase in productivity since converting to computers.
3. Interview an engineer, industrial designer, architect, or drafter. Ask about the uses and advantages of computer graphics in his/her field of work and prepare a report to your class.

Chapter 30

THE CAD WORKSTATION

The equipment used by the CAD drafter is very different than that of the drafter using traditional tools. The combination of software, the computer, and peripheral devices is referred to as the workstation. It consists basically of six components, Fig. 30-1:

1. Computer. The computer is the heart of any workstation. Its speed and memory largely determine the power of the workstation.
2. Software. The CAD program determines the available functions of the workstation. Programs range from simple drawing software to sophisticated design and analysis programs.
3. Data storage device. The data storage device holds the CAD program and drawing data.
4. Display screen. The display screen allows the drafter to view the drawing being created.
5. Input device. The input device is used to enter commands and digitize point positions.
6. Hardcopy device. The hardcopy device makes a paper or film copy of the drawing held in computer memory or stored on a data storage device.

Fig. 30-1. CAD workstation is composed of software, the computer, and peripheral devices.

COMPUTERS

The progress of computer-aided drafting and design is largely attributed to the advancement of computer technology. Early computers were large, cumbersome, and most did not have the power of today's microcomputers. The development of integrated circuits and microprocessors have led to powerful CAD systems capable of complex designs and product analysis.

A computer, by definition, simply adds, subtracts, compares, and stores data. A human is capable of the same activities; however, the computer is able to process data at high speeds. The integral elements of a computer include the central processing unit and memory.

CENTRAL PROCESSING UNIT

The central processing unit (CPU) consists of the control unit and arithmetic logic unit (ALU). The control unit directs the flow of data. It accepts data from an input device or from memory. It then sends the data to be added, subtracted, or compared to the ALU. The control unit also governs the

sequence of instructions contained in the CAD software. When the processing is complete, the control unit sends the result either back to memory or to an output device, such as a display screen or hardcopy device.

The speed of the CPU is rated by cycles and number of bits processed. A cycle is the time it takes for the computer to process an instruction. Most modern computers are rated in MHz (megahertz) or million cycles completed in one second. A typical microcomputer has a cycle rate of 4 to 20 MHz.

The amount of information that a computer can work with at one time is rated in bits. The bit is the basic unit of all digital computer operation. The topic of digital electronics is too lengthy to discuss here. Simply know that eight bits make a byte, which represents one character, such as a letter, number, or symbol. An eight bit computer can process one character at a time. A 16 bit computer can process two characters at one time. A 32 bit computer, the largest today, can process four characters at one time. Thus, the more bits the computer can process at one time, the faster it is.

MEMORY

The computer's memory stores information to be processed. Data is held in one or more microchips. All computers use at least two types of memory: Read Only Memory (ROM) and Random Access Memory (RAM).

READ ONLY MEMORY

Read only memory (ROM) contains data and instructions which are not changed by the user. The data is used when the computer is first turned on. The computer may also look to ROM for specific functions, such as outputting data to the display screen. The data in ROM is not lost when the computer is turned off. It is permanently held on one or two microchips in the computer.

RANDOM ACCESS MEMORY

Random access memory (RAM) holds informaton while you are using the computer. It must retain instructions used by the CAD program. It may also hold your drawing while you are creating or editing entities.

RAM is rated in kilobytes (KB) or megabytes (MB). A kilobyte is 1024 bytes, or characters, of information. A megabyte is 1,024,000 bytes. The amount of memory your computer has is very important. Sophisticated CAD programs depend on large amounts of memory. Also, a very large drawing might require considerable memory storage.

CATEGORIES OF COMPUTERS

Computers are classified by their speed, memory, and size of programs they can run. The three categories are mainframe, minicomputer, and microcomputers. The distinction between these computers has blurred because of advances in computer technology. Modern microcomputers can perform functions once limited to mainframe and minicomputers.

MAINFRAME COMPUTERS

A mainframe computer, Fig. 30-2, is capable of great speed and memory. They have 32 bit CPUs and may include hundreds or thousands of integrated circuits.

Mainframe systems are networked computers. Drafters work at remote terminals which are connected to the mainframe system. Each terminal includes an input device and display screen. Many drafters can access the functions of the CAD system simultaneously. Completed drawings are stored in a data storage device connected to the mainframe. Since the storage can be accessed by all drafters, it is called a common data base.

Mainframe computers are capable of multitasking, which is the computer's ability to run more than one program at the same time. They contain more than one CPU to handle increased demand for processing power.

MINICOMPUTER

A minicomputer, Fig. 30-3, is less powerful than a mainframe computer. Although they are 32 bit machines, minicomputers are generally smaller, slower, have less memory, and are less capable of multitasking. However, they do allow networking of several terminals.

A minicomputer is powerful enough to run even the most complex CAD program. This, compared with its low price

Fig. 30-2. Mainframe systems are capable of great speed and memory. (IBM)

Fig. 30-3. A minicomputer has similar capabilities to a mainframe, but requires less space than one. (IBM)

relative to mainframe computers, makes it a desirable alternative. Many industries use CAD systems based on minicomputers.

MICROCOMPUTERS

The category of microcomputer has emerged only in the last 10 years with the development of the microprocessor. Microcomputers are smaller, slower, and less powerful than mainframe and minicomputers, Fig. 30-4. Their advantage is their low cost compared to other computers. Another advantage is size; many are small enough to fit on a desktop.

Microcomputers are divided into two areas: microsystems and personal computers. Microsystems are 32 bit machines, based on UNIX operating systems. Two examples are SUN and APOLLO workstations. Personal computers are mostly 16 bit machines, based on Microsoft's Disk Operating System (DOS). The IBM PC line and its clones are typical personal computers. PCs have somewhat less speed and memory, and generally have mediocre display capabilities.

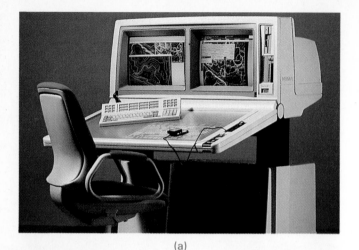

(a)

(b)

Fig. 30-4. Microcomputers. A—Microsystems are generally 32 bit machines. (Intergraph) B—PCs are 16 bit computers. (Zenith)

Microcomputers may be stand-alone systems or networked to a host computer. With a stand-alone system, each drafter uses a separate computer to perform their work. Networked PCs are connected to a host PC which contains the CAD program and drawing files. The terminal is a separate PC. Once the program and drawing is loaded into the terminal PC, it uses its own processing capability, not that of the host PC.

DATA STORAGE DEVICES

Data storage devices permanently save information. They hold both programs and data created by those programs. During a drawing session, the drafter loads the CAD program from data storage into computer memory. When finished with a drawing, the drawing data is sent to the storage device for use at a later time.

Data storage devices are classified in two ways. First, they may store data magnetically or optically. Second, they may be either disk storage or tape storage. Magnetic storage devices hold data by charging magnetic particles on the sur-

face of the tape or disk. Referring to digital electronics, the combination of charged and uncharged particles can be translated into data. A read/write head can sense (read) the charge of particles and can also change (write) the charge when saving new information. Optical storage devices use a laser to read the presence of holes on a disk. The number and sequence of holes is translated by the computer into information.

The difference between tape and disk storage is mainly speed of accessing and writing information. With tape storage, a plastic tape is passed along a read/write head, much like a cassette audio tape is played. To access information on the tape, you must wind the tape to the proper point. With disk storage, a disk is rotated and a read/write head (or laser) moves to the proper position above the rotating disk. This allows the computer to access and store data much faster.

DISK STORAGE DEVICES

Disk storage includes floppy disks, hard disks, microdisks, disk cartridges, and optical disks. Each of these devices differs in the amount of data storage.

FLOPPY DISKS

Floppy disks are 5 1/4 or 8 in. flexible plastic disks, coated with magnetic particles, and enclosed in a cardboard or plastic jacket, Fig. 30-5. The floppy disk is held in a floppy disk drive

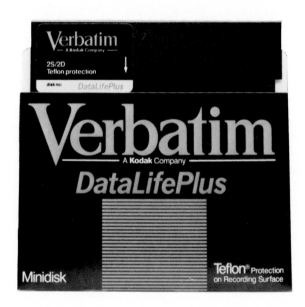

Fig. 30-5. Floppy disks. (Kodak)

which spins it at about 300 rpm. Holes cut in the jacket allow the read/write head to lower onto the disk surface to read and write information to and from the disk. The write protect notch on the jacket can be covered with an adhesive tab to prevent data from being written to the disk.

The capacity of the disk is determined by the number of sides used and the density of the magnetic coating. Most disk drives use both sides of the disk. Disks for these drives are labeled as DS, meaning double sided. Both sides of the disk are guaranteed to store data. SS (single sided) disks are only guaranteed for one side.

The density of magnetic particles on the disk may be DD (double density) or HD (high density). The DSDD (double sided, double density) disk will hold 360,000 bytes (characters) of data. DSHD (double sided, high density) disks will hold 1.2 MB of data. Floppy drives designed for DD disks will not read to or write from a HD disk.

Floppy disks must be handled very carefully. Bending the disk or writing on the jacket will damage the magnetic coating. Exposing the disk to magnetic devices (TV, stereo speaker), heat, or moisture may erase the data.

MICRODISKS

Microdisks, Fig. 30-6, are 3 1/2 in. rigid plastic disks sealed in a plastic package. Because the disk is rigid, more magnetic particles can be coated to the disk. Most will hold 1 MB of data. The disk is inserted into a microdisk device, which spins it at about 600 rpm.

HARD DISK DRIVES

Hard disk drives consist of magnetic coated aluminum platters sealed in an airtight enclosure, Fig. 30-7. Because of the number and rigidity of the platters, hard drives can store from 10 to 200 MB of data. They are used for personal computers and minicomputer CAD systems. The drive may be internal or external; however, the platters are not removable. A light on the front of the enclosure indicates when the drive is in use.

DISK CARTRIDGES

Disk cartridges are 8 in. flexible disks held in a plastic housing, Fig. 30-8 . They are a cross between floppy disks and microdisks. The cartridge is used with a disk cartridge drive.

Fig. 30-7. Hard disk drives are more durable than floppy or microdisk drives. (Seagate Technology)

Fig. 30-8. Disk cartridges may hold in excess of 20 MB. (Iomega Corporation)

Most hold in excess of 20 MB of data. This storage capacity, and the fact they are removable, make them a desirable alternative to hard disk drives.

DISK PACKS

Disk packs are used on mainframe and minicomputers. They consist of several rigid, magnetic coated disks stacked together as a unit or pack, Fig. 30-9. The pack is capable of tremendous amounts of data storage.

OPTICAL DISK STORAGE

Optical disks are plastic disks with an aluminum coating, Fig. 30-10. Holes punched, or chemically etched, in the coating are read by a laser. The number and sequence of the holes is interpreted by the computer as data. Most people are familiar with audio optical disks, more commonly referred to as compact disks.

Fig. 30-6. Microdisks are 3 1/2 in. rigid plastic disks.

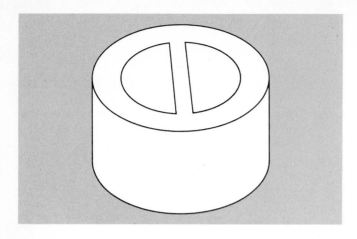

Fig. 30-9. Disk packs are composed of several rigid disks stacked together as a unit.

Tape devices are primarily used for making backups of data. They are not quick enough for data accesses while you are using the computer. A more important reason for using tape as backup is because they can hold over one gigabyte (billion bytes) of data.

There are two types of tape storage devices: reel-to-reel and tape cartridges.

REEL-TO-REEL DEVICES

Reel-to-reel tape drives consist of two removable reels, a data reel and take-up reel, which are installed on a tape transport system to wind the tape past the read/write head, Fig. 30-11. The reels vary in size. Ten to twelve in. reels with 1 in. wide tape are used on mainframe and minicomputers systems. Reel-to-reel drives for microcomputers are 6 to 8 in. diameter reels and smaller width tape.

TAPE CARTRIDGES

Tape cartridges contain both data and take-up reels in a plastic enclosure, Fig. 30-12. They are just larger than a cassette tape. Some specially made drives use videocassettes as tape cartridges.

Fig. 30-10. Optical disk storage device and drive. (IBM)

Fig. 30-11. Reel-to-reel magnetic tape. (IBM)

There are certain advantages and disadvantages of optical disks. Two advantages are that the disks can hold over 200 MB of data and are removable. The disadvantage is that once data is written to the disk, it can never be altered. This is why optical disks are primarily used for archival purposes. Drawings which will never have to be modified are kept for historical and reference use. New technology may allow data to be read and written to optical disks as easily as it is to floppy disks.

TAPE STORAGE DEVICES

Tape storage devices use a length of magnetic coated plastic tape which is wound against the read/write head to transfer data to and from the tape. Tape systems, as a rule, are slow. The read/write head cannot instantly locate any position on the tape to access data.

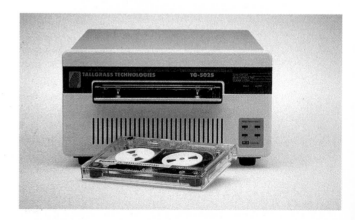

Fig. 30-12. Tape cartridges are popular as backup systems. (Tallgrass Technologies Corporation)

DISPLAY SCREENS

In traditional drafting, you create and view the drawing on paper. With CAD, you enter drawing commands and entities appear on the display, Fig. 30-13. The produced images can be modified many ways because the screen simply displays the drawing data held in memory. For example, the image can be magnified to make detail work much easier. You could also rotate an object. Instead of erasing and redrawing, as would be done with traditional tools, you simply enter a rotate command. The drawing data is changed in memory and the rotated entity is displayed.

This section will discuss the many types of display devices. Their function is the same, yet the methods used to create the image are different. The most important difference between display screens is resolution. Resolution refers to how clear the image on the screen is. Clearer images using "high resolution" monitors make drawings easier to see.

CATHODE RAY TUBE

The most common display technology is the cathode ray tube (CRT). It makes an image by propelling an electron beam against a phosphor-coated screen surface. When the beam hits the surface, individual dots of phosphor, called pixels, glow. From the drafter's point of view, the glowing pixels form images. The path of the electron beam is controlled by drawing data sent from the computer to form entities on the screen. Color is created activating three separate color pixels on the screen.

There are two technologies used to control the CRT's electron beam: raster scan and stroke writing.

RASTER SCAN DISPLAYS

In raster scan displays, the path of the CRT's electron beam is controlled to scan the phosphor-coated surface. It scans from left to right and top to bottom, much like you read this text. The beam scans continuously, over 60 times a second. This is because the phosphor pixels will only glow for a short time and must be reactivated regularly.

To form an image, the beam selectively activates phosphor pixels as it scans across the screen, Fig. 30-14. The more pixels that can be used to draw the image, the clearer the

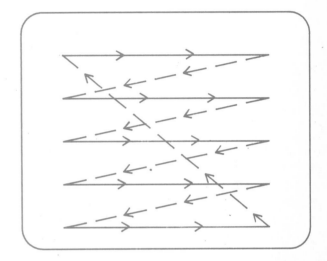

Fig. 30-14. A raster scan display scans the screen in a regular pattern.

image will be. This is referred to as resolution. Low-resolution screens have about 240 pixels horizontally and 120 vertically. Medium-resolution screens have 640 pixels horizontally and 320 vertically. Higher resolution screens are 1024 x 1024. To illustrate this example, a 1 in. horizontal line on a low-resolution screen might be made 24 pixels. A high-resolution screen might use 102.

STROKE WRITING DISPLAYS

Stroke writing displays use a different method of controlling the electron beam. The beam does not scan the screen in a regular pattern. Instead, it creates images by actually "drawing" on the screen. The beam performs much like you use a pencil to draw. For a rectangle, the beam would be controlled to draw four straight lines, Fig. 30-15. Since the phosphor will glow for only a short time, two methods are used to keep the image lit: refresh displays and direct-view storage tubes.

Refresh displays constantly "redraw" the image to prevent it from becoming dim. This is similar to scanning, yet the electron beam refreshes the image by drawing the same line pattern of the drawing. However, there is a disadvantage to this method. If there are many entities in the drawing, it takes the beam longer to refresh the image. In this case, the screen may flicker as the beam redraws a complex drawing.

To eliminate screen flicker, direct-view storage tube technology was developed. It also has the electron beam

Fig. 30-13. The display screen is used to view drawing data. (IBM)

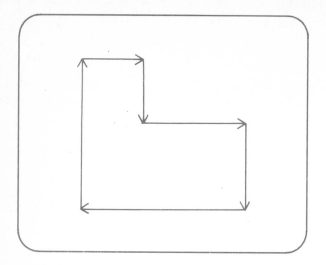

Fig. 30-15. Scanning pattern of a stroke writing display.

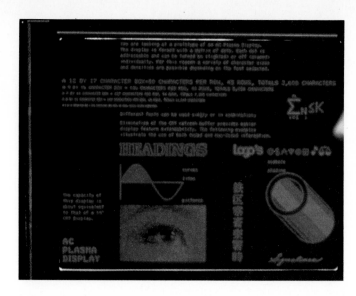

Fig. 30-16. Gas-plasma displays are characterized by their orange-red color. (IBM)

"draw" the image. However, once the beam draws the entities on the phosphor, the image is saved by flooding the screen with electrons. These electrons do not create images. They only keep pixels lit which were previously activated by the beam. The image remains constant until the screen is "redrawn." You can add entities, but none are removed until the image is regenerated.

FLAT DISPLAYS

Flat displays are thin display screens commonly seen on portable computers. They do not use cathode ray technology because the CRT requires distance to fire electrons at the screen. Instead, they produce the image near the surface of the screen. This is done using three different technologies: gas plasma, electroluminescent, and liquid crystal.

GAS-PLASMA DISPLAYS

Gas-plasma displays are made of a thin glass enclosure containing a low-pressure neon or neon/argon gas. A grid of wires runs through the enclosure. When two crossing wires are energized, the gas at their intersection glows. The glowing gas appears as a single dot, comparable to the pixel on a CRT. By energizing different wires, an image is created by a number of glowing gas pixels, Fig. 30-16. However, gas-plasma displays are monochrome (one color) which limits their use with CAD. It is also a medium-resolution display, with a typical 17 in. diagonal, 3 in. thick display having 960 horizontal and 768 vertical electrodes which creates 737,280 pixels.

ELECTROLUMINESCENT DISPLAYS

Electroluminescent displays work much like gas-plasma displays, except that instead of gas, they use a luminescent chemical. Pixels are made by the intersection of two energized wires which pass along the chemical. These displays are also monochrome.

LIQUID CRYSTAL DISPLAYS

Liquid crystal displays (LCD) are common on portable computers requiring very small, flat screens, Fig. 30-17. They require very little power and can be battery driven. You likely have seen LCD displays on digital watches and hand-held TVs.

Fig. 30-17. Liquid crystal display (LCD). (IBM)

The method of producing an image with an LCD display is done with liquid crystals and a grid of wires, called electrodes. When the wires are not energized, the crystals lay flat. To the viewer, no image is seen. However, when two intersecting wires are energized, the crystals stand up. To the viewer, this is seen as a black dot. By selectively energizing wires, the combinations of dots form the image. Unlike gas-plasma and electroluminescent displays, LCD devices can be color. The LCD image is filtered through color lenses to give the illusion of color.

DUAL SCREENS

Dual screens are used for CAD systems needing one display for text and one for graphics. When analysis is done, the results are displayed on the text display, while the design is shown on the graphics display. Another use for dual screens is separating communications text from graphics. Communications text refers to commands used to create entities. The sequence of commands used can be viewed and recorded while also viewing the design image on the graphics display.

INPUT DEVICES

Unlike traditional drafting, you cannot "touch" a CAD drawing. Instead, you use an input device to select commands and pick point positions on the display screen. Actually, the input devices lets you interact with the computer, entering instructions, and receiving some type of response.

When drawing, a cursor is displayed on the screen to show where the input device will act. For example, when drawing a line you would use the input device to move the cursor to the position of line's first endpoint. Pressing a select button on the input device chooses that location.

There are three general categories of input devices: keyboards, digitizing tablets, and cursor control devices. Keyboards are used to type in text and enter commands. Digitizing tablets are used to move the cursor across the screen, enter commands, and trace existing drawings. Cursor control devices also move the cursor across the screen and are used to enter commands. However, you cannot trace existing drawings with cursor control devices.

KEYBOARDS

There are two types of keyboards used with CAD. The most common is the alphanumeric keyboard, much like a typewriter, and simply called a keyboard by most drafters. The second is the function keyboard.

An alphanumeric keyboard, Fig. 30-18, is always used with a CAD system, even when another input device is attached. A typical keyboard has characters for letters, numbers, and symbols. It also has special keys which, when pressed, perform a series of commands which otherwise would have to be entered separately. These are called function keys. On a microcomputer, they are labeled F1, F2, up to F10. Alphanumeric keyboards are used to type in text and enter commands by typing a letter or function key.

A function keyboard, Fig. 30-19, has a number of keys which perform specific commands of the CAD system. They are not used to enter text; it is used in conjunction with an alphanumeric keyboard.

DIGITIZING TABLET

A digitizing tablet is a flat, rectangular plastic pad with a tracking device attached by a cable, Fig. 30-20. Sandwiched in the plastic tablet is a matrix of hundreds of evenly spaced wires. A specific location is determined by each intersection

Fig. 30-18. Alphanumeric keyboard is used to input data as well as control the screen cursor. (IBM)

Fig. 30-19. Function keyboard, left, may be used instead of function keys. (CADAM)

Fig. 30-20. Digitizing tablets are available in sizes ranging from 9 x 9 in. to 36 x 48 in. (Numonics)

of wires. To sense the intersection of wires, the tracking device is moved across the pad. The tracking device may be a stylus or a puck. As the tracking device is moved across the surface of the pad, the screen cursor also moves.

To select a point position, you first move the tracking device across the pad until the cursor is properly positioned on the display screen. Then you press down on the stylus or press a select button on the puck to select that location.

Digitizing tablets are often used to turn existing paper drawings into CAD drawing files. The paper drawing is taped to the surface of the pad. Then by selecting CAD commands and picking point positions of the existing drawing, you create a CAD drawing. This process is now being automated using scanners. The paper drawing is placed on a scanner drum or under a camera. The camera detects the lines on the drawing and converts this information into computer data. The data can then be loaded in as a CAD drawing file and edited as you would a drawing originally made using CAD.

CURSOR CONTROL DEVICES

Cursor control devices allow you to move the cursor across the screen and select point locations. However, unlike a digitizing tablet, they are not designed to trace an existing drawing. The advantage of cursor control devices over digitizing tablets is reduced cost.

MOUSE

A mouse is a hand-held device rolled on top of a flat surface, Fig. 30-21. A ball, wheels, or a light sensor on the bottom of the mouse detect direction and distance of move-

ment. When moving the mouse, the cursor on the display screen also moves. To select a point position, such as the end-point of a line, press the select button on the mouse. There may be more than one button. Those other than the select button will invoke certain CAD commands.

Mice are very common input devices because they are easy to use and inexpensive. However, you cannot trace an existing drawing with a mouse because there is no way to locate the mouse precisely over a point on a paper drawing.

JOYSTICK

Although not as popular as mice, joysticks are still used on many CAD systems. The joystick consists of a small box with a movable shaft, Fig. 30-22. Pushing the shaft in a direction causes the screen cursor to also move in that direction. Buttons on the joystick allow you to select point positions.

Fig. 30-22. Joysticks can move the cursor in any direction. (CH Products)

TRACKBALL

A trackball is basically an upside-down mouse. It consists of a rectangular base with a ball set in the top, Fig. 30-23. Moving the ball with the palm of your hand causes the screen cursor to move. Buttons on the device are used to pick point locations and to select commands from a screen menu (Chapter 31).

LIGHT PEN

The light pen is different from other cursor control devices because you hold it against the display screen, Fig. 30-24. The screen senses the position of the pen and places the cursor directly under the pen tip.

A light pen is efficient because you can point exactly to the position or entity on the screen. There is a pressure-sensitive tip on the pen which, when pressed, enters a point location. One drawback to the light pen is it is very tiring to use. You must keep your hand lifted to the screen.

Fig. 30-21. A mouse is a hand-held cursor control device. (Summagraphics Corporation)

Fig. 30-23. Trackball used to control the position of the cursor and pick points. (Fulcrum Computer Products)

(a)

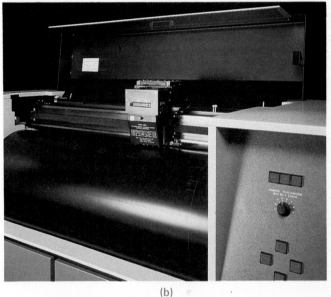

(b)

Fig. 30-24. A light pen is used to draw directly on CRT screens. (IBM)

HARDCOPY DEVICES

A CAD drawing is held in computer memory or stored on a data storage device until you are ready for a paper or film hardcopy. Then you use commands of the CAD software to send the drawing data to a hardcopy device. The device duplicates the drawing held in memory by creating an image on paper or on photographic film.

There are many types of hardcopy devices. They vary in the quality of generated drawing and the speed in which the hardcopy is complete.

PEN PLOTTERS

A pen plotter, Fig. 30-25, most closely resembles how a drafter draws by hand. The plotter moves a pen over paper

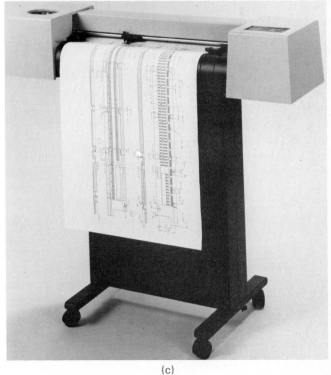

(c)

Fig. 30-25. Pen plotters. A—Flatbed plotter. (Houston Instrument) B—Drum plotter. (Gerber Scientific, Inc.) C—Microgrip plotter. (Hewlett-Packard Co.)

to create the image. The paper may also move, depending on the type of pen plotter. After drawing an entity, the pen carriage picks up the pen, relocates it, and sets it down to draw another entity.

Pen plotters may use a single pen or may be capable of holding four to twenty pens. Multipen plotters automatically select one of the many pens to draw a different lineweight or color.

Pen plotters are classified as flatbed, drum, or microgrip.

FLATBED PEN PLOTTERS

With flatbed plotters, the paper is held on a flat surface by tape, vacuum, or electrostatic charge. A pen carriage moves the pen over the paper in two directions. The dimensions of a flatbed plotter refer to the maximum size paper it can hold.

DRUM PEN PLOTTERS

Drum plotters have the paper attached to a drum. The pen moves across the drum to produce vertical (Y axis) lines on the drawing. The drum rotates to produce horizontal (X axis) lines on the drawing. Simultaneous movement of both the pen and the drum permits curves and angled lines. The paper size that can be used on a drum plotter is limited by the width and circumference of the drum.

MICROGRIP PEN PLOTTERS

Microgrip plotters resemble drum plotters in that X axis direction is produced by moving the paper and Y axis direction is produced by moving the pen. However, small rubber rollers grip the paper at the edges rather than having the paper attached to the circumference of the drum.

ELECTROSTATIC PLOTTER

Electrostatic plotters, Fig. 30-26, are much like photocopying machines. They selectively place electrostatic charges on paper which then attract ink. The charges are made by a writing head which passes back and forth across the paper. As many as 400 electrostatic "dots" are placed per inch. The paper is then fed through toner, consisting of ink suspended in a liquid, where the ink is attracted to the paper to form the image.

The position of the charges is determined by a process called rasterization. A processor in the plotter converts the drawing data into series of dots which the plotter can reproduce.

There are three types of electrostatic plotters: multi-pass web fed, single-pass web fed, and drum. Each is capable of multicolor plots. A multi-pass web fed plotter feeds paper from a roll, against the writing head, through the toner, and draws the paper up through a take-up reel. To produce different colors, the paper must be rewound, the color of toner changed, and the paper fed through the process again. A single-pass web fed plotter feeds the paper against four writing heads and four different color toners all in one pass. In drum plotters, the paper is attached to a cylinder which rotates the plotter against the writing head and through the toner. A different color toner is used for each rotation to produce multicolor plots.

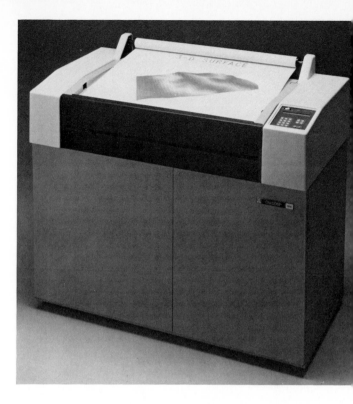

Fig. 30-26. Electrostatic plotters use a rasterization process to convert the lines of a drawing into a series of dots which are printed. (Calcomp, Inc.)

INK JET PLOTTERS

Ink jet plotters, Fig. 30-27, form images by propelling individual droplets of ink onto paper. The droplets are guided in flight to hit the paper in the proper position. Ink jet plotters are capable of producing multicolor images by using different colors of ink.

LASER PLOTTERS

Laser plotters, Fig. 30-28, use a beam of light to create an image. The beam scans across the surface of a photosensitive belt mounted on a drum. An electrical charge is applied to the belt by the beam. The drum then rolls the belt through a toner bath. Toner is attracted to the belt and then transferred to a sheet of paper.

THERMAL PLOTTERS

Thermal plotters melt ink on a separate ribbon or transfer sheet and transfer the hot ink to the paper to produce the image, Fig. 30-29. Ribbons or sheets that carry several colors are deposited in several layers to produce a multicolor image.

DOT MATRIX PLOTTING

The dot matrix printer is often used for printing letters and other business documents. In computer-aided design, the dot matrix printer is used as a plotter to produce rough check plots,

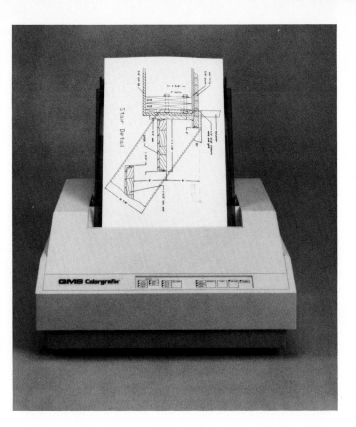

Fig. 30-27. Ink jet plotters propel ink droplets onto the plotting medium to form images. (QMS, Inc.)

Fig. 30-28. Laser plotters are a new technology in output devices. (Quadram)

Fig. 30-30. During the design, a check plot is made to verify the status of the project. A dot matrix plot is lower in quality, but much less expensive and time consuming than other hardcopy devices.

PHOTO PLOTTER

Photo plotters produce drawings by exposing photographic paper against a CRT. The dry silver-coated paper darkens in those areas where lines are present on the CRT. It is a poor reproduction and used primarily as a rough plot.

Fig. 30-29. Thermal plotters are primarily used to produce presentation materials and business graphics. (Calcomp, Inc.)

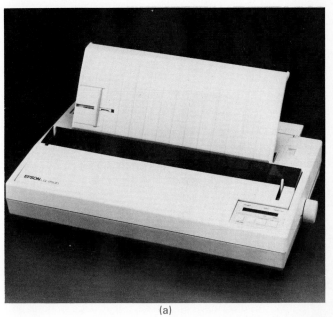

(a)

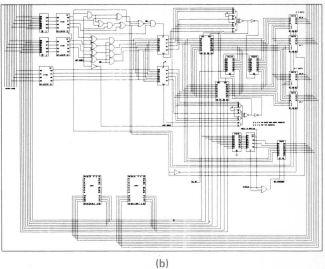

(b)

Fig. 30-30. A—Dot matrix printer is a commonly used hardcopy device. (Epson American, Inc.) B—Hardcopy produced with a dot matrix printer.

Fig. 30-31. CRT image recorders expose photographic film against a high-quality display. (Polaroid)

Fig. 30-32. COM devices utilize a laser to form images on microfilm. (3M)

CRT IMAGE RECORDER

Cathode ray tube (CRT) image recorders are an improvement over lower quality photo plotters. CRT image recorders expose photographic film against a color CRT unit. The image can be made either on slides or into photographs, Fig. 30-31.

COM DEVICES

Computer output-to-film (COM) devices produce hardcopies by directing a laser onto microfilm. The laser is controlled to "draw" on the film, much like a pen plotter. The film image is inserted into an aperture card which is coded with data related to the drawing, Fig. 30-32. Most large engineering firms use COM devices because it is easier to store aperture cards than to store full-size paper prints.

QUESTIONS FOR DISCUSSION

1. Describe the difference between read only memory and random access memory.
2. Summarize the differences between mainframe computers, minicomputers, and microcomputers.
3. Explain why disk storage devices access data much faster than tape storage devices.
4. Define "resolution" and the impact it has on the displayed image.
5. Classify the three types of input devices.
6. Name the input device which can be used to input existing drawings by tracing.
7. Explain the difference between an alphanumeric keyboard and a function keyboard.
8. Identify the three models of pen plotters.
9. Why do most large engineering firms use COM devices?

ACTIVITIES

Review current literature or visit a local computer graphics vendor to find answers to the following questions:
1. What is the highest display resolution currently available?
2. What type of hardcopy device is advertised most often?
3. How do features of personal computer (microcomputer) CAD programs differ from larger systems?
4. What is the range of storage capacities for different data storage devices?

Chapter 31

CONSTRUCTING AND EDITING DRAWINGS

Computer-aided drafting and design systems offer a wide variety of functions and features. Although no two systems offer exactly the same capabilities, the basic structure and commands sequences among them are similar. The main difference is the names used to represent drawing and editing commands. This chapter covers those drafting and detailing functions which are common among a wide variety of programs.

LOGGING ON TO THE SYSTEM

A drafter needing to use the CAD program must first progress through a series of startup steps, called a logon procedure. For a minicomputer system, this may require turning on the terminal and entering a password. With a microcomputer, the drafter usually types in the program name and enters directly into the system. If several microcomputers are networked to a common server, the drafter may have to proceed through additional steps to access the network. The exact procedure is different for every drafting department. Once you have successfully logged onto the system, the CAD program's main screen will appear, Fig. 31-1.

Fig. 31-1. The main screen provides the CAD operator with commands and options. (TEKTRONIX, Inc.)

CAD COMMANDS

When using a computer, each instruction you give is considered a command. Likewise, when using the CAD program, each direction you enter to create or edit a line, circle, or other object is a command. In its simplest form, a command is a short word which, when entered, achieves some function.

To create and edit a drawing, you select functions, usually by entering a series of commands, called command syntax. For example, to draw a horizontal line, you might have to select the commands DRAW LINE HORIZONTAL, in that order. The command structure described in this chapter is treated in a general manner and is intended to cover a number of systems.

MENUS

In most cases, you must know which command to select. The computer cannot anticipate your next move. Remember that the computer is only a machine, not a drafter. Therefore, it is your responsibility to select commands to create drawings that adhere to drafting standards. To help you identify available drawing functions, the computer supplies menus. A menu is a list of commands displayed on the screen or printed on a digitizing tablet overlay, Fig. 31-2. Besides being a helpful reference, menus prevent you from having to type in an entire command.

Commands are selected in one of three ways: typing on the keyboard, picking from the screen, or picking from a digitizing tablet. On most systems, you can use the keyboard along with the menu to select commands. However, it is often easier to "pick" the command you need from a screen menu or tablet menu.

SCREEN MENUS

A screen menu shows available commands on the display screen. This area, known as the menu pick area, might be found along the top or side of the screen. It surrounds the drawing display window, in which you see the drawing. To select a command from the screen menu, move the screen cursor (seen as crosshairs) to the menu pick area by moving your mouse, puck, or stylus. Locating the cursor in the menu pick area usually highlights single commands. Highlight

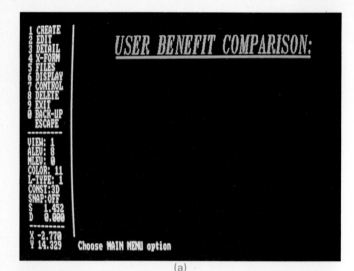

(a)

(b)

Fig. 31-2. Menus. A—Screen menu. (CADKEY, Inc.) B—Digitizing tablet overlay. (CAD Technologies)

the command you need and press the pick button on the input device.

Since there are often more than 100 commands, only so many can be displayed at one time. Thus, screen menus are divided into levels. Main menu commands are initially shown on the screen. By selecting one of these commands, a series of subcommands, or options, appear. Using this technique of "nesting," all commands can be accessed. The main menu commands might appear in a bar or as a list alongside the screen. After selecting one of these, another menu replaces the main menu. The second menu may also appear to pull-down from the command or even pop up in the middle of the screen. In addition to command names, some screen menus show functions as icons, or small pictures that represent the command.

TABLET MENUS

A tablet menu is a plastic overlay placed on a digitizing tablet. The overlay is divided into command areas and an area in which moving your puck or stylus controls the screen cursor. The tablet menu provides instant access to commands, without having to pick a series of "nested" commands as you

might on a screen menu. On the other hand, a tablet menu cannot contain all commands. Therefore, you will still rely on the screen menu or typed command entry to access all functions.

COMMON CAD FUNCTIONS

The standard functions found on drafting programs can be grouped into several categories. They are defined as follows:

FILE MANAGEMENT COMMANDS

File management commands allow you to begin, save, and load drawings from a data storage device.

DRAWING SETUP COMMANDS

Drawing setup commands allow you to determine your drawing area and units of measurement (English, metric).

DRAWING COMMANDS

Drawing commands add graphic entities—lines, circles, arcs, curves, polygons, and ellipses—to the drawing.

DRAWING AIDS

Drawing aids help you precisely locate position on the drawing. Most often, the screen cursor moves freely about the drawing as you move the input device. To select exact location, select a drawing aid which "snaps" the cursor to an object or grid dot.

EDITING COMMANDS

Editing commands allow you to change drawn entities. Moving, copying, and changing color or linetype are typical editing tasks. Editing commands also allow the drafter to construct a complex drawing by copying and manipulating just a few objects.

DISPLAY CONTROLS

Display controls determine what part and how much of the drawing is shown in the display screen.

DIMENSIONING COMMANDS

Dimensioning commands let you automatically place linear and radial dimensions. You simply have to choose the points to dimension.

TEXT COMMANDS

Text commands place notes and lettering on the drawing.

HATCH AND FILL COMMANDS

Hatch commands place section lines and graphic patterns.

LAYER COMMANDS

Layer commands allow you to separate different parts of the drawing into levels, much like using overlays.

INQUIRY COMMANDS

Inquiry commands let you ask questions about your drawing, such as distance, perimeter, area, and drawing status.

SYMBOL LIBRARY COMMANDS

Standard drafting symbols can be stored and inserted on as many drawings as needed. This prevents having to draw any symbol twice.

PLOTTING COMMANDS

Plotting commands allow you to make a hardcopy of the drawing stored on disk or in memory.

HELP COMMANDS

Most systems offer on-screen help. If you forget how to use a function, select the help option.

FILE MANAGEMENT

Each time you use a computer, you are working with files. A file is a group of related data. The instructions which make up a software program are considered a file. When working with a CAD system, you create, save, and load drawing files. This process of manipulating drawing files is called file management and includes the following functions.

STARTING A NEW DRAWING

After loading the CAD system, you typically have a choice between beginning a new drawing or editing an existing drawing. Some systems require that you select a specific command from a main menu to begin a new drawing.

SAVING THE CURRENT DRAWING

At some point in time, you will want to save information added to a new drawing. Do this frequently, even while working. This ensures that in the event of a power failure or system crash your drawing remains intact.

LOADING A DRAWING

Loading recalls a previous drawing for continued work. Remember to save the drawing again when finished to update any changes made.

LISTING DRAWING FILES

A listing shows the drawings stored on the current data storage device. It may also show the number of entities and time spent on the drawing, Fig. 31-3.

ERASING DRAWING FILES

Erase drawings with caution because you cannot "unerase" the drawing back into existence. Delete a drawing from the data storage device only with the supervisor's or instructor's permission.

MERGING DRAWING FILES

The merge command allows you to combine another drawing with the current drawing on screen. You might use this command to insert a title block and border stored as a separate drawing.

COORDINATE SYSTEMS

Each line, circle, arc, or other entity you add to a drawing is located by certain points. A line is defined by its two endpoints. A circle is defined by its center point and a point along the circumference. A square is located by its four corner points. To precisely locate entities, all CAD programs use a standard point location system, called the Cartesian coordinate system.

CARTESIAN COORDINATE SYSTEM

The Cartesian coordinate system consists of two axes, Fig. 31-4. The horizontal axis is the X axis. The vertical axis is the Y axis. The intersection of these two axes is the origin. The location of any point can be determined by measuring the distance, in units of measurement, from the origin along both the X and Y axes. These two values are called the X and Y coordinates, or coordinate pair. Each axis is divided into equal units. Each unit may refer to inches, feet, or metric measurements, such as meters or millimeters. The unit of measurement along the X and Y axes is the same.

```
Drawings pathname: \vcad53\v2d\draw\
```

drawing name	date last modified	time last modified	cumulative drawing time	object total	symbol total
Mi04001	10-Jan-1989	9:04 am	13 hrs 6 mins	635	3
Mi04007	10-Jan-1989	9:16 am	11 hrs 9 mins	592	59
Mi00006	24-Nov-1988	2:17 pm	8 hrs 10 mins	293	18
Mi03221	2-May-1988	5:14 pm	5 hrs 53 mims	1094	27
Mi04412	2-Dec-1988	3:15 pm	13 hrs 5 mins	687	0
Pd17AB24	22-Aug-1988	4:33 pm	7 hrs 2 mins	2948	0
Pd17AA43	3-Mar-1989	10:42 am	21 hrs 7 mins	3301	32

```
There is a total of 7 files.
Press [Enter] to continue.
```

Fig. 31-3. A directory of files may help you determine the most recent version of a drawing.

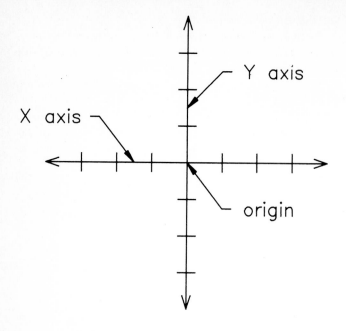

Fig. 31-4. The Cartesian coordinate system.

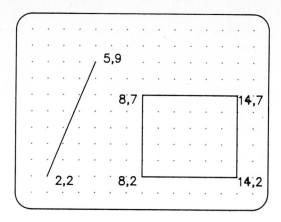

Fig. 31-5. Absolute coordinates are measured from the origin.

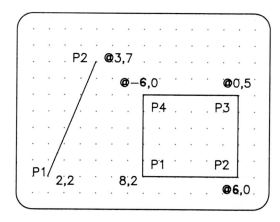

Fig. 31-6. Relative coordinates define a point from a previous point.

Coordinates for absolute point locations can be positive or negative. This depends on their location in reference to the origin. Movement to the left of the origin results in a negative X value. Movement below the origin results in a negative Y coordinate value. Looking at the view of your display screen, the origin of the axes is generally the bottom-left corner or center of the display screen.

When entering point locations to draw an entity, you can use either the absolute or relative mode.

ABSOLUTE COORDINATES

Absolute coordinates refer to exact point locations measured from the origin. The line and rectangle shown in Fig. 31-5 are defined by absolute coordinates.

RELATIVE COORDINATES

Relative coordinates define distance from a previous point. For example, suppose you have entered the first endpoint of a line at 1,1. Typing the relative coordinate @3,4 places the second endpoint three units to the right and four units above the first endpoint. (The ''@'' symbol is used by one system to designate a relative coordinate entry.) The line and rectangle shown in Fig. 31-6 are defined by relative coordinates.

POLAR COORDINATES

When using a polar coordinate entry method, points are located by angle and distance from a previous point. For example, suppose you have entered the first point of a line as absolute 1,1. You could enter the coordinate @14<30 for the second endpoint to draw a line 14 units long at a 30 degree angle, Fig. 31-7. (The ''@'' and ''<'' symbols are used by one system to designate distance and angle.) Polar coordinates are typically relative, meaning that the distance is measured from the previous point.

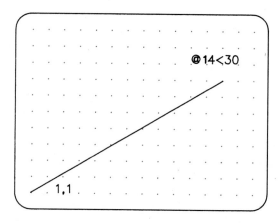

Fig. 31-7. An angle and distance are used to define polar coordinates.

COORDINATE DISPLAY

The coordinate display is a readout of your current position on the drawing. You may not always type in coordinates. Often you move the screen cursor around the drawing to location position. The coordinate display, located in a status line, shows the current position of the cursor, Fig. 31-8. It may

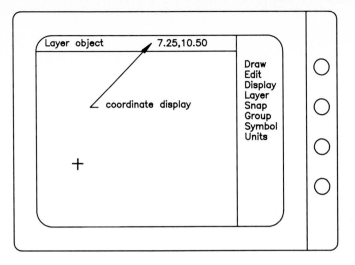

Fig. 31-8. The coordinate display indicates the position of the cursor on the display screen.

show absolute position on the drawing, relative position from the previous point, or polar position from the previous point.

DRAWING SETUP

When beginning a new drawing, you must specify certain parameters, such as unit of measurement, drawing area, and drawing scale. This process is called drawing setup. The CAD system will assign default values to each parameter.

Actually, there are a number of parameters which can be set before or during a drawing session. These might include linetype, text height, and drawing color. This section focuses on only those that should be considered before adding the first entity.

UNITS OF MEASUREMENT

The unit of measurement is the distance assigned to each unit of the coordinate system. Units may be inches, feet, millimeters, meters, or those you define. Once you set the unit of measurement as either English or metric, it is not wise to switch to the other. However, it is acceptable to change the notation, such as from inches in fractional format to inches in decimal format.

The unit resolution is the number of decimal places used to calculate dimensions you enter. Having more digits behind the decimal point results in more accurate coordinate entry.

DRAWING AREA

In traditional drafting, your drawing area is limited by the size of the sheet of paper. With CAD, your drawing area may or may not be limited. Most systems allow you to set the size of the drawing area as large as necessary. Since you are using ''real world'' measurements, the drawing area for a house plan might be 80 ft. wide by 40 ft. high.

The drawing area is defined by specifying the upper-right corner. (The lower-left corner is typically 0,0, the origin.) For

the house plan just mentioned, you would set the upper-right corner of the drawing area at 80,40.

Some systems base the drawing area on standard drafting paper sizes. In this case, you must select a sheet size large enough to contain your drawing. The drawing area is then confined to the sheet's dimensions. Sheet sizes are limiting. You obviously cannot place a 40 ft. line on a D-size (36 in. wide) sheet; therefore, you must select a drawing scale.

DRAWING SCALE

Drawing scale describes the relationship of drawing measurements to ''real life'' measurements. For example, a 1 ft. line on the drawing may only plot out to be 1/4 in. Scale is needed especially for CAD systems which offer a limited drawing area. Drawing scale will reduce the size of an entered line by some factor. However, the coordinate display and coordinates you enter should still be full size.

CONSTRUCTING GRAPHIC ENTITIES

The ''substance'' of any CAD drawing is the geometry. The drawing you create consists of graphic entities, including points, lines, polygons, circles, and arcs.

DRAWING POINTS

Points mark an exact coordinate position. They are helpful as a reference for placing other entities. After entering the DRAW POINT sequence or appropriate command, enter coordinates or pick the location on screen with your input device. A dot, or plus mark (+) should appear, Fig. 31-9.

DRAWING LINES

Lines are geometric entities defined by their two endpoints. The points may be entered as absolute or relative to another point. You can enter coordinates or pick location on screen. Entering the command sequence, such as DRAW LINE, brings up a list of options. Each of these allow you to draw a line using different ways of selecting position. Some lines are placed in relation to an existing line. The options you might choose are as follows:

DRAWING A SINGLE LINE
The simplest method to draw a line is by entering coordinates or picking location for the two endpoints. As you move the screen cursor away from the first point, a blinking or dotted ''rubberband'' line may stretch out from the first endpoint. This allows you to better place the line.

DRAWING LINES AT AN ANGLE
Lines at an angle can best be drawn by typing in the second point as a polar coordinate. Pick the first endpoint using the screen cursor.

DRAWING A SERIES OF CONNECTED LINES
With connected lines, the second endpoint of a previous line becomes the first endpoint of another line, Fig 31-10(a).

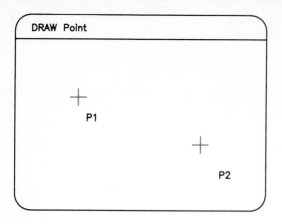

Fig. 31-9. Placing points on the display screen.

Usually this is the default option of the DRAW LINE command. After picking the first two endpoints, continue to pick the second endpoints for successive line segments.

DRAWING A SERIES OF SINGLE LINES

To draw a series of single lines, you usually must select a special option. This allows you to continue to digitize first and second endpoints without leaving the DRAW LINE command. See Fig. 31-10(b).

DRAWING HORIZONTAL AND VERTICAL LINES

Choose the HORIZONTAL and VERTICAL options after DRAW LINE to draw lines parallel to the X and Y axes, Fig. 31-10(c) and (d). After picking the first endpoint, the line will extend either horizontal or vertical depending on the option you selected. Some systems have an ORTHOGONAL mode which limits any drawn line to either horizontal or vertical. You do not need to select individual horizontal or vertical options.

DRAWING LINES AT AN ANGLE TO A GIVEN LINE

To draw a line at an angle to a given line you must first pick the existing line. Then enter an angle, pick the first endpoint, and enter a line length, Fig. 31-10(e). Some systems do this a bit differently by having you pick both endpoints on screen. As you locate the second endpoint, though, the line extends only at the entered angle.

DRAWING A LINE PARALLEL TO A GIVEN LINE

Drawing parallel lines is easy with CAD. Simply pick the existing line, then pick the endpoints of the new line. The first endpoint of the new line determines the distance away from the existing line. See Fig. 31-10(f). Some systems have you type in the distance between the two lines.

DRAWING A LINE NORMAL TO A GIVEN LINE

To draw a normal (perpendicular) line, pick the existing line, then the first endpoint of the new line. As you locate the second endpoint, the line extends at 90 degrees to the existing line, Fig. 31-10(g). Some systems have you first pick one endpoint, then the line to which the new line should be normal.

DRAWING TANGENT LINES

New lines can be drawn tangent to circles, arcs, and curves. After selecting the TANGENT option of the LINE command, pick the circle or arc to which the new line should be tangent. Then pick a second endpoint at the appropriate location. The second endpoint might also be made tangent by choosing the TANGENT option again, Fig. 31-10(h).

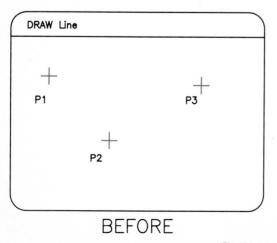

BEFORE

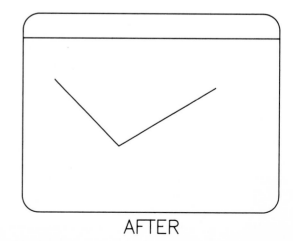

AFTER

(a)

Fig. 31-10. Drawing lines. (Continued)

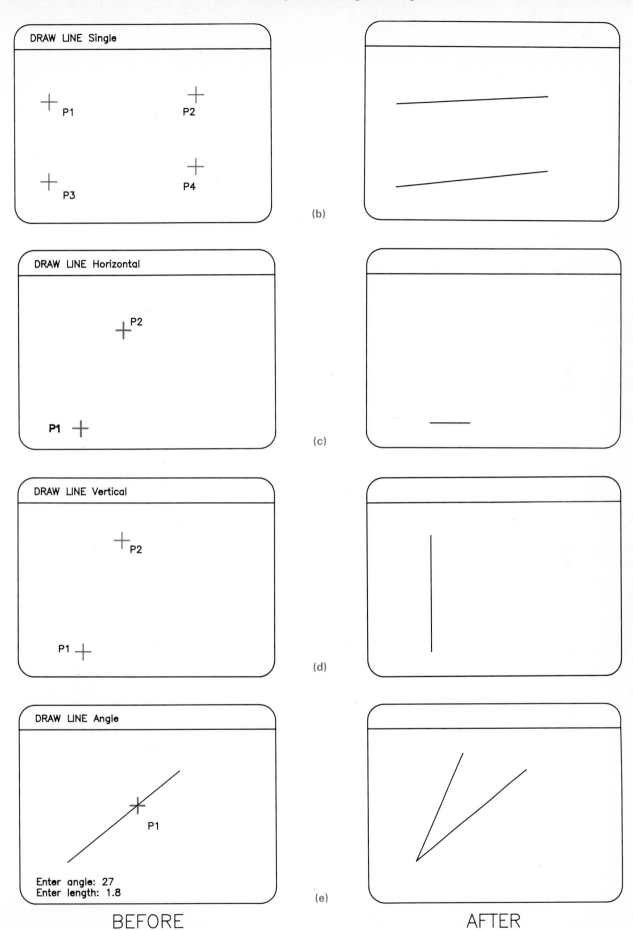

DRAW LINE Single

P1   P2

P3   P4

(b)

DRAW LINE Horizontal

P2

P1

(c)

DRAW LINE Vertical

P2

P1

(d)

DRAW LINE Angle

P1

Enter angle: 27
Enter length: 1.8

(e)

BEFORE   AFTER

Fig. 31-10. (Continued) Drawing lines.

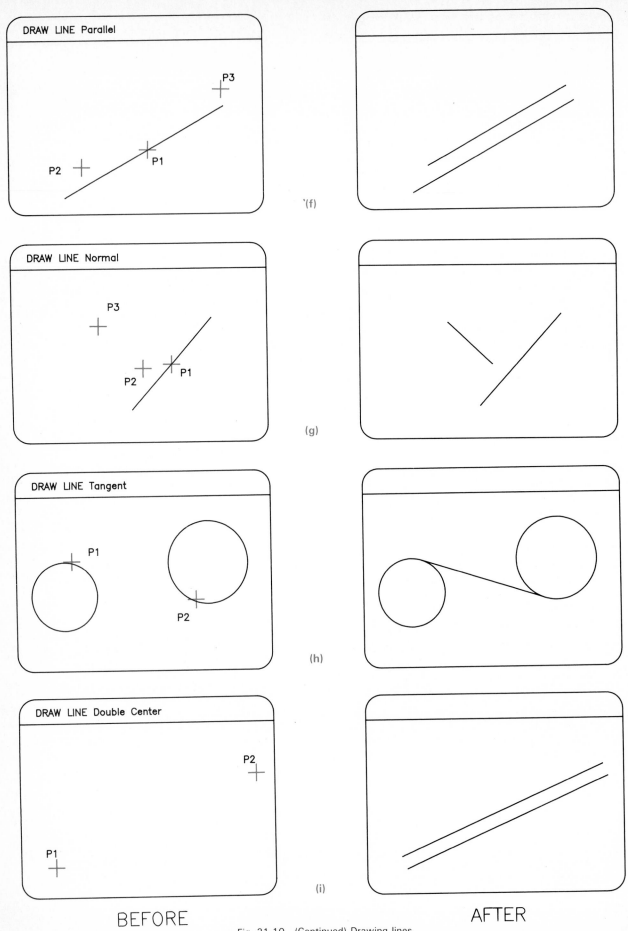

BEFORE   AFTER

Fig. 31-10. (Continued) Drawing lines.

Some systems offer the PARALLEL and NORMAL options with the TANGENT command. This allows you to place a line both tangent to a circle and parallel or perpendicular to an existing line.

DRAWING DOUBLE LINES

Double lines are parallel lines drawn at the same time. When entering or picking the endpoints, the double lines can be placed left, right, or center justified in relation to points you digitize. This command is especially useful in architectural drafting for drawing walls on the floor plan. See Fig. 31-10(i).

DRAWING POLYGONS

Any polygon can be drawn by enclosing a shape using several line segments. CAD systems offer additional specialized polygon drawing options for rectangles and regular polygons.

DRAWING A RECTANGLE

The RECTANGLE command allows you to digitize the opposite corners of a rectangle. This is much easier than drawing four individual line segments. As you locate the second corner, a temporary rectangle may appear to stretch out from the first corner to help you pick the size, Fig. 31-11.

DRAWING A REGULAR POLYGON

Regular polygons would be difficult to draw if it were not for the POLYGON command. This command generally offers three options with which to draw the polygon:
1. Center and vertex.
2. Center and edge.
3. Vertex and opposite vertex.

Before picking the two points, you must enter the number of sides for the polygon. The results of using the three above options to draw a hexagon (enter "6" sides) are shown in Fig. 31-12. As you locate the second point, the polygon may stretch out and rotate with the screen cursor to help you locate position.

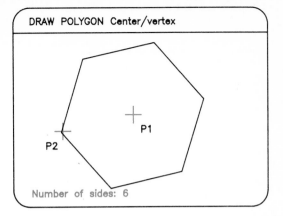

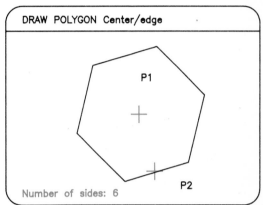

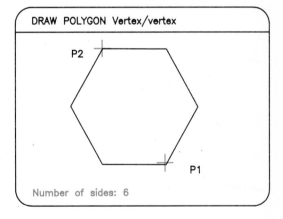

Fig. 31-12. Drawing a polygon.

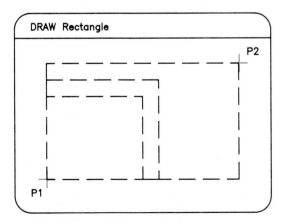

Fig. 31-11. After digitizing the first corner, a temporary rectangle "rubberbands" to allow you to pick the size.

DRAWING CIRCLES

A circle can be drawn by several methods. You should choose the appropriate method for the drawing situation. Selecting the proper commands, usually DRAW CIRCLE, brings up a list of options. Each of these options allow you to draw a circle using different methods of selecting circle attributes, such as center, radius, diameter, and circumference. See Fig. 31-13.

DRAWING A CIRCLE BY CENTER POINT AND RADIUS

After selecting the RADIUS or other appropriate option, type in the radius, and then pick or enter coordinates for the center location. Some systems allow you to pick the center point and then ''drag'' the circle to its size. In this manner, you can visually see the circle in place.

DRAWING A CIRCLE BY CENTER POINT AND DIAMETER

After selecting the DIAMETER or other appropriate option, type in the radius, and then pick or enter coordinates for the center location.

DRAWING A CIRCLE BY TWO POINTS

When drawing a circle by two points (2P option), you are actually picking two points along the circumference of the circle.

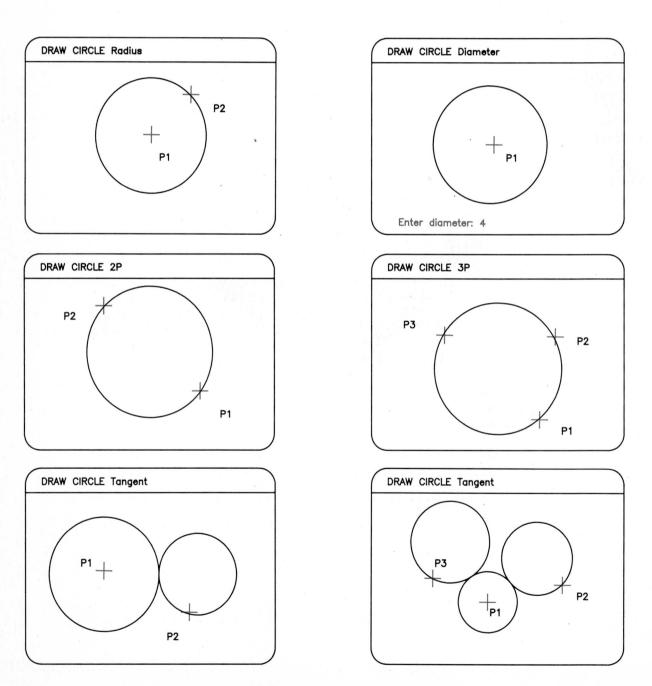

Fig. 31-13. A circle can be drawn using a variety of techniques.

DRAWING A CIRCLE THROUGH THREE POINTS

In certain situations, you must place a circle which touches three other entities. The 3P (three point) option allows you to pick three points. The circle is calculated and drawn to pass through each point.

DRAWING CIRCLES TANGENT TO OTHER ENTITIES

Just as you can draw a line tangent to a circle, you can also draw a circle tangent to a line, arc, or another circle. You can construct circles tangent to one item or to several items, depending on your system. Usually, you must pick the circle center and then the entity to which the circle should be tangent. Some systems allow you to pick two entities to which the circle should be tangent. The computer calculates the size of the circle.

DRAWING ARCS

Arcs are partial circles which can be defined by their center point, start point, endpoint, radius, and included angle. Selecting the proper commands, usually DRAW ARC, brings up a list of options. Each of these allow you to draw an arc using different aspects of an arc. Arcs are typically drawn counterclockwise; thus, pick your start and endpoints carefully. See Fig. 31-14 for an explanation of the options that follow.

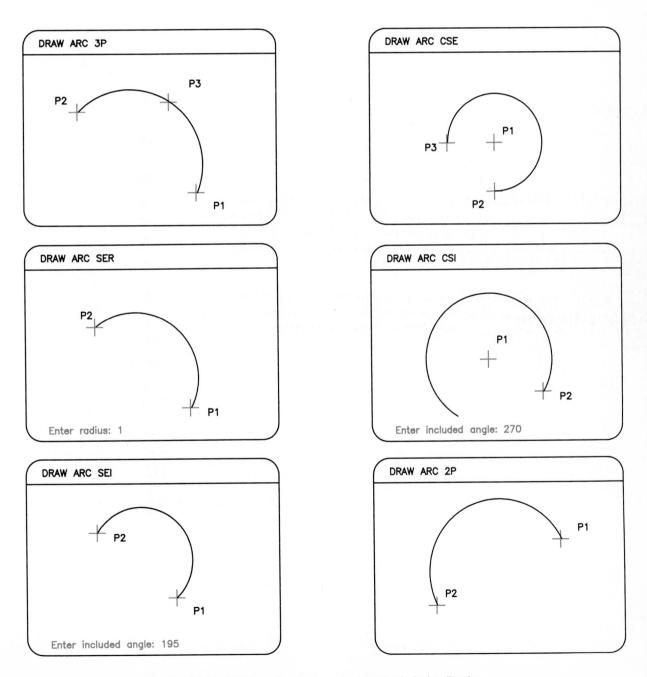

Fig. 31-14. Arcs are generally drawn in a counterclockwise direction.

THREE POINTS ON THE ARC
  Pick the start point, endpoint, and point along the arc.
CENTER, START POINT, ENDPOINT, (CSE)
  Pick the center, start point, and endpoint.
START POINT, ENDPOINT, AND RADIUS (SER)
  Type in the radius, then pick the start point and endpoint.
CENTER, START POINT, INCLUDED ANGLE (CSI)
  Type in the included angle, then pick the center point and start point.
START POINT, ENDPOINT, INCLUDED ANGLE (SEI)
  Type in the included angle. Pick start and endpoint.
TWO POINTS
  Pick the two points indicating the diameter of the arc. This results in a 180 degree arc, or semicircle.

DRAWING SPLINES

  Splines are smooth curves which pass through a series of points. The SPLINE command will either make a smooth curve out of existing connected lines or request that you pick points to create the curve, Fig. 31-15.

DRAWING ELLIPSES

  Ellipses can be drawn by several methods. The first is to locate the two axes by selecting the endpoints of one axis and one endpoint of the other axis. The second is to place the ellipse's center, then give one endpoint of each axis. A third is to pick the ellipse major axis endpoints and then enter a viewing angle. These options are all shown in Fig. 31-16.

SELECTING LINETYPE, LINEWIDTH, AND COLOR

  To comply with standard line conventions, CAD systems allow you to select the linetype and linewidth. This might be done by selecting LINETYPE and then entering HIDDEN. On some systems, the available linetypes and linewidths are shown on the screen. Simply move the cursor over the desired linetype or linewidth and pick it.
  CAD systems with color display screens allow you to assign colors to entities. For example, all hidden lines could be drawn

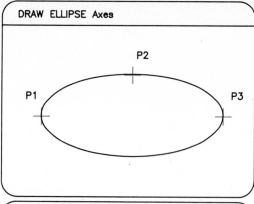

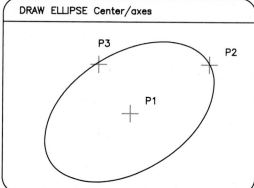

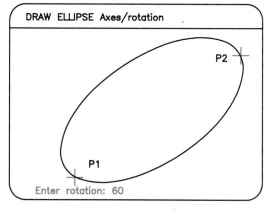

Fig. 31-16. An ellipse can be drawn by identifying the major and minor axes, the center point and an endpoint on each of the axes, or endpoints of the major axis and a viewing angle.

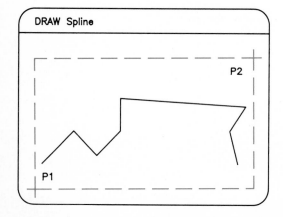

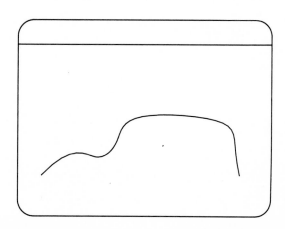

Fig. 31-15. Creating a spline from connected lines.

red, visible lines white, and dimensions blue. Separating items by color helps you to identify different parts of the drawing.

DRAWING AIDS

Drawing aids are helpful functions which allow you to locate position on the screen. The three common types are discussed here: grids, object snap, and construction lines.

GRID

A grid is a pattern of dots on the screen used much like graph paper in traditional drafting, Fig. 31-17. When adding entities, use the grid as a reference to help you locate position. The grid is not a permanent part of the drawing. Even if the grid is displayed during plotting, the grid dots are not plotted.

It is often difficult to position the cursor precisely over a grid point. The screen resolution may be too low to clearly see the dots or the cursor may float or jump. To help you, CAD systems offer snap. Grid snap locks the drawing cursor to grid dots as it moves across the screen.

You can set the X-axis and Y-axis spacing of the grid dots at any value. Set the spacing to the smallest entity size you will draw. They can be reset any number of times to meet your needs.

Grids do not have to be displayed. The grid can be turned on when needed and off when it hinders viewing the drawing. Even when not displayed, grid snap can be in effect.

Several CAD programs allow you to rotate the grid. This is helpful when drawing an auxiliary or other rotated view. Enter the angle of the grid as needed.

OBJECT SNAP

Your ability to precisely pick point locations depends on the resolution of the display and the accuracy of the input device.

However, suppose you want to locate the endpoint of a line at the corner of a rectangle. This might be difficult to do. What happens when you need to locate an endpoint at the center point of a circle? How do you find the center point? This is done with object snap.

Object snap allows you to precisely locate position on an existing entity. After entering a drawing command, select one of the object snap options. They might be found alongside the screen or in a menu. Available object snap options, shown in Fig. 31-18, allow you to place points at:

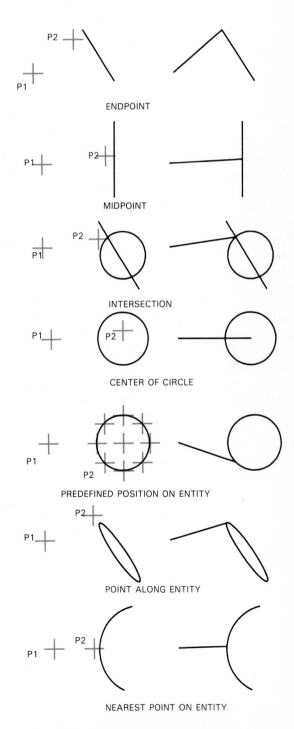

Fig. 31-18. Object snap.

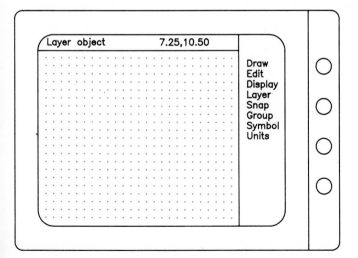

Fig. 31-17. Grids are similar to graph paper in traditional drafting.

1. The endpoint of a line, curve, or arc.
2. The midpoint of a line.
3. The intersection of two entities.
4. The center of a circle or arc.
5. Predefined positions on an entity.
6. Any point along an entity.
7. The nearest point on an entity.

CONSTRUCTION LINES

Construction lines are temporary lines placed for reference; they are not a permanent part of the drawing. Construction lines can be drawn horizontally, vertically, or at an angle. Simply enter the angle and digitize the base point, Fig. 31-19. This is the pivot around which the construction line is rotated. After placing a construction line, you can "snap" to it using any of the object snap options.

EDITING A DRAWING

Editing commands allow you to modify, erase, and manipulate previously drawn entities. Editing is the primary function of the drafter. In fact, many complex drawings are made by manipulating just a few drawn objects. CAD systems provide a variety of editing functions to suit every situation. You should be able to make the desired change using one of the commands found in this section.

When editing, you will be required to select objects to edit. You might pick an individual item, pick several items, or group objects to edit by picking corners of a window around them. The way you select items depends on the command. Usually a message will appear at the bottom of the screen telling you how to select items to perform the editing function.

ERASING AND UNERASING ENTITIES

The ERASE or DELETE command removes an entity or group of objects from the drawing. You might select entities individually, by digitizing corners of a window around them, or by entering the LAST option to erase the last drawn item. See Fig. 31-20.

CAD systems are forgiving. If you accidentally erase the wrong object(s), select the UNERASE command, often called OOPS, to restore the last erased object or a number of erased objects.

MOVING ENTITIES

Moving objects requires that you first select the object to be moved, pick a reference point, and then pick the new location. When you pick a single item, it may "drag" along with

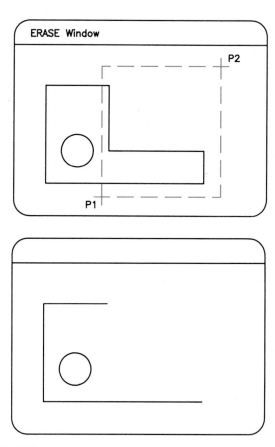

Fig. 31-20. A window may be used to indicate the entities of a drawing that are to be deleted.

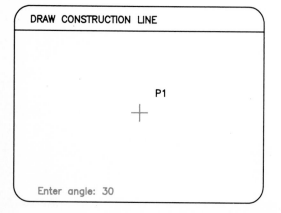

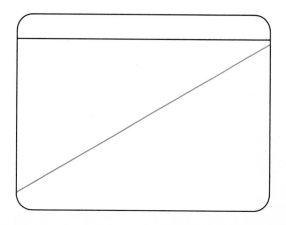

Fig. 31-19. Construction lines are temporary reference lines.

the cursor to the new position, Fig. 31-21.

A suboption of the MOVE command allows you to move a vertex of a polygon. Lines which start or end at that point are shortened or lengthened, Fig 31-22.

COPYING ENTITIES

Copying is much like moving entities, except that the original entities remain unchanged and the copy is positioned in place, Fig. 31-23.

The copy command also includes array functions. An array is a pattern of copies, placed in a rectangular or circular design. When making a rectangular array, you must select the object(s), enter the number of copies in the X and Y directions, and enter the distance between copies, Fig. 31-24. When making a circular array, you must select the object(s), digitize a pivot point, enter the number of copies, and enter the angle between copies, Fig. 31-25.

MIRRORING ENTITIES

Mirroring creates a reflected image of one or several entities on the other side of a mirror line, Fig. 31-26. A suboption allows you to delete the original entities after mirroring them.

The mirroring function is especially helpful for symmetrical objects. Draw one half and then mirror it to create the entire object.

ROTATING ENTITIES

Rotating allows you to revolve an entity or group of entities around a pivot point. Select the object(s) to be rotated, enter a rotation angle, and pick a pivot point. The objects are rotated about the pivot point at the specified angle, Fig. 31-27.

TRIMMING AN ENTITY

The TRIM command is used to shorten a line, curve, or other entity to its intersection with an existing entity. You must digitize the reference (boundary) object and the entity to be trimmed, Fig. 31-28. Check the user's manual for the digitizing order.

EXTENDING ENTITIES

The EXTEND command is the opposite of TRIM. It allows you to lengthen an entity to meet with another reference entity, Fig. 31-29.

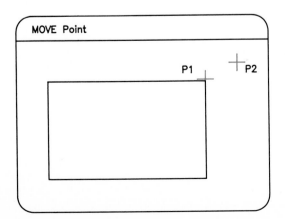

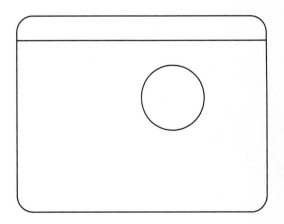

Fig. 31-21. Moving an object by "dragging" it to the new position.

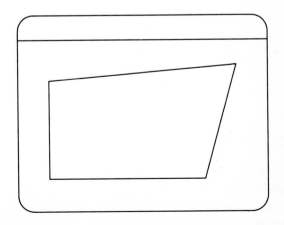

Fig. 31-22. Moving the vertex of a polygon affects all lines which start or end at that point.

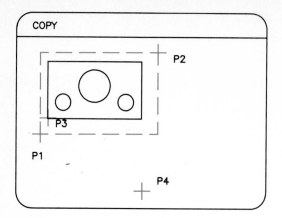
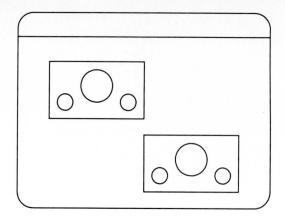

Fig. 31-23. The COPY command.

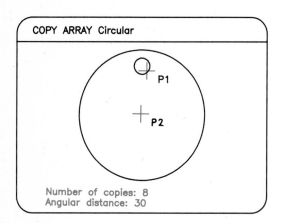
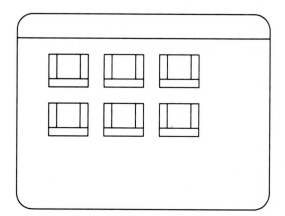

Fig. 31-24. A rectangular array created from a single object.

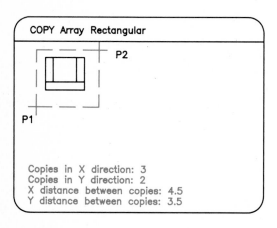
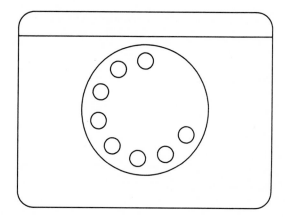

Fig. 31-25. A circular array.

STRETCHING AN OBJECT

The STRETCH command allows you to move a selected portion of the drawing while retaining all connections between entities. The portion of the object to stretch is chosen by picking corners of a window around it. Then pick a reference point and new position for the reference point, Fig. 31-30.

SCALING OBJECTS

The SCALE command allows you to enlarge or reduce the size of an entity or group of entities. After selecting items to scale, select a magnification factor, Fig. 31-31. Some systems allow you to scale objects unproportionately (different X and Y direction scaling factors).

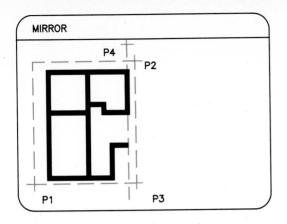

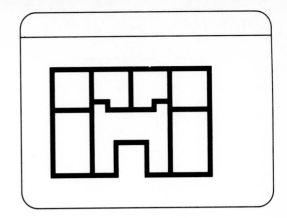

Fig. 31-26. The MIRROR command is commonly used to create symmetrical objects.

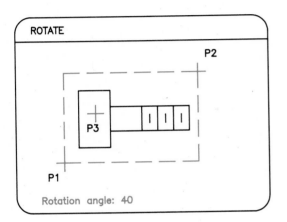

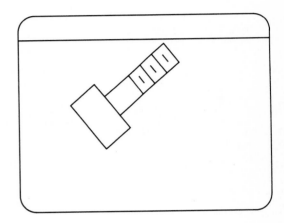

Fig. 31-27. The ROTATE command.

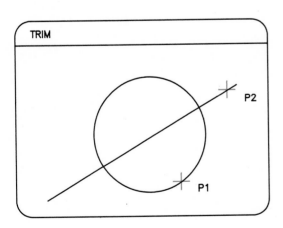

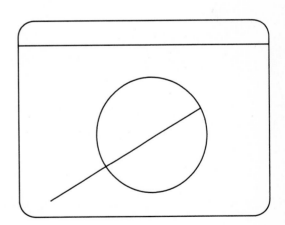

Fig. 31-28. The TRIM command shortens an entity to an intersection with another entity.

JOINING ENTITIES

The JOIN command allows you to combine two entities. It may be easier to edit or change entities joined together rather than edit them separately. To do so, the two items must end at a common meeting point. (This can also be done by trimming each.)

BREAKING AN ENTITY

The BREAK command allows you to remove a section of a line, circle, or complex object. This can be done by selecting two points on the entity, breaking a fenced portion, or selecting two objects, Fig. 31-32. When breaking with a fence, all portions of entities which are within the fence are

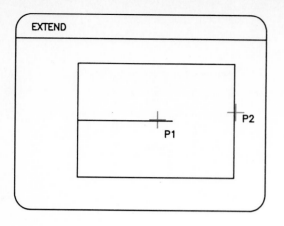

Fig. 31-29. The EXTEND command allows you to lengthen a line to meet another entity.

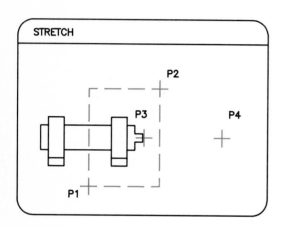

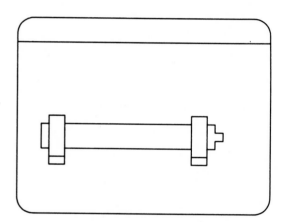

Fig. 31-30. The STRETCH command.

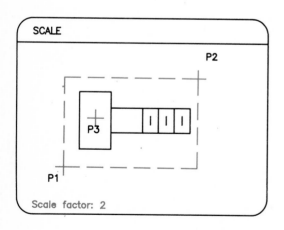

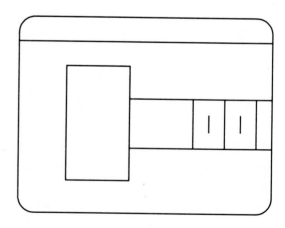

Fig. 31-31. Reductions and enlargements of an entity or group of entities is performed with the SCALE command.

removed. Breaking by object is a method of breaking a portion of two objects where they intersect.

EXPLODING AN ENTITY

The EXPLODE command breaks an entity into its component parts. A polygon, dimension, or symbol is converted into its component lines. Each item can then be edited individually. The exploded entity will not change in appearance.

CREATING FILLETS

The FILLET command creates fillets and rounds of a specified radius. Although there are several different options

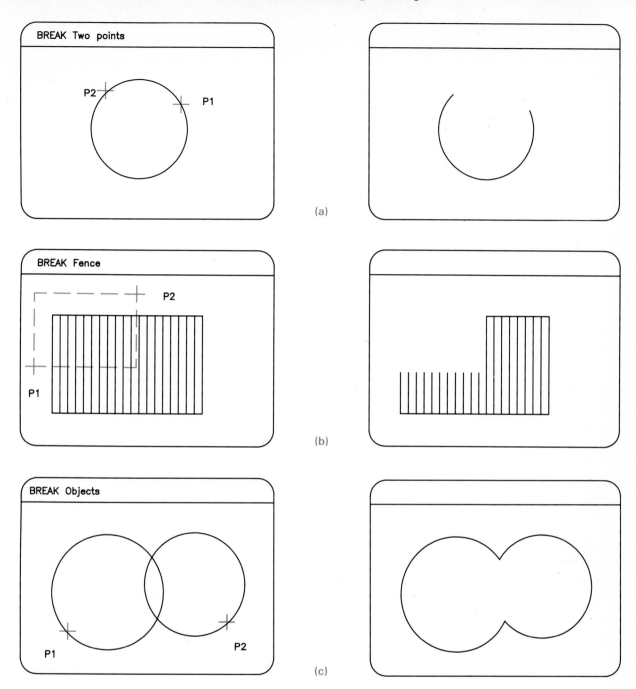

Fig. 31-32. The BREAK command. A—Breaking by two points. B—Breaking a fenced portion. C—Breaking by selecting two objects.

offered by the fillet command, the simplest technique is to pick the two lines to be filleted. The fillet made is an arc tangent to both entities, Fig. 31-33. Most systems trim the entities to the fillet arc. The two entities do not have to intersect. The fillet command will extend lines to meet the fillet arc.

CREATING CHAMFERS

The CHAMFER command connects two lines with a chamfer. The chamfer is placed equidistant from the inter-section of the lines unless you enter different chamfer distances, Fig. 31-34.

CHANGING ENTITY PROPERTIES

Properties refer to linetype, linewidth, layer, and color of lines, circles, arcs, and other entities. You are able to change these properties using editing functions, usually found under EDIT PROPERTIES. Simply enter the property to change, select the entities to change, and enter a new value, Fig. 31-35.

DISPLAY CONTROLS

Display controls are commands which determine how much and what parts of your drawing are shown on the display

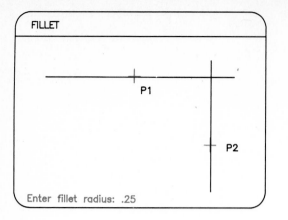

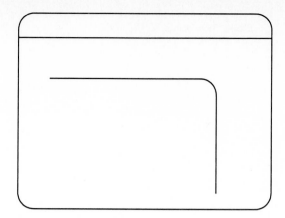

Fig. 31-33. Fillets and rounds are created using the FILLET command.

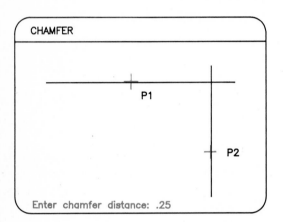

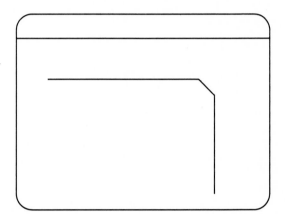

Fig. 31-34. The CHAMFER command.

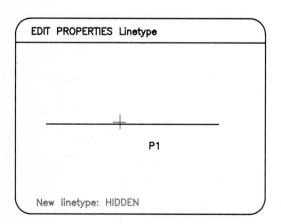

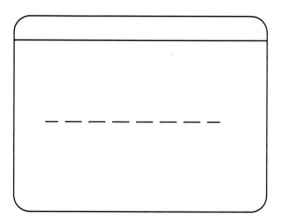

Fig. 31-35. Entity properties, such as linetype, can be changed using editing functions.

screen. Remember that your drawing is held as data in computer memory. Therefore, the CAD program can manipulate the data to show you the entire drawing or only a part of it.

ZOOM COMMANDS

ZOOM options determine how much of the drawing you see on the display screen. They do not change the actual size

of the drawing. All object sizes and shapes stay the same size. They are simply magnified or reduced in size temporarily for better viewing of the drawing. Fig. 31-36 illustrates two of the ZOOM options.

ZOOM WINDOW

This option magnifies a portion of the drawing within a

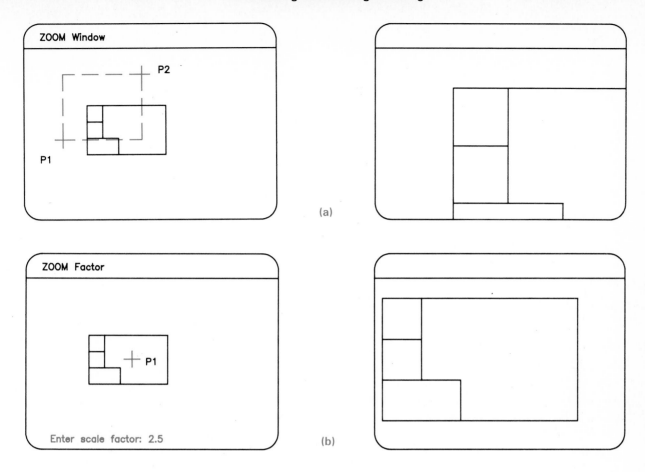

Fig. 31-36. The ZOOM command. A—Zooming in on an area within a window. B—Enlarging an object using a specified magnifying factor.

window you specify. The windowed section enlarges to fill the screen.

ZOOM OUT

This option reduces the current view to fit within a window you specify. The rest of the drawing is redrawn on the screen accordingly.

ZOOM FACTOR

This option allows you to enlarge or reduce the drawing by a magnifying factor. You must also select a center to become the center of the magnified view. Entering a magnifying factor greater than one makes the drawing appear larger. A value less than one reduces the size so that more of the drawing fits on the display.

ZOOM BASE

This option returns the displayed view to the full drawing area. No matter how large your drawing area is (even 1,000,000 ft. wide), it is displayed in its entirety.

ZOOM FULL

This option calculates how big the drawing should be to just fit on the display. It is different than the ZOOM BASE in that it does not show the drawing area defined by the boundaries. Instead, it calculates the smallest screen window which contains your entire drawing.

PANNING

When a portion of the drawing is magnified, it might be necessary to see an object which is "just off" the screen. This is done by "panning" across the enlarged view. Pick two points. The view then moves the distance between the two points in the direction of the second point, Fig. 31-37.

Panning is accomplished on some systems by specifying which direction to pan. For example, you might choose PAN LEFT, PAN RIGHT, PAN UP, or PAN DOWN. The windowed section of the drawing then moves that direction.

DISPLAY LAST

The DISPLAY LAST command sequence returns the screen to the previous view of the drawing. You might use this to return to a reduced view after having "zoomed in" to do detail work.

SAVE AND RECALL VIEW

Using DISPLAY SAVE and DISPLAY RECALL functions, you can save and recall current views of the drawing for future

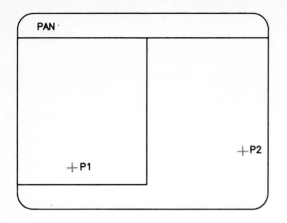

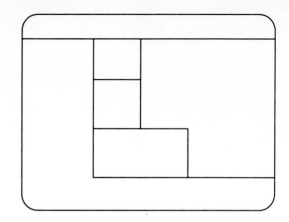

Fig. 31-37. Panning across a drawing to see a part of an object which is off the screen.

reference. Save commonly used views so that it is easier to return to these portions of the drawing for more work.

REDRAW

When editing, holes from deleted entities may be left. The REDRAW renews the view of the drawing, clearing off the current display and then redrawing the same view. This "cleans up" the drawing.

DRAWING PROBLEMS

Using standard drafting practices and a CAD system, create a drawing for each of the problems shown in Fig. 31-38. The drawing name should be P31-(problem number). For example, name Problem 1 as P31-1. Set up the drawing using decimal inches as the unit of measurement. Enter a drawing area or select a sheet size large enough to contain the drawing. Set grids and use other drawing aids to your advantage. Do not add dimensions or text to the drawing at this time.

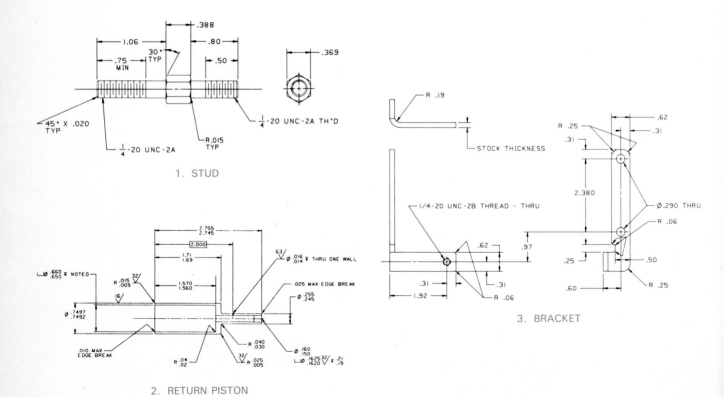

Fig. 31-38. Computer-aided drafting problems. (Continued)

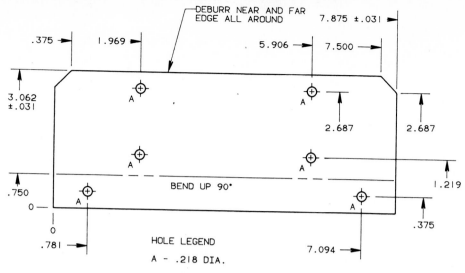

DEBURR NEAR AND FAR
EDGE ALL AROUND

7.875 ±.031

.375
1.969
5.906
7.500

3.062
±.031

2.687

2.687

A

A

BEND UP 90°

1.219

.750

A

A

.375

0

0

HOLE LEGEND

.781

7.094

A – .218 DIA.

4. BATTERY BRACKET

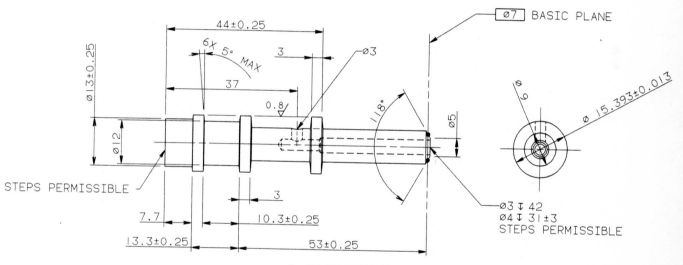

44±0.25

Ø7 BASIC PLANE

Ø13±0.25

6X 5° MAX

3

Ø3

37

0.8

Ø12

118°

Ø5

Ø 15.393±0.013

STEPS PERMISSIBLE

3

Ø3 ↧ 42
Ø4 ↧ 31±3
STEPS PERMISSIBLE

7.7

10.3±0.25

13.3±0.25

53±0.25

5. SHUTTLE VALVE

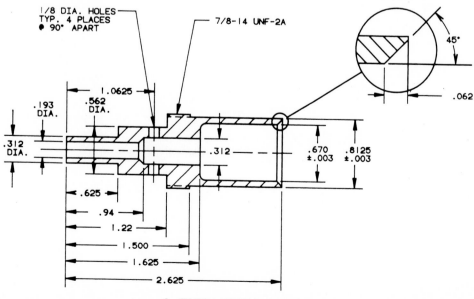

1/8 DIA. HOLES
TYP. 4 PLACES
90° APART

7/8-14 UNF-2A

45°

1.0625

.562
DIA.

.062

.193
DIA.

.312
DIA.

.312

.670
±.003

.8125
±.003

.625

.94

1.22

1.500

1.625

2.625

6. TORCH FITTING ADAPTER

Fig. 31-38. (Continued) Computer-aided drafting problems.

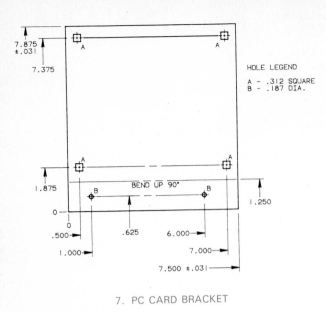

7. PC CARD BRACKET

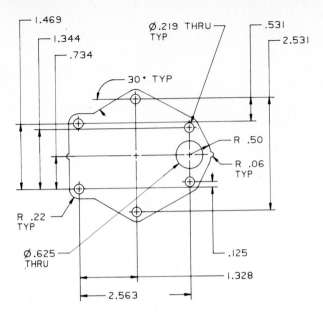

8. DIAPHRAGM

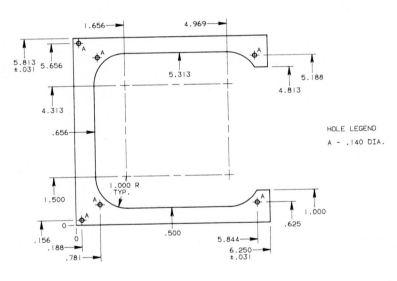

9. COVER BRACKET

Fig. 31-38. (Continued) Computer-aided drafting problems.

Chapter 32

DETAILING AND ADVANCED CAD FUNCTIONS

This chapter covers CAD functions which reduce many of the tedious tasks found in traditional drafting. Since a great amount of time typically is spent detailing engineering drawings, CAD systems automate much of this process. Automatic functions for dimensioning, adding section lines, and placing text decrease the amount of drawing time required. In addition, this chapter introduces some advanced functions, such as layering, using symbol libraries, and inquiry functions.

LAYERS

Layering commands allow you to separate different parts of the drawing into levels, much like using several layers of film in overlay drafting. This is an important concept to grasp before you begin adding text and dimensions to a drawing.

Traditionally, information is separated on different sheets of vellum or film. For example, in architectural drafting, the floor plan is drawn on one sheet, the foundation plan on another, and the plumbing plan on still another. This is done with CAD by specifying different layers, or levels. Each layer can be thought of as a plastic overlay. Related details of a drawing are drawn on the same layer while unrelated details are placed on separate layers. This allows the drafter to organize information such as object views, dimensions, text, and specifications, Fig. 32-1.

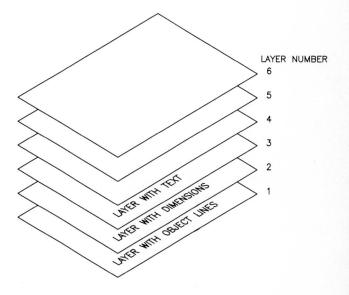

Fig. 32-1. Layers allow you to efficiently organize information about a drawing.

ACTIVE LAYER
When drawing, you are adding entities to only one layer. This is called the active or current layer. For example, suppose you are currently adding object lines but plan to now add dimensions. You may wish to choose a new active layer on which to draw the dimensions.

DISPLAYING LAYERS
All or only selected layers can be displayed. For example, when drawing a plumbing plan, you might show the floor plan, but ''hide'' layers containing electrical, foundation, and other plans. Entities on layers turned off (not displayed) are not deleted; they are just invisible. They appear again when you turn the layer back on.

LAYER NUMBERS AND NAMES
When specifying an active layer, or layers to display, you most often enter numbers. However, several systems allow

you to assign names. For example, the layer containing dimensions might be layer 10, or be named DIMEN. Names make it easier for you to identify layers.

ERASING LAYERS
Erasing entities on a layer should be done with caution. The function, often named LAYER ERASE, deletes all entities on the specified layer. Once erased, they generally cannot be brought back.

CHANGING THE LAYER OF ENTITIES
You may want to change the layer associated with an entity. For example, suppose you accidentally added a dimension to the layer containing only object views. The LAYER CHANGE function allows you to change the layer associated with the dimension to the proper value.

LAYER COLOR AND LINETYPE
Linetypes and colors may be assigned to layers. Thus, any entity drawn on that layer automatically assumes the default color and linetype. Most systems allow you to individually change the entity color and linetype if necessary.

ADDING TEXT

Text commands allow the drafter to place labels, specifications, notes, and other textual items on the drawing. There are many benefits to adding text with a CAD system compared to manual lettering. With CAD, you simply specify the text style (features), type in the text string (line of text), and digitize position on the drawing. In addition, computer-generated text can be standardized. Companies often determine what style of text is to be used. All drawings then follow those guidelines. This reduces the chance of drafters choosing their own style of lettering and increases drawing readability.

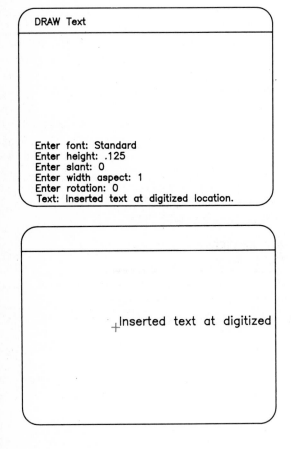

Fig. 32-2. Placing text on a drawing.

The steps taken to place text on a drawing, as shown in Fig. 32-2, are much the same among CAD systems. Although the exact command names may differ slightly, the procedure usually is:

1. Select the DRAW TEXT command.
2. Specify the various text features. Options include text font, height, slant, width, and rotation.
3. Type in the text to be placed on the drawing.
4. Select the text justification. This determines the position of the text related to the location point you digitize.
5. Digitize location on the drawing.

SELECTING TEXT FEATURES

Text features affect the appearance of the lettering on the drawing. As shown in Fig. 32-3, they include the following alternatives.

FONT

The font refers to the style of the text. It may look like the characters printed in this text, consist of simple line strokes, or be very fancy. Bold fonts are used for titles and section labels. Fancy fonts are used by graphic artists or for business graphics and presentations. For drafting, choose a font that is easy to read, usually composed of simple line strokes.

HEIGHT

The height is the distance from the bottom to top of a text character. It is usually based on the height of capital letters. Lower case letters will be smaller than the set height. Choose a height compatible with school or company standards.

WIDTH

Changing the text height automatically adjusts the width proportionately. However, you might also set the width. A larger width value makes the text look expanded; a smaller width value makes the text look condensed. Some systems use an aspect value. This is a ratio of width to height. For example, entering an aspect value of "2" means width of a text character will be twice its height.

SPACING

The spacing is the distance from the bottom of one line to the bottom of the next. It is generally just more than the text

TEXT FEATURES

FONT	HEIGHT	WIDTH	SPACING	SLANT	ROTATION
Font 1 Font 2 𝕱𝖔𝖓𝖙 3	1/16" text 1/8" text 1/4" text	Condensed text Expanded text	Variable line spacing between rows of text strings.	0 degree slant 15 degree slant 30 degree slant -30 degree slant	No rotation 15 degree rotation 30 degree rotation

Fig. 32-3. Text features.

height. A spacing option is common on systems which allow you to type more than one text string (line of text) at a time.

SLANT

The slant is the angle of each individual character in the text string. You can make text look italic by using a 15 degree slant angle. A zero (0) degree slant angle makes text characters vertical.

ROTATION

Rotation is the angle for entire text strings. Rotation is measured from horizontal in a counterclockwise manner. If you enter a negative rotation angle, the text string is rotated clockwise.

Text features are usually set to default values when you start the CAD system. Default values are parameters built into the software which the computer assigns until you change them. For example, the default text height may be .125 in. You may not wish to use the default values, and will set them to school or company standards during drawing setup. If you change the text style, text previously entered is not affected by the newly set features.

SELECTING JUSTIFICATION

Justification determines how the text string will be placed in relation to the location point you digitize. You may be required to select a justification before or after typing in the text. Justification options, illustrated in Fig. 32-4, include:

LEFT JUSTIFIED

The left side of the text string is placed at the digitized location point.

RIGHT JUSTIFIED

The right side of the text string is placed at the digitized location point.

CENTER JUSTIFIED

The text string is evenly spaced on both the left and right sides of the digitized location point.

MIDDLE JUSTIFIED

The text is spaced vertically and horizontally around the digitized location point.

VERTICAL TEXT

The text string extends vertically downward from the digitized location point. This is not rotation because all of the text characters remain horizontal.

ALIGNED TEXT

The aligned justification allows you to position the text string by digitizing two base points. The text is placed at the angle between the points, left justified from the first point selected.

TYPING IN TEXT

Most systems have you enter individual text strings, one at a time. The line of text, or text string, is considered a single unit. Thus, if you edit the text string by moving, copying, or deleting, you affect the entire string. Systems which allow you to enter multiple lines consider each line, not the entire paragraph, as a text string.

SELECTING POSITION

Once you have selected features, justification, and have entered the text string, digitize the location on the drawing.

EDITING TEXT

Standard editing commands—MOVE, COPY, ROTATE, etc.—affect a text string just as they do graphic entities. However, to revise the text, you must select a special EDIT TEXT command. Digitize the text string to edit, and make the necessary changes to the text string or its features.

DIMENSIONING

Dimensioning commands let you automatically place dimension measurements on the drawing. In traditional drafting, dimension lines are laid out by hand and measurements are made using a scale. With CAD, the drafter needs only to digitize several points. The computer then adds the proper dimension lines, extension lines, leaders, and also calculates

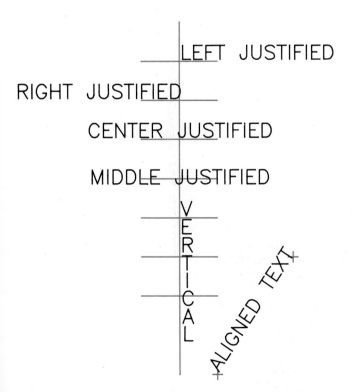

Fig. 32-4. Text justification determines how a text string is placed in relation to the insertion point.

and places the measurement. The drafter may have to accept or reject the dimension, a function which allows him/her the chance to make changes before the dimension is placed permanently.

Dimensions typically are placed on a separate layer using a different color. This distinguishes them from object lines. Set the active layer and color before dimensioning.

There are four basic methods used to dimension geometric shapes. These include linear, angular, radial (diameter and radius), and leader dimensions. Remember that drafting standards apply equally to computer-generated dimensions as they do to manual drafting. Standards are set by altering the dimensioning parameters.

DIMENSIONING PARAMETERS

When placing dimensions, there are a number of dimensioning parameters available. These are set before adding dimensions to the drawing. Common parameters, as illustrated in Fig. 32-5, include the following.

BREAK

The BREAK option determines whether the measurement is placed above or within a break in the dimension line.

LEADING ZERO

The ZERO option determines whether a zero precedes measurements less than one.

UNIDIRECTIONAL/ALIGNED SYSTEM

The ALIGN option determines whether the measurement remains horizontal or aligns with the direction of the dimension line.

TERMINATOR

The ARROW or TERMINATING SYMBOL option sets whether an arrow, tick mark, or dot is used at the intersection of dimension and extension lines.

EXTENSION LINE OFFSET

The EXTENSION LINE OFFSET option determines at what distance the extension line begins from the object view.

TOLERANCES

The TOLERANCE option allows you to automatically add plus/minus or limit tolerances.

Some systems include as many as 30 different variables you might set. These might include parameters for dual dimensioning, text size, scale factor, and arrow size. Check your user's manual to research these options.

LINEAR DIMENSIONING

Linear dimensioning commands note straight distances. They include measuring horizontal and vertical distance, and distance along an entity (called an aligned dimension). To place the dimension, digitize two points or select a line. The system

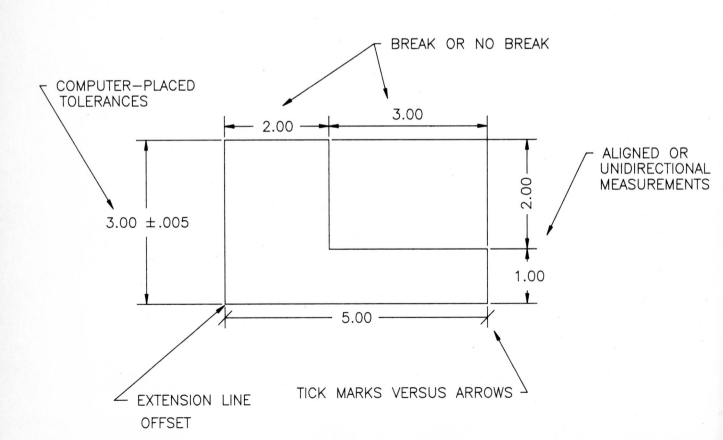

Fig. 32-5. Dimensioning parameters.

will prompt you for the placement of the dimension line. The dimension line, extension lines, and measurement is then placed automatically. This process is shown in Fig. 32-6. The system may allow you to edit the dimension text before entering it on the drawing.

Most systems automatically place the measurement outside the extension lines if the dimension will not fit. However, some systems have you digitize the third point (to indicate the dimension line) outside of the dimension area. The measurement will be placed outside the actual measured dimension and the arrows will point in, Fig. 32-7.

Linear dimensioning also includes chained (continued) and baseline (datum) dimensioning functions. These require that

you digitize several locations to measure. Then digitize the position of the first dimension line. The appropriate dimensions are added automatically as shown in Fig. 32-8.

ANGULAR DIMENSIONING

Angular dimensions measure the angle between two non-parallel lines. Two methods are used by CAD systems for dimensioning angular dimensions: between entities and three point.

ANGULAR DIMENSIONING BETWEEN ENTITIES

With this method, select the two lines which form the

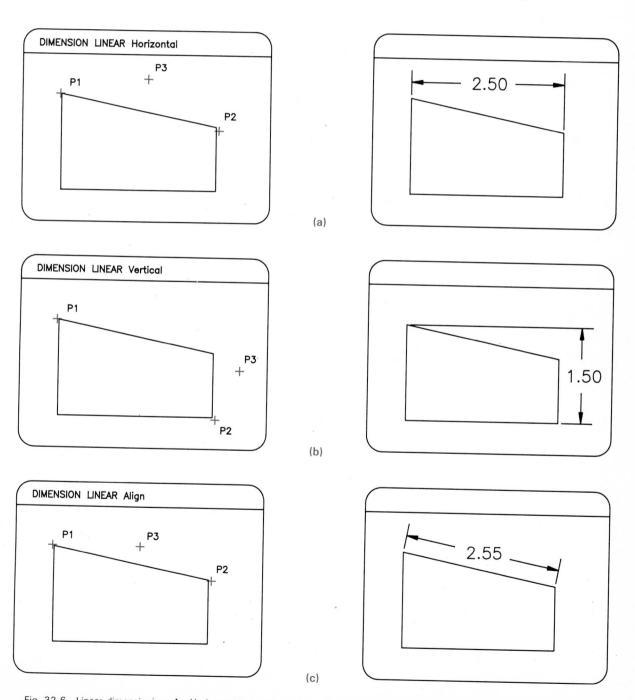

Fig. 32-6. Linear dimensioning. A—Horizontal linear dimensions. B—Vertical linear dimensions. C—Aligned linear dimensions.

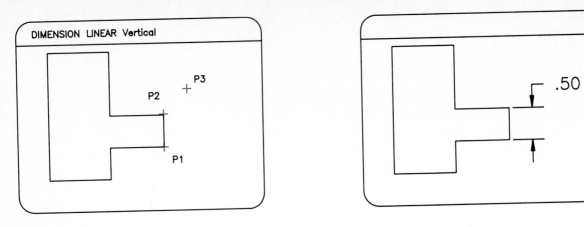

Fig. 32-7. Placing dimensions which are too large to fit within the dimension area.

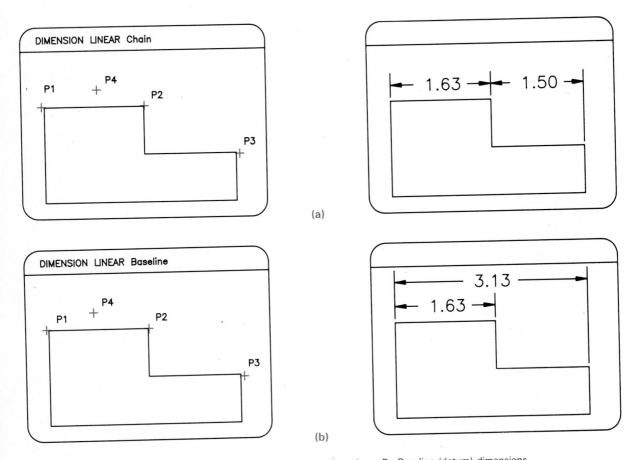

(a)

(b)

Fig. 32-8. Linear dimensioning. A—Chained dimensions. B—Baseline (datum) dimensions.

angle. Then digitize the location of the dimension line. See Fig. 32-9.

ANGULAR DIMENSIONING USING THREE POINTS

This method requires that you digitize four points, Fig. 32-10. The points are the vertex, two points which form the angle, and position of the dimension line. The vertex might be a corner, intersection of two lines, or some other point. It depends on what you are going to dimension. The two end-points might be the endpoints of two intersecting lines.

RADIAL DIMENSIONING

RADIAL DIMENSIONING includes diameter dimensioning and radius dimensioning. The steps are simple. Select the DIMENSION RADIUS or DIMENSION DIAMETER command, and digitize the circle or arc to dimension. The dimension line and measurement are placed automatically, Fig. 32-11.

Diameter and radius dimensioning can be done with or without a leader. Without a leader, the measurement is placed within the circle or arc. With a leader, the measurement

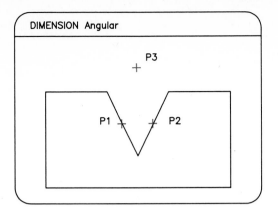

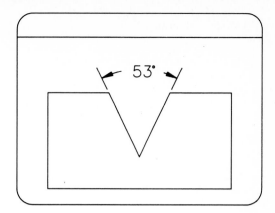

Fig. 32-9. Angular dimensioning between entities.

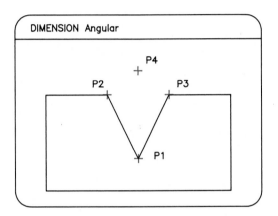

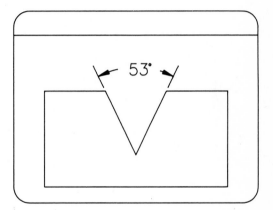

Fig. 32-10. Angular dimensioning using three points.

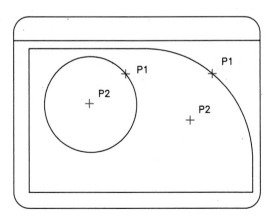

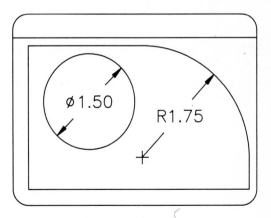

Fig. 32-11. Radial dimensioning includes diameter and radius dimensioning.

is placed outside the feature at the end of a leader pointing toward the feature. Placement of the leader is done one of two ways.

With semiautomatic dimensioning, you digitize the measurement either inside or outside the circle or arc. If you digitize inside the curve, the measurement is placed inside. If you digitize outside the feature, a leader is used. Then you must digitize the location of the measurement. See Fig. 32-12.

CAD systems which use automatic dimensioning determine whether or not to use a leader. They check whether the diameter measurement will fit in the circle. If so, the measurement is centered within a dimension line inside the circle and the dimension is complete. If the measurement will not fit, the system asks for the placement length of the leader outside the circle or arc.

Many systems add the proper symbol, diameter (ϕ) or radius

Drafting for Industry

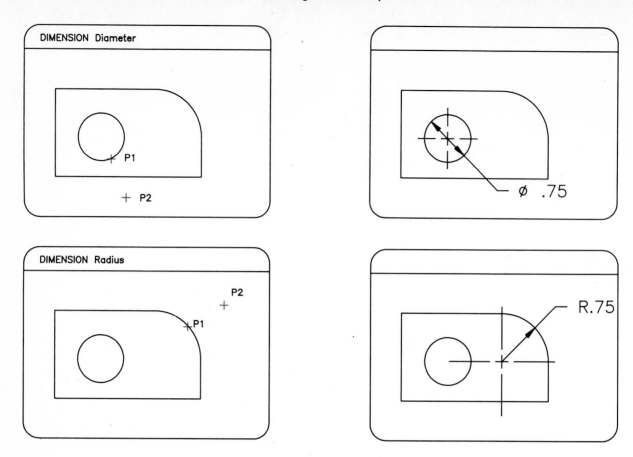

Fig. 32-12. Semiautomatic dimensioning.

(R), preceding the measurement. If not, you must add the feature manually. This can be done when the system allows you to edit the measurement before it is placed with the dimension.

LEADERS

Leaders automatically added when dimensioning arcs and circles have been mentioned. A majority of CAD systems also offer a separate command to add leaders. After entering the

DIMENSION LEADER command, digitize two points. The first marks the feature to which the leader applies. The second marks the location of the note. Then type in the text. The leader, arrow, and note are then added to the drawing. See Fig. 32-13.

COMPUTER-PLACED TOLERANCES

Tolerance values typically can be placed automatically with the dimension measurement. You must specify whether

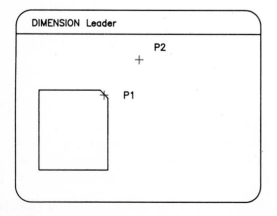

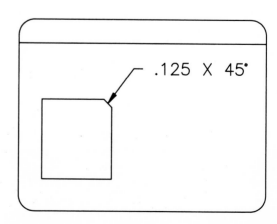

Fig. 32-13. Placing leaders on a drawing. Point P1 indicates the feature to which the leader is to be applied. Point P2 indicates the location of the note.

plus/minus or limit tolerances are used, and the tolerance values. A plus/minus tolerance is appended to the dimension. With a limit tolerance, the computer calculates and shows the upper and lower limits.

HATCHING SECTION VIEWS

One of the most tedious tasks in drafting is adding section lines on a section view. CAD makes drawing section views quicker by providing automatic hatching (section lining). The steps taken to hatch an area, after selecting the HATCH command, include selecting the pattern and digitizing the boundary.

SELECT THE HATCH PATTERN

A number of standard hatch patterns for various materials are included with most CAD systems. You might enter the pattern name, such as ANSI31 (American National Standards Institute 31), or select the pattern from a screen menu, Fig. 32-14. Systems with no preset hatch patterns request that you enter a user-defined pattern. Enter the linetype, angle, and spacing between section lines.

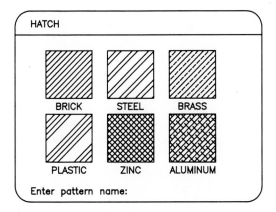

Fig. 32-14. Placing hatch patterns with CAD does not require the tedious, time-consuming procedure used with traditional equipment.

DEFINE A HATCH BOUNDARY

The hatch boundary is the lines, circles, and other entities which border the area to be hatched. Once you select these items, the hatch pattern fills the area within the boundary, Fig. 32-15. Hatching may take several seconds, especially if the pattern is complex or area is large.

Entities which form the boundary are chosen individually or by digitizing corners of a window around them. On some systems, the area must totally be enclosed. In addition, several CAD programs require that the entities forming the hatch boundary meet, but not extend beyond the boundary area. On these systems, hatching entities which do not form a perfectly enclosed boundary may cause unpredictable results. Edit entities which fall beyond the boundary area using BREAK or TRIM commands.

INQUIRY COMMANDS

Inquiry commands let you ask questions about your drawing, such as distance, perimeter, area, and drawing status. Unlike manual drafting, with CAD you cannot place a scale on the display screen to measure distance. This is done with a select group of commands.

ENTITY

The INQUIRE ENTITY function lists information about a single object. Digitize the entity in question. The returned list may include: length, diameter, radius, color, linetype, and location.

DISTANCE

To measure distance, select the INQUIRE DISTANCE function and digitize two points on the drawing. These may be at the endpoints of a line or any distance across the drawing area.

ANGLE

To measure the angle formed by two nonintersecting lines, select the INQUIRE ANGLE function and digitize the lines in question. The angle given measures the angle between the first line digitized and the second line digitized.

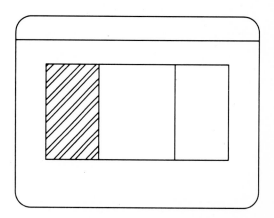

Fig. 32-15. The hatch pattern fills the area within the hatch boundary.

AREA

Area is the amount of surface enclosed by a circle, ellipse, polygon, or irregular shape composed of several entities. You must digitize entities which totally enclose the shape to be calculated. The INQUIRE AREA command will result in error message if an attempt is made to measure the area of a region not totally enclosed.

DRAWING STATUS

Selecting the INQUIRE STATUS function displays an assortment of information about the current drawing.
This might include:

1. Drawing boundaries.
2. Extent of boundaries used by drawing.
3. Portion of drawing shown on display.
4. Grid setting.
5. Current layer.
6. Layers displayed.
7. Current color.
8. Current linetype.
9. Amount of memory left (for microcomputers).
10. Amount of data storage space left (for microcomputers).
11. Elapsed drawing time.

CREATING AND USING SYMBOL LIBRARIES

Symbols are widely used in industry to represent standard parts or assemblies in diagrams, schematics, and drawings. With manual techniques, symbols are drawn by hand or by using a plastic template. This is not only time consuming, but the quality may vary and there may not be a template made for a company's unique symbols. With CAD, symbols are drawn once and saved. They can then be inserted as many times as needed on an infinite number of drawings. Some CAD programs come equipped with standard symbols for mechanical, architectural, and electrical drawings.

INSERTING A SYMBOL

Fig. 32-16 and the steps to follow are a typical procedure used to place a symbol on a drawing.

1. Select the active library that contains the symbol you need. A symbol library is a special directory on the data storage device where similar symbol types—mechanical, electrical, architectural, welding, etc.—are stored together. See Fig. 32-17. You can choose symbols from only the current library, called the active library.

2. Select the symbol you need. A prompt may ask for the symbol name or number, or a screen palette showing the available symbols could appear. In addition, with some CAD systems, you can pick symbols from the tablet menu.

3. Enter values for the symbol scale (size factor) or rotation angle. These options are not editing commands. They alter only the copy of the symbol being inserted. The functions allow you to resize, rotate, or mirror the symbol before it is added. The default position and size for the symbol stored on disk remain the same.

4. Digitize the symbol's location on the screen.

DRAWING AND STORING SYMBOLS

You are not limited to using symbols that come with your system. Creating company-specific symbols for commonly used parts and assemblies may be one of your main tasks. The steps taken to create and store symbols vary slightly among systems. Some CAD programs do not distinguish a drawing from a symbol. (Thus you can insert any existing drawing as a symbol into the current drawing.) However, most systems require that you use a special series of commands that save a group of entities in a symbol library. Fig. 32-18 and the steps to follow show the method to store a symbol.

1. Begin a new drawing and draw the symbol to size.
2. Select the LIBRARY SYMBOL ADD function.
3. Enter the name of the symbol library where the symbol is to be stored.
4. Enter the name or number for the symbol.
5. Select by digitizing the entities individually or by windowing the entities that make up the symbol.
6. Digitize an insertion point, which determines where the symbol is placed relative to the location point you pick when later inserting the symbol. The insertion point also becomes the reference point for rotating and scaling.

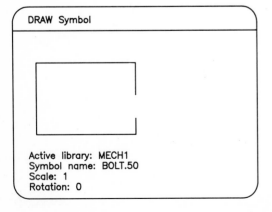

DRAW Symbol

Active library: MECH1
Symbol name: BOLT.50
Scale: 1
Rotation: 0

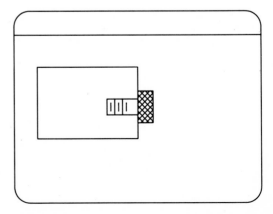

Fig. 32-16. Computer-generated symbols reduce the amount of drawing time required to produce a drawing.

Fig. 32-17. Symbol libraries. (SoftSource, Inc.)

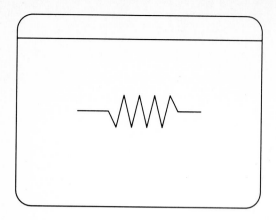

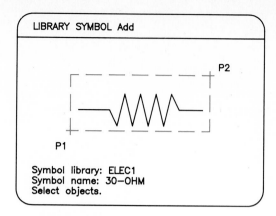

Fig. 32-18. Adding a symbol to the library.

This method has an advantage over systems which use entire drawings as symbols. Suppose that while making a drawing, you find that you need a certain shape many times. Rather than using the COPY command, follow the above procedure to save the shape as a symbol. Then continue on the drawing, inserting the shape wherever needed.

EDITING SYMBOLS

Symbols are considered to be a single item. Thus, if you select the ERASE command and pick the symbol, all entities which make up the symbol are erased. To edit the symbol, you must either:

1. Redraw and resave the symbol. The REPLACE function updates all instances of the old symbol with the revised symbol. In addition, all future instances of the symbol reflect the changes made.
2. Select the EXPLODE command to break the symbol into its individual entities. This is done when you need to alter a symbol for that particular situation only.

SYMBOL ATTRIBUTES

Symbols can be assigned attribute information which can be tabulated later when the drawing is complete. Attributes might include part number, name, material, size, weight, cost, and manufacturer. The printed report lists totals for each symbol type and can therefore determine the total amount of parts or materials required in the design.

PLOTTING DRAWINGS

Plotting is the process of making a hardcopy print of the drawing stored in computer memory. At any time, a drafter may make a check plot for an update of the drawing status. Check plots (also called screen dumps) are produced by devices which are fast, but often lack quality—such as dot matrix printers, Fig. 32-19(a). Final plots are high quality prints for reproducing and later distributing the design. Most industries continue to use pen plotters to produce final plots. Pen plotters have high resolution, quality linework, and the

ability to plot on any medium, Fig. 32-19(b).

The plotting procedure includes entering plot specifications, inserting the paper and proper pens, and selecting the command to begin the plot.

ENTERING PLOT SPECIFICATIONS

Plot specifications are a series of values which determine three things: what part of the drawing is plotted, how it appears on the paper, and to what scale it is plotted. These values include paper size, plot area, plot rotation, plot scale, and pen values. Each of these values is necessary to produce a quality plot. The CAD system is programmed with default specifications which it uses unless you enter new values.

The SIZE option allows you to choose among paper sizes supported by your plotter. The AREA option determines what part of the drawing to plot. You can plot the entire drawing, digitize a window around the portion to plot, or specify a previously saved view to plot. The ROTATION option determines whether the plot is made in a landscape (longest paper dimension horizontal) or portrait format (longest paper dimension vertical), Fig. 32-20. The SCALE option determines the plotted drawing size as a ratio to the actual size of the drawing. This is a very powerful function. It allows you to use full

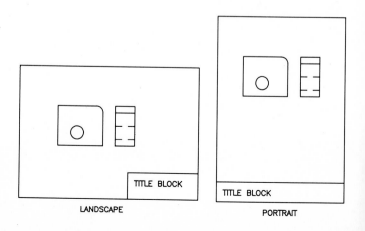

Fig. 32-20. Landscape and portrait formats.

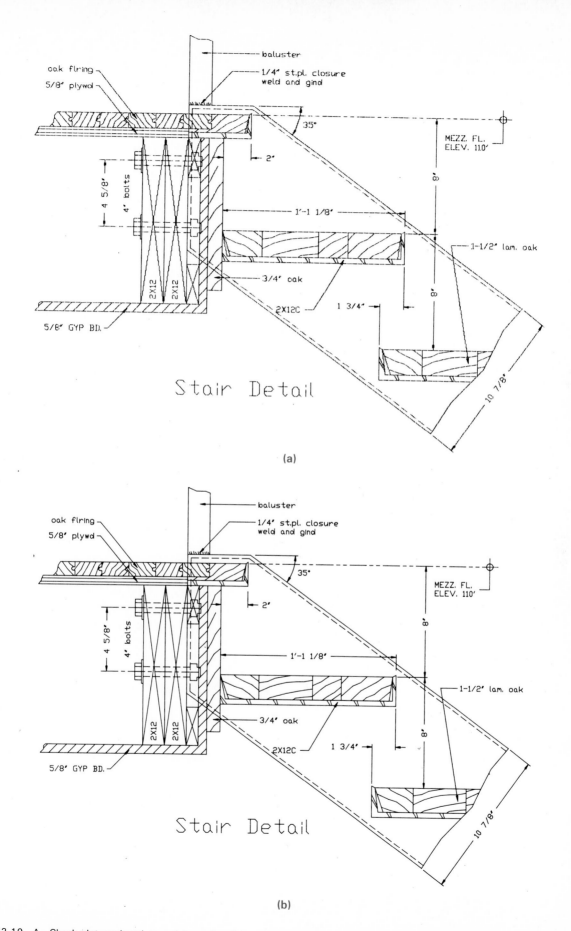

Fig. 32-19. A—Check plot produced on a dot matrix printer. (Autodesk, Inc.) B—Final plot produced on a pen plotter. (Autodesk, Inc.)

size dimensions when drawing, yet plot to an exact scale so that the drawing fits on the chosen paper size. The final plot specification, pen values, are needed for pen plotters. Pen specifications determine which pens plot which colored, numbered, or layered entities, and may include pen pressure, pen speed, and pen width. Pen specifications are often shown as a pen table for you to edit, Fig. 32-21.

```
Entity       Pen Line Pen     Entity  Pen Line Pen
Color        No. Type Speed   Color   No. Type Speed
1 (red)      1   0    8       9       1   0    8
2 (yellow)   1   0    8       10      1   0    8
3 (green)    1   0    8       11      1   0    8
4 (cyan)     1   0    8       12      1   0    8
5 (blue)     1   0    8       13      1   0    8
6 (magenta)  1   0    8       14      1   0    8
7 (white)    1   0    4       15      1   0    8
8            1   0    8       10      1   0    8

Line types: 0 = continuous line   Pen speed codes:
            1 = ...............
            2 = " "  " " " • "     Inches/Second:
            3 = ___ ___ ___ ___     1, 2, 4, 8,
            4 = ___ ___ ___ ___    16, 24, 32
            5 = ___  ___  ___
            6 = ___ ___ ___ ___    Cm/Second:
            7 = ___ ___ ___ ___     3, 5, 10, 20,
            8 = ___  ___  ___      40, 60, 80
```

Fig. 32-21. Plotter pen specifications.

PLOTTING THE DRAWING

Once the specifications are entered, the data can be sent to the plotter. First prepare the plotter by loading the paper, inserting pens, and readying the plotter. Check the plotter's operating manual for the proper way to align the paper and insert pens into the gripper, rack, or carousel. To ready the plotter, make sure that the command panel on the front of the plotter indicates the remote function. This means that data is received from the computer rather than locally. Local commands allow you to move the paper and pen by pressing buttons on the control panel of the plotter.

Once the plotter is prepared, select the START or PLOT function to begin the routing. With single-pen plotters, the plotter stops intermittently to have you insert additional pens. Multipen plotters automatically select pens in numerical order according to the pen specifications previously mentioned.

When the plot is finished, remove the paper and tightly recap all pens.

CHECK PLOTS

The previous discussion focused on pen plotters. This is because check plots generally ignore most of the specifications needed for pen plots. Most often, the entire drawing or view shown on the screen is dumped directly to the device.

PLOTTING SUPPLIES

Plotting supplies include media, pens, adaptors, and cleaning items. Carefully consider the compatibility of the media and pen type before you plot. Inks are specially made for film, vellum, gloss bond, and transparency products. Choose pen types based on the quality of plot you need and medium used. Tungsten-tip, steel-tip, and ceramic-tip liquid ink pens are best. Following in order are disposable liquid ink, plastic tip, and fiber tip pens. Note whether an adaptor must be screwed to the pen before it is inserted in the plotter. Make test plots of portions of the drawing if you are not sure how the pens will draw on the medium.

TROUBLESHOOTING

Plotting problems will arise no matter how carefully you select pens for the medium. Troubles may be caused by worn-out pens, unfit plot speeds, improper pen pressure, or environmental problems. Check your plotter manual for causes and solutions for common pen plotting problems.

DRAWING PROBLEMS

Using standard drafting practices and a CAD system, create a drawing for each of the problems in Fig. 32-22. The drawing name should be P32-(problem number). For example, name Problem 1 as P32-1. Set up the drawing using the proper unit of measurement. Enter a drawing area or select a sheet size large enough to contain the drawing. Set grids and use other drawing aids, as necessary, to your advantage. Separate object views, dimensions, and text using different layers. Follow the text and dimensioning standards used to create the problem drawing. Create symbols for geometric tolerancing frames and symbols used on the drawing. Add text with the tolerancing symbols as necessary. After completing the design, plot the entire drawing on the largest paper size supported by your plotter.

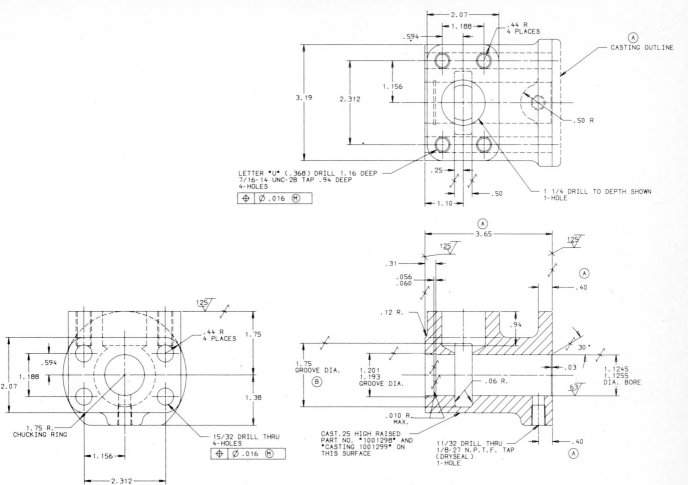

1. OUTLET REGULATOR VALVE BODY

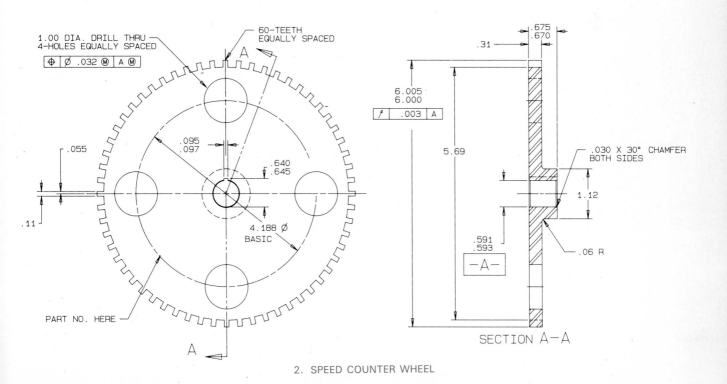

2. SPEED COUNTER WHEEL

Fig. 32-22. Computer-aided drafting problems. (Continued)

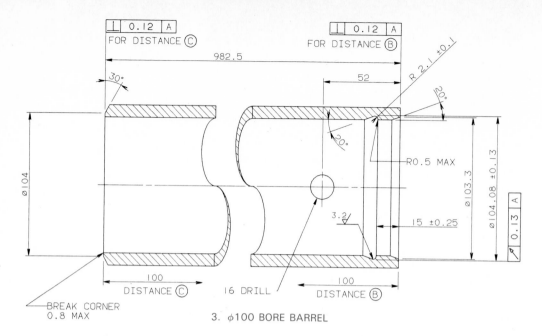

3. φ100 BORE BARREL

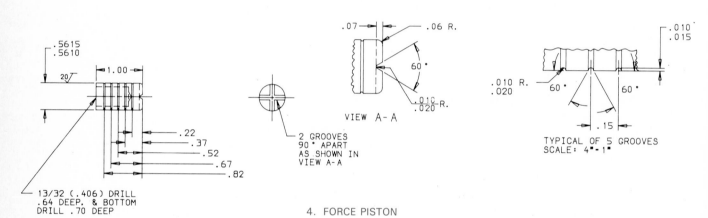

4. FORCE PISTON

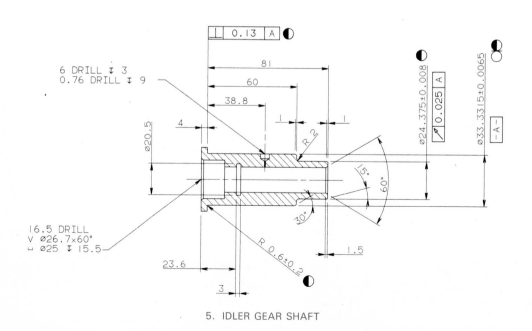

5. IDLER GEAR SHAFT

Fig. 32-22. (Continued) Computer-aided drafting problems.

REFERENCE SECTION

STANDARD ABBREVIATIONS
FOR USE ON DRAWINGS

A

Abrasive	ABRSV
Accessory	ACCESS
Accumulator	ACCUMR
Acetylene	ACET
Actual	ACT
Actuator	ACTR
Addendum	ADD
Adhesive	ADH
Adjust	ADJ
Advance	ADV
Aeronautic	AERO
Alclad	CLAD
Alignment	ALIGN
Allowance	ALLOW
Alloy	ALY
Alteration	ALT
Alternate	ALT
Alternating Current	AC
Aluminum	AL
American National Standards Institute	ANSI
American Wire Gage	AWG
Ammeter	AMM
Amplifier	AMPL
Anneal	ANL
Anodize	ANOD
Antenna	ANT
Approved	APPD
Approximate	APPROX
Arrangement	ARR
Asbestos	ASB
As Required	AR
Assemble	ASSEM
Assembly	ASSY
Attenuation, Attenuator	ATTEN
Audio Frequency	AF
Automatic	AUTO
Automatic Frequency Control	AFC
Automatic Gain Control	AGC
Auxiliary	AUX
Average	AVG

B

Babbit	BAB
Base Line	BL
Battery	BAT
Bearing	BRG
Beat-Frequency Oscillator	BFO
Bend Radius	BR
Bevel	BEV
Bill of Material	B/M
Blueprint	BP or B/P
Bolt Circle	BC
Bracket	BRKT
Brass	BRS
Brazing	BRZG
Brinell Hardness Number	BHN
Bronze	BRZ
Brown & Sharpe (Gage)	B&S
Burnish	BNH
Bushing	BUSH

C

Cabinet	CAB
Calculated	CACL
Cancelled	CANC
Capacitor	CAP
Capacity	CAP
Carburize	CARB
Case Harden	CH
Casting	CSTG
Cast Iron	CI
Cathode-Ray Tube	CRT
Center	CTR
Center to Center	C to C
Centigrade	C
Centimeter	CM
Centrifugal	CENT
Chamfer	CHAM
Check Valve	CV
Chrome Vanadium	CR VAN
Circuit	CKT
Circular	CIR
Circumference	CIRC
Clearance	CL
Clockwise	CW
Closure	CLOS
Coated	CTD
Cold-Drawn Steel	CDS
Cold-Rolled Steel	CRS
Color Code	CC
Commercial	COMM
Concentric	CONC
Condition	COND
Conductor	CNDCT
Contour	CTR
Control	CONT
Copper	COP
Counterbore	CBORE
Counterclockwise	CCW
Counter-Drill	CDRILL
Countersink	CSK
Coupling	CPLG
Cubic	CU
Cylinder	CYL

D

Datum	DAT
Decimal	DEC
Decrease	DECR
Degree	DEG
Detail	DET
Detector	DET
Developed Length	DL
Developed Width	DW
Deviation	DEV
Diagonal	DIAG
Diagram	DIAG
Diameter	DIA
Diameter Bolt Circle	DBC
Diametral Pitch	DP
Dimension	DIM
Direct Current	DC
Disconnect	DISC
Double-Pole Double-Throw	DPDT
Double-Pole Single-Throw	DPST
Dowel	DWL
Draft	DFT
Drafting Room Manual	DRM
Drawing	DWG
Drawing Change Notice	DCN
Drill	DR
Drop Forge	DF
Duplicate	DUP

E

Each	EA
Eccentric	ECC
Effective	EFF
Electric	ELEC
Electrolytic	ELCTLT
Enclosure	ENCL
Engine	ENG
Engineer	ENGR
Engineering	ENGRG
Engineering Change Order	ECO
Engineering Order	EO
Equal	EQ
Equivalent	EQUIV
Estimate	EST

F

Fabricate	FAB
Fillet	FIL
Finish	FIN
Finish All Over	FAO
Fitting	FTG
Fixed	FXD
Fixture	FIX
Flange	FLG
Flat Head	FHD
Flat Pattern	F/P
Flexible	FLEX
Fluid	FL
Forged Steel	FST

Forging	FORG
Furnish	FURN
G	
Gage	GA
Gallon	GAL
Galvanized	GALV
Gasket	GSKT
Generator	GEN
Grind	GRD
Ground	GRD
H	
Half-Hard	1/2H
Handle	HDL
Harden	HDN
Head	HD
Heat Treat	HT TR
Hexagon	HEX
High Carbon Steel	HCS
High Frequency	HF
High Speed	HS
Horizontal	HOR
Hot-Rolled Steel	HRS
Hour	HR
Housing	HSG
Hydraulic	HYD
Hydrostatic	HYDRO
I	
Identification	IDENT
Impregnate	IMPG
Inch	IN
Inclined	INCL
Include, Including, Inclusive	INCL
Increase	INCR
Independent	INDEP
Indicator	IND
Information	INFO
Inside Diameter	ID
Installation	INSTL
Intermediate Frequency	IF
International Standards Organization	ISO
Interrupt	INTER
J	
Joggle	JOG
Junction	JCT
K	
Keyway	KWY
L	
Laboratory	LAB
Lacquer	LAQ
Laminate	LAM
Left Hand	LH
Length	LG
Letter	LTR
Limited	LTD
Limit Switch	LS
Linear	LIN
Liquid	LIQ
List of Material	L/M
Long	LG
Low Carbon	LC
Low Frequency	LF
Low Voltage	LV
Lubricate	LUB
M	
Machine(ing)	MACH
Magnaflux	M
Magnesium	MAG
Maintenance	MAINT
Major	MAJ
Malleable	MALL
Malleable Iron	MI
Manual	MAN
Manufacturing (ed, er)	MFG
Mark	MK
Master Switch	MS
Material	MATL
Maximum	MAX
Measure	MEAS
Mechanical	MECH
Medium	MED
Meter	MTR
Middle	MID
Military	MIL
Millimeter	MM
Minimum	MIN
Miscellaneous	MISC
Modification	MOD
Mold Line	ML
Motor	MOT
Mounting	MTG
Multiple	MULT
N	
Nickel Steel	NS
Nomenclature	NOM
Nominal	NOM
Normalize	NORM
Not to Scale	NTS
Number	NO.
O	
Obsolete	OBS
Opposite	OPP
Oscillator	OSC
Oscilloscope	SCOPE
Ounce	OZ
Outside Diameter	OD
Over-All	OA
P	
Package	PKG
Parting Line (Castings)	PL
Parts List	P/L
Pattern	PATT
Piece	PC
Pilot	PLT
Pitch	P
Pitch Circle	PC
Pitch Diameter	PD
Plan View	PV
Plastic	PLSTC
Plate	PL
Pneumatic	PNEU
Port	P
Positive	POS
Potentiometer	POT
Pounds Per Square Inch	PSI
Pounds Per square Inch Gage	PSIG
Power Amplifier	PA
Power Supply	PWR SPLY
Pressure	PRESS
Primary	PRI
Process, Procedure	PROC
Product, Production	PROD
Q	
Quality	QUAL
Quantity	QTY
Quarter-Hard	1/4H
R	
Radar	RDR
Radio	RAD
Radio Frequency	RF
Radius	RAD or R
Ream	RM
Receptacle	RECP
Reference	REF
Regular	REG
Regulator	REG
Release	REL
Required	REQD
Resistor	RES
Revision	REV
Revolutions Per Minute	RPM
Right Hand	RH
Rivet	RIV
Rockwell Hardness	RH
Round	RD
S	
Schedule	SCH
Schematic	SCHEM
Screw	SCR

STANDARD ABBREVIATIONS
(Continued)

Screw Threads

American National Coarse	NC
American National Fine	NF
American National Extra Fine	NEF
American National 8 Pitch	8N
American Standard Taper Pipe	NTP
American Standard Straight Pipe	NPSC
American Standard Taper (Dryseal)	NPTF
American Standard Straight (Dryseal)	NPSF
Unified Screw Thread Coarse	UNC
Unified Screw Thread Fine	UNF
Unified Screw Thread Extra Fine	UNEF
Unified Screw Thread 8 Thread	8UN

Section	SECT
Sequence	SEQ
Serial	SER
Serrate	SERR
Sheathing	SHTHG
Sheet	SH
Silver Solder	SILS
Single-Pole Double-Throw	SPDT
Single-Pole Single-Throw	SPST
Society of Automotive Engineers	SAE
Solder	SLD

Solenoid	SOL
Speaker	SPKR
Special	SPL
Specification	SPEC
Spot Face	SF
Spring	SPG
Square	SQ
Stainless Steel	SST
Standard	STD
Steel	STL
Stock	STK
Support	SUP
Switch	SW
Symbol	SYM
Symmetrical	SYM
System	SYS

T

Tabulate	TAB
Tangent	TAN
Tapping	TAP
Technical Manual	TM
Teeth	T
Television	TV
Temper	TEM
Temperature	TEM
Tensile Strength	TS
Thick	THK
Thread	THD
Through	THRU
Tolerance	TOL
Tool Steel	TS
Torque	TOR
Total Indicator Reading	TIR
Transceiver	XCVR
Transformer	XFMR
Transistor	XSTR

Transmitter	XMTR
True Involute Form	TIF
Tungsten	TU
Typical	TYP

U

Ultra-High Frequency	UHF
Unit	U
Universal	UNIV
Unless Otherwise Specified	UOS

V

Vacuum	VAC
Vacuum Tube	VT
Variable	VAR
Vernier	VER
Vertical	VERT
Very High Frequency	VHF
Vibrate	VIB
Video	VD
Void	VD
Volt	V
Volume	VOL

W

Washer	WASH
Watt	W
Watt Hour	WH
Wattmeter	WM
Weatherproof	WP
Weight	WT
Wide, Width	W
Wire Wound	WW
Wood	WD
Wrought Iron	WI

Y

Yield Point (PSI)	YP
Yield Strength (PSI)	YS

ELECTRICAL AND ELECTRONICS REFERENCE DESIGNATIONS
(Partial List)

Amplifier	AR	Inductor	L
Antenna	E	Jack	J
Assembly	A	Lamp	DS
Attenuator	AT	Loudspeaker	LS
Audible Signaling Device	DS	Meter	M
Ballast Tube or Lamp	RT	Microphone	MK
Battery	BT	Motor	B
Bell	DS	Oscilloscope	M
Buzzer	DS	Pickup	PU
Capacitor	C	Plug	P
Circuit Breaker	CB	Potentiometer	R
Clock	M	Power Supply	PS
Coil	L	Receiver, Radio	RE
Computer	A	Recorder, Sound	A
Connector, Plug	P	Regulator, Voltage	VR
Connector, Receptacle	J	Relay	K
Contact, Electrical	E	Reproducer, Sound	A
Counter, Electrical	M	Resistor	R
Crystal Unit, Piezoelectric	Y	Rheostat	R
Delay Line	DL	Semiconductor Diode	CR
Electron Tube	V	Socket	X
Filter	FL	Switch	S
Fuse	F	Terminal Board	TB
Generator	G	Transformer	T
Handset	HS	Transistor	Q
Hardware	H	Transmitter, Radio	TR
Indicator	DS		

Courtesy of American National Standards Institute, Reference Designations for Electrical and Electronics Parts and Equipment, Y32.16-1968.

METRIC – INCH EQUIVALENTS

INCHES		MILLI-METERS	INCHES		MILLI-METERS
FRACTIONS	DECIMALS		FRACTIONS	DECIMALS	
	.00394	.1	15/32	.46875	11.9063
	.00787	.2		.47244	12.00
	.01181	.3	31/64	.484375	12.3031
1/64	.015625	.3969	1/2	.5000	12.70
	.01575	.4		.51181	13.00
	.01969	.5	33/64	.515625	13.0969
	.02362	.6	17/32	.53125	13.4938
	.02756	.7	35/64	.546875	13.8907
1/32	.03125	.7938		.55118	14.00
	.0315	.8	9/16	.5625	14.2875
	.03543	.9	37/64	.578125	14.6844
	.03937	1.00		.59055	15.00
3/64	.046875	1.1906	19/32	.59375	15.0813
1/16	.0625	1.5875	39/64	.609375	15.4782
5/64	.078125	1.9844	5/8	.625	15.875
	.07874	2.00		.62992	16.00
3/32	.09375	2.3813	41/64	.640625	16.2719
7/64	.109375	2.7781	21/32	.65625	16.6688
	.11811	3.00		.66929	17.00
1/8	.125	3.175	43/64	.671875	17.0657
9/64	.140625	3.5719	11/16	.6875	17.4625
5/32	.15625	3.9688	45/64	.703125	17.8594
	.15748	4.00		.70866	18.00
11/64	.171875	4.3656	23/32	.71875	18.2563
3/16	.1875	4.7625	47/64	.734375	18.6532
	.19685	5.00		.74803	19.00
13/64	.203125	5.1594	3/4	.7500	19.05
7/32	.21875	5.5563	49/64	.765625	19.4469
15/64	.234375	5.9531	25/32	.78125	19.8438
	.23622	6.00		.7874	20.00
1/4	.2500	6.35	51/64	.796875	20.2407
17/64	.265625	6.7469	13/16	.8125	20.6375
	.27559	7.00		.82677	21.00
9/32	.28125	7.1438	53/64	.828125	21.0344
19/64	.296875	7.5406	27/32	.84375	21.4313
5/16	.3125	7.9375	55/64	.859375	21.8282
	.31496	8.00		.86614	22.00
21/64	.328125	8.3344	7/8	.875	22.225
11/32	.34375	8.7313	57/64	.890625	22.6219
	.35433	9.00		.90551	23.00
23/64	.359375	9.1281	29/32	.90625	23.0188
3/8	.375	9.525	59/64	.921875	23.4157
25/64	.390625	9.9219	15/16	.9375	23.8125
	.3937	10.00		.94488	24.00
13/32	.40625	10.3188	61/64	.953125	24.2094
27/64	.421875	10.7156	31/32	.96875	24.6063
	.43307	11.00		.98425	25.00
7/16	.4375	11.1125	63/64	.984375	25.0032
29/64	.453125	11.5094	1	1.0000	25.4001

MILLIMETER – INCH CONVERSIONS

EXAMPLE:

CONVERT 2468.135 MILLIMETRES TO INCHES.

2000.	mm =	78.74016 IN.
460.	mm =	18.11024 IN.
8.13	mm =	.32008 IN.
.005	mm =	.00020 IN.
2468.135	mm =	97.17068 IN.

ROUND OFF PRACTICE:

THE TOTAL TOLERANCE APPLIED TO A MILLIMETRE DIMENSION SHALL BE THE BASIS FOR THE ACCURACY IN ROUNDING OFF DIMENSIONS AND TOLERANCES CONVERTED TO INCHES. TOTAL TOLERANCE VALUES AND REQUIRED ACCURACY FOR ROUNDING OFF ARE SHOWN IN THE FIGURES BELOW.

TOTAL TOLERANCE IN MILLIMETRES		CONVERTED VALUE IN INCHES SHALL BE ROUNDED TO
AT LEAST	LESS THAN	
0.0000	0.10	4 PLACES (.0001)
0.10	1.0	3 PLACES (.001)
1.0	--	2 PLACES (.01)

WHEN FIRST DIGIT DROPPED IS:	THE LAST DIGIT RETAINED IS:
LESS THAN 5	UNCHANGED
MORE THAN 5	INCREASED BY 1
5 FOLLOWED ONLY BY ZEROS	UNCHANGED IF EVEN
	INCREASED BY 1 IF ODD

EXAMPLE:

CONVERT 48.25 ±0.25 MILLIMETRES TO INCHES

CONVERT THE DIMENSION

40.	mm =	1.57480 IN.
8.25	mm =	.32480 IN.
48.25	mm =	1.89960 IN.

CONVERT THE TOLERANCE

±0.25 mm = .00984 IN.

CONVERTED DIM & TOL

1.89960 ±.00984 IN.

ROUND OFF TO 3 PLACES BASED ON TOTAL TOLERANCE OF 0.50 mm

ROUNDED OFF DIM & TOL

1.900 ±.010 IN.

MICRO-METRE (MICRON)	MICRO-INCH	MICRO-METRE (MICRON)	MICRO-INCH
0.025	1	0.40	16
0.050	2	0.50	20
0.075	3	0.63	25
0.100	4	0.80	32
0.125	5	1.00	40
0.15	6	1.25	50
0.20	8	1.6	63
0.25	10	2.0	80
0.32	13	2.5	100

CONVERSION OF SURFACE TEXTURE DESIGNATIONS.

EXAMPLE:

1.6√ MICROMETRES = 63√ MICROINCHES

INCREMENTS OF 1000 MILLIMETRES 1000 - 9000 MILLIMETRES

1000	2000	3000	4000	5000	6000	7000	8000	9000
39.37008	78.74016	118.11024	157.48031	196.85039	236.22047	275.59055	314.96063	354.33071

INCREMENTS OF 10 MILLIMETRES 0 - 1090 MILLIMETRES

	0	10	20	30	40	50	60	70	80	90
0	0	.39370	.78740	1.18110	1.57480	1.96850	2.36220	2.75591	3.14961	3.54331
100	3.93701	4.33071	4.72441	5.11811	5.51181	5.90551	6.29921	6.69291	7.08661	7.48031
200	7.87402	8.26772	8.66142	9.05512	9.44882	9.84252	10.23622	10.62992	11.02362	11.41732
300	11.81102	12.20472	12.59843	12.99213	13.38583	13.77953	14.17323	14.56693	14.96063	15.35433
400	15.74803	16.14173	16.53543	16.92913	17.32283	17.71654	18.11024	18.50394	18.89764	19.29134
500	19.68504	20.07874	20.47244	20.86614	21.25984	21.65354	22.04724	22.44094	22.83465	23.22835
600	23.62205	24.01575	24.40945	24.80315	25.19685	25.59055	25.98425	26.37795	26.77165	27.16535
700	27.55906	27.95276	28.34646	28.74016	29.13386	29.52756	29.92126	30.31496	30.70866	31.10236
800	31.49606	31.88976	32.28346	32.67717	33.07087	33.46457	33.85827	34.25197	34.64567	35.03937
900	35.43307	35.82677	36.22047	36.61417	37.00787	37.40157	37.79528	38.18898	38.58268	38.97638
1000	39.37008	39.76378	40.15748	40.55118	40.94488	41.33858	41.73228	42.12598	42.51969	42.91339

INCREMENTS OF .01 MILLIMETRES 0 - 10.09 MILLIMETRES

	.00	.01	.02	.03	.04	.05	.06	.07	.08	.09
0	0	.00039	.00079	.00118	.00157	.00197	.00236	.00276	.00315	.00354
.1	.00394	.00433	.00472	.00512	.00551	.00591	.00630	.00669	.00709	.00748
.2	.00787	.00827	.00866	.00906	.00945	.00984	.01024	.01063	.01102	.01142
.3	.01181	.01220	.01260	.01299	.01339	.01378	.01417	.01457	.01496	.01535
.4	.01575	.01614	.01654	.01693	.01732	.01772	.01811	.01850	.01890	.01929
.5	.01969	.02008	.02047	.02087	.02126	.02165	.02205	.02244	.02283	.02323
.6	.02362	.02402	.02441	.02480	.02520	.02559	.02598	.02638	.02677	.02717
.7	.02756	.02795	.02835	.02874	.02913	.02953	.02992	.03031	.03071	.03110
.8	.03150	.03189	.03228	.03268	.03307	.03346	.03386	.03425	.03465	.03504
.9	.03543	.03583	.03622	.03661	.03701	.03740	.03780	.03819	.03858	.03898
1.0	.03937	.03976	.04016	.04055	.04094	.04134	.04173	.04213	.04252	.04291
1.1	.04331	.04370	.04409	.04449	.04488	.04528	.04567	.04606	.04646	.04685
1.2	.04724	.04764	.04803	.04843	.04882	.04921	.04961	.05000	.05039	.05079
1.3	.05118	.05157	.05197	.05236	.05276	.05315	.05354	.05394	.05433	.05472
1.4	.05512	.05551	.05591	.05630	.05669	.05709	.05748	.05787	.05827	.05866
1.5	.05906	.05945	.05984	.06024	.06063	.06102	.06142	.06181	.06220	.06260
1.6	.06299	.06339	.06378	.06417	.06457	.06496	.06535	.06575	.06614	.06654
1.7	.06693	.06732	.06772	.06811	.06850	.06890	.06929	.06969	.07008	.07047
1.8	.07087	.07126	.07165	.07205	.07244	.07283	.07323	.07362	.07402	.07441
1.9	.07480	.07520	.07559	.07598	.07638	.07677	.07717	.07756	.07795	.07835
2.0	.07874	.07913	.07953	.07992	.08031	.08071	.08110	.08150	.08189	.08228
2.1	.08268	.08307	.08346	.08386	.08425	.08465	.08504	.08543	.08583	.08622
2.2	.08661	.08701	.08740	.08780	.08819	.08858	.08898	.08937	.08976	.09016
2.3	.09055	.09094	.09134	.09173	.09213	.09252	.09291	.09331	.09370	.09409
2.4	.09449	.09488	.09528	.09567	.09606	.09646	.09685	.09724	.09764	.09803
2.5	.09843	.09882	.09921	.09961	.10000	.10039	.10079	.10118	.10157	.10197
2.6	.10236	.10276	.10315	.10354	.10394	.10433	.10472	.10512	.10551	.10591
2.7	.10630	.10669	.10709	.10748	.10787	.10827	.10866	.10906	.10945	.10984
2.8	.11024	.11063	.11102	.11142	.11181	.11220	.11260	.11299	.11339	.11378
2.9	.11417	.11457	.11496	.11535	.11575	.11614	.11654	.11693	.11732	.11772
3.0	.11811	.11850	.11890	.11929	.11969	.12008	.12047	.12087	.12126	.12165
3.1	.12205	.12244	.12283	.12323	.12362	.12402	.12441	.12480	.12520	.12559
3.2	.12598	.12638	.12677	.12717	.12756	.12795	.12835	.12874	.12913	.12953
3.3	.12992	.13031	.13071	.13110	.13150	.13189	.13228	.13268	.13307	.13346
3.4	.13386	.13425	.13465	.13504	.13543	.13583	.13622	.13661	.13701	.13740
3.5	.13780	.13819	.13858	.13898	.13937	.13976	.14016	.14055	.14094	.14134
3.6	.14173	.14213	.14252	.14291	.14331	.14370	.14409	.14449	.14488	.14528
3.7	.14567	.14606	.14646	.14685	.14724	.14764	.14803	.14843	.14882	.14921
3.8	.14961	.15000	.15039	.15079	.15118	.15157	.15197	.15236	.15276	.15315
3.9	.15354	.15394	.15433	.15472	.15512	.15551	.15591	.15630	.15669	.15709
4.0	.15748	.15787	.15827	.15866	.15906	.15945	.15984	.16024	.16063	.16102
4.1	.16142	.16181	.16220	.16260	.16299	.16339	.16378	.16417	.16457	.16496
4.2	.16535	.16575	.16614	.16654	.16693	.16732	.16772	.16811	.16850	.16890
4.3	.16929	.16969	.17008	.17047	.17087	.17126	.17165	.17205	.17244	.17283
4.4	.17323	.17362	.17402	.17441	.17480	.17520	.17559	.17598	.17638	.17677
4.5	.17717	.17756	.17795	.17835	.17874	.17913	.17953	.17992	.18031	.18071
4.6	.18110	.18150	.18189	.18228	.18268	.18307	.18346	.18386	.18425	.18465
4.7	.18504	.18543	.18583	.18622	.18661	.18701	.18740	.18780	.18819	.18858
4.8	.18898	.18937	.18976	.19016	.19055	.19094	.19134	.19173	.19213	.19252
4.9	.19291	.19331	.19370	.19409	.19449	.19488	.19528	.19567	.19606	.19646
5.0	.19685	.19724	.19764	.19803	.19843	.19882	.19921	.19961	.20000	.20039
5.1	.20079	.20118	.20157	.20197	.20236	.20276	.20315	.20354	.20394	.20433
5.2	.20472	.20512	.20551	.20591	.20630	.20669	.20709	.20748	.20787	.20827
5.3	.20866	.20906	.20945	.20984	.21024	.21063	.21102	.21142	.21181	.21220
5.4	.21260	.21299	.21339	.21378	.21417	.21457	.21496	.21535	.21575	.21614
5.5	.21654	.21693	.21732	.21772	.21811	.21850	.21890	.21929	.21969	.22008
5.6	.22047	.22087	.22126	.22165	.22205	.22244	.22283	.22323	.22362	.22402
5.7	.22441	.22480	.22520	.22559	.22598	.22638	.22677	.22717	.22756	.22795
5.8	.22835	.22874	.22913	.22953	.22992	.23031	.23071	.23110	.23150	.23189
5.9	.23228	.23268	.23307	.23346	.23386	.23425	.23465	.23504	.23543	.23583
6.0	.23622	.23661	.23701	.23740	.23780	.23819	.23858	.23898	.23937	.23976
6.1	.24016	.24055	.24094	.24134	.24173	.24213	.24252	.24291	.24331	.24370
6.2	.24409	.24449	.24488	.24528	.24567	.24606	.24646	.24685	.24724	.24764
6.3	.24803	.24843	.24882	.24921	.24961	.25000	.25039	.25079	.25118	.25157
6.4	.25197	.25236	.25276	.25315	.25354	.25394	.25433	.25472	.25512	.25551
6.5	.25591	.25630	.25669	.25709	.25748	.25787	.25827	.25866	.25906	.25945
6.6	.25984	.26024	.26063	.26102	.26142	.26181	.26220	.26260	.26299	.26339
6.7	.26378	.26417	.26457	.26496	.26535	.26575	.26614	.26654	.26693	.26732
6.8	.26772	.26811	.26850	.26890	.26929	.26969	.27008	.27047	.27087	.27126
6.9	.27165	.27205	.27244	.27283	.27323	.27362	.27402	.27441	.27480	.27520
7.0	.27559	.27598	.27638	.27677	.27717	.27756	.27795	.27835	.27874	.27913
7.1	.27953	.27992	.28031	.28071	.28110	.28150	.28189	.28228	.28268	.28307
7.2	.28346	.28386	.28425	.28465	.28504	.28543	.28583	.28622	.28661	.28701
7.3	.28740	.28780	.28819	.28858	.28898	.28937	.28976	.29016	.29055	.29094
7.4	.29134	.29173	.29213	.29252	.29291	.29331	.29370	.29409	.29449	.29488
7.5	.29528	.29567	.29606	.29646	.29685	.29724	.29764	.29803	.29843	.29882
7.6	.29921	.29961	.30000	.30039	.30079	.30118	.30157	.30197	.30236	.30276
7.7	.30315	.30354	.30394	.30433	.30472	.30512	.30551	.30591	.30630	.30669
7.8	.30709	.30748	.30787	.30827	.30866	.30906	.30945	.30984	.31024	.31063
7.9	.31102	.31142	.31181	.31220	.31260	.31299	.31339	.31378	.31417	.31457
8.0	.31496	.31535	.31575	.31614	.31654	.31693	.31732	.31772	.31811	.31850
8.1	.31890	.31929	.31969	.32008	.32047	.32087	.32126	.32165	.32205	.32244
8.2	.32283	.32323	.32362	.32402	.32441	.32480	.32520	.32559	.32598	.32638
8.3	.32677	.32717	.32756	.32795	.32835	.32874	.32913	.32953	.32992	.33031
8.4	.33071	.33110	.33150	.33189	.33228	.33268	.33307	.33346	.33386	.33425
8.5	.33465	.33504	.33543	.33583	.33622	.33661	.33701	.33740	.33780	.33819
8.6	.33858	.33898	.33937	.33976	.34016	.34055	.34094	.34134	.34173	.34213
8.7	.34252	.34291	.34331	.34370	.34409	.34449	.34488	.34528	.34567	.34606
8.8	.34646	.34685	.34724	.34764	.34803	.34843	.34882	.34921	.34961	.35000
8.9	.35039	.35079	.35118	.35157	.35197	.35236	.35276	.35315	.35354	.35394
9.0	.35433	.35472	.35512	.35551	.35591	.35630	.35669	.35709	.35748	.35787
9.1	.35827	.35866	.35906	.35945	.35984	.36024	.36063	.36102	.36142	.36181
9.2	.36220	.36260	.36299	.36339	.36378	.36417	.36457	.36496	.36535	.36575
9.3	.36614	.36654	.36693	.36732	.36772	.36811	.36850	.36890	.36929	.36969
9.4	.37008	.37047	.37087	.37126	.37165	.37205	.37244	.37283	.37323	.37362
9.5	.37402	.37441	.37480	.37520	.37559	.37598	.37638	.37677	.37717	.37756
9.6	.37795	.37835	.37874	.37913	.37953	.37992	.38031	.38071	.38110	.38150
9.7	.38189	.38228	.38268	.38307	.38346	.38386	.38425	.38465	.38504	.38543
9.8	.38583	.38622	.38661	.38701	.38740	.38780	.38819	.38858	.38898	.38937
9.9	.38976	.39016	.39055	.39094	.39134	.39173	.39213	.39252	.39291	.39331
10.0	.39370	.39409	.39449	.39488	.39528	.39567	.39606	.39646	.39685	.39724

INCREMENTS OF .001 MILLIMETRES .001 - .009 MILLIMETRES

.001	.002	.003	.004	.005	.006	.007	.008	.009
.00004	.00008	.00012	.00016	.00020	.00024	.00028	.00031	.00035

(Caterpillar Tractor Co.)

INCH – MILLIMETER CONVERSIONS

EXAMPLE:

CONVERT 135.7924 INCHES TO MILLIMETRES.

100.	IN. =	2540.0000 mm
35.	IN. =	889.0000 mm
.792	IN. =	20.1168 mm
.0004	IN. =	.01016 mm
135.7924	IN. =	3449.12696 mm

INCREMENTS OF 100 INCHES 100 TO 900 INCHES

100	200	300	400	500	600	700	800	900
2540.0000	5080.0000	7620.0000	10160.0000	12700.0000	15240.0000	17780.0000	20320.0000	22860.0000

INCREMENTS OF 1 INCH 1 TO 109 INCHES

	0	1	2	3	4	5	6	7	8	9
0	0	25.4000	50.8000	76.2000	101.6000	127.0000	152.4000	177.8000	203.2000	228.6000
10	254.0000	279.4000	304.8000	330.2000	355.6000	381.0000	406.4000	431.8000	457.2000	482.6000
20	508.0000	533.4000	558.8000	584.2000	609.6000	635.0000	660.4000	685.8000	711.2000	736.6000
30	762.0000	787.4000	812.8000	838.2000	863.6000	889.0000	914.4000	939.8000	965.2000	990.6000
40	1016.0000	1041.4000	1066.8000	1092.2000	1117.6000	1143.0000	1168.4000	1193.8000	1219.2000	1244.6000
50	1270.0000	1295.4000	1320.8000	1346.2000	1371.6000	1397.0000	1422.4000	1447.8000	1473.2000	1498.6000
60	1524.0000	1549.4000	1574.8000	1600.2000	1625.6000	1651.0000	1676.4000	1701.8000	1727.2000	1752.6000
70	1778.0000	1803.4000	1828.8000	1854.2000	1879.6000	1905.0000	1930.4000	1955.8000	1981.2000	2006.6000
80	2032.0000	2057.4000	2082.8000	2108.2000	2133.6000	2159.0000	2184.4000	2209.8000	2235.2000	2260.6000
90	2286.0000	2311.4000	2336.8000	2362.2000	2387.6000	2413.0000	2438.4000	2463.8000	2489.2000	2514.6000
100	2540.0000	2565.4000	2590.8000	2616.2000	2641.6000	2667.0000	2692.4000	2717.8000	2743.2000	2768.6000

INCREMENTS OF .001 INCH .001 TO 1.009 INCHES

	.000	.001	.002	.003	.004	.005	.006	.007	.008	.009
0	0	.0254	.0508	.0762	.1016	.1270	.1524	.1778	.2032	.2286
.01	.2540	.2794	.3048	.3302	.3556	.3810	.4064	.4318	.4572	.4826
.02	.5080	.5334	.5588	.5842	.6096	.6350	.6604	.6858	.7112	.7366
.03	.7620	.7874	.8128	.8382	.8636	.8890	.9144	.9398	.9652	.9906
.04	1.0160	1.0414	1.0668	1.0922	1.1176	1.1430	1.1684	1.1938	1.2192	1.2446
.05	1.2700	1.2954	1.3208	1.3462	1.3716	1.3970	1.4224	1.4478	1.4732	1.4986
.06	1.5240	1.5494	1.5748	1.6002	1.6256	1.6510	1.6764	1.7018	1.7272	1.7526
.07	1.7780	1.8034	1.8288	1.8542	1.8796	1.9050	1.9304	1.9558	1.9812	2.0066
.08	2.0320	2.0574	2.0828	2.1082	2.1336	2.1590	2.1844	2.2098	2.2352	2.2606
.09	2.2860	2.3114	2.3368	2.3622	2.3876	2.4130	2.4384	2.4638	2.4892	2.5146
.10	2.5400	2.5654	2.5908	2.6162	2.6416	2.6670	2.6924	2.7178	2.7432	2.7686
.11	2.7940	2.8194	2.8448	2.8702	2.8956	2.9210	2.9464	2.9718	2.9972	3.0226
.12	3.0480	3.0734	3.0988	3.1242	3.1496	3.1750	3.2004	3.2258	3.2512	3.2766
.13	3.3020	3.3274	3.3528	3.3782	3.4036	3.4290	3.4544	3.4798	3.5052	3.5306
.14	3.5560	3.5814	3.6068	3.6322	3.6576	3.6830	3.7084	3.7338	3.7592	3.7846
.15	3.8100	3.8354	3.8608	3.8862	3.9116	3.9370	3.9624	3.9878	4.0132	4.0386
.16	4.0640	4.0894	4.1148	4.1402	4.1656	4.1910	4.2164	4.2418	4.2672	4.2926
.17	4.3180	4.3434	4.3688	4.3942	4.4196	4.4450	4.4704	4.4958	4.5212	4.5466
.18	4.5720	4.5974	4.6228	4.6482	4.6736	4.6990	4.7244	4.7498	4.7752	4.8006
.19	4.8260	4.8514	4.8768	4.9022	4.9276	4.9530	4.9784	5.0038	5.0292	5.0546
.20	5.0800	5.1054	5.1308	5.1562	5.1816	5.2070	5.2324	5.2578	5.2832	5.3086
.21	5.3340	5.3594	5.3848	5.4102	5.4356	5.4610	5.4864	5.5118	5.5372	5.5626
.22	5.5880	5.6134	5.6388	5.6642	5.6896	5.7150	5.7404	5.7658	5.7912	5.8166
.23	5.8420	5.8674	5.8928	5.9182	5.9436	5.9690	5.9944	6.0198	6.0452	6.0706
.24	6.0960	6.1214	6.1468	6.1722	6.1976	6.2230	6.2484	6.2738	6.2992	6.3246
.25	6.3500	6.3754	6.4008	6.4262	6.4516	6.4770	6.5024	6.5278	6.5532	6.5786
.26	6.6040	6.6294	6.6548	6.6802	6.7056	6.7310	6.7564	6.7818	6.8072	6.8326
.27	6.8580	6.8834	6.9088	6.9342	6.9596	6.9850	7.0104	7.0358	7.0612	7.0866
.28	7.1120	7.1374	7.1628	7.1882	7.2136	7.2390	7.2644	7.2898	7.3152	7.3406
.29	7.3660	7.3914	7.4168	7.4422	7.4676	7.4930	7.5184	7.5438	7.5692	7.5946
.30	7.6200	7.6454	7.6708	7.6962	7.7216	7.7470	7.7724	7.7978	7.8232	7.8486
.31	7.8740	7.8994	7.9248	7.9502	7.9756	8.0010	8.0264	8.0518	8.0772	8.1026
.32	8.1280	8.1534	8.1788	8.2042	8.2296	8.2550	8.2804	8.3058	8.3312	8.3566
.33	8.3820	8.4074	8.4328	8.4582	8.4836	8.5090	8.5344	8.5598	8.5852	8.6106
.34	8.6360	8.6614	8.6868	8.7122	8.7376	8.7630	8.7884	8.8138	8.8392	8.8646
.35	8.8900	8.9154	8.9408	8.9662	8.9916	9.0170	9.0424	9.0678	9.0932	9.1186
.36	9.1440	9.1694	9.1948	9.2202	9.2456	9.2710	9.2964	9.3218	9.3472	9.3726
.37	9.3980	9.4234	9.4488	9.4742	9.4996	9.5250	9.5504	9.5758	9.6012	9.6266
.38	9.6520	9.6774	9.7028	9.7282	9.7536	9.7790	9.8044	9.8298	9.8552	9.8806
.39	9.9060	9.9314	9.9568	9.9822	10.0076	10.0330	10.0584	10.0838	10.1092	10.1346
.40	10.1600	10.1854	10.2108	10.2362	10.2616	10.2870	10.3124	10.3378	10.3632	10.3886
.41	10.4140	10.4394	10.4648	10.4902	10.5156	10.5410	10.5664	10.5918	10.6172	10.6426
.42	10.6680	10.6934	10.7188	10.7442	10.7696	10.7950	10.8204	10.8458	10.8712	10.8966
.43	10.9220	10.9474	10.9728	10.9982	11.0236	11.0490	11.0744	11.0998	11.1252	11.1506
.44	11.1760	11.2014	11.2268	11.2522	11.2776	11.3030	11.3284	11.3538	11.3792	11.4046
.45	11.4300	11.4554	11.4808	11.5062	11.5316	11.5570	11.5824	11.6078	11.6332	11.6586
.46	11.6840	11.7094	11.7348	11.7602	11.7856	11.8110	11.8364	11.8618	11.8872	11.9126
.47	11.9380	11.9634	11.9888	12.0142	12.0396	12.0650	12.0904	12.1158	12.1412	12.1666
.48	12.1920	12.2174	12.2428	12.2682	12.2936	12.3190	12.3444	12.3698	12.3952	12.4206
.49	12.4460	12.4714	12.4968	12.5222	12.5476	12.5730	12.5984	12.6238	12.6492	12.6746
.50	12.7000	12.7254	12.7508	12.7762	12.8016	12.8270	12.8524	12.8778	12.9032	12.9286
.51	12.9540	12.9794	13.0048	13.0302	13.0556	13.0810	13.1064	13.1318	13.1572	13.1826
.52	13.2080	13.2334	13.2588	13.2842	13.3096	13.3350	13.3604	13.3858	13.4112	13.4366
.53	13.4620	13.4874	13.5128	13.5382	13.5636	13.5890	13.6144	13.6398	13.6652	13.6906
.54	13.7160	13.7414	13.7668	13.7922	13.8176	13.8430	13.8684	13.8938	13.9192	13.9446
.55	13.9700	13.9954	14.0208	14.0462	14.0716	14.0970	14.1224	14.1478	14.1732	14.1986
.56	14.2240	14.2494	14.2748	14.3002	14.3256	14.3510	14.3764	14.4018	14.4272	14.4526
.57	14.4780	14.5034	14.5288	14.5542	14.5796	14.6050	14.6304	14.6558	14.6812	14.7066
.58	14.7320	14.7574	14.7828	14.8082	14.8336	14.8590	14.8844	14.9098	14.9352	14.9606
.59	14.9860	15.0114	15.0368	15.0622	15.0876	15.1130	15.1384	15.1638	15.1892	15.2146
.60	15.2400	15.2654	15.2908	15.3162	15.3416	15.3670	15.3924	15.4178	15.4432	15.4686
.61	15.4940	15.5194	15.5448	15.5702	15.5956	15.6210	15.6464	15.6718	15.6972	15.7226
.62	15.7480	15.7734	15.7988	15.8242	15.8496	15.8750	15.9004	15.9258	15.9512	15.9766
.63	16.0020	16.0274	16.0528	16.0782	16.1036	16.1290	16.1544	16.1798	16.2052	16.2306
.64	16.2560	16.2814	16.3068	16.3322	16.3576	16.3830	16.4084	16.4338	16.4592	16.4846
.65	16.5100	16.5354	16.5608	16.5862	16.6116	16.6370	16.6624	16.6878	16.7132	16.7386
.66	16.7640	16.7894	16.8148	16.8402	16.8656	16.8910	16.9164	16.9418	16.9672	16.9926
.67	17.0180	17.0434	17.0688	17.0942	17.1196	17.1450	17.1704	17.1958	17.2212	17.2466
.68	17.2720	17.2974	17.3228	17.3482	17.3736	17.3990	17.4244	17.4498	17.4752	17.5006
.69	17.5260	17.5514	17.5768	17.6022	17.6276	17.6530	17.6784	17.7038	17.7292	17.7546
.70	17.7800	17.8054	17.8308	17.8562	17.8816	17.9070	17.9324	17.9578	17.9832	18.0086
.71	18.0340	18.0594	18.0848	18.1102	18.1356	18.1610	18.1864	18.2118	18.2372	18.2626
.72	18.2880	18.3134	18.3388	18.3642	18.3896	18.4150	18.4404	18.4658	18.4912	18.5166
.73	18.5420	18.5674	18.5928	18.6182	18.6436	18.6690	18.6944	18.7198	18.7452	18.7706
.74	18.7960	18.8214	18.8468	18.8722	18.8976	18.9230	18.9484	18.9738	18.9992	19.0246
.75	19.0500	19.0754	19.1008	19.1262	19.1516	19.1770	19.2024	19.2278	19.2532	19.2786
.76	19.3040	19.3294	19.3548	19.3802	19.4056	19.4310	19.4564	19.4818	19.5072	19.5326
.77	19.5580	19.5834	19.6088	19.6342	19.6596	19.6850	19.7104	19.7358	19.7612	19.7866
.78	19.8120	19.8374	19.8628	19.8882	19.9136	19.9390	19.9644	19.9898	20.0152	20.0406
.79	20.0660	20.0914	20.1168	20.1422	20.1676	20.1930	20.2184	20.2438	20.2692	20.2946
.80	20.3200	20.3454	20.3708	20.3962	20.4216	20.4470	20.4724	20.4978	20.5232	20.5486
.81	20.5740	20.5994	20.6248	20.6502	20.6756	20.7010	20.7264	20.7518	20.7772	20.8026
.82	20.8280	20.8534	20.8788	20.9042	20.9296	20.9550	20.9804	21.0058	21.0312	21.0566
.83	21.0820	21.1074	21.1328	21.1582	21.1836	21.2090	21.2344	21.2598	21.2852	21.3106
.84	21.3360	21.3614	21.3868	21.4122	21.4376	21.4630	21.4884	21.5138	21.5392	21.5646
.85	21.5900	21.6154	21.6408	21.6662	21.6916	21.7170	21.7424	21.7678	21.7932	21.8186
.86	21.8440	21.8694	21.8948	21.9202	21.9456	21.9710	21.9964	22.0218	22.0472	22.0726
.87	22.0980	22.1234	22.1488	22.1742	22.1996	22.2250	22.2504	22.2758	22.3012	22.3266
.88	22.3520	22.3774	22.4028	22.4282	22.4536	22.4790	22.5044	22.5298	22.5552	22.5806
.89	22.6060	22.6314	22.6568	22.6822	22.7076	22.7330	22.7584	22.7838	22.8092	22.8346
.90	22.8600	22.8854	22.9108	22.9362	22.9616	22.9870	23.0124	23.0378	23.0632	23.0886
.91	23.1140	23.1394	23.1648	23.1902	23.2156	23.2410	23.2664	23.2918	23.3172	23.3426
.92	23.3680	23.3934	23.4188	23.4442	23.4696	23.4950	23.5204	23.5458	23.5712	23.5966
.93	23.6220	23.6474	23.6728	23.6982	23.7236	23.7490	23.7744	23.7998	23.8252	23.8506
.94	23.8760	23.9014	23.9268	23.9522	23.9776	24.0030	24.0284	24.0538	24.0792	24.1046
.95	24.1300	24.1554	24.1808	24.2062	24.2316	24.2570	24.2824	24.3078	24.3332	24.3586
.96	24.3840	24.4094	24.4348	24.4602	24.4856	24.5110	24.5364	24.5618	24.5872	24.6126
.97	24.6380	24.6634	24.6888	24.7142	24.7396	24.7650	24.7904	24.8158	24.8412	24.8666
.98	24.8920	24.9174	24.9428	24.9682	24.9936	25.0190	25.0444	25.0698	25.0952	25.1206
.99	25.1460	25.1714	25.1968	25.2222	25.2476	25.2730	25.2984	25.3238	25.3492	25.3746
1.00	25.4000	25.4254	25.4508	25.4762	25.5016	25.5270	25.5524	25.5778	25.6032	25.6286

INCREMENTS OF .0001 INCH .0001 TO .0009 INCH

.0001	.0002	.0003	.0004	.0005	.0006	.0007	.0008	.0009
.00254	.00508	.00762	.01016	.01270	.01524	.01778	.02032	.02286

ROUND OFF PRACTICE:

THE TOTAL TOLERANCE APPLIED TO AN INCH DIMENSION SHALL BE THE BASIS FOR THE ACCURACY IN ROUNDING OFF DIMENSIONS AND TOLERANCES CONVERTED TO MILLIMETRES. TOTAL TOLERANCE VALUES AND REQUIRED ACCURACY FOR ROUNDING OFF ARE SHOWN IN THE FIGURES BELOW.

TOTAL TOLERANCE IN INCHES		CONVERTED VALUE IN MILLIMETRES SHALL BE ROUNDED TO
AT LEAST	LESS THAN	
.0000	.0004	4 PLACES (.0001)
.0004	.004	3 PLACES (.001)
.004	.04	2 PLACES (.01)
.04	--	1 PLACES (.1)

WHEN FIRST DIGIT DROPPED IS:	THE LAST DIGIT RETAINED IS:
LESS THAN 5	UNCHANGED
MORE THAN 5	INCREASED BY 1
5 FOLLOWED ONLY BY ZEROS	UNCHANGED IF EVEN
	INCREASED BY 1 IF ODD

EXAMPLE:

CONVERT 3.655 ±.002 INCHES TO MILLIMETRES.

CONVERT THE DIMENSION

3.	IN. =	76.2000 mm
.655	IN. =	16.6370 mm
3.655	IN. =	92.8370 mm

CONVERT THE TOLERANCE

±.002 IN. = ±.0508 mm

CONVERTED DIM & TOL

92.8370 ±.0508 mm

ROUND OFF TO 2 PLACES BASED ON TOTAL TOLERANCE OF .004 INCH.

ROUNDED OFF DIM & TOL

92.84 ±.05 mm

MICRO-INCH	MICRO-METRE (MICRON)	MICRO-INCH	MICRO-METRE (MICRON)
1	0.025	16	0.40
2	0.050	20	0.50
3	0.075	25	0.63
4	0.100	32	0.80
5	0.125	40	1.00
6	0.15	50	1.25
8	0.20	63	1.6
10	0.25	80	2.0
13	0.32	100	2.5

CONVERSION OF SURFACE TEXTURE DESIGNATIONS.

EXAMPLE:

63√ MICROINCHES = 1.6√ MICROMETRES

(Caterpillar Tractor Co.)

RUNNING AND SLIDING FITS

Limits are in thousandths of an inch.

Limits for hole and shaft are applied algebraically to the basic size to obtain the limits of size for the parts.

Data in bold face are in accordance with ABC agreements.

Symbols H5, g5, etc., are Hole and Shaft designations used in ABC System

Nominal Size Range Inches Over	To	Class RC 1 Limits of Clearance	Standard Limits Hole H5	Standard Limits Shaft g4	Class RC 2 Limits of Clearance	Standard Limits Hole H6	Standard Limits Shaft g5	Class RC 3 Limits of Clearance	Standard Limits Hole H7	Standard Limits Shaft f6	Class RC 4 Limits of Clearance	Standard Limits Hole H8	Standard Limits Shaft f7
0	− 0.12	0.1 / 0.45	+ 0.2 / 0	− 0.1 / − 0.25	0.1 / 0.55	+ 0.25 / 0	− 0.1 / − 0.3	0.3 / 0.95	+ 0.4 / 0	− 0.3 / − 0.55	0.3 / 1.3	+ 0.6 / 0	− 0.3 / − 0.7
0.12	− 0.24	0.15 / 0.5	+ 0.2 / 0	− 0.15 / − 0.3	0.15 / 0.65	+ 0.3 / 0	− 0.15 / − 0.35	0.4 / 1.2	+ 0.5 / 0	− 0.4 / − 0.7	0.4 / 1.6	+ 0.7 / 0	− 0.4 / − 0.9
0.24	− 0.40	0.2 / 0.6	+ 0.25 / 0	− 0.2 / − 0.35	0.2 / 0.85	+ 0.4 / 0	− 0.2 / − 0.45	0.5 / 1.5	+ 0.6 / 0	− 0.5 / − 0.9	0.5 / 2.0	+ 0.9 / 0	− 0.5 / − 1.1
0.40	− 0.71	0.25 / 0.75	+ 0.3 / 0	− 0.25 / − 0.45	0.25 / 0.95	+ 0.4 / 0	− 0.25 / − 0.55	0.6 / 1.7	+ 0.7 / 0	− 0.6 / − 1.0	0.6 / 2.3	+ 1.0 / 0	− 0.6 / − 1.3
0.71	− 1.19	0.3 / 0.95	+ 0.4 / 0	− 0.3 / − 0.55	0.3 / 1.2	+ 0.5 / 0	− 0.3 / − 0.7	0.8 / 2.1	+ 0.8 / 0	− 0.8 / − 1.3	0.8 / 2.8	+ 1.2 / 0	− 0.8 / − 1.6
1.19	− 1.97	0.4 / 1.1	+ 0.4 / 0	− 0.4 / − 0.7	0.4 / 1.4	+ 0.6 / 0	− 0.4 / − 0.8	1.0 / 2.6	+ 1.0 / 0	− 1.0 / − 1.6	1.0 / 3.6	+ 1.6 / 0	− 1.0 / − 2.0
1.97	− 3.15	0.4 / 1.2	+ 0.5 / 0	− 0.4 / − 0.7	0.4 / 1.6	+ 0.7 / 0	− 0.4 / − 0.9	1.2 / 3.1	+ 1.2 / 0	− 1.2 / − 1.9	1.2 / 4.2	+ 1.8 / 0	− 1.2 / − 2.4
3.15	− 4.73	0.5 / 1.5	+ 0.6 / 0	− 0.5 / − 0.9	0.5 / 2.0	+ 0.9 / 0	− 0.5 / − 1.1	1.4 / 3.7	+ 1.4 / 0	− 1.4 / − 2.3	1.4 / 5.0	+ 2.2 / 0	− 1.4 / − 2.8
4.73	− 7.09	0.6 / 1.8	+ 0.7 / 0	− 0.6 / − 1.1	0.6 / 2.3	+ 1.0 / 0	− 0.6 / − 1.3	1.6 / 4.2	+ 1.6 / 0	− 1.6 / − 2.6	1.6 / 5.7	+ 2.5 / 0	− 1.6 / − 3.2
7.09	− 9.85	0.6 / 2.0	+ 0.8 / 0	− 0.6 / − 1.2	0.6 / 2.6	+ 1.2 / 0	− 0.6 / − 1.4	2.0 / 5.0	+ 1.8 / 0	− 2.0 / − 3.2	2.0 / 6.6	+ 2.8 / 0	− 2.0 / − 3.8
9.85	− 12.41	0.8 / 2.3	+ 0.9 / 0	− 0.8 / − 1.4	0.7 / 2.8	+ 1.2 / 0	− 0.7 / − 1.6	2.5 / 5.7	+ 2.0 / 0	− 2.5 / − 3.7	2.2 / 7.2	+ 3.0 / 0	− 2.2 / − 4.2
12.41	− 15.75	1.0 / 2.7	+ 1.0 / 0	− 1.0 / − 1.7	0.7 / 3.1	+ 1.4 / 0	− 0.7 / − 1.7	3.0 / 6.6	+ 2.2 / 0	− 3.0 / − 4.4	2.5 / 8.2	+ 3.5 / 0	− 2.5 / − 4.7
15.75	− 19.69	1.2 / 3.0	+ 1.0 / 0	− 1.2 / − 2.0	0.8 / 3.4	+ 1.6 / 0	− 0.8 / − 1.8	4.0 / 8.1	+ 2.5 / 0	− 4.0 / − 5.6	2.8 / 9.3	+ 4.0 / 0	− 2.8 / − 5.3
19.69	− 30.09	1.6 / 3.7	+ 1.2 / 0	− 1.6 / − 2.5	1.6 / 4.8	+ 2.0 / 0	− 1.6 / − 2.8	5.0 / 10.0	+ 3.0 / 0	− 5.0 / − 7.0	5.0 / 13.0	+ 5.0 / 0	− 5.0 / − 8.0
30.09	− 41.49	2.0 / 4.6	+ 1.6 / 0	− 2.0 / − 3.0	2.0 / 6.1	+ 2.5 / 0	− 2.0 / − 3.6	6.0 / 12.5	+ 4.0 / 0	− 6.0 / − 8.5	6.0 / 16.0	+ 6.0 / 0	− 6.0 / −10.0
41.49	− 56.19	2.5 / 5.7	+ 2.0 / 0	− 2.5 / − 3.7	2.5 / 7.5	+ 3.0 / 0	− 2.5 / − 4.5	8.0 / 16.0	+ 5.0 / 0	− 8.0 / −11.0	8.0 / 21.0	+ 8.0 / 0	− 8.0 / −13.0
56.19	− 76.39	3.0 / 7.1	+ 2.5 / 0	− 3.0 / − 4.6	3.0 / 9.5	+ 4.0 / 0	− 3.0 / − 5.5	10.0 / 20.0	+ 6.0 / 0	−10.0 / −14.0	10.0 / 26.0	+10.0 / 0	−10.0 / −16.0
76.39	−100.9	4.0 / 9.0	+ 3.0 / 0	− 4.0 / − 6.0	4.0 / 12.0	+ 5.0 / 0	− 4.0 / − 7.0	12.0 / 25.0	+ 8.0 / 0	−12.0 / −17.0	12.0 / 32.0	+12.0 / 0	−12.0 / −20.0
100.9	−131.9	5.0 / 11.5	+ 4.0 / 0	− 5.0 / − 7.5	5.0 / 15.0	+ 6.0 / 0	− 5.0 / − 9.0	16.0 / 32.0	+10.0 / 0	−16.0 / −22.0	16.0 / 42.0	+16.0 / 0	−16.0 / −26.0
131.9	−171.9	6.0 / 14.0	+ 5.0 / 0	− 6.0 / − 9.0	6.0 / 19.0	+ 8.0 / 0	− 6.0 / −11.0	18.0 / 38.0	+12.0 / 0	−18.0 / −26.0	18.0 / 50.0	+20.0 / 0	−18.0 / −30.0
171.9	−200	8.0 / 18.0	+ 6.0 / 0	− 8.0 / −12.0	8.0 / 22.0	+10.0 / 0	− 8.0 / −12.0	22.0 / 48.0	+16.0 / 0	−22.0 / −32.0	22.0 / 63.0	+25.0 / 0	−22.0 / −38.0

(ANSI)   Continued

RUNNING AND SLIDING FITS (Continued)

Limits are in thousandths of an inch.

Limits for hole and shaft are applied algebraically to the basic size to obtain the limits of size for the parts

Data in bold face are in accordance with ABC agreements

Symbols H8, e7, etc., are Hole and Shaft designations used in ABC System

Class RC 5			Class RC 6			Class RC 7			Class RC 8			Class RC 9			Nominal Size Range Inches	
Limits of Clearance	Hole H8	Shaft e7	Limits of Clearance	Hole H9	Shaft e8	Limits of Clearance	Hole H9	Shaft d8	Limits of Clearance	Hole H10	Shaft c9	Limits of Clearance	Hole H11	Shaft	Over	To
0.6 / 1.6	+0.6 / −0	−0.6 / −1.0	0.6 / 2.2	+1.0 / −0	−0.6 / −1.2	1.0 / 2.6	+1.0 / 0	−1.0 / −1.6	2.5 / 5.1	+1.6 / 0	−2.5 / −3.5	4.0 / 8.1	+2.5 / 0	−4.0 / −5.6	0	0.12
0.8 / 2.0	+0.7 / −0	−0.8 / −1.3	0.8 / 2.7	+1.2 / −0	−0.8 / −1.5	1.2 / 3.1	+1.2 / 0	−1.2 / −1.9	2.8 / 5.8	+1.8 / 0	−2.8 / −4.0	4.5 / 9.0	+3.0 / 0	−4.5 / −6.0	0.12	0.24
1.0 / 2.5	+0.9 / −0	−1.0 / −1.6	1.0 / 3.3	+1.4 / −0	−1.0 / −1.9	1.6 / 3.9	+1.4 / 0	−1.6 / −2.5	3.0 / 6.6	+2.2 / 0	−3.0 / −4.4	5.0 / 10.7	+3.5 / 0	−5.0 / −7.2	0.24	0.40
1.2 / 2.9	+1.0 / −0	−1.2 / −1.9	1.2 / 3.8	+1.6 / −0	−1.2 / −2.2	2.0 / 4.6	+1.6 / 0	−2.0 / −3.0	3.5 / 7.9	+2.8 / 0	−3.5 / −5.1	6.0 / 12.8	+4.0 / −0	−6.0 / −8.8	0.40	0.71
1.6 / 3.6	+1.2 / −0	−1.6 / −2.4	1.6 / 4.8	+2.0 / −0	−1.6 / −2.8	2.5 / 5.7	+2.0 / 0	−2.5 / −3.7	4.5 / 10.0	+3.5 / 0	−4.5 / −6.5	7.0 / 15.5	+5.0 / 0	−7.0 / −10.5	0.71	1.19
2.0 / 4.6	+1.6 / −0	−2.0 / −3.0	2.0 / 6.1	+2.5 / −0	−2.0 / −3.6	3.0 / 7.1	+2.5 / 0	−3.0 / −4.6	5.0 / 11.5	+4.0 / 0	−5.0 / −7.5	8.0 / 18.0	+6.0 / 0	−8.0 / −12.0	1.19	1.97
2.5 / 5.5	+1.8 / −0	−2.5 / −3.7	2.5 / 7.3	+3.0 / −0	−2.5 / −4.3	4.0 / 8.8	+3.0 / 0	−4.0 / −5.8	6.0 / 13.5	+4.5 / 0	−6.0 / −9.0	9.0 / 20.5	+7.0 / 0	−9.0 / −13.5	1.97	3.15
3.0 / 6.6	+2.2 / −0	−3.0 / −4.4	3.0 / 8.7	+3.5 / −0	−3.0 / −5.2	5.0 / 10.7	+3.5 / 0	−5.0 / −7.2	7.0 / 15.5	+5.0 / 0	−7.0 / −10.5	10.0 / 24.0	+9.0 / 0	−10.0 / −15.0	3.15	4.73
3.5 / 7.6	+2.5 / −0	−3.5 / −5.1	3.5 / 10.0	+4.0 / −0	−3.5 / −6.0	6.0 / 12.5	+4.0 / 0	−6.0 / −8.5	8.0 / 18.0	+6.0 / 0	−8.0 / −12.0	12.0 / 28.0	+10.0 / 0	−12.0 / −18.0	4.73	7.09
4.0 / 8.6	+2.8 / −0	−4.0 / −5.8	4.0 / 11.3	+4.5 / 0	−4.0 / −6.8	7.0 / 14.3	+4.5 / 0	−7.0 / −9.8	10.0 / 21.5	+7.0 / 0	−10.0 / −14.5	15.0 / 34.0	+12.0 / 0	−15.0 / −22.0	7.09	9.85
5.0 / 10.0	+3.0 / 0	−5.0 / −7.0	5.0 / 13.0	+5.0 / 0	−5.0 / −8.0	8.0 / 16.0	+5.0 / 0	−8.0 / −11.0	12.0 / 25.0	+8.0 / 0	−12.0 / −17.0	18.0 / 38.0	+12.0 / 0	−18.0 / −26.0	9.85	12.41
6.0 / 11.7	+3.5 / 0	−6.0 / −8.2	6.0 / 15.5	+6.0 / 0	−6.0 / −9.5	10.0 / 19.5	+6.0 / 0	−10.0 / −13.5	14.0 / 29.0	+9.0 / 0	−14.0 / −20.0	22.0 / 45.0	+14.0 / 0	−22.0 / −31.0	12.41	15.75
8.0 / 14.5	+4.0 / 0	−8.0 / −10.5	8.0 / 18.0	+6.0 / 0	−8.0 / −12.0	12.0 / 22.0	+6.0 / 0	−12.0 / −16.0	16.0 / 32.0	+10.0 / 0	−16.0 / −22.0	25.0 / 51.0	+16.0 / 0	−25.0 / −35.0	15.75	19.69
10.0 / 18.0	+5.0 / 0	−10.0 / −13.0	10.0 / 23.0	+8.0 / 0	−10.0 / −15.0	16.0 / 29.0	+8.0 / 0	−16.0 / −21.0	20.0 / 40.0	+12.0 / 0	−20.0 / −28.0	30.0 / 62.0	+20.0 / 0	−30.0 / −42.0	19.69	30.09
12.0 / 22.0	+6.0 / 0	−12.0 / −16.0	12.0 / 28.0	+10.0 / 0	−12.0 / −18.0	20.0 / 36.0	+10.0 / 0	−20.0 / −26.0	25.0 / 51.0	+16.0 / 0	−25.0 / −35.0	40.0 / 81.0	+25.0 / 0	−40.0 / −56.0	30.09	41.49
16.0 / 29.0	+8.0 / 0	−16.0 / −21.0	16.0 / 36.0	+12.0 / 0	−16.0 / −24.0	25.0 / 45.0	+12.0 / 0	−25.0 / −33.0	30.0 / 62.0	+20.0 / 0	−30.0 / −42.0	50.0 / 100	+30.0 / 0	−50.0 / −70.0	41.49	56.19
20.0 / 36.0	+10.0 / 0	−20.0 / −26.0	20.0 / 46.0	+16.0 / 0	−20.0 / −30.0	30.0 / 56.0	+16.0 / 0	−30.0 / −40.0	40.0 / 81.0	+25.0 / 0	−40.0 / −56.0	60.0 / 125	+40.0 / 0	−60.0 / −85.0	56.19	76.39
25.0 / 45.0	+12.0 / 0	−25.0 / −33.0	25.0 / 57.0	+20.0 / 0	−25.0 / −37.0	40.0 / 72.0	+20.0 / 0	−40.0 / −52.0	50.0 / 100	+30.0 / 0	−50.0 / −70.0	80.0 / 160	+50.0 / 0	−80.0 / −110	76.39	100.9
30.0 / 56.0	+16.0 / 0	−30.0 / −40.0	30.0 / 71.0	+25.0 / 0	−30.0 / −46.0	50.0 / 91.0	+25.0 / 0	−50.0 / −66.0	60.0 / 125	+40.0 / 0	−60.0 / −85.0	100 / 200	+60.0 / 0	−100 / −140	100.9	131.9
35.0 / 67.0	+20.0 / 0	−35.0 / −47.0	35.0 / 85.0	+30.0 / 0	−35.0 / −55.0	60.0 / 110.0	+30.0 / 0	−60.0 / −80.0	80.0 / 160	+50.0 / 0	−80.0 / −110	130 / 260	+80.0 / 0	−130 / −180	131.9	171.9
45.0 / 86.0	+25.0 / 0	−45.0 / −61.0	45.0 / 110.0	+40.0 / 0	−45.0 / −70.0	80.0 / 145.0	+40.0 / 0	−80.0 / −105.0	100 / 200	+60.0 / 0	−100 / −140	150 / 310	+100 / 0	−150 / −210	171.9	200

End of Table

LOCATIONAL CLEARANCE FITS

Limits are in thousandths of an inch.
Limits for hole and shaft are applied algebraically to the basic size to obtain the limits of size for the parts.
Data in bold face are in accordance with ABC agreements.
Symbols H6, h5, etc., are Hole and Shaft designations used in ABC System

Nominal Size Range Inches (Over)	(To)	Class LC 1 Limits of Clearance	LC 1 Hole H6	LC 1 Shaft h5	Class LC 2 Limits of Clearance	LC 2 Hole H7	LC 2 Shaft h6	Class LC 3 Limits of Clearance	LC 3 Hole H8	LC 3 Shaft h7	Class LC 4 Limits of Clearance	LC 4 Hole H10	LC 4 Shaft h9	Class LC 5 Limits of Clearance	LC 5 Hole H7	LC 5 Shaft g6
0 —	0.12	0 / 0.45	+0.25 / −0	+0 / −0.2	0 / 0.65	+0.4 / −0	+0 / −0.25	0 / 1	+0.6 / −0	+0 / −0.4	0 / 2.6	+1.6 / −0	+0 / −1.0	0.1 / 0.75	+0.4 / −0	−0.1 / −0.35
0.12—	0.24	0 / 0.5	+0.3 / −0	+0 / −0.2	0 / 0.8	+0.5 / −0	+0 / −0.3	0 / 1.2	+0.7 / −0	+0 / −0.5	0 / 3.0	+1.8 / −0	+0 / −1.2	0.15 / 0.95	+0.5 / −0	−0.15 / −0.45
0.24—	0.40	0 / 0.65	+0.4 / −0	+0 / −0.25	0 / 1.0	+0.6 / −0	+0 / −0.4	0 / 1.5	+0.9 / −0	+0 / −0.6	0 / 3.6	+2.2 / −0	+0 / −1.4	0.2 / 1.2	+0.6 / −0	−0.2 / −0.6
0.40—	0.71	0 / 0.7	+0.4 / −0	+0 / −0.3	0 / 1.1	+0.7 / −0	+0 / −0.4	0 / 1.7	+1.0 / −0	+0 / −0.7	0 / 4.4	+2.8 / −0	+0 / −1.6	0.25 / 1.35	+0.7 / −0	−0.25 / −0.65
0.71—	1.19	0 / 0.9	+0.5 / −0	+0 / −0.4	0 / 1.3	+0.8 / −0	+0 / −0.5	0 / 2	+1.2 / −0	+0 / −0.8	0 / 5.5	+3.5 / −0	+0 / −2.0	0.3 / 1.6	+0.8 / −0	−0.3 / −0.8
1.19—	1.97	0 / 1.0	+0.6 / −0	+0 / −0.4	0 / 1.6	+1.0 / −0	+0 / −0.6	0 / 2.6	+1.6 / −0	+0 / −1	0 / 6.5	+4.0 / −0	+0 / −2.5	0.4 / 2.0	+1.0 / −0	−0.4 / −1.0
1.97—	3.15	0 / 1.2	+0.7 / −0	+0 / −0.5	0 / 1.9	+1.2 / −0	+0 / −0.7	0 / 3	+1.8 / −0	+0 / −1.2	0 / 7.5	+4.5 / −0	+0 / −3	0.4 / 2.3	+1.2 / −0	−0.4 / −1.1
3.15—	4.73	0 / 1.5	+0.9 / −0	+0 / −0.6	0 / 2.3	+1.4 / −0	+0 / −0.9	0 / 3.6	+2.2 / −0	+0 / −1.4	0 / 8.5	+5.0 / −0	+0 / −3.5	0.5 / 2.8	+1.4 / −0	−0.5 / −1.4
4.73—	7.09	0 / 1.7	+1.0 / −0	+0 / −0.7	0 / 2.6	+1.6 / −0	+0 / −1.0	0 / 4.1	+2.5 / −0	+0 / −1.6	0 / 10	+6.0 / −0	+0 / −4	0.6 / 3.2	+1.6 / −0	−0.6 / −1.6
7.09—	9.85	0 / 2.0	+1.2 / −0	+0 / −0.8	0 / 3.0	+1.8 / −0	+0 / −1.2	0 / 4.6	+2.8 / −0	+0 / −1.8	0 / 11.5	+7.0 / −0	+0 / −4.5	0.6 / 3.6	+1.8 / −0	−0.6 / −1.8
9.85—	12.41	0 / 2.1	+1.2 / −0	+0 / −0.9	0 / 3.2	+2.0 / −0	+0 / −1.2	0 / 5	+3.0 / −0	+0 / −2.0	0 / 13	+8.0 / −0	+0 / −5	0.7 / 3.9	+2.0 / −0	−0.7 / −1.9
12.41—	15.75	0 / 2.4	+1.4 / −0	+0 / −1.0	0 / 3.6	+2.2 / −0	+0 / −1.4	0 / 5.7	+3.5 / −0	+0 / −2.2	0 / 15	+9.0 / −0	+0 / −6	0.7 / 4.3	+2.2 / −0	−0.7 / −2.1
15.75—	19.69	0 / 2.6	+1.6 / −0	+0 / −1.0	0 / 4.1	+2.5 / −0	+0 / −1.6	0 / 6.5	+4 / −0	+0 / −2.5	0 / 16	+10.0 / −0	+0 / −6	0.8 / 4.9	+2.5 / −0	−0.8 / −2.4
19.69—	30.09	0 / 3.2	+2.0 / −0	+0 / −1.2	0 / 5.0	+3 / −0	+0 / −2	0 / 8	+5 / −0	+0 / −3	0 / 20	+12.0 / −0	+0 / −8	0.9 / 5.9	+3.0 / −0	−0.9 / −2.9
30.09—	41.49	0 / 4.1	+2.5 / −0	+0 / −1.6	0 / 6.5	+4 / −0	+0 / −2.5	0 / 10	+6 / −0	+0 / −4	0 / 26	+16.0 / −0	+0 / −10	1.0 / 7.5	+4.0 / −0	−1.0 / −3.5
41.49—	56.19	0 / 5.0	+3.0 / −0	+0 / −2.0	0 / 8.0	+5 / −0	+0 / −3	0 / 13	+8 / −0	+0 / −5	0 / 32	+20.0 / −0	+0 / −12	1.2 / 9.2	+5.0 / −0	−1.2 / −4.2
56.19—	76.39	0 / 6.5	+4.0 / −0	+0 / −2.5	0 / 10	+6 / −0	+0 / −4	0 / 16	+10 / −0	+0 / −6	0 / 41	+25.0 / −0	+0 / −16	1.2 / 11.2	+6.0 / −0	−1.2 / −5.2
76.39—	100.9	0 / 8.0	+5.0 / −0	+0 / −3.0	0 / 13	+8 / −0	+0 / −5	0 / 20	+12 / −0	+0 / −8	0 / 50	+30.0 / −0	+0 / −20	1.4 / 14.4	+8.0 / −0	−1.4 / −6.4
100.9 —	131.9	0 / 10.0	+6.0 / −0	+0 / −4.0	0 / 16	+10 / −0	+0 / −6	0 / 26	+16 / −0	+0 / −10	0 / 65	+40.0 / −0	+0 / −25	1.6 / 17.6	+10.0 / −0	−1.6 / −7.6
131.9 —	171.9	0 / 13.0	+8.0 / −0	+0 / −5.0	0 / 20	+12 / −0	+0 / −8	0 / 32	+20 / −0	+0 / −12	0 / 80	+50.0 / −0	+0 / −30	1.8 / 21.8	+12.0 / −0	−1.8 / −9.8
171.9 —	200	0 / 16.0	+10.0 / −0	+0 / −6.0	0 / 26	+16 / −0	+0 / −10	0 / 41	+25 / −0	+0 / −16	0 / 100	+60.0 / −0	+0 / −40	1.8 / 27.8	+16.0 / −0	−1.8 / −11.8

(ANSI)

Continued

LOCATIONAL CLEARANCE FITS (Continued)

Limits are in thousandths of an inch.

Limits for hole and shaft are applied algebraically to the basic size to obtain the limits of size for the parts.

Data in bold face are in accordance with ABC agreements.

Symbols H9, f8, etc., are Hole and Shaft designations used in ABC System

Class LC 6 Limits of Clearance	LC 6 Hole H9	LC 6 Shaft f8	Class LC 7 Limits of Clearance	LC 7 Hole H10	LC 7 Shaft e9	Class LC 8 Limits of Clearance	LC 8 Hole H10	LC 8 Shaft d9	Class LC 9 Limits of Clearance	LC 9 Hole H11	LC 9 Shaft c10	Class LC 10 Limits of Clearance	LC 10 Hole H12	LC 10 Shaft	Class LC 11 Limits of Clearance	LC 11 Hole H13	LC 11 Shaft	Nominal Size Range Inches Over — To
0.3 / 1.9	+1.0 / 0	−0.3 / −0.9	0.6 / 3.2	+1.6 / 0	−0.6 / −1.6	1.0 / 3.6	+1.6 / −0	−1.0 / −2.0	2.5 / 6.6	+2.5 / 0	−2.5 / −4.1	4 / 12	+4 / 0	−4 / −8	5 / 17	+6 / 0	−5 / −11	0 — 0.12
0.4 / 2.3	+1.2 / 0	−0.4 / −1.1	0.8 / 3.8	+1.8 / 0	−0.8 / −2.0	1.2 / 4.2	+1.8 / −0	−1.2 / −2.4	2.8 / 7.6	+3.0 / 0	−2.8 / −4.6	4.5 / 14.5	+5 / 0	−4.5 / −9.5	6 / 20	+7 / 0	−6 / −13	0.12 — 0.24
0.5 / 2.8	+1.4 / 0	−0.5 / −1.4	1.0 / 4.6	+2.2 / 0	−1.0 / −2.4	1.6 / 5.2	+2.2 / −0	−1.6 / −3.0	3.0 / 8.7	+3.5 / 0	−3.0 / −5.2	5 / 17	+6 / 0	−5 / −11	7 / 25	+9 / −0	−7 / −16	0.24 — 0.40
0.6 / 3.2	+1.6 / 0	−0.6 / −1.6	1.2 / 5.6	+2.8 / 0	−1.2 / −2.8	2.0 / 6.4	+2.8 / −0	−2.0 / −3.6	3.5 / 10.3	+4.0 / 0	−3.5 / −6.3	6 / 20	+7 / 0	−6 / −13	8 / 28	+10 / −0	−8 / −18	0.40 — 0.71
0.8 / 4.0	+2.0 / 0	−0.8 / −2.0	1.6 / 7.1	+3.5 / 0	−1.6 / −3.6	2.5 / 8.0	+3.5 / −0	−2.5 / −4.5	4.5 / 13.0	+5.0 / 0	−4.5 / −8.0	7 / 23	+8 / 0	−7 / −15	10 / 34	+12 / 0	−10 / −22	0.71 — 1.19
1.0 / 5.1	+2.5 / 0	−1.0 / −2.6	2.0 / 8.5	+4.0 / 0	−2.0 / −4.5	3.0 / 9.5	+4.0 / −0	−3.0 / −5.5	5 / 15	+6 / 0	−5 / −9	8 / 28	+10 / −0	−8 / −18	12 / 44	+16 / 0	−12 / −28	1.19 — 1.97
1.2 / 6.0	+3.0 / 0	−1.2 / −3.0	2.5 / 10.0	+4.5 / 0	−2.5 / −5.5	4.0 / 11.5	+4.5 / −0	−4.0 / −7.0	6 / 17.5	+7 / 0	−6 / −10.5	10 / 34	+12 / −0	−10 / −22	14 / 50	+18 / −0	−14 / −32	1.97 — 3.15
1.4 / 7.1	+3.5 / 0	−1.4 / −3.6	3.0 / 11.5	+5.0 / 0	−3.0 / −6.5	5.0 / 13.5	+5.0 / −0	−5.0 / −8.5	7 / 21	+9 / 0	−7 / −12	11 / 39	+14 / 0	−11 / −25	16 / 60	+22 / −0	−16 / −38	3.15 — 4.73
1.6 / 8.1	+4.0 / 0	−1.6 / −4.1	3.5 / 13.5	+6.0 / 0	−3.5 / −7.5	6 / 16	+6 / −0	−6 / −10	8 / 24	+10 / 0	−8 / −14	12 / 44	+16 / 0	−12 / −28	18 / 68	+25 / −0	−18 / −43	4.73 — 7.09
2.0 / 9.3	+4.5 / 0	−2.0 / −4.8	4.0 / 15.5	+7.0 / 0	−4.0 / −8.5	7 / 18.5	+7 / −0	−7 / −11.5	10 / 29	+12 / 0	−10 / −17	16 / 52	+18 / 0	−16 / −34	22 / 78	+28 / −0	−22 / −50	7.09 — 9.85
2.2 / 10.2	+5.0 / 0	−2.2 / −5.2	4.5 / 17.5	+8.0 / 0	−4.5 / −9.5	7 / 20	+8 / −0	−7 / −12	12 / 32	+12 / 0	−12 / −20	20 / 60	+20 / 0	−20 / −40	28 / 88	+30 / −0	−28 / −58	9.85 — 12.41
2.5 / 12.0	+6.0 / 0	−2.5 / −6.0	5.0 / 20.0	+9.0 / 0	−5 / −11	8 / 23	+9 / −0	−8 / −14	14 / 37	+14 / 0	−14 / −23	22 / 66	+22 / 0	−22 / −44	30 / 100	+35 / −0	−30 / −65	12.41 — 15.75
2.8 / 12.8	+6.0 / 0	−2.8 / −6.8	5.0 / 21.0	+10.0 / 0	−5 / −11	9 / 25	+10 / −0	−9 / −15	16 / 42	+16 / 0	−16 / −26	25 / 75	+25 / −0	−25 / −50	35 / 115	+40 / −0	−35 / −75	15.75 — 19.69
3.0 / 16.0	+8.0 / 0	−3.0 / −8.0	6.0 / 26.0	+12.0 / −0	−6 / −14	10 / 30	+12 / −0	−10 / −18	18 / 50	+20 / 0	−18 / −30	28 / 88	+30 / −0	−28 / −58	40 / 140	+50 / −0	−40 / −90	19.69 — 30.09
3.5 / 19.5	+10.0 / 0	−3.5 / −9.5	7.0 / 33.0	+16.0 / −0	−7 / −17	12 / 38	+16 / −0	−12 / −22	20 / 61	+25 / 0	−20 / −36	30 / 110	+40 / −0	−30 / −70	45 / 165	+60 / −0	−45 / −105	30.09 — 41.49
4.0 / 24.0	+12.0 / 0	−4.0 / −12.0	8.0 / 40.0	+20.0 / −0	−8 / −20	14 / 46	+20 / −0	−14 / −26	25 / 75	+30 / −0	−25 / −45	40 / 140	+50 / −0	−40 / −90	60 / 220	+80 / −0	−60 / −140	41.49 — 56.19
4.5 / 30.5	+16.0 / 0	−4.5 / −14.5	9.0 / 50.0	+25.0 / −0	−9 / −25	16 / 57	+25 / −0	−16 / −32	30 / 95	+40 / −0	−30 / −55	50 / 170	+60 / −0	−50 / 110	70 / 270	+100 / −0	−70 / −170	56.19 — 76.39
5.0 / 37.0	+20.0 / 0	−5 / −17	10.0 / 60.0	+30.0 / −0	−10 / −30	18 / 68	+30 / −0	−18 / −38	35 / 115	+50 / −0	−35 / −65	50 / 210	+80 / −0	−50 / −130	80 / 330	+125 / −0	−80 / −205	76.39 — 100.9
6.0 / 47.0	+25.0 / 0	−6 / −22	12.0 / 67.0	+40.0 / −0	−12 / −27	20 / 85	+40 / −0	−20 / −45	40 / 140	+60 / −0	−40 / −80	60 / 260	+100 / −0	−60 / −160	90 / 410	+160 / −0	−90 / −250	100.9 — 131.9
7.0 / 57.0	+30.0 / 0	−7 / −27	14.0 / 94.0	+50.0 / −0	−14 / −44	25 / 105	+50 / −0	−25 / −55	50 / 180	+80 / −0	−50 / −100	80 / 330	+125 / −0	−80 / −205	100 / 500	+200 / −0	−100 / −300	131.9 — 171.9
7.0 / 72.0	+40.0 / 0	−7 / −32	14.0 / 114.0	+60.0 / −0	−14 / −54	25 / 125	+60 / −0	−25 / −65	50 / 210	+100 / −0	−50 / −110	90 / 410	+160 / −0	−90 / −250	125 / 625	+250 / −0	−125 / −375	171.9 — 200

End of Table

LOCATIONAL TRANSITION FITS

Limits are in thousandths of an inch.

Limits for hole and shaft are applied algebraically to the basic size to obtain the limits of size for the mating parts.

Data in bold face are in accordance with ABC agreements.

"Fit" represents the maximum interference (minus values) and the maximum clearance (plus values).

Symbols H7, js6, etc., are Hole and Shaft designations used in ABC System

Nominal Size Range Inches Over	To	Class LT 1 Fit	LT 1 Hole H7	LT 1 Shaft js6	Class LT 2 Fit	LT 2 Hole H8	LT 2 Shaft js7	Class LT 3 Fit	LT 3 Hole H7	LT 3 Shaft k6	Class LT 4 Fit	LT 4 Hole H8	LT 4 Shaft k7	Class LT 5 Fit	LT 5 Hole H7	LT 5 Shaft n6	Class LT 6 Fit	LT 6 Hole H7	LT 6 Shaft n7
0	0.12	−0.10 / +0.50	+0.4 / −0	+0.10 / −0.10	−0.2 / +0.8	+0.6 / −0	+0.2 / −0.2							−0.5 / +0.15	+0.4 / −0	+0.5 / +0.25	−0.65 / +0.15	+0.4 / −0	−0.65 / +0.25
0.12	0.24	−0.15 / +0.65	+0.5 / −0	+0.15 / −0.15	−0.25 / +0.95	+0.7 / −0	+0.25 / −0.25							−0.6 / +0.2	+0.5 / −0	+0.6 / +0.3	−0.8 / +0.2	+0.5 / −0	+0.8 / +0.3
0.24	0.40	−0.2 / +0.8	+0.6 / −0	+0.2 / −0.2	−0.3 / +1.2	+0.9 / −0	+0.3 / −0.3	−0.5 / +0.5	+0.6 / −0	+0.5 / +0.1	−0.7 / +0.8	+0.9 / −0	+0.7 / +0.1	−0.8 / +0.2	+0.6 / −0	+0.8 / +0.4	−1.0 / +0.2	+0.6 / −0	+1.0 / +0.4
0.40	0.71	−0.2 / +0.9	+0.7 / −0	+0.2 / −0.2	−0.35 / +1.35	+1.0 / −0	+0.35 / −0.35	−0.5 / +0.6	+0.7 / −0	+0.5 / +0.1	−0.8 / +0.9	+1.0 / −0	+0.8 / +0.1	−0.9 / +0.2	+0.7 / −0	+0.9 / +0.5	−1.2 / +0.2	+0.7 / −0	+1.2 / +0.5
0.71	1.19	−0.25 / +1.05	+0.8 / −0	+0.25 / −0.25	−0.4 / +1.6	+1.2 / −0	+0.4 / −0.4	−0.6 / +0.7	+0.8 / −0	+0.6 / +0.1	−0.9 / +1.1	+1.2 / −0	+0.9 / +0.1	−1.1 / +0.2	+0.8 / −0	+1.1 / +0.6	−1.4 / +0.2	+0.8 / −0	+1.4 / +0.6
1.19	1.97	−0.3 / +1.3	+1.0 / −0	+0.3 / −0.3	−0.5 / +2.1	+1.6 / −0	+0.5 / −0.5	−0.7 / +0.9	+1.0 / −0	+0.7 / +0.1	−1.1 / +1.5	+1.6 / −0	+1.1 / +0.1	−1.3 / +0.3	+1.0 / −0	+1.3 / +0.7	−1.7 / +0.3	+1.0 / −0	+1.7 / +0.7
1.97	3.15	−0.3 / +1.5	+1.2 / −0	+0.3 / −0.3	−0.6 / +2.4	+1.8 / −0	+0.6 / −0.6	−0.8 / +1.1	+1.2 / −0	+0.8 / +0.1	−1.3 / +1.7	+1.8 / −0	+1.3 / +0.1	−1.5 / +0.4	+1.2 / −0	+1.5 / +0.8	−2.0 / +0.4	+1.2 / −0	+2.0 / +0.8
3.15	4.73	−0.4 / +1.8	+1.4 / −0	+0.4 / −0.4	−0.7 / +2.9	+2.2 / −0	+0.7 / −0.7	−1.0 / +1.3	+1.4 / −0	+1.0 / +0.1	−1.5 / +2.1	+2.2 / −0	+1.5 / +0.1	−1.9 / +0.4	+1.4 / −0	+1.9 / +1.0	−2.4 / +0.4	+1.4 / −0	+2.4 / +1.0
4.73	7.09	−0.5 / +2.1	+1.6 / −0	+0.5 / −0.5	−0.8 / +3.3	+2.5 / −0	+0.8 / −0.8	−1.1 / +1.5	+1.6 / −0	+1.1 / +0.1	−1.7 / +2.4	+2.5 / −0	+1.7 / +0.1	−2.2 / +0.4	+1.6 / −0	+2.2 / +1.2	−2.8 / +0.4	+1.6 / −0	+2.8 / +1.2
7.09	9.85	−0.6 / +2.4	+1.8 / −0	+0.6 / −0.6	−0.9 / +3.7	+2.8 / −0	+0.9 / −0.9	−1.4 / +1.6	+1.8 / −0	+1.4 / +0.2	−2.0 / +2.6	+2.8 / −0	+2.0 / +0.2	−2.6 / +0.4	+1.8 / −0	+2.6 / +1.4	−3.2 / +0.4	+1.8 / −0	+3.2 / +1.4
9.85	12.41	−0.6 / +2.6	+2.0 / −0	+0.6 / −0.6	−1.0 / +4.0	+3.0 / −0	+1.0 / −1.0	−1.4 / +1.8	+2.0 / −0	+1.4 / +0.2	−2.2 / +2.8	+3.0 / −0	+2.2 / +0.2	−2.6 / +0.6	+2.0 / −0	+2.6 / +1.4	−3.4 / +0.6	+2.0 / −0	+3.4 / +1.4
12.41	15.75	−0.7 / +2.9	+2.2 / −0	+0.7 / −0.7	−1.0 / +4.5	+3.5 / −0	+1.0 / −1.0	−1.6 / +2.0	+2.2 / −0	+1.6 / +0.2	−2.4 / +3.3	+3.5 / −0	+2.4 / +0.2	−3.0 / +0.6	+2.2 / −0	+3.0 / +1.6	−3.8 / +0.6	+2.2 / −0	+3.8 / +1.6
15.75	19.69	−0.8 / +3.3	+2.5 / −0	+0.8 / −0.8	−1.2 / +5.2	+4.0 / −0	+1.2 / −1.2	−1.8 / +2.3	+2.5 / −0	+1.8 / +0.2	−2.7 / +3.8	+4.0 / −0	+2.7 / +0.2	−3.4 / +0.7	+2.5 / −0	+3.4 / +1.8	−4.3 / +0.7	+2.5 / −0	+4.3 / +1.8

(ANSI)

FORCE AND SHRINK FITS

Limits are in thousandths of an inch.

Limits for hole and shaft are applied algebraically to the basic size to obtain the limits of size for the parts.

Data in bold face are in accordance with ABC agreements.

Symbols H7, s6, etc., are Hole and Shaft designations used in ABC System

Nominal Size Range Inches (Over – To)	FN 1 Limits of Interference	FN 1 Hole H6	FN 1 Shaft	FN 2 Limits of Interference	FN 2 Hole H7	FN 2 Shaft s6	FN 3 Limits of Interference	FN 3 Hole H7	FN 3 Shaft r6	FN 4 Limits of Interference	FN 4 Hole H7	FN 4 Shaft u6	FN 5 Limits of Interference	FN 5 Hole H8	FN 5 Shaft x7
0 – 0.12	0.05 / 0.5	+0.25 / –0	+0.5 / +0.3	0.2 / 0.85	+0.4 / –0	+0.85 / +0.6				0.3 / 0.95	+0.4 / –0	+0.95 / +0.7	0.3 / 1.3	+0.6 / –0	+1.3 / +0.9
0.12 – 0.24	0.1 / 0.6	+0.3 / –0	+0.6 / +0.4	0.2 / 1.0	+0.5 / –0	+1.0 / +0.7				0.4 / 1.2	+0.5 / –0	+1.2 / +0.9	0.5 / 1.7	+0.7 / –0	+1.7 / +1.2
0.24 – 0.40	0.1 / 0.75	+0.4 / –0	+0.75 / +0.5	0.4 / 1.4	+0.6 / –0	+1.4 / +1.0				0.6 / 1.6	+0.6 / –0	+1.6 / +1.2	0.5 / 2.0	+0.9 / –0	+2.0 / +1.4
0.40 – 0.56	0.1 / 0.8	+0.4 / –0	+0.8 / +0.5	0.5 / 1.6	+0.7 / –0	+1.6 / +1.2				0.7 / 1.8	+0.7 / –0	+1.8 / +1.4	0.6 / 2.3	+1.0 / –0	+2.3 / +1.6
0.56 – 0.71	0.2 / 0.9	+0.4 / –0	+0.9 / +0.6	0.5 / 1.6	+0.7 / –0	+1.6 / +1.2				0.7 / 1.8	+0.7 / –0	+1.8 / +1.4	0.8 / 2.5	+1.0 / –0	+2.5 / +1.8
0.71 – 0.95	0.2 / 1.1	+0.5 / –0	+1.1 / +0.7	0.6 / 1.9	+0.8 / –0	+1.9 / +1.4				0.8 / 2.1	+0.8 / –0	+2.1 / +1.6	1.0 / 3.0	+1.2 / –0	+3.0 / +2.2
0.95 – 1.19	0.3 / 1.2	+0.5 / –0	+1.2 / +0.8	0.6 / 1.9	+0.8 / –0	+1.9 / +1.4	0.8 / 2.1	+0.8 / –0	+2.1 / +1.6	1.0 / 2.3	+0.8 / –0	+2.3 / +1.8	1.3 / 3.3	+1.2 / –0	+3.3 / +2.5
1.19 – 1.58	0.3 / 1.3	+0.6 / –0	+1.3 / +0.9	0.8 / 2.4	+1.0 / –0	+2.4 / +1.8	1.0 / 2.6	+1.0 / –0	+2.6 / +2.0	1.5 / 3.1	+1.0 / –0	+3.1 / +2.5	1.4 / 4.0	+1.6 / –0	+4.0 / +3.0
1.58 – 1.97	0.4 / 1.4	+0.6 / –0	+1.4 / +1.0	0.8 / 2.4	+1.0 / –0	+2.4 / +1.8	1.2 / 2.8	+1.0 / –0	+2.8 / +2.2	1.8 / 3.4	+1.0 / –0	+3.4 / +2.8	2.4 / 5.0	+1.6 / –0	+5.0 / +4.0
1.97 – 2.56	0.6 / 1.8	+0.7 / –0	+1.8 / +1.3	0.8 / 2.7	+1.2 / –0	+2.7 / +2.0	1.3 / 3.2	+1.2 / –0	+3.2 / +2.5	2.3 / 4.2	+1.2 / –0	+4.2 / +3.5	3.2 / 6.2	+1.8 / –0	+6.2 / +5.0
2.56 – 3.15	0.7 / 1.9	+0.7 / –0	+1.9 / +1.4	1.0 / 2.9	+1.2 / –0	+2.9 / +2.2	1.8 / 3.7	+1.2 / –0	+3.7 / +3.0	2.8 / 4.7	+1.2 / –0	+4.7 / +4.0	4.2 / 7.2	+1.8 / –0	+7.2 / +6.0
3.15 – 3.94	0.9 / 2.4	+0.9 / –0	+2.4 / +1.8	1.4 / 3.7	+1.4 / –0	+3.7 / +2.8	2.1 / 4.4	+1.4 / –0	+4.4 / +3.5	3.6 / 5.9	+1.4 / –0	+5.9 / +5.0	4.8 / 8.4	+2.2 / –0	+8.4 / +7.0
3.94 – 4.73	1.1 / 2.6	+0.9 / –0	+2.6 / +2.0	1.6 / 3.9	+1.4 / –0	+3.9 / +3.0	2.6 / 4.9	+1.4 / –0	+4.9 / +4.0	4.6 / 6.9	+1.4 / –0	+6.9 / +6.0	5.8 / 9.4	+2.2 / –0	+9.4 / +8.0
4.73 – 5.52	1.2 / 2.9	+1.0 / –0	+2.9 / +2.2	1.9 / 4.5	+1.6 / –0	+4.5 / +3.5	3.4 / 6.0	+1.6 / –0	+6.0 / +5.0	5.4 / 8.0	+1.6 / –0	+8.0 / +7.0	7.5 / 11.6	+2.5 / –0	+11.6 / +10.0
5.52 – 6.30	1.5 / 3.2	+1.0 / –0	+3.2 / +2.5	2.4 / 5.0	+1.6 / –0	+5.0 / +4.0	3.4 / 6.0	+1.6 / –0	+6.0 / +5.0	5.4 / 8.0	+1.6 / –0	+8.0 / +7.0	9.5 / 13.6	+2.5 / –0	+13.6 / +12.0
6.30 – 7.09	1.8 / 3.5	+1.0 / –0	+3.5 / +2.8	2.9 / 5.5	+1.6 / –0	+5.5 / +4.5	4.4 / 7.0	+1.6 / –0	+7.0 / +6.0	6.4 / 9.0	+1.6 / –0	+9.0 / +8.0	9.5 / 13.6	+2.5 / –0	+13.6 / +12.0
7.09 – 7.88	1.8 / 3.8	+1.2 / –0	+3.8 / +3.0	3.2 / 6.2	+1.8 / –0	+6.2 / +5.0	5.2 / 8.2	+1.8 / –0	+8.2 / +7.0	7.2 / 10.2	+1.8 / –0	+10.2 / +9.0	11.2 / 15.8	+2.8 / –0	+15.8 / +14.0
7.88 – 8.86	2.3 / 4.3	+1.2 / –0	+4.3 / +3.5	3.2 / 6.2	+1.8 / –0	+6.2 / +5.0	5.2 / 8.2	+1.8 / –0	+8.2 / +7.0	8.2 / 11.2	+1.8 / –0	+11.2 / +10.0	13.2 / 17.8	+2.8 / –0	+17.8 / +16.0
8.86 – 9.85	2.3 / 4.3	+1.2 / –0	+4.3 / +3.5	4.2 / 7.2	+1.8 / –0	+7.2 / +6.0	6.2 / 9.2	+1.8 / –0	+9.2 / +8.0	10.2 / 13.2	+1.8 / –0	+13.2 / +12.0	13.2 / 17.8	+2.8 / –0	+17.8 / +16.0
9.85 – 11.03	2.8 / 4.9	+1.2 / –0	+4.9 / +4.0	4.0 / 7.2	+2.0 / –0	+7.2 / +6.0	7.0 / 10.2	+2.0 / –0	+10.2 / +9.0	10.0 / 13.2	+2.0 / –0	+13.2 / +12.0	15.0 / 20.0	+3.0 / –0	+20.0 / +18.0
11.03 – 12.41	2.8 / 4.9	+1.2 / –0	+4.9 / +4.0	5.0 / 8.2	+2.0 / –0	+8.2 / +7.0	7.0 / 10.2	+2.0 / –0	+10.2 / +9.0	12.0 / 15.2	+2.0 / –0	+15.2 / +14.0	17.0 / 22.0	+3.0 / –0	+22.0 / +20.0
12.41 – 13.98	3.1 / 5.5	+1.4 / –0	+5.5 / +4.5	5.8 / 9.4	+2.2 / –0	+9.4 / +8.0	7.8 / 11.4	+2.2 / –0	+11.4 / +10.0	13.8 / 17.4	+2.2 / –0	+17.4 / +16.0	18.5 / 24.2	+3.5 / –0	+24.2 / +22.0
13.98 – 15.75	3.6 / 6.1	+1.4 / –0	+6.1 / +5.0	5.8 / 9.4	+2.2 / –0	+9.4 / +8.0	9.8 / 13.4	+2.2 / –0	+13.4 / +12.0	15.8 / 19.4	+2.2 / –0	+19.4 / +18.0	21.5 / 27.2	+3.5 / –0	+27.2 / +25.0
15.75 – 17.72	4.4 / 7.0	+1.6 / –0	+7.0 / +6.0	6.5 / 10.6	+2.5 / –0	+10.6 / +9.0	9.5 / 13.6	+2.5 / –0	+13.6 / +12.0	17.5 / 21.6	+2.5 / –0	+21.6 / +20.0	24.0 / 30.5	+4.0 / –0	+30.5 / +28.0
17.72 – 19.69	4.4 / 7.0	+1.6 / –0	+7.0 / +6.0	7.5 / 11.6	+2.5 / –0	+11.6 / +10.0	11.5 / 15.6	+2.5 / –0	+15.6 / +14.0	19.5 / 23.6	+2.5 / –0	+23.6 / +22.0	26.0 / 32.5	+4.0 / –0	+32.5 / +30.0

LOCATIONAL INTERFERENCE FITS

Limits are in thousandths of an inch.

Limits for hole and shaft are applied algebraically to the basic size to obtain the limits of size for the parts.

Data in bold face are in accordance with ABC agreements,

Symbols H7, p6, etc., are Hole and Shaft designations used in ABC System

Nominal Size Range Inches (Over – To)	LN 1 Limits of Interference	LN 1 Hole H6	LN 1 Shaft n5	LN 2 Limits of Interference	LN 2 Hole H7	LN 2 Shaft p6	LN 3 Limits of Interference	LN 3 Hole H7	LN 3 Shaft r6
0 – 0.12	0 / 0.45	+0.25 / –0	+0.45 / +0.25	0 / 0.65	+0.4 / –0	+0.65 / +0.4	0.1 / 0.75	+0.4 / –0	+0.75 / +0.5
0.12 – 0.24	0 / 0.5	+0.3 / –0	+0.5 / +0.3	0 / 0.8	+0.5 / –0	+0.8 / +0.5	0.1 / 0.9	+0.5 / –0	+0.9 / +0.6
0.24 – 0.40	0 / 0.65	+0.4 / –0	+0.65 / +0.4	0 / 1.0	+0.6 / –0	+1.0 / +0.6	0.2 / 1.2	+0.6 / –0	+1.2 / +0.8
0.40 – 0.71	0 / 0.8	+0.4 / –0	+0.8 / +0.4	0 / 1.1	+0.7 / –0	+1.1 / +0.7	0.3 / 1.4	+0.7 / –0	+1.4 / +1.0
0.71 – 1.19	0 / 1.0	+0.5 / –0	+1.0 / +0.5	0 / 1.3	+0.8 / –0	+1.3 / +0.8	0.4 / 1.7	+0.8 / –0	+1.7 / +1.2
1.19 – 1.97	0 / 1.1	+0.6 / –0	+1.1 / +0.6	0 / 1.6	+1.0 / –0	+1.6 / +1.0	0.4 / 2.0	+1.0 / –0	+2.0 / +1.4
1.97 – 3.15	0.1 / 1.3	+0.7 / –0	+1.3 / +0.8	0.2 / 2.1	+1.2 / –0	+2.1 / +1.4	0.4 / 2.3	+1.2 / –0	+2.3 / +1.6
3.15 – 4.73	0.1 / 1.6	+0.9 / –0	+1.6 / +1.0	0.2 / 2.5	+1.4 / –0	+2.5 / +1.6	0.6 / 2.9	+1.4 / –0	+2.9 / +2.0
4.73 – 7.09	0.2 / 1.9	+1.0 / –0	+1.9 / +1.2	0.2 / 2.8	+1.6 / –0	+2.8 / +1.8	0.9 / 3.5	+1.6 / –0	+3.5 / +2.5
7.09 – 9.85	0.2 / 2.2	+1.2 / –0	+2.2 / +1.4	0.2 / 3.2	+1.8 / –0	+3.2 / +2.0	1.2 / 4.2	+1.8 / –0	+4.2 / +3.0
9.85 – 12.41	0.2 / 2.3	+1.2 / –0	+2.3 / +1.4	0.2 / 3.4	+2.0 / –0	+3.4 / +2.2	1.5 / 4.7	+2.0 / –0	+4.7 / +3.5
12.41 – 15.75	0.2 / 2.6	+1.4 / –0	+2.6 / +1.6	0.3 / 3.9	+2.2 / –0	+3.9 / +2.5	2.3 / 5.9	+2.2 / –0	+5.9 / +4.5
15.75 – 19.69	0.2 / 2.8	+1.6 / –0	+2.8 / +1.8	0.3 / 4.4	+2.5 / –0	+4.4 / +2.8	2.5 / 6.6	+2.5 / –0	+6.6 / +5.0
19.69 – 30.09		+2.0 / –0		0.5 / 5.5	+3 / –0	+5.5 / +3.5	4 / 9	+3 / –0	+9 / +7
30.09 – 41.49		+2.5 / –0		0.5 / 7.0	+4 / –0	+7.0 / +4.5	5 / 11.5	+4 / –0	+11.5 / +9
41.49 – 56.19		+3.0 / –0		1 / 9	+5 / –0	+9 / +6	7 / 15	+5 / –0	+15 / +12
56.19 – 76.39		+4.0 / –0		1 / 11	+6 / –0	+11 / +7	10 / 20	+6 / –0	+20 / +16
76.39 – 100.9		+5.0 / –0		1 / 14	+8 / –0	+14 / +9	12 / 25	+8 / –0	+25 / +20
100.9 – 131.9		+6.0 / –0		2 / 18	+10 / –0	+18 / +12	15 / 31	+10 / –0	+31 / +25
131.9 – 171.9		+8.0 / –0		4 / 24	+12 / –0	+24 / +16	18 / 38	+12 / –0	+38 / +30
171.9 – 200		+10.0 / –0		4 / 30	+16 / –0	+30 / +20	24 / 50	+16 / –0	+50 / +40

(ANSI)

DRILL SIZE DECIMAL EQUIVALENTS – METRIC AND INCH

NUMBER AND LETTER DRILLS.

Drill No.	Frac	Deci	Drill No.	Frac	Deci	Drill No.	Frac	Deci
80		.0135	28		.140	S		.348
79		.0145		9/64	.141		23/64	.359
	1/64	.0156	27		.144	T		.358
78		.0160	26		.147		3/8	.375
77		.0180	25		.150	U		.368
76		.0200	24		.152	V		.377
75		.0210	23		.154	W		.386
74		.0225		5/32	.156		25/64	.391
73		.0240	22		.157	X		.397
72		.0250	21		.159	Y		.404
71		.0260	20		.161		13/32	.406
70		.0280	19		.166	Z		.413
69		.0292	18		.170		27/64	.422
68		.0310		11/64	.172		7/16	.438
	1/32	.0313	17		.173		29/64	.453
67		.0320	16		.177		15/32	.469
66		.0330	15		.180		31/64	.484
65		.0350	14		.182		1/2	.500
64		.0360	13		.185		33/64	.516
63		.0370		3/16	.188		17/32	.531
62		.0380	12		.189		35/64	.547
61		.0390	11		.191		9/16	.562
60		.0400	10		.194		37/64	.578
59		.0410	9		.196		19/32	.594
58		.0420	8		.199		39/64	.609
57		.0430	7		.201		5/8	.625
56		.0465		13/64	.203		41/64	.641
	3/64	.0469	6		.204		21/32	.656
55		.0520	5		.206		43/64	.672
54		.0550	4		.209		11/16	.688
53		.0595	3		.213		45/64	.703
	1/16	.0625		7/32	.219		23/32	.719
52		.0635	2		.221		47/64	.734
51		.0670	1		.228		3/4	.750
50		.0700	A		.234		49/64	.766
49		.0730		15/64	.234		25/32	.781
48		.0760	B		.238		51/64	.797
	5/64	.0781	C		.242		13/16	.813
47		.0785	D		.246		53/64	.828
46		.0810		1/4	.250		27/32	.844
45		.0820	E		.250		55/64	.859
44		.0860	F		.257		7/8	.875
43		.0890	G		.261		57/64	.891
42		.0935		17/64	.266		29/32	.906
	3/32	.0938	H		.266		59/64	.922
41		.0960	I		.272		15/16	.938
40		.0980	J		.277		61/64	.953
39		.0995	K		.281		31/32	.969
38		.1015		9/32	.281		63/64	.984
37		.1040	L		.290		1	1.000
36		.1065	M		.295			
	7/64	.1094		19/64	.297			
35		.1100	N		.302			
34		.1110		5/16	.313			
33		.1130	O		.316			
32		.116	P		.323			
31		.120		21/64	.328			
	1/8	.125	Q		.332			
30		.129	R		.339			
29		.136		11/32	.344			

METRIC DRILLS.

MM	DEC.	MM	DEC.	MM	DEC.	MM	DEC.
1.	.0394	3.2	.1260	6.3	.2480	9.5	.3740
1.05	.0413	3.25	.1280	6.4	.2520	9.6	.3780
1.1	.0433	3.3	.1299	6.5	.2559	9.7	.3819
1.15	.0453	3.4	.1339	6.6	.2598	9.75	.3839
1.2	.0472	3.5	.1378	6.7	.2638	9.8	.3858
1.25	.0492	3.6	.1417	6.75	.2657	9.9	.3898
1.3	.0512	3.7	.1457	6.8	.2677	10.	.3937
1.35	.0531	3.75	.1476	6.9	.2717	10.5	.4134
1.4	.0551	3.8	.1496	7.	.2756	11.	.4331
1.45	.0571	3.9	.1535	7.1	.2795	11.5	.4528
1.5	.0591	4.	.1575	7.2	.2835	12.	.4724
1.55	.0610	4.1	.1614	7.25	.2854	12.5	.4921
1.6	.0630	4.2	.1654	7.3	.2874	13.	.5118
1.65	.0650	4.25	.1673	7.4	.2913	13.5	.5315
1.7	.0669	4.3	.1693	7.5	.2953	14.	.5512
1.75	.0689	4.4	.1732	7.6	.2992	14.5	.5709
1.8	.0709	4.5	.1772	7.7	.3031	15.	.5906
1.85	.0728	4.6	.1811	7.75	.3051	15.5	.6102
1.9	.0748	4.7	.1850	7.8	.3071	16.	.6299
1.95	.0768	4.75	.1870	7.9	.3110	16.5	.6496
2.	.0787	4.8	.1890	8.	.3150	17.	.6693
2.05	.0807	4.9	.1929	8.1	.3189	17.5	.6890
2.1	.0827	5.	.1968	8.2	.3228	18.	.7087
2.15	.0846	5.1	.2008	8.25	.3248	18.5	.7283
2.2	.0866	5.2	.2047	8.3	.3268	19.	.7480
2.25	.0886	5.25	.2067	8.4	.3307	19.5	.7677
2.3	.0906	5.3	.2087	8.5	.3346	20.	.7874
2.35	.0925	5.4	.2126	8.6	.3386	20.5	.8071
2.4	.0945	5.5	.2165	8.7	.3425	21.	.8268
2.45	.0965	5.6	.2205	8.75	.3445	21.5	.8465
2.5	.0984	5.7	.2244	8.8	.3465	22.	.8661
2.6	.1024	5.75	.2264	8.9	.3504	22.5	.8858
2.7	.1063	5.8	.2283	9.	.3543	23.	.9055
2.75	.1083	5.9	.2323	9.1	.3583	23.5	.9252
2.8	.1102	6.	.2362	9.2	.3622	24.	.9449
2.9	.1142	6.1	.2402	9.25	.3642	24.5	.9646
3.	.1181	6.2	.2441	9.3	.3661	25.	.9843
3.1	.1220	6.25	.2461	9.4	.3701		

TAP DRILL SIZES FOR UNIFIED STANDARD SCREW THREADS

Screw Thread Major Diameter	Threads Per Inch	Tap Drill Size Or Number	Screw Thread Major Diameter	Threads Per Inch	Tap Drill Size Or Number
0	80	3/64	3/8	16	5/16
1	64	53		24	Q
	72	53	7/16	14	U
2	56	50		20	25/64
	64	50	1/2	13	27/64
3	48	47		20	29/64
	56	45	9/16	12	31/64
4	40	43		18	33/64
	48	42	5/8	11	17/32
5	40	38		18	37/64
	44	37	3/4	10	21/32
6	32	36		16	11/16
	40	33	7/8	9	49/64
8	32	29		14	13/16
	36	29	1	8	7/8
10	24	25		12	59/64
	32	21	1 1/8	7	63/64
12	24	16		12	1 3/64
	28	14	1 1/4	7	1 7/64
1/4	20	7		12	1 11/64
	28	3	1 3/8	6	1 7/32
5/16	18	F		12	1 19/64
	24	I	1 1/2	6	1 11/32
				12	1 27/64

TAP DRILL SIZES FOR ISO METRIC THREADS

Nominal Size mm	Coarse Pitch mm	Coarse Tap Drill mm	Fine Pitch mm	Fine Tap Drill mm
10	1.5	8.5	1.25	8.75
12	1.75	10.25	1.25	10.50
14	2	12.00	1.5	12.50
16	2	14.00	1.5	14.50
18	2.5	15.50	1.5	16.50
20	2.5	17.50	1.5	18.50
22	2.5	19.50	1.5	20.50
24	3	21.00	2	22.00
27	3	24.00	2	25.00

Nominal Size mm	Coarse Pitch mm	Coarse Tap Drill mm	Fine Pitch mm	Fine Tap Drill mm
1.4	0.3	1.1	—	—
1.6	0.35	1.25	—	—
2	0.4	1.6	—	—
2.5	0.45	2.05	—	—
3	0.5	2.5	—	—
4	0.7	3.3	—	—
5	0.8	4.2	—	—
6	1.0	5.0	—	—
8	1.25	6.75	1	7.0

Sizes Primary	Sizes Secondary	Basic Major Diameter	Series with graded pitches — Coarse UNC	Fine UNF	Extra fine UNEF	4UN	6UN	8UN	12UN	16UN	20UN	28UN	32UN	Sizes
0		0.0600	—	80	—	—	—	—	—	—	—	—	—	0
	1	0.0730	64	72	—	—	—	—	—	—	—	—	—	1
2		0.0860	56	64	—	—	—	—	—	—	—	—	—	2
	3	0.0990	48	56	—	—	—	—	—	—	—	—	—	3
4		0.1120	40	48	—	—	—	—	—	—	—	—	—	4
5		0.1250	40	44	—	—	—	—	—	—	—	—	—	5
6		0.1380	32	40	—	—	—	—	—	—	—	—	UNC	6
8		0.1640	32	36	—	—	—	—	—	—	—	—	UNC	8
10		0.1900	24	32	—	—	—	—	—	—	—	—	UNF	10
	12	0.2160	24	28	32	—	—	—	—	—	—	UNF	UNEF	12
1/4		0.2500	20	28	32	—	—	—	—	—	UNC	UNF	UNEF	1/4
5/16		0.3125	18	24	32	—	—	—	—	—	20	28	UNEF	5/16
3/8		0.3750	16	24	32	—	—	—	—	UNC	20	28	UNEF	3/8
7/16		0.4375	14	20	28	—	—	—	—	16	UNF	UNEF	32	7/16
1/2		0.5000	13	20	28	—	—	—	—	16	UNF	UNEF	32	1/2
9/16		0.5625	12	18	24	—	—	—	UNC	16	20	28	32	9/16
5/8		0.6250	11	18	24	—	—	—	12	16	20	28	32	5/8
	11/16	0.6875	—	—	24	—	—	—	12	16	20	28	32	11/16
3/4		0.7500	10	16	20	—	—	—	12	UNF	UNEF	28	32	3/4
	13/16	0.8125	—	—	20	—	—	—	12	16	UNEF	28	32	13/16
7/8		0.8750	9	14	20	—	—	—	12	16	UNEF	28	32	7/8
	15/16	0.9375	—	—	20	—	—	—	12	16	UNEF	28	32	15/16
1		1.0000	8	12	20	—	—	UNC	UNF	16	UNEF	28	32	1
	1 1/16	1.0625	—	—	18	—	—	8	12	16	20	28	—	1 1/16
1 1/8		1.1250	7	12	18	—	—	8	UNF	16	20	28	—	1 1/8
	1 3/16	1.1875	—	—	18	—	—	8	12	16	20	28	—	1 3/16
1 1/4		1.2500	7	12	18	—	—	8	UNF	16	20	28	—	1 1/4
	1 5/16	1.3125	—	—	18	—	—	8	12	16	20	28	—	1 5/16
1 3/8		1.3750	6	12	18	—	UNC	8	UNF	16	20	28	—	1 3/8
	1 7/16	1.4375	—	—	18	—	6	8	12	16	20	28	—	1 7/16
1 1/2		1.5000	6	12	18	—	UNC	8	UNF	16	20	28	—	1 1/2
	1 9/16	1.5625	—	—	18	—	6	8	12	16	20	—	—	1 9/16
1 5/8		1.6250	—	—	18	—	6	8	12	16	20	—	—	1 5/8
	1 11/16	1.6875	—	—	18	—	6	8	12	16	20	—	—	1 11/16
1 3/4		1.7500	5	—	—	—	6	8	12	16	20	—	—	1 3/4
	1 13/16	1.8125	—	—	—	—	6	8	12	16	20	—	—	1 13/16
1 7/8		1.8750	—	—	—	—	6	8	12	16	20	—	—	1 7/8
	1 15/16	1.9375	—	—	—	—	6	8	12	16	20	—	—	1 15/16
2		2.0000	4½	—	—	—	6	8	12	16	20	—	—	2
	2 1/8	2.1250	—	—	—	—	6	8	12	16	20	—	—	2 1/8
2 1/4		2.2500	4½	—	—	—	6	8	12	16	20	—	—	2 1/4
	2 3/8	2.3750	—	—	—	—	6	8	12	16	20	—	—	2 3/8
2 1/2		2.5000	4	—	—	UNC	6	8	12	16	20	—	—	2 1/2
	2 5/8	2.6250	—	—	—	4	6	8	12	16	20	—	—	2 5/8
2 3/4		2.7500	4	—	—	UNC	6	8	12	16	20	—	—	2 3/4
	2 7/8	2.8750	—	—	—	4	6	8	12	16	20	—	—	2 7/8
3		3.0000	4	—	—	UNC	6	8	12	16	20	—	—	3
	3 1/8	3.1250	—	—	—	4	6	8	12	16	—	—	—	3 1/8
3 1/4		3.2500	4	—	—	UNC	6	8	12	16	—	—	—	3 1/4
	3 3/8	3.3750	—	—	—	4	6	8	12	16	—	—	—	3 3/8
3 1/2		3.5000	4	—	—	UNC	6	8	12	16	—	—	—	3 1/2
	3 5/8	3.6250	—	—	—	4	6	8	12	16	—	—	—	3 5/8
3 3/4		3.7500	4	—	—	UNC	6	8	12	16	—	—	—	3 3/4
	3 7/8	3.8750	—	—	—	4	6	8	12	16	—	—	—	3 7/8
4		4.0000	4	—	—	UNC	6	8	12	16	—	—	—	4
	4 1/8	4.1250	—	—	—	4	6	8	12	16	—	—	—	4 1/8
4 1/4		4.2500	—	—	—	4	6	8	12	16	—	—	—	4 1/4
	4 3/8	4.3750	—	—	—	4	6	8	12	16	—	—	—	4 3/8
4 1/2		4.5000	—	—	—	4	6	8	12	16	—	—	—	4 1/2
	4 5/8	4.6250	—	—	—	4	6	8	12	16	—	—	—	4 5/8
4 3/4		4.7500	—	—	—	4	6	8	12	16	—	—	—	4 3/4
	4 7/8	4.8750	—	—	—	4	6	8	12	16	—	—	—	4 7/8
5		5.0000	—	—	—	4	6	8	12	16	—	—	—	5
	5 1/8	5.1250	—	—	—	4	6	8	12	16	—	—	—	5 1/8
5 1/4		5.2500	—	—	—	4	6	8	12	16	—	—	—	5 1/4
	5 3/8	5.3750	—	—	—	4	6	8	12	16	—	—	—	5 3/8
5 1/2		5.5000	—	—	—	4	6	8	12	16	—	—	—	5 1/2
	5 5/8	5.6250	—	—	—	4	6	8	12	16	—	—	—	5 5/8
5 3/4		5.7500	—	—	—	4	6	8	12	16	—	—	—	5 3/4
	5 7/8	5.8750	—	—	—	4	6	8	12	16	—	—	—	5 7/8
6		6.0000	—	—	—	4	6	8	12	16	—	—	—	6

(ANSI)

ISO METRIC SCREW THREAD STANDARD SERIES

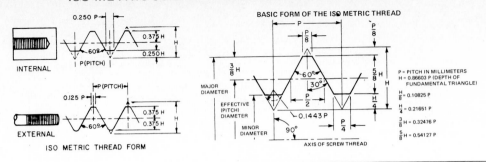

Nominal Size Diam. (mm) Column a			Series With Graded Pitches		Pitches (mm) Series With Constant Pitches												Nominal Size Diam. (mm)
1	2	3	Coarse	Fine	6	4	3	2	1.5	1.25	1	0.75	0.5	0.35	0.25	0.2	(mm)
0.25			0.075	—	—	—	—	—	—	—	—	—	—	—	—	—	0.25
0.3			0.08	—	—	—	—	—	—	—	—	—	—	—	—	—	0.3
	0.35		0.09	—	—	—	—	—	—	—	—	—	—	—	—	—	0.35
0.4			0.1	—	—	—	—	—	—	—	—	—	—	—	—	—	0.4
	0.45		0.1	—	—	—	—	—	—	—	—	—	—	—	—	—	0.45
0.5			0.125	—	—	—	—	—	—	—	—	—	—	—	—	—	0.5
	0.55		0.125	—	—	—	—	—	—	—	—	—	—	—	—	—	0.55
0.6			0.15	—	—	—	—	—	—	—	—	—	—	—	—	—	0.6
	0.7		0.175	—	—	—	—	—	—	—	—	—	—	—	—	—	0.7
0.8			0.2	—	—	—	—	—	—	—	—	—	—	—	—	—	0.8
	0.9		0.225	—	—	—	—	—	—	—	—	—	—	—	—	—	0.9
1			0.25	—	—	—	—	—	—	—	—	—	—	—	—	0.2	1
	1.1		0.25	—	—	—	—	—	—	—	—	—	—	—	—	0.2	1.1
1.2			0.25	—	—	—	—	—	—	—	—	—	—	—	—	0.2	1.2
	1.4		0.3	—	—	—	—	—	—	—	—	—	—	—	—	0.2	1.4
1.6			0.35	—	—	—	—	—	—	—	—	—	—	—	—	0.2	1.6
	1.8		0.35	—	—	—	—	—	—	—	—	—	—	—	—	0.2	1.8
2			0.4	—	—	—	—	—	—	—	—	—	—	—	0.25	—	2
	2.2		0.45	—	—	—	—	—	—	—	—	—	—	—	0.25	—	2.2
2.5			0.45	—	—	—	—	—	—	—	—	—	—	0.35	—	—	2.5
3			0.5	—	—	—	—	—	—	—	—	—	—	0.35	—	—	3
	3.5		0.6	—	—	—	—	—	—	—	—	—	—	0.35	—	—	3.5
4			0.7	—	—	—	—	—	—	—	—	—	0.5	—	—	—	4
	4.5		0.75	—	—	—	—	—	—	—	—	—	0.5	—	—	—	4.5
5			0.8	—	—	—	—	—	—	—	—	—	0.5	—	—	—	5
		5.5	—	—	—	—	—	—	—	—	—	—	0.5	—	—	—	5.5
6			1	—	—	—	—	—	—	—	—	0.75	—	—	—	—	6
		7	1	—	—	—	—	—	—	—	—	0.75	—	—	—	—	7
8			1.25	1	—	—	—	—	—	—	1	0.75	—	—	—	—	8
		9	1.25	—	—	—	—	—	—	—	1	0.75	—	—	—	—	9
10			1.5	1.25	—	—	—	—	—	1.25	1	0.75	—	—	—	—	10
		11	1.5	—	—	—	—	—	—	—	1	0.75	—	—	—	—	11
12			1.75	1.25	—	—	—	—	1.5	1.25	1	—	—	—	—	—	12
	14		2	1.5	—	—	—	—	1.5	1.25b	1	—	—	—	—	—	14
		15	—	—	—	—	—	—	1.5	—	1	—	—	—	—	—	15
16			2	1.5	—	—	—	—	1.5	—	1	—	—	—	—	—	16
		17	—	—	—	—	—	—	1.5	—	1	—	—	—	—	—	17
	18		2.5	1.5	—	—	—	2	1.5	—	1	—	—	—	—	—	18
20			2.5	1.5	—	—	—	2	1.5	—	1	—	—	—	—	—	20
	22		2.5	1.5	—	—	—	2	1.5	—	1	—	—	—	—	—	22
24			3	2	—	—	—	2	1.5	—	1	—	—	—	—	—	24
		25	—	—	—	—	—	2	1.5	—	1	—	—	—	—	—	25
		26	—	—	—	—	—	—	1.5	—	1	—	—	—	—	—	26
	27		3	2	—	—	—	2	1.5	—	1	—	—	—	—	—	27
		28	—	—	—	—	—	2	1.5	—	1	—	—	—	—	—	28
30			3.5	2	—	—	(3)	2	1.5	—	1	—	—	—	—	—	30
		32	—	—	—	—	—	2	1.5	—	—	—	—	—	—	—	32
	33		3.5	2	—	—	(3)	2	1.5	—	—	—	—	—	—	—	33
		35c	—	—	—	—	—	—	1.5	—	—	—	—	—	—	—	35c
36			4	3	—	—	—	2	1.5	—	—	—	—	—	—	—	36
		38	—	—	—	—	—	—	1.5	—	—	—	—	—	—	—	38
	39		4	3	—	—	—	2	1.5	—	—	—	—	—	—	—	39
		40	—	—	—	—	3	2	1.5	—	—	—	—	—	—	—	40
42			4.5	3	—	4	3	2	1.5	—	—	—	—	—	—	—	42
	45		4.5	3	—	4	3	2	1.5	—	—	—	—	—	—	—	45

a Thread diameter should be selected from columns 1, 2 or 3; with preference being given in that order.
b Pitch 1.25 mm in combination with diameter 14 mm has been included for spark plug applications.
c Diameter 35 mm has been included for bearing locknut applications.
The use of pitches shown in parentheses should be avoided wherever possible.
The pitches enclosed in the bold frame, together with the corresponding nominal diameters in Columns 1 and 2, are those combinations which have been established by ISO Recommendations as a selected "coarse" and "fine" series for commercial fasteners. Sizes 0.25 mm through 1.4 mm are covered in ISO Recommendation R 68 and, except for the 0.25 mm size, in AN Standard ANSI B1.10.

(ANSI)

INCH — METRIC THREAD COMPARISON

INCH SERIES			METRIC			
Size	Dia. (In.)	TPI	Size	Dia. (In.)	Pitch (MM)	TPI (Approx)
			M1.4	.055	.3 / .2	85 / 127
#0	.060	80				
			M1.6	.063	.35 / .2	74 / 127
#1	.073	64 / 72				
			M2	.079	.4 / .25	64 / 101
#2	.086	56 / 64				
			M2.5	.098	.45 / .35	56 / 74
#3	.099	48 / 56				
#4	.112	40 / 48				
			M3	.118	.5 / .35	51 / 74
#5	.125	40 / 44				
#6	.138	32 / 40				
			M4	.157	.7 / .5	36 / 51
#8	.164	32 / 36				
#10	.190	24 / 32				
			M5	.196	.8 / .5	32 / 51
			M6	.236	1.0 / .75	25 / 34
1/4	.250	20 / 28				
5/16	.312	18 / 24				
			M8	.315	1.25 / 1.0	20 / 25
3/8	.375	16 / 24				
			M10	.393	1.5 / 1.25	17 / 20
7/16	.437	14 / 20				
			M12	.472	1.75 / 1.25	14.5 / 20
1/2	.500	13 / 20				
			M14	.551	2 / 1.5	12.5 / 17
5/8	.625	11 / 18				
			M16	.630	2 / 1.5	12.5 / 17
			M18	.709	2.5 / 1.5	10 / 17
3/4	.750	10 / 16				
			M20	.787	2.5 / 1.5	10 / 17
			M22	.866	2.5 / 1.5	10 / 17
7/8	.875	9 / 14				
			M24	.945	3 / 2	.8.5 / 12.5
1"	1.000	8 / 12				
			M27	1.063	3 / 2	8.5 / 12.5

(Standard Pressed Steel Co.)

1	2	3	4	5	6	7	8	9	10		11	12
Identification		Basic Diameters			Thread Data							
Nominal Sizes (All Classes)	Threads per Inch,* n	Classes 2G, 3G, and 4G			Pitch, p	Thickness at Pitch Line, $t = p/2$	Basic Height of Thread, $h = p/2$	Basic Width of Flat, $F = 0.3707p$	Lead Angle at Basic Pitch Diameter* Classes 2G, 3G, and 4G, λ		Shear Area† Class 3G	Stress Area‡ Class 3G
		Major Diameter, D	Pitch Diameter,§ $E = D - h$	Minor Diameter, $K = D - 2h$					Deg	Min		
¼	16	0.2500	0.2188	0.1875	0.06250	0.03125	0.03125	0.0232	5	12	0.350	0.0285
5/16	14	0.3125	0.2768	0.2411	0.07143	0.03571	0.03571	0.0265	4	42	0.451	0.0474
3/8	12	0.3750	0.3333	0.2917	0.08333	0.04167	0.04167	0.0309	4	33	0.545	0.0699
7/16	12	0.4375	0.3958	0.3542	0.08333	0.04167	0.04167	0.0309	3	50	0.660	0.1022
½	10	0.5000	0.4500	0.4000	0.10000	0.05000	0.05000	0.0371	4	3	0.749	0.1287
5/8	8	0.6250	0.5625	0.5000	0.12500	0.06250	0.06250	0.0463	4	3	0.941	0.2043
¾	6	0.7500	0.6667	0.5833	0.16667	0.08333	0.08333	0.0618	4	33	1.108	0.2848
7/8	6	0.8750	0.7917	0.7083	0.16667	0.08333	0.08333	0.0618	3	50	1.339	0.4150
1	5	1.0000	0.9000	0.8000	0.20000	0.10000	0.10000	0.0741	4	3	1.519	0.5354
1 1/8	5	1.1250	1.0250	0.9250	0.20000	0.10000	0.10000	0.0741	3	33	1.751	0.709
1 ¼	5	1.2500	1.1500	1.0500	0.20000	0.10000	0.10000	0.0741	3	10	1.983	0.907
1 3/8	4	1.3750	1.2500	1.1250	0.25000	0.12500	0.12500	0.0927	3	39	2.139	1.059
1 ½	4	1.5000	1.3750	1.2500	0.25000	0.12500	0.12500	0.0927	3	19	2.372	1.298
1 ¾	4	1.7500	1.6250	1.5000	0.25000	0.12500	0.12500	0.0927	2	48	2.837	1.851
2	4	2.0000	1.8750	1.7500	0.25000	0.12500	0.12500	0.0927	2	26	3.301	2.501
2 ¼	3	2.2500	2.0833	1.9167	0.33333	0.16667	0.16667	0.1236	2	55	3.643	3.049
2 ½	3	2.5000	2.3333	2.1667	0.33333	0.16667	0.16667	0.1236	2	36	4.110	3.870
2 ¾	3	2.7500	2.5833	2.4167	0.33333	0.16667	0.16667	0.1236	2	21	4.577	4.788
3	2	3.0000	2.7500	2.5000	0.50000	0.25000	0.25000	0.1853	3	19	4.786	5.27
3 ½	2	3.5000	3.2500	3.0000	0.50000	0.25000	0.25000	0.1853	2	48	5.73	7.50
4	2	4.0000	3.7500	3.5000	0.50000	0.25000	0.25000	0.1853	2	26	6.67	10.12
4 ½	2	4.5000	4.2500	4.0000	0.50000	0.25000	0.25000	0.1853	2	9	7.60	13.13
5	2	5.0000	4.7500	4.5000	0.50000	0.25000	0.25000	0.1853	1	55	8.54	16.53

* All other dimensions are given in inches.
§ British: Effective Diameter.
† Per inch length of engagement of the external thread in line with the minor diameter crests of the internal thread. Computed from this formula: Shear Area = $\pi K_n[0.5 + h \tan 14\frac{1}{2}° (E_s - K_n)]$. Figures given are the minimum shear area based on max K_n and min E_s.
‡ Figures given are the minimum stress area based on the mean of the minimum minor and pitch diameters of the external thread.

(ANSI)

SQUARE SCREW THREADS

SIZE	THREADS PER INCH	SIZE	THREADS PER INCH	SIZE	THREADS PER INCH
3/8	12	7/8	5	2	2 1/2
7/16	10	1	5	2 1/4	2
1/2	10	1 1/8	4	2 1/2	2
9/16	8	1 1/4	4	2 3/4	2
5/8	8	1 1/2	3	3	1 1/2
3/4	6	1 3/4	2 1/2	3 1/4	1 1/2

(ANSI)

STANDARD TAPER PIPE THREADS

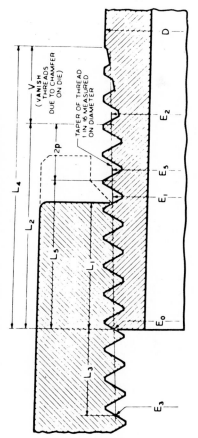

Basic Dimensions of USA (American) Standard Taper Pipe Thread, NPT[1]

(ANSI)

Nominal[8] Pipe Size	Outside Diameter of Pipe, D	Threads per inch, n	Pitch of Thread, p	Pitch Diameter at beginning of External Thread, E_0	Handtight Engagement Length[2], L_1 In.	Handtight Engagement Length[2], L_1 Thds.	Dia[3], E_1	Effective Thread, External Length[4], L_2 In.	Effective Thread, External Length[4], L_2 Thds.	Dia, E_2
1	2	3	4	5	6	7	8	9	10	11
1/16	0.3125	27	0.03704	0.27118	0.160	4.32	0.28118	0.2611	7.05	0.28750
1/8	0.405	27	0.03704	0.36351	0.1615	4.36	0.37360	0.2639	7.12	0.38000
1/4	0.540	18	0.05556	0.47739	0.2278	4.10	0.49163	0.4018	7.23	0.50250
3/8	0.675	18	0.05556	0.61201	0.240	4.32	0.62701	0.4078	7.34	0.63750
1/2	0.840	14	0.07143	0.75843	0.320	4.48	0.77843	0.5337	7.47	0.79179
3/4	1.050	14	0.07143	0.96768	0.339	4.75	0.98887	0.5457	7.64	1.00179
1	1.315	11.5	0.08696	1.21363	0.400	4.60	1.23863	0.6828	7.85	1.25630
1 1/4	1.660	11.5	0.08696	1.55713	0.420	4.83	1.58338	0.7068	8.13	1.60130
1 1/2	1.900	11.5	0.08696	1.79609	0.420	4.83	1.82234	0.7235	8.32	1.84130
2	2.375	11.5	0.08696	2.26902	0.436	5.01	2.29627	0.7565	8.70	2.31630
2 1/2	2.875	8	0.12500	2.71953	0.682	5.46	2.76216	1.1375	9.10	2.79062
3	3.500	8	0.12500	3.34062	0.766	6.13	3.38850	1.2000	9.60	3.41562
3 1/2	4.000	8	0.12500	3.83750	0.821	6.57	3.88881	1.2500	10.00	3.91562
4	4.500	8	0.12500	4.33438	0.844	6.75	4.38712	1.3000	10.40	4.41562
5	5.563	8	0.12500	5.39073	0.937	7.50	5.44929	1.4063	11.25	5.47862
6	6.625	8	0.12500	6.44609	0.958	7.66	6.50597	1.5125	12.10	6.54062
8	8.625	8	0.12500	8.43359	1.063	8.50	8.50003	1.7125	13.70	8.54062
10	10.750	8	0.12500	10.54531	1.210	9.68	10.62094	1.9250	15.40	10.66562
12	12.750	8	0.12500	12.53281	1.360	10.88	12.61781	2.1250	17.00	12.66562
14 OD	14.000	8	0.12500	13.77500	1.562	12.50	13.87262	2.2500	18.00	13.91562
16 OD	16.000	8	0.12500	15.76250	1.812	14.50	15.87575	2.4500	19.60	15.91562
18 OD	18.000	8	0.12500	17.75000	2.000	16.00	17.87500	2.6500	21.20	17.91562
20 OD	20.000	8	0.12500	19.73750	2.125	17.00	19.87031	2.8500	22.80	19.91562
24 OD	24.000	8	0.12500	23.71250	2.375	19.00	23.86094	3.2500	26.00	23.91562

[1] The basic dimensions of the USA (American) Standard Taper Pipe Thread are given in inches to four or five decimal places. While this implies a greater degree of precision than is ordinarily attained, these dimensions are the basis of gage dimensions and are so expressed for the purpose of eliminating errors in computations.

[2] Also length of thin ring gage and length from gaging notch to small end of plug gage.

[3] Also pitch diameter at gaging notch (handtight plane.)

[4] Also length of plug gage.

[5] The length L_5 from the end of the pipe determines the plane beyond which the thread form is incomplete at the crest. The next two threads are complete at the root. At this plane the cone formed by the crests of the thread intersects the cylinder forming the external surface of the pipe. $l_5 = l_2 - 2p$.

[6] Given as *information* for use in selecting tap drills. (See Appendix E.)

[7] Military Specification MIL–P–7105 gives the wrench makeup as three threads for 3 in. and smaller. The E_3 dimensions are as follows: Size 2.5 in. 2.69009 and size 3 in. 3.31719.

[8] Designated, for example, as ⅜ NPT or 0.675 NPT.

SQUARE BOLTS

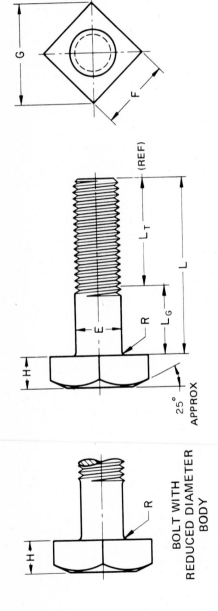

BOLT WITH
REDUCED DIAMETER
BODY

Dimensions of Square Bolts

Nominal Size or Basic Product Dia		E Body Dia.	F Width Across Flats			G Width Across Corners			H Height				R Radius of Fillet			L_T Thread Length For Bolt Lengths		
		Max	Basic	Max	Min	Max	Min	Basic	Max	Min	Max	Min	6 in. and shorter Basic	Over 6 in. Basic
1/4	0.2500	0.260	3/8	0.375	0.362	0.530	0.498	11/64	0.188	0.156	0.03	0.01	0.750	1.000
5/16	0.3125	0.324	1/2	0.500	0.484	0.707	0.665	13/64	0.220	0.186	0.03	0.01	0.875	1.125
3/8	0.3750	0.388	9/16	0.562	0.544	0.795	0.747	1/4	0.268	0.232	0.03	0.01	1.000	1.250
7/16	0.4375	0.452	5/8	0.625	0.603	0.884	0.828	19/64	0.316	0.278	0.03	0.01	1.125	1.375
1/2	0.5000	0.515	3/4	0.750	0.725	1.061	0.995	21/64	0.348	0.308	0.03	0.01	1.250	1.500
5/8	0.6250	0.642	15/16	0.938	0.906	1.326	1.244	27/64	0.444	0.400	0.06	0.02	1.500	1.750
3/4	0.7500	0.768	1 1/8	1.125	1.088	1.591	1.494	1/2	0.524	0.476	0.06	0.02	1.750	2.000
7/8	0.8750	0.895	1 5/16	1.312	1.269	1.856	1.742	19/32	0.620	0.568	0.06	0.02	2.000	2.250
1	1.0000	1.022	1 1/2	1.500	1.450	2.121	1.991	21/32	0.684	0.628	0.09	0.03	2.250	2.500
1 1/8	1.1250	1.149	1 11/16	1.688	1.631	2.386	2.239	3/4	0.780	0.720	0.09	0.03	2.500	2.750
1 1/4	1.2500	1.277	1 7/8	1.875	1.812	2.652	2.489	27/32	0.876	0.812	0.09	0.03	2.750	3.000
1 3/8	1.3750	1.404	2 1/16	2.062	1.994	2.917	2.738	29/32	0.940	0.872	0.09	0.03	3.000	3.250
1 1/2	1.5000	1.531	2 1/4	2.250	2.175	3.182	2.986	1	1.036	0.964	0.09	0.03	3.250	3.500

(ANSI)

Note: L_G is the grip gaging length (nominal bolt length minus the basic thread length L_T).
L_T is the basic thread length and is the distance from the extreme end of the bolt to the last complete (full form) thread.
Bold type indicates products unified dimensionally with British and Canadian standards.
For additional requirements see ANSI B18.2.1

HEX BOLTS

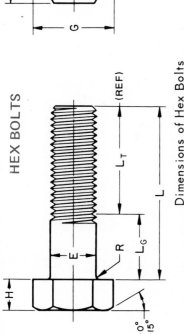

$30° \begin{smallmatrix} +0° \\ -15° \end{smallmatrix}$

L_G is the grip gaging length (nominal bolt length minus the basic thread length L_T).

Dimensions of Hex Bolts

Nominal Size or Basic Product Dia		E Body Dia Max	F Width Across Flats			G Width Across Corners		H Height			R Radius of Fillet		L_T Thread Length For Bolt Lengths	
			Basic	Max	Min	Max	Min	Basic	Max	Min	Max	Min	6 in. and Shorter Basic	Over 6 in. Basic
1/4	0.2500	0.260	7/16	0.438	0.425	0.505	0.484	11/64	0.188	0.150	0.03	0.01	0.750	1.000
5/16	0.3125	0.324	1/2	0.500	0.484	0.577	0.552	7/32	0.235	0.195	0.03	0.01	0.875	1.125
3/8	0.3750	0.388	9/16	0.562	0.544	0.650	0.620	1/4	0.268	0.226	0.03	0.01	1.000	1.250
7/16	0.4375	0.452	5/8	0.625	0.603	0.722	0.687	19/64	0.316	0.272	0.03	0.01	1.125	1.375
1/2	0.5000	0.515	3/4	0.750	0.725	0.866	0.826	11/32	0.364	0.302	0.03	0.01	1.250	1.500
5/8	0.6250	0.642	15/16	0.938	0.906	1.083	1.033	27/64	0.444	0.378	0.06	0.02	1.500	1.750
3/4	0.7500	0.768	1 1/8	1.125	1.088	1.299	1.240	1/2	0.524	0.455	0.06	0.02	1.750	2.000
7/8	0.8750	0.895	1 5/16	1.312	1.269	1.516	1.447	37/64	0.604	0.531	0.06	0.02	2.000	2.250
1	1.0000	1.022	1 1/2	1.500	1.450	1.732	1.653	43/64	0.700	0.591	0.09	0.03	2.250	2.500
1 1/8	1.1250	1.149	1 11/16	1.688	1.631	1.949	1.859	3/4	0.780	0.658	0.09	0.03	2.500	2.750
1 1/4	1.2500	1.277	1 7/8	1.875	1.812	2.165	2.066	27/32	0.876	0.749	0.09	0.03	2.750	3.000
1 3/8	1.3750	1.404	2 1/16	2.062	1.994	2.382	2.273	29/32	0.940	0.810	0.09	0.03	3.000	3.250
1 1/2	1.5000	1.531	2 1/4	2.250	2.175	2.598	2.480	1	1.036	0.902	0.09	0.03	3.250	3.500
1 3/4	1.7500	1.785	2 5/8	2.625	2.538	3.031	2.893	1 5/32	1.196	1.054	0.12	0.04	3.750	4.000
2	2.0000	2.039	3	3.000	2.900	3.464	3.306	1 11/32	1.388	1.175	0.12	0.04	4.250	4.500
2 1/4	2.2500	2.305	3 3/8	3.375	3.262	3.897	3.719	1 1/2	1.548	1.327	0.19	0.06	4.750	5.000
2 1/2	2.5000	2.559	3 3/4	3.750	3.625	4.330	4.133	1 21/32	1.708	1.479	0.19	0.06	5.250	5.500
2 3/4	2.7500	2.827	4 1/8	4.125	3.988	4.763	4.546	1 13/16	1.869	1.632	0.19	0.06	5.750	6.000
3	3.0000	3.081	4 1/2	4.500	4.350	5.196	4.959	2	2.060	1.815	0.19	0.06	6.250	6.500
3 1/4	3.2500	3.335	4 7/8	4.875	4.712	5.629	5.372	2 3/16	2.251	1.936	0.19	0.06	6.750	7.000
3 1/2	3.5000	3.589	5 1/4	5.250	5.075	6.062	5.786	2 5/16	2.380	2.057	0.19	0.06	7.250	7.500
3 3/4	3.7500	3.858	5 5/8	5.625	5.437	6.495	6.198	2 1/2	2.572	2.241	0.19	0.06	7.750	8.000
4	4.0000	4.111	6	6.000	5.800	6.928	6.612	2 11/16	2.764	2.424	0.19	0.06	8.250	8.500

Note: L_G is the grip gaging length (nominal bolt length minus the basic thread length L_T).
L_T is the basic thread length and is the distance from the extreme end of the bolt to the last complete (full form) thread.
Bold type indicates products unified dimensionally with British and Canadian standards.
For additional requirements see ANSI B18.2.1.

(ANSI)

FINISHED HEX BOLTS

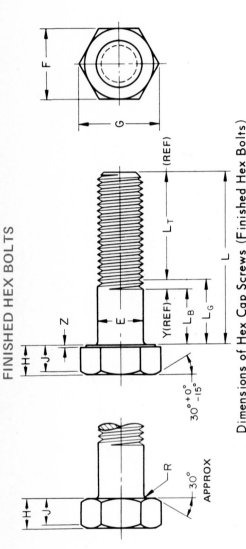

Dimensions of Hex Cap Screws (Finished Hex Bolts)

Nominal Size or Basic Product Dia		E Body Dia		F Width Across Flats			G Width Across Corners		H Height			J Wrenching Height	L_T Thread Length For Screw Lengths		Y Transition Thread Length For Screw Lengths		Z Runout of Bearing Surface FIR
		Max	Min	Basic	Max	Min	Max	Min	Basic	Max	Min	Min	6 in. and Shorter Basic	Over 6 in. Basic	6 in. and Shorter Max	Over 6 in. Max	Max
1/4	0.2500	0.2500	0.2450	7/16	0.438	0.428	0.505	0.488	5/32	0.163	0.150	0.106	0.750	1.000	0.400	0.650	0.010
5/16	0.3125	0.3125	0.3065	1/2	0.500	0.489	0.577	0.557	13/64	0.211	0.195	0.140	0.875	1.125	0.417	0.667	0.011
3/8	0.3750	0.3750	0.3690	9/16	0.562	0.551	0.650	0.628	15/64	0.243	0.226	0.160	1.000	1.250	0.438	0.688	0.012
7/16	0.4375	0.4375	0.4305	5/8	0.625	0.612	0.722	0.698	9/32	0.291	0.272	0.195	1.125	1.375	0.464	0.714	0.013
1/2	0.5000	0.5000	0.4930	3/4	0.750	0.736	0.866	0.840	5/16	0.323	0.302	0.215	1.250	1.500	0.481	0.731	0.014
9/16	0.5625	0.5625	0.5545	13/16	0.812	0.798	0.938	0.910	23/64	0.371	0.348	0.250	1.375	1.625	0.750	0.750	0.015
5/8	0.6250	0.6250	0.6170	15/16	0.938	0.922	1.083	1.051	25/64	0.403	0.378	0.269	1.500	1.750	0.773	0.773	0.017
3/4	0.7500	0.7500	0.7410	1 1/8	1.125	1.100	1.299	1.254	15/32	0.483	0.455	0.324	1.750	2.000	0.800	0.800	0.020
7/8	0.8750	0.8750	0.8660	1 5/16	1.312	1.285	1.516	1.465	35/64	0.563	0.531	0.378	2.000	2.250	0.833	0.833	0.023
1	1.0000	1.0000	0.9900	1 1/2	1.500	1.469	1.732	1.675	39/64	0.627	0.591	0.416	2.250	2.500	0.875	0.875	0.026
1 1/8	1.1250	1.1250	1.1140	1 11/16	1.688	1.631	1.949	1.859	11/16	0.718	0.658	0.461	2.500	2.750	0.929	0.929	0.029
1 1/4	1.2500	1.2500	1.2390	1 7/8	1.875	1.812	2.165	2.066	25/32	0.813	0.749	0.530	2.750	3.000	0.929	0.929	0.033
1 3/8	1.3750	1.3750	1.3630	2 1/16	2.062	1.994	2.382	2.273	27/32	0.878	0.810	0.569	3.000	3.250	1.000	1.000	0.036
1 1/2	1.5000	1.5000	1.4880	2 1/4	2.250	2.175	2.598	2.480	15/16	0.974	0.902	0.640	3.250	3.500	1.000	1.000	0.039
1 3/4	1.7500	1.7500	1.7380	2 5/8	2.625	2.538	3.031	2.893	1 3/32	1.134	1.054	0.748	3.750	4.000	1.100	1.100	0.046
2	2.0000	2.0000	1.9880	3	3.000	2.900	3.464	3.306	1 7/32	1.263	1.175	0.825	4.250	4.500	1.167	1.167	0.052
2 1/4	2.2500	2.2500	2.2380	3 3/8	3.375	3.262	3.897	3.719	1 3/8	1.423	1.327	0.933	4.750	5.000	1.167	1.167	0.059
2 1/2	2.5000	2.5000	2.4880	3 3/4	3.750	3.625	4.330	4.133	1 17/32	1.583	1.479	1.042	5.250	5.500	1.250	1.250	0.065
2 3/4	2.7500	2.7500	2.7380	4 1/8	4.125	3.988	4.763	4.546	1 11/16	1.744	1.632	1.151	5.750	6.000	1.250	1.250	0.072
3	3.0000	3.0000	2.9880	4 1/2	4.500	4.350	5.196	4.959	1 7/8	1.935	1.815	1.290	6.250	6.500	1.250	1.250	0.079

(ANSI)

Note: L_B is the body length from the underside of the head to the last scratch of thread.
L_G is the grip gaging length (nominal bolt length minus the basic thread length L_T).
L_T is the basic thread length and is the distance from the extreme end of the bolt to the last complete (full form) thread.
Bold type indicates products unified dimensionally with British and Canadian standards.
For additional requirements see ANSI B18.2.1.

SQUARE NUTS

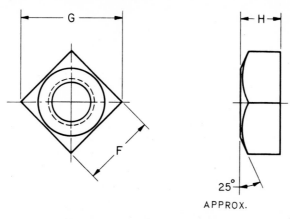

25°
APPROX.

Dimensions of Square Nuts

Nominal Size or Basic Major Dia of Thread		F Width Across Flats			G Width Across Corners		H Thickness		
		Basic	Max	Min	Max	Min	Basic	Max	Min
1/4	0.2500	7/16	0.438	0.425	0.619	0.584	7/32	0.235	0.203
5/16	0.3125	9/16	0.562	0.547	0.795	0.751	17/64	0.283	0.249
3/8	0.3750	5/8	0.625	0.606	0.884	0.832	21/64	0.346	0.310
7/16	0.4375	3/4	0.750	0.728	1.061	1.000	3/8	0.394	0.356
1/2	0.5000	13/16	0.812	0.788	1.149	1.082	7/16	0.458	0.418
5/8	0.6250	1	1.000	0.969	1.414	1.330	35/64	0.569	0.525
3/4	0.7500	1 1/8	1.125	1.088	1.591	1.494	21/32	0.680	0.632
7/8	0.8750	1 5/16	1.312	1.269	1.856	1.742	49/64	0.792	0.740
1	1.0000	1 1/2	1.500	1.450	2.121	1.991	7/8	0.903	0.847
1 1/8	1.1250	1 11/16	1.688	1.631	2.386	2.239	1	1.030	0.970
1 1/4	1.2500	1 7/8	1.875	1.812	2.652	2.489	1 3/32	1.126	1.062
1 3/8	1.3750	2 1/16	2.062	1.994	2.917	2.738	1 13/64	1.237	1.169
1 1/2	1.5000	2 1/4	2.250	2.175	3.182	2.986	1 5/16	1.348	1.276
See Notes	8	3							

(ANSI)

HEX FLAT NUTS AND HEX FLAT JAM NUTS

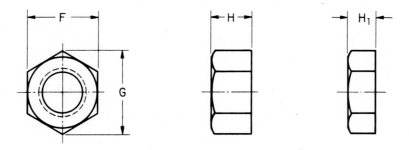

Dimensions of Hex Flat Nuts and Hex Flat Jam Nuts

Nominal Size or Basic Major Dia of Thread		F Width Across Flats			G Width Across Corners		H Thickness Hex Flat Nuts			H₁ Thickness Hex Flat Jam Nuts		
		Basic	Max	Min	Max	Min	Basic	Max	Min	Basic	Max	Min
1 1/8	1.1250	1 11/16	1.688	1.631	1.949	1.859	1	1.030	0.970	5/8	0.655	0.595
1 1/4	1.2500	1 7/8	1.875	1.812	2.165	2.066	1 3/32	1.126	1.062	3/4	0.782	0.718
1 3/8	1.3750	2 1/16	2.062	1.994	2.382	2.273	1 13/64	1.237	1.169	13/16	0.846	0.778
1 1/2	1.5000	2 1/4	2.250	2.175	2.598	2.480	1 5/16	1.348	1.276	7/8	0.911	0.839
See Notes	10	4										

(ANSI)

HEX NUTS AND HEX JAM NUTS

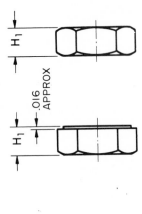

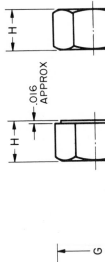

Dimensions of Hex Nuts and Hex Jam Nuts

Nominal Size or Basic Major Dia of Thread		F Width Across Flats			G Width Across Corners		H Thickness Hex Nuts			H₁ Thickness Hex Jam Nuts			Runout of Bearing Face, FIR Max Hex Nuts Specified Proof Load		Jam Nuts All Strength Levels
		Basic	Max	Min	Max	Min	Basic	Max	Min	Basic	Max	Min	Up to 150,000 psi	150,000 psi and Greater	
1/4	0.2500	7/16	0.438	0.428	0.505	0.488	7/32	0.226	0.212	5/32	0.163	0.150	0.015	0.010	0.015
5/16	0.3125	1/2	0.500	0.489	0.577	0.557	17/64	0.273	0.258	3/16	0.195	0.180	0.016	0.011	0.016
3/8	0.3750	9/16	0.562	0.551	0.650	0.628	21/64	0.337	0.320	7/32	0.227	0.210	0.017	0.012	0.017
7/16	0.4375	11/16	0.688	0.675	0.794	0.768	3/8	0.385	0.365	1/4	0.260	0.240	0.018	0.013	0.018
1/2	0.5000	3/4	0.750	0.736	0.866	0.840	7/16	0.448	0.427	5/16	0.323	0.302	0.019	0.014	0.019
9/16	0.5625	7/8	0.875	0.861	1.010	0.982	31/64	0.496	0.473	5/16	0.324	0.301	0.020	0.015	0.020
5/8	0.6250	15/16	0.938	0.922	1.083	1.051	35/64	0.559	0.535	3/8	0.387	0.363	0.021	0.016	0.021
3/4	0.7500	1 1/8	1.125	1.088	1.299	1.240	41/64	0.665	0.617	27/64	0.446	0.398	0.023	0.018	0.023
7/8	0.8750	1 5/16	1.312	1.269	1.516	1.447	3/4	0.776	0.724	31/64	0.510	0.458	0.025	0.020	0.025
1	1.0000	1 1/2	1.500	1.450	1.732	1.653	55/64	0.887	0.831	35/64	0.575	0.519	0.027	0.022	0.027
1 1/8	1.1250	1 11/16	1.688	1.631	1.949	1.859	31/32	0.999	0.939	39/64	0.639	0.579	0.030	0.025	0.030
1 1/4	1.2500	1 7/8	1.875	1.812	2.165	2.066	1 1/16	1.094	1.030	23/32	0.751	0.687	0.033	0.028	0.033
1 3/8	1.3750	2 1/16	2.062	1.994	2.382	2.273	1 11/64	1.206	1.138	25/32	0.815	0.747	0.036	0.031	0.036
1 1/2	1.5000	2 1/4	2.250	2.175	2.598	2.480	1 9/32	1.317	1.245	27/32	0.880	0.808	0.039	0.034	0.039

Note: Bold type indicates products unified dimensionally with British and Canadian standards. For additional requirements see ANSI B18.2.1.

(ANSI)

SLOTTED FLAT COUNTERSUNK HEAD CAP SCREWS

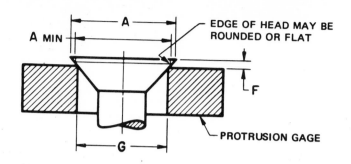

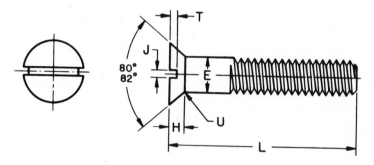

Dimensions of Slotted Flat Countersunk Head Cap Screws

Nominal Size[1] or Basic Screw Diameter		E Body Diameter		A Head Diameter		H[2] Head Height	J Slot Width		T Slot Depth		U Fillet Radius	F[3] Protrusion Above Gaging Diameter		G[3] Gaging Diameter
		Max	Min	Max, Edge Sharp	Min, Edge Rounded or Flat	Ref	Max	Min	Max	Min	Max	.Max	Min	
1/4	0.2500	0.2500	0.2450	0.500	0.452	0.140	0.075	0.064	0.068	0.045	0.100	0.046	0.030	0.424
5/16	0.3125	0.3125	0.3070	0.625	0.567	0.177	0.084	0.072	0.086	0.057	0.125	0.053	0.035	0.538
3/8	0.3750	0.3750	0.3690	0.750	0.682	0.210	0.094	0.081	0.103	0.068	0.150	0.060	0.040	0.651
7/16	0.4375	0.4375	0.4310	0.812	0.736	0.210	0.094	0.081	0.103	0.068	0.175	0.065	0.044	0.703
1/2	0.5000	0.5000	0.4930	0.875	0.791	0.210	0.106	0.091	0.103	0.068	0.200	0.071	0.049	0.756
9/16	0.5625	0.5625	0.5550	1.000	0.906	0.244	0.118	0.102	0.120	0.080	0.225	0.078	0.054	0.869
5/8	0.6250	0.6250	0.6170	1.125	1.020	0.281	0.133	0.116	0.137	0.091	0.250	0.085	0.058	0.982
3/4	0.7500	0.7500	0.7420	1.375	1.251	0.352	0.149	0.131	0.171	0.115	0.300	0.099	0.068	1.208
7/8	0.8750	0.8750	0.8660	1.625	1.480	0.423	0.167	0.147	0.206	0.138	0.350	0.113	0.077	1.435
1	1.0000	1.0000	0.9900	1.875	1.711	0.494	0.188	0.166	0.240	0.162	0.400	0.127	0.087	1.661
1 1/8	1.1250	1.1250	1.1140	2.062	1.880	0.529	0.196	0.178	0.257	0.173	0.450	0.141	0.096	1.826
1 1/4	1.2500	1.2500	1.2390	2.312	2.110	0.600	0.211	0.193	0.291	0.197	0.500	0.155	0.105	2.052
1 3/8	1.3750	1.3750	1.3630	2.562	2.340	0.665	0.226	0.208	0.326	0.220	0.550	0.169	0.115	2.279
1 1/2	1.5000	1.5000	1.4880	2.812	2.570	0.742	0.258	0.240	0.360	0.244	0.600	0.183	0.124	2.505

[1] Where specifying nominal size in decimals, zeros preceding decimal and in the fourth decimal place shall be omitted.
[2] Tabulated values determined from formula for maximum H, Appendix III.
[3] No tolerance for gaging diameter is given. If the gaging diameter of the gage used differs from tabulated value, the protrusion will be affected accordingly and the proper protrusion values must be recalculated using the formulas shown in Appendix II.

(ANSI)

SLOTTED ROUND HEAD CAP SCREWS

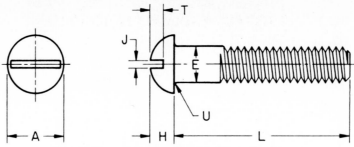

Dimensions of Slotted Round Head Cap Screws

Nominal Size[1] or Basic Screw Diameter	E Body Diameter		A Head Diameter		H Head Height		J Slot Width		T Slot Depth		U Fillet Radius	
	Max	Min	Max	Min	Max	Min	Max	Min	Max	Min	Max	Min
1/4 0.2500	0.2500	0.2450	0.437	0.418	0.191	0.175	0.075	0.064	0.117	0.097	0.031	0.016
5/16 0.3125	0.3125	0.3070	0.562	0.540	0.245	0.226	0.084	0.072	0.151	0.126	0.031	0.016
3/8 0.3750	0.3750	0.3690	0.625	0.603	0.273	0.252	0.094	0.081	0.168	0.138	0.031	0.016
7/16 0.4375	0.4375	0.4310	0.750	0.725	0.328	0.302	0.094	0.081	0.202	0.167	0.047	0.016
1/2 0.5000	0.5000	0.4930	0.812	0.786	0.354	0.327	0.106	0.091	0.218	0.178	0.047	0.016
9/16 0.5625	0.5625	0.5550	0.937	0.909	0.409	0.378	0.118	0.102	0.252	0.207	0.047	0.016
5/8 0.6250	0.6250	0.6170	1.000	0.970	0.437	0.405	0.133	0.116	0.270	0.220	0.062	0.031
3/4 0.7500	0.7500	0.7420	1.250	1.215	0.546	0.507	0.149	0.131	0.338	0.278	0.062	0.031

[1]Where specifying nominal size in decimals, zeros preceding decimal and in the fourth decimal place shall be omitted. (ANSI)

SLOTTED FILLISTER HEAD CAP SCREWS

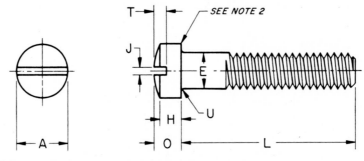

Dimensions of Slotted Fillister Head Cap Screws

Nominal Size[1] or Basic Screw Diameter	E Body Diameter		A Head Diameter		H Head Side Height		O Total Head Height		J Slot Width		T Slot Depth		U Fillet Radius	
	Max	Min	Max	Min	Max	Min	Max	Min	Max	Min	Max	Min	Max	Min
1/4 0.2500	0.2500	0.2450	0.375	0.363	0.172	0.157	0.216	0.194	0.075	0.064	0.097	0.077	0.031	0.016
5/16 0.3125	0.3125	0.3070	0.437	0.424	0.203	.0.186	0.253	0.230	0.084	0.072	0.115	0.090	0.031	0.016
3/8 0.3750	0.3750	0.3690	0.562	0.547	0.250	0.229	0.314	0.284	0.094	0.081	0.142	0.112	0.031	0.016
7/16 0.4375	0.4375	0.4310	0.625	0.608	0.297	0.274	0.368	0.336	0.094	0.081	0.168	0.133	0.047	0.016
1/2 0.5000	0.5000	0.4930	0.750	0.731	0.328	0.301	0.413	0.376	0.106	0.091	0.193	0.153	0.047	0.016
9/16 0.5625	0.5625	0.5550	0.812	0.792	0.375	0.346	0.467	0.427	0.118	0.102	0.213	0.168	0.047	0.016
5/8 0.6250	0.6250	0.6170	0.875	0.853	0.422	0.391	0.521	0.478	0.133	0.116	0.239	0.189	0.062	0.031
3/4 0.7500	0.7500	0.7420	1.000	0.976	0.500	0.466	0.612	0.566	0.149	0.131	0.283	0.223	0.062	0.031
7/8 0.8750	0.8750	0.8660	1.125	1.098	0.594	0.556	0.720	0.668	0.167	0.147	0.334	0.264	0.062	0.031
1 1.0000	1.0000	0.9900	1.312	1.282	0.656	0.612	0.803	0.743	0.188	0.166	0.371	0.291	0.062	0.031

[1]Where specifying nominal size in decimals, zeros preceding decimal and in the fourth decimal place shall be omitted.
[2]A slight rounding of the edges at periphery of head shall be permissible provided the diameter of the bearing circle is equal to no less than 90 per cent of the specified minimum head diameter.

(ANSI)

SLOTTED HEADLESS SET SCREWS

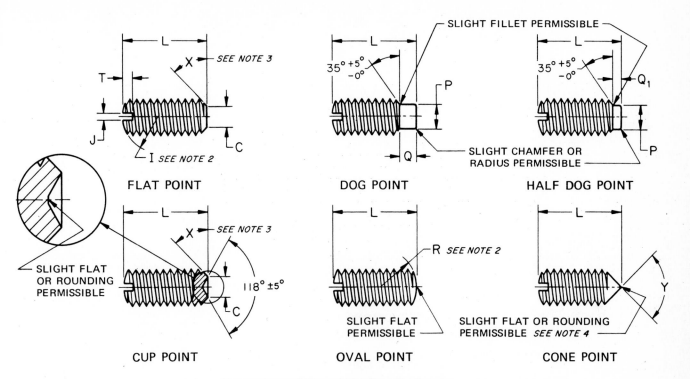

FLAT POINT   DOG POINT   HALF DOG POINT

CUP POINT   OVAL POINT   CONE POINT

Dimensions of Slotted Headless Set Screws

Nominal Size[1] or Basic Screw Diameter		I^2 Crown Radius	J Slot Width		T Slot Depth		C Cup and Flat Point Diameters		P Dog Point Diameters		Q Point Length Dog		Q_1 Point Length Half Dog		R^2 Oval Point Radius	Y Cone Point Angle $90° \pm 2°$ For These Nominal Lengths or Longer; $118° \pm 2°$ For Shorter Screws
		Basic	Max	Min	Max	Min	Max	Min	Max	Min	Max	Min	Max	Min	Basic	
0	0.0600	0.060	0.014	0.010	0.020	0.016	0.033	0.027	0.040	0.037	0.032	0.028	0.017	0.013	0.045	5/64
1	0.0730	0.073	0.016	0.012	0.020	0.016	0.040	0.033	0.049	0.045	0.040	0.036	0.021	0.017	0.055	3/32
2	0.0860	0.086	0.018	0.014	0.025	0.019	0.047	0.039	0.057	0.053	0.046	0.042	0.024	0.020	0.064	7/64
3	0.0990	0.099	0.020	0.016	0.028	0.022	0.054	0.045	0.066	0.062	0.052	0.048	0.027	0.023	0.074	1/8
4	0.1120	0.112	0.024	0.018	0.031	0.025	0.061	0.051	0.075	0.070	0.058	0.054	0.030	0.026	0.084	5/32
5	0.1250	0.125	0.026	0.020	0.036	0.026	0.067	0.057	0.083	0.078	0.063	0.057	0.033	0.027	0.094	3/16
6	0.1380	0.138	0.028	0.022	0.040	0.030	0.074	0.064	0.092	0.087	0.073	0.067	0.038	0.032	0.104	3/16
8	0.1640	0.164	0.032	0.026	0.046	0.036	0.087	0.076	0.109	0.103	0.083	0.077	0.043	0.037	0.123	1/4
10	0.1900	0.190	0.035	0.029	0.053	0.043	0.102	0.088	0.127	0.120	0.095	0.085	0.050	0.040	0.142	1/4
12	0.2160	0.216	0.042	0.035	0.061	0.051	0.115	0.101	0.144	0.137	0.115	0.105	0.060	0.050	0.162	5/16
1/4	0.2500	0.250	0.049	0.041	0.068	0.058	0.132	0.118	0.156	0.149	0.130	0.120	0.068	0.058	0.188	5/16
5/16	0.3125	0.312	0.055	0.047	0.083	0.073	0.172	0.156	0.203	0.195	0.161	0.151	0.083	0.073	0.234	3/8
3/8	0.3750	0.375	0.068	0.060	0.099	0.089	0.212	0.194	0.250	0.241	0.193	0.183	0.099	0.089	0.281	7/16
7/16	0.4375	0.438	0.076	0.068	0.114	0.104	0.252	0.232	0.297	0.287	0.224	0.214	0.114	0.104	0.328	1/2
1/2	0.5000	0.500	0.086	0.076	0.130	0.120	0.291	0.270	0.344	0.334	0.255	0.245	0.130	0.120	0.375	9/16
9/16	0.5625	0.562	0.096	0.086	0.146	0.136	0.332	0.309	0.391	0.379	0.287	0.275	0.146	0.134	0.422	5/8
5/8	0.6250	0.625	0.107	0.097	0.161	0.151	0.371	0.347	0.469	0.456	0.321	0.305	0.164	0.148	0.469	3/4
3/4	0.7500	0.750	0.134	0.124	0.193	0.183	0.450	0.425	0.562	0.549	0.383	0.367	0.196	0.180	0.562	7/8

[1] Where specifying nominal size in decimals, zeros preceding decimal and in the fourth decimal place shall be omitted.

[2] Tolerance on radius for nominal sizes up to and including 5 (0.125 in.) shall be plus 0.015 in. and minus 0.000, and for larger sizes, plus 0.031 in. and minus 0.000. Slotted ends on screws may be flat at option of manufacturer.

[3] Point angle X shall be 45° plus 5°, minus 0°, for screws of nominal lengths equal to or longer than those listed in Column Y, and 30° minimum for screws of shorter nominal lengths.

[4] The extent of rounding or flat at apex of cone point shall not exceed an amount equivalent to 10 per cent of the basic screw diameter.

(ANSI)

SQUARE HEAD SET SCREWS

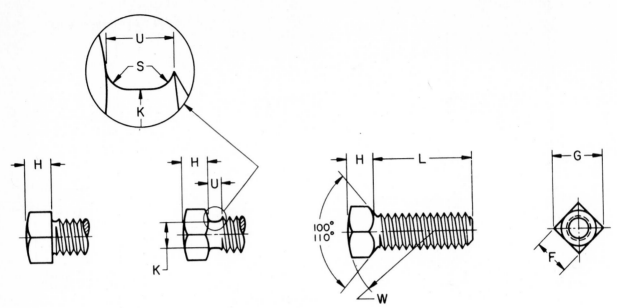

OPTIONAL HEAD CONSTRUCTIONS

Dimensions of Square Head Set Screws

Nominal Size[1] or Basic Screw Diameter		F Width Across Flats		G Width Across Corners		H Head Height		K Neck Relief Diameter		S Neck Relief Fillet Radius	U Neck Relief Width	W Head Radius
		Max	Min	Max	Min	Max	Min	Max	Min	Max	Min	Min
10	0.1900	0.188	0.180	0.265	0.247	0.148	0.134	0.145	0.140	0.027	0.083	0.48
1/4	0.2500	0.250	0.241	0.354	0.331	0.196	0.178	0.185	0.170	0.032	0.100	0.62
5/16	0.3125	0.312	0.302	0.442	0.415	0.245	0.224	0.240	0.225	0.036	0.111	0.78
3/8	0.3750	0.375	0.362	0.530	0.497	0.293	0.270	0.294	0.279	0.041	0.125	0.94
7/16	0.4375	0.438	0.423	0.619	0.581	0.341	0.315	0.345	0.330	0.046	0.143	1.09
1/2	0.5000	0.500	0.484	0.707	0.665	0.389	0.361	0.400	0.385	0.050	0.154	1.25
9/16	0.5625	0.562	0.545	0.795	0.748	0.437	0.407	0.454	0.439	0.054	0.167	1.41
5/8	0.6250	0.625	0.606	0.884	0.833	0.485	0.452	0.507	0.492	0.059	0.182	1.56
3/4	0.7500	0.750	0.729	1.060	1.001	0.582	0.544	0.620	0.605	0.065	0.200	1.88
7/8	0.8750	0.875	0.852	1.237	1.170	0.678	0.635	0.731	0.716	0.072	0.222	2.19
1	1.0000	1.000	0.974	1.414	1.337	0.774	0.726	0.838	0.823	0.081	0.250	2.50
1 1/8	1.1250	1.125	1.096	1.591	1.505	0.870	0.817	0.939	0.914	0.092	0.283	2.81
1 1/4	1.2500	1.250	1.219	1.768	1.674	0.966	0.908	1.064	1.039	0.092	0.283	3.12
1 3/8	1.3750	1.375	1.342	1.945	1.843	1.063	1.000	1.159	1.134	0.109	0.333	3.44
1 1/2	1.5000	1.500	1.464	2.121	2.010	1.159	1.091	1.284	1.259	0.109	0.333	3.75

[1]Where specifying nominal size in decimals, zeros preceding decimal and in the fourth decimal place shall be omitted.

(ANSI)

SQUARE HEAD SET SCREWS (Continued)

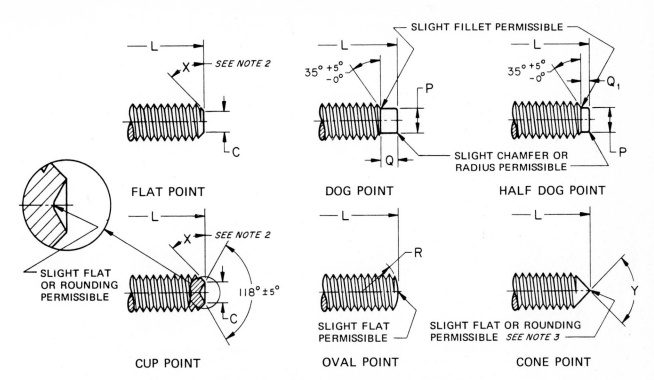

FLAT POINT   DOG POINT   HALF DOG POINT

SLIGHT FLAT OR ROUNDING PERMISSIBLE

CUP POINT   OVAL POINT   CONE POINT

Dimensions of Square Head Set Screws (continued)

Nominal Size[1] or Basic Screw Diameter		C		P		Q		Q₁		R	Y
		Cup and Flat Point Diameters		Dog and Half Dog Point Diameters		Point Length				Oval Point Radius +0.031 -0.000	Cone Point Angle 90° ±2° For These Nominal Lengths or Longer; 118° ±2° For Shorter Screws
						Dog		Half Dog			
		Max	Min	Max	Min	Max	Min	Max	Min		
10	0.1900	0.102	0.088	0.127	0.120	0.095	0.085	0.050	0.040	0.142	1/4
1/4	0.2500	0.132	0.118	0.156	0.149	0.130	0.120	0.068	0.058	0.188	5/16
5/16	0.3125	0.172	0.156	0.203	0.195	0.161	0.151	0.083	0.073	0.234	3/8
3/8	0.3750	0.212	0.194	0.250	0.241	0.193	0.183	0.099	0.089	0.281	7/16
7/16	0.4375	0.252	0.232	0.297	0.287	0.224	0.214	0.114	0.104	0.328	1/2
1/2	0.5000	0.291	0.270	0.344	0.334	0.255	0.245	0.130	0.120	0.375	9/16
9/16	0.5625	0.332	0.309	0.391	0.379	0.287	0.275	0.146	0.134	0.422	5/8
5/8	0.6250	0.371	0.347	0.469	0.456	0.321	0.305	0.164	0.148	0.469	3/4
3/4	0.7500	0.450	0.425	0.562	0.549	0.383	0.367	0.196	0.180	0.562	7/8
7/8	0.8750	0.530	0.502	0.656	0.642	0.446	0.430	0.227	0.211	0.656	1
1	1.0000	0.609	0.579	0.750	0.734	0.510	0.490	0.260	0.240	0.750	1 1/8
1 1/8	1.1250	0.689	0.655	0.844	0.826	0.572	0.552	0.291	0.271	0.844	1 1/4
1 1/4	1.2500	0.767	0.733	0.938	0.920	0.635	0.615	0.323	0.303	0.938	1 1/2
1 3/8	1.3750	0.848	0.808	1.031	1.011	0.698	0.678	0.354	0.334	1.031	1 5/8
1 1/2	1.5000	0.926	0.886	1.125	1.105	0.760	0.740	0.385	0.365	1.125	1 3/4

[1] Where specifying nominal size in decimals, zeros preceding decimal and in the fourth decimal place shall be omitted.
[2] Point angle X shall be 45° plus 5°, minus 0°, for screws of nominal lengths equal to or longer than those listed in Column Y, and 30° minimum for screws of shorter nominal lengths.
[3] The extent of rounding or flat at apex of cone point shall not exceed an amount equivalent to 10 per cent of the basic screw diameter.

SLOTTED FLAT COUNTERSUNK HEAD MACHINE SCREWS

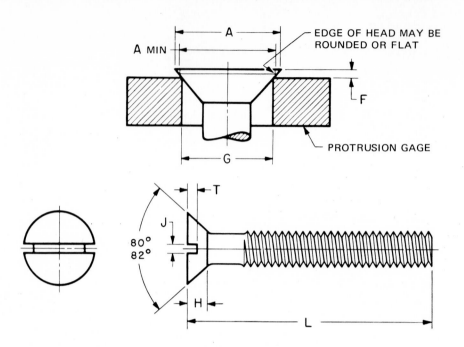

Dimensions of Slotted Flat Countersunk Head Machine Screws

Nominal Size[1] or Basic Screw Diameter		L[2] These Lengths or Shorter are Undercut.	A Head Diameter		H[3] Head Height	J Slot Width		T Slot Depth		F[4] Protrusion Above Gaging Diameter		G[4] Gaging Diameter
			Max, Edge Sharp	Min, Edge Rounded or Flat								
					Ref	Max	Min	Max	Min	Max	Min	
0000	0.0210	—	0.043	0.037	0.011	0.008	0.004	0.007	0.003	*	*	*
000	0.0340	—	0.064	0.058	0.016	0.011	0.007	0.009	0.005	*	*	*
00	0.0470	—	0.093	0.085	0.028	0.017	0.010	0.014	0.009	*	*	*
0	0.0600	1/8	0.119	0.099	0.035	0.023	0.016	0.015	0.010	0.026	0.016	0.078
1	0.0730	1/8	0.146	0.123	0.043	0.026	0.019	0.019	0.012	0.028	0.016	0.101
2	0.0860	1/8	0.172	0.147	0.051	0.031	0.023	0.023	0.015	0.029	0.017	0.124
3	0.0990	1/8	0.199	0.171	0.059	0.035	0.027	0.027	0.017	0.031	0.018	0.148
4	0.1120	3/16	0.225	0.195	0.067	0.039	0.031	0.030	0.020	0.032	0.019	0.172
5	0.1250	3/16	0.252	0.220	0.075	0.043	0.035	0.034	0.022	0.034	0.020	0.196
6	0.1380	3/16	0.279	0.244	0.083	0.048	0.039	0.038	0.024	0.036	0.021	0.220
8	0.1640	1/4	0.332	0.292	0.100	0.054	0.045	0.045	0.029	0.039	0.023	0.267
10	0.1900	5/16	0.385	0.340	0.116	0.060	0.050	0.053	0.034	0.042	0.025	0.313
12	0.2160	3/8	0.438	0.389	0.132	0.067	0.056	0.060	0.039	0.045	0.027	0.362
1/4	0.2500	7/16	0.507	0.452	0.153	0.075	0.064	0.070	0.046	0.050	0.029	0.424
5/16	0.3125	1/2	0.635	0.568	0.191	0.084	0.072	0.088	0.058	0.057	0.034	0.539
3/8	0.3750	9/16	0.762	0.685	0.230	0.094	0.081	0.106	0.070	0.065	0.039	0.653
7/16	0.4375	5/8	0.812	0.723	0.223	0.094	0.081	0.103	0.066	0.073	0.044	0.690
1/2	0.5000	3/4	0.875	0.775	0.223	0.106	0.091	0.103	0.065	0.081	0.049	0.739
9/16	0.5625	—	1.000	0.889	0.260	0.118	0.102	0.120	0.077	0.089	0.053	0.851
5/8	0.6250	—	1.125	1.002	0.298	0.133	0.116	0.137	0.088	0.097	0.058	0.962
3/4	0.7500	—	1.375	1.230	0.372	0.149	0.131	0.171	0.111	0.112	0.067	1.186

[1] Where specifying nominal size in decimals, zeros preceding decimal and in the fourth decimal place shall be omitted.
[2] Screws of these lengths and shorter shall have undercut heads as shown in Table 5.
[3] Tabulated values determined from formula for maximum H, Appendix V.
[4] No tolerance for gaging diameter is given. If the gaging diameter of the gage used differs from tabulated value, the protrusion will be affected accordingly and the proper protrusion values must be recalculated using the formulas shown in Appendix I.
*Not practical to gage.

(ANSI)

CROSS RECESSED FLAT COUNTERSUNK HEAD MACHINE SCREWS

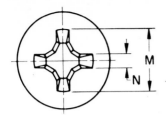

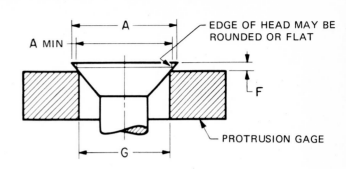

This type of recess has a large center opening, tapered wings, and blunt bottom, with all edges relieved or rounded.

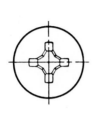

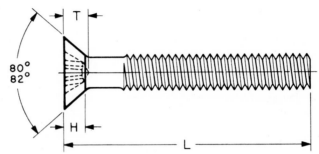

Dimensions of Type I Cross Recessed Flat Countersunk Head Machine Screws

Nominal Size[1] or Basic Screw Diameter		L[2] These Lengths or Shorter are Undercut	A Head Diameter		H[3] Head Height	M Recess Diameter		T Recess Depth		N Recess Width	Driver Size	Recess Penetration Gaging Depth		F[4] Protrusion Above Gaging Diameter		G[4] Gaging Diameter
			Max, Edge Sharp	Min, Edge Rounded or Flat	Ref	Max	Min	Max	Min	Min		Max	Min	Max	Min	
0	0.0600	1/8	0.119	0.099	0.035	0.069	0.056	0.043	0.027	0.014	0	0.036	0.020	0.026	0.016	0.078
1	0.0730	1/8	0.146	0.123	0.043	0.077	0.064	0.051	0.035	0.015	0	0.044	0.028	0.028	0.016	0.101
2	0.0860	1/8	0.172	0.147	0.051	0.102	0.089	0.063	0.047	0.017	1	0.056	0.040	0.029	0.017	0.124
3	0.0990	1/8	0.199	0.171	0.059	0.107	0.094	0.068	0.052	0.018	1	0.061	0.045	0.031	0.018	0.148
4	0.1120	3/16	0.225	0.195	0.067	0.128	0.115	0.089	0.073	0.018	1	0.082	0.066	0.032	0.019	0.172
5	0.1250	3/16	0.252	0.220	0.075	0.154	0.141	0.086	0.063	0.027	2	0.075	0.052	0.034	0.020	0.196
6	0.1380	3/16	0.279	0.244	0.083	0.174	0.161	0.106	0.083	0.029	2	0.095	0.072	0.036	0.021	0.220
8	0.1640	1/4	0.332	0.292	0.100	0.189	0.176	0.121	0.098	0.030	2	0.110	0.087	0.039	0.023	0.267
10	0.1900	5/16	0.385	0.340	0.116	0.204	0.191	0.136	0.113	0.032	2	0.125	0.102	0.042	0.025	0.313
12	0.2160	3/8	0.438	0.389	0.132	0.268	0.255	0.156	0.133	0.035	3	0.139	0.116	0.045	0.027	0.362
1/4	0.2500	7/16	0.507	0.452	0.153	0.283	0.270	0.171	0.148	0.036	3	0.154	0.131	0.050	0.029	0.424
5/16	0.3125	1/2	0.635	0.568	0.191	0.365	0.352	0.216	0.194	0.061	4	0.196	0.174	0.057	0.034	0.539
3/8	0.3750	9/16	0.762	0.685	0.230	0.393	0.380	0.245	0.223	0.065	4	0.225	0.203	0.065	0.039	0.653
7/16	0.4375	5/8	0.812	0.723	0.223	0.409	0.396	0.261	0.239	0.068	4	0.241	0.219	0.073	0.044	0.690
1/2	0.5000	3/4	0.875	0.775	0.223	0.424	0.411	0.276	0.254	0.069	4	0.256	0.234	0.081	0.049	0.739
9/16	0.5625	—	1.000	0.889	0.260	0.454	0.431	0.300	0.278	0.073	4	0.280	0.258	0.089	0.053	0.851
5/8	0.6250	—	1.125	1.002	0.298	0.576	0.553	0.342	0.316	0.079	5	0.309	0.283	0.097	0.058	0.962
3/4	0.7500	—	1.375	1.230	0.372	0.640	0.617	0.406	0.380	0.087	5	0.373	0.347	0.112	0.067	1.186

[1] Where specifying nominal size in decimals, zeros preceding decimal and in the fourth decimal place shall be omitted.

[2] Screws of these lengths and shorter shall have undercut heads as shown in Table 6.

[3] Tabulated values determined from formula for maximum H, Appendix V.

[4] No tolerance for gaging diameter is given. If the gaging diameter of the gage used differs from tabulated value, the protrusion will be affected accordingly and the proper protrusion values must be recalculated using the formulas shown in Appendix I.

(ANSI)

SLOTTED AND COUNTERSUNK HEAD MACHINE SCREWS

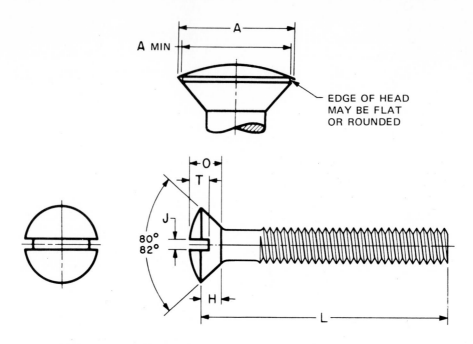

Dimensions of Slotted Oval Countersunk Head Machine Screws

Nominal Size[1] or Basic Screw Diameter		L[2] These Lengths or Shorter are Undercut	A Head Diameter		H[3] Head Side Height	O Total Head Height		J Slot Width		T Slot Depth	
			Max, Edge Sharp	Min, Edge Rounded or Flat	Ref	Max	Min	Max	Min	Max	Min
00	0.0470	—	0.093	0.085	0.028	0.042	0.034	0.017	0.010	0.023	0.016
0	0.0600	1/8	0.119	0.099	0.035	0.056	0.041	0.023	0.016	0.030	0.025
1	0.0730	1/8	0.146	0.123	0.043	0.068	0.052	0.026	0.019	0.038	0.031
2	0.0860	1/8	0.172	0.147	0.051	0.080	0.063	0.031	0.023	0.045	0.037
3	0.0990	1/8	0.199	0.171	0.059	0.092	0.073	0.035	0.027	0.052	0.043
4	0.1120	3/16	0.225	0.195	0.067	0.104	0.084	0.039	0.031	0.059	0.049
5	0.1250	3/16	0.252	0.220	0.075	0.116	0.095	0.043	0.035	0.067	0.055
6	0.1380	3/16	0.279	0.244	0.083	0.128	0.105	0.048	0.039	0.074	0.060
8	0.1640	1/4	0.332	0.292	0.100	0.152	0.126	0.054	0.045	0.088	0.072
10	0.1900	5/16	0.385	0.340	0.116	0.176	0.148	0.060	0.050	0.103	0.084
12	0.2160	3/8	0.438	0.389	0.132	0.200	0.169	0.067	0.056	0.117	0.096
1/4	0.2500	7/16	0.507	0.452	0.153	0.232	0.197	0.075	0.064	0.136	0.112
5/16	0.3125	1/2	0.635	0.568	0.191	0.290	0.249	0.084	0.072	0.171	0.141
3/8	0.3750	9/16	0.762	0.685	0.230	0.347	0.300	0.094	0.081	0.206	0.170
7/16	0.4375	5/8	0.812	0.723	0.223	0.345	0.295	0.094	0.081	0.210	0.174
1/2	0.5000	3/4	0.875	0.775	0.223	0.354	0.299	0.106	0.091	0.216	0.176
9/16	0.5625	—	1.000	0.889	0.260	0.410	0.350	0.118	0.102	0.250	0.207
5/8	0.6250	—	1.125	1.002	0.298	0.467	0.399	0.133	0.116	0.285	0.235
3/4	0.7500	—	1.375	1.230	0.372	0.578	0.497	0.149	0.131	0.353	0.293

[1] Where specifying nominal size in decimals, zeros preceding decimal and in the fourth decimal place shall be omitted.
[2] Screws of these lengths and shorter shall have undercut heads as shown in Table 24.
[3] Tabulated values determined from formula for maximum H, Appendix V.

(ANSI)

SLOTTED PAN HEAD MACHINE SCREWS

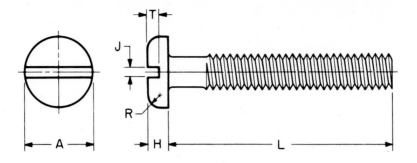

Dimensions of Slotted Pan Head Machine Screws

Nominal Size[1] or Basic Screw Diameter		A Head Diameter		H Head Height		R Head Radius	J Slot Width		T Slot Depth	
		Max	Min	Max	Min	Max	Max	Min	Max	Min
0000	0.0210	0.042	0.036	0.016	0.010	0.007	0.008	0.004	0.008	0.004
000	0.0340	0.066	0.060	0.023	0.017	0.010	0.012	0.008	0.012	0.008
00	0.0470	0.090	0.082	0.032	0.025	0.015	0.017	0.010	0.016	0.010
0	0.0600	0.116	0.104	0.039	0.031	0.020	0.023	0.016	0.022	0.014
1	0.0730	0.142	0.130	0.046	0.038	0.025	0.026	0.019	0.027	0.018
2	0.0860	0.167	0.155	0.053	0.045	0.035	0.031	0.023	0.031	0.022
3	0.0990	0.193	0.180	0.060	0.051	0.037	0.035	0.027	0.036	0.026
4	0.1120	0.219	0.205	0.068	0.058	0.042	0.039	0.031	0.040	0.030
5	0.1250	0.245	0.231	0.075	0.065	0.044	0.043	0.035	0.045	0.034
6	0.1380	0.270	0.256	0.082	0.072	0.046	0.048	0.039	0.050	0.037
8	0.1640	0.322	0.306	0.096	0.085	0.052	0.054	0.045	0.058	0.045
10	0.1900	0.373	0.357	0.110	0.099	0.061	0.060	0.050	0.068	0.053
12	0.2160	0.425	0.407	0.125	0.112	0.078	0.067	0.056	0.077	0.061
1/4	0.2500	0.492	0.473	0.144	0.130	0.087	0.075	0.064	0.087	0.070
5/16	0.3125	0.615	0.594	0.178	0.162	0.099	0.084	0.072	0.106	0.085
3/8	0.3750	0.740	0.716	0.212	0.195	0.143	0.094	0.081	0.124	0.100
7/16	0.4375	0.863	0.837	0.247	0.228	0.153	0.094	0.081	0.142	0.116
1/2	0.5000	0.987	0.958	0.281	0.260	0.175	0.106	0.091	0.161	0.131
9/16	0.5625	1.041	1.000	0.315	0.293	0.197	0.118	0.102	0.179	0.146
5/8	0.6250	1.172	1.125	0.350	0.325	0.219	0.133	0.116	0.197	0.162
3/4	0.7500	1.435	1.375	0.419	0.390	0.263	0.149	0.131	0.234	0.192

[1] Where specifying nominal size in decimals, zeros preceding decimal and in the fourth decimal place shall be omitted.

(ANSI)

SLOTTED FILLISTER HEAD MACHINE SCREWS

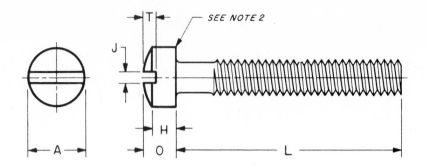

Dimensions of Slotted Fillister Head Machine Screws

Nominal Size[1] or Basic Screw Diameter		A Head Diameter		H Head Side Height		O Total Head Height		J Slot Width		T Slot Depth	
		Max	Min	Max	Min	Max	Min	Max	Min	Max	Min
0000	0.0210	0.038	0.032	0.019	0.011	0.025	0.015	0.008	0.004	0.012	0.006
000	0.0340	0.059	0.053	0.029	0.021	0.035	0.027	0.012	0.006	0.017	0.011
00	0.0470	0.082	0.072	0.037	0.028	0.047	0.039	0.017	0.010	0.022	0.015
0	0.0600	0.096	0.083	0.043	0.038	0.055	0.047	0.023	0.016	0.025	0.015
1	0.0730	0.118	0.104	0.053	0.045	0.066	0.058	0.026	0.019	0.031	0.020
2	0.0860	0.140	0.124	0.062	0.053	0.083	0.066	0.031	0.023	0.037	0.025
3	0.0990	0.161	0.145	0.070	0.061	0.095	0.077	0.035	0.027	0.043	0.030
4	0.1120	0.183	0.166	0.079	0.069	0.107	0.088	0.039	0.031	0.048	0.035
5	0.1250	0.205	0.187	0.088	0.078	0.120	0.100	0.043	0.035	0.054	0.040
6	0.1380	0.226	0.208	0.096	0.086	0.132	0.111	0.048	0.039	0.060	0.045
8	0.1640	0.270	0.250	0.113	0.102	0.156	0.133	0.054	0.045	0.071	0.054
10	0.1900	0.313	0.292	0.130	0.118	0.180	0.156	0.060	0.050	0.083	0.064
12	0.2160	0.357	0.334	0.148	0.134	0.205	0.178	0.067	0.056	0.094	0.074
1/4	0.2500	0.414	0.389	0.170	0.155	0.237	0.207	0.075	0.064	0.109	0.087
5/16	0.3125	0.518	0.490	0.211	0.194	0.295	0.262	0.084	0.072	0.137	0.110
3/8	0.3750	0.622	0.590	0.253	0.233	0.355	0.315	0.094	0.081	0.164	0.133
7/16	0.4375	0.625	0.589	0.265	0.242	0.368	0.321	0.094	0.081	0.170	0.135
1/2	0.5000	0.750	0.710	0.297	0.273	0.412	0.362	0.106	0.091	0.190	0.151
9/16	0.5625	0.812	0.768	0.336	0.308	0.466	0.410	0.118	0.102	0.214	0.172
5/8	0.6250	0.875	0.827	0.375	0.345	0.521	0.461	0.133	0.116	0.240	0.193
3/4	0.7500	1.000	0.945	0.441	0.406	0.612	0.542	0.149	0.131	0.281	0.226

[1] Where specifying nominal size in decimals, zeros preceding decimal and in the fourth decimal place shall be omitted.

[2] A slight rounding of the edges at periphery of head shall be permissible provided the diameter of the bearing circle is equal to no less than 90 per cent of the specified minimum head diameter.

(ANSI)

PLAIN AND SLOTTED HEX WASHER HEAD MACHINE SCREWS

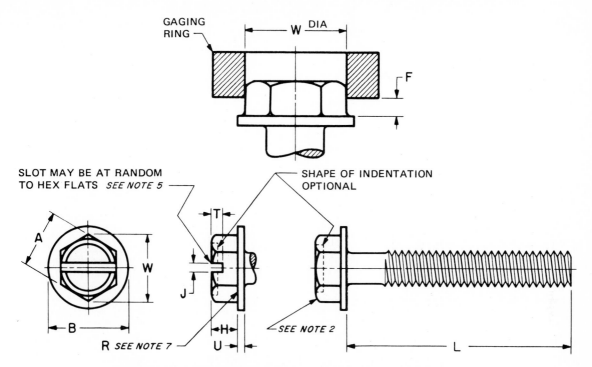

Dimensions of Plain and Slotted Hex Washer Head Machine Screws

Nominal Size[1] or Basic Screw Diameter		A^3 Width Across Flats		$W^{3,4}$ Width Across Corners	H Head Height		B Washer Diameter		U Washer Thickness		J^5 Slot Width		$T^{5,6}$ Slot Depth		F^4 Protrusion Beyond Gaging Ring
		Max	Min	Min	Max	Min	Max	Min	Max	Min	Max	Min	Max	Min	Min
2	0.0860	0.125	0.120	0.134	0.050	0.040	0.166	0.154	0.016	0.010	—	—	—	—	0.024
3	0.0990	0.125	0.120	0.134	0.055	0.044	0.177	0.163	0.016	0.010	—	—	—	—	0.026
4	0.1120	0.188	0.181	0.202	0.060	0.049	0.243	0.225	0.019	0.011	0.039	0.031	0.042	0.025	0.029
5	0.1250	0.188	0.181	0.202	0.070	0.058	0.260	0.240	0.025	0.015	0.043	0.035	0.049	0.030	0.035
6	0.1380	0.250	0.244	0.272	0.093	0.080	0.328	0.302	0.025	0.015	0.048	0.039	0.053	0.033	0.048
8	0.1640	0.250	0.244	0.272	0.110	0.096	0.348	0.322	0.031	0.019	0.054	0.045	0.074	0.052	0.058
10	0.1900	0.312	0.305	0.340	0.120	0.105	0.414	0.384	0.031	0.019	0.060	0.050	0.080	0.057	0.063
12	0.2160	0.312	0.305	0.340	0.155	0.139	0.432	0.398	0.039	0.022	0.067	0.056	0.103	0.077	0.083
1/4	0.2500	0.375	0.367	0.409	0.190	0.172	0.520	0.480	0.050	0.030	0.075	0.064	0.111	0.083	0.103
5/16	0.3125	0.500	0.489	0.545	0.230	0.208	0.676	0.624	0.055	0.035	0.084	0.072	0.134	0.100	0.125
3/8	0.3750	0.562	0.551	0.614	0.295	0.270	0.780	0.720	0.063	0.037	0.094	0.081	0.168	0.131	0.162

[1] Where specifying nominal size in decimals, zeros preceding decimal and in the fourth decimal place shall be omitted.

[2] A slight rounding of all edges and corners of the hex surfaces shall be permissible.

[3] Dimensions across flats and across corners of the head shall be measured at the point of maximum metal. Taper of sides of hex (angle between one side and the axis) shall not exceed 2 deg or 0.004 in., whichever is greater, the specified width across flats being the large dimension.

[4] The rounding due to lack of fill on all six corners of the head shall be reasonably uniform and the width across corners of the head shall be such that when a sharp ring having an inside diameter equal to the specified minimum width across corners is placed on the top of the head, the hex portion of the head shall protrude by an amount equal to, or greater than, the F value tabulated. See Appendix II for Across Corners Gaging of Hex Heads.

[5] Unless otherwise specified by purchaser, hex washer head machine screws are not slotted.

[6] Slot depth beyond bottom of indentation shall not be less than 1/3 of the minimum slot depth specified.

[7] Fillet radius R at junction of sides of hex and top of washer shall not exceed 0.15 times the basic screw diameter.

(ANSI)

PLAIN WASHERS

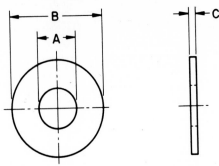

Dimensions of Preferred Sizes of Type A Plain Washers**

Nominal Washer Size***			Inside Diameter A — Basic	Tolerance Plus	Tolerance Minus	Outside Diameter B — Basic	Tolerance Plus	Tolerance Minus	Thickness C — Basic	Max	Min
—	—		0.078	0.000	0.005	0.188	0.000	0.005	0.020	0.025	0.016
—	—		0.094	0.000	0.005	0.250	0.000	0.005	0.020	0.025	0.016
—	—		0.125	0.008	0.005	0.312	0.008	0.005	0.032	0.040	0.025
No. 6	0.138		0.156	0.008	0.005	0.375	0.015	0.005	0.049	0.065	0.036
No. 8	0.164		0.188	0.008	0.005	0.438	0.015	0.005	0.049	0.065	0.036
No. 10	0.190		0.219	0.008	0.005	0.500	0.015	0.005	0.049	0.065	0.036
3/16	0.188		0.250	0.015	0.005	0.562	0.015	0.005	0.049	0.065	0.036
No. 12	0.216		0.250	0.015	0.005	0.562	0.015	0.005	0.065	0.080	0.051
1/4	0.250	N	0.281	0.015	0.005	0.625	0.015	0.005	0.065	0.080	0.051
1/4	0.250	W	0.312	0.015	0.005	0.734*	0.015	0.007	0.065	0.080	0.051
5/16	0.312	N	0.344	0.015	0.005	0.688	0.015	0.007	0.065	0.080	0.051
5/16	0.312	W	0.375	0.015	0.005	0.875	0.030	0.007	0.083	0.104	0.064
3/8	0.375	N	0.406	0.015	0.005	0.812	0.015	0.007	0.065	0.080	0.051
3/8	0.375	W	0.438	0.015	0.005	1.000	0.030	0.007	0.083	0.104	0.064
7/16	0.438	N	0.469	0.015	0.005	0.922	0.015	0.007	0.065	0.080	0.051
7/16	0.438	W	0.500	0.015	0.005	1.250	0.030	0.007	0.083	0.104	0.064
1/2	0.500	N	0.531	0.015	0.005	1.062	0.030	0.007	0.095	0.121	0.074
1/2	0.500	W	0.562	0.015	0.005	1.375	0.030	0.007	0.109	0.132	0.086
9/16	0.562	N	0.594	0.015	0.005	1.156*	0.030	0.007	0.095	0.121	0.074
9/16	0.562	W	0.625	0.015	0.005	1.469*	0.030	0.007	0.109	0.132	0.086
5/16	0.625	N	0.656	0.030	0.007	1.312	0.030	0.007	0.095	0.121	0.074
5/8	0.625	W	0.688	0.030	0.007	1.750	0.030	0.007	0.134	0.160	0.108
3/4	0.750	N	0.812	0.030	0.007	1.469	0.030	0.007	0.134	0.160	0.108
3/4	0.750	W	0.812	0.030	0.007	2.000	0.030	0.007	0.148	0.177	0.122
7/8	0.875	N	0.938	0.030	0.007	1.750	0.030	0.007	0.134	0.160	0.108
7/8	0.875	W	0.938	0.030	0.007	2.250	0.030	0.007	0.165	0.192	0.136
1	1.000	N	1.062	0.030	0.007	2.000	0.030	0.007	0.134	0.160	0.108
1	1.000	W	1.062	0.030	0.007	2.500	0.030	0.007	0.165	0.192	0.136
1 1/8	1.125	N	1.250	0.030	0.007	2.250	0.030	0.007	0.134	0.160	0.108
1 1/8	1.125	W	1.250	0.030	0.007	2.750	0.030	0.007	0.165	0.192	0.136
1 1/4	1.250	N	1.375	0.030	0.007	2.500	0.030	0.007	0.165	0.192	0.136
1 1/4	1.250	W	1.375	0.030	0.007	3.000	0.030	0.007	0.165	0.192	0.136
1 3/8	1.375	N	1.500	0.030	0.007	2.750	0.030	0.007	0.165	0.192	0.136
1 3/8	1.375	W	1.500	0.045	0.010	3.250	0.045	0.010	0.180	0.213	0.153
1 1/2	1.500	N	1.625	0.030	0.007	3.000	0.030	0.007	0.165	0.192	0.136
1 1/2	1.500	W	1.625	0.045	0.010	3.500	0.045	0.010	0.180	0.213	0.153
1 5/8	1.625		1.750	0.045	0.010	3.750	0.045	0.010	0.180	0.213	0.153
1 3/4	1.750		1.875	0.045	0.010	4.000	0.045	0.010	0.180	0.213	0.153
1 7/8	1.875		2.000	0.045	0.010	4.250	0.045	0.010	0.180	0.213	0.153
2	2.000		2.125	0.045	0.010	4.500	0.045	0.010	0.180	0.213	0.153
2 1/4	2.250		2.375	0.045	0.010	4.750	0.045	0.010	0.220	0.248	0.193
2 1/2	2.500		2.625	0.045	0.010	5.000	0.045	0.010	0.238	0.280	0.210
2 3/4	2.750		2.875	0.065	0.010	5.250	0.065	0.010	0.259	0.310	0.228
3	3.000		3.125	0.065	0.010	5.500	0.065	0.010	0.284	0.327	0.249

*The 0.734 in., 1.156 in., and 1.469 in. outside diameters avoid washers which could be used in coin operated devices.
**Preferred sizes are for the most part from series previously designated "Standard Plate" and "SAE." Where common sizes existed in the two series, the SAE size is designated "N" (narrow) and the Standard Plate "W" (wide). These sizes as well as all other sizes of Type A Plain Washers are to be ordered by ID, OD, and thickness dimensions.
***Nominal washer sizes are intended for use with comparable nominal screw or bolt sizes.

(ANSI)

REGULAR HELICAL SPRING LOCK WASHERS

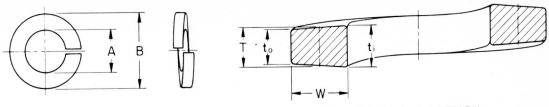

ENLARGED SECTION

Dimensions of Regular Helical Spring Lock Washers[1]

Nominal Washer Size		A Inside Diameter		B Outside Diameter	T Mean Section Thickness $\left(\frac{t_i + t_o}{2}\right)$	W Section Width
		Max	Min	Max[2]	Min	Min
No. 2	0.086	0.094	0.088	0.172	0.020	0.035
No. 3	0.099	0.107	0.101	0.195	0.025	0.040
No. 4	0.112	0.120	0.114	0.209	0.025	0.040
No. 5	0.125	0.133	0.127	0.236	0.031	0.047
No. 6	0.138	0.148	0.141	0.250	0.031	0.047
No. 8	0.164	0.174	0.167	0.293	0.040	0.055
No. 10	0.190	0.200	0.193	0.334	0.047	0.062
No. 12	0.216	0.227	0.220	0.377	0.056	0.070
1/4	0.250	0.262	0.254	0.489	0.062	0.109
5/16	0.312	0.326	0.317	0.586	0.078	0.125
3/8	0.375	0.390	0.380	0.683	0.094	0.141
7/16	0.438	0.455	0.443	0.779	0.109	0.156
1/2	0.500	0.518	0.506	0.873	0.125	0.171
9/16	0.562	0.582	0.570	0.971	0.141	0.188
5/8	0.625	0.650	0.635	1.079	0.156	0.203
11/16	0.688	0.713	0.698	1.176	0.172	0.219
3/4	0.750	0.775	0.760	1.271	0.188	0.234
13/16	0.812	0.843	0.824	1.367	0.203	0.250
7/8	0.875	0.905	0.887	1.464	0.219	0.266
15/16	0.938	0.970	0.950	1.560	0.234	0.281
1	1.000	1.042	1.017	1.661	0.250	0.297
1 1/16	1.062	1.107	1.080	1.756	0.266	0.312
1 1/8	1.125	1.172	1.144	1.853	0.281	0.328
1 3/16	1.188	1.237	1.208	1.950	0.297	0.344
1 1/4	1.250	1.302	1.271	2.045	0.312	0.359
1 5/16	1.312	1.366	1.334	2.141	0.328	0.375
1 3/8	1.375	1.432	1.398	2.239	0.344	0.391
1 7/16	1.438	1.497	1.462	2.334	0.359	0.406
1 1/2	1.500	1.561	1.525	2.430	0.375	0.422

[1] Formerly designated Medium Helical Spring Lock Washers.
[2] The maximum outside diameters specified allow for the commercial tolerances on cold drawn wire.

(ANSI)

INTERNAL TOOTH LOCK WASHERS

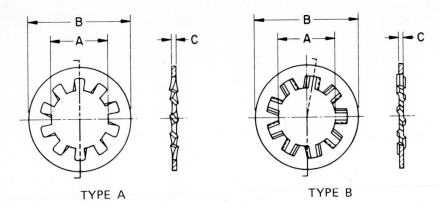

TYPE A   TYPE B

Dimensions of Internal Tooth Lock Washers

Nominal Washer Size		A Inside Diameter		B Outside Diameter		C Thickness	
		Max	Min	Max	Min	Max	Min
No. 2	0.086	0.095	0.089	0.200	0.175	0.015	0.010
No. 3	0.099	0.109	0.102	0.232	0.215	0.019	0.012
No. 4	0.112	0.123	0.115	0.270	0.255	0.019	0.015
No. 5	0.125	0.136	0.129	0.280	0.245	0.021	0.017
No. 6	0.138	0.150	0.141	0.295	0.275	0.021	0.017
No. 8	0.164	0.176	0.168	0.340	0.325	0.023	0.018
No. 10	0.190	0.204	0.195	0.381	0.365	0.025	0.020
No. 12	0.216	0.231	0.221	0.410	0.394	0.025	0.020
¼	0.250	0.267	0.256	0.478	0.460	0.028	0.023
5/16	0.312	0.332	0.320	0.610	0.594	0.034	0.028
⅜	0.375	0.398	0.384	0.692	0.670	0.040	0.032
7/16	0.438	0.464	0.448	0.789	0.740	0.040	0.032
½	0.500	0.530	0.512	0.900	0.867	0.045	0.037
9/16	0.562	0.596	0.576	0.985	0.957	0.045	0.037
⅝	0.625	0.663	0.640	1.071	1.045	0.050	0.042
11/16	0.688	0.728	0.704	1.166	1.130	0.050	0.042
¾	0.750	0.795	0.769	1.245	1.220	0.055	0.047
13/16	0.812	0.861	0.832	1.315	1.290	0.055	0.047
⅞	0.875	0.927	0.894	1.410	1.364	0.060	0.052
1	1.000	1.060	1.019	1.637	1.590	0.067	0.059
1⅛	1.125	1.192	1.144	1.830	1.799	0.067	0.059
1¼	1.250	1.325	1.275	1.975	1.921	0.067	0.059

(ANSI)

EXTERNAL TOOTH LOCK WASHERS

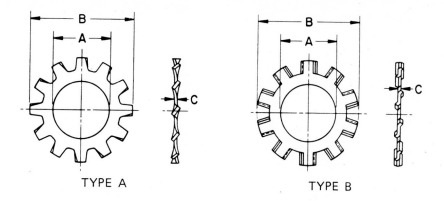

TYPE A   TYPE B

Dimensions of External Tooth Lock Washers

Nominal Washer Size		A Inside Diameter		B Outside Diameter		C Thickness	
		Max	Min	Max	Min	Max	Min
No. 3	0.099	0.109	0.102	0.235	0.220	0.015	0.012
No. 4	0.112	0.123	0.115	0.260	0.245	0.019	0.015
No. 5	0.125	0.136	0.129	0.285	0.270	0.019	0.014
No. 6	0.138	0.150	0.141	0.320	0.305	0.022	0.016
No. 8	0.164	0.176	0.168	0.381	0.365	0.023	0.018
No. 10	0.190	0.204	0.195	0.410	0.395	0.025	0.020
No. 12	0.216	0.231	0.221	0.475	0.460	0.028	0.023
$\frac{1}{4}$	0.250	0.267	0.256	0.510	0.494	0.028	0.023
$\frac{5}{16}$	0.312	0.332	0.320	0.610	0.588	0.034	0.028
$\frac{3}{8}$	0.375	0.398	0.384	0.694	0.670	0.040	0.032
$\frac{7}{16}$	0.438	0.464	0.448	0.760	0.740	0.040	0.032
$\frac{1}{2}$	0.500	0.530	0.513	0.900	0.880	0.045	0.037
$\frac{9}{16}$	0.562	0.596	0.576	0.985	0.960	0.045	0.037
$\frac{5}{8}$	0.625	0.663	0.641	1.070	1.045	0.050	0.042
$\frac{11}{16}$	0.688	0.728	0.704	1.155	1.130	0.050	0.042
$\frac{3}{4}$	0.750	0.795	0.768	1.260	1.220	0.055	0.047
$\frac{13}{16}$	0.812	0.861	0.833	1.315	1.290	0.055	0.047
$\frac{7}{8}$	0.875	0.927	0.897	1.410	1.380	0.060	0.052
1	1.000	1.060	1.025	1.620	1.590	0.067	0.059

(ANSI)

COUNTERSUNK HEAD SEMI-TUBULAR RIVETS

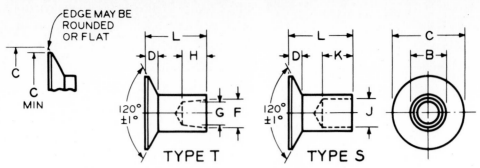

Dimensions of 120 Deg Countersunk Head Semi-Tubular Rivets
(General Purpose)

B		C		D	Type-T Taper Hole Rivets			Type-S Straight Hole Rivets		
					F	G	H	J		K
Shank Diameter		Head Diameter		Head Thickness	Hole Dia at End of Rivet		Hole Dia at Bottom of Hole	Hole Depth to Start of Apex	Hole Dia at End of Rivet		Hole Depth to Start of Apex*
		Max Sharp†	Min, Edge Rounded or Flat								
Min	Max			Ref	Min	Max	Min	Min	Min	Max	Nom
0.085	0.089	0.223	0.203	0.030	0.064	0.068	0.050	0.057	0.062	0.068	0.064
0.118	0.123	0.271	0.245	0.037	0.091	0.095	0.079	0.082	0.084	0.090	0.094
0.141	0.146	0.337	0.307	0.045	0.106	0.112	0.085	0.104	0.100	0.107	0.126
0.182	0.188	0.404	0.369	0.065	0.139	0.145	0.110	0.135	0.134	0.141	0.155
0.210	0.217	0.472	0.430	0.071	0.158	0.166	0.136	0.151	0.155	0.163	0.189
0.244	0.252	0.540	0.493	0.087	0.181	0.191	0.150	0.183	0.176	0.184	0.219

†Head diameter "Max Sharp" is a theoretical dimension determined by projection.
*For rivets having a length tolerance of ±0.015 in. or greater, the straight hole nominal depth shall be increased 0.010 in. over the depth specified.

(ANSI)

FULL TUBULAR RIVETS

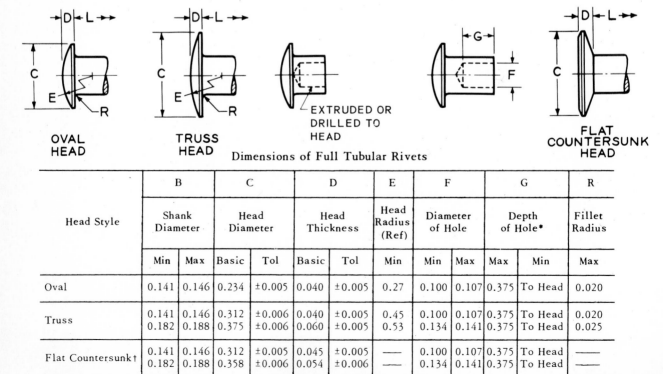

Dimensions of Full Tubular Rivets

Head Style	B		C		D		E	F		G		R
	Shank Diameter		Head Diameter		Head Thickness		Head Radius (Ref)	Diameter of Hole		Depth of Hole*		Fillet Radius
	Min	Max	Basic	Tol	Basic	Tol	Min	Min	Max	Max	Min	Max
Oval	0.141	0.146	0.234	±0.005	0.040	±0.005	0.27	0.100	0.107	0.375	To Head	0.020
Truss	0.141	0.146	0.312	±0.006	0.040	±0.005	0.45	0.100	0.107	0.375	To Head	0.020
	0.182	0.188	0.375	±0.006	0.060	±0.005	0.53	0.134	0.141	0.375	To Head	0.025
Flat Countersunk†	0.141	0.146	0.312	±0.005	0.045	±0.005	——	0.100	0.107	0.375	To Head	——
	0.182	0.188	0.358	±0.006	0.054	±0.006	——	0.134	0.141	0.375	To Head	——

*Full tubular rivets having nominal length of ⅜ in. or shorter shall be drilled or extruded to head.
†Angle of head not specified since it is assumed this type of rivet would generally be used in soft materials and therefore form its own countersink.

(ANSI)

FLAT COUNTERSUNK HEAD RIVETS

EDGE OF HEAD MAY BE ROUNDED OR FLAT
SEE NOTE 3

90° ± 2°

Dimensions of Flat Countersunk Head Rivets

Nominal Size[1] or Basic Shank Diameter	E Shank Diameter Max	Min	A Head Diameter Max[2]	Min[3]	H Head Height Ref[4]
1/16 0.062	0.064	0.059	0.118	0.110	0.027
3/32 0.094	0.096	0.090	0.176	0.163	0.040
1/8 0.125	0.127	0.121	0.235	0.217	0.053
5/32 0.156	0.158	0.152	0.293	0.272	0.066
3/16 0.188	0.191	0.182	0.351	0.326	0.079
7/32 0.219	0.222	0.213	0.413	0.384	0.094
1/4 0.250	0.253	0.244	0.469	0.437	0.106
9/32 0.281	0.285	0.273	0.528	0.491	0.119
5/16 0.312	0.316	0.304	0.588	0.547	0.133
11/32 0.344	0.348	0.336	0.646	0.602	0.146
3/8 0.375	0.380	0.365	0.704	0.656	0.159
13/32 0.406	0.411	0.396	0.763	0.710	0.172
7/16 0.438	0.443	0.428	0.823	0.765	0.186

1 Where specifying nominal size in decimals, zeros preceding decimal shall be omitted.
2 Sharp edged head. Tabulated maximum values calculated on basic diameter of rivet and 92° included angle extended to a sharp edge.
3 Rounded or flat edged irregular shaped head. See Paragraph 2.1 of General Data.
4 Head height, H, is given for reference purposes only. Variations in this dimension are controlled by the head and shank diameters and the included angle of the head.

(ANSI)

TINNERS RIVETS

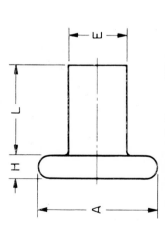

Dimensions of Tinners Rivets

Rivet Size Number[1]	E Shank Diameter Max	Min	A Head Diameter Max	Min	H Head Height Max	Min	L Rivet Length Max	Min
6 oz	0.081	0.075	0.213	0.193	0.028	0.016	0.135	0.115
8 oz	0.091	0.085	0.225	0.205	0.036	0.024	0.166	0.146
10 oz	0.097	0.091	0.250	0.230	0.037	0.025	0.182	0.162
12 oz	0.107	0.101	0.265	0.245	0.037	0.025	0.198	0.178
14 oz	0.111	0.105	0.275	0.255	0.038	0.026	0.198	0.178
1 lb	0.113	0.107	0.285	0.265	0.040	0.028	0.213	0.193
1 1/4 lb	0.122	0.116	0.295	0.275	0.045	0.033	0.229	0.209
1 1/2 lb	0.132	0.126	0.316	0.294	0.046	0.034	0.244	0.224
1 3/4 lb	0.136	0.130	0.331	0.309	0.049	0.035	0.260	0.240
2 lb	0.146	0.140	0.341	0.319	0.050	0.036	0.276	0.256
2 1/2 lb	0.150	0.144	0.311	0.289	0.069	0.055	0.291	0.271
3 lb	0.163	0.154	0.329	0.303	0.073	0.059	0.323	0.303
3 1/2 lb	0.168	0.159	0.348	0.322	0.074	0.060	0.338	0.318
4 lb	0.179	0.170	0.368	0.342	0.076	0.062	0.354	0.334
5 lb	0.190	0.181	0.388	0.362	0.084	0.070	0.385	0.365
6 lb	0.206	0.197	0.419	0.393	0.090	0.076	0.401	0.381
7 lb	0.223	0.214	0.431	0.405	0.094	0.080	0.416	0.396
8 lb	0.227	0.218	0.475	0.445	0.101	0.085	0.448	0.428
9 lb	0.241	0.232	0.490	0.460	0.103	0.087	0.463	0.443
10 lb	0.241	0.232	0.505	0.475	0.104	0.088	0.479	0.459
12 lb	0.263	0.251	0.532	0.498	0.108	0.090	0.510	0.490
14 lb	0.288	0.276	0.577	0.543	0.113	0.095	0.525	0.505
16 lb	0.304	0.292	0.597	0.563	0.128	0.110	0.541	0.521
18 lb	0.347	0.335	0.706	0.668	0.156	0.136	0.603	0.583

1 Size numbers in ounces and pounds refer to the approximate weight of 1000 rivets.

PAN HEAD RIVETS

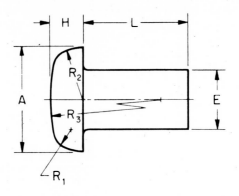

Dimensions of Pan Head Rivets

Nominal Size[1] or Basic Shank Diameter		E Shank Diameter		A Head Diameter		H Head Height		R₁ Head Corner Radius	R₂ Head Side Radius	R₃ Head Crown Radius
		Max	Min	Max	Min	Max	Min	Approx	Approx	Approx
1/16	0.062	0.064	0.059	0.118	0.098	0.040	0.030	0.019	0.052	0.217
3/32	0.094	0.096	0.090	0.173	0.153	0.060	0.048	0.030	0.080	0.326
1/8	0.125	0.127	0.121	0.225	0.205	0.078	0.066	0.039	0.106	0.429
5/32	0.156	0.158	0.152	0.279	0.257	0.096	0.082	0.049	0.133	0.535
3/16	0.188	0.191	0.182	0.334	0.308	0.114	0.100	0.059	0.159	0.641
7/32	0.219	0.222	0.213	0.391	0.365	0.133	0.119	0.069	0.186	0.754
1/4	0.250	0.253	0.244	0.444	0.414	0.151	0.135	0.079	0.213	0.858
9/32	0.281	0.285	0.273	0.499	0.465	0.170	0.152	0.088	0.239	0.963
5/16	0.312	0.316	0.304	0.552	0.518	0.187	0.169	0.098	0.266	1.070
11/32	0.344	0.348	0.336	0.608	0.570	0.206	0.186	0.108	0.292	1.176
3/8	0.375	0.380	0.365	0.663	0.625	0.225	0.205	0.118	0.319	1.286
13/32	0.406	0.411	0.396	0.719	0.675	0.243	0.221	0.127	0.345	1.392
7/16	0.438	0.443	0.428	0.772	0.728	0.261	0.239	0.137	0.372	1.500

[1] Where specifying nominal size in decimals, zeros preceding decimal shall be omitted.

(ANSI)

KEY SIZE VERSUS SHAFT DIAMETER

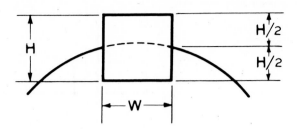

Key Size Versus Shaft Diameter

NOMINAL SHAFT DIAMETER		NOMINAL KEY SIZE			NOMINAL KEYSEAT DEPTH	
			Height, H		H/2	
Over	To (Incl)	Width, W	Square	Rectangular	Square	Rectangular
5/16	7/16	3/32	3/32		3/64	
7/16	9/16	1/8	1/8	3/32	1/16	3/64
9/16	7/8	3/16	3/16	1/8	3/32	1/16
7/8	1-1/4	1/4	1/4	3/16	1/8	3/32
1-1/4	1-3/8	5/16	5/16	1/4	5/32	1/8
1-3/8	1-3/4	3/8	3/8	1/4	3/16	1/8
1-3/4	2-1/4	1/2	1/2	3/8	1/4	3/16
2-1/4	2-3/4	5/8	5/8	7/16	5/16	7/32
2-3/4	3-1/4	3/4	3/4	1/2	3/8	1/4
3-1/4	3-3/4	7/8	7/8	5/8	7/16	5/16
3-3/4	4-1/2	1	1	3/4	1/2	3/8
4-1/2	5-1/2	1-1/4	1-1/4	7/8	5/8	7/16
5-1/2	6-1/2	1-1/2	1-1/2	1	3/4	1/2
6-1/2	7-1/2	1-3/4	1-3/4	1-1/2*	7/8	3/4
7-1/2	9	2	2	1-1/2	1	3/4
9	11	2-1/2	2-1/2	1-3/4	1-1/4	7/8
11	13	3	3	2	1-1/2	1
13	15	3-1/2	3-1/2	2-1/2	1-3/4	1-1/4
15	18	4		3		1-1/2
18	22	5		3-1/2		1-3/4
22	26	6		4		2
26	30	7		5		2-1/2

*Some key standards show 1-1/4 in. Preferred size is 1-1/2 in.

(ANSI)

PARALLEL AND TAPER KEYS

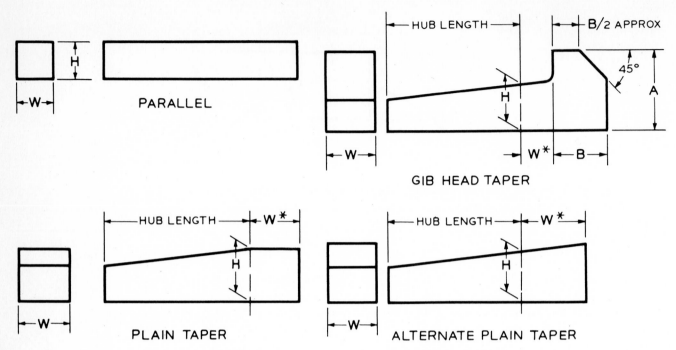

Plain and Gib Head Taper Keys Have a 1/8" Taper in 12"

Key Dimensions and Tolerances

KEY			NOMINAL KEY SIZE		TOLERANCE	
			Width, W		Width, W	Height, H
			Over	To (Incl)		
Parallel	Square	Bar Stock	— 3/4 1-1/2 2-1/2	3/4 1-1/2 2-1/2 3-1/2	+0.000 −0.002 +0.000 −0.003 +0.000 −0.004 +0.000 −0.006	+0.000 −0.002 +0.000 −0.003 +0.000 −0.004 +0.000 −0.006
		Keystock	— 1-1/4 3	1-1/4 3 3-1/2	+0.001 −0.000 +0.002 −0.000 +0.003 −0.000	+0.001 −0.000 +0.002 −0.000 +0.003 −0.000
	Rectangular	Bar Stock	— 3/4 1-1/2 3 4 6	3/4 1-1/2 3 4 6 7	+0.000 −0.003 +0.000 −0.004 +0.000 −0.005 +0.000 −0.006 +0.000 −0.008 +0.000 −0.013	+0.000 −0.003 +0.000 −0.004 +0.000 −0.005 +0.000 −0.006 +0.000 −0.008 +0.000 −0.013
		Keystock	— 1-1/4 3	1-1/4 3 7	+0.001 −0.000 +0.002 −0.000 +0.003 −0.000	+0.005 −0.005 +0.005 −0.005 +0.005 −0.005
Taper	Plain or Gib Head Square or Rectangular		— 1-1/4 3	1-1/4 3 7	+0.001 −0.000 +0.002 −0.000 +0.003 −0.000	+0.005 −0.000 +0.005 −0.000 +0.005 −0.000

*For locating position of dimension H. Tolerance does not apply.
See Table 2A for dimensions on gib heads.
All dimensions given in inches.

(ANSI)

WOODRUFF KEYS

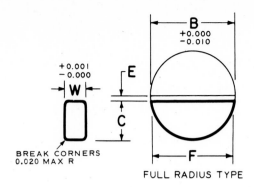

FULL RADIUS TYPE

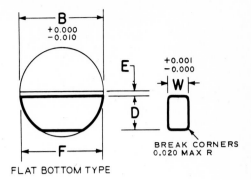

FLAT BOTTOM TYPE

Woodruff Keys

Key No.	Nominal Key Size W × B	Actual Length F +0.000-0.010	Height of Key				Distance Below Center E
			C		D		
			Max	Min	Max	Min	
202	1/16 × 1/4	0.248	0.109	0.104	0.109	0.104	1/64
202.5	1/16 × 5/16	0.311	0.140	0.135	0.140	0.135	1/64
302.5	3/32 × 5/16	0.311	0.140	0.135	0.140	0.135	1/64
203	1/16 × 3/8	0.374	0.172	0.167	0.172	0.167	1/64
303	3/32 × 3/8	0.374	0.172	0.167	0.172	0.167	1/64
403	1/8 × 3/8	0.374	0.172	0.167	0.172	0.167	1/64
204	1/16 × 1/2	0.491	0.203	0.198	0.194	0.188	3/64
304	3/32 × 1/2	0.491	0.203	0.198	0.194	0.188	3/64
404	1/8 × 1/2	0.491	0.203	0.198	0.194	0.188	3/64
305	3/32 × 5/8	0.612	0.250	0.245	0.240	0.234	1/16
405	1/8 × 5/8	0.612	0.250	0.245	0.240	0.234	1/16
505	5/32 × 5/8	0.612	0.250	0.245	0.240	0.234	1/16
605	3/16 × 5/8	0.612	0.250	0.245	0.240	0.234	1/16
406	1/8 × 3/4	0.740	0.313	0.308	0.303	0.297	1/16
506	5/32 × 3/4	0.740	0.313	0.308	0.303	0.297	1/16
606	3/16 × 3/4	0.740	0.313	0.308	0.303	0.297	1/16
806	1/4 × 3/4	0.740	0.313	0.308	0.303	0.297	1/16
507	5/32 × 7/8	0.866	0.375	0.370	0.365	0.359	1/16
607	3/16 × 7/8	0.866	0.375	0.370	0.365	0.359	1/16
707	7/32 × 7/8	0.866	0.375	0.370	0.365	0.359	1/16
807	1/4 × 7/8	0.866	0.375	0.370	0.365	0.359	1/16
608	3/16 × 1	0.992	0.438	0.433	0.428	0.422	1/16
708	7/32 × 1	0.992	0.438	0.433	0.428	0.422	1/16
808	1/4 × 1	0.992	0.438	0.433	0.428	0.422	1/16
1008	5/16 × 1	0.992	0.438	0.433	0.428	0.422	1/16
1208	3/8 × 1	0.992	0.438	0.433	0.428	0.422	1/16
609	3/16 × 1 1/8	1.114	0.484	0.479	0.475	0.469	5/64
709	7/32 × 1 1/8	1.114	0.484	0.479	0.475	0.469	5/64
809	1/4 × 1 1/8	1.114	0.484	0.479	0.475	0.469	5/64
1009	5/16 × 1 1/8	1.114	0.484	0.479	0.475	0.469	5/64

All dimensions given are in inches.

The key numbers indicate nominal key dimensions. The last two digits give the nominal diameter B in eighths of an inch and the digits preceding the last two give the nominal width W in thirty-seconds of an inch.

Example:
      No. 204 indicates a key 2/32 × 4/8 or 1/16 × 1/2.
      No. 808 indicates a key 8/32 × 8/8 or 1/4 × 1.
      No. 1212 indicates a key 12/32 × 12/8 or 3/8 × 1 1/2.

(ANSI)

WOODRUFF KEYSEATS

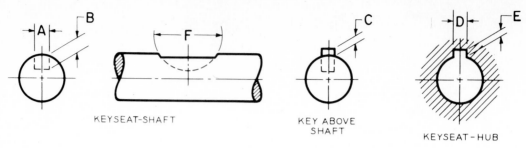

KEYSEAT-SHAFT   KEY ABOVE SHAFT   KEYSEAT-HUB

Keyseat Dimensions

Key Number	Nominal Size Key	Keyseat — Shaft					Key Above Shaft	Keyseat — Hub	
		Width A*		Depth B	Diameter F		Height C	Width D	Depth E
		Min	Max	+0.005 -0.000	Min	Max	+0.005 -0.005	+0.002 -0.000	+0.005 -0.000
202	1/16 × 1/4	0.0615	0.0630	0.0728	0.250	0.268	0.0312	0.0635	0.0372
202.5	1/16 × 5/16	0.0615	0.0630	0.1038	0.312	0.330	0.0312	0.0635	0.0372
302.5	3/32 × 5/16	0.0928	0.0943	0.0882	0.312	0.330	0.0469	0.0948	0.0529
203	1/16 × 3/8	0.0615	0.0630	0.1358	0.375	0.393	0.0312	0.0635	0.0372
303	3/32 × 3/8	0.0928	0.0943	0.1202	0.375	0.393	0.0469	0.0948	0.0529
403	1/8 × 3/8	0.1240	0.1255	0.1045	0.375	0.393	0.0625	0.1260	0.0685
204	1/16 × 1/2	0.0615	0.0630	0.1668	0.500	0.518	0.0312	0.0635	0.0372
304	3/32 × 1/2	0.0928	0.0943	0.1511	0.500	0.518	0.0469	0.0948	0.0529
404	1/8 × 1/2	0.1240	0.1255	0.1355	0.500	0.518	0.0625	0.1260	0.0685
305	3/32 × 5/8	0.0928	0.0943	0.1981	0.625	0.643	0.0469	0.0948	0.0529
405	1/8 × 5/8	0.1240	0.1255	0.1825	0.625	0.643	0.0625	0.1260	0.0685
505	5/32 × 5/8	0.1553	0.1568	0.1669	0.625	0.643	0.0781	0.1573	0.0841
605	3/16 × 5/8	0.1863	0.1880	0.1513	0.625	0.643	0.0937	0.1885	0.0997
406	1/8 × 3/4	0.1240	0.1255	0.2455	0.750	0.768	0.0625	0.1260	0.0685
506	5/32 × 3/4	0.1553	0.1568	0.2299	0.750	0.768	0.0781	0.1573	0.0841
606	3/16 × 3/4	0.1863	0.1880	0.2143	0.750	0.768	0.0937	0.1885	0.0997
806	1/4 × 3/4	0.2487	0.2505	0.1830	0.750	0.768	0.1250	0.2510	0.1310
507	5/32 × 7/8	0.1553	0.1568	0.2919	0.875	0.895	0.0781	0.1573	0.0841
607	3/16 × 7/8	0.1863	0.1880	0.2763	0.875	0.895	0.0937	0.1885	0.0997
707	7/32 × 7/8	0.2175	0.2193	0.2607	0.875	0.895	0.1093	0.2198	0.1153
807	1/4 × 7/8	0.2487	0.2505	0.2450	0.875	0.895	0.1250	0.2510	0.1310
608	3/16 × 1	0.1863	0.1880	0.3393	1.000	1.020	0.0937	0.1885	0.0997
708	7/32 × 1	0.2175	0.2193	0.3237	1.000	1.020	0.1093	0.2198	0.1153
808	1/4 × 1	0.2487	0.2505	0.3080	1.000	1.020	0.1250	0.2510	0.1310
1008	5/16 × 1	0.3111	0.3130	0.2768	1.000	1.020	0.1562	0.3135	0.1622
1208	3/8 × 1	0.3735	0.3755	0.2455	1.000	1.020	0.1875	0.3760	0.1935
609	3/16 × 1 1/8	0.1863	0.1880	0.3853	1.125	1.145	0.0937	0.1885	0.0997
709	7/32 × 1 1/8	0.2175	0.2193	0.3697	1.125	1.145	0.1093	0.2198	0.1153
809	1/4 × 1 1/8	0.2487	0.2505	0.3540	1.125	1.145	0.1250	0.2510	0.1310
1009	5/16 × 1 1/8	0.3111	0.3130	0.3228	1.125	1.145	0.1562	0.3135	0.1622

*   Width A values were set with the maximum keyseat (shaft) width as that figure which will receive a key with the greatest amount of looseness consistent with assuring the key's sticking in the keyseat (shaft). Minimum keyseat width is that figure permitting the largest shaft distortion acceptable when assembling maximum key in minimum keyseat.

Dimensions A, B, C, D are taken at side intersection.

(ANSI)

STRAIGHT PINS

CHAMFERED
25°
L ± 0.012
B ± 0.010
A

SQUARE END
L ± 0.012
A

Dimensions of Straight Pins

Nominal Diameter	Diameter A Max	Diameter A Min	Chamfer B
0.062	0.0625	0.0605	0.015
0.094	0.0937	0.0917	0.015
0.109	0.1094	0.1074	0.015
0.125	0.1250	0.1230	0.015
0.156	0.1562	0.1542	0.015
0.188	0.1875	0.1855	0.015
0.219	0.2187	0.2167	0.015
0.250	0.2500	0.2480	0.015
0.312	0.3125	0.3095	0.030
0.375	0.3750	0.3720	0.030
0.438	0.4375	0.4345	0.030
0.500	0.500	0.4970	0.030

All dimensions are given in inches.

These pins must be straight and free from burrs or any other defects that will affect their serviceability.

(ANSI)

GIB HEAD KEYS

A
45°
B/2 APPROX
B
W*
H
W

Gib Head Nominal Dimensions

Nominal Key Size Width, W	SQUARE H	SQUARE A	SQUARE B	RECTANGULAR H	RECTANGULAR A	RECTANGULAR B
1/8	1/8	1/4	1/4	3/32	3/16	1/8
3/16	3/16	5/16	5/16	1/8	1/4	1/4
1/4	1/4	7/16	3/8	3/16	5/16	5/16
5/16	5/16	1/2	7/16	1/4	7/16	3/8
3/8	3/8	5/8	1/2	1/4	7/16	3/8
1/2	1/2	7/8	5/8	3/8	5/8	1/2
5/8	5/8	1	3/4	7/16	3/4	9/16
3/4	3/4	1-1/4	7/8	1/2	7/8	5/8
7/8	7/8	1-3/8	1	5/8	1	3/4
1	1	1-5/8	1-1/8	3/4	1-1/4	7/8
1-1/4	1-1/4	2	1-7/16	7/8	1-3/8	
1-1/2	1-1/2	2-3/8	1-3/4	1	1-5/8	1-1/8
1-3/4	1-3/4	2-3/4	2	1-1/2	2-3/8	1-3/4
2	2	3-1/2	2-1/4	1-1/2	2-3/8	1-3/4
2-1/2	2-1/2	4	3	1-3/4	2-3/4	2
3	3	5	3-1/2	2	3-1/2	2-1/4
3-1/2	3-1/2	6	4	2-1/2	4	3

*For locating position of dimension H.

For larger sizes the following relationships are suggested as guides for establishing A and B.

$A = 1.8 H$ $B = 1.2 H$

All dimensions given in inches.

HARDENED AND GROUND DOWEL PINS

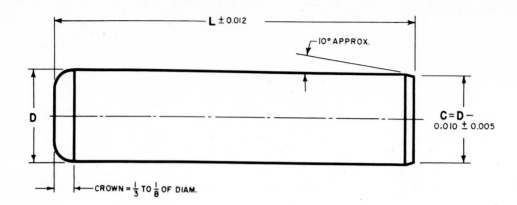

L ± 0.012

10° APPROX.

D

C = D − 0.010 ± 0.005

CROWN = ⅓ TO ⅛ OF DIAM.

Dimensions of Hardened and Ground Dowel Pins

Length, L	Nominal Diameter D									
	⅛	3/16	¼	5/16	⅜	7/16	½	⅝	¾	⅞
Diameter Standard Pins ±0.0001										
	0.1252	0.1877	0.2502	0.3127	0.3752	0.4377	0.5002	0.6252	0.7502	0.8752
Diameter Oversize Pins ±0.0001										
	0.1260	0.1885	0.2510	0.3135	0.3760	0.4385	0.5010	0.6260	0.7510	0.8760
½	X	X	X	X						
⅝	X	X	X	X						
¾	X	X	X	X	X					
⅞	X	X	X	X	X	X				
1	X	X	X	X	X	X				
1 ¼		X	X	X	X	X	X	X		
1 ½		X	X	X	X	X	X	X	X	
1 ¾		X	X	X	X	X	X	X	X	
2		X	X	X	X	X	X	X	X	X
2 ¼					X	X	X	X		
2 ½				X	X	X	X	X	X	X
3							X	X	X	X
3 ½							X	X		
4							X	X	X	X
4 ½								X	X	X
5									X	X
5 ½									X	X

All dimensions are given in inches.

These pins are extensively used in the tool and machine industry and a machine reamer of nominal size may be used to produce the holes into which these pins tap or press fit. They must be straight and free from any defects that will affect their serviceability.

(ANSI)

TAPER PINS

Dimensions of Taper Pins

Number	7/0	6/0	5/0	4/0	3/0	2/0	0	1	2	3	4	5	6	7	8	9	10
Size (Large End)	0.0625	0.0780	0.0940	0.1090	0.1250	0.1410	0.1560	0.1720	0.1930	0.2190	0.2500	0.2890	0.3410	0.4090	0.4920	0.5910	0.7060
Length, L																	
0.375	X	X															
0.500	X	X	X	X	X	X	X										
0.625	X	X	X	X	X	X	X										
0.750			X	X	X	X	X	X	X	X							
0.875					X	X	X	X	X	X							
1.000			X	X	X	X	X	X	X	X	X	X					
1.250				X		X	X	X	X	X	X	X	X				
1.500							X	X	X	X	X	X	X				
1.750								X	X	X	X	X	X	X	X		
2.000								X	X	X	X	X	X	X	X		
2.250									X	X	X	X	X	X	X	X	
2.500										X	X	X	X	X	X	X	
2.750										X	X	X	X	X	X	X	
3.000											X	X	X	X	X	X	
3.250													X	X	X	X	X
3.500													X	X	X	X	X
3.750													X	X	X	X	X
4.000															X	X	X
4.250															X	X	X
4.500															X	X	X
4.750																X	X
5.000																X	X
5.250																X	X
5.500																X	X
5.750																X	X
6.000																	X

All dimensions are given in inches.

Standard reamers are available for pins given above the line.

Pins Nos. 11 (size 0.8600), 12 (size 1.032), 13 (size 1.241), and 14 (1.523) are special sizes—hence their lengths are special.

To find small diameter of pin, multiply the length by 0.02083 and subtract the result from the large diameter.

TYPES	COMMERCIAL TYPE	PRECISION TYPE
Sizes	7/0 to 14	7/0 to 10
Tolerance on Diameter	(+0.0013, −0.0007)	(+0.0013, −0.0007)
Taper	¼ In. per Ft	¼ In. per Ft
Length Tolerance	(±0.030)	(±0.030)
Concavity Tolerance	None	0.0005 up to 1 in. long
		0.001 1¼₆ to 2 in. long
		0.002 2¹⁄₁₆ and longer

(ANSI)

SHEET METAL AND WIRE GAGE DESIGNATION

GAGE NO.	AMERICAN OR BROWN & SHARPE'S A.W.G. OR B. & S.	BIRMING-HAM OR STUBS WIRE B.W.G.	WASHBURN & MOEN OR AMERICAN S.W.G.	UNITED STATES STANDARD	MANU-FACTURERS' STANDARD FOR SHEET STEEL	GAGE NO.
0000000	- - - -	- - - -	.4900	.500	- - - -	0000000
000000	.5800	- - - -	.4615	.469	- - - -	000000
00000	.5165	- - - -	.4305	.438	- - - -	00000
0000	.4600	.454	.3938	.406	- - - -	0000
000	.4096	.425	.3625	.375	- - - -	000
00	.3648	.380	.3310	.344	- - - -	00
0	.3249	.340	.3065	.312	- - - -	0
1	.2893	.300	.2830	.281	- - - -	1
2	.2576	.284	.2625	.266	- - - -	2
3	.2294	.259	.2437	.250	.2391	3
4	.2043	.238	.2253	.234	.2242	4
5	.1819	.220	.2070	.219	.2092	5
6	.1620	.203	.1920	.203	.1943	6
7	.1443	.180	.1770	.188	.1793	7
8	.1285	.165	.1620	.172	.1644	8
9	.1144	.148	.1483	.156	.1495	9
10	.1019	.134	.1350	.141	.1345	10
11	.0907	.120	.1205	.125	.1196	11
12	.0808	.109	.1055	.109	.1046	12
13	.0720	.095	.0915	.0938	.0897	13
14	.0642	.083	.0800	.0781	.0747	14
15	.0571	.072	.0720	.0703	.0673	15
16	.0508	.065	.0625	.0625	.0598	16
17	.0453	.058	.0540	.0562	.0538	17
18	.0403	.049	.0475	.0500	.0478	18
19	.0359	.042	.0410	.0438	.0418	19
20	.0320	.035	.0348	.0375	.0359	20
21	.0285	.032	.0317	.0344	.0329	21
22	.0253	.028	.0286	.0312	.0299	22
23	.0226	.025	.0258	.0281	.0269	23
24	.0201	.022	.0230	.0250	.0239	24
25	.0179	.020	.0204	.0219	.0209	25
26	.0159	.018	.0181	.0188	.0179	26
27	.0142	.016	.0173	.0172	.0164	27
28	.0126	.014	.0162	.0156	.0149	28
29	.0113	.013	.0150	.0141	.0135	29
30	.0100	.012	.0140	.0125	.0120	30
31	.0089	.010	.0132	.0109	.0105	31
32	.0080	.009	.0128	.0102	.0097	32
33	.0071	.008	.0118	.00938	.0090	33
34	.0063	.007	.0104	.00859	.0082	34
35	.0056	.005	.0095	.00781	.0075	35
36	.0050	.004	.0090	.00703	.0067	36
37	.0045	- - - -	.0085	.00624	.0064	37
38	.0040	- - - -	.0080	.00625	.0060	38
39	.0035	- - - -	.0075	- - - - -	- - - -	39
40	.0031	- - - -	.0070	- - - - -	- - - -	40
41	.0028	- - - -	.0066	- - - - -	- - - -	41
42	.0025	- - - -	.0062	- - - - -	- - - -	42
43	.0022	- - - -	.0060	- - - - -	- - - -	43
44	.0020	- - - -	.0058	- - - - -	- - - -	44
45	.0018	- - - -	.0055	- - - - -	- - - -	45
46	.0016	- - - -	.0052	- - - - -	- - - -	46
47	.0014	- - - -	.0050	- - - - -	- - - -	47
48	.0012	- - - -	.0048	- - - - -	- - - -	48

PRECISION SHEET METAL SET-BACK CHART

MATERIAL THICKNESS (T)

90 DEG BEND RADIUS (R)	.016	.020	.025	.032	.040	.051	.064	.072	.078	.081	.091	.102	.125	.129	.156	.162	.187	.250
1/32	.034	.039	.046	.055	.065	.081	.102	.113	.121	.125	.139							
3/64	.041	.046	.053	.062	.072	.090	.108	.119	.127	.131	.145							
1/16	.048	.053	.059	.068	.079	.093	.110	.122	.134	.138	.152							
5/64	.054	.060	.066	.075	.086	.100	.117	.127	.138	.144	.158							
3/32	.061	.066	.073	.082	.092	.107	.124	.134	.142	.146	.160							
7/64	.068	.073	.080	.089	.099	.113	.130	.141	.148	.153	.167	.181						
1/8	.075	.080	.086	.095	.106	.120	.137	.147	.155	.159	.172	.186	.216	.221				
9/64	.081	.087	.093	.102	.113	.127	.144	.154	.162	.166	.179	.193	.223	.228	.263			
5/32	.088	.093	.100	.109	.119	.134	.150	.161	.169	.173	.186	.200	.230	.235	.270	.278		
11/64	.095	.100	.107	.116	.126	.140	.157	.168	.175	.179	.192	.207	.236	.242	.277	.284	.317	
3/16	.102	.107	.113	.122	.133	.147	.164	.174	.182	.186	.199	.213	.243	.248	.283	.291	.324	.405
13/64	.108	.114	.120	.129	.140	.154	.171	.181	.189	.193	.206	.220	.250	.255	.290	.298	.330	.412
7/32	.115	.120	.127	.136	.146	.161	.177	.188	.196	.199	.212	.227	.257	.262	.297	.305	.337	.419
15/64	.122	.127	.134	.143	.153	.167	.184	.195	.202	.206	.219	.233	.263	.269	.304	.311	.344	.426
1/4	.129	.134	.140	.149	.160	.174	.191	.201	.209	.213	.226	.240	.270	.275	.310	.318	.351	.432
17/64	.135	.141	.147	.156	.166	.181	.198	.208	.216	.220	.233	.247	.277	.282	.317	.325	.357	.439
9/32	.142	.147	.154	.163	.173	.187	.204	.215	.223	.226	.239	.254	.284	.289	.324	.332	.364	.446
5/16	.156	.161	.167	.176	.187	.201	.218	.228	.236	.240	.253	.267	.297	.302	.337	.345	.378	.459
11/32	.169	.174	.181	.190	.200	.214	.231	.242	.250	.253	.266	.281	.311	.316	.351	.359	.391	.473
3/8	.183	.188	.194	.203	.214	.228	.245	.255	.263	.267	.280	.294	.324	.329	.364	.372	.404	.486

(STOCK THICKNESS) T

Z = SET-BACK ALLOWANCE FROM CHART

DEVELOPED LENGTH = X + Y − Z

The constants in the table are multiplied by the diameter of the bolt-hole pitch circle to obtain the longitudinal and lateral adjustments of the right-angle slides of the jig borer, in boring equally spaced holes. While holes may be located by these right-angular measurements, an auxiliary rotary table provides a more direct method. With a rotary table, the holes are spaced by precise angular movements after adjustment to the required radius.

MULTIPLY VALUES SHOWN BY DIAMETER OF PITCH CIRCLE.

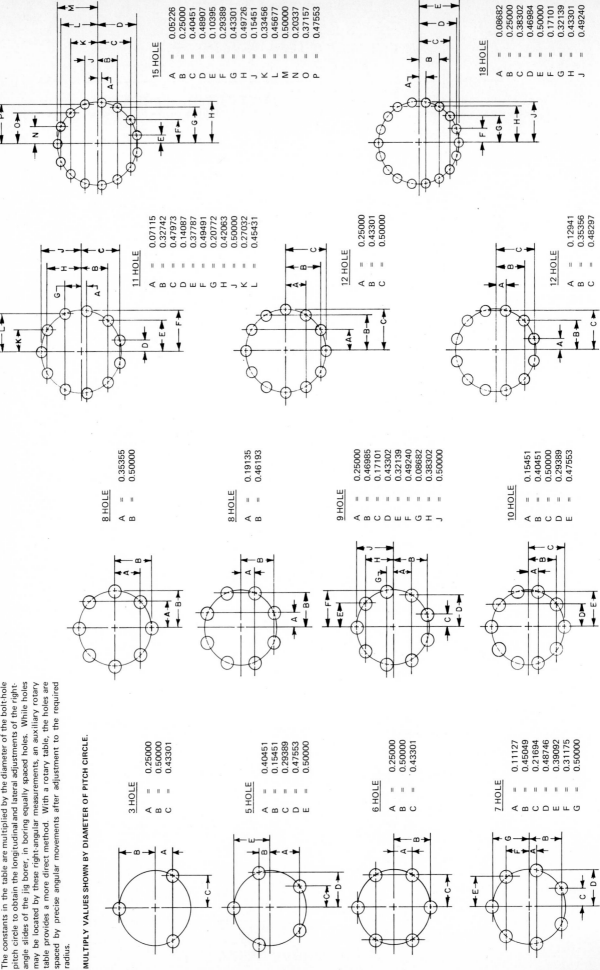

15 HOLE

A = 0.05226
B = 0.25000
C = 0.40451
D = 0.48907
E = 0.10395
F = 0.29389
G = 0.43301
H = 0.49726
J = 0.15451
K = 0.33456
L = 0.45677
M = 0.50000
N = 0.20337
O = 0.37157
P = 0.47553

18 HOLE

A = 0.08682
B = 0.25000
C = 0.38302
D = 0.46984
E = 0.50000
F = 0.17101
G = 0.32139
H = 0.43301
J = 0.49240

11 HOLE

A = 0.07115
B = 0.32742
C = 0.47973
D = 0.14087
E = 0.37787
F = 0.49491
G = 0.20772
H = 0.42063
J = 0.50000
K = 0.27032
L = 0.45431

12 HOLE

A = 0.25000
B = 0.43301
C = 0.50000

12 HOLE

A = 0.12941
B = 0.35356
C = 0.48297

8 HOLE

A = 0.35355
B = 0.50000

8 HOLE

A = 0.19135
B = 0.46193

9 HOLE

A = 0.25000
B = 0.46985
C = 0.17101
D = 0.43302
E = 0.32139
F = 0.49240
G = 0.08682
H = 0.38302
J = 0.50000

10 HOLE

A = 0.15451
B = 0.40451
C = 0.50000
D = 0.29389
E = 0.47553

3 HOLE

A = 0.25000
B = 0.50000
C = 0.43301

5 HOLE

A = 0.40451
B = 0.15451
C = 0.29389
D = 0.47553
E = 0.50000

6 HOLE

A = 0.25000
B = 0.50000
C = 0.43301

7 HOLE

A = 0.11127
B = 0.45049
C = 0.21694
D = 0.48746
E = 0.39092
F = 0.31175
G = 0.50000

GEOMETRIC TOLERANCING SYMBOLS

SYMBOL FOR:	ANSI Y14.5	ISO
STRAIGHTNESS	—	—
FLATNESS	▱	▱
CIRCULARITY	○	○
CYLINDRICITY	⌭	⌭
PROFILE OF A LINE	⌒	⌒
PROFILE OF A SURFACE	⌓	⌓
ALL AROUND—PROFILE	↖⊙	NONE
ANGULARITY	∠	∠
PERPENDICULARITY	⊥	⊥
PARALLELISM	//	//
POSITION	⊕	⊕
CONCENTRICITY/COAXIALITY	◎	◎
SYMMETRY	NONE	=
CIRCULAR RUNOUT	*↗	↗
TOTAL RUNOUT	*↗↗	↗↗
AT MAXIMUM MATERIAL CONDITION	Ⓜ	Ⓜ
AT LEAST MATERIAL CONDITION	Ⓛ	NONE
REGARDLESS OF FEATURE SIZE	Ⓢ	NONE
PROJECTED TOLERANCE ZONE	Ⓟ	Ⓟ
DIAMETER	∅	∅
BASIC DIMENSION	30	30
REFERENCE DIMENSION	(30)	(30)
DATUM FEATURE	–A–	*▰ OR *▱ A
DATUM TARGET	06/A1	06/A1
TARGET POINT	✕	✕
DIMENSION ORIGIN	⊕→	NONE
FEATURE CONTROL FRAME	⊕ ∅0.5Ⓜ A B C	⊕ ∅0.5Ⓜ A B C
CONICAL TAPER	▷	▷
SLOPE	◺	◺
COUNTERBORE/SPOTFACE	⌴	NONE
COUNTERSINK	⌵	NONE
DEPTH/DEEP	↧	NONE
SQUARE (SHAPE)	□	□
DIMENSION NOT TO SCALE	15	15
NUMBER OF TIMES/PLACES	8X	8X
ARC LENGTH	⌒105	NONE
RADIUS	R	R
SPHERICAL RADIUS	SR	NONE
SPHERICAL DIAMETER	S∅	NONE

*MAY BE FILLED IN

ELECTRICAL AND ELECTRONICS DIAGRAMS*

2.5 Battery

The long line is always positive, but polarity may be indicated in addition. Example:

2.5.1 Generalized direct-current source

2.5.2 One cell

2.5.3 Multicell

2.7 Oscillator
Generalized Alternating-Current Source

2.14 Thermal Element
Thermomechanical Transducer

Actuating device, self-heating or with external heater. (Not operated primarily by ambient temperature.) See item 9.1 for fuses, one-time devices. See item 4.30.5 for thermally operated relay.

OR

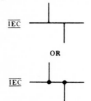

3.1.6.3 Application: junction of connected paths, conductors, or wires

OR

OR ONLY IF REQUIRED
BY LAYOUT CONSIDERATIONS

For microwave circuits, the type of coupling, power-division proportions, reflection coefficients, plane of junction, etc., may be indicated if desired.

3.1.8.4 Application: shielded 2-conductor cable with shield grounded

3.1.9‡ Coaxial cable, recognition symbol; coaxial transmission path; radio-frequency cable Ⓕ

NOTE 3.1.9A: If necessary for clarity, an outer-conductor connection shall be made to the symbol.

NOTE 3.1.9B: If the coaxial structure is not maintained, the tangential line shall be drawn only on the coaxial side.

3.1.9.1‡ General

See Note 3.1.9A

3.1.10 Grouping of leads

3.1.10.1 General

Bend of line indicates direction in which other end of path will be found.

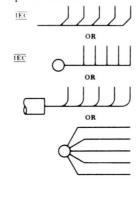

OR

OR

OR

OR

by

ited
the

The standard method of showing a contact is by a symbol indicating the circuit condition it produces when the actuating device is in the deenergized or nonoperated position. The actuating device may be of a mechanical, electrical, or other nature, and a clarifying note may be necessary with the symbol to explain the proper point at which the contact functions; for example, the point where a contact closes or opens as a function of changing pressure, level, flow, voltage, current, etc. In cases where it is desirable to show contacts in the energized or operated condition and where confusion may result, a clarifying note shall be added to the drawing.

Auxiliary switches or contacts for circuit breakers, etc, may be designated as follows:

(a) Closed when device is energized or operated position.
(b) Closed when device is in deenergized or nonoperated position.
(aa) Closed when operating mechanism of main device is in energized or operated position.
(bb) Closed when operated mechanism of main device is in deenergized or nonoperated position.

See American National Standard Manual and Automatic Station Control, Supervisory, and Associated Telemetering

Equipment, C37.2-1962, for further details.

In the parallel-line contact symbols shown below, the length of the parallel lines shall be approximately 1¼ times the width of the gap (except for symbol 4.3.7).

4.3.1 Closed contact (break)

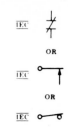

OR

OR

4.3.2 Open contact (make)

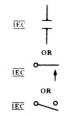

OR

OR

4.5 Operating Coil
Relay Coil Ⓕ

See also INDUCTOR; WINDING; etc (item 6.2)

NOTE 4.5A: The asterisk is not part of the symbol. Always replace the asterisk by a device designation. See, for example, ANSI C37.2-1962.

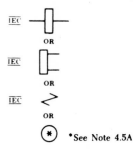

OR

OR

OR

*See Note 4.5A

4.7 Pushbutton Ⓕ , **Momentary or Spring-Return**

4.7.1 Circuit closing (make)

4.7.2 Circuit opening (break)

*Extracted from ANSI Standard Graphic Symbols for Electrical and Electronics Diagrams (ANSI Y32.3-1970), with the permission of the publisher, The Institute of Electrical and Electronics Engineers, Inc.

ELECTRICAL AND ELECTRONICS DIAGRAMS (Continued)

4.13 Selector or Multiposition Switch

The position in which the switch is shown may be indicated by a note or designation of switch position.

4.13.1 General (for power and control diagrams)

Any number of transmission paths may be shown.

OR

5.3 Connector
Disconnecting Device
Jack F̄
Plug F̄

The contact symbol is not an arrowhead. It is larger and the lines are drawn at a 90-degree angle.

5.3.1 Female contact

5.3.2 Male contact

IEC ⟶

7.3 Typical Applications

7.3.1 Triode with directly heated filamentary cathode and envelope connection to base terminal

7.3.2 Equipotential-cathode pentode showing use of elongated envelope

7.3.3 Equipotential-cathode twin triode showing use of elongated envelope and rule of item 7.2.3

7.3.4 Cold-cathode gas-filled tube

7.3.4.1 Rectifier; voltage regulator for direct-current operation

See also symbol 11.1.3.2

8.6 Typical Applications, Three- (or more) Terminal Devices

8.6.1 PNP transistor (also PNIP transistor, if omitting the intrinsic region will not result in ambiguity)

See paragraph A4.11 of the Introduction

8.6.1.1 Application: PNP transistor with one electrode connected to envelope (in this case, the collector electrode)

8.6.2 NPN transistor (also NPIN transistor, if omitting the intrinsic region will not result in ambiguity)

See paragraph A4.11 of the Introduction

8.6.8 Unijunction transistor with N-type base

See paragraph A4.11 of the Introduction

8.6.10 Field-effect transistor with N-channel (junction gate and insulated gate)

8.6.10.1 N-channel junction gate

If desired, the junction-gate symbol element may be drawn opposite the preferred source.

See paragraph A4.11 of the Introduction

12.1 Meter
Instrument

NOTE 12.1A: The asterisk is not part of the symbol. Always replace the asterisk by one of the following letter combinations, depending on the function of the meter or instrument, unless some other identification is provided in the circle and explained on the diagram.

* See Note 12.1A

A	Ammeter F̄ IEC
AH	Ampere-hour meter
C	Coulombmeter
CMA	Contact-making (or breaking) ammeter
CMC	Contact-making (or breaking) clock
CMV	Contact-making (or breaking) voltmeter
CRO	Oscilloscope F̄
	Cathode-ray oscillograph
DB	DB (decibel) meter
	Audio level/meter F̄
DBM	DBM (decibels referred to 1 milliwatt) meter
DM	Demand meter
DTR	Demand-totalizing relay
F	Frequency meter F̄
GD	Ground detector
I	Indicating meter
INT	Integrating meter
μA or UA	Microammeter
MA	Milliammeter
NM	Noise meter
OHM	Ohmmeter F̄
OP	Oil pressure meter
OSCG	Oscillograph, string
PF	Power factor meter
PH	Phasemeter F̄
PI	Position indicator
RD	Recording demand meter
REC	Recording meter
RF	Reactive factor meter
SY	Synchroscope
t°	Temperature meter
THC	Thermal converter
TLM	Telemeter
TT	Total time meter
	Elapsed time meter
V	Voltmeter F̄ IEC
VA	Volt-ammeter
VAR	Varmeter F̄
VARH	Varhour meter
VI	Volume indicator
	Audio-level meter F̄
VU	Standard volume indicator
	Audio-level meter F̄
W	Wattmeter F̄ IEC
WH	Watthour meter

Standard Welding Symbols

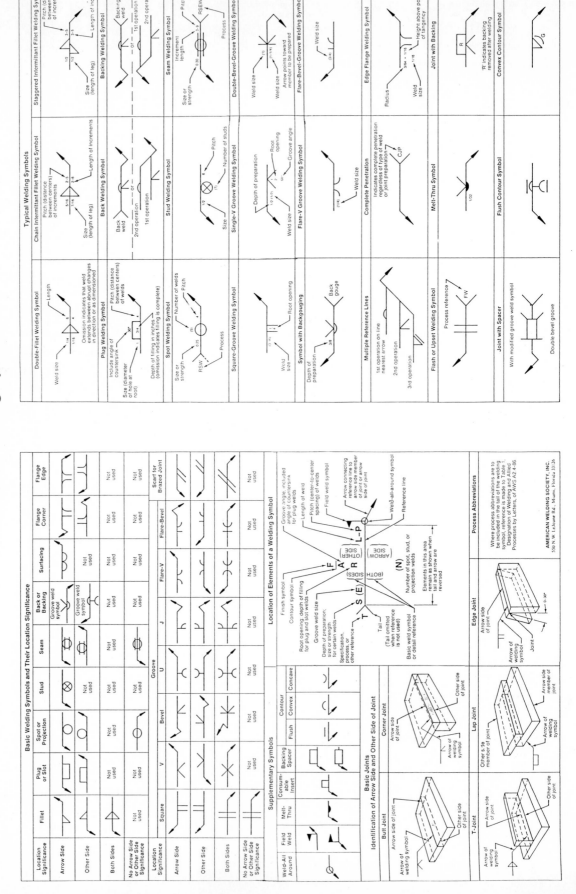

TOPOGRAPHIC MAP

VARIATIONS WILL BE FOUND ON OLDER MAPS

Hard surface, heavy duty road, four or more lanes . . . . . .

Hard surface, heavy duty road, two or three lanes . . . . . .

Hard surface, medium duty road, four or more lanes . . . .

Hard surface, medium duty road, two or three lanes . . . . .

Improved light duty road . . . .

Unimproved dirt road and trail . . . .

Dual highway, dividing strip 25 feet or less . . . .

Dual highway, dividing strip exceeding 25 feet . . . .

Road under construction . . . .

Railroad, single track and multiple track . . . .

Railroads in juxtaposition . . . .

Narrow gage, single track and multiple track . . . .

Railroad in street and carline . . . .

Bridge, road and railroad . . . .

Drawbridge, road and railroad . . . .

Footbridge . . . .

Tunnel, road and railroad . . . .

Overpass and underpass . . . .

Important small masonry or earth dam . . . .

Dam with lock . . . .

Dam with road . . . .

Canal with lock . . . .

Buildings (dwelling, place of employment, etc.) . . . .

School, church, and cemetery . . . . Cem

Buildings (barn, warehouse, etc.) . . . .

Power transmission line . . . .

Telephone line, pipeline, etc. (labeled as to type) . . . .

Wells other than water (labeled as to type) . . . . o Oil . . . . o Gas

Tanks; oil, water, etc. (labeled as to type) . . . . ● ● ● ⊘ Water

Located or landmark object; windmill . . . . o . . . 🌾

Open pit, mine, or quarry; prospect . . . . ⤫ x

Shaft and tunnel entrance . . . . ▪ . . . . Y

Horizontal and vertical control station:

Tablet, spirit level elevation . . . . BM △ 5653

Other recoverable mark, spirit level elevation . . . . △ 5455

Horizontal control station: tablet, vertical angle elevation VABM △ 9519

Any recoverable mark, vertical angle or checked elevation △3775

Vertical control station: tablet, spirit level elevation . . . . . BM × 957

Other recoverable mark, spirit level elevation . . . . × 954

Checked spot elevation . . . . × 4675

Unchecked spot elevation and water elevation . . . . × 5657 870

Boundary, national . . . .

State . . . .

County, parish, municipio . . . .

Civil township, precinct, town, barrio . . . .

Incorporated city, village, town, hamlet . . . .

Reservation, national or state . . . .

Small park, cemetery, airport, etc. . . . .

Land grant . . . .

Township or range line, United States land survey . . . . . .

Township or range line, approximate location . . . . . .

Section line, United States land survey . . . .

Section line, approximate location . . . .

Township line, not United States land survey . . . .

Section line, not United States land survey . . . .

Section corner, found and indicated . . . . + ... +

Boundary monument: land grant and other . . . . □ . . . . . . □

United States mineral or location monument . . . . ▲

Index contour . . . .   Intermediate contour . .

Supplementary contour . . . .   Depression contours . .

Fill . . . .   Cut . . . .

Levee . . . .   Levee with road . . . . . .

Mine dump . . . .   Wash . . . .

Tailings . . . .   Tailings pond . . . .

Strip mine . . . .   Distorted surface . . . . . .

Sand area . . . .   Gravel beach . . . .

Perennial streams . . . .   Intermittent streams . . .

Elevated aqueduct . . . . .   Aqueduct tunnel . . . . . .

Water well and spring . . . . .   Disappearing stream . . . . . .

Small rapids . . . .   Small falls . . . .

Large rapids . . . .   Large falls . . . .

Intermittent lake . . . .   Dry lake . . . .

Foreshore flat . . . .   Rock or coral reef . . . .

Sounding, depth curve 10   Piling or dolphin . . . . . o

Exposed wreck . . . .   Sunken wreck . . . .

Rock, bare or awash; dangerous to navigation . . . .

Marsh (swamp) . . . .   Submerged marsh . . . .

Wooded marsh . . . .   Mangrove . . . .

Woods or brushwood . .   Orchard . . . .

Vineyard . . . .   Scrub . . . .

Inundation area . . . .   Urban area . . . .

(U. S. Geological Survey)

(ANSI)

PIPE FITTINGS AND VALVES

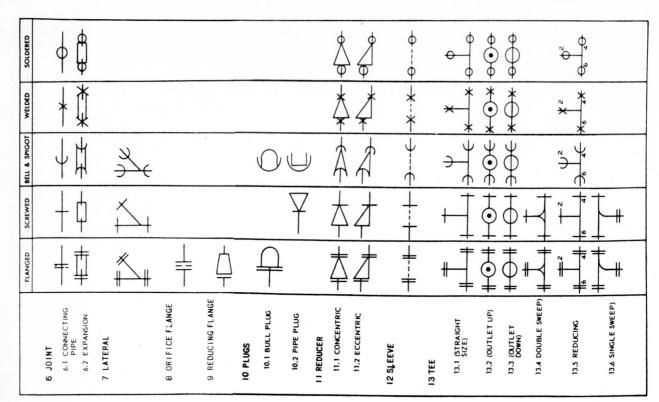

	FLANGED	SCREWED	BELL & SPIGOT	WELDED	SOLDERED
6 JOINT					
6.1 CONNECTING PIPE					
6.2 EXPANSION					
7 LATERAL					
8 ORIFICE FLANGE					
9 REDUCING FLANGE					
10 PLUGS					
10.1 BULL PLUG					
10.2 PIPE PLUG					
11 REDUCER					
11.1 CONCENTRIC					
11.2 ECCENTRIC					
12 SLEEVE					
13 TEE					
13.1 (STRAIGHT SIZE)					
13.2 (OUTLET UP)					
13.3 (OUTLET DOWN)					
13.4 DOUBLE SWEEP)					
13.5 REDUCING					
13.6 SINGLE SWEEP)					

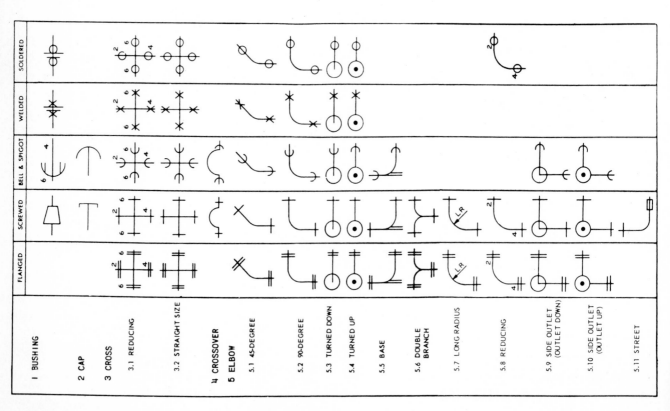

	FLANGED	SCREWED	BELL & SPIGOT	WELDED	SOLDERED
1 BUSHING					
2 CAP					
3 CROSS					
3.1 REDUCING					
3.2 STRAIGHT SIZE					
4 CROSSOVER					
5 ELBOW					
5.1 45-DEGREE					
5.2 90-DEGREE					
5.3 TURNED DOWN					
5.4 TURNED UP					
5.5 BASE					
5.6 DOUBLE BRANCH					
5.7 LONG RADIUS					
5.8 REDUCING					
5.9 SIDE OUTLET (OUTLET DOWN)					
5.10 SIDE OUTLET (OUTLET UP)					
5.11 STREET					

DRAWING SHEET LAYOUTS

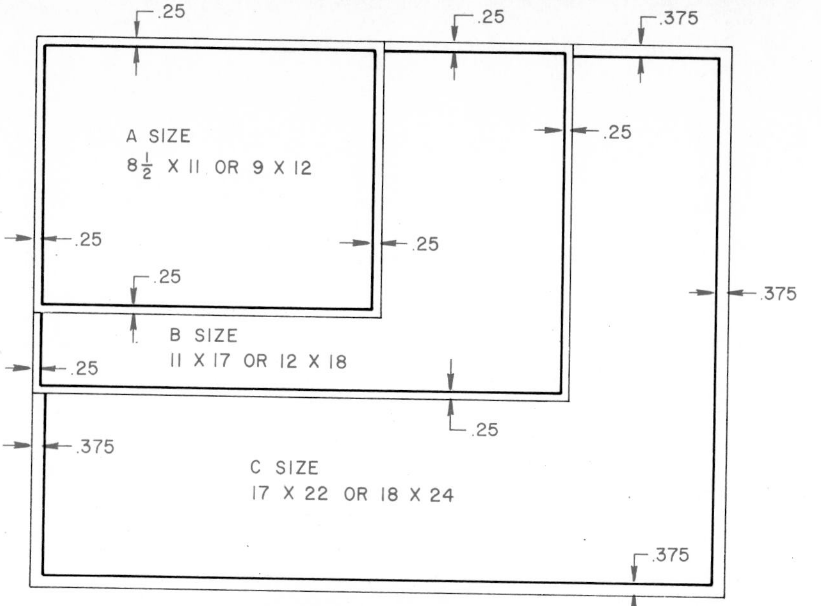

SHEET SIZE

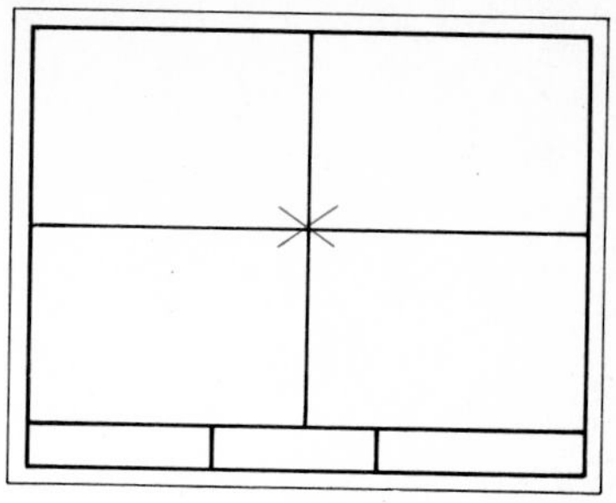

A SIZE
8½ X 11 OR 9 X 12

B SIZE
11 X 17 OR 12 X 18

C SIZE
17 X 22 OR 18 X 24

.25 .25 .375 .25 .25 .25 .25 .25 .375 .375 .375 .375

A-4

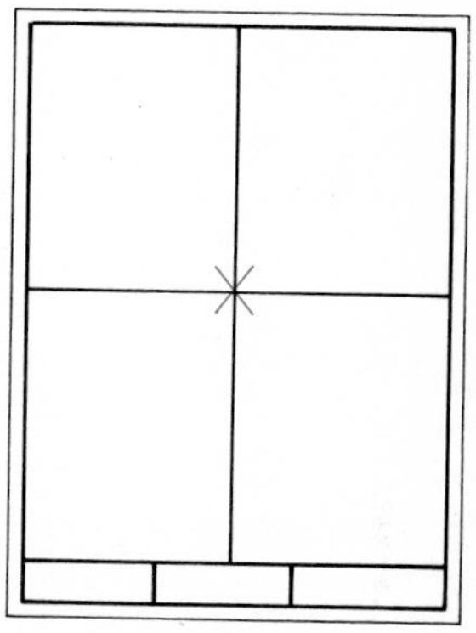

A-5

LAYOUT I

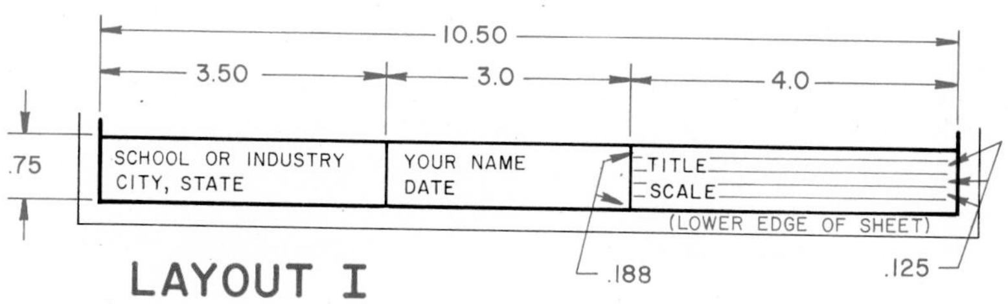

10.50
3.50 3.0 4.0
.75

| SCHOOL OR INDUSTRY CITY, STATE | YOUR NAME DATE | TITLE SCALE |

(LOWER EDGE OF SHEET)
.188 .125

LAYOUT II

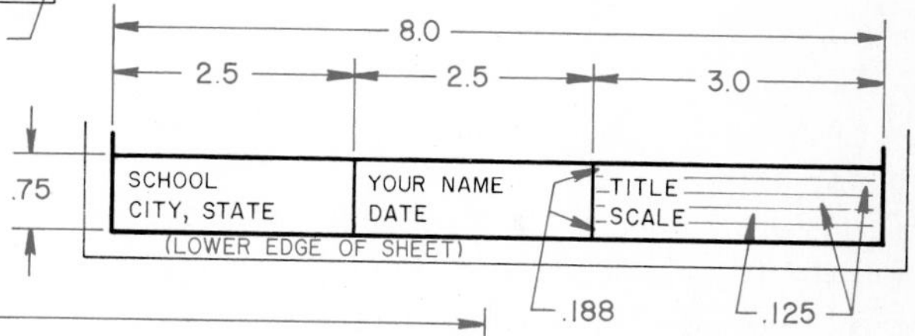

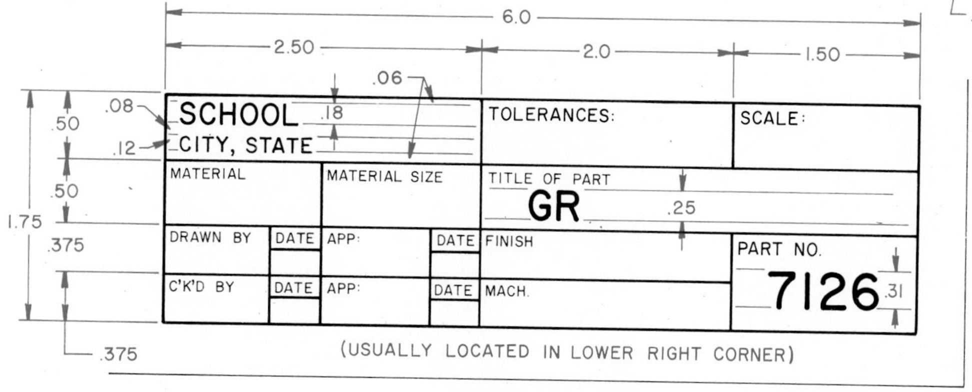

8.0
2.5 2.5 3.0
.75

| SCHOOL CITY, STATE | YOUR NAME DATE | TITLE SCALE |

(LOWER EDGE OF SHEET)
.188 .125

LAYOUT III

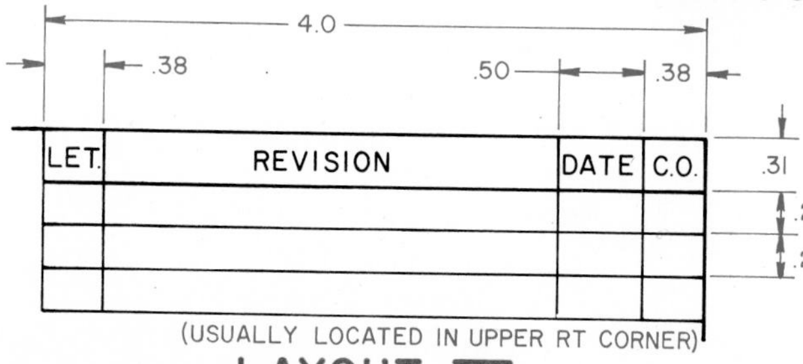

6.0
2.50 2.0 1.50
.08 .06 .18
.50 .12
.50
1.75
.375
.375

SCHOOL CITY, STATE		TOLERANCES:	SCALE:
MATERIAL	MATERIAL SIZE	TITLE OF PART **GR** .25	
DRAWN BY	DATE	APP:	DATE	FINISH	PART NO. **7126** .31
C'K'D BY	DATE	APP:	DATE	MACH.	

(USUALLY LOCATED IN LOWER RIGHT CORNER)

LAYOUT IV

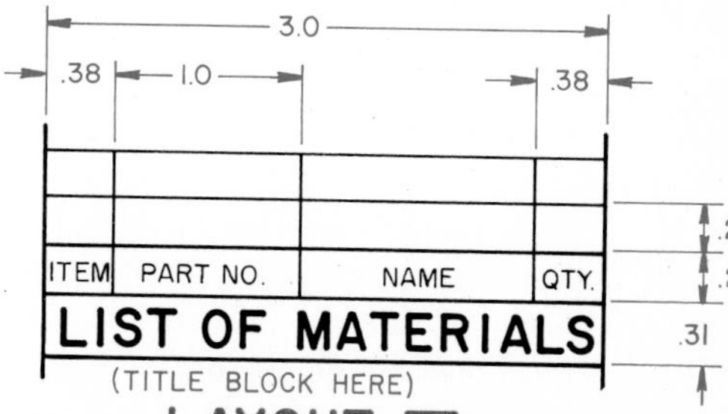

4.0
.38 .50 .38
.31
.25
.25

LET.	REVISION	DATE	C.O.

(USUALLY LOCATED IN UPPER RT CORNER)

LAYOUT V

3.0
.38 1.0 .38
.25
.25
.31

ITEM	PART NO.	NAME	QTY.
LIST OF MATERIALS			

(TITLE BLOCK HERE)

These suggested sheet layouts are recommended for use with the various sizes of drafting sheets. Layouts I and II are especially suited to A size sheets in the horizontal position or vertical. Layouts III, IV and V are suggested title block, change block and materials block for larger size sheets. A-4 and A-5 illustrate the manner of dividing an "A" size sheet into four sections horizontally and vertically.

GLOSSARY

ABRASIVE: A substance such as emery, aluminum oxide and diamond that cuts material softer than itself. It may be used in loose form, mounted on cloth, paper or bonded in a wheel.

ABSCISSA: The X-coordinate of a point, i.e., its horizontal distance from the Y-axis measured parallel to the X-axis; the horizontal axis of a graph or chart.

ABSOLUTE SYSTEM: A system of numerically controlled machining that measures all coordinates from a fixed point of origin or zero point. Also known as point-to-point N/C machining.

ACTIVE DEVICE: An electronic element capable of gain or control, such as a transistor or vacuum tube.

ACUTE ANGLE: An angle less than 90 degrees.

ADDENDUM: The radial distance between the pitch circle and the top of the tooth.

AI (Artificial intelligence): Information placed in the computer's memory that would cause the computer to make decisions normally associated with human intelligence.

ALLOWANCE: The intentional difference in the dimensions of mating parts to provide for different classes of fits.

ALLOY: A mixture of two or more metals fused or melted together to form a new metal.

AMMONIA: A colorless gas, NH_3, used in the development process of diazo and sepia prints.

ANNEAL: To soften metals by heating to remove internal stresses caused by rolling and forging.

ANODIZE: The process of protecting aluminum by oxidizing in an acid bath using a direct current.

APPROXIMATE: Describes a value that is nearly but not exactly correct or accurate.

ARBOR: A shaft or spindle for holding cutting tools.

ARC: A part of a circle.

ASSEMBLY DRAWING: A drawing showing the working relationship of the various parts of a machine or structure as they fit together.

ASYMTOTE: A straight line that is the limit of a tangent to a curve as the point of contact moves off to infinity.

AUTOPOSITIVE: A print made on paper or film by means of a positive-to-positive silver type emulsion.

AUXILIARY VIEW: An additional view of an object, usually of a surface inclined to the principal surfaces of the object to provide a true size and shape view.

AXES: Plural of axis.

AXIS: An imaginary line around which parts rotate or are regularly arranged.

AXONOMETRIC: One of several forms of one plane projection giving the pictorial effect of perspective with the possibility of measuring the principal planes directly.

BACKLASH: The play (lost motion) between moving parts, such as threaded shaft and nut or the teeth of meshing gears.

BASE-LINE DIMENSIONING: A system of dimensioning where as many features of a part as is functionally practical are located from a common set of datums.

BASIC DIMENSION: A theoretically exact value used to describe the size, shape or location of a feature.

BASIC SIZE: That size from which the limits of size are derived by the application of allowances and tolerances.

BEND ALLOWANCE: The amount of sheet metal required to make a bend over a specific radius.

BISECT: To divide into two equal parts.

BLANKING: A stamping operation in which a press uses a die to cut blanks from flat sheets or strips of metal.

BLOWBACK: To enlarge or make an enlargement of an image; an enlarged print made from a micro-image.

BLUEPRINT: A copy of a drawing.

BORING: Enlarging a hole to a specified dimension by use of a boring bar. May be done on a lathe, jig bore, boring machine or mill.

BOSS: A small local thickening of the body of a casting or forging to allow more thickness for a bearing area or to support threads.

BRAZE: To join two metal parts by adhesion with heat and a filler material of zinc and copper alloy.

BROACH: A tool for removing metal by pulling or pushing it across the work. The most common use is producing irregular hole shapes such as squares, hexagons, ovals or splines.

BURNISH: To smooth or polish metal by rolling or sliding tool over surface under pressure.

BURR: The ragged edge or ridge left on metal after a cutting operation.

BUSHING: A metal lining which acts as a bearing between rotating parts such as a shaft and pulley. Also used on jigs to guide cutting tool.

CAD (Computer-aided design or design-drafting): The design, analysis, testing and documentation of graphic solutions to problems.

Glossary

CALLOUT: A note on the drawing giving a dimension, specification or machine process.

CAM: A rotating or sliding device used to convert rotary motion into intermittent or reciprocating motion.

CAM (Computer-aided manufacturing): A plan for utilizing numerical controlled machines to perform a wide variety of manufacturing processes.

CARBURIZE: Heating of low-carbon steel for a period of time to a temperature below its melting point in carbonaceous solids, liquids or gases, then cooling slowly in preparation for heat treating.

CASE HARDENING: The process of hardening ferrous alloys so that the surface layer or case is made much harder than the interior core.

CASTING: An object made by pouring molten metal in a mold.

CHAIN DIMENSIONING: Successive dimensions that extend from one feature to another, rather than each originating at a datum. Tolerances accumulate with chain dimensions unless the note, "Tolerances do not accumulate," is placed on the drawing.

CHAMFER: A bevel on an external edge or corner, usually at 45 degrees.

CIM (Computer-integrated manufacturing): The full automation of all facets of industry to produce a product including design, documentation, materials handling, machine processes, quality assurance, storing/shipping, management and marketing.

CIRCUIT: The various connections and conductors of a specific device; the path of electron flow from the source through components and connections and back to its source.

CIRCUIT DIAGRAM: A line drawing using graphic symbols or pictorial views to show the complete path of flow in a hydraulic or electronic system.

CIRCULAR PITCH: The length of the arc along the pitch circle between the center of one gear tooth to the center of the next.

CLOCKWISE: Rotation in the same direction as hands of a clock.

CNC (Computer numerical control): The programming and control of an NC processor for production machine tools. A numerical code is produced from a CAD drawing.

COMPUTER: An electronic machine capable of making logical decisions under the control of programs.

COMPUTER GRAPHICS: The process of designing industrial products and the production of graphic documents with the aid of the computer and related input and output devices.

CONCENTRIC: Having a common center as circles or diameters.

CONJUGATE DIAMETERS: Two diameters are conjugate when each is parallel to the tangents at the extremities of the other.

CONTEMPORARY: Belonging to the present time period.

CONTOUR: A profile outline.

COORDINATE DIMENSIONING: A type of rectangular datum dimensioning in which all dimensions are measured from two or three mutually perpendicular datum planes. All dimensions originate at a datum and include regular extension and dimension lines and arrowheads.

COUNTERBORE: The enlargement of the end of a hole to a specified diameter and depth.

COUNTERSINK: The chamfered end of a hole to receive a flat head screw.

DASH NUMBER: A number preceded by a dash after the drawing number that indicates right or left-hand parts as well as neutral parts and/or detail and assembly drawings. The coding is usually special to a particular industry.

DATUM: Points, lines, planes, cylinders and the like, assumed to be exact for purposes of computation from which the location or geometric relationship (form) of features of a part may be established.

DEDENDUM: The radial distance between the pitch circle and the bottom of the tooth.

DELINEATION: To represent pictorially: a chart, a diagram or a sketch.

DESIGN SIZE: The size of a feature after an allowance for clearance has been applied and tolerances have been assigned.

DETAIL DRAWING: A drawing of a single part that provides all the information necessary in the production of that part.

DEVIATION: The variance from a specified dimension or design requirement.

DIAGRAM: A figure or drawing which is marked out by lines; a chart or outline.

DIAMETER: The length of a straight line passing through the center of a circle and terminating at the circumference on each end.

DIAMETRAL: Of a diameter; forming a diameter.

DIAZO: Diazo material is either a film or paper sensitized by means of azo dyes used for photocopying.

DIE: A tool used to cut external threads by hand or machine. Also a tool used to impart a desired shape to a piece of metal.

DIE-CASTING: A method of casting metal dies of a die-casting machine. Also the part formed by die-casting.

DIE STAMPING: A piece cut out by a die.

DIMENSION: Measurements given on a drawing such as size and location.

DNC (Direct numerical control): The control of machine tools by a computer coupled directly to the machine.

DNC (Distributive numerical control): The control of machine tools by a main host computer that controls several intermediate computers coupled to certain machine tools, robots or inspection stations.

DOCUMENT, DOCUMENTATION: Those pieces of paper, film or tapes for N/C machining which describe the engineer's idea in physical terms and tell manufacturing or construction personnel what to make.

DOWEL PIN: A pin which fits into a hole in an abutting piece to prevent motion or slipping, or to ensure accurate location of assembly.

DRAFT: The angle or taper on a pattern or casting that permits easy removal from the mold or forming die.

EAVES: The part of the roof that extends beyond and over the walls of a building.

ECCENTRIC, ECCENTRICITY: Not having the same center; off center.

EFFECTIVE THREAD: The complete thread and that portion

of the incomplete thread having fully formed roots but having crests not fully formed.

EFFECTIVITY: The serial number(s) of an aircraft, machine, assembly or part on which a drawing change applies. The change may be indicated as an effective date and would apply on that date forward.

ELLIPSE: A circle viewed at an angle.

EXPERT SYSTEMS: A branch of artificial intelligence placed in the computer's memory expressed as "rules" of human expertise.

EXTRUSION: Metal which has been shaped by forcing it in the hot or cold state through dies of the desired shape.

FASTENER: A mechanical device for holding two or more bodies in definite positions with respect to each other.

FEA (Finite element analysis): A CAD method for analyzing engineering problems involving static, dynamic and thermal stressing of parts.

FEATURE: A portion of a part, such as a diameter, hole, keyway or flat surface.

FERROUS: Metals that have iron as their base material.

FILLET: A concave intersection between two surfaces to strengthen the area.

FINISH: General finish requirements such as paint, chemical or electroplating rather than surface texture or roughness. (See surface texture.)

FIT: The clearance or interference between two mating parts.

FIXTURE: A device used to position and hold a part in a machine tool. It does not guide the cutting tool.

FLANGE: An edge or collar fixed at an angle to the main part or web as an I-beam.

FLAT PATTERN: A layout showing true dimensions of a part before bending. May be actual size pattern on polyester film for shop use.

FMS (Flexible manufacturing system): A production system of highly automated and computer controlled machines, assembly cells, robots, inspection equipment, materials and storage systems, capable of processing a wide variety of similar products.

FORGING: Metal shaped under pressure with or without heat.

FORM TOLERANCING: The permitted variation of a feature from the perfect form indicated on the drawing.

FUNCTIONAL DRAWING: A drawing which uses a minimum number of views, details and dimensions, and yet maintains accuracy.

FUSION WELD: The intimate mixing of molten metals.

GAGE: The thickness of sheet metal by number.

GENERATION: The blowback made from a microfilm of an original drawing is a first-generation print. A blowback made from a microfilm of this first-generation print is a second-generation print, etc. The term is used to express the quality required in an original drawing to be microfilmed — usually one capable of producing a clearly readable fourth-generation print.

GEOMETRIC DIMENSIONING AND TOLERANCING: A means of dimensioning and tolerancing a drawing with respect to the actual function or relationship of part features which can be most economically produced. It includes positional and form dimensioning and tolerancing.

GHOSTING: A smudged area on a reproduction copy of a drawing caused by a damaged surface due to erasing or mishandling of the intermediate.

GRAPHICS: The art and science of calculating and drawing objects and diagrams, especially those related to mathematics, engineering and industry; pertaining to the use of graphics, diagrams, charts, symbols and drawings.

GROUP TECHNOLOGY: A philosophy of organizing parts, to be manufactured, into families of parts and produced in cells of machine tools.

GUSSET: A small plate used in reinforcing assemblies.

HARDNESS TEST: Techniques used to measure the degree of hardness of heat-treated materials.

HEAT TREATMENT: The application of heat to metals to produce desired qualities of hardness, toughness and/or softness. (See anneal.)

HEXAGON: A polygon having six angles and six sides.

HOBBING: A special gear cutting process. The gear blank and hob rotate together as in mesh during the cutting operation.

HONE: A method of finishing a hole or other surface to a precise tolerance by using a spring loaded abrasive block and rotary motion.

HORIZONTAL: Parallel to the horizon.

HUMAN FACTORS: Characteristic dimension of humans such as reach, body form and size, and comfort factors. Used in the design of products.

IDENTIFICATION CODE: A number assigned by the Federal Government to industries doing contractual work for the government.

INCLINED: A line or plane at an angle to a horizontal line or plane.

INCREMENTAL SYSTEM: A system of numerically controlled machining that always refers to the preceding point when making the next movement. Also known as continuous path or contouring method of N/C machining.

INDICATOR: A precision measuring instrument for checking the trueness of work.

INSEPARABLE ASSEMBLY: A component permanently assembled by welding, bonding, riveting, potting, etc.

INSOLUBLE: Incapable of being dissolved.

INTEGRATED CIRCUIT: A complete electronic circuit composed of various electronic devices fabricated on a common substrate. Usually very small in size.

INTERCHANGEABILITY: The condition that assures the universal exchange or mutual substitution of units or parts of a mechanism or an assembly.

INTERMEDIATE: A translucent reproduction made on vellum, cloth or film from an original drawing to serve in place of the original for making other prints.

INVOLUTE: A spiral curve generated by a point on a chord as it unwinds from a circle or a polygon.

ISOMETRIC DRAWING: A pictorial drawing of an object so positioned that all three axes make equal angles with the picture plane and measurements on all three axes are made to the same scale.

JIG: A device used to hold a part to be machined and positions and guides the cutting tool.

JIT (Just-In-Time) Manufacturing: A concept of manufacturing in which the aim is to reduce work-in-progress by reducing lead times, inventories (raw materials, parts, and finished producd to be stored), and setup times to an absolute minimum.

JOGGLE: An offset in the face of a part which has an adjacent flange.

KERF: The slit or channel left by a saw or other cutting tool.

KEY: A small piece of metal (usually a pin or bar) used to prevent rotation of a gear or pulley on a shaft.

KEYSEAT: The slot machined in a shaft for square and flat keys.

KEYSLOT: The slot machined in a shaft for Woodruff type keys.

KEYWAY: The slot machined in a hub for all types of keys.

KNURL: The process of rolling depressions in the surface of an object.

LAP: To finish a surface with a very fine abrasive impregnated in a soft metal.

LASER: Light Amplification by Stimulated Emission of Radiation. A device for producing light by emission of energy stored in a molecular or atomic system when stimulated by an input signal.

LIMITS: The extreme permissible dimensions of a part resulting from the application of a tolerance.

MACHINING CENTER: A grouping of high precision, multifunctional machines for purposes of performing specific operations without the necessity of scheduling secondary operations.

MAGNAFLUX: A nondestructive inspection technique that makes use of a magnetic field and magnetic particles to locate internal flaws in ferrous metal parts.

MAXIMUM MATERIAL CONDITION (MMC): When a feature contains the maximum amount of material; that is, minimum hole diameter and maximum shaft diameter.

METRICIZE: To convert a unit other than metric to its metric (SI) equivalent.

MILL: To remove metal with a rotating cutting tool on a milling machine.

MISMATCH: The variance between depths of machine cuts on a given surface.

MONOMASTERS: Semicomplete drawing of simple parts which picture a common shape and/or combination of features ready for the addition of dimensions and other specific data by the user. For example, the drawing of a compression spring with a form for completion of data for a specific size.

MULTIVIEW PROJECTION: Two or more views of an object as projected upon the picture plane in orthographic projection.

NEGATIVE PRINT: A print, usually on opaque material, which is opposite to the original drawing; that is, light lines on a dark background.

NEXT ASSEMBLY: The next object or machine on which the part or subassembly is to be used.

NOMINAL SIZE: A general classification term used to designate size of a commercial product.

NONFERROUS: Metals not derived from an iron base or an iron alloy base, such as aluminum, magnesium and copper.

NORMALIZING: A process in which ferrous alloys are heated and then cooled in still air to room temperatures to restore the uniform grain structure free of strains caused by cold working or welding.

NUMERICAL CONTROL: A system of controlling a machine or tool by means of numeric codes which direct commands to control devices attached or built into the machine or tool.

OBLIQUE DRAWING: A pictorial drawing of an object so drawn that one of its principal faces is parallel to the plane of projection, and is projected in its true size and shape. The third set of edges is oblique to the plane of projection. at some convenient angle.

OBTUSE ANGLE: An angle larger than 90 degrees.

OCTAGON: A polygon having eight angles and eight sides.

ORDINATE: The Y-coordinate of a point, i.e., its vertical distance from the X-axis measured parallel to the Y-axis; the vertical axis of a graph or chart.

ORDINATE DIMENSIONING: A type of rectangular datum dimensioning in which all dimensions are measured from two or three mutually perpendicular datum planes. Datum planes are indicated as zero coordinates and dimensions are shown on extension lines without the use of dimension lines or arrowheads. Sometimes called arrowless dimensioning.

ORTHOGRAPHIC PROJECTION: A projection on a picture plane formed by perpendicular projectors from the object to the picture plane. Third angle projection is used in the United States and first angle projection in most countries outside the United States.

PARALLEL: Having the same direction, such as two lines which, if extended, would never meet. Lines everywhere equally distant.

PARALLEL CIRCUIT: A circuit which contains two or more paths for electrons supplied by a common voltage source.

PART PROGRAMMING: The description of a part that is to be machined by numerical control or for the development of precision artwork for printed circuit boards.

PASSIVE DEVICE: An electronic element incapable of gain or control, such as a resistor or capacitor.

PENTAGON: A polygon having five angles and five sides.

PERPENDICULAR: Lines or planes at a right angle to a given line or plane.

PERSPECTIVE DRAWING: A pictorial drawing in which receding lines converge at vanishing points on the horizon. It is the most natural of all pictorial drawings.

PICKLE: The removal of stains and oxide scales from parts by immersion in an acid solution.

PHOTODRAWING: A photograph of a drawing or an object on which additions, changes or dimensions have been drawn.

PILOT: A protruding diameter on the end of a cutting tool designed to fit in a hole and guide the cutter in machining the area around the hole.

PILOT HOLE: A small hole used to guide a cutting tool for making a larger hole. Also used to guide a drill of larger size.

PINION: The smaller of two mating gears.

PITCH: The distance from a point on one thread to a corresponding point on the next thread; the slope of a surface or roof.

PLAN VIEW: The top view of an object.

POLYGON: A plane geometric figure with three or more sides.

POSITIONAL TOLERANCING: The permitted variation of a feature from the exact or true position indicated on the drawing.

POSITIVE PRINT: A print which is similar to the original drawing; that is, dark lines on a light background.

POTASSIUM DICHROMATE: An orange-red crystalline compound used in developing blueprints with a dark blue background.

PRECISION: The quality or state of being precise or accurate; mechanical exactness.

PRISM: A solid whose bases or ends are any congruent and parallel polygons and whose sides are parallelograms.

PRISMATIC: Pertaining to or like a prism.

PROCESS SPECIFICATION: A description of the exact procedures, materials and equipment to be used in performing a particular operation such as a milling operation or spray painting.

PROGRAM (Computer): A set of step-by-step instructions telling the computer to solve a problem with the information input to it or contained in memory.

PROJECT: To extend from one point to another.

PROPORTION: Proper relation between things or parts.

QUENCHING: Cooling metals rapidly by immersing them in liquids or gases.

RADIUS: The straight-line distance from the center of a circle or arc to its circumference.

REAMING: To finish a drilled hole to a close tolerance.

REFERENCE DIMENSION: Used only for information purposes and does not govern production or inspection operations.

REGARDLESS OF FEATURE SIZE (RFS): The condition where tolerance of position or form must be met irrespective of where the feature lies within its size tolerance.

RELEASE NOTICE: Authorization indicating the drawing has been cleared for use in production.

RENDERING: Finishing a drawing to give it a realistic appearance; a representation.

RESISTANCE WELDING: The process of welding metals by using the resistance of the metals to the flow of electricity to produce the heat for fusion of the metals.

RESISTOR: A component containing resistance to flow of an electric current.

SANDBLAST: The process of removing surface scale from metal by blowing a grit material against it at very high air pressure.

SCALE: A measuring device with graduations for laying off distances. Used to draw objects full, reduced or to an enlarged size. Also refers to size an object is drawn, such as full-size, half-size or twice-size.

SCHEMATIC: Diagram of an electrical or electronic circuit showing electrical connections and identification of various components.

SECTIONAL VIEW: A view of an object obtained by the imaginary cutting away of the front portion of the object to show the interior detail.

SEMS: A generic term for screw and washer assemblies.

SEPIA: A yellow-brown print, sometimes called vandykes, made directly from the original tracing. These are the most common negatives used for making positive prints or duplicate intermediates. The negative is printed in reverse and from it the positive print is made. The negative is contacted face to face with the positive to be printed.

SERIES CIRCUIT: A circuit which contains only one possible path for electrons through the circuit.

SERRATIONS: Condition of a surface or edge having notches or sharp teeth.

SHAKES: Wooden shingles that have been hand split; usually thick and uneven.

SHIM: A piece of thin metal used between mating parts to adjust their fit.

SOLENOID: A coil of wire carrying an electric current possessing the characteristics of a magnet.

SOLUTION: The answer to a problem; a homogeneous molecular mixture of two or more substances.

SPECIFICATIONS: A written set of instructions with a proposed set of plans, giving all necessary information not shown on the blueprints such as quality, manufacturer's name and manner in which work is to be conducted.

SPLINE: A raised area on a shaft (external) or hub (internal) designed to fit into a recessed area of a mating part.

SPOTFACE: A machined circular spot on the surface of a part to provide a flat bearing surface for a screw, bolt, nut, washer or rivet head.

SPOT WELD: A resistance type weld that joins pieces of metal by welding separate spots rather than a continuous weld.

STABLE: Refers to drafting media whose dimensional characteristics remain constant (or relatively so) with changes of temperature and humidity.

STAGGERED: To arrange in an offset fashion.

STRESS RELIEVING: To heat a metal part to a suitable temperature and hold that temperature for a determined time, then gradually cool it in air. This treatment reduces the internal stresses induced by casting, quenching, machining, cold working or welding.

STRETCHOUT: A flat pattern development for use in laying out, cutting and folding lines on flat stock, such as paper or sheet metal, to be formed into a useful object (a container, air duct or funnel).

SUBSTRATE: The base material on which an integrated circuit is fabricated. Its function is primarily mechanical support but may also serve an electrical function.

SUPERHETERODYNE: A radio receiver in which the incoming signal is converted to a fixed intermediate frequency before detecting the audio signal component.

SUPERSEDENCE: The replacing of one part by another. A part that has been replaced is said to be superseded.

SURFACE TEXTURE: The roughness, waviness, lay and flaws of a surface.

SURVEYING: The process or occupation of determining location of boundaries of a part of the earth's surface.

SYMBOL: A letter, character or schematic design representing a unit or component.

TABULAR DIMENSION: A type of rectangular datum dimensioning in which dimensions from mutually perpendicular datum planes are listed in a table on the drawing instead of on the pictorial portion.

Glossary

TANGENT: A line drawn to the surface of an arc or circle so that it contacts the arc or circle at only one point.

TAP: A rotating tool used to produce internal threads by hand or machine.

TECHNICIAN: One who works in a technical field as a member of the engineering support team.

TEMPERING: Creating ductility and toughness in metal by heat treatment process.

TEMPLATE: A pattern or guide.

TENSILE STRENGTH: The maximum load (pull) a piece supports without breaking or failure.

TOLERANCE: The total amount of variation permitted from the design size of a part.

TOPOGRAPHY: The detailed description and analysis of the features of a relatively small area, district or locality.

TORQUE: The rotational or twisting force in a turning shaft.

TRADITIONAL: The customs and style from an earlier period.

TRAMMEL: An instrument consisting of a straightedge with two adjustable fixed points for drawing curves and ellipses.

TRANSLUCENT: That which passes light but diffuses it so that objects are not identifiable.

TRUE POSITION: The BASIC or theoretically exact position of a feature.

TRUNCATED: Having the apex, vertex or end cut off by a plane.

TUMBLING: The process of removing rough edges from parts by placing them in a rotating drum that contains abrasive stones, liquid and a detergent.

TYPICAL (TYP): This term, when associated with any dimension or feature, means the dimension or feature applies to the locations that appear to be identical in size and configuration unless otherwise noted.

UNIFORM: Having the same form or character; unvarying.

VERNIER SCALE: A small movable scale attached to a larger fixed scale, for obtaining fractional subdivisions of the fixed scale.

VERTEX (pl-vertices): The highest point of something; the top; the summit.

VERTICAL: Perpendicular to the horizon.

WIP: Work-in-progress.

WORKING DRAWINGS: A set of drawings which provide details for the production of each part and information for the correct assembly of the finished product.

ACKNOWLEDGMENTS

The author wishes to express his appreciation to colleagues in the Division of Technology, Arizona State University for helpful suggestions and critical review of various sections of the text: Walter Beeler, Robert P. Benzinger, Dr. Russ G. Biekert, Dr. Thomas A. Kanneman, Lyle McCurdy and Dr. Z.A. Prust; to Brian L. Duelm, technical writer, for the four chapters (Chapters 29-32) on computer-aided drafting and design; to Donal Hay for basic work on the chapter on electrical and electronics drafting; and to Dr. Walter E. Burdette for encouragement with this undertaking. Appreciation is also due to Dr. Richard A. Froese, John R. Walker, Dean M. Odegaard and Fern Rook for counsel on selection of content and technical editing.

The author would like to acknowledge the assistance of the many industries who contributed greatly to the value of this text. Gratitude is also expressed to Staedtler, Inc. for the cover photographs. In addition to those industries contributing illustrations noted in the various sections of the text, the following industries contributed drawings and technical manuals which were most helpful.

Allen-Bradley
Allied Research Associates, Inc.
Allied Structural Steel Company
Allis-Chalmers
American Hoist
American LaFrance
American Meter Company
American Standard
AMF Western Tool Division
Anaconda American Brass Company.
Autodesk, Inc.
Babcock & Wilcox
Barber-Colman Company
Barber-Greene Company
Bell Helicopter Company
Bell Laboratories
Bethlehem Steel Corporation
Black & Decker Manufacturing Company
Boeing Company
Bowmar Instrument Corporation
Burroughs Corporation
CADAM, Inc.
Cadillac Gage Company
CADKEY, INC.
CAD Technologies
Calcomp, Inc., A Lockheed Company
California Redwood Association
Caterpillar Tractor Company
CH Products
Cincinnati Milacron
Cities Service Oil Co.
Clausing Corporation
Clayton Manufacturing Company
Cummins Engine Company, Inc.
Cushman
Data Products
John Deere
DeLaval Turbine Inc.
Dresser Industries, Inc.
Eastman Kokak Co.
Eaton Corporation
Ebasco Services Incorporated

Epson American, Inc.
Fairchild Hiller
Ford Motor Company
Teledyne Farris Engineering
Faultless Division, Bliss & Laughlin Industries
Fulcrum Computer Products
Garrett Turbine Engine Co.
Gemcor Drivmatic Division
General Motors Technical Center
Gerber Scientific, Inc.
Giddings & Lewis Machine Tool Company
Globe Tool & Engineering Company
Bennie M. Gonzales, Inc. Architects
Gould, Incorporated
Heil Company
Hewlett-Packard Co.
Hyster Company
Ingersoll Milling Machine Company
Ingersoll-Rand Company
Intergraph
International Business Machines Corporation
International Harvester Company
Iomega Corporation
Johnson & Bassett, Inc.
Kodak
Koehring HPM Division
Landcadd, Inc.
Lodge & Shipley
Mack Trucks, Inc.
Martin Marietta Corporation
McDonnell Aircraft Company
Massey-Ferguson Inc.
McQuay, Inc.
Miller Electric Mfg. Co.
Monarch Machine Tool Company
Motorola Inc.
G. W. Murphy Industries
Muskegon Tool Industries, Inc.
Nash Engineering Company
National Acme
NC Microproducts, Inc.
Niagara Machine & Tool Works

Numonics
Omark Industries
Otis Elevator Company
Outboard Marine Corporation
Philco Ford
Polaroid
Procter & Gamble Company
QMS, Inc.
Quadram
Red Cedar Shingle and Handsplit Shake Bureau
Reliance Electric Company
Sanders Associates, Inc.
Seagate Technology
Simmons Machine Tool Corporation
Simmonds Precision
SKF Industries, Inc.
SoftSource
Southern Forest Products Association
Sperry Flight Systems Division
Standard Pressed Steel Co.
Summagraphics Corporation
Sundstrand Aviation
Tallgrass Technologies Corporation
TEKTRONIX, Inc.
Texas Instruments Incorporated
Thompson Electric Welder Co.
3M Company
Timken Company
Tinius Olsen Testing Machine Co., Inc.
Torin Corporation
Toro Manufacturing Corporation
The Trane Company
Twin Disc Incorporated
United Engine & Machine Co.
United States Steel Corporation
A. VanWegenen
Versatec
Warner & Swasey Co.
Waukesha Foundry Company, Inc.
Waukesha Motor Company
Westinghouse Electric Corporation
Wisconsin Bridge & Iron Co.
Zenith

Items which are credited ANSI have been extracted from AMERICAN NATIONAL STANDARD DRAFTING PRACTICES, DIMENSIONING AND TOLERANCING FOR ENGINEERING DRAWINGS (ANSI Y14.5M-1982) with the persmission of the publisher, The American Society of Mechanical Engineers, United Engineering Center, 345 East 47th Street, New York, NY 10017.

INDEX

Index